Energy, Environment, and Sustainability

Series Editor

Avinash Kumar Agarwal, Department of Mechanical Engineering, Indian Institute of Technology Kanpur, Kanpur, Uttar Pradesh, India

AIMS AND SCOPE

This books series publishes cutting edge monographs and professional books focused on all aspects of energy and environmental sustainability, especially as it relates to energy concerns. The Series is published in partnership with the International Society for Energy, Environment, and Sustainability. The books in these series are edited or authored by top researchers and professional across the globe. The series aims at publishing state-of-the-art research and development in areas including, but not limited to:

- Renewable Energy
- Alternative Fuels
- Engines and Locomotives
- Combustion and Propulsion
- Fossil Fuels
- Carbon Capture
- Control and Automation for Energy
- Environmental Pollution
- Waste Management
- Transportation Sustainability

Review Process

The proposal for each volume is reviewed by the main editor and/or the advisory board. The chapters in each volume are individually reviewed single blind by expert reviewers (at least four reviews per chapter) and the main editor.

Ethics Statement for this series can be found in the Springer standard guidelines here https://www.springer.com/us/authors-editors/journal-author/journal-author-helpdesk/before-you-start/before-you-start/1330#c14214

Kiran Bala · Tonmoy Ghosh · Vivek Kumar ·
Pritam Sangwan
Editors

Harnessing Microbial Potential for Multifarious Applications

Springer

Editors
Kiran Bala
Department of Biosciences and Biomedical Engineering
Indian Institute of Technology Indore
Indore, Madhya Pradesh, India

Vivek Kumar
Centre for Rural Development and Technology
Indian Institute of Technology Delhi
New Delhi, Delhi, India

Tonmoy Ghosh
Department of Biosciences and Biomedical Engineering
Indian Institute of Technology Indore
Indore, Madhya Pradesh, India

Institute for Water and Wastewater Technology
Durban University of Technology
Durban, South Africa

Pritam Sangwan
Centre for Fire, Explosive and Environment Safety (CFEES)
Defence Research and Development Organisation (DRDO)
New Delhi, Delhi, India

ISSN 2522-8366 ISSN 2522-8374 (electronic)
Energy, Environment, and Sustainability
ISBN 978-981-97-1151-2 ISBN 978-981-97-1152-9 (eBook)
https://doi.org/10.1007/978-981-97-1152-9

This Springer imprint is published by the registered company Springer Nature Singapore Pte Ltd.
The registered company address is: 152 Beach Road, #21-01/04 Gateway East, Singapore 189721, Singapore

Preface

The world is gravitating towards sustainability due to population explosion and the consequent pressures on its limited resources. The United Nations' Sustainable Development Goals (SDGs) also advocate moving towards a more holistic living through responsible consumption, sanitation and using clean energy, amongst others. The potent role of microorganisms cannot be ignored in our quest to minimize waste, and maximize efficiency in waste management and clean energy production. Without these unseen little warriors by our side, our planet runs the risk of being converted into a waste dump.

The International Society for Energy, Environment and Sustainability (ISEES) was founded at the Indian Institute of Technology Kanpur (IIT Kanpur), India, in January 2014 to spread knowledge and awareness and to catalyse research activities in the fields of Energy, Environment, Sustainability, and Combustion. Society's goal is to contribute to the development of clean, affordable, and secure energy resources and a sustainable environment for society, spread knowledge in the areas mentioned above, and create awareness about the environmental challenges the world is facing today. The unique way adopted by ISEES was to break the conventional silos of specialisations (engineering, science, environment, agriculture, biotechnology, materials, fuels, etc.) to tackle the problems related to energy, environment, and sustainability holistically. This is evident by the participation of experts from all fields to resolve these issues. The ISEES is involved in various activities, such as conducting workshops, seminars, conferences, etc., in the domains of its interests. The society also recognises the outstanding works of young scientists, professionals, and engineers for their contributions by conferring awards in various categories.

The Seventh International Conference on 'Sustainable Energy and Environmental Challenges' (VII-SEEC) was organised under the auspices of ISEES from December 16–18, 2022, at the Indian Institute of Technology, Banaras Hindu University, Varanasi (IIT-BHU), India. This conference provided a platform for discussions between eminent scientists and engineers from several countries, including India, the USA, Spain, Poland, Austria, the Czech Republic, and Korea. At this conference, eminent international speakers presented their views on energy, combustion, emissions, and alternative energy resources for sustainable development and a cleaner environment. The conference presented two high-voltage plenary talks by Dr. Ajay Kumar, Former Union Defense Secretary, GoI and Dr. S. S. V. Ramakumar, Director (R&D), Indian Oil.

The conference included 29 technical sessions on energy and environmental sustainability topics, including two plenary talks, 15 keynote talks, 80 invited talks from prominent scientists, and 118 contributed talks by students and researchers. The conference included technical sessions on advanced engine technologies, air pollution monitoring, anaerobic digestion, combustion and flames, air pollution control, biodegradation of toxic chemicals, energy and exergy, desalination and wastewater treatment, environmental bioengineering, pollution and climate change: challenges and priorities, alternative transportation fuels and materials, emerging environmental contaminants, sustainable processing of biomass, human health and environmental sustainability, sprays and atomisation, solid waste: challenges and mitigation, solid waste management, sustainable food and agri biotechnology, modelling and simulations, renewable energy technologies, bioremediation, biofuels and biorefineries, engine emissions and control, cleaner technologies for pollution mitigation, microbial processes, energy and environment, coal and biomass gasification, and environmental challenges mitigation.

About 250+ participants and speakers attended the conference, where 14 ISEES books published by Springer, under a dedicated series, "Energy, environment and sustainability", were released. This conference laid the roadmap for technology development, opportunities and challenges in the Energy, Environment and Sustainability domain. These topics are relevant for the country and the world in the present context. We acknowledge the support from various funding agencies and organisations for the successful conduct of the VII-SEEC, where the idea of these books germinated. We would therefore gratefully like to acknowledge IIT BHU (Special thanks to Prof. Akhilendra Pratap Singh); SERB, Government of India; Department of Scientific and Industrial Research (DSIR) (Special thanks to Dr. Vipin Shukla) and our publishing partner Springer (Special thanks to Swati Mehershi).

The editors would like to express their sincere gratitude to a large number of authors from all over the world for submitting their high-quality work on time and revising it appropriately at short notice. We want to express our special gratitude to our prolific reviewers: Dr. D. V. N. Sudheer, Dr. Imran Pancha, Dr. Shireesh Srivastava, Dr. Hamendra Singh Parmar, Dr. Chetan Paliwal, Dr. Bikash Kumar, Dr. Anand Krishna Singh, Dr. G. N. Nikhil, Dr. Lopamudra Ray, Dr. Paramasivan Balasubramanian, Prof. Ghada Abd-Elmonsef Mahmoud, Prof. Archana Tiwari, Dr. Bharti, Dr. Sanjiv Kumar Mishra, Dr. Lakhvinder Singh, Dr. Somvir Bajar, Dr. Rachna Bhateria, Dr. Rajasri Yadavalli, Dr. Vamsi Bharadwaj S. V., and Dr. Vishal Anand who reviewed chapters of this monograph and provided their valuable suggestions to improve them.

This book is based on the broad themes of bioremediation, waste mitigation, alternate energy, and healthier living choices where microbes have a crucial role to play. It includes microbial remediation processes covering wastewater, mine drainage, explosives, and plastic waste. It also includes a few chapters on the utility of microbes to produce bioplastics, enzymes, vegan neutraceutical and pigments, and biofuels.

Chapters include recent results and are focussed on current trends in processes and applications of microbes in bioremediation in various types of pollution as well as their potential to produce important biomolecules like biopolymers, pigments, nutrceuticals, and enzymes. After reading this book, readers will have an in-depth understanding about the role of microbes in pollution mitigation and the ways in which they do so. Moreover, it will also present a snapshot of the current trends of using microbial enzymes, biofuels, pigments, and nutraceuticals and the reasons for moving towards them. We hope the book greatly interests the professionals and post-graduate students involved in microbial bioremediation, biodegradable plastics, and microbial alternatives to vegan pharmaceutical and pigments.

Indore, India — Kiran Bala
Durban, South Africa — Tonmoy Ghosh
New Delhi, India — Vivek Kumar
New Delhi, India — Pritam Sangwan

Contents

Part I General

1 **Introduction to Harnessing Microbial Potential for Multifarious Applications** 3
Kiran Bala, Tonmoy Ghosh, Vivek Kumar, and Pritam Sangwan

Part II Microbial Bioremediation

2 **Microbial Remediation Technologies for Hazardous Metals in Soil, Sediments, and Water** 11
Krishna Kumar, Neha Rani Bhagat, and Vijay Kumar Bharti

3 **Microbial Remediation Technologies for Mining Waste Management** 31
Pradeep W. Samarasekere

4 **Manganese Contaminant: Revolutionizing In-Situ Microbial-Assisted Remediation Approach** 69
P. Kamleshiya

5 **Microbial Solution for Mitigating Plastic Pollution: From Environmental Hazard to Sustainable Management** 101
Suyash Devgan, Sparsh Singh, Abhilansh Pandey, and Rashmi Mathur

6 **Combatting Plastic Pollution in Natural Environments: Innovative Solutions Offered by Microbes** 137
Sushree Swagatika Mohanty, Jayashree Maharana, and Sony Pandey

7 Biodegradation of 2,4,6-Trinitrotoluene (TNT) by the Microbes and Their Synergistic Interactions 177
Jyoti Lamba, Dinesh Bhardwaj, Shalini Anand, Jayanti Dutta, and Pramod Kumar Rai

Part III Microbial Bioplastics

8 Bioplastic: Unravelling the Sustainable Approach for Petroleum Plastic 205
Manisha Mohapatra and Sanjana Singh

9 Biodegradable Plastics: Environmental Friendly Alternative to Petroleum Products 235
Mrinal Kashyap and Palak Saket

10 Greener Approach Towards Sustainable Green Plastics Through Eco-Friendly Upstream and Downstream Processing 253
Sonam Dubey, Shubhangi Parmar, Vijay Jagdish Upadhye, Anupama Shrivastav, Freny Shah, Niranjan Mishra, Pooja Bachani, and Sandhya Mishra

11 Seaweed-Based Biodegradable Packaging: A Green Alternative for a Plastic-Free Future 271
Silvia Lomartire and Ana M. M. Gonçalves

Part IV Waste-To-Wealth

12 Zero-Waste Processes Utilizing Microbes 299
Moffat Mutebele, Subbaiya Ramasamy, Todd Johnson, Manikandan Sivasubramanian, Kalirajan Arunachalam, Vinod Kumar Nathan, and George Mukupa

13 Biofuel Production from Microbial Sources: Advances and Challenges 323
Boddu Sandeep, Utkarsh Baranwal, and Ashootosh Mandpe

14 Microbes as Environmental Allies: From Wastewater Treatment to Sustainable Biofuel Production 359
Anuj Sharma, Praveen Sharma, Bansal Deepak, and Sharma Mona

Part V Miscellaneous Applications of Microbial Metabolites

15 From Pond to Pill: Microalgae's Role in Vegan Pharmaceuticals' 379
Himanshu Jain and Neeraj K. Aggarwal

16 Microbial Pigments as Vegan Colors for Food and Pharmaceuticals: A Sustainable Approach 419
Sarda Ruhil and Kalpana Nagpal

17 Microbial Biosensors for Environmental Quality Monitoring: Recent Advances and Future Outlook 439
Sompreeti Paul, Sindhoora Lakshmi, Dakshitha Akula, and Aswani Thekkangil

18 Biosensor: Application in Environmental Management 455
Arti Bhatnagar, Jamson Masih, and Ranjit Kumar

19 Metabolic Materials from Cyanobacteria and Microalgae: Biotechnology, Biochemistry and Biotherapeutics 489
Lokitha Paduvetnaya, Kamath H. Venkatesh, and Harishkumar Madhyastha

20 Novel and Conventional Uses of Fungi in Products and Services ... 519
H. Jayathunga, M. Gunasekara, M. Kavimalee, I. Jayamanna, H. K. S. de Zoysa, D. K. Hettiarachchi, and T. C. Bamunuarachchige

21 Cellulases from Psychrophilic and Psychrotrophic Microorganisms and their Potential Applications 549
L. S. Mamatha Bhanu and Soumya Chatterjee

Editors and Contributors

About the Editors

Dr. Kiran Bala is working at IIT Indore. She received her Masters and Ph.D. in environmental sciences in 2003 and 2007, respectively. She has various fellowship/awards, such as DBT women scientist, Dr. D. S. Kothari Post-doctoral Fellow award, DST INSPIRE Faculty Fellow Award and DST Early Career Research Award in her credit. In 2022, she got INYAS (Indian National Young Academy of Sciences) membership for the next five years. She has recently been elected as the core committee member of INYAS for two years. She has more than 65 papers published in journals of international repute in her credit. Currently, her research group consists of 10 Ph.D. students, two postdoc fellows and few dissertation students. Dr. Bala's research is broadly focused on the development of an integrated technology which takes nutrients from wastewater, carbon dioxide from industrial flue gas and gives us biofuels, a renewable source of energy. She is a member of scientific bodies, like BRSI and ISEES, etc. She is the Editorial Board member of several journals of repute.

Dr. Tonmoy Ghosh is currently a postdoctoral fellow at the Institute for Water and Wastewater Technology, Durban (IWWT Durban). Dr. Ghosh received his Ph.D. from the Academy of Scientific and Innovative Research (AcSIR). Prior to joining IWWT Durban, Dr. Ghosh was a National Post Doctoral Fellow at the Indian Institute of Technology Indore. He has also worked at CSIR-Central Salt and Marine Chemicals Research Institute, India and the International Centre for Genetic Engineering and Biotechnology, India. Dr. Ghosh mainly works in the field of microalgal biotechnology for alternative fuels and niche chemicals. His research interests lie in unravelling the effect of chemical modulators on microalgal and cyanobacterial growth and high-value metabolites, such as phycobiliproteins, polyunsaturated fatty acids, and tocopherols. His doctoral work was chiefly concerned with developing analytical applications for phycobiliproteins using their fluorescence as a tool. Till date, Dr. Ghosh has 28 publications in internationally recognized journals with close to 2000 citations and an h-index of 19; he also has four book chapters and three granted international patents to his credit. He is also the life member of professional organizations, such as The Biotech Research Society of India (BRSI) and the Association of Microbiologists of India (AMI).

Prof. Vivek Kumar is Professor in CRDT at Indian Institute of Technology, Delhi. Dr. Kumar is the Co-coordinator of one of the major programme of the Government of India, Unnat Bharat Abhiyan (UBA) that aims to utilize knowledge institutions of the country to help villages. Dr. Kumar is coordinating Subject Expert Group activities under UBA and also coordinating UBA collaboration with other government departments and organizations for fast racking rural livelihood. Dr. Kumar is a water system and clean technology expert for several industrial sectors and works in the area of process audit, process integration, implementation cleaner production projects, new greener product development, community engagement, circular economy and sustainability analysis of technologies. He is also actively involved in several projects of pollution prevention and control including Charter preparation and implementation especially in Paper, and textile, industries in Ganga Basin. He is the member of several Central Pollution Control Board's committees working for pollution control in process industries. He is currently working on the low footprint domestic sewage treatment systems for cold regions and rural areas, micro plastic pollution in the water and soil environment. His focuses also on natural packaging materials and valorization of biomass and industrial waste.

Kumar has published around 180 articles in journals, conference proceedings, books (as chapters). He has completed more than 35 R&D and consultancy projects and currently implementing five R&D and six consultancy projects.

Dr. (Mrs.) Pritam Sangwan is currently working as Scientist 'E' and Joint Director at Centre for Fire, Explosive and Environment Safety (CFEES), Defence Research and Development Organization (DRDO), New Delhi. Before joining DRDO she has also worked as an Assistant Professor with Gurgaon College of Engineering, Gurgaon.

She has got wide experience in the areas of remediation of Explosive contaminated sites, Solid/Hazardous Waste Management, Waste water treatment, Bioremediation and Environmental Impact Assessment. She has been awarded many fellowships/awards during her carrier. She has been awarded as 'Scientist of the Year-2018' award by Secretary, Department of Defence (DoD) and Chairman DRDO for her path breaking research and development. She was also awarded Merit Scholarship from Kurukshetra University, Research Fellowship (URF) from Guru Jambheshwar University of Science and Technology, Hisar. She was also awarded NFP fellowship by UNESCO for a course on Waste Management and Engineering at UNESCO-IHE Institute for Water Education in Delft, The Netherlands. She was also awarded Roll of Honor by Kurukshetra University, Kurukshetra.

She has published many research papers and book chapters in National and International Journals of repute. She also has many technical reports, guidelines, and a patent to her credit. She has also presented many papers in various conferences at national and international level. She is the life time member of Biotech research society of India (BRSI), Trivandrum and High Energy Materials society of India (HEMSI), Chandigarh.

Contributors

Neeraj K. Aggarwal Department of Microbiology, Kurukshetra University, Kurukshetra, Haryana, India

Dakshitha Akula Department of Biotechnology, School of Applied Science, REVA University, Bangalore, India

Shalini Anand Centre for Fire, Explosive and Environment Safety (CFEES), Defence Research Development Organization, DRDO, Timarpur, Delhi, India

Kalirajan Arunachalam Department of Biotechnology, Saveetha School of Engineering, Saveetha Institute of Medical and Technical Sciences (SIMATS), Thandalam, Chennai, Tamil Nadu, India

Pooja Bachani Amity University, Mumbai, India

Kiran Bala Department of Biosciences and Biomedical Engineering, Indian Institute of Technology Indore, Madhya Pradesh, India

T. C. Bamunuarachchige Department of Bioprocess Technology, Faculty of Technology, Rajarata University of Sri Lanka, Mihintale, Sri Lanka

Utkarsh Baranwal Department of Civil Engineering, Indian Institute of Technology Indore, Simrol, Indore, Madhya Pradesh, India

Neha Rani Bhagat DRDO-Defence Institute of High Altitude Research (DIHAR), Ladakh, UT, India

L. S. Mamatha Bhanu Department of Biotechnology, Yuvaraja's College, University of Mysore, Manasagangotri, Mysuru, Karnataka, India

Dinesh Bhardwaj Centre for Fire, Explosive and Environment Safety (CFEES), Defence Research Development Organization, DRDO, Timarpur, Delhi, India

Vijay Kumar Bharti DRDO-Defence Institute of High Altitude Research (DIHAR), Ladakh, UT, India

Arti Bhatnagar Department of Chemistry, Dayalbagh Educational Institute, Agra, India

Soumya Chatterjee Biodegradation Technology Division, Defence Research Laboratory, DRDO, Tezpur, Assam, India

H. K. S. de Zoysa Department of Bioprocess Technology, Faculty of Technology, Rajarata University of Sri Lanka, Mihintale, Sri Lanka

Bansal Deepak JBM Group, Gurugram, Haryana, India

Suyash Devgan Dr. B. R. Ambedkar Center for Biomedical Research, University of Delhi, New Delhi, Delhi, India

Sonam Dubey Applied Phycology and Biotechnology Division, CSIR-Central Salt and Marine Chemicals Research Institute, Bhavnagar, India

Jayanti Dutta Human Resource Development Centre, Panjab University, Chandigarh, India

Tonmoy Ghosh Department of Biosciences and Biomedical Engineering, Indian Institute of Technology Indore, Indore, Madhya Pradesh, India;
Institute for Water and Wastewater Technology, Durban University of Technology, Steve Biko Campus, Berea, Durban, South Africa

Ana M. M. Gonçalves Department of Life Sciences, University of Coimbra, MARE–Marine and Environmental Sciences Centre/ARNET-Aquatic Research Network, Coimbra, Portugal;
Department of Biology and CESAM, University of Aveiro, Aveiro, Portugal

M. Gunasekara Department of Bioprocess Technology, Faculty of Technology, Rajarata University of Sri Lanka, Mihintale, Sri Lanka

D. K. Hettiarachchi Department of Biological Sciences, Faculty of Applied Sciences, Rajarata University of Sri Lanka, Mihintale, Sri Lanka

Himanshu Jain Department of Microbiology, Kurukshetra University, Kurukshetra, Haryana, India

I. Jayamanna Department of Bioprocess Technology, Faculty of Technology, Rajarata University of Sri Lanka, Mihintale, Sri Lanka

H. Jayathunga Department of Bioprocess Technology, Faculty of Technology, Rajarata University of Sri Lanka, Mihintale, Sri Lanka

Todd Johnson Department of Biological Sciences, School of Mathematics and Natural Sciences, The Copperbelt University, Kitwe, Zambia

P. Kamleshiya Department of Zoology, Govt. J.S.T.P.G College Balaghat, R.S.S. University Chindwara, Chhindwara, M.P, India

Mrinal Kashyap Porter School of Environment and Earth Sciences, Tel Aviv University, Tel Aviv, Israel

M. Kavimalee Department of Bioprocess Technology, Faculty of Technology, Rajarata University of Sri Lanka, Mihintale, Sri Lanka

Krishna Kumar DRDO-Defence Institute of High Altitude Research (DIHAR), Ladakh, UT, India

Ranjit Kumar Department of Chemistry, Dayalbagh Educational Institute, Agra, India

Vivek Kumar Centre for Rural Development and Technology, Indian Institute of Technology Delhi, Hauz Khas, New Delhi, India

Sindhoora Lakshmi Department of Biotechnology, School of Applied Science, REVA University, Bangalore, India

Jyoti Lamba Department of Environment Studies, Panjab University, Chandigarh, India;
Centre for Fire, Explosive and Environment Safety (CFEES), Defence Research Development Organization, DRDO, Timarpur, Delhi, India

Silvia Lomartire Department of Life Sciences, University of Coimbra, MARE–Marine and Environmental Sciences Centre/ARNET-Aquatic Research Network, Coimbra, Portugal

Harishkumar Madhyastha Division of Cardiovascular Physiology, Faculty of Medicine, University of Miyazaki, Miyazaki, Japan

Jayashree Maharana Environment and Sustainability Department, CSIR-Institute of Minerals and Materials Technology, Bhubaneswar, Odisha, India; Academy of Scientific and Innovative Research (AcSIR), Ghaziabad, India

Ashootosh Mandpe Department of Civil Engineering, Indian Institute of Technology Indore, Simrol, Indore, Madhya Pradesh, India

Jamson Masih Department of Chemistry, Wilson College, Mumbai, India

Rashmi Mathur Department of Botany, Sri Aurobindo College, University of Delhi, New Delhi, Delhi, India

Niranjan Mishra Applied Phycology and Biotechnology Division, CSIR-Central Salt and Marine Chemicals Research Institute, Bhavnagar, India

Sandhya Mishra Applied Phycology and Biotechnology Division, CSIR-Central Salt and Marine Chemicals Research Institute, Bhavnagar, India

Sushree Swagatika Mohanty Environment and Sustainability Department, CSIR-Institute of Minerals and Materials Technology, Bhubaneswar, Odisha, India

Manisha Mohapatra Department of Biotechnology, Trident Academy of Creative Technology (TACT), Infocity, Chandrasekharpur, Odisha, India

Sharma Mona Department of Environmental Studies, School of Interdisciplinary and Applied Sciences, Central University of Haryana, Mahendergarh, Haryana, India

George Mukupa Department of Biotechnology, Saveetha School of Engineering, Saveetha Institute of Medical and Technical Sciences (SIMATS), Thandalam, Chennai, Tamil Nadu, India

Moffat Mutebele Department of Biological Sciences, School of Mathematics and Natural Sciences, The Copperbelt University, Kitwe, Zambia

Kalpana Nagpal Amity Institute of Pharmacy, Amity University Uttar Pradesh, Noida, India

Vinod Kumar Nathan School of Chemical and Biotechnology, SASTRA Deemed to Be University, Thanjavur, Tamil Nadu, India

Lokitha Paduvetnaya NITTE (Deemed to be University), NMAM Institute of Technology, Department of Biotechnology, Nitte, Karkala, Karnataka, India

Abhilansh Pandey Wallace H. Coulter Department of Biomedical Engineering, Georgia Tech and Emory University, Atlanta, GA, USA

Sony Pandey Environment and Sustainability Department, CSIR-Institute of Minerals and Materials Technology, Bhubaneswar, Odisha, India;
Academy of Scientific and Innovative Research (AcSIR), Ghaziabad, India

Shubhangi Parmar Department of Food and Bioproduct Sciences, University of Saskatchewan, Saskatoon, SK, Canada

Sompreeti Paul Department of Biotechnology, School of Applied Science, REVA University, Bangalore, India

Pramod Kumar Rai Centre for Fire, Explosive and Environment Safety (CFEES), Defence Research Development Organization, DRDO, Timarpur, Delhi, India

Subbaiya Ramasamy Department of Biological Sciences, School of Mathematics and Natural Sciences, The Copperbelt University, Kitwe, Zambia

Sarda Ruhil Raj Kumar Goel Institute of Technology (Pharmacy), Uttar Pradesh, Ghaziabad, India

Palak Saket Department of Biosciences and Biomedical Engineering, Indian Institute of Technology Indore, Indore, India

Pradeep W. Samarasekere Center for Advanced Materials and Smart Manufacturing, Faculty of Computing and Technology, University of Kelaniya, Kelaniya, Sri Lanka

Boddu Sandeep Department of Civil Engineering, Indian Institute of Technology Indore, Simrol, Indore, Madhya Pradesh, India

Pritam Sangwan Centre for Fire, Explosive and Environment Safety, Defence Research and Development Organization, Brig. S. K. Mazumdar Marg, Timarpur, Delhi, India

Freny Shah Applied Phycology and Biotechnology Division, CSIR-Central Salt and Marine Chemicals Research Institute, Bhavnagar, India

Anuj Sharma Department of Environmental Science and Engineering, Guru Jambheshwar University of Science and Technology, Hisar, Haryana, India

Praveen Sharma Department of Environmental Science and Engineering, Guru Jambheshwar University of Science and Technology, Hisar, Haryana, India

Anupama Shrivastav Faculty of Life Health, and Allied Sciences, ITM Vocational University, Vadodara, India

Sanjana Singh Department of Biotechnology, Trident Academy of Creative Technology (TACT), Infocity, Chandrasekharpur, Odisha, India

Sparsh Singh Dr. B. R. Ambedkar Center for Biomedical Research, University of Delhi, New Delhi, Delhi, India

Manikandan Sivasubramanian Department of Chemistry and Biology, School of Natural and Applied Sciences, Mulungushi University, Kabwe, Zambia

Aswani Thekkangil Department of Biotechnology, School of Applied Science, REVA University, Bangalore, India

Vijay Jagdish Upadhye Center of Research for Development (CR4D), Parul Institute of Applied Sciences (PIAS), Parul University (DSIR-SIRO Recognized), Vadodara, Gujarat, India

Kamath H. Venkatesh NITTE (Deemed to be University), NMAM Institute of Technology, Department of Biotechnology, Nitte, Karkala, Karnataka, India

Part I
General

Chapter 1
Introduction to Harnessing Microbial Potential for Multifarious Applications

Kiran Bala, Tonmoy Ghosh, Vivek Kumar, and Pritam Sangwan

At its Sustainable Development Summit 2015, the United Nations formally adopted the 17 Sustainable Development Goals (SDGs) agreed upon by the 193 member states (United Nations 2015). In the words of the then Secretary General of the United Nations, Mr. Ban Ki-moon, these SDGs "are a roadmap to ending global poverty, building a life of dignity for all and leaving no one behind. It is also a clarion call to work in partnership and intensify efforts to share prosperity, empower people's livelihoods, ensure peace and heal our planet for the benefit of this and future generations." These goals were necessitated due to the massive anthropogenic activities and the harm they are causing to the environment. In response, researchers the world over have recognized that efforts will have to be speeded up in order for them to have some effect on global problems. Consequently, research has intensified on various pressing problems currently facing the world; some of the major ones include environmental pollution by plastics and heavy metals, depleting fossil fuel reserves and increasing CO_2 due to their excessive use, and unsustainable manufacturing practices for high-value commodities.

K. Bala (✉) · T. Ghosh
Department of Biosciences and Biomedical Engineering, Indian Institute of Technology Indore, Khandwa Road, Simrol 453552, Madhya Pradesh, India
e-mail: kiranb@iiti.ac.in

T. Ghosh
Institute for Water and Wastewater Technology, Durban University of Technology, Steve Biko Campus, Berea 4001, Durban, South Africa

V. Kumar
Centre for Rural Development and Technology, Indian Institute of Technology Delhi, Hauz Khas, New Delhi 110016, India

P. Sangwan
Centre for Fire, Explosive and Environment Safety, Defence Research and Development Organization, Brig. S. K. Mazumdar Marg, Timarpur, Delhi 110054, India

K. Bala et al. (eds.), *Harnessing Microbial Potential for Multifarious Applications*, Energy, Environment, and Sustainability, https://doi.org/10.1007/978-981-97-1152-9_1

While addressing these challenges as well as nurturing innovations, the researchers have underscored the remarkable role of the versatile microbial world, ranging from viruses to algae, that has revolutionized industrial development. The genetic and biochemical diversity of these microbes empowers them to serve across a diverse spectrum of fields such as biotechnology, agriculture, environmental remediation, and therapeutics (Qian et al. 2020; Martínez et al. 2022; Priya et al. 2022; Raj et al. 2022; Xu et al. 2023; Narayanan et al. 2023; Madhavan et al. 2023; Syed et al. 2023). This book stems from an effort to put together the critical research advances made in utilizing these microbes to solve the problems detailed above. This introduction is envisaged as a guide to the reader as to the importance of the book and what to expect within. The book has been divided into five sections: (I) Introduction, (II) Microbial Bioremediation, (III) Microbial Bioplastics, (IV) Waste-to-wealth, and (V) Miscellaneous Applications of Microbial Metabolites.

The section on microbial bioremediation covers an important theme which is currently posing a burning question worldwide: pollution mitigation. During the recent COVID-19 pandemic, single-use plastic was the major pollutant responsible for environmental pollution. Researchers have detected the presence of microplastics, formed due to degradation of single-use plastics, in Earth's least inhabited place: Antarctica (Rivas et al. 2022). The widespread and irresponsible use of masks, gloves, and sanitization and protective gear has led to a situation where microplastics have even invaded our bodies (Yang et al. 2023). Additionally, heavy metal pollution has been a traditional culprit in industrial belts for a long time (Ulutaş 2022; Chen et al. 2023). Microbes have shown the ability to biosequester and contain the spread of heavy metals as well as degrade microplastics through biological means (Sreedevi et al. 2022; Sun et al. 2023). The section is dedicated to the elucidation of current advances made in the field of microbial bioremediation for pollution mitigation under various situations. Another niche where microbes play an important role is the degradation of explosives and related contaminants (Meda et al. 2021). The dedicated section will take the readers through the cutting-edge research on plastic, metals, and explosive remediation using microbes and will also detail about the mechanisms and pathways that are responsible.

Further, in order to satiate the human need for polymers and save our environment in the process, research has been directed towards the source of biodegradable polymers with similar physical and chemical properties as that of the chemically synthesized ones. Biopolymers from algae and bacteria find numerous applications in food, packaging, medical, and pharmaceutical sectors (Gupta et al. 2022; Zaccarelli et al. 2023; Rahman et al. 2023; Tabassum et al. 2023; Babaremu et al. 2023). The section on microbial bioplastics takes the readers through the advancements in the field of microbial biopolymers, especially biodegradable plastics like polyhydroxyalkanoates, and their future applications. It will also have a look at sustainable technologies currently being explored for their extraction and refinement, for example, using ionic liquids in place of organic solvents. Together, these technologies strive to make the realm of biopolymers both sustainable and practical.

In order to be sustainable, a process needs to be not just biodegradable but also generate minimal waste. Unfortunately, human activities have been devoid of such thoughts until recently, when the consequences of pollution and climate change have become more tangible. The section on waste-to-wealth essentially strives to educate the prospective reader on processes and techniques for waste utilization and creating meaningful returns. Wastewater has traditionally been one area which has attracted attention for its utility as a carbon source for microbes and the opportunity to generate biofuels. The dual advantage of waste and CO_2 mitigation has prompted researchers to explore ways to utilize these waste streams for microbial cultivation (Satpati et al. 2023; Umetani et al. 2023; Kandasamy et al. 2023; Vickram et al. 2023). This section will be dedicated to the generation of resources from waste streams and the researches that have been carried out in this direction.

Recently, we have been witnessing a shift towards greener and healthier eating alternatives. The vegan option has become more visible now in the last few years; people are preferring microbial alternatives to chemical ones for their daily consumables. In order to educate the prospective reader on these developments, the final section of this book is dedicated to microbial metabolites and their applications. Cyanobacteria are ancient prokaryotes thought responsible for creating the oxygen-rich atmosphere on Earth (Kasting and Siefert 2002). They are the microbes responsible for our oceans being the largest CO_2 sink in the world (Noreña-Caro and Benton 2018). Recent research has focused attention on cyanobacterial metabolites for their applications in diverse fields (Ahmed et al. 2023; Nawaz et al. 2023). Moreover, fungi have been investigated for their potential for enzyme production and have wide applications in agriculture, food and feed, pharmaceuticals, textiles, and paper industries (Asadollahzadeh et al. 2023; El-Ghonemy et al. 2023). Cellulases are one class of enzymes with wide-ranging applications in agriculture, textiles, packaging, paper and pulp, and biofuel industries (Saharan et al. 2023; Korsa et al. 2023; Ranjan et al. 2023). The section is, thus, dedicated to recent advances in such metabolites from microbes and their potential applications. It will talk about the detailed biochemistry of these metabolites, many of whom are important for the microbe's own survival, and will enable the readers to gain a deeper insight into the potential applications these microbes still conceal within their microscopic sizes. As discussed by the authors, these tiny, microscopic organisms hold enormous potential to propel mankind and Earth towards a more holistic, sustainable, and better future.

References

Ahmed S, Alam W, Aschner M et al (2023) Marine cyanobacterial peptides in neuroblastoma: search for better therapeutic options. Cancers 15. https://doi.org/10.3390/cancers15092515

Asadollahzadeh M, Mohammadi M, Lennartsson PR (2023) Fungal biotechnology. In: Taherzadeh MJ, Ferreira JA, Pandey A (eds) Current developments in biotechnology and bioengineering. Elsevier, pp 31–66

Babaremu K, Oladijo OP, Akinlabi E (2023) Biopolymers: a suitable replacement for plastics in product packaging. Adv Ind Eng Polym Res 6:333–340. https://doi.org/10.1016/j.aiepr.2023.01.001

Chen D, Wang X, Luo X et al (2023) Delineating and identifying risk zones of soil heavy metal pollution in an industrialized region using machine learning. Environ Pollut 318:120932. https://doi.org/10.1016/j.envpol.2022.120932

El-Ghonemy DH, Ali TH, Selim MS (2023) Extracellular β-d-fructofuranosidase from a novel Aspergillus sp. DHE1 with high potential for biotechnological applications: Purification and biochemical characterization. Biocat Agricult Biotech 48:102644. https://doi.org/10.1016/j.bcab.2023.102644

Gupta I, Cherwoo L, Bhatia R, Setia H (2022) Biopolymers: implications and application in the food industry. Biocatal Agric Biotechnol 46:102534. https://doi.org/10.1016/j.bcab.2022.102534

Kandasamy S, Narayanan M, Raja R et al (2023) The current state of algae in wastewater treatment and energy conversion: a critical review. Curr Opin Environ Sci Health 33:100469. https://doi.org/10.1016/j.coesh.2023.100469

Kasting JF, Siefert JL (2002) Life and the evolution of Earth's atmosphere. Science 296:1066–1068. https://doi.org/10.1126/science.1071184

Korsa G, Konwarh R, Masi C et al (2023) Microbial cellulase production and its potential application for textile industries. Annals Microb 73:13. https://doi.org/10.1186/s13213-023-01715-w

Madhavan A, Arun KB, Alex D et al (2023) Microbial production of nutraceuticals: metabolic engineering interventions in phenolic compounds, poly unsaturated fatty acids and carotenoids synthesis. J Food Sci Technol 60:2092–2104. https://doi.org/10.1007/s13197-022-05482-5

Martínez GM, Pire C, Martínez-Espinosa RM (2022) Hypersaline environments as natural sources of microbes with potential applications in biotechnology: the case of solar evaporation systems to produce salt in Alicante County (Spain). Curr Res Microbial Sci 3:100136. https://doi.org/10.1016/j.crmicr.2022.100136

Meda A, Sangwan P, Bala K (2021) In-vessel composting of HMX and RDX contaminated sludge using microbes isolated from contaminated site. Environ Pollut 285:117394. https://doi.org/10.1016/j.envpol.2021.117394

Narayanan M, Ali SS, El-Sheekh M (2023) A comprehensive review on the potential of microbial enzymes in multipollutant bioremediation: mechanisms, challenges, and future prospects. J Environ Manage 334:117532. https://doi.org/10.1016/j.jenvman.2023.117532

Nawaz T, Gu L, Fahad S et al (2023) A comprehensive review of the therapeutic potential of cyanobacterial marine bioactives: unveiling the hidden treasures of the sea. Food Energy Sec 12:e495. https://doi.org/10.1002/fes3.495

Noreña-Caro D, Benton MG (2018) Cyanobacteria as photoautotrophic biofactories of high-value chemicals. J CO_2 Utiliz 28:335–366. https://doi.org/10.1016/j.jcou.2018.10.008

Priya A, Dutta K, Daverey A (2022) A comprehensive biotechnological and molecular insight into plastic degradation by microbial community. J Chem Technol Biotechnol 97:381–390. https://doi.org/10.1002/jctb.6675

Qian X, Chen L, Sui Y et al (2020) Biotechnological potential and applications of microbial consortia. Biotechnol Adv 40:107500. https://doi.org/10.1016/j.biotechadv.2019.107500

Rahman MZ, Rahman M, Mahbub T et al (2023) Advanced biopolymers for automobile and aviation engineering applications. J Polym Res 30:106. https://doi.org/10.1007/s10965-023-03440-z

Raj SKH, Nair RA (2022) Anti-microbial and cytotoxic activity of ZzAMP, a serine protease inhibitor (SPI) with nutraceutical potential from rhizomes of medicinal plant, Zingiber zerumbet. Nat Prod Res 36:6375–6380. https://doi.org/10.1080/14786419.2022.2032046

Ranjan R, Rai R, Bhatt SB, Dhar P (2023) Technological road map of cellulase: a comprehensive outlook to structural, computational, and industrial applications. Biochem Eng J 198:109020. https://doi.org/10.1016/j.bej.2023.109020

Rivas ML, Albion I, Bernal B et al (2022) The plastic pandemic: COVID-19 has accelerated plastic pollution, but there is a cure. Sci Total Environ 847:157555. https://doi.org/10.1016/j.scitotenv.2022.157555

Saharan V, Tushir S, Singh J et al (2023) Application of MOGA-ANN tool for the production of cellulase and xylanase using de-oiled rice bran (DORB) for bioethanol production. Biomass Conv Bioref. https://doi.org/10.1007/s13399-023-04022-1

Satpati GG, Dikshit PK, Mal N et al (2023) A state of the art review on the co-cultivation of microalgae-fungi in wastewater for biofuel production. Sci Total Environ 870:161828. https://doi.org/10.1016/j.scitotenv.2023.161828

Sreedevi PR, Suresh K, Jiang G (2022) Bacterial bioremediation of heavy metals in wastewater: a review of processes and applications. J Water Proc Eng 48:102884. https://doi.org/10.1016/j.jwpe.2022.102884

Sun X-L, Xiang H, Xiong H-Q et al (2023) Bioremediation of microplastics in freshwater environments: a systematic review of biofilm culture, degradation mechanisms, and analytical methods. Sci Total Environ 863:160953. https://doi.org/10.1016/j.scitotenv.2022.160953

Syed Z, Sogani M, Rajvanshi J, Sonu K (2023) Microbial biofilms for environmental bioremediation of heavy metals: a review. Appl Biochem Biotechnol 195:5693–5711. https://doi.org/10.1007/s12010-022-04276-x

Tabassum Z, Mohan A, Mamidi N et al (2023) Recent trends in nanocomposite packaging films utilising waste generated biopolymers: industrial symbiosis and its implication in sustainability. IET Nanobiotechnol 17:127–153. https://doi.org/10.1049/nbt2.12122

Ulutaş K (2022) Risk assessment and spatial distribution of heavy metal in street dusts in the densely industrialized area. Environ Monit Assess 194:99. https://doi.org/10.1007/s10661-022-09762-7

Umetani I, Sposób M, Tiron O (2023) Indigenous green microalgae for wastewater treatment: nutrient removal and resource recovery for biofuels and bioproducts. BioEnergy Res. https://doi.org/10.1007/s12155-023-10611-9

United Nations (2015) Transforming Our World: The 2030 Agenda for Sustainable Development

Vickram AS, Saravanan A, Senthil Kumar P et al (2023) An integrated approach to the sustainable development and production of biofuel from biopolymers and algal biomass derived from wastewater. Fuel 349:128691. https://doi.org/10.1016/j.fuel.2023.128691

Xu G, Zhao S, Liu J, He J (2023) Bioremediation of organohalide pollutants: progress, microbial ecology, and emerging computational tools. Curr Opin Environ Sci Health 32:100452. https://doi.org/10.1016/j.coesh.2023.100452

Yang Y, Xie E, Du Z et al (2023) Detection of various microplastics in patients undergoing cardiac surgery. Environ Sci Technol 57:10911–10918. https://doi.org/10.1021/acs.est.2c07179

Zaccarelli A, Remaggi G, Elviri L (2023) 3D printed biopolymers for medical applications and devices. In: Mehrpouya M, Vahabi H (eds) Additive manufacturing of biopolymers. Elsevier, pp 303–329

Part II
Microbial Bioremediation

Chapter 2
Microbial Remediation Technologies for Hazardous Metals in Soil, Sediments, and Water

Krishna Kumar, Neha Rani Bhagat, and Vijay Kumar Bharti

2.1 Introduction

Metals, along with organic pollutants, have become common pollutants worldwide in developed and developing countries, and sometimes they are part of the natural geogenic cycle and affect many soil and water bodies (Mazumder et al. 2021). These metals are known as "hazardous metals" (HMs) when they accumulate in higher concentrations and further infiltrate into air, soil, and water through various geological processes and anthropogenic sources such as agricultural practices involving soil treatment with impure mineral phosphate fertilizer, irrigation of farmland with polluted water, soil pollution, waste treatment plants, industrial waste, and domestic sewage, as shown in Fig. 2.1 (Giri et al. 2021; Charan et al. 2023). After that, HMs become a severe threat to living organisms in the environment, depending on their concentration and exposure duration. These metals are relatively easily absorbed by plant roots, contaminating the food chain and resulting in bioaccumulation in the human and animal bodies, causing hazardous consequences for humans and animals (Yu and Tang 2022). Indeed, various reports have stated that people in developing nations are more likely to be exposed to these HMs and are majorly known to be through water, soil, and sediment pollution, which can further affect the environment and living beings (Eqani et al. 2016). Arsenic, lead, cadmium, and nickel are some common HMs that are substantial environmental polluters and, are discussed briefly in this chapter.

Arsenic (As) is one of the hazardous metals which contaminate soil, water, and food, and its long-term exposure could cause various health issues, including cancer. Lead (Pb) is found in old paint, contaminated soil, water pipes, and consumer products, which are highly toxic to the nervous system, kidneys, and other organs. Further, Metal Nickel (Ni), is extensively utilized in various industrial processes as well as

K. Kumar · N. R. Bhagat · V. K. Bharti (✉)
DRDO-Defence Institute of High Altitude Research (DIHAR), Leh-194101, Ladakh, UT, India
e-mail: vijaybharti.dihar@gov.in

K. Bala et al. (eds.), *Harnessing Microbial Potential for Multifarious Applications*, Energy, Environment, and Sustainability, https://doi.org/10.1007/978-981-97-1152-9_2

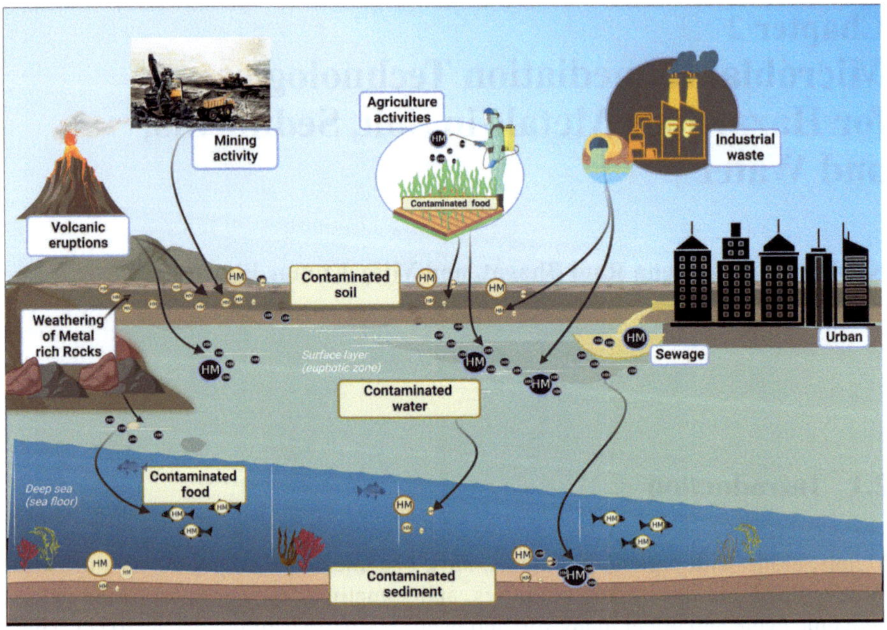

Fig. 2.1 Diagrammatic representation of the source of hazardous metals in the environment HM: hazardous metals

distributed in the environment, air, water, or soil through natural sources and anthropogenic activity (Li et al. 2016; Hayyat et al. 2020), and its high levels could lead to various adverse effects on the environment and living beings (Garai et al. 2021). Lastly, Cadmium (Cd), another toxic metal, is often present in batteries, pigments, and some industrial processes, which can cause damage to the kidneys, lungs, and bones of humans and animals (Genchi et al. 2020; Shah 2020; Abo-Alkasem et al. 2023).

Therefore, ameliorating the hazardous impact of these heavy metals on human and animal health is very important for a healthy society and human–animal welfare. In this retrospect, Contaminant's source removal, neutralization of metals, absorption, filtration, adsorption, etc., are essential means of remediation for these metals to protect the environment and inhibit their biomagnifications in the food chain. Other methods are land farming, bio-composting, biopiles, and bioreactors (Sayqal and Ahmed 2021). Other than Physical, chemical and biological remediation technologies are also introduced depending on the types of metals, level of contaminants, sources, and medium. However, due to various shortcomings of the physical and chemical remediation process, the biological degradation of HMs to levels below the concentration limits established by regulatory authorities, in recent days, microbial bioremediation of hazardous metals is gearing up importance. Under this chapter, microbial technologies, i.e., in-situ and ex-situ technologies, the microbes, viz. bacteria, fungus, and algae involved and their mechanisms in transforming the toxic form of metals present in the environment (soil, water, and sediments) into non-toxic or less hazardous as detailed in the following subheadings. Moreover, various factors affecting and challenges in existing microbial technologies are also discussed in this chapter to help further develop more efficient microbial technologies in the remediation of HMs from contaminated sources.

2.2 Impact of Hazardous Metals on the Living Organisms

Hazardous metals have a direct relationship with environmental pollution concerning their potentiality to cause toxicity in the biota, as they accumulate in the food chain at the primary producer level, followed by consumption at the consumer level, and become hazardous when they accumulate in higher concentration (Garai et al. 2021). Consequently, these metals pose several hazards to the health of plants, animals, and humans, as detailed in the following subheadings.

2.2.1 Effect of HMs Exposure on Plant Health

Hazardous metals could harm plants directly and indirectly by impacting the plant metabolism and growth through the inhibition of cytoplasmic enzymes and damage to cell structures from oxidative stress, replacement of vital nutrients at cation exchange sites in plants, the killing of the rhizoplane/rhizosphere soil microbes, inhibition of enzyme activities etc. (Djingova et al. 2000; Blaylock et al. 2000; Assche and Clijsters 1990; Taiz and Zeiger 2002). For instance, high levels of metal cadmium harm the plants through chlorosis, growth inhibition, browning of the root tips, and, ultimately, death (Mohanpuria et al. 2007; Guo et al. 2008; Asati et al. 2016). Lead (Pb) is another common hazardous metal that harms plant morphology, growth, and photosynthetic processes, causing plant mortality. Lead poisoning could also hinder early seedling growth in soybean, rice, maize, barley, tomato, and legumes, while at the same time, negatively affect the roots–stem elongation and leaf enlargement in Allium species, Barley, and Radish (Asati et al. 2016). Furthermore, nickel toxicity levels in the soil also impair nutrient imbalance, cell membrane function, plasma membrane lipid content, H-ATPase activity, and water balance in plants, impacting plant growth and metabolism (Asati et al. 2016; Pandey and Sharma 2002; Gajewska et al. 2006).

2.2.2 Effect of HMs Exposure on Animal Health

Hazardous metals constantly contaminate animals through high ambient air concentrations close to emission sources or contaminated fodder and drinking water. These metals are poisonous because individual heavy metals exhibit different toxicities based on the amount absorbed, length of exposure, species, gender, and environmental and dietary factors. These metals exposure can cause mutagenicity, carcinogenicity, teratogenicity, immunosuppression, poor physical condition, and reduced reproduction in the animals. For instance, Pb and Cd residues were found in various livestock species of animals, e.g., dairy cows, growing pigs, goats, sheep, and laying hens. Arsenic toxicity affects the gastrointestinal system, nervous system, inhibits rumen bacteria growth, which leads to weight loss, capricious appetite, conjunctively and mucosal erythematic lesions, mouth ulceration, and low milk yield in cattle, goitre in rats, as well as anaemia, liver and kidney damage and skin damage in animals (Raikwar et al. 2008). So, their long-term exposure to animals revealed an accumulation of Pb in the bone and both of these heavy metals in the liver and kidney, which could pose adverse health effects in the animals (Pandey and Madhuri 2014). Furthermore, the consumption of the products of these livestock animals also poses hazards to humans. On the other hand, in aquatic animals, these hazardous metals could impair their health and reproductive ability (Garai et al. 2021; Shahjahan et al. 2022). Moreover, the creatures such as worms, crustaceans, and insects in the benthic environment also get threatened with hazardous metal concentrations, which could result in their death or bioaccumulation. When larger animals feed on these contaminated aquatic animals, the toxic metals are taken into their bodies, moving up the food chain to human beings. So, these metals are critical environmental pollutants affecting the food chain through biomagnifications.

2.2.3 *Effect of HMs Exposure on Human Health*

These hazardous metals can enter a person's body by eating, inhaling, or skin absorption or exposure to the contaminated environment and the workplace (Afrin et al. 2015). For instance, workers in mining and manufacturing industries are exposed to heavy metals for longer and have considerably higher levels of the heavy metals Ni, Cd, and Pb (Mahmood et al. 2015). Most people also ingest these heavy metals through exposure to contaminated food and water. Growing human foods in heavy metal-contaminated medium causes these elements to bioaccumulate in the human food chains, eventually making their way into the human body. The hazardous metals diminish the body's primary antioxidants, especially those enzymes and antioxidants with thiol groups (-SH) and increase the reactive oxygen species (hydroxyl radical (OH-), superoxide radical (O_2-), and hydrogen peroxide (H_2O_2) in larger quantities, causing oxidative stress and other related diseases. In particular, nickel has many adverse effects on humans and causes allergies, mitochondrial dysfunctions, neurotoxicity, contact dermatitis, headaches, gastrointestinal manifestations, lung cancer, kidney, and cardiovascular diseases owing to the inhalation of contaminated air (Lu et al. 2005) and cadmium poisoning causes itai-itai disease. Cd and Pb are particularly hazardous to the renal cortex in humans (Wilk et al. 2017; Ercal et al. 2001). Further, arsenic poisoning leads to various multisystemic diseases like myocardial injury, cardiac arrhythmias, cardiomyopathy, Toxic polyneuropathy, peripheral neuropathy, skin alterations, genotoxicity, mutagenicity, etc. (Ozturk et al. 2021).

2.3 Conventional Technologies for Remediation of Hazardous Metals

Various technologies so far have been developed to reduce the pollution of soil, water, and sediments with hazardous metals. In the conventional approaches, some examples of physical and chemical technologies utilized to remove the HMs are presented in Fig. 2.2.

Although these conventional methods are promising in terms of separation, however, their cost of investment, long-term operation costs, use of toxic chemicals,

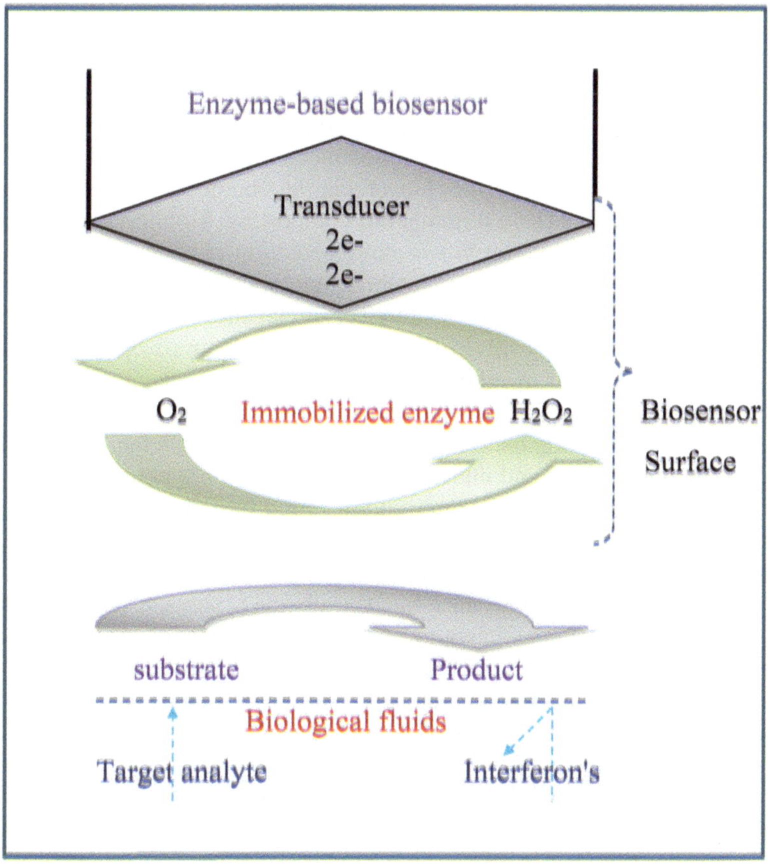

Fig. 2.2 Flow chart of conventional methods for hazardous metal remediation

ineffectiveness at low metal levels, disposal, and further storage or solid handling costs are the major drawbacks. Thus, there is a demand for cost-efficient, environment friendly alternatives, i.e., plants or microbial agents to remove heavy metals from soil, water, and sediments. In view of this, microbial remediation of hazardous metal contamination has presently been established as an economical, eco-friendly, and efficient technology for clean-up of the hazardous metal-contaminated environment (soil/water/sediment) compared to other techniques.

2.4 Microbial Technologies for Remediation of Hazardous Metals

Microbial remediation refers to decreasing the availability of HMs in the environment using indigenous or exotic microbes, e.g., bacteria, fungi, and algae. Nowadays, various microbial agents have drawn the attention of researchers and academician because of their significant potential in the removal, detoxification, and transformation of HM and protection of the surrounding environment (Jeyakumar et al. 2023; Riseh et al. 2022; Sreedevi et al. 2022; Tarekegn et al. 2020). Microbial technologies have recently been considered an advanced technique for removing HMs from contaminated sources, i.e., water, soil, and sediment. These microbial-based methods involve mainly *in-situ* (biosparging, bioventing, bioaugmentation, and biostimulation) and *ex-situ* methods (land farming, bio-composting, biopiles, and bioreactors) (Sayqal and Ahmed 2021). The *in-situ* bioremediation methods treat the contamination at the site, while the *ex-situ* method involves excavating and treating the contaminated source before returning it to its original site.

In *in-situ*, biosparging involves the injection of air or oxygen into contaminated groundwater or soil to influence the growth of indigenous microorganisms that naturally feed on pollutants to accelerate their degradation process (Vidali 2001). However, bioventing focuses on the remediation of pollutants in the contaminated soil by injecting air and nutrients. This technique promotes the biodegradation of contaminants and reduces their volatilization into the atmosphere (Chipasa and Mędrzycka 2006). Furthermore, bioaugmentation involves introducing selected strains/ mixed cultures of microorganisms with highly efficient degrading capability into the contaminated site to enhance the degradation process, especially when the indigenous microbes are insufficient to tackle the contamination (Stroo et al. 2013). On the contrary, biostimulation is a broader approach that aims to enhance the growth and activity of indigenous microorganisms. Nutrients, such as nitrogen and phosphorus, are introduced into the contaminated area to support the microbial population's growth and metabolism. This approach effectively accelerates the degradation of pollutants by capitalizing on the existing microbial community's capabilities (Sayqal and Ahmed 2021).

In the *ex-situ* methods, Land farming is a method that involves spreading polluted soil over a specific area and promoting microbial activity through regular aeration and cultivation, facilitating the remediation of HMs. Moreover, composting is based on mixing contaminated soil with non-hazardous waste to create a favourable environment for microbial growth and encourage HMs breakdown (Sivakumar et al. 2014). Further, Biopilling is considered a hybrid of both composting and land farming methods to create a conducive environment for microbial activity to transform or immobilize the HMs, rendering them less harmful over time (Sayqal and Ahmed 2021). On the other hand, a bioreactor is defined as an enclosed vessel system in which the excavated contaminated sample is introduced along with the enzymes, microbes, nutrients, and controlled environment, thereby enhancing the microbial activity and facilitates HMs remediation (Sayqal and Ahmed 2021; Vidali 2001).

All of these technologies are based on microorganisms. Therefore, various microorganisms (bacteria, fungi, and algae) involved and their mechanisms for removing HMs contamination from soil, water, and sediment are detailed in the following sections and Table 2.1.

2.4.1 Microorganisms Involved in Hazardous Metals Remediation

2.4.1.1 Bacteria

Among all microbes, bacteria can more widely clean-up the HMs because of their small size, quick growth rate, and easy cultivation (Yin et al. 2018). In fact, the dead cell biomass of bacteria could also display a high potential for HMs remediation (Sreedevi et al. 2022). These bacteria remove HMs primarily by means of adsorption on bacterial cell surfaces, wherein negatively charged carboxyl, amino, phosphoryl, and sulpho-groups on bacterial cell walls act as potential ion exchange sites and metal sinks (Sreedevi et al. 2022). Further, several reports on biosorption capacity of HMs with respect to the bacterial phyla revealed that Phylum Proteobacteria, Actinobacteria, and Firmicutes showed efficient biosorption for hazardous metals (As, Cd, Ni, and Pb), where Actinobacteria was the most efficient phylum for the biosorption of Ni ions, however, Phylum Firmicutes was the most efficient for biosorption of Cadmium and Lead, and Phylum Firmicutes and Proteobacteria for the removal of Arsenic (Hussein et al. 2004; Qamar et al. 2017; Kanamarlapudi et al. 2018; Redha 2020). Moreover, As per our literature survey, various bacterial species involved in HMs remediation are *Alcaligenes sp., Bacillus firmus, Bacillus licheniformis, Enterobacter sp., Escherichia coli, Micrococcus luteus, Pseudomonas sp., Salmonella typhi, Bacillus subtilis, Bacillus megaterium; Flavobacterium, Micrococcus sp.; Enterobacter sp., Arthrobactor sp., Azatobactor sp., Burkholderia sp., Streptomyces sp., Stenotrophomonas sp., Thiobacillus sp., Talaromyces sp., Rhodobacter sphaeroides, Desulfovibriode sulfuricans,* and some *sulphate-reducing bacteria* group as well as many actinomycetes, which has immense potential to remediate HMs from contaminated resources (Sreedevi et al. 2022; Al-Garni et al. 2010; Puyen et al. 2012; Jin et al.

Table 2.1 Hazardous metals and the specific microbes, their removal efficiency, and mechanism for hazardous metal remediation

Hazardous metal	Microbial species	% Removal	Bioremediation process	References
Lead	*Cellulosimicrobium sp.* KX710177	About 99% (Remediation)	Biosorption	Bharagava and Mishra 2018)
	Burkholderia sp. Z-90	24–52% (Bioleaching)	Bioleaching and Flocculation	Yan et al. 2018)
	P. aeruginosa	7–62% (Adsorption)	Adsorption	Elouzi et al. 2012)
	Pseudomonas sp.	NA	Biosorption	Huang et al. 2013)
	*Bacillus sp.*MSI 54	75–99%	Co-precipitation	Ravindran et al. 2020)
	Oceanobacillus profundus	54–97% (Metal sorption)	Biosorption	Abd El-Motaleb et al. 2020)
	Klebsiella sp. USl2D	97%	Biosorption	Orji et al. 2021)
	Streptomyces sp.	83%	Biosorption	Hamdan et al. 2021)
Cadmium	*Burkholderia* sp. Z-90	24–52% (Bioleaching)	Bioleaching and Flocculation	Yan et al. 2018)
	P. Aeruginosa	7–62% (Adsorption)	Adsorption	Elouzi et al. 2012)
	Pseudomonas sp. LkSO6	NA	Biosorption	Huang et al. 2013)
	Bacillus sp. MSI 54	75–99%	Co-precipitation	Ravindran et al. 2020)
	Bacillus sp. HIP3	Up to 13%	Metal chelation	Md Badrul Hisham et al. 2019)
	Citrobacter freeundii MG81234.1	40–80% Precipitation	Precipitation	Gomaa and El-Meihy 2019)
	Alcaligenes sp. MMA	Nearly 70% (Accumulation)	Bioaccumulation	Sodhi et al. 2020)
	Chlorella sp. (immobilized)	>92%	Biosorption	Shen et al. 2018)
	C. Sorokiniana	65% (metal accumulation)	Bioaccumulation	León-Vaz et al. 2021)

(continued)

Table 2.1 (continued)

Hazardous metal	Microbial species	% Removal	Bioremediation process	References
	Scenedesmus acutus, Chlorella vulgaris, Chlorella pyrenoidosa, Spirogyra hyaline, Fischerella sp., Lyngbya spiralis, Tolypothrix tenuis, Stigonema sp., Phormidium molle, Lyngbya heironymusii, Gloeocapsa sp., Oscillatoria jasorvensis, and *Nostoc sp.*	–	–	Priya et al. 2014)
Nickel	*Acinetobacter sp.* B9	69% removal	Biosorption	Bhattacharya and Gupta 2013)
	Enterobacter asburiae KUNi5	NA	Metal resistance	Paul and Mukherjee 2016)
	Bacillus sp. KL1	NA	Biosorption	Taran et al. 2019)
	Oscillatoria sp., Spirogyra sp., Anabaena variabilis, Aulosira sp., Nostoc muscorum, Oscillatoria sp., and *Westiellopsis sp.*	–	–	Priya et al. 2014)
Arsenic	*Burkholderia sp.* Z-90	24–52% (Bioleaching)	Bioleaching & Flocculation	Yan et al. 2018)
	Maugeotia genuflexa	57–90%	Biosorption	Leong and Chang 2020)
	Bacillus sp. KL1	NA	Biosorption	Taran et al. 2019)
	Bacillus macerans, Bacillus megaterium, and *Corynebacterium vitarumen*	NA	Bioaccumulation	Tyagi et al. 2018)
	Spirogyra hyaline	9.8%	Biosorption	Kumar and Oommen 2012)
	Maugeotiag enuflexa	57.48%	Biosorption	Sari et al. 2011)

2018; Wang et al. 2014; Igiri et al. 2018; Bhattacharya et al. 2018; Bhatt and Maheshwari 2020; Sher et al. 2020; Sharma and Adholeya 2012; Liu et al. 2020; Manoj et al. 2020; Zhang et al. 2020). Furthermore, In a recent study, a range of bacterial species also demonstrated remarkable proficiency in metal removal such as *Sporosarcina ginsengisoli* exhibited noteworthy efficacy in mitigating soil arsenic concentrations; *Klebsiella planticola, Bacillus cereus* strain XMCr-6, and *Bacillus cereus* were identified as potent agents for both soil and water Cd reduction; *Leclercia adecarboxylata* and *Kocuria flava* displayed remarkable capabilities for Pb contaminated soil; *Desulfovibrio desulfuricans* and *Bacillus licheniformis* were evident for soil and water Ni removal, and *Rhodopseudomonas palustris* stood out for its proficiency in Co from water and soil (Sayqal and Ahmed 2021).

In addition to indigenous microbial strains, some genetically engineered microbes are also developed for highly efficient removal of HMs, such as *Escherichia coli, Deinococcus radiodurans, Corynebacterium glutamicum, Mesorhizobium huakui,* and *Rhodopseudomonas palustris* were applied in the sites contaminated with HMs for better remediation (Diep et al. 2018).

2.4.1.2 Fungi

In microbial remediation, fungi also play an essential role in various HMs remediation capacities in soil, water, and sediments. Fungi can live under high heavy metal concentrations and accumulate both micronutrients and heavy metals. These fungi have been widely used to adsorb heavy metal ions and exhibit high metal uptake ability through chitin-chitosan complex, glucuronic acid, phosphate, polysaccharides in/on cells of fungi as well as the ionizable sites and various functional groups including amine, carboxyl, hydroxyl, phosphate and sulphhydryl groups (Yin et al. 2018). Some of the examples of these fungi involved in HMs remediation are *Rhizopus arrhizus* (for Cd contamination); *Aspergillus niger* (for Pb and Ni contamination in soil and sediments); *Lepiotahystrix, Aspergillus niger, Aspergillus terreus, Fusarium, Mucor, Penicillum, and Trichoderma longibrachiatum, Verticillium sp.,* and *Coprinopsis atramentaria* (for Cd and Pb) (Sreedevi et al. 2022; Tarekegn et al. 2020; Liu et al. 2020; Manoj et al. 2020; Dursun et al. 2003; Zeng et al. 2015; Kariuki et al. 2017).

2.4.1.3 Algae

Nowadays, algae have been reported to have a high absorbing capacity to hazardous metal ions and can effectively remove them from contaminated sources i.e., soil, sediments, and water, known as phycoremediation (Priya et al. 2014; Ahmed et al. 2021; Sun et al. 2021). In comparison to other microbes, there are several advantages associated with using Microalgae as metal remediators (1) They have a high growth rate, (2) they exhibit high metal binding capacities to heavy metals, (3) capacity to tolerate stressful conditions, (4) cost-effective, and (5) they are non-pathogenic, as they do not cause any diseases or illnesses in humans or animals and are environment friendly (Priya et al. 2014). Microalgae utilize waste materials as their nutritional sources and, consequently, enzymatically degrade or accumulate these hazardous metal pollutants on their surface through the presence of various peptides, polysaccharides, or lipids on their cell walls. Moreover, these cell walls contain various functional groups, viz. amino, hydroxyl, carboxyl, and sulphate, which serve as metal binding sites and facilitate HMs removal (Priya et al. 2014). For instance, an algae *Fucus vesiculosus* was found to be highly proficient in adsorbing the Lead from a contaminated source (Barquilha et al. 2017). Additionally, algae *Saccharina fusiforme and Saccharina japonica* were evident in improved elimination of Cd from an aqueous solution. The algal species, i.e., *Palmaria palmate, Spirulina maxima, Spirogyra hyaline, Cystoseira barbata, Cladophora sp., Chara aculeolata, Nitella opaca, Oscillatoria sp., Spirogyra sp., Anabaena variabilis, Aulosira sp., Nostoc muscorum, Oscillatoria sp., Westiellopsis sp, Chlorella pyrenoidosa, Spirogyra hyaline, Ulva lactuca,* and *Chlorella vulgaris* are some of the known algal species involved in HMs remediation, respectively (Tarekegn et al. 2020; Priya et al. 2014; Sooksawat et al. 2013; Ibrahim et al. 2018). As a result, these studies show that microalgae are highly effective at dealing with hazardous metal toxicity and bioremediation.

2.4.2 Microbial Mechanisms for Hazardous Metals Remediation

Microbes use various bioremediation pathways/processes to aid in the removal/conversion of HMs high toxic levels to low toxic level forms from soil, water, and sediments, viz. biosorption, biomineralization, bioreduction, bioleaching, biotransformation, bioaccumulation, bioleaching, precipitation, etc. as illustrated in Fig. 2.3.

Biosorption/bioaccumulation is the most widely used biotechnique to remove metal ions from effluent, among which pollutants could be adsorbed onto the walls of microbial cells (Podstawczyk et al. 2015; Li et al. 2017). In biosorption process, microorganisms interact with various bioresources acting as biosorbents. These

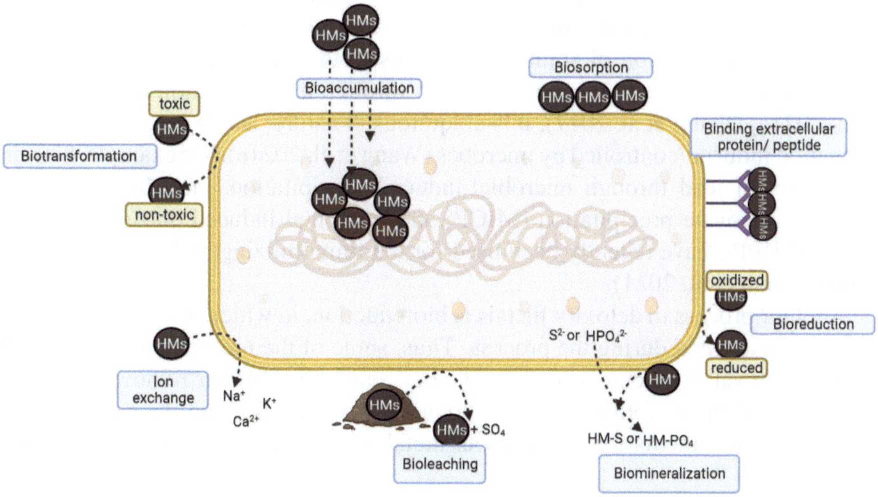

Fig. 2.3 Microbial mechanisms of removal/detoxification of hazardous metals (HMs)

biosorbents adsorb and gather soluble heavy metals from contaminated sources, particularly when the concentrations of these metals are low to moderate (Wang et al. 2021a). This mechanism encompasses the attachment of heavy metals (HMs) to a biological matrix, achieved through physical interactions like electrostatic forces and chemical interactions involving ion or proton displacement, complexation, or chelation. Conversely, bioaccumulation constitutes a dynamic metabolic process in which microorganisms actively internalize HMs within their intracellular compartments, facilitated by importer complexes through the lipid bilayer. Subsequently, proteins and peptide ligands play a role in sequestering these metals (Wang et al. 2021a, b).

On the other hand, an alternative process involves biotransformation, wherein hazardous metals undergo various reactions such as oxidation, reduction, methylation, and demethylation. These transformations lead to the conversion of these metals into less harmful forms. This intricate reaction can take place within cellular vacuoles, at the cell surface, and even in the extracellular milieu (Tarekegn et al. 2020).

Further, Bioleaching has recently been developed to recycle certain metals in treated leachate or residues by leaching them from solid wastes (Zeng et al. 2016; Dunbart 2017). The method described here is not only successful in extracting metals in their insoluble forms from secondary sources, but it also helps to reduce the volume of sludge that must be disposed of. Furthermore, it serves as a preventive measure against the unintentional discharge of metals into the surrounding environment (Wang et al. 2021a). Biological leaching methods find significant application within the realm of biohydrometallurgy, wherein microbiological processes are harnessed to facilitate the extraction of metals from both sulphide and oxide minerals. Currently, these mechanisms are utilized in leaching copper, gold, and uranium in industrial settings. Nonetheless, their potential extends beyond these examples, promising the removal of more hazardous metals from contaminated sources.

In addition, biomineralization refers to a phenomenon in which microorganisms facilitate the conversion of soluble metal ions in the water into either amorphous or crystalline deposits of mineral materials through a precipitation process (Wang et al. 2021a; Dhami et al. 2017). It is ubiquitous in nature, leading to mineral crystal formation induced/controlled by microbes (Wang et al. 2021a). For example, biomineralization of lead through microbial-induced precipitation, including microbial-induced carbonate precipitation (MICP) and microbial-induced phosphate precipitation (MIPP), have recently been used for the immobilizing and detoxification of Pb (II) (Shan et al. 2021).

Another process to detoxify metals is bioreduction, in which the valence of metal ions is transformed during the process. Thus, some of the mechanisms of microbes for the removal of arsenic, cadmium, nickel, and lead and their removal efficiency are discussed in the following table. Some of the microbes and their mechanisms involved in the remediation of hazardous metals are mentioned in Table 2.1 (Jayaram et al. 2022).

2.4.3 Factors Affecting Bioremediation of Hazardous Metals

Since the successful implementation of bioremediation at a particular site depends primarily on the physicochemical characterization and type of pollutant, it is also necessary to understand the nutrients, microbial constituents, and their interactions as the factors having a major role in influencing the bioremediation of HMs. This helps decide which technique should be suitably applied for successful decontamination. Among all the various factors, nutrients, including nitrogen, phosphate, sulphur, iron, and potassium, play a crucial role in contaminated settings by promoting and sustaining robust microbial growth, cellular metabolism, and microorganism proliferation. These nutrients are essential for sustaining life and facilitate the production of enzymes by microbes, which are crucial for the degradation of pollutants. Furthermore, the financial sustainability of bioremediation is contingent upon the remediation cost, emphasizing the significance of maintaining low costs (Sayqal and Ahmed 2021). Furthermore, the characteristics of pollutants, such as their physical state (solid, semisolid, liquid, or volatile), can have an impact on the process. These pollutants can vary in their toxicity, being either hazardous or non-toxic, and can be classified as organic or inorganic compounds viz. heavy metals, polycyclic aromatic hydrocarbons, pesticides, and chlorinated solvents, which could diminish their efficient remediation process (Sayqal and Ahmed 2021). Thus, the characterization of the contaminated region is also of the utmost importance, as it can exert an influence on the efficacy of bioremediation measures. Additionally, the consideration of pH, temperature, and various physicochemical variables also holds significance in the context of bioremediation. The careful determination of the optimal range for these parameters can significantly impact the rate and extent of bioremediation since it directly affects microbial growth and, consequently, the removal of pollutants (Sayqal and Ahmed 2021). Furthermore, moisture content, specifically water, is another crucial determinant for the effectiveness of biological growth and bioremediation processes. At last, the microbial variety capable of bioremediating many contaminants, including *Pseudomonas, Aeromonas, Acinetobacter, Mycobacterium, Flavobacteria,* etc., is of significant interest that could also positively influence the broad-spectrum removal efficiency of Hazardous metals from contaminated sites using microbial technologies (Sayqal and Ahmed 2021).

2.4.4 Challenges to Microbial Technologies and Future Aspects

Even though these remediation technologies are eco-friendly and cost-effective in nature, however, some limitations still persist with this technology, such as heavy metal toxicity to microorganisms, the influence of environmental factors on the bioremediation process, poor leaching rate, time-consuming, low metal removal efficiency rate, testing limited upto laboratory level (Wang et al. 2021a; Das and Dash 2014;

Verma et al. 2021; Sánchez-Castro et al. 2023). Out of all these factors, the major drawback in microbial technologies for HMs removal is the environmental factors, viz. temperature, pH, organic acids, and humic acids, which can alter the transformation, transportation, valence state of heavy metals and, consequently, their bioavailability for microorganisms in the contaminated sources (Tarekegn et al. 2020).

Furthermore, the products of bioremediation sometimes become more toxic than the original compound (Sayqal and Ahmed 2021). In fact, sometimes, even if the compounds are biodegradable, their downstream processing and degradation further generate toxic material, or even sometimes, a particular bacterial strain functional at one site may not successfully work on different sites (Vishwakarma et al. 2020). Indeed, there are some differences in the bioremediation technologies used for water, soil, and sediments. Unlike water, bioremediation in soil cannot be simulated in a laboratory to estimate the real field impact.

These limitations highlight the need for further studies to look for more biological solutions for the bioremediation of heavy metal contamination from different environmental systems. To counter these limitations, recent advances in genetic engineering and the acceptability of genetically engineered microorganisms/biocatalysts for environmental remediation have proven more efficient than natural microorganisms (Das and Dash 2014). In addition, hybrid/integrated biotechnologies, computational biology, and multi-omics approaches are also gaining attention for efficient HMs bioremediation (Diep et al. 2018).

2.5 Conclusions

On the basis of these preceding discussions, it could be stated that microbes definitely have the potential to increase the efficiency of HMs removal from soil, water, and sediments. However, more scientific research should be encouraged and promoted about the toxicity of heavy metals, their trophic level transfer, and their effect on the environment. This will help improve existing technologies by understanding the physical, chemical, and biological behaviour of microorganisms in the presence of HMs in soil and water environments. Overall, this chapter elucidates that there is still ample opportunity for developing this sector through fostering higher HMs tolerance in microbes and evaluating their feasibility as well as their application of on-site treatment.

Acknowledgements All the authors would like to express their gratitude towards DRDO-Defence Institute of High Altitude Research (DIHAR) and Director, DIHAR, for providing support and facilities to complete this work. The authors would also like to acknowledge Biorender.com for helping to create the diagrams.

References

Mazumder P, Haq, I, Das, A, Kalamdhad A (2021) Microbial remediation of soil and water metal contaminants. In: Shah M, Rodriguez-Couto S (eds), vol 2021, Elsevier, pp 523–536. Doi: https://doi.org/10.1016/B978-0-12-822503-5.00006-0

Giri A, Bharti VK, Kalia S, Kumar B, Chaurasia OP (2021) Health-risk assessment of heavy metals through cow milk consumption in trans-Himalayan high altitude region. Biol Trace Elem Res 199:4572–4581

Charan G, Bharti VK, Giri A, Kumar P (2023) Evaluation of physico-chemical and heavy metals status in irrigation, stagnant, and Indus River water at the trans-Himalayan region. Discover Water 3:3. https://doi.org/10.1007/s43832-023-00027-z

Yu Z, Tang J (2022) Quality improvement in vegetable greenhouse by cadmium pollution remediation. J Food Qual 2022:8335753

Eqani S, Khalid R, Bostan N, Saqib Z, Mohmand J, Rehan M, Ali N, Katsoyiannis I, Shen H (2016) Human lead (Pb) exposure via dust from different land use settings of Pakistan: a case study from two urban mountainous cities. Chemosphere 155:259–265

Mitra S, Chakraborty AJ, Tareq AM, Emran TB, Nainu F, Khusro A, Simal-Gandara J (2022) Impact of heavy metals on the environment and human health: Novel therapeutic insights to counter the toxicity. J King Saud Univ Sci 34:101865

Sarker A, Al Masud MA, Deepo DM, Das K, Nandi R, Ansary MWR, Islam T (2023) Biological and green remediation of heavy metal contaminated water and soils: a state-of-the-art review. Chemosphere 332:138861

Li S, Xu S, Niu Q et al (2016) Lutein alleviates arsenic-induced reproductive toxicity in male mice via Nrf2 signaling. Hum Exp Toxicol 35:491–500. https://doi.org/10.1177/0960327115595682

Hayyat MS, Adnan M, Awais M, Bilal HM, Khan B, Rahman HA (2020) Effect of heavy metal (Ni) on plants and soil: a review. Int J Appl Res 6:313–318

Garai P, Banerjee P, Mondal P, Saha NC (2021) Effect of heavy metals on fishes: toxicity and bioaccumulation. J Clin Toxicol S18:1–10

Genchi G, Sinicropi MS, Lauria G, Carocci A, Catalano A (2020) The effects of cadmium toxicity. Int J Env Res Public Health 17:3782. https://doi.org/10.3390/ijerph17113782

Shah MP (Ed) (2020) Microbial bioremediation & biodegradation. Springer, pp IX-550. https://doi.org/10.1007/978-981-15-1812-6

Abo-Alkasem MI, Hassan NH, Abo Elsoud MM (2023) Microbial bioremediation as a tool for the removal of heavy metals. Bull Natl Res Cent 47:1–15. https://doi.org/10.1186/s42269-023-01006-z

Sayqal A, Ahmed OB (2021) Advances in heavy metal bioremediation: an overview. Appl Bionics Biomech 2021:1609149

Djingova R, Kuleff I (2000) Instrumental techniques for trace analysis. In: Vernet JP (ed) Vol 4. Elsevier, London, pp 137–185

Blaylock MJ, Huang JW (2000) Phytoextraction of metals in phytoremediation of toxic metals: using plants to clean up the environment. In: Raskin I, Ensley BD (eds) 2000. Wiley, NY, USA, pp 53–70

Assche F, Clijsters H (1990) Effects of metals on enzyme activity in plants. Plant Cell Environ 24:1–15

Taiz L, Zeiger E (2002) Photosynthesis: physiological and ecological considerations. Plant Physiol 9:172–174

Mohanpuria P, Rana NK, Yadav SK (2007) Cadmium induced oxidative stress influence on glutathione metabolic genes of Camella sinensis (L.). O Kuntze. Environ Toxicol 22:368–374

Guo J, Dai X, Xu W, Ma M (2008) Over-expressing GSHI and AsPCSI simultaneously increases the tolerance and accumulation of cadmium and arsenic in Arabidopsis thaliana. Chemosphere 72:1020–1026

Asati A, Pichhode M, Nikhil K (2016) Effect of heavy metals on plants: an overview. Int J Appl Inn Engg Manage 5:56–66

Pandey N, Sharma CP (2002) Effect of heavy metals Co+2, Ni+2, and Cd+2 on growth and metabolism of cabbage. Plant Sci 163:753–758

Gajewska E, Sklodowska M, Slaba M, Mazur J (2006) Effect of nickel on antioxidative enzymes activities, proline and chlorophyll contents in wheat shoots. Biol Planta 50:653–659

Raikwar MK, Kumar P, Singh M, Singh A (2008) Toxic effect of heavy metals in livestock health. Vet World 1:28

Pandey G, Madhuri S (2014) Heavy metals causing toxicity in animals and fishes. Res J Anim Vet Fishery Sci 2:17–23

Shahjahan M, Taslima K, Rahman MS, Al-Emran M, Alam SI, Faggio C (2022) Effects of heavy metals on fish physiology-a review. Chemosphere 300:134519. https://doi.org/10.1016/j.chemosphere.2022.134519

Afrin R, Mia MY, Ahsan MA, Akbor A (2015) Concentration of heavy metals in available fish species (bain, Mastacembelus armatus; taki, Channa punctatus and bele, Glossogobius giuris) in the Turag river, Bangladesh. Pak J Sci Ind Res Series B: Biological Sci 58:104–110

Mahmood Q, Wang J, Pervez A, Meryem SS, Waseem M, Ullah Z (2015) Health risk assessment and oxidative stress in workers exposed to welding fumes. Toxicol Environ Chem 97:634–639

Lu H, Shi X, Costa M et al (2005) Carcinogenic effect of nickel compounds. Mol Cell Biochem 279:45–67. https://doi.org/10.1007/s11010-005-8215-2

Wilk A, Kalisińska E, Danuta IK, Maciej R, Jacek R, Kazimierz C, Marcin S, Natalia ŁWA, Kalisinska E, Kosik-Bogacka DI (2017) Cadmium, lead and mercury concentrations in pathologically altered human kidneys. Env Geochem Health 39:889–899

Ercal N, Gurer-Orhan H, Aykin-Burns N (2001) Toxic metals and oxidative stress Part I: mechanisms involved in metal-induced oxidative damage. Curr Topics Med Chem 1:529–539

Ozturk M, Metin M, Altay V, Bhat RA, Ejaz M, Gul A, Kawano T (2021) Arsenic and human health: genotoxicity, epigenomic effects, and cancer signaling. Biol Trace Elem Res 200:988–1001. https://doi.org/10.1007/s12011-021-02719-w

Gunatilake SK (2015) Methods of removing heavy metals from industrial wastewater. Methods 1:14

Kumar M, Seth A, Singh AK, Rajput MS, Sikandar M (2021) Remediation strategies for heavy metals contaminated ecosystem: a review. Environ Sustain Indic 12:100155

Razzak SA, Faruque MO, Alsheikh Z, Alsheikhmohamad L, Alkuroud D, Alfayez A, Hossain MM (2022) A comprehensive review on conventional and biological-driven heavy metals removal from industrial wastewater. Environ Adv 7:100168

Jeyakumar P, Debnath C, Vijayaraghavan R, Muthuraj M (2023) Trends in bioremediation of heavy metal contaminations. Environ Eng Res 28:220631

Riseh RS, Vazvani MG, Hajabdollahi N, Thakur VK (2022) Bioremediation of heavy metals by Rhizobacteria. Appl Biochem Biotechnol 195:4689–4711

Sreedevi PR, Suresh K, Jiang G (2022) Bacterial bioremediation of heavy metals in wastewater: a review of processes and applications. J Water Proc Engg 48:102884

Tarekegn MM, Salilih FZ, Ishetu AI, Yildiz F (2020) Microbes used as a tool for bioremediation of heavy metal from the environment. Microbial technologies for heavy metal remediation: efect of process conditions and current. Food Sci Technol 6:1783174

Vidali M (2001) Bioremediation. an overview. Pure Appl Chem 73:1163–1172

Chipasa K, Mędrzycka K (2006) Behaviour of lipids in biological wastewater treatment processes. J Inds Microbiol Biotechnol 33:635–645

Stroo HF, Leeson A, Ward CH (2013) Bioaugmentation for groundwater remediation. In: Stroo H, Leeson A, Ward C (eds) 2013. Springer, New York, pp 117–140

Sivakumar D, Kandaswamy A, Gomathi V, Rajeshwaran R, Murugan N (2014) Bioremediation studies on reduction of heavy metals toxicity. Pollution Res 33:553–558

Yin K, Wang Q, Lv M, Chen L (2018) Microorganism remediation strategies towards heavy metals. Chem Engg J 360:1553–1563. https://doi.org/10.1016/j.cej.2018.10.226

Hussein H, Ibrahim SF, Kandeel K, Moawad H (2004) Biosorption of heavy metals from waste water using Pseudomonas sp. Elect J Biotechnol 7:30–37

Qamar N, Rehman Y, Hasnain S (2017) Arsenic resistant and plant growth promoting Firmicutes and γ-Proteobacteria species from industrially polluted irrigation water and corresponding cropland. J Appl Microbiol 123:748–758. https://doi.org/10.1111/jam.13535

Kanamarlapudi SLRK, Chintalpudi VK, Muddada S (2018) Application of biosorption for removal of heavy metals from wastewater. Biosorption 18:70–116

Redha AA (2020) Removal of heavy metals from aqueous media by biosorption. Arab J Basic Appl Sci 27:183–193

Al-Garni SM, Ghanem KM, Ibrahim AS (2010) Biosorption of mercury by capsulated and slime layer forming Gram-ve bacilli from an aqueous solution. Afr J Biotechnol 9:6413–6421

Puyen ZM, Villagrasa E, Maldonado J, Diestra E, Esteve I, Sole A (2012) Biosorption of lead and copper by heavy-metal tolerant Micrococcus luteus DE2008. Bioresour Technol 126:233–237

Jin Y, Luan Y, Ning Y, Wang L (2018) Effects and mechanisms of microbial remediation of heavy metals in soil: a critical review. Appl Sci 8:1336

Wang T, Sun H, Mao H, Zhang Y, Wang C, Zhang Z, Wang B, Sun L (2014) The immobilization of heavy metals in soil by bioaugmentation of a UV-mutant Bacillus subtilis 38 assisted by NovoGro biostimulation and changes of soil microbial community. J Hazard Mater 278:483–490

Igiri BE, Okoduwa SIR, Idoko GO, Akabuogu EP, Adeyi AO, Ejiogu IK (2018) Toxicity and bioremediation of heavy metals contaminated ecosystem from tannery wastewater: a review. J Toxicol 2018:2568038. https://doi.org/10.1155/2018/2568038

Bhattacharya A, Naik SN, Khare SK (2018) Harnessing the biomineralization ability of urease producing Serratia marcescens and Enterobacter cloacae EMB19 for remediation of heavy metal cadmium (II). J Environ Manage 215:143–152

Bhatt K, Maheshwari DK (2020) Insights into zinc-sensing metalloregulator “Zur” deciphering mechanism of zinc transportation in Bacillus spp. by modeling, simulation and molecular docking. J Biomol Struct Dyn 14:1–16

Sher S, Hussain SZ, Rehman A (2020) Phenotypic and genomic analysis of multiple heavy metal resistant Micrococcus luteus strain AS2 isolated from industrial waste water and its potential use in arsenic bioremediation. Appl Microbiol Biotechnol 104:2243–2254

Sharma S, Adholeya A (2012) Hexavalent chromium reduction in tannery efuent by bacterial species isolated from tannery effluent contaminated soil. J Environ Sci Technol 5:142–154

Liu S, Yang B, Liang Y, Xiao Y, Fang J (2020) Prospect of phytoremediation combined with other approaches for remediation of heavy metal-polluted soils. Environ Sci Pollution Res 27:16069–16085

Manoj SR, Karthik C, Kadirvelu K, Arulselvi PI, Shanmugasundaram T, Bruno B, Rajkumar M (2020) Understanding the molecular mechanisms for the enhanced phytoremediation of heavy metals through plant growth promoting rhizobacteria: a review. J Environ Manag 254:109779

Zhang Z, Si R, Lv J, Ji Y, Chen W, Guan W, Cui Y, Zhang T (2020) Effects of extracellular polymeric substances on the formation and methylation of mercury sulfide nanoparticles. Environ Sci Technol 54(13):8061–8071

Diep P, Mahadevan R, Yakunin AF (2018) Heavy metal removal by bioaccumulation using genetically engineered microorganisms. Front Bioeng Biotechnol 6:157

Dursun AY, Uslu G, Cuci Y, Aksu Z (2003) Bioaccumulation of copper (II), lead (II) and chromium (VI) by growing Aspergillus niger. Process Biochem 38:1647–1651

Zeng X, Wei S, Sun L, Jacques DA, Tang J, Lian M, Xu Z (2015) Bioleaching of heavy metals from contaminated sediments by the Aspergillus niger strain SY1. J Soils Sediments 15:1029–1038

Kariuki Z, Kiptoo J, Onyancha D (2017) Biosorption studies of lead and copper using rogers mushroom biomass Lepiota hystrix. S Afr J Chem Eng 23:62–70

Priya M, Gurung N, Mukherjee K, Bose S (2014) Microalgae in removal of heavy metal and organic pollutants from soil. Microbial Biodeg Biorem 2014:519–537. https://doi.org/10.1016/b978-0-12-800021-2.00023-6

Ahmed DAEA, Gheda SF, Ismail GA (2021) Efficacy of two seaweeds dry mass in bioremediation of heavy metal polluted soil and growth of radish (Raphanus sativus L.) plant. Environ Sci Pollut Res 28:12831–12846. https://doi.org/10.1007/s11356-020-11289-8

Sun W, Cheng K, Sun KY, Ma X (2021) Microbially mediated remediation of contaminated sediments by heavy metals: a critical review. Curr Poll Rep 7:201–212. https://doi.org/10.1007/s40726-021-00175-7

Barquilha CER, Cossich ES, Tavares CRG, Silva EA (2017) Biosorption of nickel (II) and copper(II) ions in batch and fixed-bed columns by free and immobilized marine algae Sargassum sp. J Cleaner Prod 150:58–64

Sooksawat N, Meetam M, Kruatrachue M, Pokethitiyook P, Nathalang K (2013) Phytoremediation potential of charophytes: bioaccumulation and toxicity studies of cadmium, lead and zinc. J Environ Sci 25:596–604

Ibrahim WM, Abdel Aziz YS, Hamdy SM, Gad NS (2018) Comparative study for biosorption of heavy metals from synthetic wastewater by different types of marine algae. J Bioremed Biodegrad 9:1–425

Podstawczyk D, Witek-Krowiak A, Dawiec A, Bhatnagar A (2015) Biosorption of copper (II) ions by flax meal: empirical modelling and process optimization by response surface methodology (RSM) and artificial neural network (ANN) simulation. Ecol Eng 83:364–379

Li D, Xu X, Yu H, Han X (2017) Characterization of pb2+ biosorption by psychrotrophic strain Pseudomonas sp. I3 isolated from permafrost soil of Mohe wetland in Northeast China. J Environ Manag 196:8–15

Wang S, Liu T, Xiao X et al (2021a) Advances in microbial remediation for heavy metal treatment: a mini review. J Leather Sci Eng 3:1–10. https://doi.org/10.1186/s42825-020-00042-z

Wang Y, Selvamani V, Yoo IK, Kim TW, Hong SH (2021b) A novel strategy for the microbial removal of heavy metals: cell-surface display of peptides. Biotechnol Bioprocess Eng 26:1–9

Zeng J, Gou M, Tang Y, Li G, Sun Z, Kida K (2016) Effective bioleaching of chromium in tannery sludge with an enriched sulfur-oxidizing bacterial community. Bioresour Technol 218:859e866

Dunbart WS (2017) Biotechnology and the mine of tomorrow. Trends Biotechnol 35:79e89

Dhami NK, Quirin ME, Mukherjee A (2017) Carbonate biomineralization and heavy metal remediation by calcifying fungi isolated from karstic caves. Ecol Eng 103:106–117

Shan B, Hao R, Xu H, Li J, Li Y, Xu X, Zhang J (2021) A review on mechanism of biomineralization using microbial-induced precipitation for immobilizing lead ions. Environ Sci Poll Res 28:30486–30498. https://doi.org/10.1007/s11356-021-14045-8

Jayaram S, Ayyasamy PM, Aiswarya KP, Devi MP, Rajakumar S (2022) Mechanism of microbial detoxification of heavy metals: a review. J Pure Appl Microbiol 16:1562–1574

Ahemad M, Kibret M (2013) Recent trends in microbial biosorption of heavy metals: a review. Biochem Mol Biol 1:19–26

Jing R, Kjellerup BV (2018) Biogeochemical cycling of metals impacting by microbial mobilization and immobilization. J Environ Sci 66:146–154

Bharagava RN, Mishra S (2018) Hexavalent chromium reduction potential of Cellulosimicrobium sp. isolated from common effluent treatment plant of tannery industries. Ecotoxicol Environ Safety 147:102–109

Yan X, Liu M, Zhong J, Guo J, Wu W (2018) How human activities affect heavy metal contamination of soil and sediment in a long-term reclaimed area of the Liaohe River Delta, North China. Sustainability 10:338

Elouzi AA, Akasha AA, Elgerbi AM, El-Baseir M, El Gammudi BA (2012) Removal of heavy metals contamination by bio-Surfactants (Rhamnolipids). J Chem Pharma Res 4:4337–4341

Huang L, Pu X, Pan JF, Wang B (2013) Heavy metal pollution status in surface sediments of Swan Lake lagoon and Rongcheng Bay in the northern Yellow Sea. Chemosphere 93:1957–1964

Ravindran A, Sajayan A, Priyadharshini GB, Selvin J, Kiran GS (2020) Revealing the efficacy of thermostable biosurfactant in heavy metal bioremediation and surface treatment in vegetables. Front Microbiol 11:493383

Abd El-Motaleb M, El-Sabbagh S, Mohamed W, Wafy K (2020) Biosorption of Cu2+, Pb2+ and Cd2+ from wastewater by dead biomass of Streptomyces cyaneus Kw42. Int J Curr Microbiol Appl Sci 9:422–435

Orji OU, Awoke JN, Aja PM, Aloke C, Obasi OD, Alum EU, Oka GO (2021) Halotolerant and metalotolerant bacteria strains with heavy metals biorestoration possibilities isolated from Uburu Salt Lake, Southeastern. Nigeria. Heliyon 7:e07512

Hamdan AM, Abd-El-Mageed H, Ghanem N (2021) Biological treatment of hazardous heavy metals by Streptomyces rochei ANH for sustainable water management in agriculture. Sci Rep 11:9314

Md Badrul Hisham NH, Ibrahim MF, Ramli N and Abd-Aziz S (2019) Production of biosurfactant produced from used cooking oil by Bacillus sp. HIP3 for heavy metals removal. Mol 24: 2617

Gomaa EZ, El-Meihy RM (2019) Bacterial biosurfactant from Citrobacter freundii MG812314. 1 as a bioremoval tool of heavy metals from wastewater. Bull Natl Res Cent 43:1–14

Sodhi KK, Kumar M, Singh DK (2020) Multi-metal resistance and potential of Alcaligenes sp. MMA for the removal of heavy metals. SN Appl Sci 2:1–13

Shen Y, Zhu W, Li H, Ho SH, Chen J, Xie Y, Shi X (2018) Enhancing cadmium bioremediation by a complex of water-hyacinth derived pellets immobilized with Chlorella sp. Biores Technol 257:157–163

León-Vaz A, León R, Giráldez I, Vega JM, Vigara J (2021) Impact of heavy metals in the microalga Chlorella sorokiniana and assessment of its potential use in cadmium bioremediation. Aquatic Toxicol 239:105941

Bhattacharya A, Gupta A (2013) Evaluation of Acinetobacter sp. B9 for Cr (VI) resistance and detoxification with potential application in bioremediation of heavy-metals-rich industrial wastewater. Environ Sci Poll Res 20:6628–6637

Paul A, Mukherjee SK (2016) Enterobacter asburiae KUNi5, a nickel resistant bacterium for possible bioremediation of nickel contaminated sites. Pol J Microbiol 65:115–118

Taran M, Fateh R, Rezaei S, Gholi MK (2019) Isolation of arsenic accumulating bacteria from garbage leachates for possible application in bioremediation. Iran J Microbiol 11:60–66

Leong YK, Chang JS (2020) Bioremediation of heavy metals using microalgae: recent advances and mechanisms. Biores Technol 303:122886

Tyagi D, Tyagi S, Narayan RK (2018) Bacterial bioremediation of arsenic from contaminated groundwater and soil. Eur J Biotechnol Biosci 6:45–51

Kumar N, Oommen C (2012) Removal of heavy metals by biosorption using freshwater alga Spirogyra hyalina. J Environ Biol 33:27

Sari A, Uluozlü D and Tüzen M (2011) Equilibrium, thermodynamic and kinetic investigations on biosorption of arsenic from aqueous solution by algae (Maugeotia genuflexa)

Das S, Dash HR (2014) Microbial bioremediation. Microbial Biodeg Biorem 1–21. https://doi.org/10.1016/b978-0-12-800021-2.00001-7

Verma S, Bhatt P, Verma A, Mudila H, Prasher P and Rene ER (2021) Microbial technologies for heavy metal remediation: effect of process conditions and current practices. Clean Technol Environ Policy 1–23

Sánchez-Castro I, Molina L, Prieto-Fernández MÁ and Segura A (2023) Past, present and future trends in the remediation of heavy-metal contaminated soil-Remediation techniques applied in real soil-contamination events. Heliyon

Vishwakarma GS, Bhattacharjee G, Gohil N, Singh V (2020) Current status, challenges and future of bioremediation. Bioremed Poll 403–415. https://doi.org/10.1016/b978-0-12-819025-8.00020-x

Chapter 3
Microbial Remediation Technologies for Mining Waste Management

Pradeep W. Samarasekere

3.1 Introduction

Minerals are an essential part of human civilization, and mineral exploration, mining, and extraction are pivotal for providing valuable resources for human activities. Mining plays an important part in industrial and technical advancements by providing raw materials for industry. Almost all industrial sectors rely on mined minerals, which serve as the foundations of industrial operations, technological advancements, and so on. Coal, iron, copper, tin, and lead are common industrial minerals, whereas gold, platinum, silver, titanium, and other valuable metals are considered precious metals. Rare earth elements, a category of key elements required for high-tech and green-energy applications, are also extracted through mines. As economies grow and demand for minerals increases, mining activities become an increasingly important aspect of supporting global economies (Kesler and Simon 2015; Deshmukh 2010). However, mining operations, despite their economic benefits, have a significant environmental cost as well as threats to health of humans and other living organisms. Mining operations causes numerous issues, ranging from pit mining, which destroys habitats, to final mineral extraction stages, which generate large amounts of chemical waste. Other major challenges include deforestation, ecosystem destruction, endangering plant and animal species, altering landscapes, soil erosion, affecting agricultural productivity, and water pollution. Among the many pollutants and toxins created by mining, dust, and emissions from mining activities, particularly in open-pit mining and smelting processes are considered extremely dangerous to the health of living beings. These air pollutants contain hazardous particles and gases such as sulfur dioxide and nitrogen oxides, and prolonged exposure to these pollutants can cause respiratory problems as well as other severe health issues in the surrounding

P. W. Samarasekere (✉)
Center for Advanced Materials and Smart Manufacturing, Faculty of Computing and Technology, University of Kelaniya, Kelaniya, Sri Lanka
e-mail: pradeeps@kln.ac.lk

K. Bala et al. (eds.), *Harnessing Microbial Potential for Multifarious Applications*, Energy, Environment, and Sustainability, https://doi.org/10.1007/978-981-97-1152-9_3

population. Apart, mining regularly contaminates local water sources with heavy metals, chemicals including acid and alkaline, and sediments, posing a serious threat to both aquatic life and endangering nearby communities that rely on water bodies for sustenance (Punia and Siddaiah 2017; Pitchaiah 2017; Hilson 2002; Ghose 2004).

This chapter primarily focuses on the challenges associated with mine waste management and the opportunity to address the issues using microbiological approaches. Mining waste includes a wide range of waste products generated during extraction and processing, the majority of which are hazardous solids, sludge, slags, gases, and mining tailing. Mine tailings are the materials left over after valuable minerals have been extracted, and they frequently contain dangerous elements and chemicals, whereas slag is a byproduct of smelting ore to purify metals, and it can contain heavy metals and toxins. Waste rock, the non-value part of the ore mined alongside valuable minerals, can also be hazardous to the environment. The large amount of waste generated is staggering, resulting in considerable environmental degradation and contamination, thus, appropriate waste management is of the utmost importance for mitigating the negative consequences caused by mining industrial activities (Fig. 3.1).

The major part of mineral processing and extraction is based on chemical processing, often resulting in a significant amount of mining waste containing large quantities of harsh chemicals, including acid waste. Additionally, the amounts of heavy metal concentrates also have a significant impact on the environment, making the mineral industry an environmentally destructive industry. Acid wastes, including sulfuric, nitric, and halogenic acids, are common environmental destructors, while

Fig. 3.1 Environments impacted by mine waste and examples of the mining sites **a** Shamokin Creek, Pennsylvania polluted by high levels of metals from abandoned mine drainage. **b** Acid mine drainage at the Davis Mine, Massachusetts. **c** Acid mine drainage in Portugal **d** Kingston Fossil Plant fly ash spill, Tennessee **e** Kajaran copper-molybdenum open-pit mine in southern Armenia **f** Mining worker in the sulfur mine at the edge of Ijen's Crater Lake, Indonesia. Image source (**a**–**f**): Wikimedia (https://commons.wikimedia.org/)

heavy metals such as lead, mercury, cadmium, uranium, etc., are often toxic pollutants released into the environment in heavy concentrations. These chemicals are usually considered toxic for living organisms and can cause significant contamination of water and soil bodies. Even though such chemical releases are unavoidable with current mining activities, researchers are often investigating remediation methods to limit the adverse impact these chemicals are causing to the environment (Mudd 2010; Kuyucak and Akcil 2013; Johnson and Hallberg 2005; Agboola et al. 2020; Aznar-Sánchez et al. 2018; Acharya and Kharel 2020; Moore and Luoma 1990; Rebello et al. 2021; Rodríguez-Galán et al. 2019; Sengupta 2021).

Many challenges faced by the mining industry require actions at the international and national regulatory levels to oversee mining waste management and process innovation and technologies to the challenges. Regulations aim to manage the disposal and treatment of mining waste and the innovative process remediation technologies for minimizing environmental damages. These regulations have evolved over time, becoming increasingly stringent and comprehensive. However, their effectiveness in reducing environmental consequences remains a source of concern. Striking a balance between economic interests and environmental conservation remains a challenging task that requires ongoing examination and adaptation of these regulatory frameworks and innovative technologies to protect our world for future generations. Among the many methods researched for the remediation of heavy metal and acid mine drainage-contaminated mining sites, microbial remediation methods are often considered a powerful and environmentally friendly approach to treating mining waste. Given the diversity of microbes in the environment, there are often many options that can be found to treat the diverse amount of chemical contaminants found in mining waste sites. Due to their ability to adapt to harsh environments and their extraordinary metabolic abilities, microorganisms can be effectively utilized in processing mining wastes. The different abilities and behaviors of microorganisms provide a variety of interactions with different contaminants (Bernoth et al. 2000; Bosecker 1999; Das and Ghosh 2022; Godoy-Faúndez et al. 2015; Gupta and Joia 2016).

1. Bioremediation: Bioremediation is a common process that utilizes the abilities of microorganisms to disintegrate or neutralize pollutants. Extensive research has been conducted on bacteria that can transform poisonous forms of metals such as lead, cadmium, and arsenic into less dangerous metallic forms. They are also effectively utilized for the breakdown of organic contaminants like hydrocarbons through enzymatic routes (Chakraborty et al. 2012; Anekwe and Isa 2023; Sánchez-Andrea et al. 2012).
2. Bioaccumulation or Biosorption: Bioaccumulation or biosorption is a process where certain microbes accumulate or bind heavy metals onto their surfaces. Through this process, contaminants can be concentrated in biomasses, facilitating the easy removal of contaminants from the sites (Das 2010; Park et al. 2010; Ortiz-Castillo et al. 2021).
3. Microorganism-Catalyzed Chemical Reactions: Some microorganisms have the ability to catalyze chemical reactions that convert toxic or harmful contaminants

into less toxic or harmful forms of chemicals. For instance, certain bacteria can oxidize sulfides found in mining tailings, thereby lowering the discharge of acidic substances responsible for acid mine drainage (AMD) (Rawlings and Johnson 2007; Sand et al. 2001; Coker 2016).

It is fascinating to see the diversity of microorganisms found in the harsh conditions of mine sites. Consequently, it provides many opportunities and approaches for utilizing microorganisms in mining waste remediation. Extreme conditions such as high metal concentrations, low pH levels, and a lack of nutrients are common characteristics of mining tailings. This makes the mining environment uninhabitable for most life forms. However, many microbial species, including a wide range of bacteria, fungi, and archaea, have been identified to have evolved and survived in these harsh conditions.

1. Extremophiles: Many microorganisms in mining waste can endure and succeed in harsh environments, commonly known as extremophiles. Acidophilic bacteria, for example, can withstand very low pH levels, a characteristic of mining tailings affected by acid mine drainage (AMD) (Valdes et al. 2009; Jiang et al. 2007).
2. Microorganisms that are resistant to metals: Some bacteria and fungi living in mining tailings have evolved defenses against and detoxification systems for the high quantities of heavy metals present in the tailings. Bioremediation efforts heavily depend on these metal-resistant microorganisms, as heavy-metal contamination is a common issue that needs to be addressed to prevent environmental pollution caused by metal mining waste (Coelho et al. 2020; Mendez et al. 2008).
3. Synergistic connections: It is fascinating to observe the synergistic connections that microbial communities create, wherein one species contributes products or by-products, or waste products that help another species grow and function. If used strategically, these synergistic interactions can significantly improve the overall effectiveness of mining waste cleanup (Tang et al. 2021; Gadd 2010) (Fig. 3.2).

3.1.1 Role of Bacteria, Archaea, and Fungi in Metal Sequestration

Microorganisms such as bacteria, fungi, and archaea have a remarkable ability to interact with heavy metals, with some being metal resistant and others having the ability to sequester metals while interacting with metal-rich environments. Genera like *Pseudomonas*, *Cupriavidus*, and *Bacillus* are capable of metal sequestering. Archaea, such as extremophiles, demonstrate resistance to high metal concentrations, making them valuable bioremediation candidates in extreme environments. Fungi such as *Aspergillus*, *Trichoderma*, and *Penicillium* are often researched as candidates for potential heavy metal treatments, as they can accumulate heavy metals within their mycelia, allowing mycoremediation of heavy metal contaminations (Dopson and Lindström 1999; Nies 1999; Bosecker 1997).

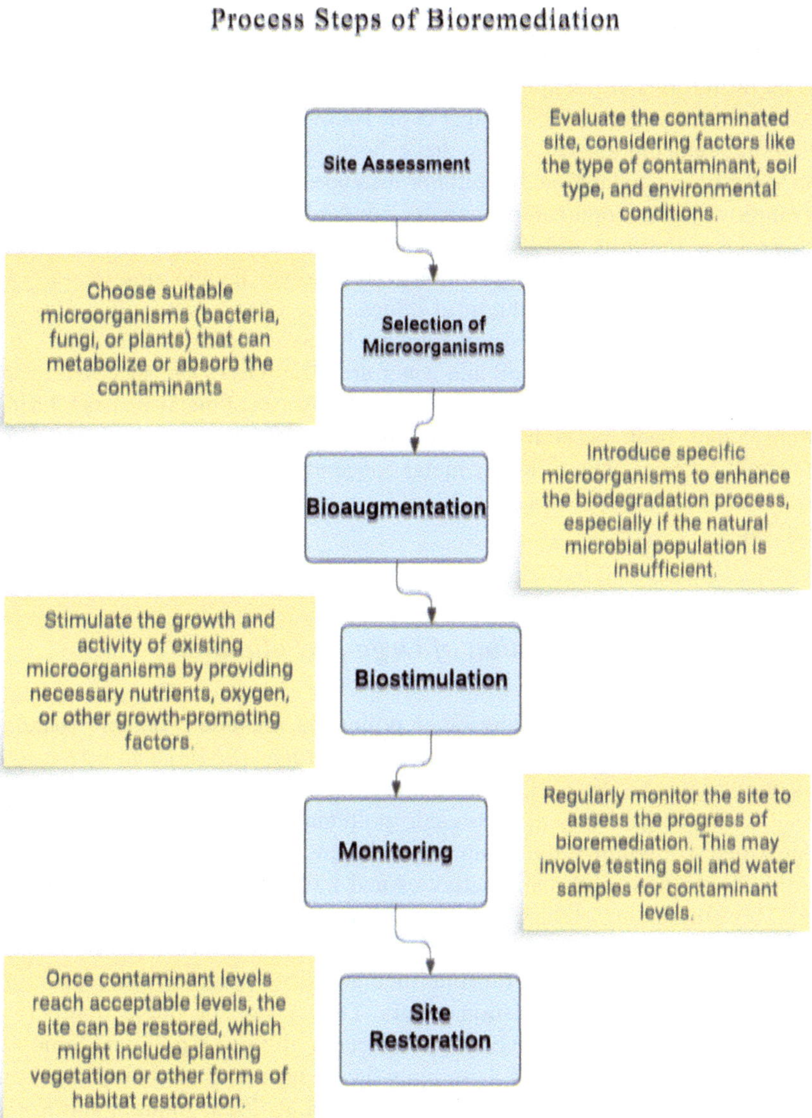

Fig. 3.2 Simplified infographic illustrating the comprehensive steps involved in bioremediation, a sustainable environmental solution for cleaning up mine waste-contaminated sites

Microbial metal resistance and reduction mechanisms are often complex and diverse, involving both passive and active mechanisms. Active processes involve the efflux of metal ions through specialized transporters or molecular pumps to reduce intracellular metal concentrations. Passive mechanisms often involve extracellular metal binding through cell wall structures such as lipopolysaccharides and exopolysaccharides, inhibiting the entry of metals into the cell. Additionally, some microorganisms have the capability to reduce metal cations and convert toxic forms to less harmful forms or precipitate them into easily removable forms such as metal sulfates, oxides, and carbonates (Singer et al. 2000; Barkay et al. 2003).

Microbial remediation techniques have been successfully employed in many scenarios for heavy metal remediation. For instance, bioremediation has been effectively employed in the removal of mercury-contaminated sites, where bacteria like *Pseudomonas aeruginosa* and *Bacillus cereus* can be used for the treatment process. The process for treating mercury with bacteria involves methylating mercury into less toxic forms, mitigating its harmful effects. The use of *Aspergillus niger* and *Trichoderma viride* to treat environmental toxin metals such as chromium, lead, and cadmium in contaminated soil and water are examples of the successful use of bioremediation for heavy-metal contamination. Consortia of metal-resistant bacteria and archaea have been researched for the remediation of abandoned mining sites to reduce significant amounts of heavy metal concentrates in these mines and restore the ecosystem in the areas (Das et al. 2016; Kapahi and Sachdeva 2019; Amobonye et al. 2023).

3.1.2 Microbial Degradation of Organic Pollutants

In addition to heavy metals, many other pollutants, including organic pollutants such as hydrocarbons, polycyclic aromatic hydrocarbons (PAHs), and various other organic toxins, may be present in mining waste. By utilizing the metabolic capacities of specific microorganisms, these organic pollutants can be effectively degraded and detoxified to minimize their impact on the environment. Bacterial and fungal consortia, due to their enzymatic activities and metabolic routes, demonstrate the potential for eliminating organic pollutants. Bacteria frequently dominate the early phases of degradation by breaking down complex organic structures into simpler intermediate structures. The resulting intermediates serve as carbon sources for fungi, where the completion of degradation occurs. This synergistic interaction between bacteria and fungi leads to the overall effectiveness of organic pollutant removal in contaminated environments such as mining sites (Das and Chandran 2010; Leitão 2009; Shahriari Moghadam et al. 2014; Vaksmaa et al. 2023; Johnsen and Karlson 2004).

Organic contaminants are broken down by bacteria along specific metabolic pathways. For instance, enzymatic processes such as deoxygenation, hydrogenation, and ring cleavage are used to break down PAHs. Certain bacteria, including *Pseudomonas*, *Mycobacterium*, and *Burkholderia*, are well-documented for their ability to break down PAHs. Bacteria and fungi can also break down hydrocarbons through co-metabolism, aerobic respiration, and anaerobic respiration. When these contaminants are broken down, byproducts like carbon dioxide and water are produced (Wiegel and Wu 2000; Wu et al. 2022a; Furukawa 2000).

Additionally, the development of modern research approaches such as genetic engineering has created new opportunities for utilizing microorganisms' capabilities in organic pollution cleanup. Scientific research on altering the genomes of bacteria and fungi is leading to the improvement of their ability to degrade contaminants or introduce new enzymes. In relation to mining waste remediation, genetically modified fungi provide more effective and focused approaches to waste cleanup. These genetically modified fungi show promise in degrading resistant contaminants like chlorinated chemicals (Malik et al. 2023; Rafeeq et al. 2023; Fetzner and Lingens 1994).

3.1.3 Microbial Detoxification of Toxic Chemicals

Various toxic chemicals and synthetic compounds, in addition to heavy metals and organic contaminants, pose serious threats to the environment and public health. Strategies for microbial detoxification have great potential to change and immobilize these harmful chemicals.

3.1.3.1 Enzymatic Transformation of Toxic Compounds by Microorganisms

The transformation of toxic chemicals into less harmful forms by microorganisms usually occurs through diverse enzymatic reactions. For instance, dehalogenation processes have been reported in bacteria like *Rhodococcus*, breaking down chlorinated solvents such as trichloroethylene. Some bacteria can detoxify pesticides and herbicides into much harmless forms by breaking down their chemical structures through complex enzymatic processes. Algae are identified as effective detoxifiers of harmful chemicals through sequestration due to their rapid growth rates and ability to absorb various chemicals. Some microalgae have proven to absorb radioactive materials, heavy metals, and organic pollutants from polluted water sources. There is hope for cleaning up polluted aquatic habitats using the phytoremediation method, in which algae are employed to remove and concentrate toxins (Rai 2008; Singh and Walker 2006; Ntougias et al. 2016; Klaus-Joerger et al. 2001).

3.1.3.2 Potential of Microbial-Mediated Nanoparticles in Chemical Remediation

Emerging research fields such as microbial-mediated nanoparticles carry the potential to create novel, long-lasting solutions to problems involving chemical contamination in the environment. It has been demonstrated that some microbes can produce nanoparticles with unique properties that can be used in chemical remediation. These nanoparticles, often referred to as biogenic nanoparticles, have the ability to immobilize and change hazardous chemical forms into less dangerous forms. For instance, silver nanoparticles produced by bacteria like *Bacillus licheniformis* have demonstrated the potential for waste remediation through their antibacterial capabilities (Moldovan and David 2020; Sintubin et al. 2009; Dold 2003). The value of microbial diversity as a tool for addressing the complex environmental problems caused by mining waste cannot be overestimated. Microorganisms are capable of a wide range of remediation operations, including the immobilization of heavy metals, the breakdown of complex organic contaminants, and the conversion of harmful substances into harmless forms. It is important to continue researching on microbial remediation, given the possibility of revealing a variety of approaches that could fundamentally revise how we approach environmental cleaning. Microbial diversity serves as an excellent example of nature's creativity. By unlocking the full potential of microbial remediation solutions through devoted research, innovation, and a commitment to sustainable practices, we have the unparalleled opportunity to pave the way for a cleaner, healthier, and more peaceful interaction with the environment.

3.2 Factors Affecting Microbial Activity in Mining Wastes

Complex environmental conditions encountered in mining waste-polluted areas pose particular difficulties for microbial activity. Inhibitory chemicals and interactions between microorganisms and their physical and chemical surroundings can greatly impact microbial activity in these environments. In order to harness the maximum potential of microbial diversity in remediating mining waste sites, it is essential to understand these factors to build efficient bioremediation strategies. Various methods, such as complexation-promoted reactions (complexolysis), acid-promoted reactions (acidolysis), and processes based on redox reactions (redoxolysis), are often used to facilitate biological activities in mining waste.

3.2.1 Complexolysis

This involves the release of metals from solid materials, utilizing organic acids and siderophores made by microorganisms. In this mechanism, microorganisms do not directly interact with the elements but generate organic acids that dissolve the metals, freeing them from the solid matrix. Microorganisms such as *Acidithiobacillus ferrooxidans*, *Enterobacter aerogenes*, *Pseudomonas putida*, and *Aspergillus niger* are among the common microbes that have shown to effectively mobilize metals bound within the complex solid matrices in minerals and mine wastes through syntrophic processes (Johnson and Hallberg 2007; Newsome and Falagán 2021; Sand and Gehrke 2006).

3.2.2 Redoxolysis

This mechanism involves microorganisms that produce catalytic chemicals which can control the oxidation potential of the solution. The efficacy and speed of treating the effluent are influenced by the types of metals and other chemical constituents and their oxidation states. *Acidithiobacillus thiooxidans*, *Acidithiobacillus ferrooxidans*, and *Leptospirillum ferrooxidans* are among the common microorganisms involved in redoxolysis of mining wastes (Schippers and Sand 2013, 1999; Sánchez-Andrea et al. 2011).

3.2.3 Acidolysis

This mechanism involves the microorganisms that dissolve the complex metal matrices through the production of acids. The dissolving of mine wastes by sulfur-oxidizing bacteria is an example in which the oxidation of sulfide to sulfuric acid happens with the involvement of bacteria such as *Alicyclosbacillus disulfidooxidans*, *Acidithiobacillus ferrooxidans*, and *Acidithiobacillus thiooxidans* (Nies 1999; Hallberg et al. 2010; Dopson and Johnson 2012).

3.3 Physicochemical Parameters

Physicochemical characteristics of mining wastes have a significant impact on the microbial populations present in the environment and their activity. Depending on the type of mining operation, the type of minerals extracted, and the conditions of surrounding environment, these characteristics might vary greatly. Among the several factors that could affect microbial activities, pH, temperature, and redox potential are considered as key physicochemical factors affecting microbial activities.

3.3.1 pH

Depending on the composition of minerals, quantity of minerals, content, and weathering processes, the pH of mining wastes can range from highly acidic to alkaline. Extreme pH levels can significantly reduce the diversity and activity of microorganisms. Acidophilic bacteria, such as *Acidithiobacillus ferrooxidans*, are essential for the bioleaching of metals because they can survive in extremely acidic settings. Alkaliphilic bacteria, on the other hand, may function in alkaline conditions, assisting in pH stabilization and the elimination of some pollutants (Foti et al. 2007; Cavicchioli et al. 2011).

3.3.2 Temperature

The temperature of mining wastes can vary widely, impacting the growth and metabolic rates of microorganisms. Temperature affects the solubility of metals, the reaction kinetics, and the growth and metabolism of microorganisms, making it an essential and determining factor in bio-based technologies. The type of microorganisms that have been utilized and their tolerance to temperature determine the process temperatures. Extreme temperatures, however, can also prevent microbial activity, and some microbes may go dormant until circumstances are more hospitable (Noll et al. 2020; Noike et al. 1983).

3.3.3 Redox Potential

The availability of electron acceptors and donors has a direct impact on a system's redox potential or its capacity for oxidation–reduction reactions. Alternative electron acceptors, such as nitrate or sulfate, may be used by microbes in anaerobic conditions, such as wet mine wastes, for respiration. Changes in redox potential might influence the activity of several microbial metabolic pathways, impacting the system's overall remediation capability (Kupka et al. 2009; Konopka et al. 1999). Nutrient Availability and Limitations. In mining wastes, nutrient availability significantly impacts microbial development and activity. The natural balance of nutrients is frequently upset by mining activities, creating both nutrient-rich and nutrient-poor ecosystems. Carbon, nitrogen, and phosphorus are crucial for the development and metabolism of microorganisms. Energy is produced by carbon sources like organic matter, which also acts as the foundation for microbial biomass. Proteins, nucleic acids, and other biological structures all depend on phosphorus and nitrogen to function. The availability of these nutrients in mining wastes can vary greatly, affecting the diversity and activity of the microorganisms. Nutrient supplementation, using techniques like bioaugmentation or the addition of organic amendments, can improve microbial activity and the efficacy of bioremediation. Micronutrients such as trace metals and vitamins play a crucial role as cofactors in enzyme activities. However, high levels of certain trace metals in mining wastes may impede microbial activity. Conversely, a deficiency in vital micronutrients can restrict microbial activity and growth, impairing their capacity to properly break down mining waste (Rajeev et al. 2021; Johnson and Hallberg 2003).

3.3.4 *Salinity and Osmolarity*

High salinity and osmolarity in some mining waste settings can inhibit microbial activities, especially those affected by seawater or brine. High salinity can potentially damage microbial membranes and restrict metabolic activities. However, halotolerant microorganisms can withstand osmotic stress and adapt to high-salinity environments. In salty conditions, some halophilic bacteria, such as Halomonas species, demonstrate the ability to sequester metals and degrade organic pollutants. Osmotic stress can interfere with cellular functions and protein structures, having a detrimental effect on microbial activity. For microorganisms to survive and function in saline mining waste conditions, they must be able to balance osmotic pressure (Dopson and Holmes 2014; Hassen et al. 1998).

3.3.5 *Inhibitory Substances*

Many inhibitory substances are commonly found in mining wastes, negatively impacting microbial activity. To develop efficient bioremediation solutions, it is essential to comprehend the effects of these substances.

3.3.5.1 Heavy Metals

Heavy metals are the most commonly found inhibitory substances in mining wastes, damaging bacterial cell membranes, denaturing proteins, and obstructing enzymatic processes, inhibiting microbial growth and activity. Several microbes have evolved metal-resistance mechanisms, such as metal efflux pumps, metal-binding proteins, and metal sequestration, to survive in environments with metal contamination. Multiple heavy metals can either positively or negatively impact microbial activity. Their synergistic interactions may increase the toxicity of some, whereas competition for the same resistance mechanisms may decrease the toxicity of others (Imron et al. 2019; Kebede et al. 2021).

3.3.5.2 Organic Pollutants

Mining wastes often contain a variety of organic pollutants that can be toxic to microorganisms. Microbial decomposition is often influenced by the bioavailability of organic contaminants. Some organic contaminants are hard to break down by microbes and are recalcitrant. These persisting substances can inhibit microbial activity, particularly in situations where the essential metabolic pathways are not easily accessible (Seo et al. 2009; Sigel et al. 2005).

3.4 Co-Culturing and Microbial Synergism

Microbial co-culturing and synergistic interactions between microorganisms are promising approaches to enhance microbial activity and boost bioremediation effectiveness. Certain microbial consortia can degrade complex waste contaminants by utilizing the complementing metabolic abilities of diverse species. For instance, bacteria can start the breakdown of hydrocarbons, resulting in intermediary chemicals that fungi can then metabolize. This synergistic interaction can accelerate the overall pace at which organic pollutants degrade (Gao and Wu 2023; Abo-Alkasem et al. 2023).

To enhance bioremediation through synergistic approaches, consortium engineering has been utilized to build effective and robust microbial communities. Consortium engineering involves the deliberate selection and fusion of microorganisms. Researchers can modify microbial consortia to fit particular mining waste conditions by choosing species with specific pollutant-degrading capabilities and metal resistance. Bioaugmentation, the addition of particular microorganisms, is used to improve the ability of existing microbial communities to remediate, while biostimulation uses the supply of nutrients or electron acceptors to promote the development and activity of local microorganisms. Both strategies might be used to encourage efficient pollutant degradation at mining waste sites and overcome barriers (Anekwe and Isa 2023; Mosa et al. 2016; Wang 2018; Thiruvenkatachari and Su 2017).

Efficient bioremediation solutions require an understanding of the many parameters influencing microbial activity in mining wastes. The potential of microbial variety to remediate mining waste environments is greatly influenced by physicochemical factors, inhibitory chemicals, and microbial interactions.

3.5 Optimization of Microbial Remediation Processes in Mining Wastes

The environmental conditions in mining waste generating sites often present a complex and challenging environment for microbes to survive. Thus, various optimization strategies need to be employed to increase the effectiveness and efficiency of bioremediation methods. These approaches delve into critical areas of enhancing microbial remediation, such as proper bioreactor design and operation, effective molecular approaches for monitoring and measuring microbial populations, and efficient ways to improve microbial adaptability and evolution.

3.5.1 Bioreactor Design and Operation

Bioreactors offer controlled environments for microbial remediation that enable the analysis of critical parameters to enhance microbial activity and waste treatment. The selection of the appropriate bioreactor for the purpose depends on the specific requirements of the remediation process and the characteristics and properties of the mining waste. There are three main types of bioreactors used for this purpose.

3.5.1.1 Aerobic Bioreactors

Aerobic bioreactors provide oxygen-rich conditions necessary for aerobic microorganisms engaged in the oxidation of metals and the breakdown of organic pollutants. These can be built as stirred-tank bioreactors, packed-bed bioreactors, or biofilm reactors, based on the requirements for oxygen transport effectiveness and mixing efficiency (Johnson and Hallberg 2005; Dopson and Lindström 1999; Costa et al. 2022).

3.5.1.2 Anaerobic Bioreactors

These devices create oxygen-free environments ideal for anaerobic bacteria that reduce metals and decompose some organic contaminants, such as chlorinated compounds. These bioreactors can be built as up-flow anaerobic sludge blankets (UASB), fixed-bed, or fluidized-bed reactors (Willow and Cohen 2003; Morello et al. 2017).

3.5.1.3 Hybrid Bioreactors

Hybrid bioreactors combine aerobic and anaerobic zones to accommodate microorganisms with various redox requirements. These bioreactors can be useful when treating mining waste with diverse pollutant profiles (Crisostomo et al. 2016; Kiani et al. 2022; Perez et al. 2020). The choice of bioreactor operational mechanism is based on waste treatment operations and mining waste properties. Continuous processing and batch processing are the operational mechanisms used in bioreactors.

3.5.1.4 Continuous Processes

Continuous processes involve a continuous inflow of waste to the reactor and continuous outflow of treated effluent. This mechanism provides microbial communities with steady environmental conditions, resulting in consistent pollution removal rates. However, continuous processes could be more challenging to manage and might require more advanced monitoring and control systems (Hille et al. 1999; Luka et al. 2018).

3.5.1.5 Batch Processes

In contrast, batch processes fill the bioreactor with polluted waste, allowing the microbial remediation process to take place over a period of time, and then empty the reactor after the processing time. Batch operations offer more precise control, but pollutant removal rates may vary over time (Robles-González et al. 2008; Sun et al. 2019).

The ideal microbial remediation procedures in mining waste require effective monitoring and control systems. Real-time measurement of parameters, including pH, temperature, dissolved oxygen, and pollutant concentrations, is essential for continuous monitoring during operations. Using these data, microbial activity and pollutant degradation can be optimized instantaneously. Furthermore, automated control systems can be employed to modify operational conditions automatically based on feedback from online monitoring. For instance, controlling the concentration of dissolved oxygen can create optimal aerobic conditions for pollutant breakdown.

3.6 Molecular Techniques in Monitoring and Assessing Microbial Communities

Understanding the complex dynamics of microbial communities based on molecular methods enables researchers to delve into intricate microbe interactions. The development of novel molecular techniques has provided more sophisticated approaches for tracking and evaluating the behavior of microbial communities, tapping into their potential for ecological stability and biotechnological advancements. Molecular methods provide crucial insights into the composition and functional potential of microbial communities participating in mining waste processing.

3.6.1 DNA/RNA-Based Analysis for the Profiling of Microbial Communities

Using DNA/RNA-based techniques such as amplicon sequencing and metatranscriptomics, it is possible to characterize the makeup of microbial communities and the expression of their genes in mining waste environments (Heyde et al. 2020; Větrovský and Baldrian 2013).

3.6.1.1 Amplicon Sequencing

Amplicon sequencing focuses on specific genetic markers, such as internal transcribed spacer (ITS) regions for fungi or 16S rRNA genes for bacteria and archaea. This method reveals the diversity and relative abundance of many microbial species (Arif et al. 2021; Lemes et al. 2021; Poretsky et al. 2014; Nkuna et al. 2022; Hua et al. 2015).

3.6.1.2 Metatranscriptomics

Metatranscriptomics involves sequencing mRNA transcripts from microbial communities, providing information on gene expression involved in pollution breakdown and active metabolic pathways (Lehembre et al. 2013; Mukherjee and Reddy 2020; Zhang et al. 2017).

3.6.1.3 Remediation Monitoring Using Transcriptomics and Metagenomics

Researchers have the ability to monitor variations in gene expression and assess the potential functions of microbial populations throughout remediation by employing transcriptomic and metagenomic techniques.

1. **Transcriptomics**: Transcriptomic analysis focuses on RNA transcripts produced by microbial communities, identifying activated metabolic pathways in response to pollutants or other environmental factors (Alcasabas et al. 2020; Chen et al. 2015).
2. **Metagenomics**: Metagenomics entails direct sequencing of DNA from environmental samples, revealing information about microbial capacities present, including genes encoding enzymes involved in pollution breakdown (Li et al. 2018, 2022; Zhang et al. 2021) (Fig. 3.3).

3.6.1.4 Stable Isotope Probing (SIP) for Monitoring Microbiological Activity

Stable isotope probing (SIP) is a robust molecular method used to track and evaluate the activity of microbial communities in various conditions and environments. This method relies on microbes' ability to incorporate specific isotopically tagged chemical compounds into their cellular structures while performing specific metabolic functions. SIP operates on the premise that microbes metabolizing the isotopically labeled substrates would leave an identifiable "isotope fingerprint" in their biomolecules, distinguishing them from inactive or differently active organisms. Monitoring can be achieved by incorporating these isotopes into the DNA, RNA,

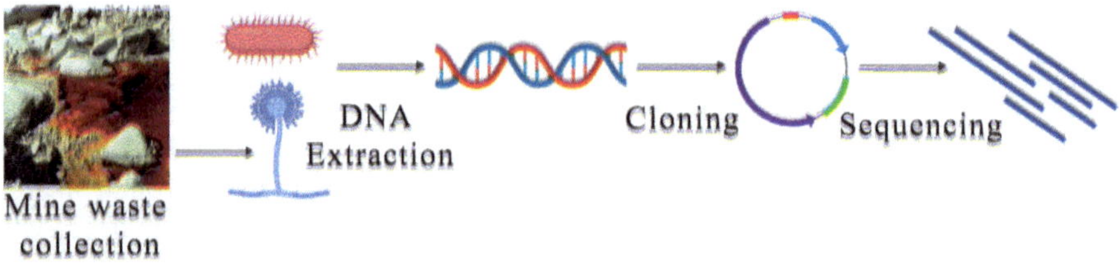

Fig. 3.3 Metagenomic Analysis Workflow: This diagram illustrates the step-by-step process of metagenomic analysis, a sophisticated approach to studying microbial communities in diverse environments. Steps involved; **a** Sample Collection: Gather diverse environmental samples. **b** DNA Extraction: Isolate genetic material from samples. **c** Cloning: Fragment DNA for sequencing. **d** Sequencing: Use high-throughput methods for DNA sequencing and Data Processing to remove artifacts and ensure data quality

proteins, and other biomolecules of active microbial populations through the addition of a substrate enriched with stable isotopes, often carbon-13 (^{13}C) or nitrogen-15 (^{15}N). Cutting-edge analytical methods such as Isotope Ratio Mass Spectrometry (IRMS) or High-throughput Sequencing can be employed to identify interactions within the microbial community. In the bioremediation process, isotopically tagged substrates are introduced into the bioreactor, and stable isotopes from these substrates are incorporated into the cellular biomass of microorganisms that actively metabolize them. SIP enables the monitoring of specific microorganisms engaged in the oxidation of organic contaminants or the sequestration of heavy metals. This method assists in optimizing bioremediation strategies by identifying the key participants in the remediation process (Zhang et al. 2020; Guri et al. 2012; Gabay 2022; Wu et al. 2022b).

3.7 Enhancing Microbial Adaptation and Evolution

At the intersection of biology, technology, and ecological knowledge, improving microbial adaptation and evolution has emerged as an exciting prospect. Researchers enable microbes to adapt quickly to new settings, overcome obstacles, and unleash unique functionalities through methodologies like adaptive laboratory evolution (ALE), microbial bioprospecting, and synthetic biology techniques. These techniques accelerate the appearance of advantageous features, such as improved toxin resistance or increased production of valuable substances, by subjecting microbial populations to controlled selective pressures or adding genetic alterations. This control over microbial development can revolutionize industries, including processes involved in mining waste remediation.

3.7.1 *Improved Remediation via Adaptive Laboratory Evolution (ALE)*

Adaptive Laboratory Evolution (ALE) is a revolutionary strategy that employs evolution to enhance the effectiveness and efficiency of environmental remediation procedures. Microbial populations in ALE undergo regulated laboratory settings over generations, selecting strains with superior ability to break down specific pollutants or toxins. Through this iterative process, microorganisms evolve and adapt, acquiring genetic changes and enhancing their metabolic pathways for target molecule breakdown. Beneficial mutations develop providing improved pollutant-degrading capabilities. ALE accelerates natural evolution, developing microbial consortia with enhanced bioremediation capability (Abashina and Vainshtein 2023; Sana et al. 2021).

3.7.2 *Microbial Bioprospecting for Novel Pollutant-Degrading Strains*

Microbial Bioprospecting is a revolutionary approach in the search for long-term solutions to pollution-related problems. This novel method entails investigating harsh habitats with the prospect of finding previously undetected microorganisms with outstanding capacities for pollution degradation. Finding microbial strains with specific enzymatic pathways that facilitate the breakdown of resistant contaminants is the main objective of the search process. Frequently studied habitats involve extreme environments such as hydrothermal vents, contaminated soils, and industrial effluents. High-throughput screening techniques, coupled with functional assays, allow researchers to identify microorganisms with specific pollutant-degrading abilities, some of which may possess unique capabilities for pollutant degradation. This microbial hunt offers a possible arsenal of natural bioremediation agents and helps comprehend the extraordinary flexibility of microbes. These unique pollutant-degrading strains can be used to create eco-friendly bioremediation techniques that reduce environmental pollution, opening the path for a cleaner and more sustainable future through careful isolation, culture, and characterization (Das and Ghosh 2018; Costa et al. 2021; Tran et al. 2021).

3.7.3 *Synthetic Biology and Gene Editing for Enhanced Bioremediation*

Synthetic biology and gene editing provide transformative technologies for precisely modifying microorganisms. Synthetic biology uses genetic circuits and metabolic pathways to engineer microbes for increased pollutant affinity, faster degradation, and improved adaptation to changing environmental conditions. Gene editing methods like CRISPR-Cas9 accurately alter microbial genomes, introducing advantageous genetic alterations increasing pollutant-specific enzyme production or providing resistance to hazardous substances. Well-characterized microbes receive specific genes or gene clusters involved in pollution breakdown, enhancing their capacities (Rylott and Bruce 2020; Roldán et al. 2021; Jiménez-Díaz et al. 2022; Jaiswal et al. 2019; Capeness and Horsfall 2020; Bhattacharjee et al. 2020; Ribeiro and Prasad 2021). A multidisciplinary strategy integrating ideas from various scientific fields is necessary to optimize microbial remediation processes in mining waste landscapes successfully. Effective bioremediation results from the complex interplay of variables, including bioreactor design, operational parameters, molecular techniques for monitoring microbial communities, and creative strategies catalyzing microbial adaptation and evolution.

3.8 Advantages, Limitations, and Applicability of Microbial Techniques in Mining Wastes

Mining operations generate substantial amounts of waste, which, if mismanaged, pose significant environmental, and public health risks. Microbial strategies have emerged as beacons of hope in the field of mining waste remediation, addressing the critical need for innovative approaches. Microorganisms' inherent ability to metabolize and convert a wide array of pollutants provides a compelling solution for managing the intricacies of toxins originating from mining activities. However, it is vital to acknowledge that while microbial techniques hold immense potential, they are not without challenges. Microbial activity and success are heavily influenced by factors such as site-specific conditions, varying pollutant concentrations, and the harsh environments of mining sites. To develop efficient and enduring remediation strategies, a precise understanding of different mining waste profiles and microbiological approaches is crucial. This necessitates a comprehensive assessment of pollutant bioavailability, ecosystem interactions, and microbial adaptability, enabling the formulation of focused and sustainable strategies that leverage the innate abilities of microorganisms while navigating the complexities of each mining waste scenario.

3.8.1 Advantages of Microbial Remediation

Microbial approaches offer several advantages for the management and remediation of mining waste. Among these, cost-effectiveness stands out as a key benefit. In contrast to conventional remediation techniques like physical and chemical methods, microbial remediation relies on naturally occurring microbes, eliminating the need for expensive materials and energy-intensive procedures. Microbiological methods are also deemed more environmentally friendly and sustainable. Microorganisms utilize their inherent capabilities to trap heavy metals and break down contaminants, causing minimal environmental disturbances. Additionally, certain microbial processes generate harmless byproducts, contributing to the overall sustainability of the remediation process. Moreover, microbial methods facilitate flexible in-situ and ex-situ applications, allowing customized approaches to various mining waste sites. In-situ bioremediation addresses toxins directly at their source without excavation, making it particularly useful for large quantities like mining tailings and hard-to-reach areas where ex-situ remediation may be impractical. On the other hand, ex-situ bioremediation involves relocating polluted materials to controlled environments such as bioreactors or constructed wetlands. Ex-situ methods provide superior environmental control, enabling researchers to maximize microbial activity and contaminant degradation. By employing microbial techniques, contaminated sites can be permanently rehabilitated through long-term remediation mechanisms. Once established, microbial communities can endure and adapt to changing environmental conditions, enabling continuous pollutant degradation over extended periods. Long-term efficacy is especially crucial in managing mining waste, where many pollutants can persist for decades or even centuries. Instead of repetitive, expensive interventions, microbial approaches offer a sustainable solution for ongoing waste management and remediation.

3.9 Limitations and Challenges of Microbial Remediation

While microbiological approaches offer numerous benefits, they also present several drawbacks and challenges that need to be addressed for their effective application. One of the primary drawbacks is the comparatively slow remediation rates of microbial approaches, especially in harsh or extreme conditions. Microbial activity and pollution breakdown can be hampered by factors such as poor nutrient availability, excessive salinity, and severe pH levels. In some cases, the sluggish rates of microbial breakdown may fail to meet legal standards for cleanup or adequately control pressing environmental issues. In such circumstances, expediting the remediation process may require combining microbial treatments with other remediation procedures. The complexity of microbial communities and their responses to changing environmental conditions pose challenges in monitoring and managing microbial activities at mining waste sites. Monitoring microbial activity necessitates the use of complex molecular techniques, which can be both expensive and time-consuming. Additionally, diverse mining waste sites and the spatial distribution of microorganisms can impact the accuracy and reliability of monitoring methods. Regulating microbial activity in large-scale in-situ applications can be difficult, as maintaining stable conditions conducive to microbial activity and pollutant degradation is challenging due to frequently changing environmental conditions. Moreover, regulations and social acceptance issues may hinder the use of microbial approaches in mining waste remediation. Lack of experience of regulatory agencies with microbial remediation methods can complicate obtaining approval and permits for large-scale microbial remediation initiatives. Public perception of microbial remediation may vary, with some stakeholders expressing concerns about the risks associated with using living organisms to control wastes. Effective communication and public outreach strategies are essential to increase awareness and confidence in microbial remediation technology.

3.9.1 Applicability to Different Types of Mining and Industrial Wastes

Each type of mining and industrial waste presents unique challenges for remediation, but microbial approaches are adaptable and applicable to many instances. Mine tailings, characterized by high metal concentrations and potential acid production, require remediation. Microbial approaches, including bioleaching and metal-resistant bacteria, can mobilize and sequester heavy metals from tailings, reducing their environmental impact. Acidophiles and metal-reducing microbes can neutralize acidity and lower metal concentrations, making them valuable for treating acidic and metal-rich mine drainage. Residues from smelting and refining often contain hazardous metals such as lead, zinc, and copper. Microbial methods can stabilize these residues by promoting the precipitation of metal sulfides or converting metal ions into less hazardous forms. Unlike conventional approaches that rely on physical or chemical immobilization techniques, microbial-assisted stabilization of smelting and refining wastes is economical and environmentally friendly. Heap leach pads, used in the mining industry to extract metals from low-grade ores, can lead to the production of acid mine drainage when acidic leachate containing metals runs from the heap. Microbial approaches, especially those involving acidophiles, can remediate acid mine drainage sites by neutralizing acidity and encouraging metal precipitation. Additionally, heap leach pads can serve as locations for microbial remediation, utilizing microorganisms to expedite metal extraction and recovery, reduce harmful chemicals, and advance environmentally friendly mining techniques. Microbial approaches offer significant benefits in the rehabilitation of mining waste through their affordability, sustainability, and potential for long-term use. However, these approaches face challenges such as slow remediation rates, monitoring complexities, and social and legal acceptance issues. Despite these drawbacks, microbial approaches can efficiently manage waste and remediate various mining and industrial waste in an environmentally beneficial manner.

3.10 Synergistic Approaches in Microbial Remediation in Mining Waste

The effective remediation of mining waste sites demands the deployment of cutting-edge, efficient techniques capable of handling the complex nature of contaminants. Innovative and efficient procedures are essential due to the intricate interactions between various pollutants and the diverse characteristics found in mining waste. Synergistic approaches offer hope for meeting this pressing need. These methods successfully combine the efficiency of microbial remediation with a range of physical, chemical, and biological procedures to create a potent toolkit for addressing contamination-related difficulties. By integrating the unique qualities of several remediation modalities, these synergistic techniques have the potential for dramatically improved remediation solutions.

3.10.1 Combining Microbial Remediation with Physical and Chemical Methods

A dynamic, all-encompassing strategy that utilizes the best aspects of several remediation modalities involves combining physical and chemical techniques with microbial remediation. This integrated strategy leverages microorganisms' innate capacity to biodegrade contaminants while enhancing its effectiveness through the strategic use of chemical treatments and physical interventions. Managing this harmonious inter-play enables the remediation process to effectively target a broader spectrum of toxins and adapt to various environmental conditions. Microorganisms act as bioagents, breaking down pollutants into less dangerous components, while chemical and physical agents either enhance the amount of pollutants that microbes can degrade or expedite their removal. This collaborative approach not only increases the overall effectiveness of remediation processes but also opens up prospects for solving challenging contamination scenarios that may resist individual solutions.

3.10.2 Electrobioremediation

Electrokinetics, a technique in engineering that utilizes electrical fields to accelerate the flow of contaminants in the subsurface, can be paired with bioremediation, a process known as electrobioremediation. This combination enhances the transport and bioavailability of contaminants to microbial communities, thereby speeding up the rate of breakdown. Electrobioremediation involves placing electrodes in contaminated soil or water and applying an electric current. The electric field accelerates the movement of contaminants, making them more vulnerable to microbial populations. It promotes microbial community growth and activity by improving electron transport and nutrition availability. Additionally, the electric field accelerates the biodegradation of contaminants through microbial redox reactions. Heavy metals like lead and zinc in areas with active microbial populations are more frequently trapped and immobilized as a result (Puggioni 2022; Kaya et al. 2023; Hernández et al. 2023; Hlihor et al. 2022).

3.10.3 Phytoremediation Combined with Microbial Remediation

Phytoremediation, the use of plants to remove, stabilize, or degrade pollutants from the environment, can be combined with microbial remediation to accelerate the breakdown and transformation of specific contaminants. Plants play a crucial role in phytoremediation, where contaminants accumulate within plant cells. Interactions with plant exudates enable rhizosphere bacteria connected to plant roots to expedite the breakdown of pollutants. When microbial and phytoremediation techniques are coupled, the removal of organic contaminants and heavy metals from mining waste sites has shown promising results (Rabani et al. 2022; Mishra et al. 2015).

3.10.4 Nanotechnology-enhanced Microbial Remediation

Nanotechnology, which involves the use of nanoscale materials to enhance various processes, including environmental remediation, can improve the efficacy of microbial remediation techniques. Introducing nanoparticles enhances microbial remediation by transporting nutrients and electron acceptors to microbial populations, encouraging their activity and growth. Nanoparticles also increase the surface area for microbial adhesion and the bioavailability of pollutants. Nanoparticles such as zerovalent iron and metal oxide nanoparticles have demonstrated potential in enabling metal reduction and sequestration by microbial communities. Research in microbial remediation enhanced by nanotechnology is active, particularly for the remediation of organic contaminants and heavy metals in mining waste sites (Wong-Pinto et al. 2021; Punia et al. 2023; Okanigbe et al. 2022; Mandeep 2020; Bionanomining of Copper-Based Nanoparticles Using Pre-Processed Mineral Tailings as Precursor 2023; Chaurasia et al. 2022; Ai et al. 2020a).

3.10.5 *Microbial Bioremediation in Conjunction with Microbial Fuel Cells*

Microorganisms are used by microbial fuel cells (MFCs) to transform organic materials into electrical energy. Systems that simultaneously generate energy and remove toxins can be made self-sustaining by combining MFCs and microbial bioremediation. MFCs produce electricity through microbial redox processes that occur as organic matter are degraded. The remediation process may be operated by this energy, eliminating the need for extra sources of energy. Some microbial populations involved in pollutant degradation are stimulated by the electron transfer mechanisms in MFCs. A self-sustaining cycle is created when the remediation process simultaneously produces organic matter that fuels the MFC. The remediation of organic pollutants and the reduction of metal contaminants in mining waste sites have both been evaluated using MFCs coupled with microbial bioremediation. The method is especially useful for remote or off-grid mining waste sites since it can generate energy while pollutant degradation occurs. Electroactive microbes can be highly significant in MFC-assisted microbial bioremediation. During microbial metabolism, electroactive microorganisms can transport electrons directly to the anode of the MFC. These microbes assist in the breakdown of organic matter, producing electricity. By providing an additional electron transfer channel, electroactive microorganisms may improve microbial bioremediation. As a result, both the rate of pollutant degradation and the production of power are enhanced. These approaches can be considered highly sustainable given their additional power generation capacity (Stefanova et al. 2018; Ai et al. 2020b; Sahoo et al. 2022). The use of synergistic methods is a promising approach for the treatment of mining waste, offering potential solutions to the complex problems caused by mining pollution. Synergistic techniques will be crucial in fostering a cleaner and more sustainable mining industry, maintaining ecosystems, and safe-guarding public health as research and technology development progress.

3.11 Future Perspectives in Microbial Remediation in Mining Waste

Recent developments and research frontiers have the potential to entirely transform how we approach environmental remediation as the field of microbial remediation in mining waste develops. This chapter examines the exciting potential provided by cutting-edge technology, identifies knowledge gaps that require further research, and proposes future research routes for microbial remediation in mining waste.

a. Emerging Technologies and Research Frontiers

Application of CRISPR-Cas and Synthetic Biology. In the field of microbial treatment for mining wastes, the groundbreaking CRISPR-Cas technology has proven itself as a pioneering method. CRISPR-Cas, an innovative gene editing technique derived from the built-in defenses of some bacteria, stands for Clustered Regularly Interspaced Short Palindromic Repeats and CRISPR-associated proteins. CRISPR-Cas, first identified as a component of bacterial adaptive immunity, has been exploited and used across various organisms for precise gene editing. This cutting-edge technology enables scientists to target specific DNA sequences with unparalleled accuracy and delete, insert, or change genes with a level of precision that was previously deemed challenging. CRISPR-Cas systems empower scientists to develop microbes with specialized abilities to handle challenging contamination issues due to their remarkable accuracy in genetic editing. Using this technology, microbial strains with altered genetic makeup can be engineered to better break down specific toxins or withstand the harsh environmental conditions found in mining waste sites. The use of CRISPR-Cas in microbial remediation accelerates the remediation procedure and improves the overall efficacy and efficiency of bioremediation techniques. Additionally, through synthetic biology approaches, specialized microbial consortia with enhanced metabolic pathways for selective pollutant breakdown can be created. This strategy significantly improves the effectiveness and precision of microbial remediation procedures. However, the use of gene editing and synthetic biology in environmental remediation could raise ethical questions about potential unexpected ecological effects and the release of genetically modified organisms into the environment. Implementing solid risk assessment frameworks and adhering to ethical standards are crucial for responsible implementation (Saxena et al. 2020; Hu et al. 2022; Ding et al. 2020; Neu et al. 2010).

Microbial Biomineralization for Metal Recovery. Microbial biomineralization, a process in which microorganisms induce the formation of mineral deposits, has drawn significant attention due to its dual advantages of metal immobilization and recovery. This biological process converts soluble metal ions into insoluble solid forms through the specific metabolic capacities of microorganisms such as bacteria and fungi. These minerals encapsulate and isolate metal contaminants as they precipitate out of solution, making them less mobile and less likely to diffuse into the environment. Microorganisms' inherent capacity to direct the precipitation of metals into stable mineral matrices provides a natural remedy for the remediation of metal-contaminated environments. Beyond its environmental impact, microbial biomineralization has the potential to recover metals, aligning with sustainable resource management goals. Precious metals such as nickel, cobalt, copper, and rare earth elements can be found in the mineral deposits formed through microbial biomineralization. Moreover, the variable nature of microbial biomineralization offers a flexible method for enhancing metal recovery. Researchers can fine-tune the precipitation process by adjusting surrounding factors, microbial consortia, and nutrient inputs to produce mineral deposits enriched with specific target metals. This versatility enhances the effectiveness of metal recovery and allows for the modification of the procedure to accommodate various waste types and metal compositions. Like any emerging technology, the development of microbial biomineralization for metal recovery faces challenges. Thorough research is required into aspects such as the scalability of process, the stability of biominerals produced over time, and potential interactions with co-contaminants (Kalin et al. 2005; Begum et al. 2022; Muniruzzaman and Pedretti 2021).

Improving Predictive Models for Remediation Performance. Optimizing remediation procedures necessitates enhancing the reliability of microbial remediation performance. Multiscale modeling enhances the comprehension of microbial processes in complex ecosystems by integrating data at various scales, from the molecular to the ecosystem level. Multiscale models can forecast how microbial changes will affect the ultimate outcomes of remediation. Utilizing machine learning and artificial intelligence algorithms, patterns and connections in massive datasets can be identified, allowing for more precise predictive models. This method aids in the adaptive management of remediation initiatives and real-time decision-making. Long-term field site monitoring will improve the accuracy of models and their adaptability to various mining waste conditions (Tran et al. 2021; Zhang 2021).

Understanding Microbial Community Dynamics in Complex Environment. The dynamics of microbial communities are crucial in determining the extent to which remediation processes function. A deeper understanding of these processes is necessary for creating focused interventions. Researching microbial interactions, such as mutualism, rivalry, and cross-feeding, can help comprehend the intricate web of connections that affect community dynamics. Clarifying these connections will facilitate the design of consortia for optimal remedial performance. Additionally, omics technologies such as metagenomics, metatranscriptomics, and metabolomics can be employed to characterize microbial populations in great detail. Integrating multiomics data will demonstrate how functional capacities and community composition adapt to environmental changes. Applying ecological concepts to microbial communities at mining waste sites, such as community assembly theory and succession, can enhance understanding of how these communities respond to environmental changes.

Integration of Bioremediation with Ecological Restoration. Mining waste management becomes more comprehensive and sustainable when bioremediation and ecological restoration techniques are combined. Understanding the potential benefits and ecosystem services provided by microbial remediation can aid in planning restoration efforts for achieving beneficial ecological results. Combining microbial treatment with habitat restoration facilitates the repopulation of diverse plant and animal species in mining-affected areas. Regenerative techniques, such as adding organic matter to the soil and using microbial inoculants, can accelerate ecosystem recovery and enhance ecological resilience over the long run (Enkateswarlu et al. 2016; Sun et al. 2018; Ma 2014; Lin et al. 2021; Stierle and Stierle 2005).

In the rapidly developing field of microbial remediation in mining waste, there is significant potential to address environmental issues and advance environmentally friendly mining methods. New technologies like CRISPR-Cas gene editing, microbial biomineralization, present intriguing opportunities for improving remediation efficacy. However, careful consideration of moral and biosecurity issues is necessary for the responsible implementation of these technologies. Further research should focus on enhancing models for predicting remediation effectiveness, increasing our understanding of microbial community dynamics, and integrating bioremediation with ecological restoration techniques.

3.11.1 Real-World Case Studies of Successful Microbial Remediation Projects for Mining Waste Remediation

The real-world implementation of microbial remediation techniques in the management of mining waste has resulted in numerous examples of successful project outcomes, providing viable alternatives to the challenging issues presented by mining-related degradation of the environment. Microbial remediation has shown a great deal of promise in laboratory and pilot-scale research. However, for the effective application of these strategies at a larger scale, particularly in real-world situations, careful planning, detailed appraisal of impediments, and adherence to best practices are required. This section delves into several noteworthy case studies that demonstrate the efficacy and adaptability of microbial remediation in recovering polluted landscapes, cutting pollution, and encouraging sustainable land use practices.

3.11.2 Berkeley Pit Lake, Montana, USA

The open-pit copper mine, referred to as the Berkeley Pit Lake in Montana, USA, is an example of how microbial remediation may revolutionize the restoration of polluted ecosystems. The pit turned into a hazardous reservoir with water that was heavily acidic and metallic after the mine was shut down. The success of the bioremediation restoration was due to the addition of sulfate-reducing bacteria, which aided in the creation of metal sulfides and decreased the solubility and toxicity of the metals. These bacteria were able to change the water in the pit over time, transforming it from a harsh, caustic environment to one that could accommodate a wider variety of aquatic life. This instance shows how microbial intervention may be a key component in cleaning up even the most challenging contaminated environments (Tuovinen et al. 2018).

3.11.3 Talvivaara Nickel Mine, Finland

The Talvivaara Sotkamo Mine in Northeastern Finland is a recent commercial use case of bioleaching. The ore primarily consists of pentlandite, pyrrhotite, chalcopyrite, sphalerite, and pyrite as the primary sulfide minerals present in the low-grade black schist ore. To provide sulfate-reducing bacteria with a favorable habitat, microbial bioreactors were introduced. These microbes aided in the transformation of heavy metals and sulfates into less harmful forms, reducing the amount of contaminants released into the environment. A significant on-site pilot trial yielded positive performance data, and bioheap leaching technology was chosen for the extraction of nickel from the ore due to its favorable capital and operating costs. Mining began in 2008 with the industrial-scale bioheap. This instance demonstrates the effectiveness of using microbial processes to mitigate the negative ecological effects of mining operations (Riekkola-Vanhanen 2013; Whitehead and Prior 2005).

3.11.4 Wheal Jane Mine, United Kingdom

The Wheal Jane Mine in the United Kingdom demonstrated the benefits of combining microbiological and plant science techniques for site restoration. This research, which combined phytoremediation and microbial communities, utilized plants to stabilize soil and reduce erosion, while microbial communities helped to break down pollutants, notably heavy metals. A pilot passive treatment facility was constructed at the Wheal Jane Mine in Cornwall, UK, in 1994 as part of a larger plan to find a long-term treatment solution for acid mine drainage. This facility is the biggest experimental facility of its kind in Europe and comprises three distinct systems, each containing aerobic reed beds, anaerobic cells, and rock filters. A significant research project was undertaken to analyze the pilot plant and thoroughly examine the biological mechanisms, geochemical processes, and physical processes that regulate passive treatment systems. The Wheal Jane pilot plant presented a unique opportunity to gather information, models, knowledge, and design standards for passive treatment system design, planning, and sustainable management in the future. The effectiveness of this combined approach demonstrated the possibility of coordinating several biological processes for efficient control of mining wastes (Krümmelbein et al. 2012).

3.11.5 Lusatian Lignite Mining Region, Germany

Large-scale open-pit coal mining operations in the German Lusatian lignite mining region produced a significant amount of acid mine drainage, endangering the aquatic ecosystems. Microbial mats and biofilm-forming microorganisms were utilized on exposed mining surfaces to create biocovers. These biocovers served as organic barriers that minimized oxygen and moisture intrusion, hence reducing acid mine drainage. The success of this strategy illustrated how microbial technologies could be used to control the acidity brought on by coal mining effluent (Yin et al. 2018).

3.11.6 Dexing Copper Mine, Jiangxi Province, China

A large-scale example of the use of microbial approaches in mining is the Dexing Copper Mine (DCM) in China. The DCM used acidophilic microorganisms to speed mineral oxidation and bioleaching to recover copper from low-grade ores. This groundbreaking method improved copper recovery while significantly lowering waste production and sulfuric acid consumption. Dexing Copper Mine's process involves spraying initial leaching solution onto the top of a bio-dump-leaching, and as the solution percolates through ores, the concentration of Cu(II) increases, eventually yielding pregnant leaching solution at the bottom of the dumps. The absence of a pad and larger boulder yield in comparison to bioheap leaching are seen as key challenges in the process. A smaller bacterial population and the inherent permeability of the waste rock dump are hypothesized to impede greater extraction in DCM. The DCM's reduced heap permeability led to the identification of preferred flow as a seepage phenomenon and its impact on extraction and surface morphology. Leaching from biodump in the DCM established a waste rock dump with lower intrinsic permeability's leachability and potential for commercial gain, marking a significant milestone. The Dexing case highlights microbial bioleaching potential as a sustainable substitute for traditional mineral extraction techniques. Each of these case studies demonstrates the disruptive nature and the potential of microbial remediation technology in solving the challenging issues of mining waste management. These studies show the versatility and efficacy of using microorganisms for site restoration and environmental preservation, whether in contaminated water bodies or acidic landscapes. These real-world successes provide a promising guide for a sustainable future where the fusion of biology and technology results in harmonious relationships between people and the environment. This is especially important as the globe struggles to strike a balance between the need for economic advancement and ecological preservation. Microbial remediation will continue to redefine the mining waste management landscape through ongoing research, innovation, and collaboration and will help create a more resilient and ecologically balanced world.

3.12 Conclusion: Paving the Way for Sustainable Mining Waste Management

The exploration of microbial remediation technologies has emerged as a beacon of hope in the journey toward sustainable mining practices and environmental conservation. Throughout this chapter, we have delved into the intriguing field of microbial remediation technologies for mining waste management, emphasizing their transformational potential and effectiveness in addressing the most significant issues related to mining. Several significant trends have emerged in the ever-changing mining waste management environment, pointing to a more sustainable future. Incorporating cutting-edge technologies, such as microbial treatment, is one option. Researchers are increasingly interested in microorganisms' potential to modify contaminants, indicating a paradigm shift toward ecologically friendly solutions. The convergence of biotechnology, nanotechnology, and environmental science has created new opportunities for mining waste cleanup. At the core of microbial remediation lies the utilization of microorganisms' innate abilities to mitigate the effects of mining waste on ecosystems. This interdisciplinary field harnesses microorganisms to remove heavy metals from the environment, restore ecosystems, and degrade contaminants, paving the way for a cleaner and more sustainable future. Microbes offer a natural and sustainable method of cleaning up contaminated sites due to their versatility and wide range of metabolic capacities. The effectiveness of remediation techniques is heavily reliant on microbial diversity. A diverse array of abilities is required to biodegrade organic contaminants, immobilize heavy metals, and detoxify harmful substances through bacteria, archaea, fungi, and algae. These microorganisms possess a variety of metabolic pathways that enable them to adapt to and thrive in harsh environments, making them invaluable allies in the remediation of contaminated soil. Nature provides a blueprint for innovation that can restore balance to settings damaged by mining activities, whether it's the capacity of certain bacteria to sequester metals or the ability of algae to absorb nutrients. Microbial remediation technologies showcase remarkable adaptability. Ranging from bioleaching to biosorption, these methods offer specialized solutions that can be tailored to a variety of pollutants and environmental conditions. This adaptability empowers us to tackle the complex challenges posed by diverse mining waste scenarios, underscoring the fact that there is no one-size-fits-all approach to environmental restoration. Maintaining a commitment to ethical and responsible methods is essential for advancing in the field of microbial remediation. The potential benefits of these technologies are vast, but to avoid unforeseen consequences and to ensure that our efforts have a positive impact on both the environment and humans, they must be accompanied by rigorous research, comprehensive risk assessments, and robust regulatory frameworks. All stakeholders—researchers, entrepreneurs, legislators, and communities—are urged to respond to the potential of microbial remediation technology for managing mining waste. Despite these advances, many gaps in current scientific knowledge remain. The intricacies of natural ecosystems frequently put microbial solutions to the test. Variations in soil composition, meteorological circumstances, and the different types of contaminants present generate complex problems that necessitate nuanced solutions. Furthermore, further research is needed to determine the long-term consequences of microbial remediation on ecosystems and human health. More studies are needed to properly understand the ecological impact of these technologies and assure their long-term viability. Moreover, controversies and disagreements in the literature highlight the need for a nuanced perspective. Debates persist regarding the scalability of microbial remediation technologies. While successful at the laboratory scale, challenges arise when implementing these techniques in large, real-world mining waste sites. Questions about the economic viability, technological feasibility, and social acceptance of microbial solutions necessitate critical examination. Addressing these controversies demands not just scientific insight but also an understanding of the socio-economic contexts in which these technologies are applied. From an author's perspective, it is crucial to identify the challenges and limitations of microbial remediation technologies in mining waste management. Microbial solutions, while promising, are not a one-size-fits-all remedy. Factors such as the adaptability of microorganisms to specific contaminants, site-specific conditions, and regulatory constraints pose significant challenges. Additionally, the potential unintended consequences of introducing engineered microorganisms into natural environments raise ethical and ecological concerns. In conclusion, as we navigate the complexities of mining waste management, acknowledging these trends, and existing gaps is essential. Embracing a multidisciplinary approach, fostering collaboration, and addressing controversies will drive the field forward. Recognizing the limitations and challenges of microbial remediation technologies provides a foundation for future research, ensuring that our quest for sustainable mining practices remains both innovative and ethically sound. As authors, our responsibility lies in advocating for continued research, mindful application, and responsible implementation, guiding the trajectory of mining waste management toward a greener and more sustainable future. Together, we must work toward promoting the use of these technologies and opening the door to a greener, more sustainable future by investing in research and development, fostering collaborations, and advocating for ethical mining practices. As this chapter concludes, mining waste management is on the brink of a new era. Technologies for microbial remediation have the potential to fundamentally alter how we interact with the environment. With each experiment conducted, technological advancement achieved, and the site successfully remediated, we move a step closer to striking a balance between the needs of contemporary society and the preservation of our natural world.

Acknowledgment This research was supported by the Science and Technology Human Resource Development Project, Ministry of Higher Education, Sri Lanka, funded by the Asian Development Bank (Grant No. CRG/R2/KE6).

References

Abashina T, Vainshtein M (2023) Current trends in metal biomining with a focus on genomics aspects and attention to arsenopyrite leaching—a review. Microorganisms. 11:183

Abo-Alkasem M, Hassan N, Abo EM (2023) Microbial bioremediation as a tool for the removal of heavy metals. Bull Nat Res Centre 47:31

Acharya B, Kharel G (2020) Acid mine drainage from coal mining in the United States – an overview 588:125061

Agboola O, Babatunde D, Isaac FO, Sadiku E, Popoola P, Moropeng L (2020) A review on the impact of mining operation: Monit##oring, assessment and management. Resul Eng 8:100181

Ai C, Yan Z, Hou S, Zheng X, Zeng Z, Amanze C (2020a) Effective treatment of acid mine drainage with microbial fuel cells: an emphasis on typical energy substrates. Minerals 10:443

Ai C, Yan Z, Hou S, Zheng X, Zeng Z, Amanze C, Dai Z, Chai L, Qiu G, Zeng W (2020b) Effective treatment of acid mine drainage with microbial fuel cells: an emphasis on typical energy substrates. Minerals 10:443

Alcasabas A, Massingberd-Mundy F, Breeze B, Ruiz Pérez M, Martínez García C (2020) BIORECOVER - New bio-based technologies for recapture of critical raw materials, Johnson matthey technology review

Amobonye A, Aruwa C, Aransiola S, Omame J, Alabi T, Lalung J (2023) The Potential of Fungi in the Bioremediation of Pharmaceutically Active Compounds: a Comprehensive Review 14:1207792

Anekwe I, Isa Y (2023) Bioremediation of acid mine drainage – Review 65:1047–1075

Arif S, Schliekmann E, Hoppert M (2021) 16S rRNA amplicon sequencing of microbial biofilms from Marsberg copper mine. Germany 10:e01315-e1320

Aznar-Sánchez J, García-Gómez J, Velasco-Muñoz J, Carretero-Gómez A (2018) Mining waste and its sustainable management: advances in worldwide research. Minerals 8:284–311

Barkay T, Miller S, Summers A (2003) Bacterial Mercury Resistance from Atoms to Ecosystems. 27:355–384

Begum S, Rath S, Rath C (2022) Applications of Microbial Communities for the Remediation of Industrial and Mining Toxic Metal Waste: a Review 39:282–293

Bernoth L, Firth I, McAllister P, Rhodes S (2000) Biotechnologies for Remediation and Pollution Control in the Mining Industry 17:105–111

Bhattacharjee G, Gohil N, Singh V (2020) 14 - Synthetic biology approaches for bioremediation. In: Pandey VC, Singh V (eds), Bioremediation of pollutants. Elsevier

das Neves Vasconcellos Brandão I, Ferreira de Macedo E, Barboza de Souza Silva P, Fontana Batista A, Graciano Petroni S, Gonçalves M (2023) Bionanomining of copper-based nanoparticles using pre-processed mineral tailings as precursor 338:117804

Bosecker K (1997) Bioleaching: metal solubilization by microorganisms. FEMS Microbiol Rev 20:591–604

Bosecker K (1999) Microbial leaching in environmental clean-up programmes. Process Metallurgy. Elsevier 533–536

Capeness M, Horsfall L (2020) Synthetic biology approaches towards the recycling of metals from the environment. Biochem Soc Trans 48:1367–1378

Cavicchioli R, Charlton T, Ertan H, Omar S, Siddiqui K, Williams T (2011) Biotechnological Uses of Enzymes from Psychrophiles 4:449–460

Chakraborty R, Wu C, Hazen T (2012) Systems biology approach to bioremediation. Curr Opin Biotechnol 23:483–490

Chaurasia P, Jasuja ND, Kumar S (2022) Role of nanotechnology in microbial mediated remediation. In Review

Chen L-X, Hu M, Huang L-N, Hua Z-S, Kuang J-L, Li S-J (2015) Comparative metagenomic and metatranscriptomic analyses of microbial communities in acid mine drainage 9:1579–1592

Coelho E, Reis T, Cotrim M, Mullan T, Corrêa B (2020) Resistant fungi isolated from contaminated uranium mine in Brazil shows a high capacity to uptake uranium from water. Chemosphere 248:126068

Coker J (2016) Extremophiles and Biotechnology: Current Uses and Prospects 5:396

Costa J, de Castro K, Rodriguez R, Sancinetti G (2022) Anaerobic Reactors for the Treatment of Sulphate and Metal-Rich Wastewater: a Review. 102:923–934

Costa F, Lima É, Brito Y, de Oliveira D, da Rocha Filho G, do Nascimento L (2021) Microbial approach for valorization of mining wastes and tailings: an overview, pp 261–281

Crisostomo C, Lima F, Dias R, Cardoso V, de Resende M (2016) Joint assessment of bioreduction of chromium (VI) and of removals of both total chromium and total organic carbon (TOC) in sequential hybrid bioreactors. Water, Air, Soil Poll 227:51

Das N (2010) Recovery of precious metals through biosorption — A review 103:180–189

Das N, Chandran P (2010) Microbial degradation of petroleum hydrocarbon contaminants: an overview. Biotechnol Res Int 2011:941810

Das A, Ghosh S (2022) Role of Microorganisms in Extenuation of Mining and Industrial Wastes 39:173–175

Das S, Dash H, Chakraborty J (2016) Genetic basis and importance of metal resistant genes in bacteria for bioremediation of contaminated environments with toxic metal pollutants. Appl Microbiol Biotechnol 100:2967–2984

Das A, Ghosh S (2018) Bioleaching of manganese from mining waste materials. In: Second International Conference on Materials Science

Deshmukh D (2010) Elements of mining technology Vol. 1 (8th Edition). Denett & Company

Ding W, Zhang Y, Shi S (2020) Development and Application of CRISPR/Cas in Microbial Biotechnology 8:711

Dold B (2003) Speciation of the Most Soluble Phases in a Sequential Extraction Procedure Adapted for Geochemical Studies of Copper Sulfide Mine Waste 80:55–68

Dopson M, Holmes D (2014) Metal resistance in acidophilic microorganisms and its significance for biotechnologies. Appl Microbiol Biotechnol 98:8133–8144

Dopson M, Johnson D (2012) Biodiversity, metabolism and applications of acidophilic sulfur-metabolizing microorganisms. Environ Microbiol 14:2620–2631

Dopson M, Lindström E (1999) Potential Role of *Thiobacillus caldus* in arsenopyrite bioleaching. Appl Environ Microbiol 65:36–40

Enkateswarlu K, Nirola R, Kuppusamy S, Thavamani P, Naidu R, Megharaj M (2016) Abandoned metalliferous mines: ecological impacts and potential approaches for reclamation. Rev Environ Sci Bio/Technol 15:327–354

Fetzner S, Lingens F (1994) Bacterial dehalogenases: biochemistry, genetics, and biotechnological applications. Microbiol Rev 58:641–685

Foti M, Sorokin D, Lomans B, Mussman M, Zacharova E, Pimenov N (2007) Diversity, activity, and abundance of sulfate-reducing bacteria in saline and hypersaline Soda Lakes. Appl Environ Microbiol 73:2093–2100

Furukawa K (2000) Engineering dioxygenases for efficient degradation of environmental pollutants. Curr Opin Biotechnol 11:244–249

Gabay T (2022) Investigating the effects of phosphate mining in the Negev Desert on biological soil crust communities. Ben-Gurion University of the Negev

Gadd G (2010) Metals, minerals and microbes: geomicrobiology and bioremediation 156:609–643

Gao W, Wu C (2023) Editorial: biogeochemistry of metals in contaminated environments. Front Environ Sci 11:1187361

Ghose M (2004) Effect of opencast mining on soil fertility. J Sci Ind Res 63

Godoy-Faúndez A, Aitken D, Reyes-Bozo L, Rivera D (2015) Environmental-microbial biotechnology inside mining operations from an engineering viewpoint based on LCA. In: Sukla LB, Pradhan N, Panda S, Mishra BK (eds) Environmental microbial biotechnology. Springer International Publishing, Cham, pp 133–158

Gupta GA, Joia J (2016) Microbes as potential tool for remediation of heavy metals: a review. J Micro Biochem Technol

Guri M, Durand L, Cueff-Gauchard V, Zbinden M, Crassous P, Shillito B (2012) Acquisition of epibiotic bacteria along the life cycle of the hydrothermal shrimp Rimicaris exoculata. ISME J 6:597–609

Hallberg K, González-Toril E, Johnson D (2010) Acidithiobacillus ferrivorans, sp. nov.; facultatively anaerobic, psychrotolerant iron-, and sulfur-oxidizing acidophiles isolated from metal mine-impacted environments. Extremophiles 14:9–19

Hassen A, Saidi N, Cherif M, Boudabous A (1998) Resistance of Environmental Bacteria to Heavy Metals 64:7–15

Hernández I, Morales M, Morales F, Romero L, Camacho J (2023) Removal of heavy metals from mine tailings by in-situ bioleaching coupled to electrokinetics 117183

Hilson G (2002) The Environmental Impact of Small-Scale Gold Mining in Ghana: Identifying Problems and Possible Solutions. 168:57–72

Hlihor R, Petronela C, Gavrilescu M (2022) Removal of heavy metals from the environment by phytoremediation and microbial remediation. In El-Gendy N (ed) Sustainable solutions for environmental pollution. Wiley, pp 95–146

Hu Y, Chen Y, Xu J, Wang X, Luo S, Mao B (2022) Metagenomic discovery of novel CRISPR-Cas13 systems 8:107

Hua Z-S, Han Y-J, Chen L-X, Liu J, Hu M, Li S-J (2015) Ecological Roles of Dominant and Rare Prokaryotes in Acid Mine Drainage Revealed by Metagenomics and Metatranscriptomics 9:1280–1294

Imron M, Kurniawan S, Soegianto A (2019) Characterization of mercury-reducing potential bacteria isolated from Keputih non-active sanitary landfill leachate. Surabaya, Indonesia under Different Saline Conditions. 241:113–122

Jaiswal S, Singh D, Shukla P (2019) Gene Editing and Systems Biology Tools for Pesticide Bioremediation: a Review 10:87

Jiang H, Dong H, Yu B, Liu X, Li Y, Ji S (2007) Microbial response to salinity change in Lake Chaka, a hypersaline lake on Tibetan plateau 9:2603–2621

Jiménez-Díaz V, Pedroza-Rodríguez A, Ramos-Monroy O, Castillo-Carvajal L (2022) Synthetic biology: a new Era in hydrocarbon bioremediation. Processes 10:712

Johnsen A, Karlson U (2004) Evaluation of bacterial strategies to promote the bioavailability of polycyclic aromatic hydrocarbons. Appl Microbiol Biotechnol 63:452–459

Johnson D, Hallberg K (2007) Techniques for detecting and identifying acidophilic mineral-oxidizing microorganisms. In: Rawlings DE, Johnson DB (eds), Biomining. Berlin, Heidelberg: Springer Berlin Heidelberg, pp 237–261

Johnson D, Hallberg K (2003) The Microbiology of Acidic Mine Waters 154:466–473

Johnson D, Hallberg K (2005) Acid mine drainage remediation options: a review. Sci Total Environ 338:3–14

Kalin M, Wheeler W, Meinrath G (2005) The removal of uranium from mining waste water using algal/microbial biomass. J Environ Radioact 78:151–177

Kapahi M, Sachdeva S (2019) Bioremediation Options for Heavy Metal Pollution 9:191203

Kaya S, Ozcelikay G, Cetinkaya A, Bakirhan N, Ozkan S (2023) Emerging trends—current developments in nanobioremediation via electrochemistry. Elsevier, In Nanoremediation

Kebede G, Tafese T, Abda E, Kamaraj M, Assefa F (2021) Factors Influencing the Bacterial Bioremediation of Hydrocarbon Contaminants in the Soil: Mechanisms and Impacts 2021:9823362

Kesler S, Simon A (2015) Mineral resources. Economics and the Environment. Cambridge University Press, Cambridge

Kiani S, Lehosmaa K, Kløve B, Ronkanen A-K (2022) Nitrogen removal of mine-influenced water in a hybrid bioreactor with floating hook-moss (Warnstorfia fluitans) in cold climate conditions. Ecol Eng 177:106562

Klaus-Joerger T, Joerger R, Olsson E, Granqvist C-G (2001) Bacteria as workers in the living factory: metal-accumulating bacteria and their potential for materials science. Trends Biotechnol 19:15–20

Konopka A, Zakharova T, Bischoff M, Oliver L, Nakatsu C, Turco R (1999) Microbial biomass and activity in lead-contaminated soil. Appl Environ Microbiol 65:2256–2259

Krümmelbein J, Bens O, Raab T, Naeth M (2012) A history of lignite coal mining and reclamation practices in Lusatia, eastern Germany. Can J Soil Sci 92:53–66

Kupka D, Liljeqvist M, Nurmi P, Puhakka J, Tuovinen O, Dopson M (2009) 160:767–774

Kuyucak N, Akcil A (2013) Cyanide and Removal Options from Effluents in Gold Mining and Metallurgical Processes 50–51:13–29

Lehembre F, Doillon D, David E, Perrotto S, Baude J, Foulon J (2013) Soil Metatranscriptomics for Mining Eukaryotic Heavy Metal Resistance Genes 15:2829–2840

Leitão A (2009) Potential of Penicillium Species in the Bioremediation Field 6:1393–1417

Lemes C, Villa M, Felestrino É, Perucci L, Assis R, Cordeiro I (2021) 16S rRNA gene amplicon sequencing data of the iron quadrangle ferruginous caves (Brazil) Shows the importance of conserving this singular and threatened geosystem 13:494

Li F, Neves A, Ghoshal B, Guan L (2018) Symposium review: Mining metagenomic and metatranscriptomic data for clues about microbial metabolic functions in ruminants1. J Dairy Sci 101:5605–5618

Li JT, Jia P, Wang X-J, Ou S-N, Yang T-T, Feng S-W (2022) Metagenomic and metatranscriptomic insights into sulfate-reducing bacteria in a revegetated acidic mine wasteland 8:71

Lin Y, Jiao Y, Zhao M, Wang G, Wang D, Xiao W (2021) Ecological Restoration of Wetland Polluted by Heavy Metals in Xiangtan Manganese Mine Area 9:1702

Luka Y, Highina B, Zubairu A (2018) Bioremediation: a solution to environmental pollution-a review 7:101–109

Ma X (2014) Study on ecological restoration of Handan iron mining area. Adv Mater Res 955:2102–2105

Malik S, Bora J, Nag S, Sinha S, Mondal S, Rustagi S, Mondal S, Rustagi S et al (2023) Fungal-based remediation in the treatment of anthropogenic activities and pharmaceutical-pollutant-contaminated wastewater. Water 15:2262

Mandeep SP (2020) Microbial Nanotechnology for Bioremediation of Industrial Wastewater 11:590631

Mendez M, Neilson J, Maier R (2008) Characterization of a Bacterial Community in an Abandoned Semiarid Lead-Zinc Mine Tailing Site 74:3899–3907

Mishra S, Panda S, Pradhan N, Biswal S, Sukla L, Mishra B (2015) Microbe–mineral interactions: exploring avenues towards development of a sustainable microbial technology for coal beneficiation. In LB S, Pradhan N, Panda S, Mishra B (eds), Environmental microbial biotechnology. Cham: Springer International Publishing, pp 33–52

Moldovan B, David L (2020) Green synthesis of biogenic silver nanoparticles for efficient catalytic removal of harmful organic dyes. Nanomaterials 10:202

Moore J, Luoma S (1990) Hazardous wastes from large-scale metal extraction. A Case Study. 24:1278–1285

Morello L, Raga R, Lavagnolo M, Pivato A, Ali M, Yue D (2017) The S.An.A.® concept: Semi-aerobic, Anaerobic, Aerated bioreactor landfill. Waste Manag 67:193–202

Mosa K, Saadoun I, Kumar K, Helmy M, Dhankher O (2016) Potential biotechnological strategies for the cleanup of heavy metals and metalloids, 7

Mudd G (2010) The Environmental Sustainability of Mining in Australia: Key Mega-Trends and Looming Constraints 35:98–115

Mukherjee A, Reddy M (2020) Metatranscriptomics: an approach for retrieving novel eukaryotic genes from polluted and related environments 10:71

Muniruzzaman M, Pedretti D (2021) Mechanistic Models Supporting Uncertainty Quantification of Water Quality Predictions in Heterogeneous Mining Waste Rocks: a Review 35:985–1001

Neu M, Boukhalfa H, Merroun M (2010) Biomineralization and biotransformations of actinide materials. MRS Bull 35:849–857

Newsome L, Falagán C (2021) The microbiology of metal mine waste: bioremediation applications and implications for planetary health. GeoHealth 5:e2020GH000380

Nies D (1999) Microbial heavy-metal resistance. Appl Microbiol Biotechnol 51:730–750

Nkuna R, Ijoma G, Matambo T (2022) Applying EDTA in chelating excess metal ions to improve downstream DNA recovery from mine tailings for long-read amplicon sequencing of acidophilic fungi communities. J Fungi. 8:419

Noike T, Nakamura K, Matsumoto J (1983) Oxidation of ferrous iron by acidophilic iron-oxidizing bacteria from a stream receiving acid mine drainage. Water Res 17:21–27

Noll P, Lilge L, Hausmann R, Henkel M (2020) Modeling and Exploiting Microbial Temperature Response 8:121

Ntougias S, Polkowska Ż, Nikolaki S, Dionyssopoulou E, Stathopoulou P, Doudoumis V (2016) Bacterial community structures in freshwater polar environments of Svalbard. Microbes Environ 31:401–409

Okanigbe D, Popoola A, Malatji N, Lesufi T, Sekgobela G (2022) Bionanomining: a revised insight into processing of south Africa's complex gold ores. Springer

Ortiz-Castillo J, Mirazimi M, Mohammadi M, Dy E, Liu W (2021) The role of microorganisms in the formation, dissolution, and transformation of secondary minerals in mine rock and drainage: a review 11:1349
Park D, Yun Y-S, Park J (2010) The past, present, and future trends of biosorption. Biotechnol Bioprocess Eng 15:86–102
Perez ID, Botelho Junior AB, Aliprandini P (2020) Espinosa DCR: copper recovery from nickel laterite with high-iron content: A continuous process from mining waste. Canadian J Chem Eng 98:957–998
Pitchaiah P (2017) Impacts of sand mining on environment–a review. Intern J Geoinform Geol Sci 4:1–6
Poretsky R, Rodriguez RL, Luo C, Tsementzi D, Konstantinidis K (2014) Strengths and limitations of 16S rRNA gene amplicon sequencing in revealing temporal microbial community dynamics 9:e93827
Puggioni G (2022) Remediation of Multi-contaminated groundwater using bioelectrochemical systems
Punia A, Pratap Singh R, Singh V, Chauhan N (2023) Chapter 17 - environment sustainability with microbial nanotechnology. In Singh P, Kumar V, Bakshi M, Hussain C, Sillanpää M (eds), Environmental applications of microbial nanotechnology. Elsevier, pp 289–314
Punia A, Siddaiah N (2017) Assessment of heavy metal contamination in groundwater of khetri copper mine region. India Health Risk Asses 14:9–19
Rabani M, Hameed I, Mir T, Wani B, Gupta M, Habib A (2022) Chapter 5 - Microbial-assisted phytoremediation. In Bhat R, Tonelli F, Dar G, Hakeem K (eds), Phytoremediation. Academic Press, pp 91–114
Rafeeq H, Afsheen N, Rafique S, Arshad A, Intisar M, Hussain A (2023) Genetically engineered microorganisms for environmental remediation. Chemosphere 310:136751
Rai P (2008) Heavy Metal Pollution in Aquatic Ecosystems and Its Phytoremediation Using Wetland Plants: an Ecosustainable Approach 10:133–160
Rajeev M, Sushmitha T, Aravindraja C, Toleti S, Pandian S (2021) Exploring the Impacts of Heavy Metals on Spatial Variations of Sediment-Associated Bacterial Communities 209:111808
Rawlings D, Johnson D (2007) The microbiology of biomining: development and optimization of mineral-oxidizing microbial consortia. Microbiology 153:315–324
Rebello S, Anoopkumar A, Aneesh E, Sindhu R, Binod P, Kim S (2021) Hazardous Minerals Mining: Challenges and Solutions 402:123474
Ribeiro A, Prasad M (2021) Electrokinetic remediation for environmental security and sustainability. Wiley Online Library
Riekkola-Vanhanen M (2013) Talvivaara mining company – from a project to a mine. Miner Eng 48:2–9
Robles-González I, Fava F, Poggi-Varaldo H (2008) A Review on Slurry Bioreactors for Bioremediation of Soils and Sediments 7:5
Rodríguez-Galán M, Baena-Moreno F, Vázquez S, Arroyo-Torralvo F, Vilches L, Zhang Z (2019) Remediation of Acid Mine Drainage 17:1529–1538
Roldán M, Olaya-Abril A, Sáez L, Cabello P, Luque-Almagro V, Moreno-Vivián C (2021) Bioremediation of cyanide-containing wastes: the potential of systems and synthetic biology for cleaning up the toxic leftovers from mining. EMBO Rep 22:e53720
Rylott E, Bruce N (2020) How synthetic biology can help bioremediation. Synth Biol Synth Biomol 58, 86–95
Sahoo S, Routray SP, Lenka S, Bhuyan R, Mohanty JN (2022) CRISPR/Cas-Mediated functional gene editing for improvement in bioremediation: an emerging strategy. In Kumar V, Thakur IS (eds), Omics insights in environmental bioremediation. Springer, Singapore
Sana S, Neelam D, Gupta V, Devki D, Rahi R (2021) An overview: application of microorganisms in bio-mining of metals. Intern J Pharm Biolog Sci 11:01–08

Sánchez-Andrea AI, Rodríguez N, Amils R, Sanz J (2011) Microbial diversity in anaerobic sediments at Río Tinto, a naturally acidic environment with a high heavy metal content. Appl Environ Microbiol 77:6085–6093

Sánchez-Andrea I, Triana D, Sanz J (2012) Bioremediation of acid mine drainage coupled with domestic wastewater treatment. Water Sci Technol 66:2425–2431

Sand W, Gehrke T, Jozsa P-G, Schippers A (2001) (Bio)chemistry of bacterial leaching—direct vs. indirect bioleaching. Biohydrometallurgy and the environment-toward the mining 59:159–75

Sand W, Gehrke T (2006) Extracellular polymeric substances mediate bioleaching/biocorrosion via interfacial processes involving iron(III) ions and acidophilic bacteria. Res Microbiol 157:49–56

Saxena P, Singh N, Harish, null, Singh A, Pandey S, Thanki A (2020) Recent advances in phytoremediation using genome engineering CRISPR–Cas9 technology. Bioremediation of Pollutants. Elsevier

Schippers A, Sand W (1999) Bacterial leaching of metal sulfides proceeds by two indirect mechanisms via thiosulfate or via polysulfides and sulfur. Appl Environ Microbiol 65:319–321

Schippers VM, Sand W (2013) Progress in bioleaching: fundamentals and mechanisms of bacterial metal sulfide oxidation—part A. Appl Microbiol Biotechnol 97:7529–7541

Sengupta M (2021) Environmental impacts of mining: monitoring. Restoration and Control. CRC Press, Boca Raton, FL

Seo J-S, Keum Y-S, Li Q (2009) Bacterial degradation of aromatic compounds. Int J Environ Res Public Health 6:278–309

Shahriari Moghadam M, Ebrahimipour G, Abtahi B, Ghassempour A, Hashtroudi M (2014) Biodegradation of polycyclic aromatic hydrocarbons by a bacterial consortium enriched from mangrove sediments. J Environ Health Sci Eng 12:114

Sigel HSRM (2005) Metal ions in biological systems: biogeochemisty, Availability, and Transport of Metals in the Environment. CRC Press

Singer A, Gilbert E, Luepromchai E, Crowley D (2000) Bioremediation of polychlorinated biphenyl-contaminated soil using carvone and surfactant-grown bacteria. Appl Microbiol Biotechnol 54:838–433

Singh B, Walker A (2006) Microbial Degradation of Organophosphorus Compounds 30:428–471

Sintubin L, De Windt W, Dick J, Mast J, Van Der Ha D, Verstraete W (2009) Lactic acid bacteria as reducing and capping agent for the fast and efficient production of silver nanoparticles. Appl Microbiol Biotechnol 84:741–749

Stefanova A, Angelov A, Bratkova S, Genova P, Nikolova K (2018) Influence of electrical conductivity and temperature in a microbial fuel cell for treatment of mining waste water. Lett Soc Sci Series 3:1–7

Stierle A, Stierle D (2005) Bioprospecting in the Berkeley pit: Bio active metabolites from acid mine waste extremophiles, pp 1123–1175

Sun W, Ji B, Khoso S, Tang H, Liu R, Wang L (2018) An Extensive Review on Restoration Technologies for Mining Tailings 25:33911–33925

Sun W, Xiao E, Krumins V, Dong Y, Li B, Deng J (2019) Comparative Analyses of the Microbial Communities Inhabiting Coal Mining Waste Dump and an Adjacent Acid Mine Drainage Creek 78:651–664

Tang C, Zhong J, Lv Y, Liu X, Li Y, Zhang M (2021) Response and dynamic change of microbial community during bioremediation of Uranium Tailings by Bacillus sp. Minerals 11:967

Thiruvenkatachari R, Su S (2017) 5 - aerobic treatment of effluents from the mining industry. In: Lee D-J, Jegatheesan V, Ngo H, Hallenbeck P, Pandey A (eds) Current Developments in Biotechnology and Bioengineering. Elsevier

Tran, K., Lee, H.-M., Thai, T., Shen, J., Eyun, S.-. i ., Na, D.: Synthetically engineered microbial scavengers for enhanced bioremediation. Journal of Hazardous Materials. 419, 126516, (2021)

Tuovinen H, Pelkonen M, Lempinen J, Pohjolainen E, Read D, Solatie D (2018) Behaviour of metals during Bioheap leaching at the Talvivaara mine. Finland 8:66

Vaksmaa A, Guerrero-Cruz S, Ghosh P, Zeghal E, Hernando-Morales V, Niemann H (2023) Role of fungi in bioremediation of emerging pollutants. Front Mar Sci 10:1070905

Valdes V, Quatrini J, Hallberg R, Dopson K, Valenzuela PDT, null, Holmes D (2009) Draft genome sequence of the extremely acidophilic bacterium *Acidithiobacillus caldus* ATCC 51756 reveals metabolic versatility in the genus *Acidithiobacillus*. J Bacteriol 191:5877–5878

van der Heyde M, Bunce M, Dixon K, Wardell-Johnson G, White N, Nevill P (2020) Changes in Soil Microbial Communities in Post Mine Ecological Restoration: Implications for Monitoring Using High Throughput DNA Sequencing 749:142262

Van Hille R, Boshoff G, Rose P, Duncan J (1999) A continuous process for the biological treatment of heavy metal contaminated acid mine water. Resour Conserv Recycl 27:157–167

Větrovský T, Baldrian P (2013) Analysis of soil fungal communities by amplicon pyrosequencing: current approaches to data analysis and the introduction of the pipeline SEED. Biol Fertil Soils 49:1027–1037

Wang Y (2018) Biostimulation remediation technologies for groundwater contaminants. IGI Global

Wang X, Ma L, Wu J, Xiao Y, Tao J, Liu X (2020) Effective bioleaching of low-grade copper ores: insights from microbial cross experiments. Biores Technol 308:123273

Whitehead P, Prior H (2005) Bioremediation of Acid Mine Drainage: an Introduction to the Wheal Jane Wetlands Project 338:15–21

Wiegel J, Wu Q (2000) Microbial reductive dehalogenation of polychlorinated biphenyls. FEMS Microbiol Ecol 32:1–15

Willow M, Cohen R (2003) PH, dissolved oxygen, and adsorption effects on metal removal in anaerobic bioreactors. J Environ Qual 32:1212–1221

Wong-Pinto L-S, Mercado A, Chong G, Salazar P, Ordóñez J (2021) Biosynthesis of copper nanoparticles from copper tailings ore–an approach to the 'Bionanomining.' J Clean Prod 315:128107

Wong-Pinto L-S, null, Menzies A, Ordóñez J (2020) Bionanomining: biotechnological synthesis of metal nanoparticles from mining waste—opportunity for sustainable management of mining environmental liabilities. Appl Microb Biotech 104:1859–1869

Wu A, Yin S, Wang H, Qin W, Qiu G (2009) Technological assessment of a mining-waste dump at the Dexing copper mine, China, for possible conversion to an in situ bioleaching operation. Biores Technol 100:1931–1936

Wu Y, Jameel A, Xing X-H, Zhang C (2022b) Advanced strategies and tools to facilitate and streamline microbial adaptive laboratory evolution. Trends Biotechnol 40:38–59

Wu Z, Man Q, Niu H, Lyu H, Song H, Li R (2022) Recent advances and trends of trichloroethylene biodegradation: a critical review. 13

Yin S, Wang L, Kabwe E, Chen X, Yan R, An K (2018) Copper bioleaching in China: review and prospect. Minerals 8:32

Zhang M, Li Z, Haggblom M, Young L, He Z, Li F (2020) Characterization of nitrate-dependent As (III)-oxidizing communities in arsenic-contaminated soil and investigation of their metabolic potentials by the combination of DNA-stable isotope probing and metagenomics. Environ Sci Technol 54:7366–7377

Zhang CY (2021) Application of microbiotechnology in ecological restoration of mines. IOP conference series: earth and environmental science: IOP publishing, pp 1755–1315

Zhang T, Surampalli R, Tyagi R, Benerji S (2017) 14 - biological treatment of hazardous waste. In: Wong J, Tyagi R, Pandey A (eds) Current developments in biotechnology and bioengineering. Elsevier, pp 311–340

Zhang M, Kolton M, Li Z, Lin H, Li F, Lu G (2021) Bacteria responsible for antimonite oxidation in antimony-contaminated soil revealed by DNA-SIP coupled to metagenomics. FEMS Microbiol Ecol 97:fiab057

Chapter 4
Manganese Contaminant: Revolutionizing In-Situ Microbial-Assisted Remediation Approach

P. Kamleshiya

4.1 Introduction

Manganese (Mn) is a strategically crucial transition metal with a multifaceted array of industrial applications, from steel production to batteries, non-ferrous metallurgical, leather tanning, catalyst, firework, colouring agent, and animal feed (Rongrong et al. 2020; Silva et al. 2009). Manganese is the 12th most abundant element and the 3rd most abundant transition metal present in the earth's crust (Das et al. 2015; Volesky 1990). Its deposits mostly exist in the form of different oxides, ores, and carbonates (Das et al. 2015; Rongrong et al. 2020). Manganese is the fourth most used metal after iron, aluminium, and copper. (Bafghi et al. 2008; Silva et al. 2009). The rapid rise of anthropogenic sources such as industrialization causes increasing environmental contamination of this heavy metal pollutant (Rongrong et al. 2020; Silva et al. 2009), resulting in serious ecosystem damage and health hazards. Manganese is introduced into the environment through industrial waste and wastewater irrigation (Rahhou et al. 2023; Santamaria 2008; Sharma xxxx; Sinha and Purcell 2019). During the extraction of manganese from their ores, some of its portion is left out as leachate in the open and, via flood or wind, gets relocated from one place to another. (Pande et al. 2022; Rongrong et al. 2020). Due to hazardous health effects and environmental issues, the eradication of heavy metals and minerals via different possible means has been a topic of concern. Detoxification measures for Mn pollutants, which have been utilized for decades, involve several physical and electrochemical processes such as adsorption, ionic exchange, and membrane filtration to recover the valuable metal (Zapffe 1933). Although physical and chemical eradication methods for manganese

P. Kamleshiya (✉)
Department of Zoology, Govt. J.S.T.P.G College Balaghat, R.S.S. University Chindwara, Chhindwara, M.P, India
e-mail: kpushp99@gmail.com

K. Bala et al. (eds.), *Harnessing Microbial Potential for Multifarious Applications*, Energy, Environment, and Sustainability, https://doi.org/10.1007/978-981-97-1152-9_4

removal have been utilized for decades, these methods do not have much impact on effluent detoxification (Das et al. 2015). Formation of undesirable secondary byproducts during the process, (Su et al. 2014) low removal percentage, their high cost, and energy inefficiency impacted these physiochemical methods of oxidation. The advancement of novel molecular techniques focuses light on understanding Mn detoxifying microbes present in the environment, which led to an increase in bioremediation applications for Mn oxidation. This kind of biological approach is also seen as a novel remediation for emerging contaminants. A broad group of microbes participate in this bioremediation process, such as aerobic and anaerobic bacteria, algae, and fungi. According to the site where the process of bioremediation is held, it is generally categorized into two schemes (Rahhou et al. 2023): *ex-situ* and *in-situ* bioremediation methods. The basic strategy employed in *ex-situ* remediation for pollutant treatment is to excavate the contaminated soil from the location by microbial organisms. This type of method implies the use of bioreactors and nutritional supplements to accelerate the oxidation of environmental pollutants. The disintegration or removal of toxic dissolved pollutants by the microorganism directly at the location of pollution is referred to as the *in-situ* approach. It can be applied to treat polluted soil, groundwater, and mine effluent (Chu et al. 2011; Oleksandr, et al. 2013). This chapter mainly endeavours some microbial-assisted in-situ remediation approaches to overcome difficulties generated during diverse physiochemical treatments implemented for low-grade Mn ore. Such bioremediation methods are discussed here, including biosorption, bio-oxidation, bioaccumulation, bioleaching, and microbial-induced carbonate precipitation (MICP), in detail with their mechanisms, regulatory factors, advantages, and limitations. A comprehensive summary of combination approaches in microbe-intervened remediation and other methods is also discussed. By upgrading the potential of metal-oxidizing microorganisms, bioremediation technology can predominantly remediate Mn toxicity. The microbial species possessing manganese oxidation ability are the key factors in Mn biomining and oxidation. This chapter finally concluded with an in-depth observation of effective microbial-assisted Mn removal techniques, which will further guide the selection of appropriate operation procedures for Mn bioremediation.

4.2 Occurrence, Properties, and Distribution

Manganese is a hard, brittle, silvery-grey ferroalloy metal. This metallic element is abundant and is found in areas of metamorphic and sedimentary rock (Rongrong et al. 2020). Mineralogical investigations revealed that in nature numerous forms of manganese occur (~300 minerals), which are abundant in rock, soil, and sediments (Das et al. 2015), among them major Mn minerals found as higher oxides and minerals, mainly pyrolusite (MnO_2) (Acharya et al. 2004), birnessite (MnO_2), psilomelane [Ba, $Mn^{2+}(Mn^{4+})_8O_{16}(OH)_4$], manganite (MnOOH), romanechite ($BaMnMn_5O_{16}(OH)_4$), vernadite (MnO_2), hausmannite (Mn_3O_4), braunite (Mn, Si2O3), and Mn-nodules (Acharya et al. 2004; Hariprasad 2007; Diaconu et al.

2023a; Nour et al. 2022). Manganese nodules are mineral contents found at the bottom of the sea and are made of manganese and iron oxides. It also exists in very low concentrations in animals, plants, and other living beings. Being a transition metal, manganese has several diverse oxidation states, ranging from + 2 to + 7, with the majority being manganese oxide (Mn^{2+}/Mn^{3+}) (Aschner et al. 1999). In nature, it exists mainly in its oxidized chemical forms, such as manganese dioxide (MnO_2) or tri manganese tetraoxide (Mn_3O_4) (Post 1994). The versatile chemical nature of Mn is due to its different oxidation stages. By virtue of these states, manganese can form a variety of chemical compounds, making it applicable to many industrial processes (Michalke and Fernsebner 2014; Rongrong et al. 2020; Su et al. 2014). In groundwater, concentrations of manganese normally range between 1 and 100 mg/ L, with most values being below 10 mg/L. Typical airborne levels of manganese (in the absence of excessive pollution) range from 0.05 to 0.10 mg (Alexander 1999; Post 1994). The main foremost reserve of Mn is found in countries like South Africa and the Ukraine (>80%), China, Australia, Gabon, Brazil, India, and Mexico (Das et al. 2015; Mangood et al. 2022). Mn deposits are frequently present in the Paleoproterozoic Hotazel formation in South Africa, conserving more than 4 billion heaps of Mn ores, making the Kalahari manganese field the world's largest land-based economic reserve (Das et al. 2015; Kirschvink 1992). In India, manganese ore deposits are primarily associated with Madhya Pradesh, Maharashtra, Gujarat, Odisha, and Andhra Pradesh. Manganese Ore India Limited (MOIL) is the major manganese manufacturer, producing approximately two billion tonnes of high-grade ores annually. Manganese dioxide (MnO_2) is the main oxidized mineral of Mn (Das et al. 2015; Zhang et al. 2019; Zhu et al. 2011). In the general map of India, Madhya Pradesh stood in an important position in manganese ore production.

4.3 Manganese Biogeochemical Cycle

The biogeochemical cycle of Mn occurs through the atmosphere, hydrosphere, lithosphere, and biosphere (Ehrlich 2008). It is cycled as similarly as iron through the earth but via different redox pathways. Anthropogenic activities have affected the fluxes of manganese among the different spheres of the Earth. Manganese (Mn) circulation among the global "spheres" is arbitrated by both physiochemical and biological processes, as depicted in Fig. 4.1a and b. The entrance of manganese from the lithosphere to the hydrosphere occurs from the dissolution and erosion of bedrock in the river; in solution form, it then makes its way into the ocean. Manganese enters the ocean as dust or runoff in the form of dissolved Mn (II). It leaves the ocean via diffusion. This dissolved Mn is oxidized and reduced by organisms and then sinks to the ocean floor. As it sinks, it undergoes aggregation or scavenging. From the ocean floor, Mn is lost through burial and is inputted into the ocean cycle by sediment diffusion and hydrothermal process (Van et al. 2016). Anthropogenic influences mainly stem from mineral processing and mining for industrial use; the burning of fossil fuels alters the global Mn cycle (Howe and Malcolm 2004).

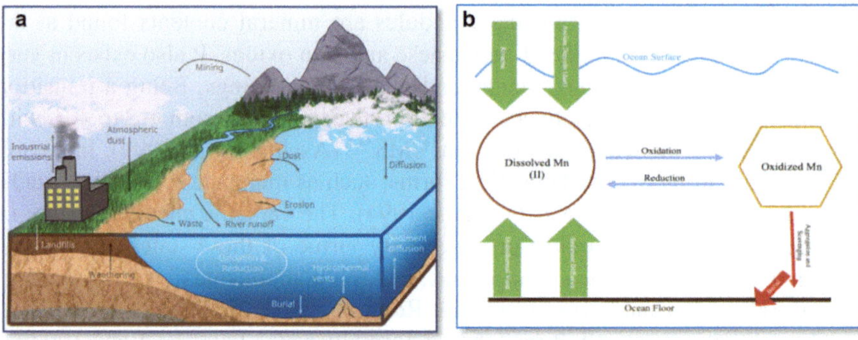

Fig. 4.1 **a** Manganese biogeochemical cycle: Adapted from https://www.frontiersin.org/journals/plant-science/articles/10.3389/fpls.2020.00300/full. **b** Stages of manganese biogeochemical cycle: Adapted from https://www.mountsinai.org/health-library/supplement/manganese

4.4 Manganese Indispensability

The Mn (II) ion exhibits a crucial role as an essential trace element in all living organisms. As a cofactor, it is involved in carbohydrate, lipid, cholesterol, and amino acid metabolism; it helps in bone formation, immune response, free radical scavenging activity, and as a component of the blood clotting cascade (Chen et al. 2018; Failed 2018). It plays an important role in fat and carbohydrate metabolism, calcium absorption, and blood sugar regulation. Manganese is required for brain functioning. Manganese functions as a cofactor for some antioxidant enzymes, such as superoxide dismutase (SOD). Naturally generated and freely wandering radical ions can start the oxidation cascade, thus damaging cell membranes and vital biomolecules like nucleic acid and protein. Therefore, it is responsible for developing various ailments such as mental disability, heart disease, and cancer.

Antioxidant enzymes, such as SOD, can help neutralize free radicals and reduce or even prevent the damage they cause (https://www.mountsinai.org/health-library/supplement/manganese; Post 1994). Mn deficiency can also lead to bone malformation, weakness, seizures, and infertility. From the perspective of self-defence, findings of a recent study show that Mn not only promotes a body's potential to protect against deoxy ribonucleic acid (DNA) viruses by increasing the sensibility of the DNA sensor cyclic GMP-AMP synthase (cGAS) and its downstream adapter protein, stimulator of interferon genes (STING), but it also helps in the inherent immune sensing of tumours and thereby accelerates the adaptive immune response against tumours (Lv et al. 2020). Mn is one of capital important micronutrients for healthy plant growth and development. It plays a key role in several crucial physiological processes in plants. It assists as a cofactor of the photosynthetic machinery in the water-splitting reaction of photosystem II (PSII) for the oxygen-evolving complex (OEC) (Cordova et al. 2012; Hakala et al. 2006). Manganese is a requisite trace element naturally available as a dietary supplement in many foods (https://www.frontiersin.org/journals/plant-science/articles/10.3389/fpls.2020.00300/full).

4.5 Manganese (Mn) Pollution and Toxicity

The environment is greatly harmed by the global expansion of the industrial revolution, unchecked anthropogenic activity, and the use of natural resources for metal mining and industry. Exposure to mine effluents in large quantities has resulted from widespread mining and industrialization. Land filling is currently an approved method of disposing of manganese solid waste (MSW), although this method has serious ecological consequences, including soil and groundwater contamination. Utilizing environmental manganese solid waste (EMSW) has thus become a problem for environmental conservation and sustainable development. Pollution of the water supply and the air is the main factor in Mn exposure (Santamaria 2008). Releasing large amounts of acid mine drainage (AMD) with high levels of manganese (Mn) imparts potentially detrimental effects on the environment. There are many sources of Mn in the environment, including weathered rocks, soil, and decaying plants. Human activities expose individuals to other sources of manganese, including the fungicides maneb and mancozeb, water purifying factors, and medical imaging. In addition, several countries, including the United States, Australia, Canada, Argentina, Russia, Bulgaria, France, New Zealand, China, and the European Union, have approved the use of the fuel additive methylcyclopentadienyl manganese tricarbonyl (MMT) (Michalke and Fernsebner 2014; Nadaska et al. 2012).

Mines, steelworks, battery factories, welding, fossil fuel combustion, and chemical engineering are among the industries with occupational exposure to Mn contamination (Das et al. 2015; Santamaria 2008). In the steel, iron, dry-cell battery, and ferromanganese manufacturing sectors, manganese exposure mostly results from inhaling manganese fumes or dusts. Manganese electrolysis produces a large amount of electrolytic Mn slag. For living beings, heavy metal pollution is of great concern as they are non-biodegradable and get accumulated easily in the food chain (Ngwenya et al. 2019). Elevated levels of manganese may have deleterious effects on the health of animals, plants, and the environment as they cause dangerous morbidity and mortality (Volesky 1990).

As per the World Health Organization (WHO) guidelines, Mn concentration in drinking water should not exceed 0.1 mg/L. Its excessive accumulation in the central nervous system (CNS) induces various ailments related to neurotoxicity, including psychiatric symptoms, Parkinson's disease, and (Zoni et al. 2012) a permanent neurological disorder known as "manganism." Mn overexposure can lead to cognitive and neurodegenerative damage and other impairments in the neuromotor system (Cordova et al. 2012; Jakkapong et al. 2019; Xiong et al. 2018). The effect of Mn on the nervous system is also influenced by its exposure route and magnitude of accumulation. Mn inhalation induces free radical generation and causes oxidative stress, thereby increasing neuronal apoptosis. In contrast, Mn ingestion seems to have a more subtle effect, altering neurochemistry, cognition, and neural chemistry. Nowadays, Mn toxicity is chiefly connected with chronic, high-level exposure of airborne Mn particulate to occupational industrial workers of manganese. On inhalation, it causes pulmonary inflammation gradually, including cough, bronchitis, pneumonitis, and impaired functioning of lung (Aschner et al. 1999; Rongrong et al. 2022). Vital biomolecules, viz., proteins, nucleic acids, lipids, and enzymes, get destructed due to intracellular free radical generation by heavy metals (HMs), thereby leading to oxidative stress. This stress results in numerous physiological problems like cell dysfunction, enzyme inhibition, aberrant protein trafficking, and misfolding, leading to interference in the vital activities of living organisms. An elevated level of Mn is responsible for immunological impairment and reproductive problems (Das et al. 2015).

Although heavy metals (HMs) such as Mn are part of the soil, when present in excess, they not only cause severe damage to the soil but also to the plants (Nour, et al. 2022). HMs are responsible for changing soil composition as well as increasing stress levels in plants, leading to crop failure. (Oleksandr et al. 2013) The symptoms of manganese toxicity in terrestrial plants vary widely between species. They include marginal chlorosis, necrotic lesions, and distorted development of the leaves. In plants, manganese tends to accumulate in the shoots rather than the roots, resulting in symptoms occurring in the leaves (Loneragan 1988). Manganese (Mn) is one of the typical metal pollutants that are found in aquatic environments. Unlike organic substances, it cannot be easily broken down into less harmful substances due to its lack of biodegradability, making it challenging to eliminate. It exists in a variety of forms, such as compounds or complexes with organic compounds, making it very persistent in the water body and only precipitating out if the pH is raised to 10.0. Mn is either absorbed by biota like fish and invertebrates or redistributed throughout the water column after being introduced to an aquatic environment (Hasan et al. 2016; Nielsen 2012) causing a variety of biological impacts that range from being necessary for living things to being deadly, respectively. Their higher concentrations could affect the entire trophic chain, including humans. Aquatic ecosystems, including aquatic flora and fauna, have the capacity to bioaccumulate metal residues from contaminated habitats (Bohu et al. 2015; Hasan et al. 2016; Yilmaz 2003), which may then enter the human food chain and cause health issues (Varol and Şen 2010). According to the physical and chemical conditions present between the sediment and the associated water column, metal accumulation in sediments occurs through processes of precipitation of specific compounds, binding fine solid particles, association with organic molecules, co-precipitation with Fe or Mn oxides, or species bounded as carbonates (Hatayama 2020). Some aquatic creatures have been put in danger by excessive Mn levels in the aquatic environment. According to a 2015 study by Fatimah et al., Mn toxicity reduces the quantity of free-living amoebae. As amoebae serve as bacterial predators, a decline in their abundance brought on by the presence of heavy metals will cause an imbalance in the natural ecosystem (Valavanidis et al. 2006).

4.6 Physiochemical Remediation of Mn

Several methods, as given below in Table 4.1, are several conventional physiochemical processes that have been utilized to remove heavy metal toxicity. These traditional methods, although excellent in nature, also show noticeable drawbacks. However, according to Hut and Puchooa (2017), the conventional method is regarded as effective for broad areas that are heavily polluted locally and for areas that are heavily contaminated with modest levels of heavy metals. Inevitably, there is a challenge for scientists and ecologists to design a novel technique that helps in the complete removal of heavy metals.

Table 4.1 Physiochemical remediation methods for manganese

Methods	Advantages	Disadvantages	References
Adsorption method	Efficient and economic strategy, adsorbent is easily assessable, good reusability of adsorbent	Generation of secondary pollutants, tends to cause large adsorbent sludge	Baoying and Kano (2015), Rongrong et al. (2022)
Electrodialysis	Relatively high productivity, Accurate regulation of kinetics	Low efficiency, Large power supply, High operating cost	Oleksandr et al. (2013), Tebo et al. (2004)
Reverse osmosis	Efficient filtration mechanism, Best method for water softening, Can remove large spectrum of contaminants	High operating cost, energy inefficient	https://www.frontiersin.org/journals/plant-science/articles/10.3389/fpls.2020.00300/full, Lei et al. (2018)
Chemical Precipitation	High removal, simple, and inexpensive	Production of large amounts of salty secondary pollutants, large amounts of sludge deposition, low solubility of metals	Alina (2020), Rongrong et al. (2022), Lei et al. (2018)
Ion exchange	High removal potential, Renewable and non-toxic, high recovery rate, simple equipment, and convenient operation and control	Excessive waste regeneration, lengthy experimental cycles, high salt consumption, ion exchange material is being susceptible to contamination	Baoying and Kano (2015), Chernikova et al. (2006), Das et al. (2015)
Membrane ultrafiltration	Low energy consumption, high efficiency, simple process	Frequent membrane replacement, Membrane fouling, cost inefficient	Nadaska et al. (2012), Yin et al. (2020)
Coagulations/ Flocculation	Economically feasible	High sludge formation and generation of large particles	Kushwaha et al. (2017), Nurlliana et al. (2017)

4.7 Microbial-Assisted Remediation of Mn

Bioremediation is a treatment technique that utilizes naturally occurring microbial diversity, either directly or indirectly, to break down toxic pollutants into less toxic or non-toxic materials. (Sharma and Bano 2018). Today, microorganism-assisted bioremediation techniques fascinate scientists' great attention, thus gaining substantial usage and emerging at an exponential rate (Priya et al. 2022; Sharma et al. 2017). It is a holistic approach where, with the aid of metabolic processes in biological organisms like bacteria, algae, fungi, actinomycetes, and plants, environmental pollutants are removed or neutralized. Microorganisms thrive in the broadest variety of settings within the Earth's biosphere (Cuaxinque-Flores et al. 2020; Hojnacka 2010; Valavanidis et al. 2006). They flourish in environments with soil, water, plants, animals, the deep sea, and icy conditions. Microorganisms are the ideal choice to escort our environment due to their sheer number and desire for a variety of pollutants. They can overcome heavy metal toxicity stress via fast mutation and evolution. With the advantage of strong metabolic potential, rapid reproduction ability, and immense types of species, microbes can efficaciously reduce the concentration and toxicity of environmental pollutants using a variety of mechanisms.

Detoxification of pollutants using microbial processes (bioremediation) has proven reliable and effective because of their remarkable benefits, such as eco-friendly nature, cost-effectiveness, high efficiency, and the absence of secondary pollutant generation (Sharma et al. 2017; Zhang et al. 2019). Microbial consortiums used to bioremediate in different environments comprise *Achromobacter, Arthrobacter, Alcaligenes, Bacillus, Corynebacterium, Pseudomonas, Flavobacterium, Mycobacterium, Nitrosomonas, Xanthobacter*, etc. (Acharya et al. 2020; Nour, et al. 2022).

Researchers have developed various bioremediation technologies that restore polluted environments. For the remediation of toxic pollutants, two different types of strategies are in use depending on the location at which the remediation process occurs (Hariprasad 2007). *Ex-situ* remediation includes removing contaminants from the subsoil and treating them at the same place or transferring them for the cleanup process to another location. In contrast, *in-situ* techniques are methods where the decontamination process is done directly at the contaminated site itself by applying the biological agents or enhancing the indigenous microbial growth by facilitating the optimal conditions for their propagation to remove contaminants (Miyata et al. 2007; Nielsen 2012). *In-situ* treatment is often chosen since it is cheaper than *ex-situ* monitoring.

4.8 In-Situ Bioremediation

Conventional in-situ bioremediation approaches involve using microbial assistance for heavy metals. Manganese pollution has been comprehensively outlined here, as shown in Fig. 4.2, focusing on biosorption, bioaccumulation, bio-oxidation, microbial-induced carbonated precipitation, bioleaching, bioventing, and disparaging, including some combination methods used to detoxify the Mn pollutant effectively, as we can summarize in Fig. 4.2. Bioremediation is considered an effective method to detoxify organic chemicals as well as to treat heavy metals and inorganic components, contrary to ex-situ (Baragaño et al. 2020; Miyata et al. 2007). The chemotactic nature of microorganisms facilitates their movement towards metal-contaminated areas, making the process safer (https://www.uobabylon.edu.iq/eprints/publication_5_10694_432.pdf; Miyata et al. 2007). Furthermore, the reduction of heavy metals at the location of contamination minimizes the cost of contaminant transportation.

4.8.1 Biosorption

The ability of microbial biomass to absorb heavy metals from wastewater via metabolically mediated or physiochemical uptake mechanisms or the use of bio-based material is known as biosorption (Nealson 1989; Santamaria 2008). Wide arrays of the microbial spectrum have been analysed for adsorption properties (Osmani et al. 2015; Santamaria 2008) with heavy metals. To increase the potential capacity and specificity of these microbial bioadsorbents, researchers also engineered the microbe with recombinant metal-binding protein ligands and peptides on the extracellular surface of microorganisms.

4.8.1.1 Mechanism

The cell surface adsorption mechanism involves the fact that certain biomass can passively concentrate and bind pollutants onto its cellular structure through the physiochemical process, viz., the binding of cell wall functional groups, viz., carboxylic group (COOH), Phosphate group (PO_4), Carbonate group(CO_3), amino group (NH_2), hydroxyl group (OH), etc. with metal ions or their mutual coordination (Nealson 1989; Sharma and Bano 2018). It is incorrect to equate biosorption and bioaccumulation because they are different phenomena. By leveraging physical interactions (viz., electrostatic interaction), chemical reactions (viz., ion or proton displacement), complexation, or chelation, particles are adsorbed to a biological matrix in the process of biosorption. Anionic moieties present on the extracellular surfaces of bacteria offer cationic HMs with binding sites when the pH is neutral.

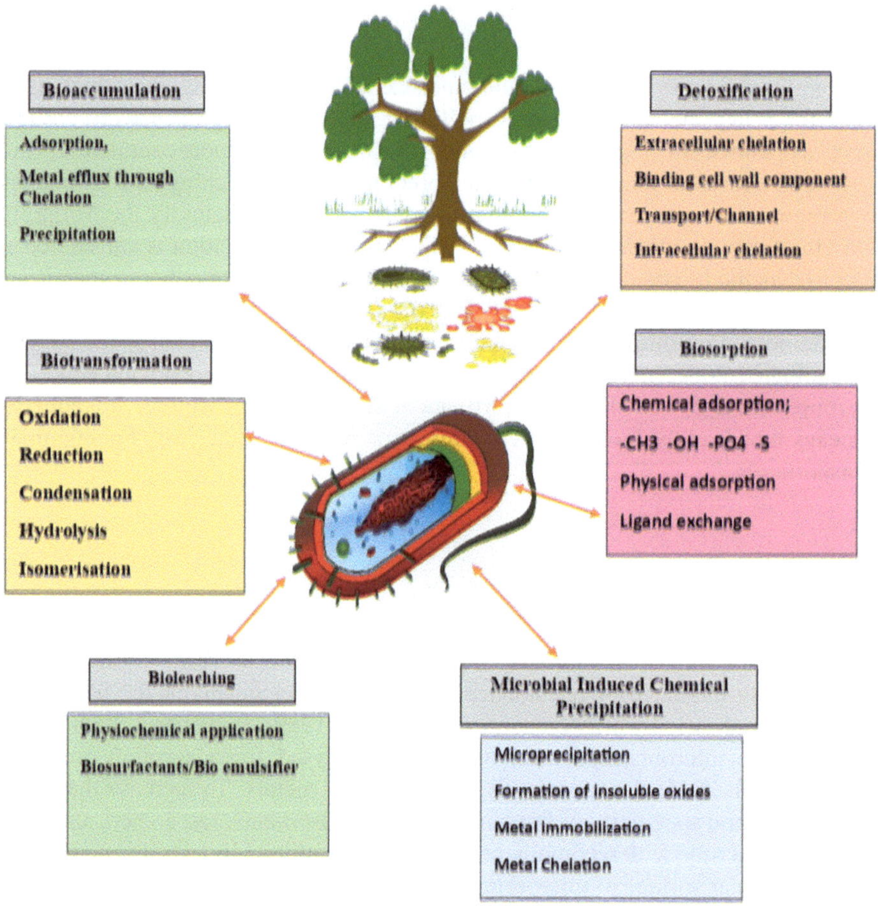

Fig. 4.2 In-situ bioremediation techniques: Adapted from https://www.uobabylon.edu.iq/eprints/publication_5_10694_432.pdf; https://pixabay.com/vectors/bacterium-nucleoid-cytoplasm-cell-307660/, CCBY-SA; https://cdn.pixabay.com/photo/2014/04/03/00/40/tree-309046_1280.png, CCBY-SA

4.8.1.2 Regulatory Factors

Since biosorption is achieved by equilibrium, there are many factors exerting an effect on Mn eradication, including pH, temperature, concentration of biomass, contact time for biosorption, and the interaction among different metallic ions (Ahalya et al. 2003; Das et al. 2011). Microorganisms take a certain amount of time to biosorb Mn. It was found that Mn biosorption increased with time. For instance, the maximum time utilized for manganese biosorption by *Saccharomyces cerevisiae* and 19.34 mg/g by *Aspergillus niger* was 20 and 60 min, respectively (Nadaska et al. 2012; Sharma and Bano 2018). With the increase in Mn and biomass concentration, the biosorption of Mn is also increased. The removal of manganese ions using a composite sorbent as the pH increased from low to high (2–7 pH) is favoured by the sorbent. Like for Mn in water, the biosorption potential of *Pseudomonas aeruginosa* rises with an increase in pH in the range of 5.34–7.82 (Nurlliana, et al. 2017). Lower Mn uptake at an acidic pH may be determined by the process in which multiple protons (H^+) in the solution compete with Mn ions for attachment to binding sites of the biomass. The Mn uptake was minimal at alkaline pH (at pH 9 or 11), which is expected since these metals typically occur as hydroxide colloids (precipitates) at alkaline pH, which results in a slower pace of biosorption (Baik et al. 2002). Consequently, the acidic environment is ideal for biosorption. At an initial metal ion concentration of 600 mg/L, it is stated that the *Bacillus cereus* strain HM-5 (a kind of bacteria) has a biosorption capability for Mn ions of up to 98.9%. (Nealson 1989; Su et al. 2013) Analysis of the ingestion of *Bacillus, Pseudomonas sp.*, and *Saccharomyces cerevisiae* demonstrates the high efficacy of removing manganese, with a specific removal rate of about 96%. Several studies have reported that temperatures in the range of 20–35 °C cannot influence the biosorption process

4.8.1.3 Advantages and Disadvantages

Compared to traditional treatment options, biosorption has several benefits, including efficiency and cost-effectiveness. Additionally, it has the potential to reduce chemical and biological sludge and regenerate itself with the ability for metal recovery (Alluri et al. 2007) and the uptake of multiple heavy metals at a time. The routine use of inexpensive adsorbents made from plant or agricultural waste products as a replacement for pricy conventional heavy metal removal techniques from wasted streams has recently been studied (Keen and Zidenberg-Cherr 1993; Nealson 1989). Saturation of active sites of metal-binding ligands and reversible sorption of metals on biomass might be drawbacks to biosorption. Additionally, biosorbents have a short lifespan since they frequently utilize decaying dead biomass, and fouling prevents the binding sites from being available.

4.8.2 Bioaccumulation

It is the gradual accumulation of xenobiotic components like heavy metals in an organism (Alexander 1999). When an organism absorbs any component faster than its elimination or loss from the body, the process of bioaccumulation takes place. When the rate of pollution clearance is outpaced by the amount of pollutant absorption, bioaccumulation takes place. Because of this, the pollutant or toxicant is trapped inside the cell structure and accumulates in it (Forrez et al. 2010). Since metals are not metabolized, their bioaccumulation in organisms is particularly useful as an exposure indicator.

4.8.2.1 Mechanism

Microorganisms accumulate heavy metals in their intracellular space during the metabolically active process of bioaccumulation via ion channels or transporter complexes. This plasma membrane transport system creates a translocation pathway across it. Once the heavy metal enters the intracellular space, it gets sequestered by the protein or peptide ligands present in the intracellular space.

4.8.2.2 Regulatory Factors

The bioaccumulation process is primarily regulated by factors such as the physio-chemical nature of metal toxicants, temperature, pH, and phase of microbial growth. It has been suggested that some macrofungal species, if present, can retain heavy metals in the mycelium, sporophore, and rhizomorphs at various phases of their growth (Tebo et al. 2004; Therdkiattikul et al. 2020). The bioaccumulation potential of manganese (Mn^{2+}), zinc (Zn^{2+}), copper (Cu^{2+}), nickle (Ni^{2+}), and cobalt (Co^{2+}) by *Bacillus circulans* strain EB1 also demonstrated variation in different growth phases, according to Yilmaz et al. (Sinha and Purcell 2019). Ozdemir et al. found that *Geobacillus thermantarcticus* and *Anoxybacillus amylolyticus* had the highest bioaccumulation capacity of Mn^{2+} during a 24-h incubation period, with respective values of 24.5 (8 h) and 28.5 (20 h) mg/g dry weight. They also found that Mn was the highest metal that was bioaccumulated by both bacteria, as opposed to cadmium (Cd^{2+}), cobalt (Co^{2+}), and copper (Cu^{2+}) (Michalak et al. 2020). This research also depicted that the metal accumulation capacity of these microbes is highly influenced by the variant stages of their growth phase, showing optimal accumulation at the end of the stationary phase (20 h). Many studies concurred that the inability of Mn^{2+} to form bonds with cell walls in an acidic environment ($pH < 5$) with high protons, such as hydroxide hydrate (H_3O^+) and proton (H^+), may lower the rate of Mn^{2+} adsorption (Hatayama 2020). Using a metallurgical waste heap as a source, Noszczynska et al. (Loneragan 1988) isolated and identified a *Pseudomonas sp.* strain with significant Mn removal capacity. Given that their cells have a large capacity and distinct process of accumulation, they hypothesized that their cells had unique mechanisms to store Mn.

4.8.3 Bio-Oxidation

The bio-oxidation method refers to processes in which the recovery of a metal is achieved by microbial decomposition of the mineral. It is one of the most practicable approaches to treating Mn pollutants in waste, sewage, and groundwater. A diverse group of microflorae known as the "manganese-oxidizing group of microorganisms" is shown phylogenetically and categorized according to their capacity to catalyze the oxidation of soluble Mn (II) to insoluble manganese oxides (Li et al. 2019). Mn-oxidizing bacteria (MnOB) oxidize the divalent Mn ions (II) to trivalent Mn ions or tetravalent Mn ions. Several isolates of bacterial species such as *Proteobacteria, Actinobacteria*, and *Firmicutes* have been found to be easily able to transform Mn (II) to either Mn (III) or Mn (IV) (Hariprasad 2007). MnOB can be isolated from diverse locations, like Mn-rich biofilms, deep sea nodules, fresh water lakes, ferromanganese deposits, and sediments. Microbial-assisted Mn oxidation is several folds faster than abiotic Mn oxidation.

4.8.3.1 Mechanism

Mn bio-oxidation follows two different pathways, which may occur collaterally (Baoying and Kano 2015; Bohu et al. 2015; Pagnanelli et al. 2004): the direct method assisted by cellular components, such as enzymes (Alluri et al. 2007; Buchman 2014; Burgstaller and Schinner 1993), and (ii) indirect, whereby Mn (II) oxidation occurs when environmental pH and redox conditions are altered by bacterial metabolites and/or microbial development, as mentioned below (https://pixabay.com/vectors/bacterium-nucleoid-cytoplasm-cell-307660/, CCBY-SA). Direct oxidation depends on MnOB to speed up the oxidation of Mn with the help of a specialized Mn oxidase or on other macromolecular components such as polysaccharides and proteins in cellular membranes that bind, collect, and adsorb Mn (Post 1994). A broad lineage of enzymes is related to Mn oxidation with multicopper oxidases (MCOs), manganese peroxidase, and lactase. Among these, MCOs are widely used in the oxidation process. (Deenan and Yen-Peng 2006; Huang et al. 2020). Previous analysis showed that many MCOs, viz., MofA, MnxG, and MoxA, function as strong Mn-oxidizing candidates. It has been demonstrated that a wide spectrum of organisms, such as fungi (Laccase), bacteria, insects, plants, and vertebrates, show the presence of multicopper oxidases. Bacteria such as *Bacillus sp. strains SG-1*, SS-1, *Psuedomonas putida* strain MnB1, GB-1 *Leptothrix* strain, and *Leptothrix discophora* strain were found to possess Mn-oxidizing activity (Diaconu et al. 2023b).

It has been proven that MCO-type enzymes are responsible for oxidizing manganese, but the accurate mechanism of action is not yet fully understood (Hatayama 2020; Priya et al. 2022). Instead, some Bacillus strains were found to have Mn oxidation capability without producing MCOs indicating that the only mechanism for oxidation is not Mn oxidase activity in every MnOB strain. Changes in pH, dissolved oxygen, and metabolites are known to have a substantial impact on the oxidation process. Bacteria use the plentiful amino acids of peptone during the exponential phase of their growth and yeast extract to create ammonium ions (NH^{4+}) in the culturing medium, which results in a rise in pH. Aside from a pH increase during the stationary phase, extracellular superoxides are important in indirect oxidation as well. Mn^{2+} oxidation by superoxide anion (O_2^-) and reactive oxygen species (ROS) has a positive thermodynamic outcome in all pertinent pH circumstances (Barboza et al. 2016; Nealson 1989).

4.8.3.2 Regulatory Factor

The pH value, oxygen concentrations, cell counts, and value of pH are all factors that impact the oxidation process. The activity of Mn oxidase often remains high in neutral or slightly alkaline circumstances (pH 6.5–8.5) (Tabiason et al. 2016). Strong alkaline and acidic conditions can therefore impede Mn oxidative action (Ozdemir et al. 2013). This does not exclude acidic environments from acknowledging MnOB's presence, either, because several bacterial species, including *Bacillus altitudinis, Frondihabitans,* and *Sphingomonas*, have been found in extreme environments. In the late logarithmic phase, the dissolved oxygen concentration is also found to be relevant to enzyme-assisted oxidation. Katsoyiannis et al. revealed that bacterial-assisted Mn detoxification involves more saturated DO (dissolved oxygen) conditions than that of iron in a two-phase ascending bed filtration unit (Hsieh et al. 2013). Studies also depict that the rate of Mn oxidation is directly proportional to biomass concentration.

4.8.3.3 Advantages and Disadvantages

A biological oxidation method to counteract metal pollution has been proven eminent, economically valuable, and does not produce any secondary pollutant. Although the treatment of Mn mine effluent using MnOB has not been completely competent due to several reasons, such as the higher manganese concentration and strong acidic waste of Mn^{2+}, which impose toxicity on MnOB, a lack of organic nutrient material in Mn ore effluents restricts the growth and biological activity of MnOB (Learman et al. 2011). Simultaneously, while processing Mn ore, the ferrous ion co-exists in pyrite and discharges into waste, consuming the oxygen present and leading to insufficient dissolved oxygen (DO), thereby decreasing the rate of oxidation (https://cdn.pixabay.com/photo/2014/04/03/00/40/tree-309046_1280.png, CCBY-SA; https://www.biorender.com/).

4.8.4 Microbial-Induced Carbonate Precipitation (MICP)

A biogeochemical technique known as microbially induced calcium carbonate precipitation (MICP) causes calcium carbonate to precipitate when different types of microbial metabolism have taken place in the soil matrix (And and Tebo 1994). There are various mechanisms that have been observed through which microorganisms can initiate calcium carbonate precipitation, including urea hydrolysis, sulphate reduction, denitrification, ammonification, iron reduction, and photosynthesis (Cho et al. 2019; Dhami 2013). To fill and repair cracks in stone and concrete materials, produce self-healing cement, prevent building leaks, reduce the occurrence of sandy soil liquefaction, and prevent soil erosion of sand or soil slope, sand dike erosion, piping, and other diseases, the technology uses mineralizing bacteria found in nature to induce mineral components (such as calcium carbonate, etc.).

4.8.4.1 Mechanism

Currently, urease-producing, sulphate-reducing, denitrifying, oxidizing, and other types of bacteria make up nearly most of the known microbial mineralizing bacteria. Urease-producing bacteria are most common among them, and they have received the most attention due to their low cost, ease of extraction and cultivation, good mineralization and cementation impacts, and simplicity of reaction process control. Bacterial mineralization can be explained succinctly as follows: Bacterial metabolism involves the adsorption of Ca^{2+} from the surrounding fluid by the cell surface's negative charge. Following the addition of urea, the bacterial cell urease hydrolyzes the urea to produce $CO_3{}^{2-}$ and NH^{4+}. The formation of ammonium ions causes an increase in pH. Since the alkaline condition is favourable for carbonates to get conjugated with divalent heavy metals such as Cu^{2+} and Mn2 + , they eventually induce the interaction of plasma and Ca^{2+} with $CO_3{}^{2-}$ to create calcium carbonate crystals on the cell surface. After some time, calcium carbonate had completely encircled the whole cell (Hojnacka 2010).Rhodochrosite ($MnCO_3$) precipitation, or CaMn $(CO_3)_2$ synthesized when Mn replaces Ca^{2+} in the crystal lattice of calcite, thus leading to the conversion of Mn from soluble to insoluble mineral forms (Chugh et al. 2022).

Equations (4.1)–(4.6) show the biochemical reactions in the process of MICP driven by urea hydrolysis (Cho et al. 2019).

$$CO(NH_2)_2 + H_2O \rightarrow NH_2COOH + NH_3 \tag{4.1}$$

$$NH_2COOH + H_2O \rightarrow NH_3^+H_2CO_3 \tag{4.2}$$

$$H_2CO_3 \leftrightarrow HCO_3^- + H^+ \tag{4.3}$$

$$2NH_3 + 2H_2O \leftrightarrow 2NH_4^+ + 2OH^- \tag{4.4}$$

$$HCO_3^- + H^+ + 2NH_4 + + 2OH- \leftrightarrow CO_3^{2-} + 2NH_4^+ + 2H_2O \tag{4.5}$$

$$Cell + Ca_2^+ \rightarrow Cell - Ca_2^+ \tag{4.6}$$

$$Cell - Ca_2^+ + CO_3^{2-} \rightarrow Cell - CaCO_3 \tag{4.7}$$

Similarly, ammonification-induced Mn carbonate precipitation has also been reported. Through the pathway of amino acid metabolism, bacteria can generate CO_2 and ammonia. Further, ammonia breakdown would produce ammonium and hydroxyl ions, thereby increasing the pH.

Thus, under the influence of alkaline conditions, the dissolved CO_2 gets transformed into carbonate ions. As the solution gets completely saturated, the precipitation of Mn carbonate takes place (Equeenuddin et al. 2013).

4.8.4.2 Regulatory Factors

The pH level, effective nucleation site, and calcium ion concentration are the primary factors affecting the traditional precipitation of calcium carbonate (Alina 2020). MICP is a very intricate biological and chemical procedure. MICP is primarily influenced by the essential variables listed below: (i) bacterial species; (ii) microbial concentration (Rahhou et al. 2023); (iii) surrounding temperature; (iv) pH value; (v) calcium ion concentrate; (vi) kind of cement used (soil, sand, or other); and (vii) for the required quantity of precipitation for cementation, the precipitation of $CaCO_3$ requires an ample amount of calcium and carbonate ions (Das et al. 2015).

4.8.4.3 Advantages and Disadvantages

MICP has greater advantages in evolving the properties of rock and soil compared to conventional modification technology: its construction method is easy, it almost never produces toxic or harmful substances, it generates fewer chemical contaminants for the soil and water environment (Hazaimeh 2023; Rahhou et al. 2023), and the percentage of bacteria liquid and bonding liquid employed for the customized object is low compared with typical chemical slurry. The synthesis of micro- or nanosized particles of Mn carbonate has wide applications in the fields of catalysts and electrochemical device production. Although there is still a great deal of room for research and development, for instance, the use of MICP may be limited to engineering projects and sand embankments, and despite studies into the use of MICP for heavy metal removal, Mn is hardly ever addressed. Furthermore, there has not been any specific research on the contributing elements and features of the sediments in Mn treatments using MICP (Su et al. 2014).

4.8.5 Bioleaching

Bioleaching is the process of recovering minerals from their ores with the aid of living things. The ability of a miscellaneous group of microorganisms to naturally dissolve solid metallic waste into soluble, extractable form is known as bioleaching. Acharya et al. reported that recovery of manganese from manganese ores utilizing microorganisms can be achieved (Acharya et al. 2004, 2020) by using *penicillium citrum* on operating Mn ore. Several scholars report the bioleaching of low-grade manganese ore with *E. cloacae*, *Pseudomonas cepacia*, *Bacillus coagulans* (Dhami

2013), *Thiomonas sp.* (Silva et al. 2009) heterotrophic microorganisms, and *Enterobacter agglomerans* and MnO_2 ore bioleaching with *Aspergillus niger* (Cordova et al. 2012), *Thiobacillus thiooxidans,* and *Bacillus sp.*

4.8.5.1 Mechanism

The bioleaching of heavy metals through microbial assistance relies on the biological breakdown and oxidation of MnO_2 to liberate positively charged heavy metal ions from intractable ores. The metabolism of bacteria results in the synthesis of secretions like organic acids with a low molecular weight that efficiently degrade heavy metals in metallic leachate. Several results demonstrate that under ample nutritional conditions, microbes release organic acids that further stimulate the bioleaching of manganese (Acharya et al. 2020).

4.8.5.2 Regulatory Factors

Several scholars have depicted that the bioleaching of Mn is highly influenced by certain parameters such as particle size, temperature, pH, inoculum size, and pulp density. Their result shows that due to the accessibility of small size particles to the fungus, extraction was highest at the finest particle size of 45 microns (Acharya et al. 2004, 2020). Simultaneously, research carried out by Marshell et al. (1979) observed that the maximum activity of bioleaching occurs at pH > 5 as Mn ions get absorbed on the surface of ore particles of MnO_2 (Das et al. 2011; Kirschvink 1992). Veglio et al. and some other scholars observed that higher production of Mn^{2+} extraction is achieved with lower pulp densities. A reduction in yield might be observed due to an insufficient supply of nutrients and oxygen available for growing microbes. The manganese reduction was found to be directly proportional to the inoculum size of microbes or their growth phase. At higher concentrations of microbial inoculums, there is an increase in organic acid production, which helps in the reduction of manganese leachate. This research revealed a close relationship between the product production (manganese reduction), the microbial inoculum or growth, and the linked product. The formation of organic acids, which finally cause manganese reduction, increases with inoculum concentration (Burgstaller and Schinner 1993; Hatayama 2020). The Mn-leaching fungus *P. citrinum* was discovered to exhibit fluctuation in its leaching ability across its several growth phases, with its optimal leaching activity being noted during the logarithmic phase.

4.8.5.3 Advantages and Disadvantages

Bioleaching is a simpler, innovative method to treat metal pollutants than smelting. It requires fewer staff to operate, is an easy operation method, and has a reduced cost; therefore, it can be preferred as an economic and eco-friendly option. Besides having

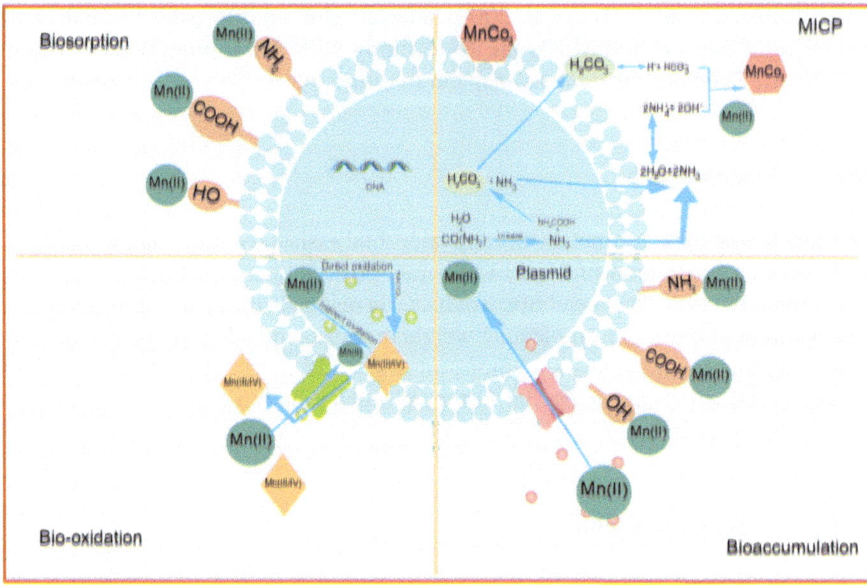

Fig. 4.3 Different mechanisms showing microbial and microalgae bioremediation: Adapted from https://www.biorender.com/

such benefits, it has several operating limitations, like that it is a time-consuming process, effective recovery may not be achieved, and it simultaneously produces large amounts of acidic or alkaline waste effluent (Fig. 4.3 and Tables 4.2 and 4.3).

4.9 Combination Approaches in Microbes—Intervened Remediation

In many cases, the treatment of heavy metals under complex conditions could not be achieved successfully. Thus, researchers are in search of several novel and efficient combination approaches to maximize the ideal removal efficacy.

4.9.1 Microalgae and MnOB

Techniques for cleaning up manganese pollution have been developed in recent years. This entails converting Mn from its soluble to its insoluble state. Non-living microbial biomass of bacteria, fungi, and algae has been used to remediate Mn. Microalgae have received attention for their propensity to bioaccumulate contaminants in an ecosystem and for their diverse dominance responses and ability to

Table 4.2 List of some microbial species depicting the removal efficiency of manganese contaminant using different remediation mechanisms

Method	Microbial species	Removal Efficacy (%)	Optimum Temp. °C	Optimum pH	References
Biosorption	*Serratia sp*	94.80	34	5–6	Cao et al. (2011), Nadaska et al. (2012)
	Bacillus cereus	60.78	35	5–6	Sharma et al. (2017), Sharma and Bano (2018)
	Saccharomyces cerevisiae	96.00	37	6–7	Baik et al. (2002), Barboza et al. (2016)
	Aspergillus niger	94.30	37	5–7	Baik et al. (2002), Barboza et al. (2016)
	Psuedomonas aeruginosa	75.60	30	5–7	Baik et al. 2002; Barboza et al. (2016)
Bioaccumulation	*Bacillus cerculan strain EB1*	93.67	35	6–7	Sharma and Bano (2018), Srivastava et al. (2016)
	Geobacillus thermatarcticus	79.34	28	<5	Dhami (2013), Marshall (1979)
	Anoxybacillus amylolyticus	82.00	20	<5	Dhami (2013), Marshall (1979)
	Pseudomonas sp.	77.58	25	5–6	Huang et al. (2020), Failed (2018)
	Papiliotrema huenov	88.36	28	Self-regulation	Huang et al. (2020), Failed (2018), Marshall (1979)

(continued)

Table 4.2 (continued)

Method	Microbial species	Removal Efficacy (%)	Optimum Temp. °C	Optimum pH	References
	Bacillus sp.	99.10	Self-regulation	6.5–8.5	https://cdn.pixabay.com/photo/2014/04/03/00/40/tree-309046_1280.png, CCBY-SA
	Leptothrix discophora	76.20	27	7	https://cdn.pixabay.com/photo/2014/04/03/00/40/tree-309046_1280.png, CCBY-SA
	Acinetobacter sp.	88.39	35	5–7	Baglin (1992), Huang et al. (2020), Mortensen et al. (2011)
Biological oxidation	*Citrobacter sp.*	69.15	35	5–7	Huang et al. (2020), Mortensen et al. (2011)
	Bacillus megaterium	57.68	37	7–8	Muynck et al. (2010)
	Bacillus cereus	73.20	37	7–8	Muynck et al. (2010)
	Bacillus thuringiensi	81.46	35	7–8	Muynck et al. (2010)
	Bacillus subtilis	72.20	37	7–8	Muynck et al. (2010)
	Clostridium	70.00	30	6–8	Mitchell and Carlos Santamarina (2005), Sharma and Bano (2018)
	Sporosarcina pasteurii strain ATCC 11,859	83.20	37	6–8	Deenan and Yen-Peng (2006), Srivastava et al. (2016)

(continued)

Table 4.2 (continued)

Method	Microbial species	Removal Efficacy (%)	Optimum Temp. °C	Optimum pH	References
	Bacillus sphaericus	96.10	36	6–8	Das et al. (2011)
	Penicillium citrium	87.23	30–35	>5	Priya et al. (2022)
	Aspergillus niger	92.25	30–36	3–7	Priya et al. (2022)
	Thiobacillus thioaxidans	48.69	30–37	3–7	De et al. (2013), Ehrlich (2008), Sharma (xxxx)
	Enterobacter agglomerans	73.83	30–38	3–7	De et al. (2013), Ehrlich (2008)
	Pseudomonas cepacia	58.39	30–39	3–7	De et al. (2013), Ehrlich (2008)
	E. cloacae	72.46	30–40	3–7	Abayomi et al. (2017)
Bioleaching	*Bacillus coagulans*	78.30	30–41	3–7	Abayomi et al. (2017; De et al. (2013), Sharma (xxxx)
	Thiomonas sp.	63.20	30–42	3–7	Abayomi et al. (2017)
	Microbacterium trichothecenolyticum	98.00	45	3–4	Abayomi et al. (2017), De et al. (2013), Sharma (xxxx)

survive in extreme environmental conditions. Taxa of microalgae like Euglenophyta, Cryptophyta, Chloraphyta, and Rhodaphyta can grow in tough conditions (Rongrong et al. 2022; Zhu et al. 2010). They are tiny, unicellular, or multicellular organisms that predominantly use photosynthetic processes. Despite being single-celled, some of them could group together to create colonies, such as filaments or spheres of the same species. Detailed data about MnOB has been described in the above literature

Table 4.3 Manganese remediation efficiency of different algae by some bioremedial techniques

Algae species	Remediation technique	Mn removal efficiency (%)	References
Chlorella vulgaris	Phycoremediation	80–90	Chugh et al. (2022), Hazaimeh (2023), Lei et al. (2018), Smythers and Perry (2019)
Spirulina platensis	Phycoremediation	69–72	Chernikova et al. (2006), Michalak et al. (2020)
Scenedesmus obliquus	Phycoremediation	58	Failed (2018), Narayanan et al. (2020)
Chlamydomonas reinhardtii	Phycoremediation	70–73	Diaconu et al. (2023a), Smythers and Perry (2019)
Ulva lactuca	Phytoremediation	45	Rahhou et al. (2023)
Gracilaria sp	Phytoremediation	55–60	Jakkapong et al. (2019), Rivas et al. (2022)
Porphyra sp.	Phytoremediation	55	Diaconu et al. (2023a)
Gelidium amansii	Phytoremediation	68	El-Naggar et al. (2018), Nour et al. (2022)

on bio-oxidation. Microalgae have the potential to scavenge free radicals because of enzymes such as superoxide dismutase (SOD), catalase (CAT), ascorbate peroxidase (APX), and glutathione peroxidase (GPX), which can also assimilate the high metal concentration (Miah et al. 2020). Thus, the living microalgae, *Pediastrum duplex* AARLG060, get remarkable attention for this type of application (Holguera et al. 2018; Lv et al. 2020).

The Mn oxidation and removal of microalgae were done by biosorption and bio-oxidation of Mn^{2+}. In an oxidation pond rich in organic waste material, a symbiotic relationship between microalgae and bacteria exists. In the presence of sunlight, algae liberate oxygen via photosynthesis. By virtue of this liberated oxygen, DO (dissolved oxygen) and pH get increased, thereby providing an optimal condition for the growth of oxidizing bacteria like MnOB, which consequently accelerates the oxidation of Mn^{2+} to Mn^{3+} or Mn^{4+}. The products of organic waste oxidized by MnOB are CO_2, H_3PO_4, water, and NH_4 which are further required for continuous oxygen production. It is to be noted that macroalgae shows bioaccumulation of Mn^{2+} through its cell surface binding group (Rongrong et al. 2020; Tebo et al. 2004). Several studies have shown that the rate of MnOB growth can be increased with the help of extracellular polymorphic substances (EPS) secreted by microalgae, concluding that mutualism between microalgae and MnOB is a very helpful approach towards Mn toxicants with a very high efficiency. Thus, it is suggested that it is a crucial step to get rid of heavy metal toxicity through this symbiotic association.

4.9.2 BioMnOx and MnOB

Oxidation of Mn $^{2+}$ transforms it into insoluble trivalent (Mn^{3+}) and tetravalent (Mn^{4+}) oxides. These oxides are also referred to as biogenic manganese oxides (BioMnOx). BioMnOx are considered foragers of trace elements, strong oxidizing agents, and prime reservoirs of organic carbon (Johnson et al. 2015; Su et al. 2014). Mn-oxide-forming bacteria are considered astonishingly skilled engineers of nanoparticles. Their enzyme-assisted mechanism is suitable for bioenergy production and bioremediation applications. They have the potential to detoxify the heavy metals and waste effluents of mines and soil (Ali et al. 2019; Osmani et al. 2015). BioMnOx forms a layered, large surface-to-volume structure of nanoparticle minerals (Holguera et al. 2018). With its strong adsorption capability, it is the strongest contender as an oxidizing agent and has been found to have a higher reactivity to oxidize metal than chemical manganese oxides (Forrez et al. 2010). An innovative effort of joint activity of manganese-oxidizing bacteria (MnOB) and bacterium with BioMnOx to detoxify the Mn pollutant comes into consideration with the help of a new highly Mn-tolerant bacterium, *Provindencia sp.* LLDRA6, which exceptionally makes possible the removal of many heavy metals, viz., Cr, Cu, Cd, Mn, and Zn, from polluted soil, water, and mine leachate (And and Tebo 1994). This type of combination strategy involving symbiotic and synergistic treatment of Mn is preferred (Table 4.4).

4.10 Conclusion

Removal of heavy metal pollution still represents a big challenge for ecologists, which is why this review sheds some light on bioremediation technology, its importance, mechanism of action, and prospects.

Mn is one of the prevalent elements that exist in the earth's crust in different oxidized forms. Being a cofactor of multiple metabolic enzymes, it is an important trace element of the physiological system. Mining, mineral processing, the release of a large amount of acid mine drainage (AMD), and unchecked anthropogenic activities pose a notable threat to the environment and public health. Its elevated level is reported to be linked with molecular damage and could also generate mental problems like Parkinson's disease and manganism. This chapter reviews the essentiality and toxicity related to manganese composites, Mn regulation, and biological techniques employed for Mn recovery. Due to limitations in physiochemical remediation technology, the low-grade Mn ore cannot be extracted completely. In the present era, a novel expertise has been established, termed "bioremediation," with an improved microbial-assisted technology for mineral recovery and removal. The use of thermophiles and archaea bacteria imposes a major contribution in this field. A miscellaneous group of microflorae involving autotrophs, acidophiles, and heterophiles is frequently applied in the biological treatment of metal pollution. For bioremediation of Mn toxicants, a variety of manganese-oxidizing bacteria have been used.

Table 4.4 Advantages and disadvantages of in-situ and ex-situ bioremediation technologies (Bafghi et al. 2008; Keen and Zidenberg-Cherr 1993; Post 1994; Yin et al. 2019)

Technologies		Advantages	Drawbacks
In-situ	Biosorption	1. The contaminants can be fully rendered totally harmless substances (e.g., Water, carbon dioxide, and ethane) 2. Often cost-effective than other remediation techniques 3. As an in-situ method there is reduced potential for cross-media transfer of contamination 4. Bioremediation "in-situ" treatment volume can be achieved, which deals with both the dissolved and the sorbed contaminant 5. In this technique there is a reduced risk of human exposure to contaminated media	1. Depending on the site, some contaminants may not be completely converted into harmless compounds 2. Some contaminants may not be biodegradable (i.e., they are recalcitrant) 3. Toxic concentrations of organic compounds or heavy metals may inhibit indigenous microorganisms' activity
	Bioaccumulation		
	Biological oxidation		
	MICP		
	Bioleaching		
Ex-situ	Biopiling	1. Moisture, heat, nutrients, oxygen, and pH can be controlled to stimulate biodegradation 2.Generates a small number of wastes 3. The soil is mixed with fillers and organic amendments such as scrap wood, hay, manure, and plant wastes and organic pollutants are transformed by microorganisms (aerobic and anaerobic conditions) into harmless byproducts derived	1. Extended treatment is the time required for phytoremediation (usually several seasons) 2. Limited groundwater depth (90–300 cm) possibility of contaminant entering the food chain through the consumption of plants by animals 3. Mass transfer problem 4. Bioavailability limitation
	Composting		
	Farming		

In this article, diverse Mn-solubilizing microbial-mediated in-situ approaches, viz., biosorption, bioaccumulation, bio-oxidation, bioleaching, and MICP, are discussed in detail. To promote bioremediation efficiency, a combination of strategies involving symbiotic and synergistic treatment of Mn is preferred. It is observed that for the biosorption method, to reduce the cost, there is a need for plant-derived, low-cost adsorbent material that can be used as a substitute for expensive adsorbents. For the bio-oxidation process, there is an urgent need to explore and identify more microbes with strong oxidation potential. Parallel to this, we could optimize the environmental conditions to improve the activity of the available bacterial strains. Furthermore, the efficacy of biological remediation is influenced by the type of soil and heavy metals since the strong interaction between the soil and heavy metals may reduce microbial absorption (Mitchell and Carlos Santamarina 2005). Hydrogen ion concentration (pH) is an important parameter that affects microbial activity and metallic properties. Unfavourable pH conditions may negatively affect microbial growth; the ionic charge present on the cell wall may also alter, thereby hampering the binding capacity between the microbe and HM, which leads to poor bio-adsorption and bioaccumulation of HM (Bai et al. 2021; Pande et al. 2022). Similarly, temperature is also an important factor that regulates optimal enzymatic and metabolic activity to achieve maximum growth. It is also observed that different phases of microbial growth also play a role in different bioremediation techniques.

In a nutshell, due to its environment-friendly nature, the application of bioremediation is a characteristically safe approach, making it a globally accepted method of waste removal. With this approach, the elimination of different pollutants could be possible at the same time. There is a low risk of contamination transferring from one location to another. Low energy consumption and cost-effectiveness make it a better method than other physiological methods.

4.11 Future Perspective

At present, there is an increasing demand for manganese and other heavy metals. A novel and pioneering technique involving microorganisms for heavy metal remediation is receiving more consideration. However for future aspects, various limitation factors must be addressed promptly for the advancement of this technique so that it is completely able to remove metal toxicants and can be commercialized.

References

Abayomi OO, Olugbenga OS, John Olatunji A, Farag Twibi M, Alftessi SA (2017) A bioleaching regeneration and recovery of spent refinery catalyst using adapted microorganisms. J Adv Chem Eng 07. https://doi.org/10.4172/2090-4568.1000158

Acharya C, Kar RN, Shukla LB, Misra VN (2004) Fungal leaching of manganese ore. Trans Indian Inst Met 57(5)

Acharya C, Shukla LB, Misra VN (2020) Environmentally-friendly bioleaching of manganese from pyrolusite: performance and mechanisms. J Clean Prod 249

Ahalya N, Ramachandra TV, Kanamadi RD (2003) Biosorption of Heavy Metals. Res J Chem Environ 7(4)

Alexander DE (1999) Bioaccumulation, bioconcentration. Biomagnification. Encyclop Earth Sci 43–44. https://doi.org/10.1007/1-4020-4494-1_31

Ali I, Peng C, Khan ZM, Naz I, Sultan M, Ali M, Abbasi IA, Islam T, Ye T (2019) Overview of microbes based fabricated biogenic nanoparticles for water and wastewater treatment. J Environ Manage 230:128–150. https://doi.org/10.1016/j.jenvman.2018.09.073

Alina P (2020) Removal of heavy metal ions from water and wastewaters by sulfur-containing precipitation agents. Water, Air, Soil Pollut 231

Alluri H, Srinivasa RR, Vijaya SS, Jayakumar SB, Suryanarayana V, Venkateshwar P (2007) Biosorption: an eco-friendly alternative for heavy metal removal. Afr J Biotech 6:2924–2931

Aschner M, Vrana KE, Zheng W (1999) Manganese uptake and distribution in the central nervous system (CNS) Neurotoxicology. 20:173–180

Asim H, Rehman F, Rafeeq H, Waqas M, Asghar A, Afsheen N, Rahdar A, Bilal M, Iqbal HMN (2022) In-situ, ex-situ, and nano-remediation strategies to treat polluted soil, water, and air – a review. Chemosphere 289:133252

MSh, Bafghi, Zakeri A, Ghasemi Z, Adeli M (2008) Reductive dissolution of manganese ore in sulfuric acid in the presence of iron metal. Hydrometallurgy 90:207–212. https://doi.org/doi.org/10.1016/j.hydromet.2007.07.003

Bai Y, Zhou Y, Gong J (2021) Physiological mechanisms of the tolerance response to manganese stress exhibited by pinus massoniana, a candidate plant for the phytoremediation of Mn-contaminated soil. Environ Sci Pollut Res 28:45422–45433. https://doi.org/10.1007/s11356-021-13912-8

Baik WY, Bae JH, Cho KM, Hartmeier W (2002) Biosorption of heavy metals using whole mold mycelia and parts thereof. Biores Technol 81:167–170

Baoying W, Kano N (2015) Adsorption of heavy metal onto the materials prepared by biomass. Biomass Production Uses

Baragaño D, Forján R, Welte L, Gallego JL (2020) Nanoremediation of as and metals polluted soils by means of graphene oxide nanoparticles. Scient Reports 10. https://doi.org/10.1038/s41598-020-58852-4

Barboza NR, Guerra-Sá R, Leão VA (2016) Mechanisms of manganese bioremediation by microbes: an overview. J Chem Technol Biotechnol 91:2733–2739

Bohu T, Santelli CM, Akob DM, Neu TR, Ciobota V, Rösch P, Popp J, Nietzsche S, Küsel K (2015) Characterization of ph dependent mn(ii) oxidation strategies and formation of a bixbyite-like phase by Mesorhizobium Australicum T-G1. Front Microbiol 6

Buchman AR (2014) Manganese. In Modern Nutrition in Health and Disease. Baltimore, MD, Lippincott Williams & Wilkins 11:238–244

Burgstaller, W. and Schinner, F. "Leaching of Metals with fungi." Journal of Biotechnology, vol.27, pp.91–116(1993).

Acharya C et al. (2002) Bioleaching of low-grade manganese ore with penicillium citrinum. Europ J Min Proc Environ Prot 3:1303–0868, 2:197–204

Cao JG, Li XM, Ouyang YZ, Zheng W, Yang Q (2011) Mn waste water treatment: biosorption of Mn by Serratia sp. J Environ Prot Ecol 12:661–666

Chen P, Bornhorst J, Aschner M (2018) Manganese metabolism in humans. Front Biosci (landmark Ed) 23:1655–1679

Chen X, Zhang D, Larson SL et al (2021) Microbially induced carbonate precipitation techniques for the remediation of heavy metal and trace element-polluted soils and water. Water Air Soil Pollut 232:268

Chernikova AA, Tsoglin LN, Markelova AG, Zorin SN (2006) Capacity of spirulina platensis to accumulate manganese and its distribution in cell. Russ J Plant Physiol 53(6):800–806. https://doi.org/10.1134/S1021443706060112

Cho G, Park Y, Hong YK, Ha DH (2019) Ion exchange: an advanced synthetic method for complex nanoparticles. Nano Converg 6:39

Chu J, Ivanov V, He J, Naeimi M, Li B, Stabnikov V (2012) Development of microbial geotechnology in Singapore. Geo-Frontiers 2011:4070–4078. https://doi.org/10.1061/41165(397)416.ISBN97807844116

Chugh A, Kumar L, Shah MP, Bharadvaja N (2022) Algal bioremediation of heavy metals: an insight into removal mechanisms, recovery of by-products, challenges, and future opportunities. Energy Nexus 7. https://doi.org/10.1016/j.nexus.2022.100129

Cordova FM, Aguiar AS, Peres TV, Lopes MW, Gonc¸alves FM (2012) In vivo manganese exposure modulates Erk, Akt and Darpp-32 in the striatum of developing rats, and impairs their motor function. PLoS One 7(3)

Cuaxinque-Flores G, Luis Aguirre-Noyola J, Hernandez-Flores G, Martinez-Romero E, Romero-Ramirez Y, Talavera Mendoza O (2020) Bioimmobilization of toxic metals by precipitation of carbonates using Sporosarcina luteola: an in vitro study and application to sulfide-bearing tailings. Sci. Total Environ 724:138124

Hariprasad D et al. (2007) Leaching of manganese ores using sawdust as a reductant. Miner. Eng

Das AP, Sukla LB, Pradhan N, Nayak S (2011) Manganese biomining: a review. Bioresour Technol 102:7381–7387

Das AP, Ghosh S, Mohanty S, Shukla LB (2015) Advances in manganese pollution and its bioremediation. Environ Microb Biotechnol Soil Biol 45:313–327. Springer International Publishing Switzerland. https://doi.org/10.1007/978-3-319-19018-1_16

De M, Willem Kim V et al. (2013) Influence of temperature on the effectiveness of a biogenic carbonate surface treatment for limestone conservation. Appl Microbiol Biotechn 97(3):1335–1347

De W Muynck, De Belie N, Verstraete W (2010) Microbial carbonate precipitation in construction materials: a review. Ecol Eng 36:118–136

Deenan S, Yen-Peng T (2006) Use of adapted aspergillus niger in the bioleaching of spent refinery processing catalyst. J Biotechnol 211:62–74

Dhami NK (2013) Biomineralization of calcium carbonate polymorphs by the bacterial strains isolated from calcareous sites. J Microbiol Biotechnol 23(5):707–714

Diaconu LI, Covaliu-Mierlă CI, Păunescu O, Covaliu LD (2023) Phytoremediation of wastewater containing lead and manganese ions using Algae. MDPI Biol 12(6):773. https://doi.org/10.3390/biology12060773

Diaconu LI, Covaliu-Mierlă CI, Păunescu O, Covaliu LD (2023) Phytoremediation of wastewater containing lead and manganese ions using Algae. Biology 12(6):773. https://doi.org/10.3390/biology12060773

Dick GJ, Torpey JW, Beveridge TJ, Tebo BA (2008) Direct identification of a bacterial Mn (II) oxidase, the multicopper oxidase MnxG, from spores of several different marine Bacillus species. Appl Environ Microbiol 74:1527–1532

Baglin EG et al. (1992) Solubilization of manganese from ores by heterotrophic micro-organisms. Hydrometallurgy

Ehrlich Henry Lutz, Newman Dianne K (2008) Geomicrobiology. CRC Press, pp 347–426.. ISBN 978-0-8493-7907-9

El-Naggar NEE, Hamouda R, Mousa I (2018) Biosorption optimization, characterization, immobilization and application of Gelidium amansii biomass for complete Pb2+ removal from aqueous solutions. Sci Reports 8(1). https://doi.org/10.1038/s41598-018-31660-7

Emerson D, Ghiorse WC (1993) Ultrastructure and chemical-composition of the sheath of Leptothrix discophora SP-6. J Bacteriol 175:7808–7818

Equeenuddin SM, Tripathy S, Sahoo P, Panigrahi M (2013) Metal behavior in sediment associated with acid mine drainage stream: role of pH. J Geochem Explor 124:230–237. https://doi.org/10.1016/j.gexplo.2012.10.010

Forrez I, Carballa M, Verbeken K, Vanhaecke L, Ternes T, Boon N, Verstraete W (2010) Diclofenac oxidation by biogenic Mn oxides. Environ Sci Technol 44:3449–3454. https://doi.org/10.1021/es9027327

Francis CA, Tebo BM (2002) Enzymatic Mn (II) oxidation by metabolically dormant spores of diverse Bacillus species. Appl Environ Microbiol 68:874–880

Gomeze Ramiraze M et al. (2020) Acidithiobacillus thiooxidans DSM 26636: an alternative for the bioleaching of metallic burrs. Catalysts 1230, 10(11). https://doi.org/10.3390/catal10111230

Guechi E, Benabdesselam S (2020) Removal of cadmium and copper from aqueous media by biosorption on cattail (Typha angustifolia) leaves: Kinetic and isotherm studies. Desalin Water Treat 173:367–382. https://doi.org/10.5004/dwt.2020.24768

Hakala M, Rantamaki S, Puputti EM, Tyystjarvi T, Tyystjarvi E (2006) Photoinhibition of manganese enzymes: insights into the mechanism of photosystem II photoinhibition. J Exp Bot 57:1809–1816

Hasan MR, Khan MZH, Khan M, Aktar S, Rahman M, Hossain F (2016) Heavy metals distribution and contamination in surface water of the Bay of Bengal coast. Environ Sci 2(1):1–12. https://doi.org/10.1080/23311843.2016.1140001

Hatayama K (2020) Mn Carbonate precipitation induced by calcite-forming bacteria. Geomicrobiol J 37:603–609

Hazaimeh M (2023) Phycoremediation of heavy metals and production of biofuel from generated algal biomass: a review. Environ Sci Poll Res

Hojnacka K (2010) Biosorption and bioaccumulation – the prospects for practical applications. Environ Int 36(3):299. https://doi.org/doi.org/10.1016/j.envint.2009.12.001

Holguera JG, Etui ID, Jensen LHS, Peña J (2018) Contaminant loading and competitive access of Pb, Zn and Mn (III) to vacancy sites in biogenic MnO2. Chem Geol 502:76–87. https://doi.org/10.1016/j.chemgeo.2018.10.020

Howe P, Malcolm H (2004) Manganese, and its compounds: environmental aspects. World Health Organization. ISBN 978-92-4-153063-7. 4

Hsieh SI, Castruita M, Malasarn D, Pellegrini M (2013) The proteome of copper, iron, zinc, and manganese micronutrient deficiency in chlamydomonas reinhardtii. Mol Cell Proteomics 12(1):65–86. https://doi.org/10.1074/mcp.M112.021840

https://www.frontiersin.org/journals/plant-science/articles/10.3389/fpls.2020.00300/full

https://www.mountsinai.org/health-library/supplement/manganese

https://www.uobabylon.edu.iq/eprints/publication_5_10694_432.pdf

https://pixabay.com/vectors/bacterium-nucleoid-cytoplasm-cell-307660/, CCBY-SA

https://cdn.pixabay.com/photo/2014/04/03/00/40/tree-309046_1280.png, CCBY-SA

https://www.biorender.com/

Huang HM, Zhao YL, Xu ZG, Ding Y, Zhou XM, Dong M (2020) A high Mn (II)-tolerance strain, Bacillus thuringiensis HM7, isolated from Mn ore and its biosorption characteristics. PeerJ 8:24

Chu J, Ivanov V, He J, Naeimi M, Li B, Stabnikov V (2011) Development of microbial geotechnology in Singapore. Geo-Frontiers, American Society of Civil Engineers, Dallas, Texas, United States, pp 4070–4078

Mitchell JK, Carlos Santamarina J (2005) Biological considerations in geotechnical engineering. J Geotech Geoenviron Eng 131(10):1222–1233

Jakkapong T et al. (2019) Remediation of manganese-contaminated coal-mine water using biosorption and bio-oxidation by the microalga pediastrum duplex (AARLG060): a laboratory-scale feasibility study. Front Microbiol Sec Microbiotech 10. https://doi.org/10.3389/fmicb.2019.02605

Johnson K et al (2015) Towards a mechanistic understanding of carbon stabilization in manganese oxides. Nat Commun 6:7628

Katsoyiannis IA, Zouboulis AI (2004) Biological treatment of Mn (II) and Fe (II) containing groundwater: Kinetic considerations and product characterization. Water Res 38:1922–1932

Keen CL, Zidenberg-Cherr S (1993) Manganese. Food Technology and Nutrition, Academic Press Copyright, Encyclopaedia of Food Science

Kirschvink JL (1992) The proterozoic biosphere: a multidisciplinary study. Cambridge University Press, Cambridge, pp 51–52. (1992), ISBN 0521366151

Koul A, Sharma K, Shah MP (2022) Phycoremediation: A sustainable alternative in wastewater treatment (WWT) regime. Environ Technol Innov 25:102040. https://doi.org/10.1016/j.eti.2021.102040

Kushwaha A, Rani R, Kumar S, Thomas T, David AA, Ahmed M (2017) A new insight to adsorption and accumulation of high lead concentration by exopolymer and whole cells of lead-resistant bacterium Acinetobacter junii L. Pb1 isolated from coal mine dump. Environ Sci Pollut Res 24:10652–10661. https://doi.org/10.1007/s11356-017-8752-8

Learman DR, Wankel SD, Webb SM, Martinez N, Madden AS, Hansel CM (2011) Coupled biotic–abiotic Mn (II) oxidation pathway mediates the formation and structural evolution of biogenic Mn oxides. Geochim Cosmochim Acta 75:6048–6063

Lee Y, Tebo BM (1994) Cobalt (II) oxidation by the marine manganese (II)-oxidizing bacillus sp. Strain SG-1. Applied and Environmental Microbiology, American Society for Microbiology, pp 2949–2957

Lei et al (2018) Application of reverse osmosis in purifying drinking water. https://www.researchgate.net/publication/325559205_Application_of_reverse_osmosis_in_purifying_drinking_water/citation/download

Li D, Li R, Ding Z, Ruan X, Luo J, Chen J, Zheng J, Tang J (2020) Discovery of a novel native bacterium of Providencia sp. with high biosorption and oxidation ability of Mn for bioleaching of heavy metal contaminated soils. Chemosphere 241:125039. https://doi.org/10.1016/j.chemosphere.2019.125039

Li L, Yang X (2018) The essential element manganese, oxidative stress, and metabolic diseases: links and interactions. Oxid Med Cell Longev

Li YC, Xu Z, Ma HQ, Hursthouse AS (2019) Removal of Mn (II) from acid mine wastewater: a review of the challenges and opportunities with special emphasis on Mn-oxidizing bacteria and microalgae. Water 11:2493

Loneragan JF (1988) Distribution and movement of manganese in plants. In Manganese in soils and plants, pp 113–124

Lv MZ, Chen MX, Zhang R, Zhang, Liu et al. (2020) Mn is critical for antitumor immune responses via cGAS-STING and improves the efficacy of clinical immunotherapy. Cell Res 30:966–979

Mangood AH, Abdelfattah I, El-Saied FA, Mansour MZ (2022) Removal of heavy metals from polluted aqueous media using berry leaf. Int J Environ Anal Chem 1–17. https://doi.org/10.1080/03067319.2021.1928102

Marshall KC (1979) Biogeochemistry of manganese minerals. In Biogeochemical cycling of mineral forming elements (Trudinger, P.A. and Swaine, D.J. Eds.), Elsevier, Amsterdam. Chapter 5, 253–292

Miah MR, Ijomone OM, Okoh COA, Ijomone OK, Akingbade GT, Ke T, Krum B, Martins AD, Akinyemi A, Aranoff N et al. (2020) The effects of Mn overexposure on brain health. Neurochem Int 135104688. https://doi.org/10.1016/j.neuint.2020.104688

Michalak M, Mironiuk M, Godlewska K, Trynda J, Marycz K (2020) Arthrospira (Spirulina) platensis: an effective biosorbent for nutrients. Process Biochem 88:129–138. https://doi.org/10.1016/j.procbio.2019.10.004

Michalke B, Fernsebner K (2014) New insights into manganese toxicity and speciation. J Trace Elem Med Biol 28:106–116

Miyata N, Tani Y, Sakata M, Iwahori K (2007) Microbial Mn oxide formation and interaction with toxic metal ions. J Biosci Bioeng 104:1–8

Mortensen BM, Haber MJ, DeJong JT, Caslake LF, Nelson (2011) Effects of environmental factors on microbial induced calcium carbonate precipitation. J Appl Microb 111(2):338–349. https://doi.org/10.1111/j.1365-2672.2011.05065. x.PMID21624021

Nadaska G, Lesny J, Michalik I (2012) Environmental aspect of manganese chemistry, 1–16

Narayanan M et al. (2020) In-Situ and Ex-Situ phycoremediation competence of innate scenedesmus Sp. On polluted river water, 235. Date of Publication. https://doi.org/10.21203/rs.3.rs-95696/v1

Nealson KH (2006) National Research Council (NRC). Recommended dietary allowances, 10th.“ Washington DC: National Academic Press, 1989. Manganese-oxidizing bacteria. Prokaryotes 5:222–231

Ngwenya S, Guyo U, Zinyama NP, Chigondo F, Nyamunda BC, Muchanyereyi N (2019) Response surface methodology for optimization of Cd (II) adsorption from wastewaters by fabricated tartaric acid-maize tassel magnetic hybrid sorbent. Biointerface Res Appl Chem 3996–4005. https://doi.org/10.33263/briac94.996005

Nielsen FH (2012) Manganese, Molybdenum, Boron, Chromium, and Other Trace Elements. In John W Erdman Jr. IAM, Steven H Zeisel (eds), Present Knowledge in Nutrition. 10th ed: Wiley-Blackwell, 586–607

Noszczynska M, Lakomy K, Nowacki K, Piotrowska-Seget Z (2020) A high Mn-tolerant Pseudomonas sp. strain isolated from Metallurgical waste heap can be a tool for enhancing Mn removal from contaminated soil. Appl Sci 10:5717

Nour E et al. (2022) Inhibition of antioxidant enzyme activities enhances carotenogenesis in microalga Dactylococcus dissociatus MT1. Front Bioeng Biotechnol Sec Industrial Biotechnol 10. https://doi.org/10.3389/fbioe.2022.1014604

Nurlliana I et al. (2017) Review on adsorption of heavy metal in wastewater by using geopolymer. MATEC Web Conf 97(1):01023

Oleksandr P et al. (2013) Electrospark method in galvanic wastewater treatment for heavy metal removal. J Appl Sci 15(10)

Osmani M, Bani A, Hoxha B (2015) Heavy metals and ni phytoextractionin in the metallurgical area soils in Elbasan. Albanian J. Agric. Sci. 1:414

Ozdemir S, Kilinc E, Nicolaus B, Poli A (2013) Resistance and bioaccumulation of Cd2+, Cu2+, Co2+ and Mn2+ by thermophilic bacteria, Geobacillus thermantarcticus and Anoxybacillus amylolyticus. Ann Microbiol 63:1379–1385

Pagnanelli F et al. (2004) Leaching of low-grade manganese ores by using nitric acid and glucose: optimization of the operating conditions. Hydrometallurgy 75, 1(4):157–167. https://doi.org/10.1016/j.hydromet.2004.07.007

Pande V, Pande S, Sati D et al. (2022) Microbial interventions in bioremediation of heavy metal contaminants in agroecosystem. Front Microbiol Sec Microbiotechn 13. https://doi.org/10.3389/fmicb.2022.824084

Post JE (1994) Manganese oxide minerals: crystal structures and economic and environmental significance. Proc Natl Acad Sci USA 96:3447–4345

Priya AK, Gnanasekaran L, Dutta K, Rajendran S, Balakrishnan D, Soto-Moscoso M (2022) Biosorption of heavy metals by microorganisms: evaluation of different underlying mechanisms. Chemosphere 307:12. https://doi.org/10.1016/j.chemosphere.2022.135957

Quaing N, Lin Jin et al. (2021) Removal of Mn (II) by a nitrifying bacterium Acinetobacter sp. AL-6: efficiency and mechanisms. Environ Sci Pollut Res Int 28(24):31218–31229. https://doi.org/10.1007/s11356-021-12764-6

Rahhou A, Layachi M, Akodad M, El Ouamari N, Rezzoum NE (2023) The bioremediation potential of Ulva lactuca (Chlorophyta) causing green tide in marchica lagoon (NE Morocco, Mediterranean Sea): biomass, heavy metals, and health risk assessment. Water 15(7):1310. https://doi.org/10.3390/w15071310

Rivas J, Piña F, Araya M (2022) Heavy metal depuration steps for gracilaria chilensis in outdoor culture systems. Molecules 27(20):6832. https://doi.org/10.3390/molecules27206832

Rongrong Wu, Fangting YX, Li Chongjing S (2022) Manganese pollution and its remediation: a review of biological removal and promising combination strategies. Microorganisms. MDPI 10(12):2411. https://doi.org/10.3390/microorganisms10122411

Santamaria AB (2008) Manganese exposure, essentiality & toxicity. Indian J Med Res 28:484–500

Sharma B, Sarkar A, Singh P, Singh RP (2017) Agricultural utilization of biosolids: a review on potential effects on soil and plant grown. Waste Manag 64:117–132. https://doi.org/10.1016/j.wasman.2017.03.002

Sharma I. Trace metals in the environment—new approaches and recent advances bioremediation techniques for polluted environment: concept, advantages, limitations and prospects. https://doi.org/10.5772/intechopen.90453

Sharma P, Bano A et al (2018) Recent advancements in microbial-assisted remediation strategies for toxic contaminants. Clean Chem Eng 2(11):667. https://doi.org/10.1016/j.clce.2022.100020

Silva RMP, Rodriguez AA, De Oca J, Moreno DC (2009) Biosorption of chromium, copper, Mn and zinc by Pseudomonas aeruginosa AT18 isolated from a site contaminated with petroleum. Bioresour Technol 100:1533–1538. https://doi.org/10.1016/j.biortech.2008.06.057

Sinha MK, Purcell W (2019) Reducing agents in the leaching of manganese ores: a comprehensive review. Hydrometallurgy 187:168–186. https://doi.org/10.1016/j.hydromet.2019.05.021

Smythers AL, Perry NL, et al. (2019) Chlorella vulgaris bioaccumulates excess manganese up to 55× under photomixotrophic conditions. Springer Algal Research, 43. https://doi.org/10.1016/j.algal.2019.101641

Srivastava V, de Araujo ASF, Vaish B, Bartelt-Hunt S, Singh P, Singh RP (2016) Biological response of using municipal solid waste compost in agriculture as fertilizer supplement. Rev Environ Sci Biotechnol 15:677–696. https://doi.org/10.1007/s11157-016-9407-9

Su J, Bao P, Bai T, Deng L, Wu H, Liu F, He J, Cot A (2013) A multicopper oxidase from Bacillus pumilus WH4, exhibits manganese-oxidase activity. PLoS ONE 8(4):60573

Su JM, Deng L, Huang LB, Guo SJ, Liu F, He J (2014) Catalytic oxidation of Mn (II) by multicopper oxidase CueO and characterization of the biogenic Mn oxide. Water Res 56:304–313

Tabiason J et al (2016) Manganese removal from drinking water sources. Curr Pollut Reports 2:168–177

Tebo BM et al (2004) Biogenic manganese oxides: properties and mechanisms of formation. Annu Rev Earth Planet Sci 32:287–328

Therdkiattikul N, Ratpukdi T, Kidkhunthod P, Chanlek N, Siripattanakul-Ratpukdi S (2020) Mn-contaminated groundwater treatment by novel bacterial isolates: kinetic study and mechanism analysis using synchrotron-based techniques. Sci Rep 10:12. https://doi.org/10.1038/s41598-020-70355-w

Vakilchap F et al (2016) Role of Aspergillus niger in recovery enhancement of valuable metals from produced red mud in Bayer process. Biores Technol 218:991–998. https://doi.org/10.1016/j.biortech.2016.07.059

Valavanidis A, Vlahogianni T, Dassenakis M, Scoullos M (2006) Molecular biomarkers of oxidative stress in aquatic organisms in relation to toxic environmental pollutants. Ecotoxicol Environ Saf 64(2):178–189

Van H, Marco Dutay Jean-C, Middag et al. (2016) Manganese in the world ocean: a first global model (PDF) . https://doi.org/10.5194/bg-2016-282

Varol M, Şen B (2010) Assessment of nutrient and heavy metal contamination in surface water and sediments of the upper Tigris River, Turkey. CATENA 92:1–10. https://doi.org/10.1016/j.catena.2011.11.011

Vivanov V, Chu J, Stabnikov V (2015) Basics of construction microbial biotechnology, in Biotechnologies and Biomimetics for Civil Engineering, pp 21–56, Springer International Publishing, Cham

Volesky B (1990) Biosorption of heavy metals. Florida: CRC Press. ISBN 978–0849349171

Wang D, Li H, Wei Z, Wang X, Hu F (2006) Effect of earthworms on the phytoremediation of zinc-polluted soil by ryegrass and Indian mustard. Biol Fertil Soils 43(1):120–123

Xiong S, Li X, Liu P, Hao S, Hao F, Yin Z, Liu J (2018) Recovery of manganese from low-grade pyrolusite ore by reductively acid leaching process using lignin as a low-cost reductant. Miner Eng 125:126–132

Xu ZG, Ding Y, Huang HM, Wu L, Zhao YL, Yang GY (2019) Biosorption characteristics of Mn (II) by Bacillus cereus Strain HM-5 Isolated from soil contaminated by Mn Ore. Pol J Environ Stud 28:463–472. https://doi.org/10.15244/pjoes/84838

Han Y et al. (2013) Sulfur-oxidizing bacteria dominate the microbial diversity shift during the pyrite and low-grade pyrolusite bioleaching process. J Biosci Bioeng

Yilmaz EI (2003) Metal tolerance and biosorption capacity of Bacillus circulans strain EB1. Res Microbiol 154:409–415

Yin X, Zhang ZJ, Ma HY, Venkateswaran S, Hsiao BS (2020) Ultra-fine electrospun nanofibrous membranes for multicomponent wastewater treatment: filtration and adsorption. Sep Purif Technol 242:116794

Yin K, Wang QN, Lv M, Chen LX (2019) Microorganism remediation strategies towards heavy metals. Chem Eng J 360:1553–1563

Zapffe C (1933) The history of manganese in water supplies and methods for its removal. J Am Water Works Assoc 25(5):655–676

Zhang Y, Tang YK, Qin ZY, Luo PH, Ma Z, Tan MY, Kang HY, Huang ZN (2019) A novel Mn oxidizing bacterium-Aeromonas hydrophila strain DS02: Mn (II) oxidization and biogenic Mn oxides generation. J Hazard Mater 367:539–545

Zhou D, Kim DG, Ko SO (2015) Heavy metal adsorption with biogenic Mn oxides generated by Pseudomonas putida strain MnB1. J Ind Eng Chem 24:132–139

Zhu G, Zhao Y, Cheng Z (2010) Thermal analysis and kinetic modeling of manganese oxide ore reduction using biomass straw as reductant. Hydrometallurgy 105:96–102

Zhu FK, Qu L, Fan WX, Qiao MY, Hao HL, Wang XJ (2011) Assessment of heavy metals in some wild edible mushrooms collected from Yunnan Province, China. Environ Monit Assess 179:191–199

Zoni S, Bonettia G, Lucchinia R (2012) Olfactory functions at the intersection between environmental exposure to manganese and Parkinsonism. J Trace Elem Med Biol 26:179–182

Chapter 5
Microbial Solution for Mitigating Plastic Pollution: From Environmental Hazard to Sustainable Management

Suyash Devgan, Sparsh Singh, Abhilansh Pandey, and Rashmi Mathur

5.1 Introduction

The middle of the last century witnessed the manufacturing of a weapon for large-scale destruction “The Atomic Bomb”. It also witnessed the manufacturing and development of another weapon, capable of large-scale destruction. Though the previous century was unfortunate to have witnessed the power of the previous weapon—The Atomic bomb. It is this century that will witness the destruction from the latter, “Plastics”. The term “plastic” originated from the Greek word “plastikos”. The root word of which is “plasso”, meaning “to mold” or “to shape”. The word “plastikos” is generally referred to as something that can be molded. The common modern-day usage of the word plastic is to denote any class of substances that can be molded into various shapes and sizes (OECD 2022). Ever since its entry into the market, we as a society have only developed more dependency on this material. So much so that to have envisioned a world of the present, without plastic or synthetic organic polymers seems impossible (Geyer et al. 2017). It is not surprising that the past three decades have witnessed a plastic revolution, and plastic consumption has almost quadrupled since. The plastic industry is one of the most significant contributors to greenhouse gas (GHG) emissions, with data suggesting 3.4% of global GHG emissions in 2019 (OECD 2022). A May 2019 report from the Centre for International environmental

S. Devgan · S. Singh
Dr. B. R. Ambedkar Center for Biomedical Research, University of Delhi, New Delhi, Delhi 110007, India

A. Pandey
Wallace H. Coulter Department of Biomedical Engineering, Georgia Tech and Emory University, Atlanta, GA 30332, USA

R. Mathur (✉)
Department of Botany, Sri Aurobindo College, University of Delhi, New Delhi, Delhi 110017, India
e-mail: rmathur_botany@aurobindo.du.ac.in

K. Bala et al. (eds.), *Harnessing Microbial Potential for Multifarious Applications*, Energy, Environment, and Sustainability, https://doi.org/10.1007/978-981-97-1152-9_5

law suggested the plastic production and incineration could emit 2.8 gigatons of CO_2 per year, releasing as many emissions as 615 five-hundred-megawatt coal plants, by 2050 (OECD 2022; Home 2022). It is due to their lightweight, easy availability, low production cost, long life, ease of molding, packaging, convenient handling, and durability that plastic is an indispensable component of our daily lives (Geyer et al. 2017). It is these facts in combination that make the dealing of the problem in hand even tougher and plastic as a poisoned chalice.

The problem with plastics is not confined to their degradation and manufacturing. Only 9% of plastic waste is recycled, 50% turns into landfill, and 22% evades waste management and instead finds its way to unregulated dumping sites (Home 2022). It undergoes open pit burning, or enters terrestrial and aquatic environments, particularly in economically disadvantaged nations (Kale et al. 2015; Geyer et al. 2017). In this fast-moving world, the race to produce something new every day creates dump yards of waste. Data published by OECD (Organisation for Economic Co-operation and Development) in the Global Plastics Outlook Database reported that 353 million tons of global plastic waste generated from 2000 to 2019 has more than doubled. Roughly 66% of plastic waste originates from plastics that have life spans shorter than five years, comprising 40% from packaging, 12% from consumer goods, and 11% from clothing and textiles. An average annual growth rate of 4% is observed by urbanization, increased population, and lifestyle changes with the generation of millions of tons of waste (OECD 2022; Home 2022). Fossil fuel-based plastics emit greenhouse gases throughout their lifecycle from production to landfilling (International Bottled Water Association 2020). A significant portion of the plastics produced enters and remains in marine ecosystems, including shorelines, seabeds, water columns, and the surface of the world's oceans. According to a study in 2019, a total of 6.1 Million tons of plastic waste escaped into aquatic habitats and a total of 1.7 million tons of plastic garbage made its way into seas (Tarun et al. 2019). This makes the unacquainted marine biodiversity vulnerable to the threats of plastic. The Major pathways contributing to marine pollution include river systems, atmospheric transport, beach littering, aquaculture, shipping, and fishing activities (Geyer et al. 2017; International Bottled Water Association 2020). It might come as an alarming fact to a few readers that an estimated 30 million tons of plastic trash is floating around in the oceans and another 109 million tons floating around in rivers. The Great Pacific Garbage Patch (GPGP) covers an estimated surface area of 1.6 million square kilometers, an area twice the size of Texas or three times the size of France. 80,000 tons of plastic float in the GPGP, equivalent to 500 jumbo jets (Home 2022). Despite the possibility of greatly reducing the poorly managed plastic trash, the currently accumulated plastics in our rivers and oceans will continue to reside there for decades to come. The effects of which will not be limited to the marine flora and fauna, but will directly affect the biosphere as a whole. Given the alarming rate of plastic pollution, urgent monitoring and management solutions are needed. Therefore, it is crucial to treat and eliminate plastic waste permanently from the environment as soon as possible (Geyer et al. 2017).

Studies have revealed the existence of diverse microbial communities associated with plastic waste, both in natural environments and in specialized treatment systems.

These microbial communities have shown promising results in plastic degradation. They exhibit synergistic interactions and metabolic pathways that contribute to the breakdown of plastic materials. Understanding the composition, dynamics, and functional potential of these microbial communities is vital for developing effective bioremediation strategies. Microbes capable of degrading plastic have been extensively studied, and numerous studies point to several species having been identified with the ability to break down various types of plastic. Furthermore, the identification of plastic-degrading genes and enzymes within the microbial genomes has opened new possibilities for engineering microbial strains with enhanced plastic degradation capabilities. This chapter navigates the intricate landscape of plastic pollution, offering an examination of potential microbes for remedy and genetic advancements to combat this global menace. Our goal is to provide a comprehensive understanding of the role microorganisms play in transforming plastic pollution into an opportunity for sustainable practices and environmental stewardship.

5.2 Environmental Consequences of Plastics

Plastic is ubiquitous in our world. Surprisingly, a single water bottle purchased is never recycled as a whole. The use of different types of plastics for the manufacturing of different day-to-day useful items like bottles, makes it impossible to recycle them as a unit. They must be separated into different resin types. The plastic lids/ caps both are made from polypropylene while the plastic bottles are made from polyethylene terephthalate (PET). The manual sorting of plastic is an inordinately slow process (International Bottled Water Association 2020). Therefore, to sort plastic with precision, advanced methods are employed involving AI and deep learning, wherein machines are trained to distinguish plastic types and other substances. As of now, crystal polystyrene, acrylonitrile butadiene styrene (ABS), polypropylene, and high-impact polystyrene (HIPS) are electrical plastic waste, separated using the Machine Learning technique. It is particularly effective for sorting dark plastics that are otherwise difficult to detect using conventional near-infrared techniques due to the deleterious effects of radiation absorption. Machine learning and AI continue to advance, and many new approaches to sorting plastic innovation will open up in the future (Tarun et al. 2019).

The early months of 2020, recorded a worldwide decrease in plastic production (0.3%) and in plastic usage (2.2%) as a result of a disruption of the supply chain during the global lockdown. However, the demand for single-use plastic masks, PPE kits, PCR apparatus, etc. generated 1.6 million tons of plastic each day during COVID-19. Widespread use of face masks, disposable gloves, and PPE increased the rise in plastic and municipal solid waste. A major contributor was single-use face masks and shields thrown daily at the rate of 3.4 billion. The utilization of the RT-PCR test to detect SARS-CoV-2 resulted in the creation of 15,000 tons of plastic waste, a majority of which underwent incineration and was reduced to ashes (Yousefi et al. 2021). Apart from that, a boost in the packaging industry was observed suddenly in October when most of the countries eased the lockdown restrictions. As per the findings of Graulich and coworkers (2021), there was a decrease of 227 kilotons in plastic usage in the time period from April–October 2020. However, during the same period, e-commerce trade witnessed an extra increment of 11.5–17.5 kilotons of plastics. (Graulich et al. 2021).

Microplastics are classified as plastic particles with diameters smaller than 5 mm that extensively disperse across aquatic, terrestrial, and atmospheric environments (Lechthaler et al. 2021). They are largely present in cleaning and cosmetic products like face scrub/ face wash, toothpaste, and synthetic clothes laundry. Some examples of primary and secondary microplastic pollutants, like the abrasion of tires while driving and the breakdown of large plastic objects such as fishing nets, plastic bags, and bottles respectively. When fibrous microplastics are inhaled, they have the potential to persist in the lungs and may pose health risks due to associated contaminants such as dyes and plasticizers, including carcinogenic and mutagenic effects, according to Gasperi and coworkers (2018) and Wong and associates (2020). The microparticles of <1 mm have the potential to enter the food chain and cause accumulation in living tissues. Due to unmanaged recycling, the presence of microplastics has been reported in samples from human lungs, urine, blood, and even breast milk (Gasperi et al. 2018; Wong et al. 2020). To minimize the release of microplastics and microfibers during washing, an effective solution is to utilize filter washing bags. These bags, such as the Guppyfriend Washing Bag, are specifically designed to hold plastics or other textiles during the washing process. The bag captures any residues or particles that may be shed from the items being washed. Once the washing is complete, the captured residues can be easily removed from the bag and thrown into the waste bin. The filter washing bags are woven from polyamide, chosen for their durability, and have a user-friendly design, ensuring maximum convenience for users (Lebreton et al. 2017).

Various efforts are being made to reduce the mismanagement of plastic waste. The Ocean Cleanup Interceptor 2.0, is one of the initiatives that incorporate autonomous systems to remove plastic waste from rivers and oceans. Equipped with barriers, it guides floating plastic to a collection vessel, aimed at capturing plastic waste leaked from rivers and water bodies before reaching the ocean. The vessel operates autonomously by solar energy and has a storage capacity of up to 50 cubic meters of waste (Graulich et al. 2021).

5.3 Classification of Plastics

Plastics are synthesized through the process of polymerization, wherein small molecules are chemically bonded together to form larger polymer chains. This synthesis allows for the creation of a wide range of plastic materials with various properties and applications.

5.3.1 *Based on Thermal Properties*

Plastics are categorized into two classes based on their thermal properties: thermoplastics and thermosetting polymers.

- **Thermoplastics**-Thermoplastics refer to a category where materials are molded multiple times without undergoing any significant chemical changes in their composition when heated. For example, polypropylene (PP), polytetrafluoroethylene (PTFE), polystyrene (PS), polyethylene (PE), and polyvinyl chloride (PVC) as enlisted in Table 5.1—most common plastics used in everyday life with molecular weights ranging from 20,000 to 500,000 atomic mass units (amu). In thermoplastics, macromolecules are organized in a linear manner, creating chains where different molecules and atoms are linked together within carbon chains. The additional polymerization involves the formation of linear macromolecules, and the opening of double bonds achieved through free radical mechanisms (Zeenat et al. 2021).
- **Thermosetting polymers**—Thermosetting polymers undergo a chemical reaction wherein plastics can be melted and cast into a rigid, irreversible structure. They can undergo slight chemical changes upon heating, causing the monomers to transform into an unfusible mass (Ghosh et al. 2013). Examples of thermosetting polymers include formaldehyde and polyurethanes are formed by step-growth polymerization, under favorable conditions. During each step of the process, the release of H_2O and HCl occurs as byproducts, facilitating the intermolecular condensation of bi-functional molecules (Zeenat et al. 2021).

5.3.2 *Based on Plastic Degradability*

Plastics can be classified into three categories based on their degradability, comprising photodegradable, bio-based, and compostable bioplastics.

- **Photodegradable bioplastics**—In certain types of plastics, the polymer backbone is linked to light-sensitive groups and prolonged exposure to UV radiation can result in the disintegration of the polymeric structure. However, once the supply of radiation is stopped, further degradation ceases. It is important to note that plastics in landfills, where sunlight is absent, do not undergo natural degradation. Nonetheless, there is a possibility of artificial photodegradation processes occurring, which could result in the emission of hazardous volatile organic compounds (VOCs), linked to the environmental degradation of plastic debris (Lomonaco et al. 2020).
- **Bio-based bioplastics**—Plastics where all carbon content is derived from renewable resources such as forestry and agricultural sources is called bio-based bioplastics. Examples of such resources include starch, corn, soybean, and cellulose. Polylactic acid (PLA), polyhydroxyalkanoates (PHAs), starch plastics, cellulose esters, and protein-based plastics are commonly known bio-based plastics in terms

Table 5.1 Most common plastics used in everyday life & their general properties

S No	Plastics	Abbreviation	Tm (°C)	Decomposition time (yrs)	Application	Structure	Recycling codes
1	Polyethylene terephthalate	PET	260	5–10	Carbonated drinks bottles, microwavable packaging, peanut butter jars	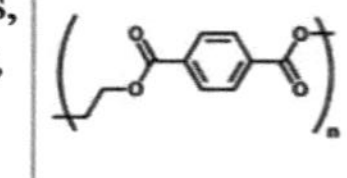	1 PET
2	High-density polyethylene	HDPE	200–300	100	Detergent bottles and milk jugs	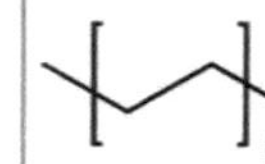	2 HDPE
3	Polyvinyl chloride	PVC	100–260	Never	Plumbing pipes, shower curtains, window frames, flooring, and guttering	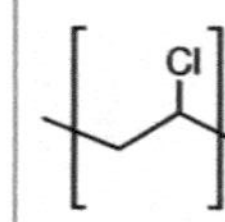	3 PVC
4	Low- density polyethylene	LDPE	160–260	500–1,000	Outdoor furniture, siding, floor tiles, shower curtains, clamshell packaging	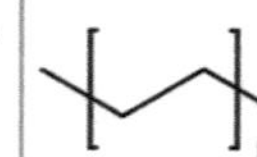	4 LDPE

(continued)

Table 5.1 (continued)

S No	Plastics	Abbreviation	Tm (°C)	Decomposition time (yrs)	Application	Structure	Recycling codes
5	Polypropylene	PP	130	20–30	Bottle caps, drinking straws, yogurt containers, appliances, car bumpers, plastic pressure pipe	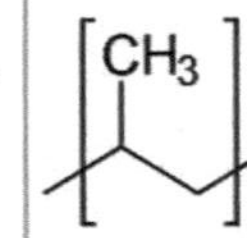	PP
6	Polystyrene	PS	240	50	Packaging foam, food containers, plastic tableware, disposable cups, plates, cutlery and CD	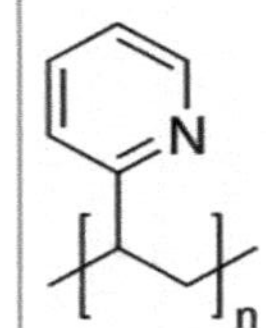	PS
7	Other plastic	Other	8–20 (soft)	Most of plastic (never), polylactic acid (6 months)	Polycarbonate (PC) used for CD, traffic eyeglasses, riot shields, security windows,		OTHER

of their productivity and renewability in the marketplace. It has the potential to address the sustainability concerns and environmental issues associated with the manufacturing and disposal of synthetic plastics (Zeenat et al. 2021).

- **Compostable bioplastics**—During the composting process, specific conditions are necessary to facilitate the breakdown of materials. On the other hand, biodegradable products naturally decompose without requiring specific settings. In the case of biologically decomposed plastics, no toxic substances are left behind (Meereboer et al. 2020). The composting rate for these plastics is similar to that of different compostable materials. Plastics are labeled as bio-compostable based on their complete disintegration, biodegradability, and ecological toxicity level as determined through standardized testing (Meereboer et al. 2020).

5.4 Degradation Pathways of Polymers

5.4.1 Photo Oxidative Degradation

The process in which polymer degradation occurs by the amount of light absorbed is called photodegradation or photo-oxidation degradation. UV radiation from the sun, ranging in wavelength from 290 to 400 nm, cleaves the C-C bond easily and also cuts down the lifespan of the polymeric material. When exposed to photoirradiation, the soft segments within the polymeric region degrade to generate functional groups like propyl, ester, and aldehyde (Fig. 5.1) (Rånby 1989).

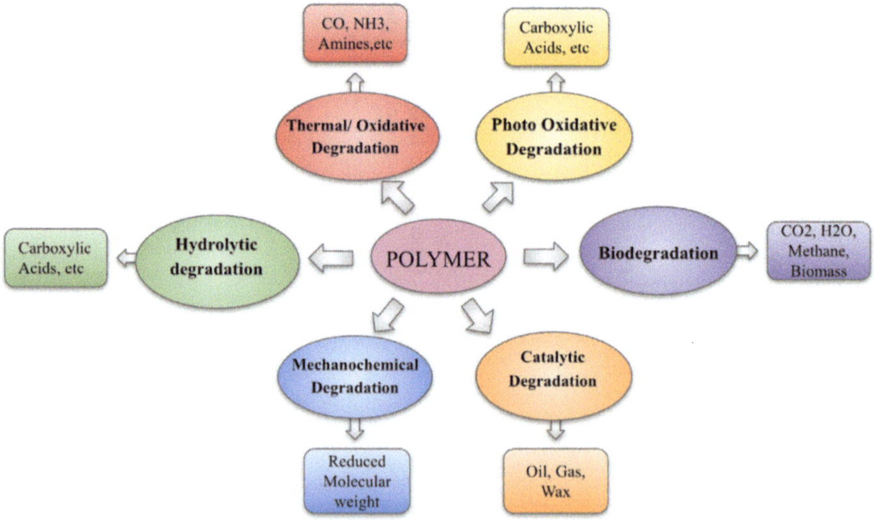

Fig. 5.1 Types of degradation pathways of polymer

5.4.2 *Thermal/Oxidative Degradation*

When plastic degrades either by accident or via a depolymerization reaction, it is called thermal degradation. The peroxide and ether link has weak bonds due to imperfection in the chain and starts depolymerization. For Example, Polyethylene (PE) and Polymethylmethacrylate (PMMA) decompose at high temperature and generate monomers. In the case of photochemical degradation and thermal degradation, both are kind of the same, the only difference they have is in the initial steps. During thermal degradation, reactions take place throughout the entire polymer, whereas in photochemical degradation, reactions are limited to the polymer's surface (Meereboer et al. 2020; Zeenat et al. 2021).

5.4.3 *Catalytic Degradation*

The transformation of waste polymers into hydrocarbons through catalysis is an area of significant interest. The catalytic degradation of polyolefins results in the production of oils and gases. This method of degradation not only enhances the quality of the obtained products from plastic pyrolysis but also offers the potential to obtain targeted outcomes. Various catalysts have been employed for polymer degradation, such as Pt-Co, Pt-Mo supported via SiO_2, transition metal catalysts (chromium, nickel, molybdenum, cobalt, and ferrous) with the addition of Al_2O_3 and SiO_2, non-zeolite catalysts, zeolite, and zeolite catalysts. In the case of polypropylene (PP), the degradation process operates through the mechanism of free radical, utilizing Fe or activated C as the catalyst (Gimouhopoulos et al. 2000; Lin and Yen 2005; Sekine and Fujimoto 2003).

5.4.4 *Mechano-chemical Degradation*

In this degradation process, polymer chains undergo fragmentation due to ultrasonic irradiation and mechanical stress (Li et al. 2005). This process leads to an increase in the number of branches in long polymer chains due to chain-side radical reactions. Additionally, the width of the weight distribution function of molecules decreases, indicating a correlation between crosslinks and ruptures, and the concentration of double bonds also undergoes changes (Striegel 2003). During the mechanochemical degradation of polymethylmethacrylate (PMMA), molecules of nitroxide serve as terminating agents of the chain, generating radicals known as macro radicals, utilized in radical-polymerization reactions. Moreover, when air is present polyvinyl chloride's molecular weight decreases through mechanochemical dichlorination employing different powders of oxide such as Al_2O_3, SiO_2, Fe_2O_3, and CaO (Schmidt-Naake et al. 2002).

5.4.5 *Hydrolytic Degradation*

Hydrolysis, a significant degradation process, occurs when susceptible bonds within a polymer chain react with water molecules, leading to the fragmentation of the chains into smaller segments. The degradation is influenced by various factors, including the chemical reactivity of polymer bonds, diffusion rates of water and other reactants and products, interactions between polymer and water, as well as the presence of ions in the testing environment. All of these elements play a role in the hydrolysis process (Lyu and Untereker 2009).

5.5 Plastic Waste Management

5.5.1 *Greenhouse Gas Emissions During the Lifecycle of Plastic*

Plastic pollution is one of the greatest threats our planet faces today and our planet is choking on plastic. The production of synthetic plastics relies on fossil fuels as feedstock, and these fuels are converted into monomers through a process that consumes significant amounts of energy. The greenhouse gas (GHG) emissions in the economy are dependent on the energy and factor intensity of different economic sectors, as well as their emission intensity during the manufacturing process. Plastics have the tendency to emit greenhouse gases throughout their life cycle. According to the OECD Global Plastics Outlook Database, 4.5% of the total global GHG emission with 1.8 gigatons of carbon dioxide equivalent (Gt CO_2eq) were released in 2019. To accomplish the objectives outlined in the Paris Agreement, it is imperative to halt the increase in emissions associated with plastics (Home 2022).

Approximately 90% of the total greenhouse gas (GHG) emissions throughout the lifecycle of fossil fuel-based synthetic plastics originate from their conversion to and production of various products. The specific emissions from polymer production and conversion differ based on the type of polymer, ranging from 2.7 to 6.3 tons of CO_2 equivalent per ton of plastics. The recycling process results in the direct emission of 0.9 tons of CO_2 equivalent per ton of plastic (Home 2022). However, utilizing secondary plastics obtained from recycling can help avoid emissions stemming from the production of primary plastics. Conversely, sanitary-landfilling represents the method of plastic disposal with the lowest emissions intensity approach for plastic disposal, with direct emissions totaling less than 0.1 tons of CO_2 equivalent per ton of plastic. It is worth noting that sanitary-landfilling does not generate usable energy in other applications. The calculations used in estimating greenhouse gas (GHG) emissions do not currently include the impact of plastic leakage. However, a recent study conducted by Shen and coworkers (2020) utilized experimental data from Royer and associates (2018) to estimate that annual methane emissions, equivalent to approximately 2 million tons of CO_2, result from plastic leakage.

The production of plastics has wide-ranging impacts, including health effects from the release of particulate matter (PM) emissions and socioeconomic effects by providing employment opportunities and generating value-added benefits. The global plastics value chain is responsible for greenhouse gas (GHG) emissions, known as the carbon footprint of plastics. The global carbon footprint of plastics is anticipated to increase by 31% from 2015 to 2030 for a 6 °C scenario, under the IEA's projection matching the projected growth rate of plastics production while in the 2 °C scenario would result in a 10% reduction in the global carbon footprint of plastics from 2015 to 2030, despite an expected 40% increase in plastics production. Even if the 2 °C scenario is followed, coal combustion contributes more than 1/3rd of the carbon footprint of plastics by 2030. Even though emerging economies are experiencing an increasing demand for plastics, the carbon footprints of plastics still remain significantly higher on a per capita basis in high-income regions (Cabernard et al. 2022).

Does replacing plastic reduce GHG emissions? Replacing plastic with more environmentally friendly alternatives, such as biodegradable materials or recycling, can potentially reduce GHG emissions associated with the production and disposal of plastics. However, it would depend on various factors such as the specific alternative material, the manufacturing process, and the end-of-life treatment. Interestingly, seaweed can potentially be utilized to make plastic, an Indian biotech company called Zerocircle is involved in the development of plastic made from seaweed. The novel process involves the collection and preparation of seaweed, followed by the heating and mixing with a specialized solution to create a gel-like substance. Seaweed-based plastic characteristics include tensile strength, flexibility, biodegradability, and degradation by simply dissolving in water. The material exhibits excellent barrier properties, rendering it suitable for packaging perishable goods, snacks, and beverages. Zerocircle's pioneering achievement signifies a crucial step toward a more sustainable and ecologically balanced future.

5.5.2 Repurposing of Plastics

To efficiently handle the ever-increasing production and disposal of plastics, the European-Commission's Waste Framework introduced a zero-waste hierarchy (Fig. 5.2) as a viable strategy, aiming to address the management of plastics more effectively. In this framework, the improper disposal of plastics in the environment and landfilling are identified as the least desirable options due to their significant contribution to pollution and greenhouse gas emissions, offering no economic benefits. The directive emphasizes the importance of adopting alternative approaches to reduce the environmental impact of plastics and promote more sustainable waste management practices. The top priority in this hierarchy is to redesign the life cycle of plastics, enabling their reuse, recycling, repair, or composting to establish a circular economy with minimal reliance on producing new virgin plastic. Although these goals are desirable, accomplishing them will take time and necessitate global

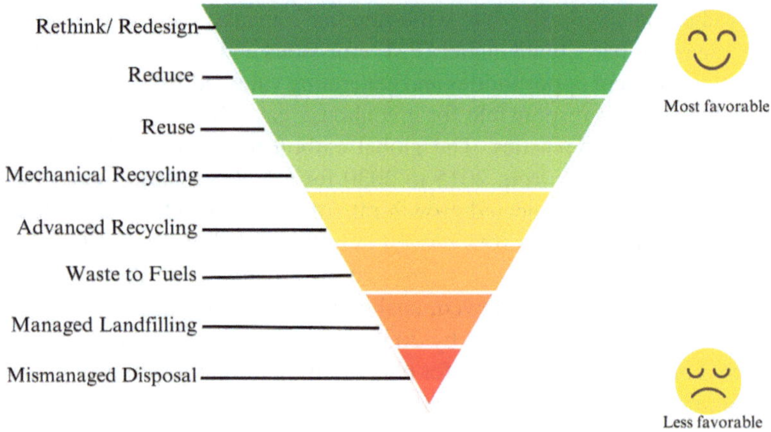

Fig. 5.2 A representation of the zero waste hierarchy of plastic waste management

cooperation among consumers, governments, and industries. During the transitional period, the most effective approaches include decreasing consumption, advocating for the reuse of products, enhancing recycling endeavors, and embracing upcycling techniques (Zhao et al. 2022).

5.5.2.1 Construction Materials

As the second-largest consumer of plastics, the construction industry accounts for approximately 19–20% of the total annual production of plastics. PP and PVC are commonly used in this sector for insulation, plumbing, and pipes while PC is utilized for finishing interiors and various other uses. These plastics are favored for their moldability, lightweight nature, insulating properties, and corrosion resistance however their load-bearing capacity is typically limited (Zhao et al. 2022).

Recently, there has been a growing interest in utilizing plastics that are recycled for construction purposes such as concrete, bricks, and interior design. PET, PP, LDPE, and melted dumped rubber and combined via different materials for the creation of a more novel type of bricks. For instance, PET bottles were thoroughly cleaned, shredded, dried in the sun, and mixed with riverbed sand at varying ratios before being formed into bricks. A study conducted to check compression strength, water absorption, and fire resistance, among other tests, showed that PET with a sand ratio of 1:2 exhibited higher compressive strength as compared to ratios of 1:3 or 1:4. An analytical comparison of bricks made from waste PP, waste rubber, and $CaCO_3$ showcased superior compressive strength (6.3 MPa), less water absorption (<5%), compared to clay bricks 3.6 N mm^{-2} and 15–20% respectively. Nevertheless, it is important to highlight that the bricks containing plastic began to melt at 350 °C, which is considerably lower than the conventional clay-based alternatives (Hopewell et al. 2009).

Aiming at converting plastic waste into plastic bricks and blocks the company called Conceptos Plastics based in Columbia creates a 100% recycled plastic construction system used to assemble houses and schools for vulnerable people in developing countries. A city called Abidjan creates enough waste to build about 45 classrooms/day (Building a sustainable future from plastic waste 2024). The other application of recycled plastics is in concrete when the slag, cement, sand, and water are mixed in a blast furnace to create lightweight aggregate concrete. The findings propose that waste plastics should not be treated as an independent category of construction material, given the significant diversity in their physical and mechanical characteristics (Zhao et al. 2022).

5.5.2.2 Liquid Refuse-Derived Fuel (RDF) Oil

The process of converting plastic waste to molten refuse-derived fuel (RDF) oil involves catalytic pyrolysis and mechanical segregation from solid waste. However, the division method is not flawless, leading to the separation of other fragmentary waste materials along with plastic waste. To achieve 100% source separation of plastic waste by the means of separation methods, the segregated waste generated is then delivered through a conveyor belt equipped with a device specializing in optical segregation. Through catalytic pyrolysis, the miscellaneous plastic waste undergoes the transformation into a valuable fuel-liquid RDF oil. All of the input material is transformed into liquid RDF, gases, with some portion of sludge. Importantly, this conversion process generates no effluent, and any remaining gases from the reactor are safely released through vents (Zhao et al. 2022; Hopewell et al. 2009).

5.6 Biodegradation

Biodegradation is a process that involves microorganisms assisting in degradation, under predefined conditions. The fate of the reaction is to a very large extent known and involves the conversion of the substrate (usually organic, could be natural or synthetic) into CH_4, CO_2, H_2O, and biomass by composting, and soil biodegradation. It is widely accepted and depends on both the physical and chemical properties of biopolymers. Multiple factors affect biodegradation such as molecular weight, composition of polymer, reduction potential, crystallinity, hydrophilicity, etc. The biodegradation process is affected by the ability of microorganisms to degrade the polymer's carbon backbone. Degradation of polymer can be classified as either Surface Erosion or Bulk Erosion. Enzymatic degradation, or biotic degradation, predominantly occurs at the surface of the polymer due to the limited ability of the enzymes, which are macroscopic particles, to infiltrate the polymer composition compared to smaller chemicals and free radicals (Zeenat et al. 2021; Meereboer et al. 2020). In contrast, abiotic degradation encompasses both bulk and surface deterioration and is commonly used as a preliminary step for biodegradation. The polymer is commonly fragmented into smaller pieces through bulk erosion, thereby enhancing the rate of surface erosion and facilitating molecular weight reduction. This process is particularly relevant when considering the potential consequences of improperly disposed plastic waste.

5.6.1 *Plastic Biodegradation Characteristics*

Water-soluble polymers can readily degrade and transform into alcohols, ketones, and acids; however, not all polymers are soluble in water. A few methods according to which plastic biodegradation is monitored are:

- Changes in Plastic surface properties,
- Alteration of Physical and Chemical Properties,
- Chemical composition-based analysis,
- Determination of the rate of oxygen consumption,
- Evaluation of the speed at which carbon dioxide is produced,
- Biomass generation indicates that microorganisms utilize plastics as their carbon source to support their growth.

5.6.2 *Factors Influencing Biodegradation of Polymers*

- The availability of functional groups that increase hydrophobicity affects biodegradation (Wang et al. 2021).
- The structural complexity, such as linear or branched configurations, influences biodegradation (Tokiwa et al. 2009).
- The type of chemical bonds present in the polymer, with amide bonds and ester bonds being more easily breakable, impacts biodegradation. The coupling within the polymer chain follows the order of urethane $<$ amide $<$ ether $<$ ester (Shams et al. 2021).
- Biodegradation is influenced by the makeup of the polymer, which depends on its monomers (Shams et al. 2021).
- The structure of the polymer, such as pellets, films, or powders, as well as its nature and physical appearance, can influence biodegradation (Meereboer et al. 2020).
- The molecular weight and density of the polymer play a role in its biodegradation (Tokiwa et al. 2009).
- The morphology of the polymer, including the amount of amorphous and crystalline regions, affects biodegradation (Meereboer et al. 2020).
- The toughness of the polymer, with soft polymers degrading faster than tough or hard ones, impacts biodegradation (Swift 1993).

5.7 Microbes and Their Role in Plastic Biodegradation

Plastic is no longer a stranger to the deepest point in the ocean, the Mariana Trench, and the highest mountain peaks in the world, Mt. Everest. Microbes are omnipresent. From highlands to the depths of oceans and skies, and space. And even in outer space. They are everywhere. The environment dictates the possibilities of finding specific kinds of microbes. The environment of residence of these microbes remains no stranger to the influences of these microbes too. Microbes, therefore, with their brief division time and omnipresence have had a larger time to interact, adapt, and evolve into newer niches in due regard to their constantly changing environment. They affect their surroundings in several ways. They are responsible for maintaining the flow of nutrients in the biosphere (Yoshida et al. 2016). For instance, they play a significant role in nutrient cycles by helping in maintaining the flow or the continuity of the cycle. The flow of Carbon is crucial to all life processes and therefore life itself. Carbon, initially thought to be a molecule of life and restricted only to the beings that possess life, can now easily be found in synthetic polymers. Carbon is an important element. Plastics, being synthetic organic compounds, pose a significant challenge for proper management. Therefore, leveraging microbes for the degradation of man-made synthetic organic compounds, as in plastics, allows nature to resolve the environmental consequences of plastics, which we as the creators of the problem, have struggled to address adequately. The disintegration of the synthetic polymer is dependent on many variables including substrate availability, morphology, molecular weight of the polymer, and surface characteristics of the material (Cassone et al. 2020). These microorganisms, unlike any man-made process, have shown an unparalleled response to the degradation of plastic keeping in mind the idea of a sustainable future. Heterogeneity in microorganisms, through their location, interactions, and adaptations has evolved to degrade different types of plastics. Figure 5.3, shows the types of plastics and the respective organism responsible for bringing about their biodegradation (Ghosh et al. 2019; Sekiguchi et al. 2011; Pischedda et al. 2019). Before, we dive into the realm of microbes and their degradation abilities. We, for the sake of readers, introduce them first to stories that led to the groundbreaking work and innovative technologies in the war against these synthetic polymers to only then talk about those discoveries in the subsequent paragraph.

In 2016, a team of researchers from Kyoto University in Japan made a groundbreaking discovery. They isolated a strain of bacteria, *Ideonella sakaiensis*, from a PET (polyethylene terephthalate) plastic recycling plant. This bacterium had a unique ability to feed on PET plastic, breaking it down into its basic components. The researchers found that these bacterial species produced two enzymes, PETase and MHETase, that collaborated harmoniously to efficiently break down PET plastic. At its active site, PETase contains a catalytic triad consisting of the amino acids Serine, Histidine, and Aspartic Acid. This triad can hydrolyze PET into monohydroxyethyl terephthalate (MHET), terephthalate, and ethylene glycol. This enters

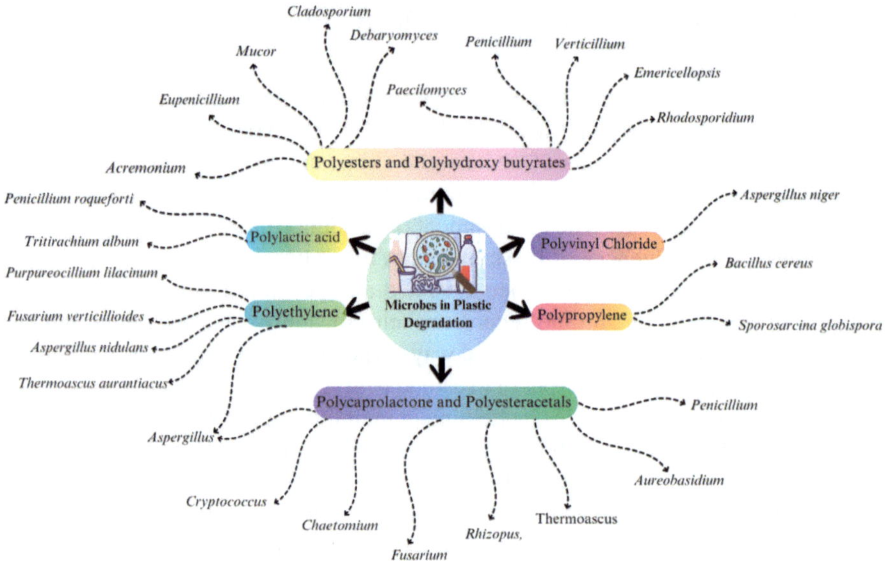

Fig. 5.3 Types of Plastics and microbes that are reported to degrade them

into the beta-oxidation pathway to produce protocatechuic acid (PCA) or protocatechuate (Yoshida et al. 2016). In 2017, a study conducted by scientists at the University of Cambridge surprised the world. Wax worms, the larvae of the greater wax moth (*Galleria mellonella*), were found to have a remarkable ability to consume and digest polyethylene, a common type of plastic used in shopping bags and packaging materials (Cassone et al. 2020).

In another study conducted in 2015, researchers at Stanford University discovered that mealworms (the larvae of darkling beetles; *Tenebrio molitor* L) could consume and digest Styrofoam, which is a notoriously difficult-to-recycle form of plastic (Ghosh et al. 2019). The researchers found that the mealworms could break down the Styrofoam into smaller fragments and convert it into biodegradable waste. Further investigations revealed that the mealworms' gut bacteria played a crucial role in aiding their ability to digest plastic. This finding offered a glimpse into the potential of harnessing natural plastic-degrading organisms to combat plastic pollution.

These discoveries, among many others, have opened a path to microbe-mediated plastic degradation, a crucial, if not the only, environmentally friendly way of plastic degradation. These real-life discovery tales demonstrate the resilience and adaptability of nature. Scientists continue to explore and harness the capabilities of plastic-degrading organisms in nature, offering hope for a more sustainable future in the battle against plastic pollution. As in the words of John Muir, the Scottish-born American naturalist, known as the Father of National Parks, "In every walk with nature, one receives far more than he seeks". Our current understanding of adaptation has hence been struck by awe by such innovations. Microbial adaptations to a material produced by humans during the last century are proof of nature watching

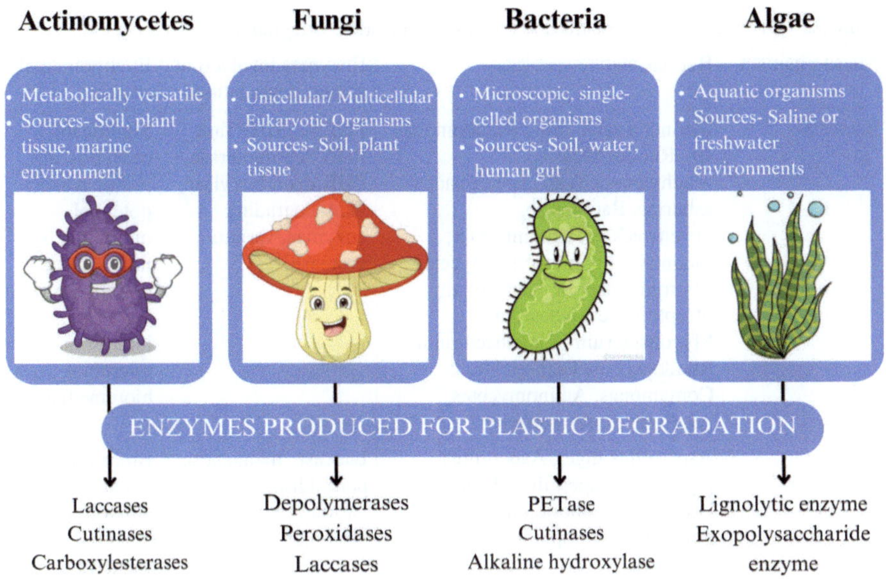

Fig. 5.4 The image depicts the different classes of organisms with the associated enzymes reported that are useful in microbial-mediated plastic degradation

us. Figure 5.4 represents a gist of the enzymes and their sources that are discovered and can be employed in plastic degradation. It is not merely adequate to be aware of the knowledge of plastic waste management. For a healthy sustainable environment, reliable solutions in terms of redesigning the current practices need to be enforced. Here, we now introduce you to microbial avengers for plastic waste management. Table 5.2 also outlines representative microbes from different phyla and the associated enzymes useful in plastic degradation (Yang et al. 2014; Nanda and Sahu 2010; Arkatkar et al. 2010).

5.7.1 Bacteria

Many studies have reported as many as 90 genera of microbes that could degrade plastics. Reports on poly(ε-caprolactone) biodegradation proposed the involvement of *Rhodococcus sp* and *Pseudomonas* along with 2 other members of phylum mycota. Jointly, they managed to degrade PCL films to more than half of their original weight in a span of a month (Ghosh et al. 2019). Therefore, *Rhodococcus sp* and *Pseudomonas* emerge as prominent biodegradation candidates. Temperature is an essential factor in biodegradation. During the degradation of biodegradable plastics such as PBS, PBSA, PLA, and polyhydroxybutyrate (PHB), *Pseudomonas, Tenacibaculum, and Alcanivorax sp* where were active at low temperatures, meanwhile no

Table 5.2 Diverse microorganisms and enzymes in plastic biodegradation

Microorganism	Representative species	Enzymes involved in plastic degradation	Environmental impact
Bacteria	Ideonella sakaiensis, Pseudomonas sp., Rhodococcus sp., Staphylococcus sp., Enterobacter asburiae, Bacillus sp., Owenweeksia, Crocinitomix, Ideonella, Thermoactinomycetes, Micromonospora, Klebsiella, Streptomyces, Nocardia, Mycobacterium, Flavobacterium, Rhodococcus, Escherichia, Comamonas, Actinomycetes, Alcaligenes, Azotobacter	PETase, MHETase, PHB depolymerase, PHB-decarboxylase, PLA-degrading enzymes, various protease	Significant reduction in plastic waste, potential mitigation of microplastic pollution, increased interest in microbial bioremediation
Fungi	Aspergillus niger, Aspergillus japonicus, Aspergillus flavus, Penicillium simplicissimum, Penicillium pinophilum	Laccase, manganese peroxidase, and various enzymes for oligomer production	Biodegradation in natural environments, potential use in wastewater treatment
Actinomycetes	Actinomadura, Microbispora, Streptomyces, Thermoactinomyces, Saccharomonospora, Amycolatopsis, Saccharothrix, Kibdelosporangium aridum	PHBV depolymerase, PHB depolymerase, various enzymes for polyester degradation	Biodegradable plastics production, potential role in soil enrichment
Algae	Spirulina, Chlorella, Microalgae	Enzymes for lowering activation energy, ligninolytic and oxidative enzymes	Potential use in wastewater treatment, carbon sequestration, and biofuel production
Insects	Galleria mellonella, Tenebrio molitor	Gut microbiota producing enzymes for plastic degradation	Insights into insect-microbe interactions, potential in waste management

degradation was observed at normal temperature (Sekiguchi et al. 2011; Pischedda et al. 2019).

5.7.1.1 *Pseudomonas* sp.

In a study on polyethylene, in comparison to other species in the study, *Pseudomonas* sp. produced the most viscous and flocculent biofilms (refer to section 6.7 for biofilms) on the surface during the degradation of polyethylene (PE). It was hypothesized that bacteria selectively utilized the medium, and depleted basal nutrients of the medium, with polyethylene acting as the readily available nutrient source (Nanda and Sahu 2010). Other studies, such as Arkatkar and associates (2010), corroborate the aforementioned results of the presence of flocculent microcolonies of *Pseudomonas* sp. and Actinomycetes sp. on the surface of polypropylene. It is also well supported during polypropylene (PP) degradation. Undoubtedly, *Pseudomonas* sp. adheres to biodegradable PE films. The degradation ability of *Pseudomonas* strains varies widely. *Pseudomonas* sp. mediated biodegradation of PE could be modulated by adjusting the hydrophobic interaction between the PE film and the organism. Certain agents, such as mineral oil, stimulate hydrophobic interactions, leading to an increase in bacterial adhesion and biofilm formation. This improved attachment hastened polymer decomposition. Tween 80, a non-ionic surfactant, inhibits biofilm formation by reducing hydrophobic interactions, thereby reducing bacterial attachment and PE degradation (Yang et al. 2014; Tribedi and Sil 2013).

5.7.1.2 *Rhodococcus* sp.

Other bacterial species besides *Pseudomonas* sp., have been found to be effective in the plastic degradation process. Gilan and coworkers (Orr et al. 2004) reported that *Rhodococcus* ruber (C208) used PE as its primary carbon source and formed a dense biofilm in a flask culture study and ultimately fed on PE as determined by weight loss analysis (Ghosh et al. 2019). According to reports, it also degrades polyolefins. Extracellular Polymeric Substrate (EPS) (are organic compounds secreted by microbes to help them in establishing interbacterial interactions, and therefore biofilm formation) of *Rhodococcus* sp. mainly comprises proteins. The addition of proteases hampers biofilm formation followed by plastic biodegradation (Gilan and Sivan 2013). *Rhodococcus* sp. could degrade PP more effectively than *Bacillus* sp. both isolated from the soil of mangroves. *Rhodococcus* sp. degraded PP with up to 1.6 times better efficiency than *Bacillus* sp in the same time interval of 40 days (Auta et al. 2018). While, in some other studies, *Rhodococcus* sp. demonstrated the lowest oxidation of PE as compared to *Pseudomonas* and *Brevibacillus* sp. (Nanda and Sahu 2010).

5.7.1.3 Other Bacteria(s)

Bacteria exhibit robust growth and demonstrate a wide range of diversity. The involvement of various genera of microbes in polymer degradation, in addition to *Pseudomonas* and *Rhodococcus* sp., has been well-documented. The report by Vatsaldutt and Anbuselvi (2014), (Anbuselvi and Pandey 2015) documented the findings that *Staphylococcus* sp. exhibited the highest level of polyethylene degradation at 52%, while *Pseudomonas* sp. accounted for 11% of the degradation, equivalent to 20%. In a study conducted by Yang and associates (2014), compelling evidence was presented regarding the participation of Enterobacter asburiae YT1 and *Bacillus* sp. YP1, which was isolated from the intestinal tracts of wax worms that ingest plastic, resulting in the degradation of PEA. Also, he stated that the analysis of PET-degrading communities revealed a high prevalence of Tenacibaculum and various other members belonging to the Flavobacteriaceae family (Yang et al. 2014). The genera Owenweeksia and Crocinitomix, which are classified under Cryomorphaceae and Bacteriodetes, respectively, exhibited significant presence in the process of PET degradation. Previous studies on the degradation of polyethylene terephthalate (PET) have also indicated the presence of Saprospiraceae, Cryomorphaceae, and Flavobacteriaceae in the process (Shah 2007). Additional bacterial species that are pertinent to the biodegradation process encompass *Ideonella, Thermoactinomycetes, Micromonospora, Klebsiella, Streptomyces, Nocardia, Mycobacterium, Flavobacterium, Rhodococcus, Escherichia, Comamonas, Actinomycetes, Alcaligenes,* and *Azotobacter.*

5.7.2 *Fungi*

Both members of bacteria and fungi have been documented as being implicated in the process of biodegradation. Bacteria are the primary colonizers during the fouling process on plastic surfaces. Upon colonization on the plastic surface, fungi are permitted to become entrapped, thereby facilitating the subsequent succession of additional species. The sharing of metabolic intermediates within the different colonizers accelerates degradation (Orr et al. 2004). The study documented the impact of fungal samples obtained from agricultural soils on the degradation process of plastic fragments that were buried within the agricultural soil, which had been previously mixed with sewage sludge. The additional examination also uncovered the presence of bacterial and fungal adherence on the plastic surface composed of low-density polyethylene (LDPE), suggesting the potential utilization of plastic as a substrate for nourishment. The fungi that were subsequently isolated consisted of species belonging to the genera *Penicillium, Fusarium,* and *Aspergillus* (Shah 2007). Most of the degradation studies on fungus-mediated biodegradation of plastics report *Aspergillus*, and *Penicillium sp.* as active degrading candidates. In contrast to bacteria, the ability of fungal species to form biofilms on polyethylene surfaces is linked to a gradual decrease in surface hydrophobicity. This suggests an inverse relationship between the biofilm formation capacity of fungal species on plastic materials and the hydrophobic nature of the surface (Ghosh et al. 2019).

5.7.2.1 *Aspergillus* sp.

Numerous species of *Aspergillus* exhibit the capability to degrade polyethylene. The findings of a research investigation focused on the isolation of fungi from sites contaminated with polyethylene revealed that the predominant organisms identified were *Aspergillus niger* and *A. japonicus*. The degradative capacity of *Aspergillus* species exhibits variation, with *A. japonicas* demonstrating a higher efficiency in degrading LDPE compared to *A. niger* within a similar incubation period of approximately one month. In vitro, the degradation of *A. japonicas* was observed to be 11.11%, while *A. niger* exhibited a degradation rate of only 5.8% (Raaman et al. 2012). In contrast, it was observed that *A. niger* exhibited the highest degradation capacity, reaching up to 38% degradation within a 60-day time frame, whereas *A. flavus* only achieved a degradation rate of 31%. Fungi have been predominantly employed for the degradation of highly durable plastics, such as low-density polyethylene (LDPE), owing to their ability to secrete hydrophobic proteins that facilitate attachment to other organisms, thereby promoting colonization (Mohan and Suresh 2015). Certain strains of *Aspergillus niger*, specifically those identified as ITCC (Indian Type Culture Collection) number 6052, have demonstrated the capability to degrade modified polyethylene. After 30 days of incubation, it was observed that thermally oxidized polyethylene experienced a weight reduction of approximately 3.44%. Additionally, there was a significant decline of 61% in the material's tensile strength. The results of the Scanning Electron Microscopy analysis revealed the presence of a highly concentrated network of biofilm on the surface of the polymer (Ghosh et al. 2019).

5.7.2.2 *Penicillium* sp.

Certain species of *Penicillium* exhibit initial enzymatic activity that is responsible for the production of short-chain oligomers, subsequently facilitating the further degradation of the polymer. Strains of the microorganism *P. simplicissimum* can synthesize enzymes such as laccase and manganese peroxidase (Ojha et al. 2017). In the course of polyhydroxybutyrate (PHB) degradation, both fungi, and bacteria secrete PHB depolymerase enzymes that facilitate the hydrolysis of PHB into 3-hydroxybutyrate monomers and short-chain oligomers. The enzyme, possessing a molecular weight of 35 kilodaltons, attaches to the polymer's surface via its substrate binding domain, thereby facilitating the catalytic reaction. The study conducted by Jyoti and coworkers (2014) (Ojha et al. 2017) demonstrated that the substance underwent further degradation and assimilation into carbon dioxide (CO_2) and water (H_2O). The production of PHB-decarboxylase is observed in two species of Penicillium, specifically *P. pinophilum* and *Penicillium* sp. (Panagiotidou et al. 2014). The degradation of PE films was observed to be enhanced within a 45-day incubation period when *Penicillium* sp. was included in a consortium consisting of *Bacillus megaterium*, *Pseudomonas mediterranea*, *Phanerochaete* sp., *Aspergillus* sp., and *Pseudomonas putida*. This enhancement was found to be more significant compared to the degradation observed in individual cultures over a 90-day incubation period (Mahalakshmi and Siddiq 2015). Pointing toward a cross-genus communication network.

5.7.2.3 *Mucor* sp.

Mucor sp. are generally brings the biodegradation of PE in a synergistic form, by being associated with other microbes. *Mucor circinilloids*, for instance, in association with *Aspergillus flavus*, isolated from municipal landfill areas showed after nine months, promising LDPE degradation when mixed with cow dung and poultry droppings (39. Pramila and Vijayaramesh 2011). The degradation potential varies depending on the type of consortia used. In some cases, consortia may decrease the degrading ability of the fungi as in a study carried out by Singh and associates (2014) where fungal consortia comprised of *A. flavus* F1, *Fusarium* sp F6, *A. japonicas* F3, showed significant biodegradation results in four weeks as compared to by *Penicillium* sp F5, *A. niger* F2, *Mucor* sp. F4 in terms of LDPE weight loss measurements (Jyoti and Gupta 2014).

5.7.3 Actinomycetes

Actinomycetes are predominantly a form of life in the soil. Staining positive to gram stains, this group of organisms is thought to be intermediates between bacteria and fungus. They are medically relevant due to an important class of medicines extracted from them; Antibiotics. Hence, studies on their metabolites as a potential for the biodegradation of plastics are not a novel idea. Aerobic microorganisms present in the natural environment were found to degrade the plastics poly (β-hydroxybutyrate) (PHB) and poly (ε-caprolactone) (PCL). A study on *Actinomadura* reported PLA-degrading enzymes (Sukkhum et al. 2009). Studies *Streptomyces* and *Micromonospora* strains were obtained and subjected to a screening process to assess their potential for the degradation of poly (ethylene succinate) (PES), poly(ε-caprolactone) (PCL), and poly(β-hydroxybutyrate) (PHB). These strains were isolated from both the upstream and downstream areas of the Touchien River in Taiwan (Nishida and Tokiwa 1993).

The study identified *Streptoverticillium kashmirense* AF1 as a microorganism present in municipal sewage that can degrade poly (3-hydroxybutyrate-co-3-hydroxyvalerate) (PHBV). The extracellular enzymes PHBV depolymerase, which are secreted by *Streptoverticillium kashmirense* AF1, were successfully purified and demonstrated the capability to degrade PHBV film. *Actinomadura, Microbispora, Streptomyces, Thermoactinomyces,* and *Saccharomonospora* are strains of thermophilic actinomycetes that possess the capability to degrade poly (ethylene succinate) (PES), poly (ε-caprolactone) (PCL), and poly (β-hydroxybutyrate) (PHB). The thermophilic actinomycetes *Microbispora rosea, Excellospora japonica,* and *E. viridilutea* exhibited the capability to degrade the aliphatic polyester known as poly(tetramethylene succinate) (Nishida and Tokiwa 1993).

Polylactic acid (PLA) is broadly used in food packaging. The production of a novel *Actinomycete, Streptomyces sp.* KKU215, capable of degrading polylactic acid (PLA) packaging, was conducted using PLA packaging as the exclusive carbon source. The highly effective strain was employed in the process of biodegrading PLA-packaging materials. The *Actinomycete strain* HT-6, which has been classified as *Amycolatopsis sp.*, has been discovered to possess the ability to degrade poly(tetramethylene carbonate) (PTMC), a polymer consisting of tetramethylene succinate (PTMS). Actinomycetes's strain demonstrates efficient polycarbonate PTMC degradation when cultivated in a liquid medium containing 150 mg of PTMC film. The process of degradation exhibits a rapid and comprehensive nature, leading to a notable augmentation in cellular proliferation (Nishida and Tokiwa 1993). The presence of microorganisms capable of degrading polylactide (PLA) in soil environments is limited. The degradation of polylactic acid (PLA) by actinomycetes was observed. The findings of this study demonstrate that a particular strain of *Amycolatopsis sp.* exhibits the capability to significantly decrease the weight of PLA film in a liquid culture environment at a temperature of 30 °C over a period of 14 days. There exist multiple species of actinomycetes that have been identified as capable of degrading Polybutylene succinate (PBS), PCL, and PLA. These species include *Amycolatopsis sp., Saccharothrix, Kibdelosporangium aridum, Actinomadura keratinilytica, Amycolatopsis thailandensis, Streptomyces bangladeshensis, and Streptomyces thermoviolaceus sub sp. Thermoviolaceus* (Nishida and Tokiwa 1993).

5.7.4 Algae

Microalgae are widely regarded as a highly promising means for the degradation of plastics. They also provide innovative alternatives to plastics. These organisms exhibit a remarkable ability to rapidly adapt to emerging ecological niches (Oberbeckmann et al. 2014). Microalgae, like other microorganisms, can colonize surfaces and create biofilms. This process, known as (Bio)fouling, paves the way for larger microorganisms like diatoms, protists, and some microscopic fungi to establish colonies (Pauli et al. 2017). According to reports, microalgae have the potential to enhance the degradation of synthetic polymers through the production of enzymes that can effectively lower the energy requirements for the cleavage of chemical bonds within the polymer structure. This occurs especially in areas where there is a high concentration of energy due to toxic systems (Chia et al. 2020). Microorganisms that have enhanced the ability to produce ligninolytic and oxidative enzymes demonstrate significantly greater proficiency in the biodegradation of polyethylene (Paladhi et al. 2022).

Spirulina and *Chlorella* due to their small size (~50 mm) and higher amounts of protein composition increase their suitability for plastic degradation without any pretreatment. A 3 step degradation process is seen in *Chlorella.* The first step is similar to *Spirulina* i.e. water evaporation, and the formation of a bimodal peak occurs as a result of the second and third steps of degradation, this happens due to degradation of Carbohydrates and proteins and a few minor thermally resistant and thermally sensitive constituents (Zeller et al. 2013). Algae attaches to the substratum by initiating the synthesis of exopolysaccharides and ligninolytic enzymes, which are crucial enzymes involved in the degradation of plastics. Microorganisms utilize plastic as a carbon and energy source, subsequently facilitating its degradation (Paladhi et al. 2022). Various microalgae in their capacity promote the biodegradation of material. The microalgae secrete enzymes that effectively decrease the activation energy and attenuate the chemical bonds within the polymer, consequently diminishing the energy demand for degradation (Zeller et al. 2013).

5.7.5 *Insects*

Members of the class Insecta, commonly known as insects, have shown the ability to participate in plastic degradation. Some insects possess enzymes and gut microbiota that enable them to break down certain types of plastics. One such example is the wax moth caterpillar, *Galleria mellonella* (Bombelli et al. 2017). Researchers have found that these caterpillars can consume and digest polyethylene, a common type of plastic used in various applications. In addition to *Galleria mellonella*, there have been studies exploring the plastic-degrading capabilities of other insects as well. For instance, certain species of mealworms; *Tenebrio molitor* (larval stages) have been observed to consume and metabolize polystyrene foam, another type of plastic. The gut microbiota of these insects plays a significant role in breaking down the plastic, converting it into simpler compounds that can be utilized by the insects as a food source (Machona et al. 2022) Research in this area is ongoing, and scientists are continually investigating the plastic-degrading abilities of various insect species. Understanding how insects can contribute to plastic biodegradation may provide valuable insights for developing sustainable waste management strategies and eco-friendly solutions to tackle plastic pollution (Bombelli et al. 2017).

5.7.6 *Anti-protein also Anti-plastic*

In 1981, William and his colleagues published their research findings on the protease-mediated degradation of plastics (Sukkhum et al. 2009). They conducted experiments to test the degradation of two types of plastics: Poly (L-lactide) and polylactic acid (PLA) were exposed to enzymatic degradation using three different proteases, specifically bromelain, pronase, and proteinase K. Following the conclusion of their experimental inquiries, the researchers ascertained that proteinase K sourced from the *Tritirachium album* demonstrated the utmost efficacy in regard to the fragmentation of polymer chains. Proteinase K was found to be effective in breaking down the amorphous section of L-PLA, which resulted in an accelerated degradation rate. However, the enzyme's efficiency decreased in the crystalline region, indicating that the protease may not be as effective in breaking down the more structured and organized parts of the polymer. The findings from this study significantly implicated the development of more environmentally friendly plastics, as they provide insight into which enzymes may be most effective in breaking down different types of polymers. This research has paved the way for further studies aimed at discovering new and more effective enzymes for the degradation of plastics (Chaignon et al. 2007; Gilan and Sivan 2013). It is now that we have even created Inotropic materials that have enzymes embedded in them and once they serve their purpose they initiate their self-degradation process. However, with plastics, it's a tough knot to break since the high temp manufacturing conditions denatured the enzymes within. Researchers then employed a very elegant but well-known class of protein molecules called chaperones that are known to protect enzymes from heat stress and bring about enzymatic reactions later and voila we have plastics that contain their degrading machinery (Zettler et al. 2013). The effect of two proteases, trypsin and proteinase K, on the biofilm formation of *R. ruber* C208 has been studied in other research. The results showed that these two proteases exerted opposite effects on the biofilm formation. The presence of trypsin resulted in a decrease in biofilm formation, whereas the addition of proteinase K led to an increase in biofilm formation. However, the reason for these different effects of the two common proteases is still under study. Both of these proteases though have a broad range of substrates that they can act on, and have a target specificity associated with bond cleavage to them.

Both proteases in one study were found to be effective in removing the biofilm produced by some *S. aureus* strains, but in other studies, proteinase K was observed to be more effective than trypsin. Prior research has demonstrated that proteinase K possesses the ability to impede the formation of biofilms and facilitate the dispersion of pre-existing biofilms. The (agr) quorum-sensing system which is the accessory gene regulator is also involved in biofilm detachment. Prior research has demonstrated that proteinase K possesses the ability to impede the formation of biofilms and facilitate the dispersion of pre-existing biofilms. The results of these studies indicate that agr-deficient *S. aureus* strains form more resilient and tough biofilms compared with wild-type strains. The authors of a study (Gilan and Sivan 2013) suggested that extracellular proteins serve different purposes in the structure and function of the biofilm matrix, and may also be involved in detachment processes through quorum-sensing mechanisms. Certain proteins identified are believed to be extracellular proteolytic enzymes functioning within the matrix. The researchers propose that this methodology has the potential to be extended to characterize other EPS polymers. Biofilms likely rely on matrix proteins, which could eventually lead to the development of an enzymatic method for treating polyolefin waste.

5.7.7 Biofilm Formation: Let's Talk About Bacteria

Biofilms are functionally, and phylogenetically diverse communities of bacteria, fungi, and algae conjointly termed as Microbial Conglomeration, attached to the surface. Biofilms provide a lot of benefits for survival and competition strategies, including bioavailability of nutrients, horizontal gene transfer, cell viability, and preventing toxic shock. Biofilms are dynamic structures that influence the plastic-microbe interaction, and the microbe-mediated breakdown of plastics. Therefore, the understanding of the dynamics of biofilms is crucial to the development of processes to enhance microbe-mediated-plastic-degradation.

Microbial communities undergo a process of aggregation on solid surfaces, resulting in the formation of a complex, dynamic structure called biofilm (Zettler et al. 2013), altering the physicochemical characteristics of plastic (Ye and Andrady 1991; Rummel et al. 2017). Surface hydrophobicity is crucial for bacterial attachment and degradation, as plastic's hydrophobic nature prompts hydrophobic interactions causing chemical alterations while contacting microbes (Ye and Andrady 1991; Gilan and Sivan 2013; Sivan 2011). Increased surface roughness facilitates microbial attachment and polymer modification, promoting degradation by enhancing the surface area-to-volume ratio (Rummel et al. 2017). Extracellular polymeric substances (EPS) from initiator microorganisms contribute to biofilm adhesion and colonization, impacting biofilm properties. Recent research findings have indicated that the hydrophobic properties of a bacterial culture can be enhanced through the process of nutrient deprivation. Consequently, this exacerbated the degradation process. Furthermore, these findings provide additional support for the increased affinity of *Rhodococcus ruber* to adhere to the polyethylene (PE) surface.

Microbial filaments infiltrate the polymer matrix, participating in the creation of biosurfactants and retaining water for up to six hours. In the conducted research, additional hydrolysis was observed to elevate conductivity (Ghosh et al. 2019). Furthermore, the release of lipophilic pigments contributes to altering the coloration of the plastic (Zettler et al. 2013). Biofouling plays a crucial role, particularly in aquatic environments, in microbial-mediated plastic degradation. This process involves adsorption, immobilization of biomass, and subsequent micro and macro-fouling. Bacteria act as primary colonizers, capturing other organisms, such as fungi, diatoms, and others (Selim et al. 2017). The metabolic activities of biomass attached to surfaces lead to polymer chain desorption, adsorption, and fragmentation, ultimately resulting in debris degradation (Harrison et al. 2011; Balasubramanian et al. 2010). The degradation of polymers by microorganisms, as illustrated in Fig. 5.5, is an area of ongoing research. Microbial polymer degradation encompasses various processes, including biodeterioration, bio-fragmentation, mineralization, and assimilation, as documented in the literature.

- **Biodeterioration**—The process under consideration is an important process that impacts the external characteristics of plastic materials, leading to alterations in their chemical composition, physical attributes, and mechanical behavior that is brought in by the activities of a living organism(s) and is termed as biodeterioration (Pathak 2017). The chemical and structural alterations are contingent upon the specific structure and composition of polymers. Environmental conditions play a significant role in driving change. Abiotic factors like temperature, UV radiation, and Oxidative stress, Contribute to the chemical and physical changes of the material. Colonization by Algae/ bacteria, which are primarily the pioneer community, influence the further degradation by attachment to the substrate(plastic) and

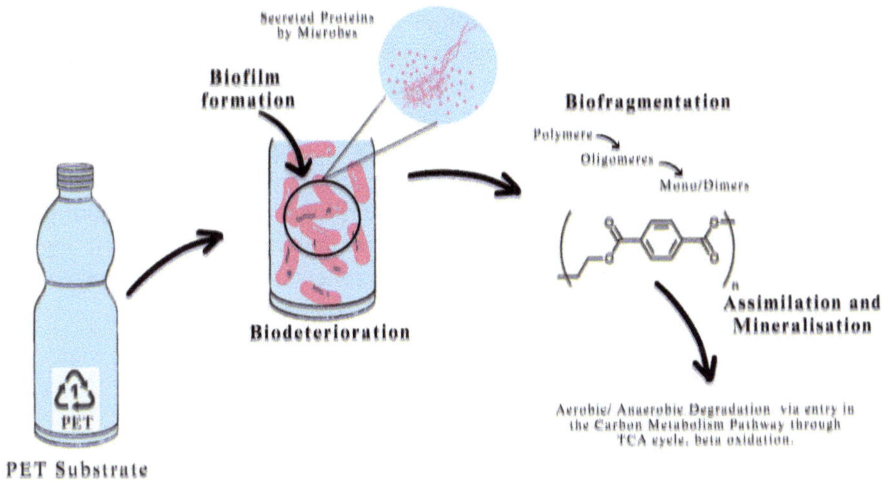

Fig. 5.5 The figure is a representative diagram of microbe-mediated plastic degradation. Plastic is broken into chunks by abiotic factors (UV radiations). Microbes colonize these chunks, leading to biofilm formation. This leads via complex communication (within the biofilm), in the secretion of enzymes, specific to the class of organism (bacteria shown here), which proceeds into plastic degradation and the utilization of the constituents into building blocks of life for the organism

the secretion of enzymes Fig. 5.5 (e.g. depolymerase, exopolysaccharides, etc.) (Rummel et al. 2017; Harrison et al. 2011).

- **Bio-fragmentation**—Following biodeterioration, distinguishes itself from fragmentation, which can occur due to abiotic factors before microbe attachment, as it involves enzymes secreted by microbes acting on plastic polymers. The most commonly studied enzymes Oxygenases, which are enzymatic proteins found in bacteria, possess the ability to cleave oxygen molecules within carbon chains. Consequently, the liberated oxygen atoms are incorporated into the carbon chain, resulting in the generation of carbon chains containing functional groups such as alcohol and peroxyl products. These modified carbon chains exhibit reduced levels of harmfulness (Pathak 2017; Dussud et al. 2018). In addition, lipases play a crucial role in facilitating the conversion of carboxylic groups, while esterases or endopeptidases are primarily responsible for the modification of amide groups (Zeenat et al. 2021)
- **Mineralization**—Bio-fragmentation process leads to the generation of oligomers, dimers, and monomers. Microbial cells are penetrated by these substances via their cell membranes. Cells are unable to accommodate and retain monomers of significant size, resulting in their exclusion from the cellular interior. The diminutive monomers that undergo intracellular movement are subject to oxidation and subsequently utilized for the generation of energy. The energy is ultimately harnessed for biomass production (Kale et al. 2015; Lucas et al. 2008).
- **Assimilation**—The ultimate reason for microbe-mediated plastic degradation sums up this step. All forms of life spend energy in finding, feeding, and secreting

enzymes responsible for the digestion or breakdown of complex molecules for the ultimate reason for carrying out their life processes. During the process of assimilation, the molecules that are generated are incorporated into the microbial cells to facilitate their complete degradation. The extracellular transportation of secondary metabolites, or their transfer to other microorganisms for subsequent degradation and utilization. Metabolism produces the oxidized products like CO_2, N_2, H_2O, and CH_4 (Lucas et al. 2008). Amorphous polymers are more prone to microbial enzyme attack than the crystalline polymers or polymeric regions within a polymer harboring both microenvironments. Therefore, an increase in crystallinity decreases polymer degradation (Krzan et al. 2006).

5.8 Synthetic Biology

A frontier discipline seated at the intersection of biology, engineering, and computer science, seeks to refashion, redesign, and repurpose the genetic and metabolic capabilities of organisms. Its approach to biological design is underpinned by a central tenet: that life is programmable and that through human ingenuity, we can imbue microbes with novel functionality to aid us in our struggles against various environmental challenges. The applications of synthetic biology are vast and eclectic, but we shall narrow our focus on its prospective role in the mitigation of plastic waste (Cheng and Lu 2012). Plastic, that ubiquitous, yet insidious manifestation of human civilization, is today found in every corner of our planet. Synthetic biology, with its toolkit of genetic manipulation, is uniquely poised to convert this crisis into an opportunity by exploiting the plastic-degrading capacities of microbes. The field is in essence, a paradigm shift, converging the boundaries between the digital realm and biological reality, where the code of life becomes a programming language to be mastered and manipulated. Synthetic biology promises to extend our humanistic imperatives of sustainable progress and environmental stewardship into the microscopic world, enabling microbes to perform feats of biochemistry that are scarcely found in nature (Andrianantoandro et al. 2006).

5.8.1 Harnessing Synthetic Biology for Sustainable Plastic Waste Management Synthetic Biology Is Emerging as a Powerful Tool in the Global Effort to Address Plastic Waste. Its Adoption Is Driven by a Multitude of Compelling Reasons

- **Adaptability and Customization**: Synthetic biology's unique strength lies in its adaptability. It empowers scientists to design custom-built biological systems tailored to the specific type of plastic waste and environmental conditions at hand (Nguyen et al. 2023).

- **Scalability**: Unlike traditional plastic waste management methods, such as recycling or physical degradation, synthetic biology offers scalability. Biological solutions can self-sustain and proliferate, presenting the potential to operate at the same magnitude as the plastic waste problem itself (French 2019).
- **Sustainability**: Synthetic biology promotes environmentally friendly solutions. Biodegradation of plastic, a key application, produces no harmful by-products and leaves no residual microplastics, reducing the environmental footprint of plastic waste (Nguyen et al. 2023).
- **Uncovering New Solutions**: The untapped genetic diversity within the microbial world holds the promise of innovative and unexpected solutions to the plastic problem. Synthetic biology can manipulate and combine genetic resources to discover new enzymatic pathways, microbial consortia, and degradation mechanisms, significantly enhancing plastic pollution mitigation (Santomartino et al. 2023).

Synthetic biology, with its versatility, scalability, sustainability, and untapped potential, stands at the forefront of a new era in waste management. It has the capacity to fundamentally reshape our relationship with plastic, paving the way for a more sustainable and plastic-free world (Andrianantoandro et al. 2006).

5.8.2 Empowering Microbes with Recombinant DNA Technology

Recombinant DNA technology, or genetic engineering, is one of the quintessential tools, empowers scientists to manipulate genetic material for enhanced plastic degradation (Cheng and Lu 2012). This technology allows for:

- **Gene Transfer**: Specific genes can be identified, isolated, and introduced into plastic-degrading microbes, amplifying their efficacy. Genes can be translocated between organisms to confer the ability to degrade resistant plastics (Cheng and Lu 2012). Take the case of bacteria known to harbor enzymes capable of breaking down PET (Polyethylene Terephthalate)—a common type of plastic used in beverage bottles. Recombinant DNA technology allows us to extract these enzymes and insert them into other organisms, scaling up the PET-degradation process and paving the way toward more effective plastic waste remediation (Yoshida et al. 2016; Andrianantoandro et al. 2006).
- **Precise Solutions**: Synthetic biology enables the creation of bespoke microbial solutions for various plastic pollutants, enhancing the precision and effectiveness of plastic waste management (Nguyen et al. 2023). Researchers identified an enzyme that is not found in significant quantities for breakdown of polystyrene. Using synthetic biology, they isolated the responsible degradation gene from these

microorganisms and inserted it into *E. coli.* The expression of this gene was optimized to enhance enzyme production. The custom-engineered bacteria were then scaled up for industrial applications. These engineered bacteria were utilized in controlled environments, such as wastewater treatment, to precisely and efficiently break down polystyrene into eco-friendly molecules (Rafeeq et al. 2022).

However, it is not merely about transferring genes, but also about optimizing their expression, control, and regulation within the host-microbe. As we delve into the artistry of synthetic biology, the scope of recombinant DNA technology broadens, encompassing not just gene transfer, but also the entire orchestra of gene expression, regulation, and feedback controls (Santomartino et al. 2023). Synthetic biology and recombinant DNA technology have thereby unfurled a new narrative of hope and possibility in our fight against plastic pollution. By leveraging the power of microbes, we are on the brink of a biological revolution that could redefine our relationship with plastics, transforming a global environmental challenge into an opportunity for sustainable development. This approach leverages the power of microbes, heralding a biological revolution that can transform a global environmental challenge into an opportunity for sustainable development (Nguyen et al. 2023).

5.9 Conclusion

The chapter offers valuable insights into the notable environmental challenges that stem from plastic waste, as well as the potential solutions that have been suggested through the application of various microbial agents. Bacteria, fungi, actinomycetes, algae, and other organisms, including insects, have exhibited notable effectiveness in the process of plastic degradation, offering a potential resolution to the escalating ecological crisis. The ability to break down persistent plastics into harmless byproducts presents a promising opportunity for waste management and conservation of the environment. In addition, the emerging field of synthetic biology offers unprecedented opportunities for revolutionizing our approach to addressing the problem of plastic waste in the future. Genetic engineering equips researchers with the capacity to manipulate the genetic composition of microorganisms. This alteration of their genetic makeup enhances their plastic degradation potential under various environmental conditions. The innovations explored in this chapter represent the early phases of wider technological progress focused on the microbial degradation of plastics. Though limited to a small scale at the moment, the understanding of biology of the microbes and the degradation processes promises to lead us to a sustainable future. By applying the principles of synthetic biology and further developing the microbial solutions discussed in this chapter, it is conceivable to envision a future in which the problem of plastic pollution is successfully eliminated. Yet, it's crucial to acknowledge our understanding of nature remains incomplete, thereby limiting our ability to reconstruct it if ravaged. Hence, recognizing the "stopping rules"—our call to halt plastic use—becomes pivotal while using synthetic biology to engineer microbes for plastic degradation. This quest demands unwavering commitment and broad-scale collaboration to bear fruit. It's a testament to our tenacity, a challenge to our ingenuity, and a plea for our unity in safeguarding our world. It is crucial to also acknowledge that the innovations being discussed are not isolated solutions, but rather significant steps toward fostering an environmentally conscious society. The effective transfer of innovations from the laboratory to widespread implementation requires collaborative efforts among scientists, governments, industries, and communities. By leveraging continuous research, advancements in technology, and a shared dedication to environmental stewardship, it is feasible to overcome the challenges posed by plastic waste and pave the way toward a more sustainable and thriving planet for forthcoming generations. This won't happen in a day, but shall happen someday.

References

Anbuselvi S, Pandey V (2015) A comparative study of LDPE degradating bacteria from polythene dumped garbage. Int J Pharm Technol ISSN

Andrianantoandro E, Basu S, Karig DK, Weiss R (2006) Synthetic biology: new engineering rules for an emerging discipline. Mol Syst Biol 2(2006):0028. https://doi.org/10.1038/msb4100073

Arkatkar A, Juwarkar AA, Bhaduri S, Uppara PV, Doble M (2010) Growth of *Pseudomonas* and *Bacillus* biofilms on pretreated polypropylene surface. Int Biodeterior Biodegradation 64(6):530–536. https://doi.org/10.1016/j.ibiod.2010.06.002

Auta HS, Emenike CU, Jayanthi B, Fauziah SH (2018) Growth kinetics and biodeterioration of polypropylene microplastics by *Bacillus* sp. and *Rhodococcus* sp. isolated from mangrove sediment. Mar Pollut Bull 127:15–21. https://doi.org/10.1016/j.marpolbul.2017.11.036

Balasubramanian V, Natarajan K, Hemambika B, Ramesh N, Sumathi CS, Kottaimuthu R, Rajesh KV (2010) High-density polyethylene (HDPE)-degrading potential bacteria from marine ecosystem of Gulf of Mannar, India. Lett Appl Microbiol 51(2):205–211. https://doi.org/10.1111/j.1472-765X.2010.02883

Bombelli P, Howe CJ, Bertocchini F (2017) Polyethylene bio-degradation by caterpillars of the wax moth *Galleria mellonella*. Curr Biol 27(8):R292–R293. https://doi.org/10.1016/j.cub.2017.02.060

Building a sustainable future from plastic waste. Conceptos Plasticos. https://conceptosplasticos.com/

Cabernard L, Pfister S, Oberschelp C et al (2022) Growing environmental footprint of plastics driven by coal combustion. Nat Sustain 5:139–148. https://doi.org/10.1038/s41893-021-00807-2

Cassone BJ, Grove HC, Elebute O, Villanueva SMP, LeMoine CMR (2020) Role of the intestinal microbiome in low-density polyethylene degradation by caterpillar larvae of the greater wax moth, Galleria mellonella. The Royal Society 287(1922). https://doi.org/10.1098/rspb.2020.0112

Chaignon P, Sadovskaya I, Ragunah Ch, Ramasubbu N, Kaplan JB, Jabbouri S (2007) Susceptibility of staphylococcal biofilms to enzymatic treatments depends on their chemical composition. Appl Microbiol Biotechnol 75(1):125–132.https://doi.org/10.1007/s00253-006-0790-y

Cheng AA, Lu TK (2012) Synthetic biology: an emerging engineering discipline. Annu Rev Biomed Eng 14:155–178. https://doi.org/10.1146/annurev-bioeng-071811-150118

Chia WY, Tang DYY, Khoo KS, Lup ANK, Chew KW (2020) Nature's fight against plastic pollution: algae for plastic biodegradation and bioplastics production. Environ Sci Ecotechnol 4:100065. https://doi.org/10.1016/j.ese.2020.100065

Dussud C, Hudec C, George M, Fabre P, Higgs P, Bruzaud S, Delort AM, Eyheraguibel B, Meistertzheim AL, Jacquin J, Cheng J, Callac N, Odobel C, Rabouille S, Ghiglione JF (2018) Colonization of non-biodegradable and biodegradable plastics by marine microorganisms. Front Microbiol 9:1571. https://doi.org/10.3389/fmicb.2018.01571

French KE (2019) Harnessing synthetic biology for sustainable development. Nat Sustain 2(4):250–252. https://doi.org/10.1038/s41893-019-0270-x

Gasperi J, Wright SL, Dris R, Collard F, Mandin C, Guerrouache M, Tassin B (2018) Microplastics in air: are we breathing it in? Curr Opin Environ Sci Health 1:1–5. https://doi.org/10.1016/j.coesh.2017.10.002

Geyer R, Jambeck JR, Law KL (2017) Production, use, and fate of all plastics ever made. Sci Adv 3(7):e1700782. https://doi.org/10.1126/sciadv.1700782

Geyer R, Jambeck JR, Law KL (2017) Production, use, and fate of all plastics ever made. Sci Adv 3(7):e1700782. https://doi.org/10.1126/sciadv.1700782.

Ghosh SK, Pal S, Ray S (2013) Study of microbes having potentiality for biodegradation of plastics. Environ Sci Pollut Res 20:4339–4355. https://doi.org/10.1007/s11356-013-1706-x

Ghosh S, Qureshi A, Purohit HJ (2019) Microbial degradation of plastics: biofilms and degradation pathways. Contam Agric Environ Health Risks Remediat 1:184–199. https://doi.org/10.26832/AESA-2019-CAE-0153-014

Gilan I, Sivan A (2013) Effect of proteases on biofilm formation of the plastic-degrading actinomycete *Rhodococcus ruber* C208. FEMS Microbiol Lett 342(1):18–23. https://doi.org/10.1111/1574-6968.12114

Gimouhopoulos K, Doulia D, Vlyssides A, Georgiou D (2000) Optimization studies of waste plastics-lignite catalytic coliquefaction. Waste Manage Res 18(4):352–357. https://doi.org/10.1177/0734242X00018004

Graulich K et al (2021) Impact of COVID-19 on single-use plastics and the environment in Europe. European Environment Agency

Harrison JP, Sapp M, Schratzberger M, Osborn AM (2011) Interactions between microorganisms and marine microplastics: a call for research. Mar Technol Soc J 45(2):12–20. https://doi.org/10.4031/MTSJ.45.2.2

Hopewell J, Dvorak R, Kosior E (2009) Plastics recycling: challenges and opportunities. Philos Trans Royal Soc b Biol Sci 364(1526):2115–2126. https://doi.org/10.1098/rstb.2008.0311

International Bottled Water Association, Bottled water packaging has the lowest environmental footprint of all packaged drinks (2020). https://bottledwater.org/packaging/#:~:text=Plastic%20bottled%20water%20containers%20are,and%20HDPE%20for%209.2%20percent

Jyoti S, Gupta KC (2014) Screening and identification of low density polyethylene (LDPE) degrading soil fungi isolated from polythene polluted sites around Gwalior city (MP). Int J Curr Microbiol App Sci 3(6):443–448

Kale SK, Deshmukh AG, Dudhare MS, Patil V (2015) Microbial degradation of plastic-a review. J Biochem Technol 6:952–961. https://doi.org/10.31838/ijpr/2021.13.01.245

Krzan A, Hemjinda S, Miertus S, Corti A, Chiellini E (2006) Standardization and certification in the area of environmentally degradable plastics. Polym Degrad Stab 91(12):2819–2833. https://doi.org/10.1016/j.polymdegradstab.2006.04.034

Lebreton LC, Van Der Zwet J, Damsteeg JW, Slat B, Andrady A, Reisser J (2017) River plastic emissions to the world's oceans. Nat Commun 8(1):15611. https://doi.org/10.1038/ncomms15611

Lechthaler S, Waldschläger K, Sandhani CG, Sannasiraj SA, Sundar V, Schwarzbauer J, Schüttrumpf H (2021) Baseline study on microplastics in Indian rivers under different anthropogenic influences. Water 13(12):1648. https://doi.org/10.3390/w13121648

Li J, Guo S, Li X (2005) Degradation kinetics of polystyrene and EPDM melts under ultrasonic irradiation. Polym Degrad Stab 89(1):6–14. https://doi.org/10.1016/j.polymdegradstab.2004.12.017

Lin YH, Yen HY (2005) Fluidised bed pyrolysis of polypropylene over cracking catalysts for producing hydrocarbons. Polym Degrad Stab 89(1):101–108. https://doi.org/10.1016/j.polymdegradstab.2005.01.0

Lomonaco T, Manco E, Corti A, La Nasa J, Ghimenti S, Biagini D, Castelvetro V (2020) Release of harmful volatile organic compounds (VOCs) from photo-degraded plastic debris: a neglected source of environmental pollution. J Hazard Mater 394:12259. https://doi.org/10.1016/j.jhazmat.2020.122596

Lucas N, Bienaime C, Belloy C, Queneudec M, Silvestre F, Nava-Saucedo JE (2008) Polymer biodegradation: mechanisms and estimation techniques–a review. Chemosphere 73(4):429–442. https://doi.org/10.1016/j.chemosphere.2008.06.064

Lyu S, Untereker D (2009) Degradability of polymers for implantable biomedical devices. Int J Mol Sci 10(9):4033–4065. https://doi.org/10.3390/ijms10094033

Machona O, Chidzwondo F, Mangoyi R (2022) *Tenebrio molitor*: possible source of polystyrene-degrading bacteria. BMC Biotechnol 22:2. https://doi.org/10.1186/s12896-021-00733-3

Mahalakshmi V, Siddiq SA (2015) Enhanced biodegradation of polyethylene by development of a consortium. Pelagia Res Libr 6(4):183–189

Meereboer KW, Misra M, Mohanty AK (2020) Review of recent advances in the biodegradability of polyhydroxyalkanoate (PHA) bioplastics and their composites. Green Chem 22(17):5519–5558. https://doi.org/10.1039/D0GC01647K

Michels J, Stippkugel A, Lenz M, Wirtz K, Engel A (2018) Rapid aggregation of biofilm-covered microplastics with marine biogenic particles. Proc R Soc B 285(1885):20181203. https://doi.org/10.1098/rspb.2018.1203

Mohan SK, Suresh B (2015) Studies on biodegradation of plastics by *Aspergillus* sp. isolated from dye effluent enriched soil. Indo Am J Pharm Sci 2(12):1636–1639

Nanda S, Sahu SS (2010) Biodegradability of polyethylene by *Brevibacillus, Pseudomonas*, and *Rhodococcus.* spp. https://api.semanticscholar.org/CorpusID:52061941

Nguyen PQ, Huang X, Collins DS, Collins JJ, Lu T (2023) Harnessing synthetic biology to enhance ocean health. Trends Biotechnol. https://doi.org/10.1016/j.tibtech.2022.12.015

Nishida H, Tokiwa Y (1993) Distribution of poly(β-hydroxybutyrate) and poly(ε-caprolactone)aerobic degrading microorganisms in different environments. J Environ Polym Degr 1:227–233. https://doi.org/10.1007/BF01458031

Oberbeckmann S, Loeder MG, Gerdts G, Osborn AM (2014) Spatial and seasonal variation in diversity and structure of microbial biofilms on marine plastics in Northern European waters. FEMS Microbiol Ecol 90(2):478–492. https://doi.org/10.1111/1574-6941.12409

Oberbeckmann S, Osborn AM, Duhaime MB (2016) Microbes on a bottle: substrate, season and geography influence community composition of microbes colonizing marine plastic debris. PLoS ONE 11(8):e0159289. https://doi.org/10.1371/journal.pone.0159289

OECD (2022) Global plastics outlook: economic drivers, environmental impacts and policy options. OECD Publishing, Paris.https://doi.org/10.1787/de747aef-en

OECD Home (2022) Plastic pollution is growing relentlessly as waste management and recycling fall short, says OECD. https://www.oecd.org/environment/plastic-pollution-is-growing-relentlessly-as-waste-management-and-recycling-fall-short.htm

Ojha N, Pradhan N, Singh S et al (2017) Evaluation of HDPE and LDPE degradation by fungus, implemented by statistical optimization. Sci Rep 7:39515. https://doi.org/10.1038/srep39515

Orr IG, Hadar Y, Sivan A (2004) Colonization, biofilm formation and biodegradation of polyethylene by a strain of *Rhodococcus* ruber. Appl Microbiol Biotechnol 65(1):97–104. https://doi.org/10.1007/s00253-004-1584-8

Paladhi AG, Vallinayagam S, Rajendran S, Rathinam V, Sharma VK (2022) Microalgae: a promising tool for plastic degradation. In Microbes and microbial biotechnology for green remediation. Elsevier, pp 575–587. https://doi.org/10.1016/B978-0-323-90452-0.00049-9

Panagiotidou E, Konidaris C, Baklavaridis A, Zuburtikudis I, Achilias D, Mitlianga P (2014) A simple route for purifying extracellular poly (3-hydroxybutyrate)-depolymerase from *Penicillium pinophilum.* Enzyme research.https://doi.org/10.1155/2014/159809

Pathak VM (2017) Review on the current status of polymer degradation: a microbial approach. Bioresour Bioprocess 4(1):1–31. https://doi.org/10.1186/s40643-017-0145-9

Pauli NC, Petermann JS, Lott C, Weber M (2017) Macrofouling communities and the degradation of plastic bags in the sea: an *in situ* experiment. Royal Soc Open Sci 4(10):170549. https://doi.org/10.1098/rsos.170549

Pischedda A, Tosin M, Degli-Innocenti F (2019) Biodegradation of plastics in soil: the effect of temperature. Polym Degrad Stab 170:109017. https://doi.org/10.1016/j.polymdegradstab.2019.109017

Pramila R, Vijay Ramesh K (2011) Biodegradation of low-density polyethylene(LDPE) by fungi isolated from municipal landfill area. J Microbiol Biotech Res 1(4):131–136. https://doi.org/10.5897/AJMR11.670

Raaman, N., Rajitha, N., Jayshree, A., & Jegadeesh, R. (2012). Biodegradation of plastic by *Aspergillus* spp. isolated from polythene polluted sites around Chennai. J Acad Indus Res 1(6):313–316

Rafeeq H, Afsheen N, Rafique S, Arshad A, Intisar M, Hussain A, Bilal M, Iqbal HM (2022) Genetically engineered microorganisms for environmental remediation. Chemosphere 310:136751. https://doi.org/10.1016/j.chemosphere.2022.136751

Rånby B (1989) Photodegradation and photo-oxidation of synthetic polymers. J Anal Appl Pyrol 15:237–247. https://doi.org/10.1016/0165-2370(89)85037-5

Royer S-J, Ferrón S, Wilson ST, Karl DM (2018) Production of methane and ethylene from plastic in the environment. In: Pardha-Saradhi P (ed) PLOS ONE, vol 13, Issue 8. Public Library of Science (PLoS), p e0200574. https://doi.org/10.1371/journal.pone.0200574

Rummel CD, Jahnke A, Gorokhova E, Kühnel D, Schmitt-Jansen M (2017) Impacts of biofilm formation on the fate and potential effects of microplastic in the aquatic environment. Environ Sci Technol Lett 4(7):258–267. https://doi.org/10.1021/acs.estlett.7b00164

Santomartino R, Averesch NJ, Bhuiyan M, Cockell CS, Colangelo J, Gumulya Y, Lehner B, McMahon S, Mohanty A, Santa Maria SR, Urbaniak C, Volger R, Yang J, Zea L (2023) Toward sustainable space exploration: a roadmap for harnessing the power of microorganisms. Nat Commun 14(1):1–11. https://doi.org/10.1038/s41467-023-37070-2

Schmidt-Naake G, Drache M, Weber M (2002) Combination of mechanochemical degradation of polymers with controlled free-radical polymerization. Macromol Chem Phys 203(15):2232–2238. https://doi.org/10.1002/1521-3935(200211)203

Sekiguchi T, Saika A, Nomura K, Watanabe T, Watanabe T, Fujimoto Y, Kanehiro H (2011) Biodegradation of aliphatic polyesters soaked in deep seawaters and isolation of poly (ε-caprolactone)-degrading bacteria. Polym Degrad Stab 96(7):1397–1403. https://doi.org/10.1016/j.polymdegradstab.2011.03.0

Sekine Y, Fujimoto K (2003) Catalytic degradation of PP with an Fe/activated carbon catalyst. J Mater Cycles Waste Manage 5:107–112. https://doi.org/10.1007/s10163-003-0091-9

Selim MS, Shenashen MA, El-Safty SA, Higazy SA, Selim MM, Isago H, Elmarakbi A (2017) Recent progress in marine foul-release polymeric nanocomposite coatings. Prog Mater Sci 87:1–32. https://doi.org/10.1016/j.pmatsci.2017.02.001

Shah AA (2007) Role of microorganisms in biodegradation of plastics (Doctoral dissertation, Quaid-i-Azam University Islamabad, Pakistan)

Shams M, Alam I, Mahbub MS (2021) Plastic pollution during COVID-19: plastic waste directives and its long-term impact on the environment. In Environmental advances, vol 5, p 100119. Elsevier BV. https://doi.org/10.1016/j.envadv.2021.100119

Shen M, Huang W, Chen M, Song B, Zeng G, Zhang Y (2020) (Micro) plastic crisis: un-ignorable contribution to global greenhouse gas emissions and climate change. J Clean Prod 254:120138. Elsevier BV. https://doi.org/10.1016/j.jclepro.2020.120138

Singh J, Gupta KC (2014) Screening and identification of low density polyethylene (LDPE) degrading soil fungi isolated from polythene polluted sites around Gwalior City (MP). Int J Current Microbiol Appl Sci 3:443–448. ISSN: 2319-7706

Sivan A (2011) New perspectives in plastic biodegradation. Curr Opin Biotechnol (3):422–6. https://doi.org/10.1016/j.copbio.2011.01.013

Striegel AM (2003) Influence of chain architecture on the mechanochemical degradation of macromolecules. J Biochem Biophys Methods 56(1–3):117–139. Elsevier BV. https://doi.org/10.1016/s0165-022x(03)00054-x

Sukkhum S, Tokuyama S, Kitpreechavanich V (2009) Development of fermentation process for PLA-degrading enzyme production by a new thermophilic Actinomadura sp. T16–1. Biotechnol Bioproc E 14:302–306. https://doi.org/10.1007/s12257-008-0207-0

Swift G (1993) Directions for environmentally biodegradable polymer research. Acc Chem Res 26(3):105–110. https://doi.org/10.1021/ar00027a005

Tarun K, Sreelakshmi K, Peeyush K (2019) Segregation of plastic and non-plastic waste using convolutional neural network. IOP conference series: materials science and engineering, vol 561/1, p 012113. https://doi.org/10.1088/1757-899X/561/1/012113

Tokiwa Y, Calabia B, Ugwu C, Aiba S (2009) Biodegradability of plastics. Int J Mol Sci 10(9):3722–3742. MDPI AG. https://doi.org/10.3390/ijms10093722

Tribedi P, Sil AK (2013) Low-density polyethylene degradation by *Pseudomonas* sp. AKS2 biofilm. Environ Sci Pollut Res 20:4146–4153. https://doi.org/10.1007/s11356-012-1378-y

Vatsal dutt P, Anbuselvi S (2014) Isolation and characterization of polythene degrading bacteria from polythene dumped garbage. Int J Pharm Sci Rev Res 25:205–206. ISSN 0976 – 044X.

Wang J, Zhao X, Wu A, Tang Z, Niu L, Wu F, Fu Z (2021) Aggregation and stability of sulfate-modified polystyrene nanoplastics in synthetic and natural waters. Environ Pollut 268:114240. https://doi.org/10.1016/j.envpol.2020.114240

Wong JKH, Lee KK, Tang KHD, Yap PS (2020) Microplastics in the freshwater and terrestrial environments: prevalence, fates, impacts, and sustainable solutions. Sci Total Environ 719:137512. https://doi.org/10.1016/j.scitotenv.2020.137512

Yang J, Yang Y, Wu WM, Zhao J, Jiang L (2014) Evidence of polyethylene biodegradation by bacterial strains from the guts of plastic-eating waxworms. Environ Sci Technol 48(23):13776–84. https://doi.org/10.1021/es504038a

Ye S, Andrady AL (1991) Fouling of floating plastic debris under Biscayne Bay exposure conditions. Mar Pollut Bull 22(12):608–613. https://doi.org/10.1016/0025-326X

Yoshida S, Hiraga K, Takehana T, Taniguchi I, Yamaji H, Maeda Y, Toyohara K, Miyamoto K, Kimura Y, Oda K (2016) A bacterium that degrades and assimilates poly(ethylene terephthalate). Science 351(6278):1196–9. https://doi.org/10.1126/science.aad6359

Yousefi M, Oskoei V, Jonidi Jafari A et al (2021) Municipal solid waste management during COVID-19 pandemic: effects and repercussions. Environ Sci Pollut Res 28:32200–32209. https://doi.org/10.1007/s11356-021-14214-9

Zeenat EA, Bukhari DA, Shamim S, Rehman A (2021) Plastics degradation by microbes: a sustainable approach. J King Saud Univ Sci 33(6):101538. Elsevier BV. https://doi.org/10.1016/j.jksus.2021.101538

Zeller MA, Hunt R, Jones A, Sharma S (2013) Bioplastics and their thermoplastic blends from Spirulina and Chlorella microalgae. J Appl Polym Sci 130(5):3263–3275. https://doi.org/10.1002/app.39559

Zettler ER, Mincer TJ, Amaral-Zettler LA (2013) Life in the "plastisphere": microbial communities on plastic marine debris. Environ Sci Technol 47(13):7137–7146. https://doi.org/10.1021/es401288x

Zhao X, Boruah B, Chin KF, Đokić M, Modak JM, Soo HS (2022) Upcycling to sustainably reuse plastics. Adv Mater 34(25):2100843. https://doi.org/10.1002/adma.202100843

Chapter 6
Combatting Plastic Pollution in Natural Environments: Innovative Solutions Offered by Microbes

Sushree Swagatika Mohanty, Jayashree Maharana, and Sony Pandey

6.1 Introduction

Plastics are a diverse set of artificial polymer materials that are immensely useful but have come to be seen as a big environmental pollutant and a manufactured evil because they become unmanageable waste at the end of their life. They are chemically diverse and referred to as "essential" materials in several spheres of social and economic growth due to their flexibility, stability, lightweight, and low manufacturing costs. The demand for plastic production is rising exponentially as the purchasing power of consumers has increased, and they are constantly looking to make life convenient. However, the amount of plastic waste humans generate, without a second thought, is overwhelming and poses a massive threat to all life on earth because now plastics have entered into the food chain in their micro and nano form. The plastic pollution problem has been unanimously recognized as nations are already gathering to chalk out a global treaty to curb plastic pollution on the lines of United Nation's Climate treaty. The most common plastics in the market are polyethylene (PE) and polypropylene (PP) because of their high demand. According to Statista, a leading market, and consumer data provider, the global PE and PP production in 2022 stands at 110.13 and 79.01 million metric tons, respectively. Out of all the different varieties of plastics produced, about 350 million metric tons of plastic is waste, but only 10% is recycled (https://www.statista.com/). The remaining non-recycled plastics get fragmented into microplastics due to UV exposure and mechanical wear and tear (Chen et al. 2020; Yang et al. 2021). Due to their small particle size, with a diameter of less than 5 mm, and widespread dispersion, microplastics are found all over the world

S. S. Mohanty · J. Maharana · S. Pandey (✉)
Environment and Sustainability Department, CSIR-Institute of Minerals and Materials Technology, Bhubaneswar, Odisha 751013, India
e-mail: sony@immt.res.in

J. Maharana · S. Pandey
Academy of Scientific and Innovative Research (AcSIR), Ghaziabad 201002, India

K. Bala et al. (eds.), *Harnessing Microbial Potential for Multifarious Applications*, Energy, Environment, and Sustainability, https://doi.org/10.1007/978-981-97-1152-9_6

in various ecosystems, such as soil, surface and groundwater, food, air, and even in the human body. They pollute the environment and act as a vehicle for other pollutants. While migrating and spreading, they produce harmful consequences. Now it is known as a fact that marine environment contains microplastic that causes a million birds and thousands of marine species to perish every year. Microplastic is also prevalent in the air, and research is underway to find out how they affect the earth's climate system. Albeit the production and disposal of plastic waste on land are enormous, microplastic research has only recently started focusing on terrestrial systems, as the soil appeared to be a long-term sink for microplastic debris (Moller et al. 2020). There are a lot of fibrous and fragmented microplastics in soils all over the world from sources such as water for irrigation, plastic mulch films, sewage sludge, compost, and air because they are exposed to artificial polymer material from urban and industrial areas (Moller et al. 2020; Zhou et al. 2020; Berg et al. 2020). With increase in soil organic carbon content and ionic strength, microplastic retention in soils rises creating a negative impact on the native microorganisms, thereby reducing the soil's potential. An array of microorganisms, spanning bacteria, fungi, bacterial consortia, and biofilms, have been categorically identified as prospective agents of MP degradation. Microplastics biodegradation results in mass loss, chemical composition changes, a loss of mechanical and molecular weight, and the production of CO_2 and CH_4. This approach holds promise, but its quantitative effects can be influenced by factors such as microbial community composition and competition, environmental conditions like temperature, pH, nutrient availability, and the type of microplastics targeted. Certain microorganisms possess the ability to degrade microplastics under controlled laboratory conditions.

Microplastics in various environments are still poorly understood due to a lack of standardization in the quantification, sample collection, processing, and analysis procedures, preventing a thorough comparison of the prevalence of microplastics found by different research groups (Huang et al. 2020; Sarker and Rashid 2013). Therefore, to correctly evaluate the presence of microplastics in different environmental matrices, it is crucial to provide an exhaustive review of the interlinkage and source-pathway of MPs, on-spot sampling, and analytical laboratory techniques used (Zhang et al. 2020).

In the following sections, the subject of this chapter will be elaborated under the five themes:

(1) Critical analysis of the possible sources or entry points for microplastics in environment.
(2) Outlines of the present state of knowledge on the dangers that microplastics may pose to humans and environment.
(3) Description of microplastic remediation and the factors affecting microplastic biodegradation.
(4) Thorough assessment of the analytical techniques used in the scientific literature for sample collection, purification, extraction, characterization, and identification of microplastics in environmental matrices.

(5) Several significant problems are explored, and recommendations are made for more study.
(6) We explore the existing research status, identification of research gaps and committing to future research on microplastics that is standardized and comparable across studies in various parts of the world.

6.1.1 Plastics—Brewing a Manmade Disaster

According to the United States Environmental Protection Agency, plastics made up more than 12% of municipal solid garbage in 2011. Less than 1% of the materials in the solid municipal garbage in the 1960s were plastics. Plastics usages and their disposal in a large scale has a negative impact on humans, plants, animals and wildlife habitat, land, streams, and oceans. The groundwater ecosystem may be impacted by the leaching of hazardous chemicals from chlorinated plastics. The air gets polluted with the burning or incineration of plastic waste and during the recycling process, along with greenhouse gas emissions that contribute to global warming. The plastics in the water bodies and oceans get brittle and break due to exposure to sunlight and contaminated water, and they eventually break into micro and nanoplastics causing health hazards to humans, plants, and animals.

Synthetic plastics were created to protect natural resources (like animal horns, ivory, plants and plant products, silkworms, etc., from which they were produced earlier), and for convenience, but with the intention of reuse because they are relatively strong, flexible, and durable material. However, slowly plastics replaced many of the daily use material, and disposable items for use and throw became a norm as it provided comfort and convenience at a very less cost. Polyethylene replaced the cloth shopping bags, glass bottles, toys, utility items in our kitchens and dining, and may parts of automobiles, appliances, and what not! Phthalates are chemical compounds that are used in a variety of items, including emulsions, inks, shoes, and toys. PVC is a type of plastic used in utility items, packaging, and containers that release dioxins, a class of the most toxic synthetic chemicals, when PVC plastic is made or burned. Some plastic water bottles have the chemical bisphenol A (BPA), which is thought to stimulate early puberty, weaken the immune system, causes cancer, and also development of diabetes and obesity. Many other such carcinogenic chemicals are used as plasticizers, additives, and fillers in these common use items on which humans have been irreversibly dependent. As a result, cancer, birth defects, genetic conditions, immune and metabolic disorders, and organ dysfunction or failures are commonly noticed these days. As attention mounts regarding the potential repercussions on both micro and macroflora and fauna, smart and efficient plastic waste management is the need of the hour with dire consequences otherwise.

Traditional plastic waste management methods often fall short in effectively removing plastic waste from terrestrial and aquatic ecosystems due to the durability and slow degradation rates of most plastics. The proliferation of environmental contamination arising from the unregulated and mismanaged utilization of plastic materials and their derivatives has become a pressing issue in diverse settings, spanning from quaint rural locales to bustling urban centres. This predicament is further exacerbated by the fact that the decomposition process of plastic bags alone stretches over a millennium. Considering the number of synthetic polymers, which is thought to be around 20–25%, in municipal solid trash in developed countries like the United States of America, or Australia (Jayasekara et al. 2005), the municipal landfills will fall short and slowly take over major parts of human habitation or the nearby wild areas. Already the water bodies close to human settlements and the far corners of the ocean are contaminated by the plastics because the plastic waste requires no permits to cross international boundaries. It is an alarming situation but the majority of human population is ignorant to its consequences and not doing enough to prevent it. Such rampant use and disposal of plastics is not only an eye sore but a serious issue that cannot, and should not, be ignored by the public and governments. Responsible use by consumers, and responsible manufacturing practices have to be inculcated to reduce the wastage. Safe disposal methods and upscaling, or conversion of plastics to other useful products, have to be worked out.

6.1.2 Microplastics in the Food Chain

The plastics or microplastics accumulated in land and aquatic systems, ingested by primary producers and predators, cause upward movement of microplastics in the food chain. The toxins/contaminants adsorbed on the surface of microplastics also get bioaccumulated in the food chain because of their small size. Numerous aquatic species have been shown to have the most frequent bits and fibres (800–1600 m) of microplastics that have been collected in the field. Trophic transfer of tiny nano/ microplastics happens when MPs are consumed by lower trophic fauna and absorbed by the flora, which then get transferred to the organisms in the higher trophic levels, thereby spreading in the food web. Invertebrates, crustaceans, and fish that are meant for human diet all now a days include microplastics. Similarly, plants have been reported to have accumulated micro and nanoplastics in their conducting tissues. Plants are eaten raw or cooked by higher animals and humans resulting in plastics in the blood streams. It is worrying to note that through the food chain, MPs pose a major health hazard but many of the health effects of bioaccumulation of MPs is still unknown on the living organism. Human life is slowly affected by microplastics both physically and chemically. Microplastics are eliminated via the human body's excretory system, which probably excretes >90% of all ingested micro and nanoplastics. The size, shape, type of polymer, and additional chemical compounds of microplastics that humans ingest influence the retention and clearance rates of microplastics. How severe the adverse effects of exposures are contingent on the kind of hazardous chemical, exposure parameters, and personal sensitivity. The disturbance of the gut microbiome is one of several potentially dangerous impacts that have been examined, along with enhanced autoimmune responses, size-related toxicity of plastic particles, chemical transfer of adsorbed chemical pollutants, and many more. The distribution and storage of toxins in the human body are better understood than the physical effects of accumulated microplastics. Surface functional groups, size, shape, surface charge, buoyancy, and hydrophobicity all contribute to the estimation of microplastic absorption. Modelling studies of mammalian systems reveal that some types of microplastics can be transported across live cells in mammalian systems through dendritic cells, or cells of the lymphatic or circulatory system and build up in different organ systems and affect immune system. Microplastics get absorbed and translocated through the endocytic pathways and persorption into the gastrointestinal or respiratory epithelium, but the effect is yet to be fully understood.

6.2 Sources

6.2.1 Source of Microplastic in Air

The primary source of airborne microplastic is synthetic cloth. When wearing, washing, and drying clothing and other fibre items, little fibres are readily torn from them. For instance, it was found that 1 g of acrylic cloth released more than 1100 fibres. Because there are more textiles indoors, there may be a greater quantity of airborne microplastics indoors than outdoors (Sridharan et al. 2021). The description of airborne microplastics also points to the possibility of alternative sources, such as the breakdown of big plastics, industrial emissions, dust re-suspension, etc. Microplastics on the ground are propelled into the air by wind, where they are carried to remote and sparsely populated locations (Hadad et al. 2005). As a result, microplastics travel a great distance through the atmosphere from urban areas to far-flung terrestrial regions. When they settle, these airborne microplastics will be a source of pollution for ecosystems on land and in water. After that, surface runoff and stormwater may resuspend the settled microplastics into the air or carry them into aquatic environments. A dynamic cycle of microplastics in the environment involves their interaction in the air, on land, and in water.

6.2.2 Source of Microplastic in Soil

There are two main categories of sources for microplastic: primary and secondary. Abrasives for cosmetic use, medicine delivery systems, and technical and industrial uses like air blasting are only a few of the purposes for which primary microplastics are specifically made. Once they enter wastewater, these microplastics which are often challenging to remove using sewage disposal techniques will eventually build in the environment. Larger plastics become secondary microplastics as a result of various, intricate environmental factors including wind, waves, temperature, and UV radiation gradually breaking them up into tiny bits. The usage of plastic items repeatedly can also lead to fragmentation and the creation of secondary microplastics. In addition, plastic emissions from moving vehicles, such as those from brakes, road markings, and tyre wear and tear, are a significant source of microplastics in the environment (Kole et al. 2017). The average annual per-person emission of microplastics from road vehicle tyres was calculated to be 0.81 kg. Apart from emissions from road traffic, aeroplane tyre debris makes for around 2% of all tyre-related emissions. The single largest source of intentionally added microplastics is the granular infill material used on artificial turf pitches that loses 0.5 to 8.0% of its blades due to the grinding action during play, UV radiation and environmental exposure, with an emission of up to 16,000 tonnes annually.

Secondary microplastics are also released into the soil from solid waste plastic debris, single-use plastic packaging, and landfills. Biological processes such as feeding, digesting, and excretion activities carried out by soil organisms also add microplastics through soil supplements, sewage sludge applied to the ground, compost made from wastewater, etc. (Chae and An 2018). Microplastics' migrates and transmitted up the trophic levels in severely contaminated soils, especially those in areas where wastewater is used in irrigation and plastic mulch film is coated, represent significant threats to the ecosystem and significantly impair soil quality (Weithmann et al. 2018; Corradini et al. 2019).

6.2.3 Source of Microplastic in Water

Marine microplastic contamination has several different origins, which may be broadly categorized as inland, maritime, and aerial sources. The most significant routes for the movement of microplastics from inland areas to the ocean are thought to be rivers. Land-based sources contribute 80% of the plastic debris that comes from coastal tourism, recreational, commercial fishing (such as the usage of plastic fishing gear, etc., contributing 18%), marine vessels, and marine industries. Plastic garbage from beach tourism is dumped directly on the sand or the place itself, as is plastic debris in urban drainage systems, sewage effluents. Ocean-based sources come from the fishing, shipping, and offshore sectors of the economy (Klein et al. 2015). Large ship emissions and leaks are regarded as a significant source of microplastics. Equipment loss and damage from farming and fishing may readily release plastic into the ocean. The world's fishing fleet is the primary offshore source, followed by marine aquaculture, trash illegally dumped from ships or offshore rigs, and a significant amount of goods originate from misplaced containers. Furthermore, airborne Microplastics are significant sources as well (Bell et al. 2016). Primary microplastics (Waller et al. 2019) are widely used in the production of fibre and textiles as well as personal care and pharmaceutical that contain abrasives and scrubs (such as toothpaste, shower gels, hand and facial cleansers), cosmetics formulations (like eye shadow, nail polish, and hair dye), and air-blasting tools. Home washing machines produce microfibre masses at a count of 1471–2121 microfibres per garment, that is typically more in front-load machines compared to top-load machines. A recent study found that textile clothing lost between 30,000 and 465,000 microfibres per square metre (or 175 and 560 microfibres per gram) (Hartline et al. 2016).

The gradual degradation of massive plastic waste owing to chemical, biological, and physical deterioration, which are correspondingly indicative of photodegradation (sunlight UV radiation), microbial species biodegradation, and mechanical abrasion (wave action), is what is often implied by secondary microplastics (Wagner et al. 2014). Fragile plastics are broken down into microplastics by mechanical damage, photodegradation, and oxidative degradation in the water. Further degradation of microplastics might result in plastic particles that are nanometre in size. Future public concern will arise from these micro and nanoplastics since they are readily absorbed in tissue structure and will have chronic effects on the marine ecosystem.

6.3 Effects

6.3.1 Effect of Microplastic in Air

The behaviour of airborne microplastics in the atmosphere and their potential impact on Earth's climate system is an emerging area of scientific inquiry. Here are some ways airborne microplastics may interact with the climate:

1. Radiative forcing: Radiative forcing refers to the influence that a factor has on the balance of incoming and outgoing energy in the Earth's atmosphere. Some studies suggest that certain types of airborne microplastics, such as black carbon particles that result from the incomplete combustion of plastics, could have a warming effect on the climate. These particles can absorb sunlight and contribute to the greenhouse effect, trapping heat in the atmosphere.
2. Cloud formation: Airborne microplastics, particularly those with specific surface properties, may act as cloud condensation nuclei (CCN). CCNs are small particles around which water droplets condense to form clouds. The presence of microplastics in the atmosphere could potentially affect cloud formation and properties, impacting weather patterns and regional climate.
3. Transport of pollutants: Airborne microplastics can adsorb and concentrate various pollutants, including toxic chemicals and heavy metals, from the surrounding environment. When carried by the wind, these microplastics may transport pollutants over long distances, potentially affecting ecosystems and human health far from the original source.

Since there is now a very low concentration of microplastics in the atmosphere, their impact on the earth's climate is currently relatively minimal. But considering that plastic trash is expected to double over the next few decades, microplastics may have a greater effect on the planet's climate system.

6.3.2 Effect of Microplastic in Soil

The nature of the soil is crucial for the migration of microplastics through natural occurrence, they alter the physical and biological properties of soil, including its structure, function, microbial diversity (Murugan et al. 2014). These changes may have an impact on plants and animals, raise questions about the quality and safety of food, and ultimately endanger human health. The existence of substantial residual plastic films in the soil that impact soil microbial activity and abundance and lower saturated hydraulic conductivity of the soil, eventually affecting soil fertility. In contrast, in the complex soil environment, physical and abiotic variables (like erosion of top layer soil) and biotic factors (like microorganisms, earthworms, and plants) that depend on the fundamental physical and chemical characteristics of soil, alter the properties of microplastics in soil (Rillig et al. 2019). However, a small number of research have focused on the modification of soil physicochemical characteristics, soil microorganisms, as well as the toxicity of microplastic pollution. The bulk of studies on microplastics are still focused on the characterization of semi-quantitative and qualitative metrics. More severe soil issues may be caused by the interactions between microplastics and the soil, which may have an unpredicted effect on the environmental behaviours of other contaminants in soil (Kasirajan and Ngouajio 2012).

6.3.3 Effect of Microplastic in Water

Effect microplastics on the marine biota is a cause for worry. Due to their micron-sized particle size and persistence in the marine ecosystem, these tiny plastic fragments are mistaken for food by a variety of marine biota, including corals, phytoplankton (Mao et al. 2018), zooplankton, sea urchins (Martinez-Gomez et al. 2017), lobsters, fish, etc., (Diepens and Koelmans 2018) and are subsequently transported to higher tropic levels. While bigger plastic causes entanglement, ingestion of small plastics or microplastics becomes fatal to marine creature. Microplastic-associated microbial biofilms may harm coral reefs by facilitating the spread of pathogens. The most important component of the marine ecosystem, planktons, are also negatively impacted by microplastics. Chlorophyll absorption is decreased as a result of microplastics getting within phytoplankton cell walls. The ratio of triglycerides and cholesterol in fish blood serum levels changes as a result of ingesting micro and nanoplastics, and the amount of cholesterol delivered to the fish's liver and muscle varies as well (Lu et al. 2016).

6.4 Impact on human Health

Some potential ways microplastics could impact human health are:

1. Ingestion: Microplastics can enter the human body through ingestion of contaminated food and water. Seafood, for example, may contain microplastics if the marine organisms have ingested them. It's possible that these particles could accumulate in the gastrointestinal tract and, in some cases, may pass into other tissues and organs (Naji et al. 2021; Karbalaei et al. 2018). The main exposure route for MPs and NPs is the gastrointestinal tract (GIT). When evaluated on a model of intestinal epithelium, the NPs showed a high potential to pass the GIT barrier. Plastic particles pose a special risk because of their capacity to translocate across the gut barrier and penetrate cells, which can allow them to enter the bloodstream and cause systemic exposure. The human GIT, which has a length of 6 to 8 m and a surface area of 200 square metres, is where most of the entering plastic particles are first exposed. The effects of MPs and NPs on GIT have been studied using a variety of intestinal barrier cell models, including monocultures and co-cultures of diverse cell types. It is noteworthy to highlight the fact that GIT is a complex element. Therefore, it is important to consider how the environment of the digestive tract—which includes pH, ions, organic substances, and enzymes—affects bioplastic. The digestion therapy reduced PS MPs and arsenic's combined toxicity, as well as the Caco-2 monolayer's cytotoxicity and transport function abnormality. However, the digestion process causes the proinflammatory effects of MPs to grow. The first physical barrier that the body encounters after ingestion is a layer of mucous gel that covers the epithelium of the intestines. Mucins serve as the structural foundation for these protective mucous gels on the surfaces of the human gastrointestinal tract, providing surface entrapment and protection while also exhibiting the right characteristics for elimination by flow.
2. Inhalation: Microplastics have been detected in the air, and individuals may inhale these particles, especially in areas with high plastic pollution. The potential health effects of inhaling microplastics are still being studied (Yurtsever and Kaya 2018).
3. Chemical exposure: Microplastics can absorb and concentrate various toxic chemicals from the environment, such as persistent organic pollutants (POPs) and heavy metals. When humans come into contact with microplastics, they might also be exposed to these chemicals, which could pose health risks. Superoxide radical (O_2) and H_2O_2 are ROS that are necessary for sustaining healthy physiological functions, but too much of them can overwhelm the body's antioxidant defences, causing oxidative stress and additional damage to DNA, proteins, and lipids. It is thought that oxidative stress is the most likely mechanism behind the toxicity that MPs and NPs cause.
4. Inflammation and immune response: Some research suggests that microplastics could trigger inflammation and immune responses in the body when ingested or inhaled. Chronic inflammation is associated with various health conditions, including cardiovascular disease and certain cancers (Hodson et al. 2017; Groh et al. 2019). Generally, inflammation plays a major role in the emergence of

several illnesses. Additionally, proinflammatory and anti-inflammatory cytokines predominated the inflammation. Changes in inflammatory factors can occur in a variety of types, including human alveolar epithelial type 1-like cells, A549 cells, and AGS cells. These impacts also depended on the size of the microplastic. The size and surface chemistry of plastic particles are connected to the complicated activation of cytokine mediators in response to MPs and NPs. The release of IL-6 and IL-8 from cells subjected to 50 nm NPs (1, 50, and 100 lg/mL) was reported to be considerably greater than that of cells treated to 100 nm NPs (1, 50, and 100 lg/mL).

5. Disruption of gut microbiota: Microplastics can alter the composition of gut microbiota. Since the gut microbiota plays a crucial role in human health, any disruption to this ecosystem could have implications for overall well-being. The body's most numerous cells, known as red blood cells (RBCs), are highly specialized. In gas exchange, detecting and signalling, preserving the pH and redox state's equilibrium, regulating vascular tone, clotting, and other functions, RBCs play a crucial role. There is proof that nanoparticles may spread throughout the body, including to the blood and internal organs. NPs may also be purposefully injected into the bloodstream in the intended medicinal use, in addition. Recent research has shown that RBCs, important blood components, are employed as drug delivery vehicles since investigations have shown that binding NPs to an RBC significantly extends their time in circulation. Thus, it is important to consider how plastic particles interact with blood constituents.

To address these concerns, efforts are being made to reduce plastic pollution and promote sustainable practices. Governments, industries, and individuals are working together to find solutions to minimize plastic waste and prevent microplastic pollution from accumulating in the environment (Fig. 6.1).

6.5 Remediation of Microplastics

Long chains of repetitive monomers make up the synthetic polymers that make up plastics. There are several different kinds of plastics, including polystyrene (PS), polyethylene terephthalate (PET), polypropylene (PP), polyethylene (PE) terephthalate (PE), and polyethylene (PE) (PE). There was always a speculation that some microbes, especially bacteria and fungi, have developed enzymes that can target and degrade particular kinds of plastics. However, it was the discovery of the plastic-eating bacteria in the Japanese plastic dump site, that changed the notion of non-biodegradability of plastics. The bacterium *Ideonella sakaiensis* 201-F6 consumes the plastic polyethylene terephthalate (PET) as a sole carbon and energy source (Yoshida et al. 2016). The discovery of two-enzyme PET depolymerization system in *I.sakaiensis* offered insights into the biological deconstruction of plastics (Kumar et al. 2020). Subsequently, PETase, MHETase, Cutinases, and Lipases are important categories of enzymes involved in the breakdown of plastics. These enzymes

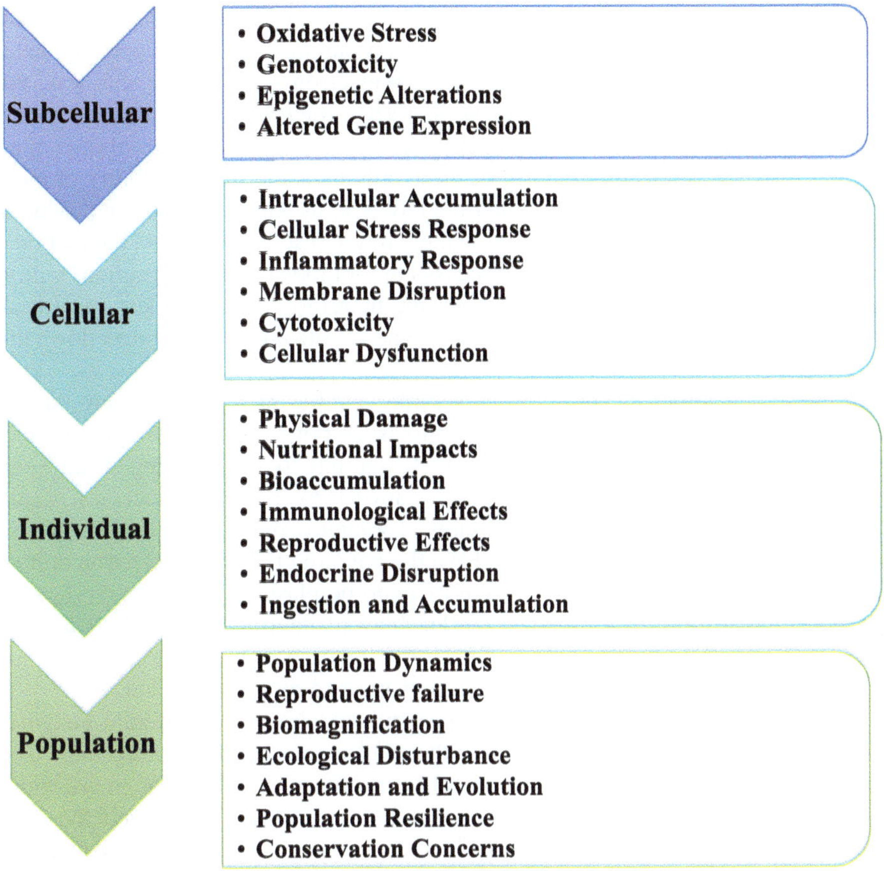

Fig. 6.1 Biological effects of microplastics on humans from population to sub cellular level

may break down a variety of polymers, including polyethylene (PE) and polypropylene. These enzymes have been found by researchers in a variety of habitats, such as plastic-contaminated regions and waste sites, where microbes have evolved to use plastic as a carbon source. Some researchers have also improved the enzymes' capacity to break down plastic in the lab using approaches like directed evolution and enzyme engineering. The enzymes break the lengthy polymer chains into smaller pieces by specifically targeting certain chemical bonds within the polymer structure. Other microbes can more easily reach these tiny bits and further metabolize them. The qualities of the polymer and the environment are the two most significant factors that influence the biodegradation pathways. Biogeochemical cycle is the basis for the synthesis and decomposition of organic matter in all aquatic environments. Microplastics and microorganisms are more likely to interact as a result of the easy degradability of small sized plastics which have been extensively introduced into the marine environment. This might result in the microbial use of microplastics or its breakdown products as a carbon source, substrate, or co-substrate. Microbial carbon utilization in water or sediment is dependent on the presence of oxygen, sulphate, nitrate, and ferric iron, which are all possible electron acceptors. In this way, the makeup of the microbial population and important microbial respiration routes are influenced by the redox environment. As a result, microbes and microplastics may interact actively, connecting the use of these polymers and the availability of electron acceptors. Microplastics and their degradation, as well as metabolic products, are transported within the aquatic environment through active or passive interactions, passed through food webs, and made accessible for biotic or abiotic reactions within aquatic biogeochemical cycles. Redox setting therefore becomes a key consideration for microplastic fate in aquatic systems, and for the microbes to break down the marine trash.

A mixture of abiotic and biotic processes may start the integration of microplastics and the intermediate breakdown products into the carbon cycle at the water's surface. Methane and other hydrocarbon gases are released into the surrounding ocean when plastic interacts with sunlight. When breaking down through abiotic processes, plastic particles produce gases in addition to dissolved organic carbon (DOC) in water. These intermediate degradation by-products have the potential to serve as electron donors for microbial metabolism. The creation of microbe-plastic aggregates serves as a biogeochemical route in addition to the leaching of intermediates and breakdown products from microplastics, which impact marine carbon cycle. Biofilm develops on the microplastics that float in the surface water and receives sunlight.

Plastic degradation is often a multi-step process involving different enzymes working collaboratively to break down the plastic into simpler compounds. The potential applications of these microbial enzymes for plastic degradation are wide-ranging. They include recycling plastic waste, reducing plastic pollution in the environment, and developing more sustainable plastic materials. It's essential to note that while significant progress has been made, there are still challenges to overcome, such as optimizing the efficiency of these enzymes and understanding their environmental impact fully.

Alcaligenes faecalis (MK517568) and *Bacillus cereus (MK517567)*, exhibited a significant degrading activity. LLDPE is broken down by 3.5%, HDPE by 5.8%, and polyester by 17.3% by *Alcaligenes faecalis (MK517568). Bacillus cereus (MK517567)* degrades Polyester by 29% and is more tolerable at 30 °C. Depending on the kinds of plastics addressed, variations in infrared spectra revealed the breakdown routes of various strains.

6.5.1 Remediation of Microplastic in Soil

Three phases make up the microbial activity that aids in the microbial biodegradation process:

1. Microbial development results in modifications to the physical, chemical, and mechanical characteristics of microplastics within or on the surface.
2. Microplastics are converted by microbial activity into oligomers and monomers.
3. 3- Through absorption, microplastic particles provide microorganisms with the energy, carbon, and nutrients they require, turning the carbon in the plastic into CO_2, H_2O, and biomass.

Microorganisms have the capacity to adapt to their surroundings and degrade various organic contaminants, including microplastic. Microplastics are used as a substrate by microorganisms in the biofilm formation process. When a biofilm forms, the structure of the microplastic material disintegrates, and fungus or bacteria secrete enzymes that aid in the particular and atypical breakdown of the microplastic particles. If a fragment weighs less than 600 kDa during biofilm assimilation, bacteria can readily take up the microplastics (Rillig et al. 2019).

Fungal mycelium may stick to microplastics' surfaces more securely and may even enter the particle's interior. By encouraging the development of chemical bonds such as carboxyl, carbonyl, and ester groups in microplastics, fungi may be able to decrease the hydrophobicity of such materials.

Microorganisms are picky in the polymers they break down, thus the enzymes they release may not be suitable for the target plastic. For instance, when inoculated alone in nutritional solution, Bacillus subtilis only decomposed PS with a weight loss of 20.0%, compared to PET's weight loss of 74.59%. Likewise, *Pseudomonas aeruginosa* only degraded PS with a weight loss of 5.0%, whereas PET showed no weight loss. Additionally, because microbes and enzymes dominate the biodegradation process, environmental elements that impact their metabolic activities, such as temperature, humidity, and light, also affect how quickly microplastics degrade (Zhang and Liu 2018).

6.5.2 Remediation of Microplastic in Water

Most of the plastic waste enter in to the ponds, lakes, rivers, and ocean. In current situation, the application of engineering tools and technology for the removal of plastics from marine water is of paramount importance, otherwise it will disturb the ocean ecosystem and cause climate change. Therefore, ideally, prior to discharge of solid waste and waste water, plastic debris should be filtered out, and wastewater should be treated in waste water treatment facilities (WWTPs). Membrane bioreactor (MBR), ultrafiltration, or microfiltration are employed to treat wastewater (Talvitie et al. 2017; Lares et al. 2018).

Currently, attention is being paid to the three R hierarchy of an integrated waste management system: reduction, reuse, and recycling. The utilization of plastic materials minimized by outlawing the use of specific plastics, such as single-use plastics. Increased recycling rates can significantly lessen the amount of plastic waste in the ocean. The steps in the difficult process of recycling plastic are as follows: managing consumer waste, separating recyclables and removing pollutants, grounding and differentiating polymers, distilling polymer and colour samples, and recycling products for manufacturing (Bing and Bloemhof 2014; Walker 2018). Direct dumping must be prevented at all costs, and various wastes should be carefully processed in recycling units. There are many plastics can be discovered in a large size of medical waste like Personal protection equipment (PPE), one-time-use masks, and covers in the present Covid-19 pandemic situation. Such wastes must be carefully managed in order to prevent future environmental pollution that will harm the ecosystem (Issac and Kandasubramanian 2021). Use of the pandemic scenario as a justification for skipping over the treatment of such wastes is not appropriate. In fact, it is important to take these medical plastic wastes even more seriously through enzymatic action, different bacterial and fungal species can break down conventional BPA-based plastics (Sarma et al. 2019; Yoshida et al. 2016). It is crucial to develop methods for microplastics biodegradation on-site using microbes or to boost natural attenuation using native microflora. Since physical and environmental factors have a significant impact on bioremediation, replicating these elements will make it simpler and more convenient to hasten the microbial degradation process (Tiwari et al. 2020). Marine sources produce plastic wastes all the way down to the ocean floor. The microorganisms that are already present in each of these niches can quickly adapt to the plastic wastes and are likely to create biofilms on their surface. There are various studies on the biodegradation of specific types of plastics to various extents by some bacteria and fungi as tabulated (Table 6.1).

A new thermophilic bacteria strain, *Bacillus sp. BCBT21,* was identified from agricultural waste being composted in Vietnam, can degrade chemically different plastic bags using extracellular hydrolase enzymes lipase, CMCase, xylanase, chitinase, and protease (Dang et al. 2018). According to a study, soil isolates of *Streptococcus*

Table 6.1 Potential microbial strains that are capable of degrading different types of plastic waste

Microorganism	Plastic type	Percentage of plastic degradation	References
Ideonella sakaiensis	PET (polyester)	~20–30%	Yoshida et al. (2016)
Pseudomonas putida	PET (polyester)	~10–20%	Bradenberg et al. (2022)
Bacillus subtilis	PLA (polylactic acid)	~30–40%	Dang et al. (2018)
Aspergillus niger	PET (polyester)	~10–15%	Albertsson et al. (1987)
Rhodococcus sp.	PVC (polyvinyl chloride)	~10–20%	Orr et al. (2004), Auta et al. (2018), Gilan et al. (2004)
Exophiala sp.	PET (polyester)	~5–10%	Giyahchi and Moghimi (2023)
Comamonas testosteroni	PS (polystyrene)	~20–30%	Fangzhi et. al. (2023)

sp., Pseudomonas sp., and Bacillus sp. of bacteria showed 23% of the degradation roughly after 30 days; and *Aspergillus sp.* and *Fusarium sp.* of fungi degraded 44% of plastic bags (Vignesh et al. 2016).

6.6 Factors Affecting Biodegradation of Microplastics

The qualities of the polymer, the type of organism, and the method of pre-treatment are some of the variables that influence biodegradation. The properties of the polymer including its flexibility, crystallinity, molecular weight, kind of functional groups, and substituted Structure, plasticizers, and other polymer-added compounds all have a significant impact when it is degraded (Gu et al. 2000; Artham and Doble 2008). Polymer biodegradability is influenced by two key variables, such as the exposure circumstances and the distinguishing characteristics polymer of abiotic and biotic conditions are additional categories for exposure biotic variables. In order to increase the accessibility of the polymer chain to moisture and microbes, the primary chain scission from photodegradation lowers the average molecular weight. Depending upon exposure conditions there are three main factors affecting degradation, they are—abiotic, biotic, and physical as elaborated below:

6.6.1 Abiotic

Moisture: Given that bacteria need water to thrive and reproduce, moisture can have a variety of effects on polymer biodegradation. Consequently, polymer degradation when there is enough moisture present, speed is increased because of rapid microbial activity (Ho et al. 1999). Furthermore, moisture-rich circumstances encourage the hydrolysis process by increasing the number of chain scission processes.

Temperature: The temperature can modify the rate of hydrolysis reactions by changing the increase and decrease condition. The softening temperature of the polymer has a considerable impact on its capacity to be broken down by enzymes. Less chance of biodegradation exists in polyester with a higher melting point. As temperature rises, potential enzymatic degradability declines. For instance, R. delemar's pure lipase effectively hydrolysed Low melting points are demonstrated by polyesters such as Polycaprolactone (PCL) (Degradation of microbial polyesters 2004; Tokiwa et al. 2009).

pH: By adjusting the acidic or basic conditions, the pH can alter the pace of hydrolysis reactions. For instance, the rate of PLA capsule hydrolysis is best at pH 5 (Auras et al. 2004; Henton et al. 2005). Various degradation by products the pH conditions are altered by polymers, and the rate of the microbial growth and the decay process.

UV Radiation: All UV can have a photochemical reaction within the polymer structure that might cause the material to deteriorate. The UVC with higher energy is the one that is most likely to harm polymers. The photons in a plastic can be excited by absorbed UV light. Free radicals can be produced when photons are excited. The presence of catalyst remnants and other impurities (for instance, oxygen and salt), even in small amounts, will frequently operate as free radical receptors even though many pure polymers cannot absorb UV radiation. These free radicals have the potential to damage the polymer bond.

6.6.2 *Biotic*

Enzyme Characteristics: Different enzymes have distinctive active sites and can biodegrade different kinds of polymers. As an illustration, straight chain polyesters made from diacid 6 to 12 C-atoms, which enzymes swiftly destroyed. A comparison between the fungi *A. flavus* and *A. niger* compared to polyesters with straight chains made from another monomer (Kale et al. 2007) and according to that research, the extracellular enzymes, a part of the PHB depolymerization process (depolymerases), a variety of mechanisms are produced by microorganisms depending on the particular depolymerase (Onodera et al. 2001).

Hydrophobicity: Due to their hydrophobicity and three-dimensional structure, plastics made from petrochemical sources are unable to decompose easily in the environment (Onodera et al. 2001). Additionally, PE's hydrophobic properties affect the creation of a microbe biofilm to slow down biodegradation (Hadad et al. 2005).

Biosurfactants: Amphiphilic substances known as biosurfactants are mostly formed on living surfaces. Due to their low toxicity and high biodegradability, the inclusion of a biosurfactant accelerates the biodegradation of polymers (both fossil and biobased) (Orr et al. 2004). Because they contain particular functional groups, biosurfactants speed up the biodegradation process and as a result, they permit the activity in extremely hot environments, Conditions for pH and salinity (Kawai et al. 2002, 2004).

6.6.3 *Physical Characteristics*

Molecular Weight: In terms of biodegradability, molecular weight is a key factor in determining many polymer qualities with an increase in molecular weight, degradability is reduced. PCL with a higher molecular weight (>4,000) was gradually compared to modest amounts that a strain of R. delemar's lipase destroyed polymer with molecular weight (Tokiwa et al. 2009). It develops an easy substrate to attack by microbial enzymes one that is low in molecular mass (Auras et al. 2004).

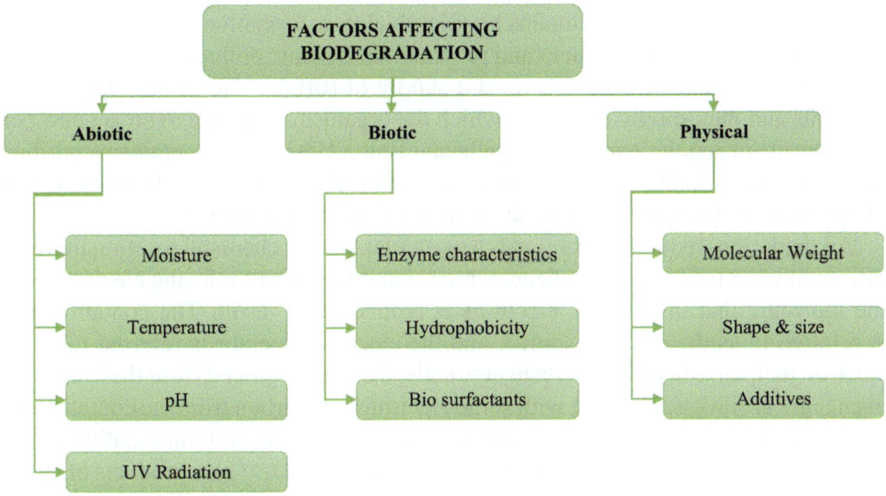

Fig. 6.2 Flow diagram of the factors affecting microplastic biodegradation

Shape and Size: The degradation process is significantly influenced by the polymer's characteristics, such as its size and form. Compared to those with a limited surface area, polymers with a wide surface area can deteriorate more quickly (Kijchavengkul and Auras 2008).

Additives: The capacity to degrade is impacted by non-polymeric impurities such as colours, waste or debris of catalysts employed for the polymerization, and additives or filler. It has been shown that as the samples lignocellulosic filler content rises, the thermal stability and ash content decrease. The main factors influencing the thermal stability of the composite system are interfacial adhesion and dispersion between the thermoplastic polymer and lignocellulosic filler (Yang et al. 2005). Similarly, metals serve as good pro-oxidants in polyolefin manufacturing of polymer sensitive to thermo-oxidative degradation (Fig. 6.2).

6.7 The Importance of Microbes in Biodegradation

The term "biodegradation process" refers to microbial activity that can use plastic as a carbon source to transform organic molecules into biogas and leftover biomass (Kerr 1994). Globally, synthetic plastics including PES, PVA, polyamide (PA), and polyurethane (PU) are the most extensively utilized materials. However, due to the nature of plastic and its physicochemical qualities, which make plastic a poor growth substrate for microorganisms, only a tiny portion of their wastes may be biodegraded.

The organisms utilize the pollutants by microbial transformation of organic contaminants for their own development and procreation. Organic pollutants have two functions for the organisms: they serve as a supply of carbon, one of the fundamental elements and they supply electrons, which the organisms can take to create new cell components to acquire energy (Chapelle and Bathe 1993). Microorganisms produce energy by accelerating chemical reactions that break chemical bonds and transport electrons away from the contamination in order to produce energy.

An oxidation–reduction reaction is a particular kind of chemical reaction in which the organic contaminant is oxidized, which is another word for losing electrons, and the molecule that obtains the electrons is reduced as a result. The contamination is referred to as the electron source, and the electron acceptor is referred to as the electron recipient. In order to create new cells, the energy gained from these electron exchanges is "invested," along with some electrons and carbon from the contaminant.

Anaerobic respiration, which is a technique used by many microbes, allows them to survive without oxygen. Nitrate (NO_3) and sulphate ($SO_4{}^{-2}$) are products of anaerobic respiration and metals like iron (Fe^{3+}) and Magnesium (Mn^{4+}) or even carbon dioxide (CO_2) can act as oxygen by absorbing electrons from the deteriorated contamination. In light of this, anaerobic respiration electron acceptors made of inorganic substances. The by-products include fresh cell matter in addition to nitrogen gas (N_2), hydrogen sulphide (H_2S), reduced nitrogen, and oxygen (O_2). Depending on the electron acceptor, several types of metals and methane (CH_4) can be found. The primary process for the deterioration of plastics is oxidative degradation. The enzymes that are both intracellular and extracellular. The microorganisms break down the polymer into monomer, dimer, and oligomer.

Both biotic and abiotic environmental conditions, play a significant role in the beginning of PE (LDPE/HDPE) degradation. When the conditions are conducive microorganisms colonize the surface of plastics and secrete the necessary enzymes, or else if it is thermodynamically advantageous they activate enzyme production as needed to metabolize PE (Albertsson et al. 1987; Balasubramanian et al. 2014).

6.8 The Significance of Enzyme in Biodegradation

Enzyme work by lowering the activation energy and turning substrate into product. Enzymes released by the bacteria break polymer chains, and the hydrolysis products are ingested and metabolized. A polypeptide moiety, such as a protein or a glycoprotein, must make up at least one portion of an enzyme. The main component of an enzyme is a catalytic site, a part of the holoenzyme. A holoenzyme includes a group and apoenzyme is prosthetic. The substratum is joined to the enzyme by a covalent or noncovalent bond through the active site bond. A collection of polypeptide chains with protein or glycoprotein moieties makes up the apoenzymes. Organismal enzymes aid in ensuring the bioremediation is a successful and environment friendly method (Karigar and Rao 2011). The enzymes responsible for the breakdown of petro-polymers like PE and PP as well as the processes behind these interactions between the enzyme and the substrate remain mainly unknown. Enzymatic processes such as hydrolysis, oxidation, and hydroxylation cause the polymer chains to break down into oligomers and monomers. Petro-plastic polymers share the same ester bonds as other polyester polymers, and esterases are enzymes that break ester bonds. Esterase comes in a variety of forms, each with unique biological properties, substrate specificities, and protein structures. Biodegradation is more likely to occur in polymers having hydrolyzable ester linkages in their backbones, such as PET and PUR, than in polymers with carbon chain backbones, such as PE, PS, PP, and PVC. The high-molecular weight polymers PET and ester-based PU are degraded by recognized enzymes, and their specific activities have been characterized.

6.8.1 *Enzymes in Bioremediation*

Microbial Oxidoreductases: Through the oxidative-reductive coupling, oxidoreductases are in charge of the detoxification and degradation of hazardous natural mixtures by various microorganisms and higher plants. The oxidoreductases take role in the detoxification of numerous synthetic organics that are found in xenobiotics or soil, such as phenolic, azo rings, and aniline compounds in the environment. The effectiveness of mushrooms like Basidiomycetes, Pleurotus, Trametes versicolor and P. chrysosporium, Oyster mushrooms, or ostreatus, are better for microbes in the soil (Khatoon et al. 2017).

Microbial Oxygenases: When using FAD/NADH/NADPH as a co-substrate, oxygenases, which are members of the oxidoreductase category, add oxygen from subatomic oxygen (O_2) to the reduced substrates. Based on the quantity of oxygenases, they are divided into two categories: monooxygenases and dioxygenases (Gewert et al. 2015).

Microbial Laccases: A collection of multi-copper oxidases can be found in the fundamental structure of laccases (p-diphenol: dioxygen oxidoreductase). Specifically, plants, insects, and bacteria create them was oxidized by several diminished phenolics, and fragrant substrates with a sub-atomic density decline that is systematic water with oxygen. Several genes produce various structures isoenzymes are produced by laccases (Chauhan et al. 2017).

Microbial Peroxidases: High oxidizing enzymes are called peroxidases (contributor: hydrogen peroxide oxidoreductases). The oxidation of coal to low molecular mass is catalysed by them. These relate to the breakdown of auxin, the organization of lignin and suberin in plants, the cross-linking of cell divider segments, a defence against pathogens, or cell expansion. Additionally, microbes perform bio-pulping, bio-bleaching, polymerization. The many fungal species contain numerous lignin and manganese peroxidase isoenzymes.

Microbial Lignin Peroxidases: The majority of the white-rot and brown-rot basidiomycetes release lignin peroxidases, which are heme proteins, as part of their metabolizing process by optionally digesting lignocellulose activity. They very effectively degrade the woodlands efficiently (Bansal et al. 2017).

Microbial Manganese Peroxidases: An extracellular heme chemical called manganese peroxidase is found in the lignin-degrading basidiomycetes organism.

Microbial Versatile Peroxidase: Versatile peroxidase (VP) is characterized by its capacity to oxidize the typical substrates of other basidiomycete peroxidases, including (i) Mn(2+), the substrate for manganese peroxidase (MnP) Mn(3+) has the ability to oxidize phenols and initiate lipid peroxidation reactions), (ii) veratryl alcohol (VA), the typical substrate for lignin peroxidase.

Hydrolases: Hydrolytic enzymes breakdown chemical links between hazardous compounds to lessen toxicity. This procedure is easily capable of breaking down pesticides like carbamate, organophosphate, and oil spill. Additionally, it catalyses alcoholysis and condensation.

Microbial Lipases: Lipase is a lipid-degrading enzyme that has been identified from microbial, animal, and plant sources. It contributes to the significant reduction of organic contaminants in the polluted ground. The lipase activity was caused the significant reduction of polluted total hydrocarbon soil (Table 6.2).

6.9 Characterization of Plastic Degradation

6.9.1 X-ray Diffraction

X-ray Diffraction (XRD) is an analytical technique that can be used to study the results of plastic degradation. XRD provides valuable information about the crystalline structure of materials, including changes that may occur due to degradation. Many plastics are semi-crystalline, meaning they consist of both crystalline and amorphous regions. During degradation, the crystalline regions of the plastic may undergo changes, leading to a reduction in crystallinity. This can be observed in XRD by a decrease in the intensity of crystalline peaks (Li et al. 2018). Sometimes the crystalline domains may become smaller or less ordered. This can result in peak broadening in the XRD pattern. Broader peaks indicate a decrease in the size of crystalline domains or an increase in their disorder. The degradation of plastic may lead to chemical changes in the material, affecting the lattice parameters of the crystalline phase. These changes can cause shifts in the positions of XRD peaks, indicating alterations in the crystal structure (Gao and Sun 2021). During degradation, chemical breakdown or reactions may occur, resulting in the formation of new phases within the plastic material. XRD can detect the appearance of new peaks corresponding to these new phases. In addition to studying the crystalline phase,

Table 6.2 Commonly used polymers—present production and recovery rates, with corresponding microorganisms/enzymes known so far for their biodegradation

Polymer	Applications	Generation (1000 tons/year)	Recovery (%)	Microorganisms	Enzyme	References
Polyethylene Terephthalate (PET/PETE)	Soft drink bottles, peanut butter jars, frequently recycled	2,860	19	Bacillus sp, Aspergillus sp., and Spirulina sp.	Cutinases and lipases	Dang et. al. (2018)
High density Polyethylene (HDPE)	Milk, water, juice, detergent, motor oil bottles; often recycled	5,890	9	Arthrobacter sp. and Pseudomonas sp	–	Balasubramanian et. al. (2010)
Polyvinyl Chloride (PVC)	Cooking oil bottles, Credit cards, household food wrap, building materials; rarely recycled, toxic if burned	1,640	0	Poliporusversicolor, Pleurotus sajor caju, Phanerochaete chrysosporium, Pleurotus ostreatus, Pleurotus sapidus, Pleurotus eryngii, Pleurotus florida	Fungal laccases and peroxidases	Chapelle and Bathe (1993)
Low density polyethylene (LDPE)	Shrink wrap, garbage and shopping bags, lids, cups, sometimes recycled	6,450	3	Penicillium pinophilium and Aspergillus niger	Laccase,esterase, and alkane hydroxylase	Albertsson et al. (1987)
Polypropylene (PP)	Snack food wrap, margarine tubs, Straw, car batteries, occasionally recycled	4,000	0.3	Bacillus sp., Brevibacillus sp., Phanerochaete chrysoporium and Rhodococcus sp.	Lipase or dehydrogenase	Auta et. al. (2018)
Polystyrene (PS)	Pharmaceuticals bottles, Styrofoam cups, clear salad bowl lids, packing peanuts; rarely recycled	2,590	0	Plesiophthalmus davidis and Serratia sp.	–	Woo et. al. (2020)
Polyethelene (PE) and Other resins	Multilayered, mixed material containers, squeezable bottles; relax recycled	5,480	0.001	Bacillus mycoides	Fungal laccases and peroxidases	Karigar and Rao (2011)

XRD can provide information about the amorphous content of the plastic. Changes in the amorphous regions can also occur during degradation and may be reflected in the XRD pattern (Gaonkar et al. 2020). In severe cases of degradation, the plastic's molecular structure may break down significantly, leading to the loss of its original crystallographic features. This can result in a more featureless or amorphous-like XRD pattern. Depending on the extent of degradation and the specific properties of the plastic, quantitative analysis of the XRD data may be performed to quantify the degree of crystallinity, crystal size, or identify the presence of specific phases.

6.9.2 FTIR

The results of plastic degradation in Fourier Transform Infrared Spectroscopy (FTIR) can reveal various changes in the material's chemical structure. FTIR is particularly useful for identifying functional groups and chemical bonds present in the plastic, which can be affected by degradation processes. By analysing the FTIR data, researchers can gain valuable insights into the chemical changes that occur during plastic degradation. This information is crucial for understanding the material's performance, durability, and potential applications in various environments.

Degradation can lead to the breaking of chemical bonds, resulting in the decrease or disappearance of characteristic peaks in the FTIR spectrum. This reduction in peak intensity indicates a loss of specific functional groups in the plastic. Degradation can also introduce new chemical entities or functional groups as a result of chain scission, oxidation, or other chemical reactions (Huang et al. 2021). New peaks may emerge in the FTIR spectrum, indicating the presence of these newly formed groups. Changes in the plastic's chemical structure due to degradation can cause shifts in the positions of specific peaks in the FTIR spectrum. These shifts are indicative of modifications in bond lengths or chemical environments. Different regions of the FTIR spectrum correspond to various types of chemical bonds and functional groups (Wright et al. 2021). Degradation may alter the intensity and shape of these absorption bands, providing insights into the nature of chemical changes in the plastic. So FTIR data is used for quantitative analysis, allowing researchers to determine the extent of degradation based on peak intensities, peak areas, or the calculation of specific chemical ratios. FTIR can also help identify the mechanisms involved in the degradation process. For example, the presence of characteristic oxidation products or degradation intermediates can provide clues about the degradation pathway. FTIR can be used to study how different environmental factors, such as exposure to heat, light, moisture, or chemicals, contribute to the degradation of plastics. Similar to other analytical techniques, FTIR analysis involves comparing the spectra of the degraded plastic with those of the original, pristine material (control). This comparison allows for a direct assessment of the changes that occurred during degradation. FTIR is used to track the degradation of plastics over time, providing valuable information about the material's stability and potential lifespan under specific conditions (Erni-Cassola et al. 2017).

6.9.3 Scanning Electron Microscopy

SEM can reveal changes in the surface morphology of the plastic due to degradation. These changes may include cracks, fractures, surface roughness, pitting, and other forms of mechanical damage. In some cases, the degradation of plastics can lead to the formation of fibrils or microfibres on the surface. SEM can capture these fibrillar structures, which are more pronounced in certain types of polymers (Ojha et al. 2017). Chemical degradation can cause erosion and etching on the surface of the plastic, leading to the loss of material or changes in surface texture. SEM can help visualize these features (Sowmya et al. 2014). During degradation, the plastic material may undergo agglomeration, where small particles or fibres clump together. SEM can identify the presence of agglomerates and their distribution on the surface. SEM can reveal the presence of contaminants or foreign particles on the surface of the degraded plastic, which may have contributed to or resulted from the degradation process (Gniadeka and Dąbrowska 2019).

6.9.4 Transmission Electron Microscopy

TEM provides detailed images of the internal structure of the plastic material, including information about the arrangement of polymer chains and crystalline regions. TEM can show changes in the microstructure, such as changes in crystalline size and orientation, due to degradation (Hadria et al. 2020). TEM can reveal evidence of chain scission, which is the breaking of polymer chains during degradation, leading to changes in the molecular arrangement. TEM can identify changes in the amorphous phase of the plastic, such as an increase in the amorphous content due to chain cleavage. TEM can provide insights into nanoscale features, such as nanoparticles, nanofibres, or nanocrystals that may form during the degradation process. TEM can help identify the presence of degradation products or new phases that have formed as a result of degradation (Kang et al. 2019). Both SEM and TEM are valuable tools for studying plastic degradation as they provide complementary information about the material's structural changes at different length scales. The combination of these techniques with other analytical methods, such as FTIR and mechanical testing, can lead to a comprehensive understanding of the degradation mechanisms and overall behaviour of the plastic material under different conditions.

6.9.5 Gel Permeation Chromatography (GPC)

Also known as Size Exclusion Chromatography (SEC), a powerful analytical technique used to study the molecular weight distribution of polymers, including plastics. GPC can provide valuable insights into the results of plastic degradation. One of the primary results of plastic degradation is a reduction in the molecular weight of the polymer chains. As the polymer chains undergo chain scission or other degradation processes, their length decreases, resulting in a decrease in the average molecular weight of the polymer. This reduction in molecular weight is reflected in the GPC analysis as a shift toward lower molecular weight values compared to the original, pristine plastic. Degradation can lead to a broader molecular weight distribution of the polymer (Biver et al. 2018). As the polymer chains break into smaller fragments, the resulting distribution of molecular weights becomes wider, showing a greater variety of chain lengths in the degraded plastic compared to the narrow distribution of the pristine material. In addition to the main peak corresponding to the degraded polymer, GPC may also reveal the presence of low molecular weight fractions, such as oligomers or monomers. These are the smaller fragments resulting from chain scission during degradation. In GPC analysis, the retention time of a polymer peak corresponds to its molecular weight. As the molecular weight decreases due to degradation, the retention time of the main peak shifts toward shorter retention times. The GPC results of the degraded plastic are typically compared with those of the original, pristine plastic (control). This comparison allows for a direct assessment of the changes that occurred in the molecular weight distribution during degradation (Schrank et al. 2022). The analysis of GPC data can provide insights into the mechanisms of plastic degradation. For example, different degradation pathways may result in distinct changes in the molecular weight distribution, helping to understand the degradation process better. GPC allows for quantitative determination of various parameters related to the molecular weight distribution, such as the number-average molecular weight, weight-average molecular weight, and polydispersity index (PDI). Changes in these parameters can be indicative of degradation effects.

6.9.6 Differential Scanning Calorimetry (DSC)

Differential Scanning Calorimetry (DSC) is a thermal analysis technique used to study the thermal behaviour of materials, including plastics. DSC can provide valuable results related to the degradation of plastics. DSC can detect changes in the melting and crystallization behaviour of plastics during degradation. The melting temperature (Tm) and the heat of fusion associated with melting can be altered as the crystalline regions of the polymer are affected by degradation processes. Degradation can lead to a decrease in the melting point of the plastic due to the reduction in molecular weight or the disruption of the crystalline structure (Capitaina et al. 2020). As the polymer chains degrade, the heat of fusion associated with the melting process may decrease due to a decrease in the size and perfection of crystalline regions. Depending on the nature of the degradation, DSC may show additional endothermic peaks corresponding to the decomposition of specific functional groups or the release of volatile degradation products during heating. In some cases, DSC can reveal exothermic peaks associated with exothermic reactions that occur during the degradation process, such as oxidation or degradation reactions. For amorphous plastics, DSC can detect changes in the glass transition temperature (Tg) due to degradation-induced changes in the molecular structure or molecular weight. In severely degraded plastics, multiple thermal transitions may be observed, indicating the presence of different degraded components with distinct thermal behaviour (Zhang et al. 2023). DSC data can be used for quantitative analysis, such as the determination of the heat of fusion, the crystallinity of the material, or the enthalpy of specific degradation processes. The analysis of DSC data can provide insights into the mechanisms of plastic degradation. Different degradation pathways may lead to distinct thermal behaviour, helping to identify the degradation process and its kinetics.

6.9.7 Thin Layer Chromatography (TLC)

Thin Layer Chromatography (TLC) is a chromatographic technique used to separate and analyse different components in a mixture. While TLC is not a typical method used to study plastic degradation directly, it can be applied in certain cases to analyse the presence of degradation products or changes in the chemical composition of the plastic material. TLC can be used to detect the presence of degradation products resulting from the breakdown of plastic molecules. These products may have different chemical properties than the original plastic material, leading to distinct spots or bands on the TLC plate. In TLC, the Retention Factor (Rf) is the ratio of the distance travelled by a compound to the distance travelled by the solvent front. During plastic degradation, changes in the chemical structure or polarity of the material may cause shifts in the Rf values of certain components, indicating alterations in their behaviour during chromatography. Plastics often contain various additives, such as stabilizers, plasticizers, or colourants. During degradation, these additives may also undergo changes, leading to the appearance of new spots or shifts in Rf values on the TLC plate. In some cases, TLC can be used for semi-quantitative analysis by comparing the intensity of spots or bands on the TLC plate. This can provide information about the relative amounts of different components present in the degraded plastic.

6.9.8 GC–MS

Gas Chromatography-Mass Spectrometry (GC–MS) is a powerful analytical technique used to study the chemical composition of materials, including plastics, and is particularly valuable for analysing the products of plastic degradation. GC–MS can identify and characterize the various degradation products formed during the breakdown of plastic molecules. The technique separates individual components in the mixture based on their volatility and molecular weight, and the mass spectrometer provides information about their molecular structure and mass. GC–MS allows for quantitative analysis of the degradation products, providing information about their relative abundance or concentration in the degraded plastic sample (Sangale et al.

2019). It is especially well-suited for detecting volatile degradation products, such as small molecules, gases, and volatile organic compounds (VOCs) released during the degradation process (Sarker and Rashid 2013). Plastics often contain various additives, such as stabilizers, plasticizers, or flame retardants. GC–MS can help identify and quantify these additives, which may undergo changes during the degradation of the plastic. The identification of degradation products and their relative abundance can provide insights into the mechanisms of plastic degradation and the pathways through which the degradation occurs. In addition to volatile compounds, GC–MS can also analyse semi-volatile and non-volatile compounds that may be formed during plastic degradation. GC–MS can help identify and quantify potentially harmful or environmentally significant degradation products, contributing to the understanding of the environmental impact of plastic degradation (Fries et al. 2013).

6.9.9 Chemiluminescence

Chemiluminescence is a unique analytical technique used to study the degradation of plastics. It involves the measurement of light emission resulting from chemical reactions occurring during degradation. While chemiluminescence is not as commonly used as other analytical methods like FTIR, SEM, TEM, or GC–MS for studying plastic degradation, it can provide valuable insights into the degradation processes. Chemiluminescence analysis can reveal the light emission profile over time as the plastic degrades. The intensity and duration of light emission can provide information about the rate and extent of the degradation process. Chemiluminescence may indicate the formation of reactive oxygen species (ROS) during plastic degradation. ROS, such as singlet oxygen or hydroxyl radicals, are often involved in oxidative degradation pathways. The presence of chemiluminescence can suggest the generation of free radicals during plastic degradation. Free radicals play a significant role in chain scission reactions, leading to the breakdown of polymer chains. The chemiluminescence patterns can provide insights into the mechanisms of plastic degradation. Different degradation pathways may result in distinct chemiluminescence behaviour. Chemiluminescence data can be correlated with results from other analytical methods, such as FTIR or GC–MS, to gain a comprehensive understanding of the degradation processes and identify the chemical changes occurring in the plastic. Chemiluminescence analysis can demonstrate the sensitivity of plastic degradation to environmental factors like temperature, humidity, or exposure to light. Depending on the chemiluminescence setup and the plastic's composition, it may be possible to detect the emission of light from specific additives that undergo chemical changes during degradation.

6.9.10 *Nuclear Magnetic Resonance (NMR) Spectroscopy*

It is a powerful analytical technique used to study the structure and chemical composition of materials, including plastics. NMR can provide valuable results related to the degradation of plastics. NMR can identify and characterize the various degradation products formed during the breakdown of plastic molecules. By comparing the NMR spectra of the degraded plastic with those of the pristine material, specific changes in chemical shifts and peak patterns can indicate the presence of degradation products. As the plastic undergoes degradation, changes in its chemical structure may occur. These changes can lead to shifts in the NMR signals, which are directly related to the local chemical environment of the nuclei within the plastic (Falkenstein et al. 2021). NMR allows for quantitative analysis, providing information about the relative abundance or concentration of different chemical species, including degradation products, in the degraded plastic sample. NMR can help identify and quantify the presence of additives, such as stabilizers, plasticizers, or other modifiers, which may undergo changes during the degradation process. NMR can selectively probe specific functional groups within the plastic, allowing for the investigation of the degradation of particular moieties or chemical bonds. By studying the changes in NMR spectra, researchers can gain insights into the mechanisms of plastic degradation and the specific chemical reactions involved. NMR can provide information about cross-linking or chain scission reactions that may occur during plastic degradation, leading to changes in the polymer structure (Nishida et al. 2018). NMR can be used for kinetic studies of plastic degradation, allowing researchers to monitor the progress of degradation over time and study the rate of degradation processes.

6.9.11 *Adenosine Triphosphate (ATP) Assays*

These are not typically used directly for determining plastic degradation. ATP assays are commonly used to quantify the amount of ATP, a molecule involved in energy transfer in living cells, as an indicator of cell viability or metabolic activity. The assay is based on the principle that ATP is present in all living cells and is rapidly degraded upon cell death or damage (Gilan et al. 2004).

6.9.12 *Protein Assay*

Some work focused on the use of protein assays and FDA hydrolysis by extracellular esterases as efficient tools for determining the colonization and biofilm formation on polyethylene surfaces. The results showed rapid colonization of polyethylene in the first 2 days of incubation, followed by a sharp decrease in biomass density. This early colonization was attributed to the utilization of mineral oil adhering to the polyethylene as a carbon source. Once the mineral oil was depleted, most of the biomass in the high-density, multilayered biofilm was washed off into the medium. It is hypothesized that a low-density population of cells capable of utilizing polyethylene as a carbon source is responsible for the biodegradation of the plastic (Gilan et al. 2004).

6.9.13 Fluorescein Diacetate (FDA) Analysis

This is a commonly used method to assess the metabolic activity and viability of microorganisms in a sample, including those involved in the degradation of plastics. The FDA assay relies on the enzymatic hydrolysis of FDA by intracellular esterases in living cells, leading to the release of fluorescein, a fluorescent compound. The fluorescence intensity is directly proportional to the metabolic activity and number of viable cells in the sample.

FDA analysis can be used to measure the metabolic activity of microorganisms present on or near the plastic surface. An increase in fluorescence indicates active microbial populations, which may be involved in the biodegradation of the plastic. By performing FDA analysis at different time points during the degradation process, researchers can track changes in microbial activity over time. This allows for the monitoring of the plastic degradation rate and the dynamics of microbial colonization. FDA analysis can provide quantitative data about the number of viable microbial cells associated with the plastic material. Comparing the fluorescence intensity of samples with known standards or controls can give insight into the relative abundance of degrading microorganisms. FDA analysis can be performed under different environmental conditions to assess the influence of temperature, pH, or nutrient availability on microbial activity and plastic degradation. By using FDA analysis, researchers can optimize conditions that promote higher microbial activity and accelerate the biodegradation of plastics. FDA analysis can be combined with other analytical methods, such as GC–MS or GPC, to correlate the metabolic activity of microorganisms with the chemical changes occurring during plastic degradation (Shabaka et al. 2019). Overall, FDA analysis is a valuable tool for studying plastic degradation as it provides insights into the microbial activity associated with the degradation process. However, it is essential to use FDA analysis in conjunction with other analytical techniques to obtain a comprehensive understanding of plastic degradation mechanisms and the involvement of microorganisms in the process (Fig. 6.3).

6.10 Molecular Methods to Study Plastic Degradation

Molecular biology plays a significant role in understanding the mechanisms and processes involved in plastic degradation. It involves studying the genetic and molecular aspects of microorganisms and enzymes responsible for breaking down plastics.

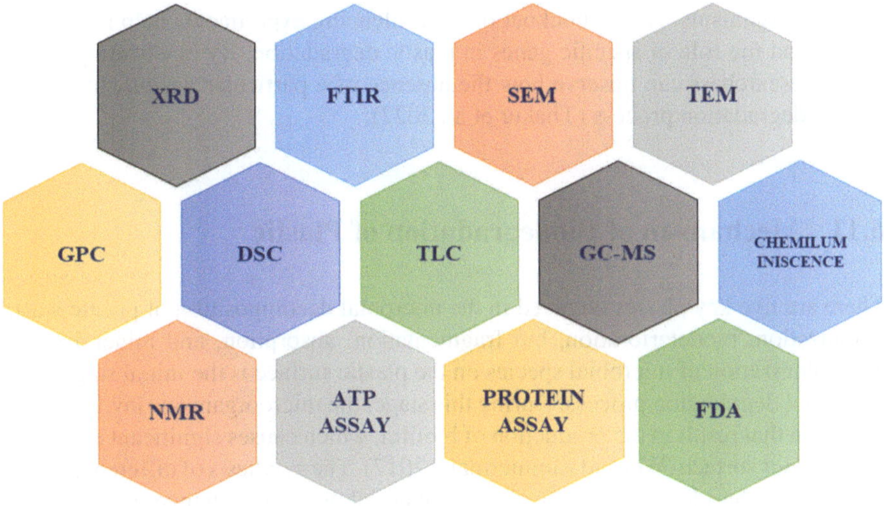

Fig. 6.3 Characterization techniques for plastic degradation at a glance

Microbial Identification: Molecular techniques, such as Polymerase Chain Reaction (PCR) and DNA sequencing, are used to identify and characterize microorganisms capable of degrading plastics. By analysing specific genetic markers or genes associated with plastic degradation, researchers can identify the microbial communities involved in biodegradation processes. Metagenomics involves the study of the genetic material obtained directly from environmental samples. In the context of plastic degradation, metagenomics can reveal the genetic diversity of microbial communities present on or near plastic surfaces (Urbanek et al. 2021). This approach allows researchers to explore the potential for plastic biodegradation in various environments. To understand the molecular mechanisms of plastic degradation, researchers analyse gene expression patterns of microorganisms during plastic degradation. Quantitative Reverse Transcription-PCR (qRT-PCR) and other gene expression profiling techniques provide insights into which genes are upregulated or downregulated in response to the presence of plastics. Molecular biology techniques are used to isolate, clone, and characterize enzymes involved in plastic degradation. These enzymes, such as esterases, lipases, and hydrolases, are responsible for breaking down the polymer chains of plastics. Understanding the structure and function of these enzymes helps in developing biotechnological approaches for enhancing plastic degradation. Genetic engineering techniques are employed to modify microorganisms or their enzymes to improve their plastic-degrading capabilities. This includes introducing genes or gene clusters from plastic-degrading organisms into other microbes to enhance their ability to degrade plastics (Malla et al. 2023). Molecular biology tools are used to elucidate the pathways involved in plastic degradation. By identifying the genes responsible for specific enzymatic reactions, researchers can construct metabolic pathways that explain the breakdown of plastics by microorganisms. Gene knockout or gene silencing experiments help researchers understand the role of specific genes in plastic degradation. By inactivating certain genes, researchers can observe how the absence of a particular enzyme affects the overall degradation process (Thakur et al. 2022).

6.11 Mechanism of Biodegradation of Plastic

There are five key phases involved in the microbial decomposition of plastic waste: colonization, biodeterioration, bio fragmentation, absorption, and mineralization. The colonization of microbial species on the plastic surface is the initial stage of the microbial degradation process. During this stage, the microorganisms involved form consortia that result in the production of biofilm, which causes significant damage to the polymer surface (Wei and Zimmermann 2017). The synthesis of different proteins and polysaccharides, which seep into the material pores and change the size of the pores, is what allows microbes to adhere to the polymer surface. As a result of the microbial attachment and activity on the plastic surface, the biodeterioration process is carried out. This step sees the surface degrading, changing the physicochemical characteristics of the plastic polymer. Regarding the activity of different microorganisms, filamentous fungi use their mycelia to penetrate the polymeric material to increase the size of the pores and induce the formation of cracks, resulting in decreased polymer resistance and durability (Bonhomme et al. 2003; Phaniendra et al. 2014). Also, the microbial penetration rate can be enhanced by the microbial secretion of extracellular substances that affect hydrophobic and hydrophilic phases.

Additionally, microorganisms that are chemo-organotrophic (using oxygen and/or chemical bonds in organic substrates as an energy source) can emit organic acids, such as gluconic, oxalic, glyoxylic, citric, oxaloacetic, glutaric, etc. Due to the simultaneous modification of the medium pH brought on by the production of acids and bases during microbial metabolic activity, surface erosion may become more pronounced (Lugauskas and Loreta Peciulyte 2003). It is also important to note that organic acids, which are required for process completion, are more effective than mineral acids at preventing biodeterioration by fixing cations. By invading the polymer material and causing an increase in holes and fractures, filamentous microorganisms (bacteria and fungi) have the potential to cause physical biodeterioration (Hakkarainen et al. 2000). However, certain microbial species, such as chemo lithotrophic bacteria that can ingest manganese and/or iron (Fe^{3+}) cations through particular proteins in the microbial cell membrane, may also execute oxidation–reduction and chemical biodeterioration. Several extracellular enzymes, including peroxidases, are involved in the process of enzymatic biodeterioration (Otsuka et al. 2003). However, some plastic polymers, including PU and PVC, are resistant to degradation (Shah et al. 2008). To get around the crystallinity of these polymers, microorganisms commonly create enzymatic groups such lipases, esterases, ureases, and proteases.

The process of fragmentation, a lytic process is used by microorganisms to degrade polymers into smaller units like monomers, dimers, and oligomers. This process involves the secretion of specific enzymes, such as oxidoreductases and hydrolases, as well as free radicals. The chemical bonds in plastic polymers are similar to those found in natural polymers, making it reasonable to assume that lignocellulose-degrading enzymes can also degrade plastic polymers (Austin et al. 2018). The synthesis of microbial endo- and exo-enzymes takes time to start, and their concentration increases over time, with activity ending when the substrate is depleted. The hydrolysis stage is influenced by various enzymes, with depolymerase enzymes catalysing reactions to break ester bonds. Due to the hydrophobic crystalline nature of polymers, multiple enzymes are involved in the chain scission reactions to transform the polymer structure effectively. Additional enzymes like mono and dioxygenases, peroxidases, and oxidases containing copper atoms play roles in increasing polymer structure polarity and catalysing specific reactions during the biodegradation process (Phaniendra et al. 2014).

On the other hand, the development of a hydroxyl function and/or a carbonyl or carboxyl group during free radical oxidation seeks to increase the molecule's polarity. Microbial attack is amplified by the increased hygroscopic characteristics. Additionally, a number of enzymes that may produce free radicals and cause oxidative stress can catalyse a number of oxidation processes, which in turn can cause chain reactions that convert and degrade polymeric structures (Gewert et al. 2015). However, because the enzymes can't interact with the interior component of these structures, the breakdown of linear polymeric substances by enzymatic processes is severely restricted. H_2O_2 may also be produced by a number of microbial degraders, including bacteria, fungus, and algae. Additionally, a variety of microbial degraders, such as bacteria, fungus, and algae, have the ability to generate H_2O_2 (Diaz et al. 2018). Regarding the oxidation capacity of algal species, Diaz et al. (2018) showed that five marine harmful bloom-forming algae (HBA) species, including *Aureococcus anophagefferens, Pseudonitzschia sp., Heterosigma akashiwo, Chattonella marina,* and *Karenia brevis,* produced extracellular superoxide and H_2O_2 (Diaz et al. 2018). When it comes to assimilating the created monomers, the microbial cell gets its energy sources from these integrated monomers, enabling the microbial cell to proliferate. Some monomers enter cells via membrane carriers that are unique to each individual. However, because of the permeability of the cell membrane, other molecules cannot be taken up. When it comes to assimilating the created monomers, the microbial cell gets its energy sources from these integrated monomers, enabling the microbial cell to proliferate. Some monomers enter cells via membrane carriers that are unique to each individual. However, because of the permeability of the cell membrane, other molecules cannot be taken up. When it comes to assimilating the created monomers, the microbial cell gets its energy sources from these integrated monomers, enabling the microbial cell to proliferate. Some monomers enter cells via membrane carriers that are unique to each individual (Ali et al. 2020).

When plastics biodegrade, they first break down into smaller subunits, which then break down further into tiny inorganic molecules like carbon dioxide and water (Andrady 1998). The breakdown products can be taken up into the cell for further metabolization after the molecular size of the synthetic polymers has been reduced to a range of 10 to 50 carbon atoms (Wei and Zimmermann 2017; Lucas et al.

2008; Restrepo-Florez et al. 2014; Khan and Majeed 2019; Brandenberg et al. 2022). Particle absorption by cells includes intricate biomolecular interactions and highly controlled systems. The biological response and absorption pathways of the particle may define its function, and the paths through which MPs and NPs enter cells are strongly related to their harmful effects. The quantity of internalized particles is connected to the toxicity brought on by MPs and NPs.

6.12 Conclusion

Microplastics have become increasingly prevalent in all our ecosystems, posing significant harm to various life forms within these ecosystems. However, with the implementation of a suitable legal framework and the utilization of advanced biotechnology, we have the potential to effectively eliminate microplastic pollution through strain engineering, protein engineering, genomic and proteomic understanding, and the production of enzyme-based products for the large-scale management of plastic waste. One promising solution lies in microbial biodegradation, a process in which microorganisms break down these harmful particles, with significant implications for addressing the global issue of plastic pollution. Microorganisms have demonstrated the remarkable ability to break down these persistent synthetic polymers through various enzymatic processes and metabolic pathways. As we delve deeper into understanding the mechanisms of microbial plastic degradation, we gain valuable insights into harnessing these natural processes for more effective and sustainable solutions. The potential benefits of microbial degradation of microplastics are multi-faceted. First and foremost, this biodegradation offers an eco-friendly and non-intrusive approach to reducing plastic waste in the environment. Unlike physical removal methods that may be challenging and costly, harnessing microbial action can enable us to tackle microplastic pollution at its source. Furthermore, the study of microbial plastic degradation has unveiled the diversity and adaptability of microorganisms in coping with environmental challenges. However, it is crucial to recognize that microbial degradation of microplastics is a complex process influenced by various factors such as plastic type, environmental conditions, and microbial communities. Thus, ongoing research is necessary to make big strides in technological innovations for optimization and scale up of the microbial enzyme production and biodegradation processes for real-world applications. In the quest to combat plastic pollution, interdisciplinary collaboration between environmental scientists, biotechnologists, and policymakers is vital. Together, we can foster innovations in biotechnology and waste management, promoting sustainable practices that leverage the power of nature to create a cleaner, healthier planet. As we continue to explore and refine this natural solution, we move closer to a future where the detrimental impact of microplastics on ecosystems and human health is mitigated, leaving a positive legacy for generations to come.

References

Albertsson AC, Andersson SO, Karlsson S (1987) The mechanism of biodegradation of polyethylene. J Polym Degrad Stab 18:73–87. https://doi.org/10.1016/0141-3910(87)90084-X

Ali SS, Kornaros M, Manni A, Sun J, El-Shanshoury AER, Kenawy E-R, Khalil MA (2020) Enhanced anaerobic digestion performance by two artificially constructed microbial consortia capable of woody biomass degradation and chlorophenols detoxification. J Hazard Mater 389:122076. https://doi.org/10.1016/j.jhazmat.2020.122076

Andrady AL (1998) Biodegradation of plastics: monitoring what happens. In: Pritchard G (eds) Plastics additives. Polymer science and technology series, 1. https://doi.org/10.1007/978-94-011-5862-6_5

Artham T, Doble M (2008) Biodegradation of aliphatic and aromatic polycarbonates. J Macromol Biosci 8(1):14–24. https://doi.org/10.1002/mabi.200700106

Auras R, Harte B, Selke S (2004) An overview of polylactides as packaging materials. J Macromol Biosci 4(9):835–864. https://doi.org/10.1002/mabi.200400043

Austin HP, Allen MD, Donohoe BS, Rorrer NA, Kearns FL, Silveira RL, Pollard BC, Dominick G, Duman R, El Omari K, Mykhaylyk V, Wagner A, Michener WE, Amore A, Skaf MS, Crowley MF, Thorne AW, Johnson CW, Woodcock HL, McGeehan JE, Beckham GT (2018) Characterization and engineering of a plastic-degrading aromatic polyesterase. Proc Natl Acad Sci U S A 115:E4350–E4357. https://doi.org/10.1073/pnas.1718804115

Auta HS, Emenike CU, Jayanthi B, Fauziah SH (2018) Growth kinetics and biodeterioration of polypropylene microplastics by Bacillus sp. and Rhodococcus sp. isolated from mangrove sediment. Pollut Bull 127:15–21. https://doi.org/10.1016/j.marpolbul.2017.11.036

Balasubramanian V, Natarajan K, Hemambika B, Ramesh N, Sumathi CS, Kottaimuthu R, Kannan VR (2010) High-density polyethylene (HDPE)-degrading potential bacteria from marine ecosystem of Gulf of Mannar, India. Lett Appl Microbiol 51(2):205–211. https://doi.org/10.1111/j.1472-765X.2010.02883.x

Balasubramanian V, Natarajan K, Rajesh KV, Perumal P (2014) Enhancement of in vitro high-density polyethylene (HDPE) degradation by physical, chemical and biological treatments. J Environ Sci Pollut Res 21:12549–12562. https://doi.org/10.1007/s11356-014-3191-2

Bansal V, Gassenhuber J, Phillips T (2017) Spectrum of mutations in monogenic diabetes genes identified from high-throughput DNA sequencing of 6888 individuals. BMC Med 15:213. https://doi.org/10.1186/s12916-017-0977-3

Bell JD, Watson RA, Ye Y (2016) Global fishing capacity and fishing effort from 1950–2012 (18, 489. Fish Fish 18:792–793. https://doi.org/10.1111/faf.12187

Benyathiar P, Kumar P, Carpenter G, Brace J, Mishra DK (2022) Polyethylene terephthalate (PET) bottle-to-bottle recycling for the beverage industry: a review. Polymers 14:2366. https://doi.org/10.3390/polym14122366

Bing X, Bloemhof JM (2014) van der Vorst.: sustainable reverse logistics network design for household plastic waste. Flex Serv Manuf J 26:119–142. https://doi.org/10.1007/s10696-012-9149-0

Biver T, Bianchi S, Carosi MR, Ceccarini A, Corti A, Manco E, Castelvetro V (2018) Selective determination of poly(styrene) and polyolefin microplastics in sandy beach sediments by gel permeation chromatography coupled with fluorescence detection. Mar Pollut Bull 136:269–275. https://doi.org/10.1016/j.marpolbul.2018.09.024

Bonhomme S, Cuer A, Delort AM, Lemaire J, Sancelme M, Scott G (2003) Environmental biodegradation of polyethylene. Polym Degrad Stab 81(3):441–452. ISSN 0141–3910. https://doi.org/10.1016/S0141-3910(03)00129-0

Brandenberg OF, Schubert OT, Kruglyak L (2022) Towards synthetic PET trophy: engineering pseudomonas putida for concurrent polyethylene terephthalate (PET) monomer metabolism and PET hydrolase expression. Microb Cell Fact 21:119. https://doi.org/10.1186/s12934-022-01849-7

Capitaina C, Ross-Jonesb J, Möhringa S, Tippkötter N (2020) Differential scanning calorimetry for quantification of polymer biodegradability in compost. Int Biodeterior Biodegrad 149:104914. https://doi.org/10.1016/j.ibiod.2020.104914

Chae Y, An YJ (2018) Current research trends on plastic pollution and ecological impacts on the soil ecosystem: a review. Environ Pollut 240:387–395. https://doi.org/10.1016/j.envpol.2018.05.008

Chapelle D, Bathe KJ (1993) The Inf-Sup Test. Comput Struct 47:537–545. https://doi.org/10.1016/0045-7949(93)90340-J

Chauhan SP, Goradia B, Saxena A (2017) Bacterial laccase: recent update on production, properties and industrial applications. J Nat Center Bio 7:1–20. https://doi.org/10.1007/s13205-017-0955-7

Chen Y, Leng Y, Liu X, Wang J (2020) Microplastic pollution in vegetable farmlands of suburb Wuhan, central China. Environ Pollut 257:113449. https://doi.org/10.1016/j.envpol.2019.113449

Corradini F, Meza P, Eguiluz R, Casado F, Huerta-Lwanga E, Geissen V (2019) Evidence of microplastic accumulation in agricultural soils from sewage sludge disposal. Sci Total Environ 671:411–420. https://doi.org/10.1016/j.scitotenv.2019.03.368

Dang TCH, Nguyen TD, Thai H, Nguyen CT, Tran HTT, Nguyen HV, Tran XB, Pham TPT, Nguyen GT, Nguyen TQ (2018) Plastic degradation by thermophilic Bacillus sp. BCBT21 isolated from composting agricultural residual in Vietnam. J Adv Nat Sci Nanosci Nanotechnol 9(1):015014. https://doi.org/10.1088/2043-6254/aaabaf

Tokiwa Y, Calabia PB (2004) Degradation of microbial polyesters. J Biotechnol Lett 26:1181–1189.https://doi.org/10.1023/b:bile.0000036599.15302.e5

Diaz JM, Plummer S, Tomas C, Alves-de-Souza C (2018) Production of extracellular superoxide and hydrogen peroxide by five marine species of harmful bloom-forming algae. J Plankton Res 40(6):667–677. https://doi.org/10.1093/plankt/fby043

Diepens NJ, Koelmans AA (2018) Accumulation of plastic debris and associated contaminants in aquatic food webs. Environ Sci Technol 52:8510–8520. https://doi.org/10.1021/acs.est.8b02515

Erni-Cassola G, Gibson MI, Thompson RC, Christie-Oleza J (2017) Lost, but found with Nile Red: a novel method for detecting and quantifying small microplastics (1 mm to 20 μm) in environmental samples. Environ Sci Technol 51(23):13641–13648. https://doi.org/10.1021/acs.est.7b04512

Falkenstein P, Wei R, Matysik J, Song C (2021) Mechanistic investigation of enzymatic degradation of polyethylene terephthalate by nuclear magnetic resonance. Methods Enzymol 648:231–252. https://doi.org/10.1016/bs.mie.2020.11.002

Fangzhi J, Ziwei J, Jiayue H, Pengfei T, Jizhe C, Wenxuan F, Chunjing Y, Chang F, Qian L (2023) Exploration of potential driving mechanisms of Comamonas testosteroni in polycyclic aromatic hydrocarbons degradation and remodelled bacterial community during co-composting. J Hazard Mater 458:132032. https://doi.org/10.1016/j.jhazmat.2023.132032

Fries E, Dekiff JH, Willmeyer J, Nuelle MT, Ebertc M, Remy D (2013) Identification of polymer types and additives in marine microplastic particles using pyrolysis-GC/MS and scanning electron microscopy. Environ Sci Process 1949–1956. https://doi.org/10.1039/c3em00214d

Gao R, Sun C (2021) A marine bacterial community capable of degrading poly(ethylene terephthalate) and polyethylene. J Hazard Mater 416:125928. https://doi.org/10.1016/j.jhazmat.2021.125928

Gaonkar AA, Murudkar VV, Deshpande VD (2020) Comparison of crystallization kinetics of polyethylene terephthalate (PET) and reorganized PET. Thermochim Acta 683:178472. https://doi.org/10.1016/j.tca.2019.178472

Gewert B, Plassmann MM, MacLeod M (2015) Pathways for degradation of plastic polymers floating in the marine environment. Environ Sci Process Impacts 17:1513. https://doi.org/10.1039/C5EM00207A

Gilan I, Hadar Y, Sivan A (2004) Colonization, biofilm formation and biodegradation of polyethylene by a strain of Rhodococcus ruber. Appl Microbiol Biotechnol 65:97–104. https://doi.org/10.1007/s00253-004-1584-8

Giyahchi M, Moghimi H (2023) Aerobic biodegradation of untreated polyester–polyether urethanes by newly isolated yeast strains Exophilia sp. NS-7 and Rhodotorula sp. NS-12. Sci Rep 13:5016. https://doi.org/10.1038/s41598-023-31639-z

Gniadeka M, Dąbrowska A (2019) The marine nano- and microplastics characterisation by SEM-EDX: the potential of the method in comparison with various physical and chemical approaches. Mar Pollut Bull 148:210–216. https://doi.org/10.1016/j.marpolbul.2019.07.067

Groh KJ, Backhaus T, Carney-Almroth B, Geueke B, Inostroza PA, Lennquist A, Leslie HA, Maffini M, Slunge D, Trasande L, Warhurst AM, Muncke J (2019) Overview of known plastic packaging-associated chemicals and their hazards. Sci Total Environ 651:3253–3268. https://doi.org/10.1016/j.scitotenv.2018.10.015

Gu JD, Ford TE, Mitton DB, Mitchell R (2000) Microbial corrosion of metals. The uhlingn corrosion hand book. 2 Edition. Whiley, New York, USA, pp 915–927

Hadad D, Geresh S, Sivan A (2005) Biodegradation of polyethylene by the thermophilic bacterium Brevibacillus borstelensis. J Appl Microbiol 98:1093–1100. https://doi.org/10.1111/j.1365-2672.2005.02553.x

Hadria HE, Gigault J, Maxit B, Grassla B, Reynaud S (2020) Nanoplastic from mechanically degraded primary and secondary microplastics for environmental assessments NanoImpact. S2452–0748(19)30115–6. https://doi.org/10.1016/j.impact.2019.100206

Hakkarainen M, Karlsson S, Albertsson AC (2000) Rapid (bio)degradation of polylactide by mixed culture of compost microorganisms–low-molecular weight products and matrix changes. Polymer 41:2331–2338. https://doi.org/10.1016/S0032-3861(99)00393-6

Hartline NL, Bruce NJ, Karba SN, Ruff EO, Sonar SU, Holden PA (2016) Microfiber masses recovered from conventional machine washing of new or aged garments. Environ Sci Technol 50(21):11532–11538. https://doi.org/10.1021/acs.est.6b03045

Henton ED, Gruber P, Lunt J, Randall J (2005) Polylactic acid technology natural fibers. J Biopolym Biocomposites. 16:527–577

Ho KLG, Pometto AL, III Hinz PN (1999) Effects of temperature and relative humidity on polylactic acid plastic degradation. J Environ Polym Degrad 7:83–92. 1021808317416

Hodson ME, Duffus-Hodson CA, Clark A, Prendergast-Miller MT, Thorpe K (2017) Plastics in agriculture: sources and impacts. L., Plastic bag derived-microplastics as a vector for metal exposure in terrestrial invertebrates. Environ Sci Technol 51:4714–4721. https://doi.org/10.1021/acs.est.7b00635

Huang Y, Liu Q, Jia W, Yan C, Wang J (2020) Agricultural plastic mulching as a source of microplastics in the terrestrial environment. Environ Pollut 260:114096. https://doi.org/10.1016/j.envpol.2020.114096

Huang D, Xu Y, Le,i F, Yu X, Ouyang Z, Chen Y, Jia H, Guo X (2021) Degradation of polyethylene plastic in soil and effects on microbial community composition. J Hazard Mater 416:126173. https://doi.org/10.1016/j.jhazmat.2021.126173

Issac MN, Kandasubramanian B (2021) Effect of microplastics in water and aquatic systems. J Environ Sci Pollut 28:19544–19562. https://doi.org/10.1007/s11356-021-13184-2

Jayasekara R, Harding I, Bowater I, Lonergan G (2005) Biodegradability of a selected range of polymers and polymer blends and standard methods for assessment of biodegradation. J Pol Env 13(3). https://doi.org/10.1007/s10924-005-4758-2

Kale G, Kijchavengkul T, Auras R, Rubino M, Selke ES, Singh PS (2007) Compostability of bioplastic packaging materials. J Macromol Biosci 7(3):255–277. https://doi.org/10.1002/mabi.200600168

Kang J, Zhou L, Duan X, Sun H, Ao Z, Wang S (2019) Degradation of cosmetic microplastics via functionalized carbon nanosprings. Matter 745–758. https://doi.org/10.1016/j.matt.2019.06.004

Karbalaei S, Hanachi P, Walker TR, Cole M (2018) Occurrence, sources, human health impacts and mitigation of microplastic pollution. Environ Sci Pollut Res 25(36):36046–36063. https://doi.org/10.1007/s11356-018-3508-7

Karigar CS, Rao SS (2011) Role of microbial enzymes in the bioremediation of pollutants: a review. Enzyme Res 2011:805187. https://doi.org/10.4061/2011/805187

Kasirajan S, Ngouajio M (2012) Polyethylene and biodegradable mulches for agricultural applications: a review Agron. Sustain Dev 32:501–529. https://doi.org/10.1007/s13593-011-0068-3

Kawai F, Watanabe M, Shibata F, Yokoyama S, Sudate S, Hayashi S (2004) Comparative study on biodegradability of polyethylene by bacteria an dfungi. Polym Degrad Stabil 86:105–114. https://doi.org/10.1016/j.polymdegradstab.2004.03.015

Kawai F, Watanabe M, Shibata M, Yokoyama S, Sudate Y (2002) Experimental analysis and numerical simulation for biodegradability of polyethylene. Polym Degrad Stab 6:129e35. https://doi.org/10.1016/S0141-3910(02)00006-X

Kerr C (1994) Reviews. J Educ 49(4):78–89. https://doi.org/10.1177/107769589404900411

Khan KA, Majeed T (2019) Biodegradation of synthetic and natural plastics by microorganisms. J Nat Appl Sci Pakistan 1(2):180–184. Thakur M, Medintz LI, Walper AS (2019) Enzymatic bioremediation of organophosphate compounds—progress and remaining challenges. J Front Bioeng Biotechnol 7:1–21. https://doi.org/10.3389/fbioe.2019.00289

Khatoon N, Pandey RK, Prajapati VK (2017) Exploring Leishmania secretory proteins to design B and T cell multi-epitope subunit vaccine using immunoinformatics approach. Sci Rep 7:8285. https://doi.org/10.1038/s41598-017-08842-w

Kijchavengkul T, Auras R (2008) Compostability of polymers. J Polym 57:793–804. https://doi.org/10.1002/pi.2420

Klein S, Worch E, Knepper TP (2015) Occurrence and spatial distribution of microplastics in river shore sediments of the rhine-main Area in Germany. Environ Sci Technol 49:6070–6076. https://doi.org/10.1021/acs.est.5b00492

Kole PJ, Löhr AJ, Van Belleghem F, Ragas A (2017) Wear and tear of tyres: a stealthy source of microplastics in the environment. Int J Environ Res Public Health 14(10):1265. https://doi.org/10.3390/ijerph14101265

Kumar M, Xiong X, He M, Tsang DCW, Gupta J, Khan E, Harrad S, Hou D, Ok YS, Bolan NS (2020) Microplastics as pollutants in agricultural soils. Environ Pollut 265:114980. https://doi.org/10.1016/j.envpol.2020.114980

Kunlere IO, Fagade OE, Nwadike BI (2019) Biodegradation of low density polyethylene (LDPE) by certain indigenous bacteria and fungi. Int J Environ Stud 76:3:428–440. https://doi.org/10.1080/00207233.2019.1579586

Lares M, Ncibi MC, Sillanpää M, Sillanpää M (2018) Occurrence, identifcation and removal of microplastic particles and fbers in conventional activated sludge process and advanced MBR technology. Water Res 133:236–246. https://doi.org/10.1016/j.watres.2018.01.049

Li J, Zhang K, Zhang H (2018) Adsorption of antibiotics on microplastics. Environ Pollut 237:460–467. https://doi.org/10.1016/j.envpol.2018.02.050

Lu Y, Zhang Y, Deng Y, Jiang W, Zhao Y, Geng J, Ding L, Ren H (2016) Uptake and accumulation of polystyrene microplastics in zebrafish (danio rerio) and toxic effects in liver. Environ Sci Technol 50:4054–4060. https://doi.org/10.1021/acs.est.6b00183

Lucas N, Bienaime C, Belloy C (2008) Polymer biodegradation: mechanisms and estimation techniques–a review. Chemosphere 73(4):429–442. https://doi.org/10.1016/j.chemosphere.2008.06.064

Lugauskas A, Levinskaitė L, Peciulyte D (2003) Micromycetes as deterioration agents of polymeric materials. Int Biodeterior Biodegrad 52. https://doi.org/10.1016/S0964-8305(03)00110-0

Malla MA, Dubey A, Kumar A, Yadav S (2023) Unlocking the biotechnological and environmental perspectives of microplastic degradation in soil-ecosystems using metagenomics 170:372–379. https://doi.org/10.1016/j.psep.2022.11.084

Mao Y, Ai H, Chen Y, Zhang Z, Zeng P, Kang L, Li W, Gu W, He Q, Li H (2018) Phytoplankton response to polystyrene microplastics: perspective from an entire growth period. Chemosphere 208:59–68. https://doi.org/10.1016/j.chemosphere.2018.05.170

Martinez-Gomez C, Leon VM, Calles S, Gomariz-Olcina M, Vethaak AD (2017) The adverse effects of virgin microplastics on the fertilization and larval development of sea urchins. Mar Environ Res 130:69–76. https://doi.org/10.1016/j.marenvres.2017.06.016

Moller JN, Loder MGJ, Laforsch C (2020) Finding microplastics in soils: a review of analytical methods. Environ Sci Technol 54:2078–2090. https://doi.org/10.1021/acs.est.9b04618

Murugan R, Beggi F, Kumar S (2014) Belowground carbon allocation by trees, understory vegetation and soil type alter microbial community composition and nutrient cycling in tropical Eucalyptus plantations. Soil Biol Biochem 76:257–267. https://doi.org/10.1016/j.soilbio.2014.05.022

Naji A, Azadkhah S, Farahani H, Uddin S, Khan FR (2021) Microplastics in wastewater outlets of Bandar Abbas city (Iran): a potential point source of microplastics into the Persian Gulf, Chemosphere, 262(128039). ISSN 0045–6535. https://doi.org/10.1016/j.chemosphere.2020.128039

Nishida M, Tanaka T, Hayakawa Y, Nishida M (2018) Solid-state nuclear magnetic resonance (NMR) and nuclear magnetic relaxation time analyses of molecular mobility and compatibility of plasticized polyhydroxyalkanoates (PHA). Polymers 10:506. https://doi.org/10.3390/polym10050506

Ojha N, Pradhan N, Singh S, Barla A, Shrivastava A, Khatua P, Rai V, Bose S (2017) Evaluation of HDPE and LDPE degradation by fungus, implemented by statistical optimization. Sci Rep 7:39515. https://doi.org/10.1038/srep39515

Onodera YK, Mukumoto H, Katsuyaya Y, Saiganji A, Tani Y (2001) Degradation of polyethylene by a fungus Penicillium Simplicissimum YK. J Polym Degrad 72(2):323–327. https://doi.org/10.1016/S0141-3910(01)00027-1

Orr IG, Hadar Y, Sivan A (2004) Colonization, biofilm formation and biodegradation of polyethylene by a strain of Rhodococcus ruber. Appl Microbiol Biotechnol 65(1):97–104. https://doi.org/10.1007/s00253-004-1584-8

Otsuka K, Pulec JL, Suzuki M (2003) Assessment of intravenous lidocaine for the treatment of subjective tinnitus. Ear Nose Throat J 82(10):781–784. https://doi.org/10.1177/014556130308201011

Phaniendra A, Jestadi DB, Periyasamy L (2014) Free radicals: properties, sources, targets, and their implication in various diseases. Indian J Clin Biochem 30(1):11–26. https://doi.org/10.1007/s12291-014-0446-0

Restrepo-Florez JM, Bassi A, Thompson MR (2014) Microbial degradation and deterioration of polyethylene–a review. Int Biodeterior Biodegrad 88:83–90. https://doi.org/10.1016/j.ibiod.2013.12.014

Rillig MC, Lehmann A, de Souza Machado AA, Yang G (2019) Microplastic effects on plants. New Phytol. https://doi.org/10.1111/nph.15794

Sangale MK, Shahnawaz M, Ade AB (2019) Gas chromatography-mass spectra analysis and deleterious potential of fungal based polythene-degradation products. Sci Rep 9:1599. https://doi.org/10.1038/s41598-018-37738-6

Sarker M, Rashid MM (2013) Thermal degradation of poly (ethylene terephthalate) waste soft drinks bottles and low density polyethylene grocery bags. Int J Sustain Energy Environ 1(3)78–86, 2327–0330

Sarma H, Nava AR, Manriquez AME, Dominguez DC, Lee WY (2019) Biodegradation of bisphenol A by bacterial consortia isolated directly from river sediments. Environ Technol Innov 14:100314. https://doi.org/10.1016/j.eti.2019.01.008

Schrank I, Möller JN, Imhof HK, Hauenstein O, Zielke F, Agarwal S, Löder MGJ, Greiner A, Laforsch C (2022) Microplastic sample purification methods-assessing detrimental effects of purification procedures on specific plastic types. Sci Total Environ 833:154824. https://doi.org/10.1016/j.scitotenv.2022.154824

Shabaka SH, Ghobashy M, Mareya RS (2019) Identification of marine microplastics in Eastern Harbor, Mediterranean Coast of Egypt, using differential scanning calorimetry. Mar Pollut Bull 142:494–550. https://doi.org/10.1016/j.marpolbul.2019.03.062

Shah AA, Hasan F, Hameed A, Ahmed S (2008) Biological degradation of plastics: a comprehensive review. Biotechnol Adv 26(3):246–265, 0734–9750. https://doi.org/10.1016/j.biotechadv.2007.12.005

Sowmya HV, Ramalingappa KM (2014) Degradation of polyethylene by Trichoderma harzianum—SEM, FTIR, and NMR analyses. Environ Monit Assess 186:6577–6586. https://doi.org/10.1007/s10661-014-3875-6

Sridharan S, Kumar M, Singh L, Bolan NS, Saha M (2021) Microplastics as an emerging source of particulate air pollution: a critical review. J Hazard Mater 418:126245. https://doi.org/10.1016/j.jhazmat.2021.126245

Talvitie J, Mikola A, Setälä O, Heinonen M, Koistinen A (2017) How well is microlitter purifed from wastewater? A detailed study on the stepwise removal of microlitter in a tertiary level wastewater treatment plant. Water Res 109:164–172. https://doi.org/10.1016/j.watres.2016.11.046

Temporiti MEE, Nicola L, Nielsen E, Tosi S (2022) Fungal enzymes involved in plastics biodegradation. Microorganisms 8 10(6):1180. https://doi.org/10.3390/microorganisms10061180

Thakur S, Mathur S, Patel S, Paital B (2022) Microplastic accumulation and degradation in environment via biotechnological approaches. Water 14(24):4053. https://doi.org/10.3390/w14244053

Tiwari N, Santhiya D, Sharma JG (2020) Microbial remediation of micro-nano plastics: current knowledge and future trends. Environ Pollut 265:115044. https://doi.org/10.1016/j.envpol.2020.115044

Tokiwa Y, Calabia PB, Ugwu UC, Aiba S (2009) Biodegradability of plastics. Int J Mol Sci 10(9):3722–3742. https://doi.org/10.3390/ijms10093722

Urbanek AK, Kosiorowska KE, Mirończuk AM (2021) Current knowledge on polyethylene terephthalate degradation by genetically modified microorganisms. Front Bioeng Biotechnol Sec Bioprocess Eng 9. https://doi.org/10.3389/fbioe.2021.771133

Van den Berg P, Huerta-Lwanga E, Corradini F, Geissen V (2020) Sewage sludge application as a vehicle for microplastics in eastern Spanish agricultural soils. Environ Pollut 261:114198. https://doi.org/10.1016/j.envpol.2020.114198

Vignesh R, Deepika RC, Manigandan P, Janani R (2016) Screening of plastic degrading microbes from various dumped soil samples. Int Res J Eng Tech 3(4):2493–2498

Wagner M, Scherer C, Alvarez-Munoz D, Brennholt N, Bourrain X, Buchinger S, Fries E, Grosbois C, Klasmeier J, Marti T et al (2014) Microplastics in freshwater ecosystems: what we know and what we need to know. Environ Sci Eur 26:12. https://doi.org/10.1186/s12302-014-0012-7

Walker TR (2018) Drowning in debris: solutions for a global pervasive marine pollution problem. Mar Pollut Bull 126:338. https://doi.org/10.1016/j.marpolbul.2017.11.039

Waller CL, Griffiths HJ, Waluda CM, Thorpe SE, Loaiza I, Moreno B, Pacherres CO, Hughes KA (2019) Microplastics in the Antarctic marine system: an emerging area of research. Sci Total Environ 2017(598):220–227. https://doi.org/10.1016/j.scitotenv.2017.03.283

Webb JS, Nixon M, Eastwood IM, Greenhalgh M, Robson GD, Handley PS (2000) Fungal colonization and biodeterioration of plasticized polyvinyl chloride. Appl Environ Microbiol 66(8):3194–3200

Wei R, Zimmermann W (2017) Microbial enzymes for the recycling of recalcitrant petroleum-based plastics: how far are we? Microb Biotechnol 10(6):1308–1322. https://doi.org/10.1111/1751-7915.12710

Weithmann N, Möller JN, Löder MGJ, Piehl S, Laforsch C, Freitag R (2018) Organic fertilizer as a vehicle for the entry of microplastic into the environment. Sci Adv 4. https://doi.org/10.1126/sciadv.aap8060

Wright RJ, Bosch R, Langille MGI, Gibson MI, Christie-Oleza JA (2021) A multi-OMIC characterisation of biodegradation and microbial community succession within the PET plastisphere. Microbiome 9:141. https://doi.org/10.1186/s40168-021-01054-5

Woo S, Song I, Cha HJ (2020 Sep) Fast and Facile Biodegradation of Polystyrene by the Gut Microbial Flora of Plesiophthalmus davidis Larvae. Appl Environ Microbiol 1:86(18):e01361–20. https://doi.org/10.1128/AEM.01361-20. PMID: 32631863; PMCID: PMC7480385

Yang HS, Wolcott M, Kim HS, Kim HJ (2005) Thermal properties of lignocellulosic filler-thermoplastic polymer bio-composites. J Therm Anal Calorim 82:157–160. https://doi.org/10.1007/s10973-005-0857-5
Yang L, Zhang Y, Kang S, Wang Z, Wu C (2021) Microplastics in soil: a review on methods, occurrence, sources, and potential risk. Sci Total Environ 146546. https://doi.org/10.1016/j.scitotenv.2021.146546
Yoshida S, Hiraga K, Takehana T, Taniguchi I, Yamaji H, Maeda Y, Toyohara K, Miyamoto K, Kimura Y, Oda K (2016) A bacterium that degrades and assimilates poly (ethylene terephthalate). Science 351(6278):1196–1199
Yurtsever M, Kaya AT (2018) A research on microplastic presence in outdoor air. Senem Çiftçi Bayraktar 89–97. https://doi.org/10.1007/978-3-319-71279-6_13
Zhang GS, Liu YF (2018) The distribution of microplastics in soil aggregate fractions in southwestern China. Sci Total Environ 642:12–20. https://doi.org/10.1016/j.scitotenv.2018.06.004
Zhang L, Xie Y, Liu J, Zhong S, Qian Y, Gao P (2020) An overlooked entry pathway of microplastics into agricultural soils from application of sludge-based fertilizers. Environ Sci Technol 54:4248–4255. https://doi.org/10.1021/acs.est.9b07905?ref=pdf
Zhang M, Zhang Y, Li C, Jing N, Shao S, Wang F, Mei H, Rogers KM, Kong X, Yuan Y (2023) Identification of biodegradable plastics using differential scanning calorimetry and carbon composition with chemometrics. J Hazard Mater 10:100260. https://doi.org/10.1016/j.hazadv.2023.100260
Zhou Y, Wang J, Zou M, Jia Z, Zhou S, Li Y (2020) Microplastics in soils: a review of methods, occurrence, fate, transport, ecological and environmental risks. Sci Total Environ 748:141368. https://doi.org/10.1016/j.scitotenv.2020.141368

Chapter 7
Biodegradation of 2,4,6-Trinitrotoluene (TNT) by the Microbes and Their Synergistic Interactions

Jyoti Lamba, Dinesh Bhardwaj, Shalini Anand, Jayanti Dutta, and Pramod Kumar Rai

7.1 Introduction

Explosive contamination is the foremost alarm in association with environmental well-being. Inappropriate discarding of remnants from military establishments and industries leads to critical disorder in terrestrial along with adjacent aquatic resources (Nagar et al. 2021, 2018). This added to the adverse consequences on the inhabitants' health (Nagar et al. 2020). Hence, supervision of explosive dumping expertly, remedially, and ecologically has become a topmost alarm.

Environmental pollution via explosives is largely by explosive manufacturing. Nevertheless, the manufacturing and practice of explosives universally are inevitable thus adding to ground contamination (Kalderis et al. 2011). Topsoil contamination with the explosive residue's precipitation is not a solitary concern. Because of the enormously unstable compounds, explosives are rapid enough to spread into the local water bodies by making their pathway through leaching and settling down into the groundwater. Further, because of the continuous conversion of the compounds, eventually, they erect their route to the neighboring freshwater source, troubling perhaps the whole watershed. Even with minimal solubility, high explosives turn out to be enthusiastically mobile on suspension (Lynch et al. 2001). Energetic compounds

J. Lamba
Department of Environment Studies, Panjab University, Chandigarh 160 014, India

J. Lamba · D. Bhardwaj · S. Anand (✉) · P. K. Rai
Centre for Fire, Explosive and Environment Safety (CFEES), Defence Research Development Organization, DRDO, Timarpur, Delhi 110054, India
e-mail: shalinianand1975@rediffmail.com

J. Dutta
Human Resource Development Centre, Panjab University, Chandigarh 160 014, India

K. Bala et al. (eds.), *Harnessing Microbial Potential for Multifarious Applications*, Energy, Environment, and Sustainability, https://doi.org/10.1007/978-981-97-1152-9_7

come up with mutagenic and cancer-causing properties turning them into a noxious threat to the biota (Lata et al. 2021).

In consideration of possible health risks to human beings and the environment, the U.S. Environmental Protection Agency (USEPA) judiciously selected numerous spots from their permissible perimeter in air, water, and soil. However, human evolution escape from the utilitarian backing of explosives is hard, but suitable practices for remediation are meant for defending the ecosystem (Chatterjee et al. 2017). In the existing scenario, environmental laws have *turned* out to be strict in the direction of well-being, economy, and decrease of contamination (Sathya et al. 2023). Henceforth for explosive remediation, green, cost-effective, and ecological technology is the need of the hour.

7.1.1 Explosives

Explosives, existing as chemical complexes comprises high oxygen and nitrogen contents, which on explosion expand to a tremor-like movement that produces a severe burden on the bordering environment, causing an explosion and leaving deadly leftovers in the environment (Singh et al. 2012). Massive volumes of hot gases for instance, "carbon dioxide, carbon monoxide, nitrogen, oxygen, and water vapor" are released immediately after an explosive explosion (i.e., at a speed of kilometer per second). The key choices of the army are HMX, RDX, and TNT, because of their energy. Although, "nitrocellulose (NC), 2,4-dinitrotoluene (DNT), nitroglycerin (NG), and nitroguanidine (NQ)" are being opted for rocket, missile, and gun propellants.

7.1.2 TNT and Its History

2,4,6-trinitrotoluene, TNT (molecular formula $C_7H_5N_3O_6$) is an enormously volatile single-ring nitroaromatic complex carrying three nitro groups. Although it was synthesized as a yellow dye by Joseph Wilbrand, a German chemist in 1863, however till 1902, TNT was not recognized as an explosive (Gad 2014).

In 1902 only, the wide casting of TNT as an explosive took place (Higson 1992) and was employed instead of picric acid by Germans against the British in armor-piercing artillery shells. At the preliminary stage of World War I, TNT was bulk-produced at the topmost level. TNT was massively manufactured together at commercial as well as at government ammunition plants throughout World War I and II (USEPA 2014) with utmost TNT manufacturing during World War II.

Generally, TNT was synthesized using the conventional nitration method which was carried out on a large scale in batches. Formation reactions are susceptible to runway exothermic reactions which are important for safety concerns. Recently it has

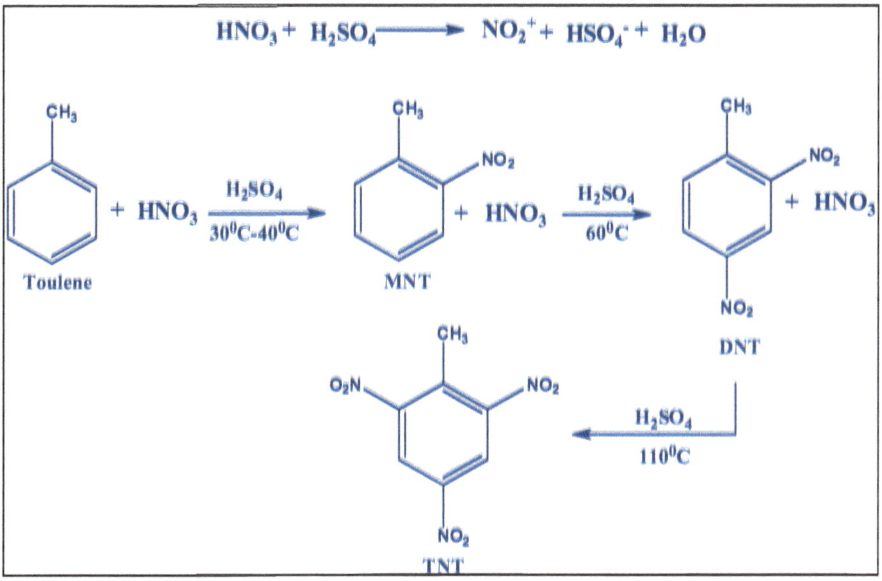

Fig. 7.1 Synthesis route of TNT

been reported TNT can also be synthesized by DNT as a reactant using flow chemistry via an electrophilic aromatic substitution mechanism as depicted in Fig. 7.1 (Kyprianou et al. 2020). Flow chemistry methods are found to be safer and can be carried out easily at higher temperatures as compared to the conventional method. During the addition of nitric acid and sulphuric acid in a nitration reaction high amount of heat is dissipated which can be easily controlled by using flow chemistry methodology by performing chemical reactions with constant controlled temperature and controlled pressure. By this methodology, high purity of products can be obtained.

Manufacturing of TNT results in the production of wastewater in large *volumes. This* wastewater contains numerous derivatives along with additional nitrobenzene by-products. Such derivatives further convert the wastewater into a dark red color, *known as* TNT red wastewater (Bui et al. 2021).

7.1.3 Fate and Transport, and Toxicity of TNT

Fate and transport of TNT are directed by the physio-chemical properties of TNT, for instance, Henry's law constant, solubility, and vapor pressure (Table 7.1). Determination of suppressed TNT on the soil surface depends upon numerous characteristics such as microbial action, incubation time, soil organic matter, and concentration of O_2 and TNT. The cause of hydrogen bond formation with H_2O is due to the nitro

Table 7.1 Physio-chemical properties of TNT

Property	Value
Chemical abstract service	118–96-7
Physical state (at room temperature)	Crystalline, odorless solid
Color	Yellow
Molecular weight (g/mol)	227.13
Water solubility at 20 °C (mg/L)	130
Octanol–water partition coefficient (Low K_{ow})	1.6
Soil organic carbon–water coefficient (Log K_{oc})	300
Boiling point (°C)	240 (explodes)
Melting point (°C)	80.1
Specific gravity	1.654
Henry's law constant (atm-m^3/mol at 20 °C)	4.57×10^{-7}

(USEPA 2014; National Center for Biotechnology Information 2023)

groups (NO_2^-) presence which further dictates solubility of TNT in aqueous medium accompanied by its affinity for exciting surfaces. This leads to TNT transportation to ground and surface water, along with the aquatic ecosystem. During fine or colloidal particle existence, TNT affinity predominantly with high cation exchange capacity upsurges.

Solubility of nitroaromatic, i.e., TNT is higher in contrast to nitroamines (like RDX and HMX), i.e., 130 mg/L, accordingly, it has the proficiency to dissolve steadily with time causing groundwater pollution. TNT proliferation in oceans has turned out to be a worldwide problem too (Xu et al. 2023). Processes such as dissolution and adsorption manage TNT passage although volatilization alleyway is insignificant due to TNT's low vapor pressure. In direct photolysis, TNT may either dissolve or stay as an unhindered product in aqueous media or sorb on a alive or degraded substance. Fundamentally, photolysis is the absorption of light energy by TNT, accompanied by its derivatives, and this practice is extremely structured via light intensity and wavelength. For TNT-polluted soils, its hydrolysis can be selected as one of the therapeutic conditionals however reaction rate and satisfactory pH situations are the main constraints. During photolysis, the primary TNT photodegradation product is 1,3,5-Trinitrobenzene (1,3,5-TNB). TNT showed elevated sorption affinity of 170 L/Kg as compared to RDX and HMX sorption affinity, i.e., 8 and 30 L/kg. Irreversible sorption is an outcome of TNT covalent bond binding on the way to soil and humic constituents further adding to the difficulty level for TNT removal. Hence, biodegradation is the lone achievable selection over hydrolysis and photolysis. The TNT breakdown amid water is evident by means of biodegradation but at a significantly gradual rate as contrasted to photolysis. TNT degradation follows a pathway involving intermediates. These intermediates yet again get blended along soil organic matter and further experience added oxidation which continuously leaches and pollutes groundwater (Pichtel 2012; Wang et al. 2002).

In accordance with EPA IRIS, 1993, TNT is categorized as a weight-of-evidence carcinogenic C (probable human carcinogenic). On 19 December 2008, intended for the Safe Drinking Water and Toxic Enforcement Act (1986), the California Office of Environmental Health Hazard Assessment listed TNT as a cancer-causing toxic chemical (USEPA 2014). The synthesis and unserviceable munitions both liberate TNT (Agency for Toxic Substances and Disease Registry (ATSDR) 1995). EPA National Priorities List (NPL) recognized TNT at over 30 sites since 2013 (EPA 2013). Bioaccumulation of TNT does not appear to take place in animals, but in plants (counting garden, aquatic, and wetland plants) it is used up and broken down. The reason that TNT has a low octanol–water partition coefficient (K_{ow}) and experimental bioconcentration factor, TNT bioconcentration is not looked ahead to high levels in the exposed aquatic organism's tissues and plants (Agency for Toxic Substances and Disease Registry (ATSDR) 1995; HSDB 2012).

During the twentieth century, over 17,000 cases of TNT toxic conditions resulting in more than 475 losses of life were recorded throughout production procedures during World War I (Bodeau 1993; USEPA 2014). The prime route source for TNT introduction is the production sites due to the manufacturing, loading, packaging, or discharging of TNT (USEPA 2014). Historical marine environment introduction to 2,4,6-trinitrotoluene (TNT) turns out as a result of residual munitions throwing away (Kundu et al. 2023). Koske et al. 2019 examined lethal and sub-lethal consequences of TNT and its intermediate products to Zebrafish embryos (*Danio rerio*) with an exposure of 120 h and the results confirmed genotoxicity in fish embryos in vivo exposure to Dust inhalation, digestion, and epidermal sorption of TNT particulates, fallouts to substantial health effects such as liver necrosis and aplastic anemia (HSDB 2012). TNT carcinogenicity concerning humans has restricted confirmations, even though in female rats, urinary bladder carcinoma has been spotted. Outcomes from animal studies exhibited injurious effects on the liver, blood, immune, and reproductive systems after the intake of TNT. The breakdown of TNT into metabolites can be realized at the minute in the liver as verified by their existence in urine samples (USEPA 2014). Malfunctioning of the liver due to extreme TNT amount in the air was observed amidst the labors employed at manufacturing spots. Some workers felt skin itchiness and got cataracts later in long-term contact with the skin and eyes (Massachusetts Military Reservation (MMR) 2001). TNT contact to biomarker responses in mussels, although effects are also documented after extended exposure even though no death ensues (Schuster et al. 2021). Evidence concerning human birth defects has not been shown yet. Even though, vital reproductive system impacts have developed by the male rats treated with high doses of TNT (HSDB 2012). TNT properties have been summarized in Table 7.1

7.1.4 TNT Degradation

With the passage of time, awareness of the toxicity and hazards related to TNT compounds has amplified, and this contributed to the development of numerous cost-effective technologies associated with TNT remediation. Various physical, chemical, and biological approaches used in the last few decades for TNT treatment are discussed in Table 7.2.

As soon as explosives and propellants waste are not appropriately disposed of; the environment along with human well-being are in danger because such waste even now contains ignition and detonation properties. Environmental pollution consequences from conservative practices such as incineration and landfilling. "Biodegradation" is a green demolition kind of technology that proposes the benefits of reasonable investment, effective outcomes, and a harmless and successful decomposition method (Li et al. 2023). One remediating methodology for contaminated soil and water is via microorganisms as agents of bioremediation, i.e., an ecologically approachable and affordable substitute (Gupta et al. 2022).

7.1.5 Biological Treatment of TNT

Pollutant degradation either into a lesser amount or non-toxic elements with living organisms is recognized as biological treatment (Alexander 1999). Such techniques put into practice algae, bacteria, fungi, or higher living beings like floras and deliver potential remediation for contaminated water and soil. In the evaluation of conservative physicochemical methods, as explained directly above, biological treatment is an idiosyncratic reduction procedure that enhances the green eradication of pollutants. Numerous aspects can interrupt the biodegradation method to say molecule's nature may be considered to be reduced (e.g., charge, molecule number, proportions, and functional group's location, solubility, and toxicity) along with the ecological conditions (pH, organic content temperature, and soil moisture). As stated by Kalderis et al. (2011), the bioremediation procedure is justifiable by being green and cost-effective. Therefore, bioremediation develops the valuable potential for explosives degradation.

7.1.5.1 Phytoremediation of TNT

Phytoremediation is a profitable, ecological process for explosive remediation. Herein, the plant's capability is used to flourish in explosive pollutant existence, to tolerate their harmfulness, and to supplementary remediate soil with the support of precise plant enzymes. The process is based on phyto-degradation, phyto-extraction, phyto-stabilization, or phyto-volatization, which is a largely discovered field along with plentiful current literature on explosive pollution remediation (Brentner et al.

Table 7.2 Physico-chemical methods

Parameters	Adsorption	Electrolysis and electrocatalytic	Photolysis	Bi-metallic reduction
Overview	Mass transfer practice wherein fluid elements are deposited upon the adsorbent surface via chemical or physical forces (electrostatic interaction, hydrogen bonding, Vander Wall forces) (Hinshaw et al. 1987) Materials employed in adsorption-based water treatment methods are activated carbon, alumina zeolites, and ion-interchange resins	The process of sorting out ions through the assistance of electrons flow from anion to cation (i.e., negative to positive) to boost interaction reactions amid two phases (Kirk–Othmer 1993) The electrochemical reduction process is potentially energy efficient and operable at ambient conditions. Transition metals were used as a catalyst to degrade TNT (Wong et al. 2023)	The chemical procedure is used to break down the molecules into reduced units all through the light absorption phenomenon. The formation of hydroxyl radicals in water is encouraged by UV radiation that results in energetic material degradation within the solution (Alnaizy and Akgerman 1999)	Zerovalent iron (ZVI) can be utilized for the remediation of groundwater. ZVI acts as a reducing agent and works through electron transfer from Fe° to the pollutant

(continued)

Table 7.2 (continued)

Parameters	Adsorption	Electrolysis and electrocatalytic	Photolysis	Bi-metallic reduction
Efficiency	In the course of the process, devoid of any modification explosives transformation proceeds from one medium to another. Activated carbon is a non-specific adsorbent that can remediate both the contaminated groundwater as well as soil	Explosives like 3-Nitro-1,2,4-triazole-5-one (NTO) and TNT has only been studied with this process (Wallace et al. 2009) which points toward the efficacy of electrolysis along with other nitroaromatic compounds	The success of the process is being stated against RDX analysis (Bordeleau et al. 2013). Mahbub and Nesterenko (2016) stated that nitroaromatic compounds breakdown through ring cleavage, and as a consequence of the nitro group loss following compounds are formed NH_2CHO, HCHO, HCOOH, N_2O, NO^{2-} and NO^{3-}	This technique was tested to examine explosive degradation (Wanaratna et al. 2006) covering TNT, RDX, NTO, and DNAN with varying concentrations (48, 50, 410, and 430 mg/l, respectively) (Koutsospyros et al. 2012)
Hindrance	Expensive, incomplete explosives degradation, retaining of untreated nitro compounds, and prerequisite for supplementary nutrients	High-cost operation and potential space necessity for the system	Involves precisely skilled individual and management facilities for hydrogen peroxide that further uplifts the running costs of the treatment	Metallic particles themselves have the potential to be lost and get discharged along the treated water which further harms the environment

2008; Burken et al. 2000; Gunning et al. 2014; Palazzo and Leggett 1986; Vila et al. 2007). A few of the commonly used diversity of plant systems in the procedures are bush bean, aquatic and wetland species, yellow nutsedge, switch grass, and hybrid polar. Combinations of plant-bacterium to phyto-remediate contaminated soil were invented through the *Pseudomonas* strain against TNT which caused TNT breakdown to amino compounds (Thenmozhi and Devasena 2019). However, this process has been limiting in plant category choice, maintenance glitches, contaminants phytotoxicity, sluggish degradation regularity, unspecified effects of the formed by-products all over the process in the accumulation with the hazard of pollutants invading the food chain restricts its usage for on-the-spot remediation (Hannink et al. 2002). In reference to this process, only explosives transformation is reported, even though in a few cases mineralization at a slow rate also happened. In conclusion, in the environment, whole remediation of explosives does not occur.

7.1.5.2 Mycoremediation of TNT

As stated by Hawari (2000), in most of the fungal biodegradation reports *Phanerochaete chrysosporium*, a white-rot fungus has been focused and reported to degrade a legion of organic contaminants through its non-specific lignin-degrading system. Few reports conclude that under aerobic conditions, white-rot fungi species such as *Phanerochaete chrysosporium* and *Stropharia* species can mineralize TNT (Esteve-Nunez et al. 2001). The reactions associated with TNT conversion via fungi are generally lignin-degrading enzymes. The utmost capable and effective pathway for the release of nitrogen from TNT is the conversion of the parental complex via hydride ions accumulation fallouts in the Meisenheimer *hydride complexes* formation along with their succeeding devastation. Between the aromatic ring, hydrogenation products, i.e., hydride-Meisenheimer complexes were spotted (Ziganshin et al. 2015). The capability of *Yarrowia lipolytica* AN-L15 to form hydride complexes due to aromatic ring breakdown followed through their de-nitration, specifies it as a probable nominee for explosive bioremediation at polluted sites. In the studies conducted subsequently by researchers, the degradation of TNT compounds by *T. viride* proceeded via biotransformation mechanisms that enable the exclusion of nitrogen and recyclization of the slashed compounds (Alothman et al. 2020). However, studies are lacking and may be steered to clearly identify the enzymes and consistent genes accountable for the ability of the fungus to degrade and remediate TNT explosives.

7.1.5.3 Composting for TNT Remediation

Composting is a traditional practice where the simple combination of polluted soil along with bulking agents (wood chips, fodder, straw) and organic substrates (cattle manure, waste from vegetables) for a specific time period supports explosive exclusion (Thenmozhi and Devasena 2019). The resultant soil and modifications combination can be distributed in the piles or windrows form. The bulking agent's role is generally to deliver porosity, aeration, and nourishment, however, the organic modification benefits in keeping an optimal carbon–nitrogen ratio in the direction to encourage biotic movement. An additional significant highlight of composting is temperature raising which assists in microorganisms' catabolic actions for explosive degradation (Chaudhary and Kim 2019).

Decontamination of TNT from polluted soils using municipal solid wastes as a feasible exercise has been investigated (Rezaei et al. 2010). A mix of TNT-contaminated soil, cow manure, and microbial suspension with 50,000 (mg/Kg) TNT presence was treated. TNT elimination (99.99%) was gained within 15 days. Complete dissipation of TNT (100 kg) in a study based on an in-vessel composting system, augmented with TNT-treating bacteria isolates and containing a combination of TNT, organic waste, dung, wood chips, was attained in an aerobic environment in 15 days (Gumuscu et al. 2015). The foremost drawback of the practice is the episodic maintenance of the arrangement (Thenmozhi and Devasena 2019). Composting studies deliver an idea of employing soil microbes for explosives remediation afterward isolation and characterization.

7.1.5.4 Microbial Remediation of TNT

Microbes flaunt an exceptional competence to alter nearly all compounds, viz, both organic and inorganic, for their evolution and hence can inhabit numerous ecological niches (Timmis and Pieper 1999). Microbes have the capability to perform in any circumstances (Mishra et al. 2001) which contributes to establishing microbes as a striking contender for bioremediation methods. Furthermore, microorganisms perform a vital character in biogeochemical cycles along with ecosystem sustainability. In the 1960s, it was tracked down that microbes are capable of breaking down xenobiotic contaminants introduced into the atmosphere (Singleton 1994). The microbial conversion of xenobiotic contaminants can occur either in oxygenic or anoxygenic situations. Nevertheless, among widely held cases, oxygen molecules contribute to the primary conversion of xenobiotic compounds (aliphatic and aromatic) (Sinha et al. 2009). Microbes have been verified to be valuable in countless food, industrial, medicinal, and pharmaceutical sectors (Kumar et al. 2023). Amid numerous microbe groups, bacteria have been considered as utmost effectual and prevailing in the natural bioremediation procedure. Beneath both the circumstances (aerobic and anaerobic), bacteria have established a line of attack for gaining energy effectively from every single compound via electron acceptors (ferric ions, sulfate, nitrate, etc.). A huge range of microorganisms exists, that participates in the aerobic conversion of xenobiotic contaminants. Excess research is accessible for remediation of explosive *ex-situ* (Singh et al. 2012) but then again, the foremost task lies in its conversion to ground level, as maximum isolation misses the mark to continue in real contaminated sites. The foremost advantage of employing an isolate from the contaminated site lies in its capability to continue in situ (Cupples 2013). *Stenotrophomonas* strain SG1's aptitude to convert TNT efficiently beneath aerobic and anaerobic circumstances along with and deprived of supplementary carbon sources and ammonium chloride was tried and the study confirms the TNT elimination aptitude of the strain (Gupta et al. 2023).

Throughout the last century, military and civilian purposes became the chief reason for explosive contamination of the environment due to which visible effects were seen on human and environmental health which further contributes to making it an important pollutant (Anand et al. 2016). There is a necessity for distinctive biological practices for the operative remediation of these pollutants. Microbes are significant players in bioremediation as they can produce enzymes that will ignite degradation reactions and consume pollutant substances as carbon and energy sources for their reproduction and metabolism and to bring out several conversions' responses. An ample amount of trinitrotoluene (TNT) has been allied with previous and current military events universally. Considering the toxicity of this contaminant an investigation with bacterial combinations separated from explosive-soiled location was evaluated in TNT-spiked loam strata for its efficacy if degradation. A reduction in the TNT amount (30%) within a span of 5 months was estimated. A part of TNT can be degraded in strata however the rest of it got attached or positioned on spots that are unreachable to degrading strains. However, the presence of microbial consortia containing specialized microbes enhanced the capabilities of specific degraders, enhancing the extent of mineralization (Robertson and Jjemba 2005). Cabrera et al. (2020) isolated Pseudomonas sp. TNT3 from Deception Island and found that the sp. possesses enzymes that contribute to a diversity of molecular functions associated with xenobiotic removal. The capabilities of marine microbiomes are less explored against TNT. Kundu et al. 2023 studied the potential of *Colwellia, Thalossospira,* and *Methylophaga* and observed that the capacity of these microbes can be enhanced with the addition of Methanol as it boosted the whole degradation process by delivering supplementary energy along with the establishment of syntrophic relationship among methanol-utilizing and TNT-utilizing bacteria. This is suggestive of the utilization of suitable groupings of microbial strains for enhancing TNT release and its consequent mineralization in a struggle to biotreat spots adulterated with this pollutant in a more effective manner.

The microbial remediation method directs to the remediation technology that practices microorganisms by means of precise functions to break down pollutants from harmful to harmless substances all through their particular metabolism within the suitable environment (Xin-feng et al. 2023).

7.2 Bacterial Degradation of TNT

TNT degradation degree and the resultant products may differ conditional on the microorganism type along with the culture conditions. Although, the pathway for de-nitration is beneficial toward microorganisms due to the decline in TNT's electrophilic nature, that further supports them in utilizing substrates, i.e., dioxygenases and mono- or dinitrotoluenes (Kim and Song 2003; Martin et al. 1997). To various aerobic and anaerobic microorganisms, explosive compounds behave as a carbon and nitrogen source (Hawari 2000). A bacterial consortium from the carefully chosen three strains was investigated for TNT deprivation and related degradation metabolites. The consortium changed TNT to 2-amino-4,6-dinitrotoluene, 4-amino-2,6-dinitrotoluene, and an unidentified compound of treatment (Gök et al. 2018). TNT degradation may necessitate collective exertions of numerous enzymes triggering varied metabolic pathways in microbes. Prokaryotic transcriptomics analysis of *Buttiauxella* sp. S19-1 subsequent to TNT acquaintance discovered the increased activity of the gene protocatechuic acid 3, 4-dioxygenase (P34O) and amplified the degradation of both TNT and ADNT catalyzing the ring cleavage of TNT transitional products (Xu et al. 2021). Afterward, the involvement of monooxygenase (MO) in the conversion of 4-hydroxybenzaldehyde to 4-hydroxybenzoate was established (Xu et al. 2023). Simplified steps for TNT mineralization are depicted in Fig. 7.2.

7.2.1 *TNT Degradation in an Aerobic Environment*

Aerobic biodegradation is an exercise of accumulating oxygen in saturated soil and groundwater to upsurge the quantity and strength of indigenous microorganisms to accomplish biodegradation. Oxygen is considered by numerous as a key growth-limiting factor for hydrocarbon-degrading bacteria. *Pseudomonas* can nurture TNT as the sole source of nitrogen. *P. savastanoi* is capable of transforming TNT into 2,4-DNT. In glucose enhanced medium, during degradation production of 4-Amino-2,6-dinitrotoluene (4-A-2,6-DNT) and 2-Amino-4,6-dinitrotoluene (2-ADNT) increases whereas TNT declines via de-nitration. Reports suggest, 4-Amino-2-nitroso-6-nitrotoluene (4A-2NOC-6-NT) is one of the concluding products ensuing aerobic pathway among the *Bacillus sp.*, *Pseudomonas aeruginosa,* and *Staphylococcus sp.* cultures (Esteve-Nunez et al. 2001). *P. aeruginosa* too can break down amino-di-nitro-toluene into polar compounds. A decrease in a nitro group initiates aerobic degradation. The nitro-group stays simpler to break once the N-atom merges its elevated electronegativity and partial positive charge (Claus 2014). By one or two nitro group reductions, aerobic bacteria generally convert the TNT molecule via non-specific NAD(P)H-dependent nitroreductase to hydroxylamine. Hydroxylamine further gets converted to amino groups by nitro reductases, and 2,4 amino-2-hydroxylamine-6-nitrotoluene or 2 amino-4,6-dinitrotoluene forms. In addition, with 2,4 amino-2 hydroxylamino-6-nitrotoluene, supplementary complexes also formed

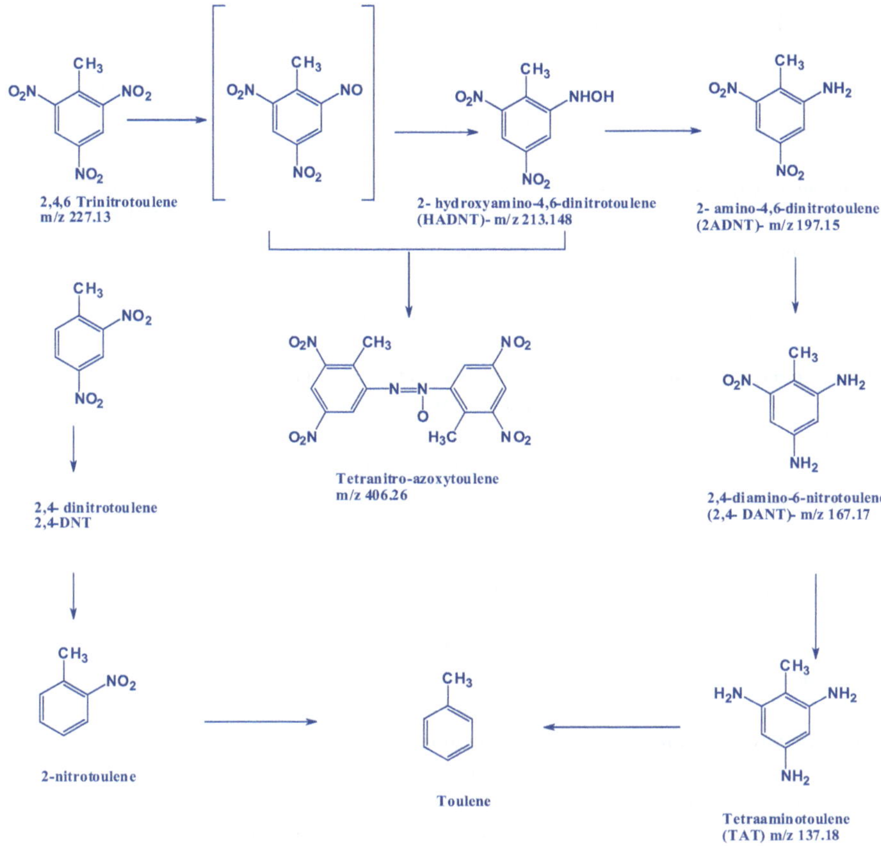

Fig. 7.2 Steps of TNT degradation

during the reduction procedure. Nevertheless, among a few incidents, in the existence of oxygen, TNT forms react to produce recalcitrant azoxytetranitrotoluene (Esteve-Nunez et al. 2001).

Aerobic bacteria can solitarily reduce 2 nitro groups, whereas the third involves anaerobic bacteria (McFarlan and Sara 2016). Once the electrons confirm the detachment of TNT aromatic ring via electronegative nitro groups, the nucleus turns out to be electrophilic (Claus 2014), and in a few cases these fallouts into non-aromatic structures formation like Meisenheimer complex, along with dinitrotoluenes formation (Esteve-Nunez et al. 2001; HaÏdour and Ramos 1996). Bacterial strains, such as *Rhodococcus erythropolis* and *Mycobacterium sp.*, are capable of producing Meisenheimer complexes (McFarlan and Sara 2016). A list of several aerobically involved microorganisms in TNT degradation is presented in Table 7.3.

Table 7.3 Testified bacteria and mechanisms projected for TNT degradation and conversion in prevailing aerobic environments

Bacteria	Metabolism	References
Pseudomonas sp. strain FR2	DANT formation	McCormick et al. (1976)
Pseudomonas sp. strain CBS3	DANT formation	Schackmann and Müller (1991)
Pseudomonas sp. clone A	TNT de-nitration capability	Duque et al. (1993)
Mycobacterium ***sp. strain HL4NT-1***	Meisenheimer complex formation	Vorbeck et al. (1994)
Pseudomonas aeruginosa MA101	Conversion into very polar compounds	Alvarez et al. (1995)
Pseudomonas fluorescens	Conversion to ADNT, DANTs, and 4-acetamide-2-amino-6- nitrotoluene	Gilcrease and Murphy (1995)
Pseudomonas sp.	Usage of TNT as a solitary nitrogen source along with the transfiguration to ADNT	Jones et al. (1995)
Enterobacter **sp.**	Conversion to unknown polar compounds, mineralization, and half assimilation to lipids	Vanderberg et al. (1995)
Pseudomonas pseudoalcaligenes	Nitro and nitrite group release from 2,4-dihydroxylamino-6-nitrotoluene	Fiorella and Spain (1997)
Pseudomonas savastanoi	De-nitration to nitrite and reduction of nitro group to amino	Martin et al. (1997)
Enterobacter cloacae **PB2**	Formation of Meisenheimer complex along with the release of nitrite	French et al. (1998 Aug)
Pseudomonas aeruginosa	Transformation of Nitrite to 2-amino-4-nitrotoluene	Kalafut et al. (1998)
Rhodococcus erythropolis	The evolution of picric acid catalyzes ring hydrogenation	Vorbeck et al. (1998)
Pseudomonas fluorescens and pseudomonas fluorescens B3468	An NADPH protein ignites various conversions and produces H-TNT and H_2-TNT	Pak et al. (2000)
Pseudomonas sp. strain iibx	Initial authorization of [^{14}C]-TNT mineralization	Esteve-Nunez et al. (2001)
Stenotrophomonas maltophilia	TNT as the sole nitrogen source	Kim and Song (2003)
Methylobacterium sp.	Degrade TNT	Aken et al. (2004)
P. Putida JLR11	Nitroreductase enzymes such as *pnra* and *pnrb*	Caballero et al. (2005)
Klebsiella **sp. CI**	Break down TNT	Kim and Song (2005)
Raoultellaterrigena strain HB	Formation of amino products along with dinitrotoluene and nitrobenzenes	Claus et al. (2007)

(continued)

Table 7.3 (continued)

Bacteria	Metabolism	References
K. Palustris RS32, R. Opacus IG, Acinobacter sp. VT11	Entire transformation	Solyanikova et al. (2011)
Psuedarthrobacter chlorophenolicus	Complete TNT degradation to TCA cycle	Lamba et al. (2021)

7.2.2 TNT Degradation in an Anaerobic Environment

Anaerobic bacteria likewise, Archaea, *Methanococcus sp, Clostridium sp.,* and *Desulfovibrio sp.* have the capability to entirely transform TNT to TAT (Crawford 1995). Species like *Clostridium* are capable of anaerobically reducing TNT (Ederer et al. 1997). Species like *Desulfovibrio* can break down TNT in anoxic circumstances and further use it as the exclusive nitrogen source. Species like *Clostridium* and *Desulfovibrio* are proficient to ignite the overall transformation, i.e., from TNT to 2,4,6-triaminotoluene (TAT) and this transformation is considered as one of the concluding stages of TNT degradation. Numerous intermediary reactions arise previously to the conversion to TAT, for instance, the DANTs to TAT reduction, which requires pyruvate and H_2 to continue. As per Preuss et al. (1993), this is the step that regulates the speed of the entire procedure. Table 7.4 delivers a complete list of several anaerobic microorganisms involved in TNT degradation.

Anaerobic microbial processes remain proficient in the speedy degradation of TNT at low redox potential (Lewis et al. 1997). Through TNT aerobic degradation, the nitro groups uninterruptedly transform to nitroso, afterward hydroxylamino, and finally to amino groups. The concluding complexes are similar to the aerobic degradation products such as 2-ADNT and 4-A-2,6-DNT. Both "4-ADNT and 2-ADNT" are transformed to "DANT" through "non-specific NAD(P)H nitroreductase. Afterward via dissimilatory sulfite reductase TAT is formed, and DANT is transmuted to DAHAT through hydrogenase or carbon monoxide dehydrogenase". Furthermore, "DANT gets transferred directly to TAT". TAT, uninterruptedly gets transformed to 2,4,6-tri hydroxytoluene, afterward "4-hydroxytoluene (*p*-Cresol)" forms. The *p*-Cresol is an intermediate metabolite of the toluene degradation pathway and heads toward the concluding conversions of the path to succinyl CoA and Acetyl-CoA production.

7.2.3 Simultaneous Aerobic and Anaerobic TNT Degradation

In both mixed or undefined culture systems, TNT mineralization has been initiated, however scanty literature is available on mineralization via pure bacterial systems in aerobic or anaerobic circumstances (Kalderis et al. 2011). The best TNT degradation rate is given by nitrate in contrast with sulfate and CO_2 (Boopathy et al. 1993). In the

Table 7.4 Reported bacteria and mechanisms projected for TNT degradation and conversion in anaerobic environments

Bacteria	Metabolism	References
Clostridium pasterianum	TNT to TAT breakdown	McCormick et al. (1976)
Veillonella alkalescens	Break TNT via H_2 to toluenetriamine	McCormick et al. (1976)
***Desulfovibrio* sp. Strain B**	TNT, a nitrogen source, ends in toluene as a hypothetical intermediate	Boopathy and Kulpa (1992)
***Desulfovibrio* sp.**	TNT is enacted as a sole nitrogen source and ends in the transformation to TAT	Preuss et al. (1993)
***Methanococcus* sp. Strain B**	TNT to DANT reduction	Boopathy and Kulpa (1994)
***Clostridium bifermentans* CYS-1**	TNT reduces through 4-ADNT and 2,4DANT to aliphatic polar compounds	Shin and Crawford (1995)
Clostridium sordelii	TNT to TAT reduction	Ederer et al. (1997)
***Clostridium bifermentans* LJP-1**	TNT to TAT transformation besides phenolic compounds	Lewis et al. (1997)
Escherichia coli	Conversion to TAT	Lewis et al. (1997)
***Lactobacillus* sp.**	Change to TAT	Lewis et al. (1997)
Serratia marcescens	Degrades TNT	Montpas et al. (1997)
***Desulfovibrio* sp.**	Transmutes into DANT and TAT along with the association of TNT with cell biomass	Drzyzga et al. (1998)
Clostridium acetobutylicum	TNT to TAT	Cai et al. (2013)

existence of aerobic and anaerobic situations, TNT is transmuted to amino derivatives via non-specific NAD(P)H-dependent nitroreductase as stated by Kalderis et al. (2011). Maximum TNT-degrading bacteria practice solitary one initial metabolic pathway; still, some bacteria can practice concurrently two diverse pathways for TNT degradation (Haïdour and Ramos 1996; Vorbeck et al. 1994). Production of reduced intermediates, DANT, 2-hydroxyl-4, 6-aminodinitrotoluene, 2-ADNT, 4-ADNT, isomers, and azo and azoxy dimers in one of the metabolic pathways is a successive reduction of electrons (two numbers) (Lewis et al. 1997).

Another pathway is a straight decline of two numbers of electrons of the ring through hydride accumulation, this reaction products comprised of hydride and dihydride Meisenheimer complex, along with protonated tautomers (Esteve-Nunez et al. 2001). Disaggregating the two preliminary biochemical paths in aerobic degradation, the initial stage of TNT degradation is its conversion to 4-HDNAT or 2-HDNT, depending upon the basis of the bacteria, mediated via nitroreductase/dihydropteridine reductase, N-ethylmaleimide reductase, and nitrobenzene nitroreductase. Such outcomes are pursued with the formation of 4-A-2,6-DNT or 2-ADNT through nitro reductases (Serrano-González et al. 2018). Along with this, during the aerobic degradation of 4-HDNAT and 2-HDNT supplementary compounds such as tetranitro-azoxybenzene also form. The concluding aerobic route conversions led to 2A-4-NT, 4A-2NOC-6-NT, and 2,6-DA-4-NT formation, afterward 4-A-2,6-DNT and 2-ANDT are produced. Through anaerobic bacteria, 4-A-2,6-DNT and 2-ADNT get converted to DANT (Lamba et al. 2022). Supplementary DANT transmutes into DAHAT, followed by TAT formation through dissimilatory sulfite reductase. Alternatively, DANT can precisely break down into TAT via *C. bifermentans, C. sordelli,* and *C. sporogenes.* TAT, consecutively degraded to 2,4,6-trihydroxytoluene and formerly to *p*-Cresol. The *p*-Cresol is a transitional metabolite of toluene degradation; its formation during TNT degradation is the reduction consequence of hydroxide bonds from 2,4,6-trihydroxytoluene aromatic ring. Further, *p*-Cresol gets transmuted to 4-hydroxybenzaldehyde, along with oxidation of the methyl group to 4-hydroxybenzoate formation (Perlman 1972). 4-hydroxybenzoate sequentially gets oxidized to Benzoyl-CoA, which went through sequence of reactions: initially it transformed into 2, 3-epoxybenzoyl-CoA, along with the conversion to 3,4-dehydroadipyl-CoA-Semialdehyde, and further 3, 4-dehydroadipyl-CoA to lastly produce succinyl and Acetyl-CoA (McFarlan and Sara 2016). Succinyl and acetyl-CoA are vibrant mediates of the TCA cycle.

7.3 Metabolic and Co-metabolic Treatment of 2-Methyl-1,3,5-Trinitrobenzene

The consequence of associated constituents on TNT biotransformation may be cooperative or incompatible and may be dependent on the microbes present, their digestion, and the category and amount of the co-substrate. In a study, the vanishing of TNT was happening side by side with the disappearance of the main substrate (glucose) along with the strong and fast growth of the consortia. The use of a quantum of culture with no co-substrate drove the enhanced TNT transformation in aqueous media opined the standing of handling the isolate's proliferation and survival, and its' associated constituents facilitated co-metabolic biotransformation competencies, execution of bioremediation procedures for hazardous waste clean-up (Tharakan and Gordon 1999) (Fig. 7.3).

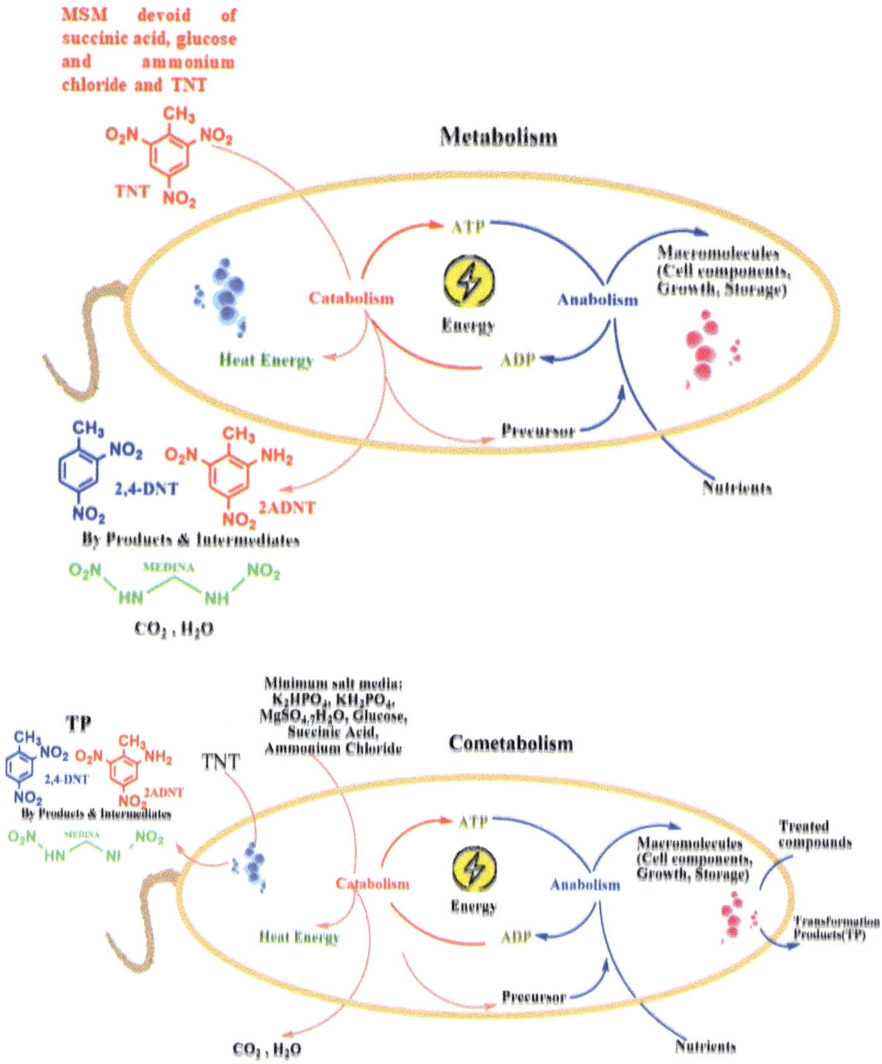

Fig. 7.3 Metabolism and cometabolism of TNT

7.4 Case Studies for Onsite Application

Considering the importance of contamination of TNT at US military sites, an investigation was carried out by a team of researchers for the development of microbial isolates having the potential to degrade amalgams of nitrobenzene derivatives and spread the treatment potential to TNT as well for onsite field applications. A thorough elucidation of routes and enzymes, regulation systems, and identification of the persuading molecules was done with a vision to develop augmented recombinant isolate proficient in TNT degradation (Spain et al. 2023). Biotransformation of TNT waste, which did not detonate due to some flaw, by microbial strains *Pseudomonas putida GG04 and* Bacillus *SF* was investigated. The complete biotransformation of TNT moieties upon transfer in liquid media was observed after 5 days. The inoculated strains broke TNT consecutively into secondary metabolites, buildup of which was efficiently prohibited by adding guaiacol and catechol during the biotransformation process (Nyanhongo et al. 2009).

Treatment solution for the groundwater pollutants at a previous explosive manufacturing facility in the UK has been provided by Regenesis (https://regenesis.com/wp-content/uploads/2020/06/CS_REG_JaS60088_3DME_ROx_ORCA_UK.pdf, @ REGENESIS JaS60088). Numerous model trials were commenced to examine the ideal degradation pathway of mixed TNT, DNT, and MNT. It was found that the in situ TNT degradation pathways follow both aerobic and anaerobic mineralization leading to a considerable reduction in pollutant concentrations.

7.5 Conclusion

The presence *of 2,4,6-trinitrotoluene (TNT) in the atmosphere necessitates the remediation of contaminated sites. Although various established physical and chemical abiotic technologies exist, bioremediation has emerged as a more environmentally* friendly alternative. Effective execution of remediation techniques at a specific location is influenced by the contaminant's characterization and form. Understanding these aspects aids in determining whether ex situ or in situ practices should be employed for efficient remediation.

A synergistic aerobic–anaerobic system can break down and mineralize the explosive to a bigger level than a specific method. Holdups to the gathering of a TNT treatment pathway included enzyme specificity, absence of enzyme regulation, and the reactivity of pathway metabolites that led to alternation in the degradation route. With a fundamental understanding of the microbial processes involved in the degradation of TNT, insight into the enzymes involved and their regulation will help in developing improved strategies for the degradation of TNT contaminants. This knowledge forms the foundation for metabolic engineering, paving the way for the development of innovative biochemical strategies for TNT degradation. The chapter discussed the contamination of the environment caused by explosives and the need for effective remediation technologies. It highlights the potential of microbes to degrade TNT and presents various approaches such as microbial strains in isolation, consortium, and immobilized conditions. Microbial bioremediation is presented as a green substitute for TNT disposal, emphasizing its efficiency and ecological benefits. This will further provide the basis for metabolic engineering to develop a new biochemical strategy for the degradation of TNT. Furthermore, leveraging microbes (fungi and enzymes) can assist in degrading the resulting molecules of NT degradation. Ongoing research focuses on advanced microbial approaches to mitigate the risk of TNT contamination effectively. As we advance in our understanding of microbial interaction and enzymatic mechanisms, we move closer to suitable and effective strategies for minimizing the risk of TNT adulterations. The bioremediation approach shows promise for effective and eco-friendly TNT remediation. The future aspects for TNT remediation such as microbial optimization, environmental application, and synergistic approaches should focus on enhancing the efficacy of microbial consortia to maximize TNT degradation rate and implementing biodegradation in the real world, i.e., field scale will validate its effectiveness. These future aspects help in the advancement in the field of biodegradation of TNT, ensuring suitable and efficient approaches to mitigate TNT contamination.

References

Agency for Toxic Substances and Disease Registry (ATSDR) (1995) Toxicological Profile for 2,4,6-Trinitrotoluene (TNT). www.atsdr.cdc.gov/toxprofiles/TP.asp?id=677&tid=125

Alexander M (1999) Biodegradation and bioremediation, 2nd. Academic Press, London, England

Alnaizy R, Akgerman A (1999) Oxidative treatment of high explosives contaminated wastewater. Water Res 33(9):2021–2030. https://doi.org/10.1016/S0043-1354(98)00424-2

Alothman ZA, Bahkali AH, Elgorban AM, Al-Otaibi MS, Ghfar AA, Gabr SA, Ahmed AYBH (2020) Bioremediation of explosive TNT by trichoderma viride. Molecules 25(6):1393. https://doi.org/10.3390/molecules25061393

Alvarez MA, Kitts CL, Botsford JL, Unkefer PJ (1995) *Pseudomonas aeruginosa* strain MA01 aerobically metabolizes the amino-di-nitrotoluenes produced by 2,4,6-Trinitrotoluene nitro group reduction. Can J Microbiol 41:984–991. http://www.ncbi.nlm.nih.gov/pubmed/7497356

Anand S, Celin SM, Bhalla R, Sahai S, Hooda L (2016) Bioremediation potential of *Pseudomonas sp.* isolated from a 2,4,6-trinitrotoulene contaminated site. Int J Environ Waste Manag 18(2):145. https://doi.org/10.1504/IJEWM.2016.080401

Bodeau D (1993) Occupational health: the soldier and the industrial base, Chapter 9, Military energetic materials: explosives and propellant

Boopathy R, Kulpa CF (1992) Trinitrotoluene (TNT) as a sole nitrogen source for a sulfate-reducing bacterium *Desulfovibrio sp.* (B strain) isolated from an anaerobic digester. Curr Microbiol 25(4):235–241. http://www.ncbi.nlm.nih.gov/pubmed/1368976

Boopathy R, Kulpa CF (1994) Biotransformation of 2, 4, 6-trinitrotoluene (TNT) by a *Methanococcus sp* (strain B) isolated from a lake sediment. Can J Microbiol 40(4):273–278. http://www.ncbi.nlm.nih.gov/pubmed/8039051

Boopathy R, Wilson M, Kulpa CF (1993) Anaerobic removal of 2,4,6-trinitrotoluene (TNT) under different electron accepting conditions: laboratory study. Water Environ Res 65(3): 271–275. http://www.jstor.org/stable/25044298

Bordeleau G, Martel R, Ampleman G, Thiboutot S (2013) Photolysis of RDX and nitroglycerin in the context of military training ranges. Chemosphere 93(1):14–19. https://doi.org/10.1016/j.chemosphere.2013.04.048

Brentner LB, Mukherji ST, Merchie KM, Yoon JM, Schnoor JL, Aken BV (2008) Expression of glutathione S-transferases in poplar trees (Populustrichocarpa) exposed to 2, 4, 6-trinitrotoluene (TNT). Chemosphere 73(5):657–662. https://doi.org/10.1016/j.chemosphere.2008.07.059

Bui ND, Minh TT (2021) Investigation of TNT red wastewater treatment technology using the combination of advanced oxidation processes. Sci Total Environ 756:143852. https://doi.org/10.1016/j.scitotenv.2020.143852

Burken JG, Shanks JV, Thompson PL (2000) Phytoremediation and plant metabolism of explosives and nitroaromatic compounds. In: Spain JC, Hughes JB, Knackmuss H (eds) Biodegradation of nitroaromatic compounds and explosives. CRC Press, Boca Raton, Florida, pp 239–276

Caballero A, Lázaro JJ, Ramos JL, Esteve-Núñez A (2005) PnrA, a new nitroreductase-family enzyme in the TNT-degrading strain *Pseudomonas putida* JLR11. Environ Microbiol 7:1211–1219. https://doi.org/10.1111/j.1462-2920.2005.00801.x

Cabrera ÁM, Márquez LS, Osorio IM, Castro-Nallar E, González-Nilo DF, Pérez-Donoso MJ (2020) Biotransformation of 2,4,6-trinitrotoluene by Pseudomonas sp. TNT3 isolated from Deception Island, Antarctica. Environ Pollut 262:113922. https://doi.org/10.1016/j.envpol.2020.113922

Cai X, Servinsky M, Kiel J, Sund C, Bennett GN (2013) Analysis of redox responses during TNT transformation by *Clostridium acetobutylicum* ATCC 824 and mutants exhibiting altered metabolism. Appl Microbiol Biotechnol 97:4651–4663. https://doi.org/10.1007/s00253-012-4253-3

Chatterjee S, Deb U, Datta S, Walther C, Gupta DK (2017) Common explosives (TNT, RDX, HMX) and their fate in the environment: emphasizing bioremediation. Chemosphere 184:438–451. https://doi.org/10.1016/j.chemosphere.2017.06.008

Chaudhary DK, Kim J (2019) New insights into bioremediation strategies for oil-contaminated soil in cold environments. Int Biodeterior Biodegradation 142:58–72. https://doi.org/10.1016/j.ibiod.2019.05.001

Claus H, Bausinger T, Lehmler I, Perret N, Fels G, Dehner U, Preuß J, König H (2007) Transformation of 2,4,6- trinitrotoluene (TNT) by *Raoultella*. Biodegradation 18(1):27–35. https://doi.org/10.1007/s10532-005-9033-7

Claus H (2014) Microbial degradation of 2,4,6- Trinitrotoluene in vitro and in natural environments, In: Singh SN (ed) Biological remediation of explosive residues, environmental science and engineering. Springer International Publishing, pp 15–38. https://doi.org/10.1007/978-3-319-01083-0_2

Crawford R (1995) The microbiology and treatment of nitroaromatic compounds. Curr Opin Biotechnol 6(1995):329–336. https://doi.org/10.1016/0958-1669(95)80055-7

Cupples AM (2013) RDX degrading microbial communities and the prediction of microorganisms responsible for RDX bioremediation. Int Biodeterior Biodegradation 85:260–270. https://doi.org/10.1016/j.ibiod.2013.08.002

Drzyzga O, Bruns-Nagel D, Gorontzy T, Blotevogel KH, Gemsa D (1998) Mass balance studies with 14C-labeled 2,4,6-trinitrotoluene (TNT) mediated by an Anaerobic *desulfovibrio species* and an Aerobic *serratia species*. Curr Microbiol 37:380–386. http://www.ncbi.nlm.nih.gov/pubmed/9806975

Duque E, Haidour A, Godoy F, Ramos LJ (1993) Construction of a Pseudomonas hybrid strain that mineralizes 2,4,6- trinitrotoluene. J Bacteriol 175(8):2278–2283. https://doi.org/10.1128/jb.175.8.2278-2283.1993

Ederer MM, Lewis TA, Crawford RL (1997) 2, 4, 6-Trinitrotoluene (TNT) transformation by clostridia isolated from a munition-fed bioreactor: comparison with non-adapted bacteria. J Ind Microbiol Biotechnol 18(2–3):82–88. https://doi.org/10.1038/sj.jim.2900257

EPA (2013) Superfund information systems, superfund site information. www.epa.gov/superfund/sites/cursites/

Esteve-Nunez A, Caballero A, Ramos JL (2001) Biological degradation of 2,4,6-trinitrotoluene. Microbiol Mol Biol Rev 65(3):335–352. https://doi.org/10.1128/mmbr.65.3.335-352.2001

Fiorella PD, Spain JC (1997) Transformation of 2,4,6-trinitrotoluene by *Pseudomonas pseudoalcaligenes* JS52. Appl Microbiol Biotechnol 63:2007–2015. https://doi.org/10.1128/aem.63.5.2007-2015.1997

French CE, Nicklin S, Bruce NC (1998) Aerobic degradation of 2,4,6-trinitrotoluene by Enterobacter cloacae PB2 and by pentaerythritol tetranitrate reductase. Appl Environ Microbiol 64(8):2864–2868. https://doi.org/10.1128/AEM.64.8.2864-2868.1998.PMID:9687442;PMCID:PMC106784

Gad CS (2014) Trinitrotoluene, Encyclopedia of. Toxicology 4:855–857. https://doi.org/10.1016/b978-0-12-386454-3.00957-X

Gilcrease PC, Murphy VG (1995) Bioconversion of 2,4-diamino-6-nitrotoluene to a novel metabolite under anoxic and aerobic conditions. Appl Microbiol Biotechnol 61:4209–4214. https://doi.org/10.1128/aem.61.12.4209-4214.1995,CorpusID:12700897

Gök ZG, Inal M, Yiğitoğlu M (2018) Effective degradation of 2,4,6-trinitrotoluene (TNT) with a bacterial consortium developed from high TNT-degrading bacteria Isolated from TNT-contaminated soil. Hacettepe J Biol Chem 45(3):445–455. https://doi.org/10.15671/HJBC.2018.252

Gumuscu B, Cekmecelioglu D, Tekinay T (2015) Complete dissipation of 2,4,6-trinitrotoluene by in-vessel composting. RSC Adv 5(64):51812–51819. https://doi.org/10.1039/C5RA07997G

Gunning V, Tzafestas K, Sparrow H, Johnston EJ, Brentnall AS, Potts JR, Rylott EL, Bruce NC (2014) Arabidopsis glutathione transferases U24 and U25 exhibit a range of detoxification activities with the environmental pollutant and explosive, 2, 4, 6-trinitrotoluene. Plant Physiol 165(2):854–865. https://doi.org/10.1104/pp.114.237180

Gupta S, Siebner H, Ramanathan G, Ronen Z (2022) Inhibition effect of 2,4,6-trinitrotoluene (TNT) on RDX degradation by *rhodococcus* strains isolated from contaminated soil and water. Environ Pollut 311:2022. https://doi.org/10.1016/j.envpol.2022.120018

Gupta S, Goel SS, Siebner H, Ronen Z, Ramanathan G (2023) Transformation of 2, 4, 6-trinitrotoluene by *Stenotrophomonas* strain SG1 under aerobic and anaerobic conditions. Chemosphere 311(1).https://doi.org/10.1016/j.chemosphere.2022.137085

Haïdour A, Ramos JL (1996) Identification of products resulting from the biological reduction of 2,4,6-trinitrotoluene, 2,4-dinitrotoluene, and 2,6-dinitrotoluene by *Pseudomonas spp.* Environ Sci Technol 30(7):2365–2370. https://doi.org/10.1021/es950824u

Hannink NK, Rosser SJ, Bruce NC (2002) Phytoremediation of explosives. Crit Rev Plant Sci 21(5):511–538. https://doi.org/10.1080/0735-260291044340

Hawari J (2000) Biodegradation of RDX and HMX: from basic research to field application In: biodegradation of nitroaromatic by compounds and explosives. Biodegrad Nitroaromatic Compd Explos 277–310. https://doi.org/10.1201/9781420032673.ch11

Higson FK (1992) Microbial degradation of nitroaromatic compounds. Adv Appl Microbiol 37:1–19. https://doi.org/10.1016/s0065-2164(08)70250-8

Hinshaw GD, Fansk CB, Fiscus DE, Sorenson SA (1987) Granular activated carbon (GAC) system performance and capabilities and optimization, Report AMXTH-TE-GR87111, U.S. Army Toxic and Hazardous Materials Agency, Aberdeen Proving Ground, MD, U.S

HSDB (2012) 2,4,6-Trinitrotoluene (TNT), www.atsdr.cdc.gov/toxprofiles/TP.asp?id=677&tid=125. Hazardous Substances Data Bank (HSDB)

Jones AM, Greer CW, Hawari J, Ampleman G, Lavigne J (1995) Biodegradability of selected highly energetic pollutants under aerobic conditions. http://www.osti.gov/scitech/biblio/490985-biodegradability-selected-highly-energetic-pollutants-under-aerobic-conditions

Kalafut T, Wales ME, Rastogi VK, Naumov PR, Zaripova KS, Wild RJ (1998) Biotransformation patterns of 2,4,6-trinitrotoluene by aerobic bacteria. Curr Microbiol 36:45–54. https://doi.org/10.1007/s002849900278

Kalderis D, Juhasz LA, Boopathy R, Comfort S (2011) Soils contaminated with explosives: environmental fate and evaluation of state-of-the-art remediation processes (IUPAC Technical Report). Pure Appl Chem 83(7):1407–1484. https://doi.org/10.1351/PAC-REP-10-01-05

Kao C-M, Lin B-H, Chen S-C, Wei S-F, Chen C-C, Yao C-L, Chien C-C (2016) Biodegradation of trinitrotoluene (TNT) by indigenous microorganisms from TNT-contaminated soil, and their application in TNT bioremediation. Bioremediat J 1–9.https://doi.org/10.1080/10889868.2016.1148007

Kim HY, Song HG (2003) Transformation and mineralization of 2,4,6-trinitrotoluene by the white rot fungus *Irpex lacteus*. Appl Microbiol Biotechnol 61:150–156. https://doi.org/10.1007/s00253-002-1211-5

Kim HY, Song HG (2005) Purification and characterization of NAD(P)H-dependent nitroreductase I from *Klebsiella sp.* C1 and enzymatic transformation of 2,4,6-trinitrotoluene. Appl Microbiol Biotechnol 68:766–773. https://doi.org/10.1007/s00253-005-1950-1

Kirk-Othmer RE (1993) Kirk-Othmer encyclopedia of chemical technology, chlorocarbons and chlorohydrocarbons-CSUB 2/SUB to combustion technology

Koske D, Goldenstein IN, Kammann U (2019) Nitroaromatic compounds damage the DNA of zebrafish embryos (*Danio rerio*). Aquat Toxicol 217. https://doi.org/10.1016/j.aquatox.2019.105345

Koutsospyros A, Pavlov J, Fawcett J, Strickland D, Smolinski B, Braida W (2012) Degradation of high energetic and insensitive munitions compounds by Fe/Cu bimetal reduction. J Hazard Mater 219:75–81. https://doi.org/10.1016/j.jhazmat.2012.03.048

Kumar V, Sehgal R, Gupta R (2023) Microbes and wastewater treatment. In: Shah MP, Rodriguez-Couto S, Nadda AK, Daverey A (eds) Development in wastewater treatment research and processes. Elsevier, pp 239–255. https://doi.org/10.1016/B978-0-323-88505-8.00010-3

Kundu K, Landuyt VJ, Mattelin V, Martin B, Neyts M, Parmentier K, Boon N (2023) Enhanced removal of warfare agent tri-nitro-toluene by a *Methylophaga*-dominated microbiome. Mar Pollut Bull 190. https://doi.org/10.1016/j.marpolbul.2023.114866

Kyprianou D, Berglund M, Emma G, Rarata G, Anderson D, Diaconu G, Exarchou V (2020) Synthesis of 2,4,6-trinitrotoluene (TNT) using flow chemistry. Molecules 25:3586. https://doi.org/10.3390/molecules25163586

Lamba J, Anand S, Dutta J, Chatterjee S, Nagar S, Celin SM, Rai PK (2021) Study on aerobic degradation of 2,4,6-trinitrotoluene (TNT) using *Pseudarthrobacter chlorophenolicus* collected from the contaminated site. Environ Monit Assess 193(2):80. https://doi.org/10.1007/s10661-021-08869-7

Lamba J, Anand S, Dutta J, Rai PK (2022) 2,4,6-trinitrotoluene (TNT) degradation by *Indiicoccus explosivorum* (S5-TSA-19). Arch Microbiol 204:447. https://doi.org/10.1007/s00203-022-03057-8

Lata K, Kushwaha A, Ramanathan G (2021) Bacterial enzymatic degradation and remediation of 2,4,6-trinitrotoluene. In: Das S, Dash HR (eds) Microbial and natural macromolecules. Academic Press, pp 623–659.https://doi.org/10.1016/B978-0-12-820084-1.00024-7

Lewis TA, Ederer MM, Crawford RL, Crawford DL (1997) Microbial transformation of 2, 4, 6-trinitrotoluene. J Ind Microbiol Biotechnol 18(2):89–96. https://doi.org/10.1038/sj.jim.2900258

Li C, Li W, Ding Y, Zhou J, Ziao Z (2023) Progress and future of green explosives and propellants. Chem Energy Sour Chin Sci Bull 68(25):3311–3321. https://doi.org/10.1360/TB-2023-0203

Lynch JC, Myers KF, Brannon JM et al (2001) Effects of pH and temperature on the aqueous solubility and dissolution rate of 2,4,6-Trinitrotoluene (TNT), Hexahydro-1,3,5-trinitro-1,3,5-triazine (RDX), and Octahydro-1,3,5,7-tetranitro-1,3,5,7-tetrazocine (HMX). J Chem Eng Data 46(6):1549–1555. https://doi.org/10.1021/je0101496

Mahbub P, Nesterenko PN (2016) Application of photo degradation for remediation of cyclic nitramine and nitroaromatic explosives. Royal Soc Chem (great Britain) 6(81):77603–77621. https://doi.org/10.1039/C6RA12565D

Martin LJ, Comfort DS, Shea JP, Drijber AR, Kokjohn AT (1997) Denitration of 2,4,6-trinitrotoluene by *Pseudomonas savastanoi*. Can J Microbiol 43:447–455. https://doi.org/10.1139/m97-063

Massachusetts Military Reservation (MMR) (2001) Impact area groundwater study program, chemical fact sheet–TNT, Fact Sheet 2001–05

McCormick NG, Feeherry FE, Levinson HS (1976) Microbial transformation of 2,4,6-trinitrotoluene and other nitroaromatic compounds. Appl Environ Microbiol 31:949–958. https://doi.org/10.1128/aem.31.6.949-958.1976

McFarlan G, Sara Y (2016) 2,4,6-Trinitrotoluene degradation pathway. http://eawag-bbd.ethz.ch/tnt/tnt_map.html

Mishra V, Lal R, Srinivasan S (2001) Enzymes and operons mediating xenobiotic degradation in bacteria. Crit Rev Microbiol 27:133–166. https://doi.org/10.1080/20014091096729

Montpas S, Samson J, Langlois E, Lei J, Piché Y, Chênevert R (1997) Degradation of 2,4,6-trinitrotoluene by *Serratia marcescens*. Biotech Lett 19(3):291–294. https://doi.org/10.1023/A:1018326228448

Nagar S, Anand S, Chatterjee S, Rawat DC, Lamba J, Rai KR (2021) A review of toxicity and biodegradation of octahydro-1,3,5,7-tetranitro-1,3,5,7-tetrazocine (HMX) in the environment. Environ Technol Innov 23(2–3):101750. https://doi.org/10.1016/j.eti.2021.101750

Nagar S, Shaw AK, Anand S, Celin SM, Rai PK (2018) Aerobic biodegradation of HMX by *Planomicrobium flavidum.* 3 Biotech 8(11):45. https://doi.org/10.1007/s13205-018-1479-5

Nagar S, Shaw AK, Anand S, Celin SM, Rai PK (2020) Biodegradation of octogen and hexogen by *Pelomonas aquatica* strain WS2-R2A-65 under aerobic condition. Environ Technol 1–22.https://doi.org/10.1080/09593330.2020.1812731

National Center for Biotechnology Information (2023) PubChem compound summary for CID 8376, 2,4,6-trinitrotoluene. Accessed 27 July 2023

Nyanhongo GS, Aichernig N, Ortner M, Steiner W, Guebitz GM (2009) Incorporation of 2,4,6-trinitrotoluene (TNT) transforming bacteria into explosive formulations. J Hazard Mater 165(1–3):285–290. https://doi.org/10.1016/j.jhazmat.2008.09.107

Pak JW, Knoke KL, Noguera DR, Fox BG, Chambliss GH (2000) Transformation of 2,4,6-trinitrotoluene by purified xenobiotic reductase B from *Pseudomonas fluorescens* I-C. Appl Environ Microbiol 66(11):4742–4750. https://doi.org/10.1128/aem.66.11.4742-4750.2000

Palazzo AJ, Leggett DC (1986) Effect and disposition of TNT in a terrestrial plant. J Environ Qual 15(1):49. https://doi.org/10.2134/jeq1986.00472425001500010012x

Perlman D (1972) Advances in applied microbiology. Academic Press

Pichtel J (2012) Distribution and fate of military explosives and propellants in soil: a review. Appl Environ Soil Sci 1–33.https://doi.org/10.1155/2012/617236

Preuss A, Fimpel J, Diekert G (1993) Anaerobic transformation of 2,4,6-trinitrotoluene (TNT). Arch Microbiol 159:345–353. https://doi.org/10.1007/BF00290917

REGENESIS JaS60088. https://regenesis.com/wp-content/uploads/2020/06/CS_REG_JaS60088_3DME_ROx_ORCA_UK.pdf, @. Accessed 16 Aug 2023

Rezaei MR, Abdoli MA, Karbassi A, Baghvand A, Khalilzadeh R (2010) Bioremediation of TNT contaminated soil by composting with municipal solid wastes. Soil Sedim Contam Int J 19(4):504–514. https://doi.org/10.1080/15320383.2010.486049

Robertson, B. K., & Jjemba, P. K. (2005). Enhanced bioavailability of sorbed 2, 4, 6-trinitrotoluene (TNT) by a bacterial consortium. *Chemosphere, 58*(3), 263–270. https://doi.org/10.1016/j.chemosphere.2004.08.080

Sathya R, Arasu VM, Al-Dhabi AN, Vijayaraghavan P, Ilavenil S, Rejiniemon ST (2023) Towards sustainable wastewater treatment by biological methods–a challenges and advantages of recent technologies. Urban Clim 47. https://doi.org/10.1016/j.uclim.2022.101378

Schackmann A, Müller R (1991) Reduction of nitroaromatic compounds by different Pseudomonas species under aerobic conditions. Appl MicrobiolBiotechnol 34.https://doi.org/10.1007/BF00169355

Schuster R, Strehse JS, Ahvo A, Turja R, Maser E, Bickmeyer U, Lehtonen KK, Brenner M (2021) Exposure to dissolved TNT causes multilevel biological effects in Baltic mussels (*Mytilus spp.*) Mar Environ Res 167:105264. https://doi.org/10.1016/j.marenvres.2021.105264

Serrano-González YM, Chandra R, Castillo-Zacarias C, Robledo-Padilla F, Rostro-Alanis JM, Parra-Saldivar R (2018) Biotransformation and degradation of 2,4,6-trinitrotoluene by microbial metabolism and their interaction. Def Technol 14(2):151–164. https://doi.org/10.1016/j.dt.2018.01.004

Shin CY, Crawford DL (1995) Biodegradation of trinitrotoluene (TNT) by a strain of *Clostridium bifermentans.* U.S. Department of Energy Office of Scientific and Technical Information. http://www.osti.gov/scitech/biblio/474237

Singh B, Kaur J, Singh K (2012) Microbial remediation of explosive waste. Crit Rev Microbiol 38(2):152–167. https://doi.org/10.3109/1040841x

Singleton I (1994) Microbial metabolism of xenobiotics: fundamental and applied research. J Chem Tech Biotechno 59:9–23. https://doi.org/10.1002/jctb.280590104

Sinha S, Chattopadhyay P, Pan I, Chatterjee S, Chanda P, Bandyopadhyay D, Das K, Sen SK (2009) Microbial transformation of xenobiotics for environmental bioremediation. Afr J Biotech 8(22):6016. https://doi.org/10.5897/AJB09.740

Solyanikova IP, Baskunov BP, Baboshin MA, Saralov AI, Golovleva LA (2011) Detoxification of high concentrations of trinitrotoluene by bacteria. Appl Microbiol Biotechnol 48:21–27. https://doi.org/10.1134/S0003683812010152

Spain PR, Johnson J, Glenn R (2023) Bacterial degradation of DNT and TNT mixtures, CU 1212. Final technical report, 31 OCT 2005. Accessed 24 Aug 2023

Tharakan JP, Gordon JA (1999) Cometabolic biotransformation of trinitrotoluene (TNT) supported by aromatic and non-aromatic co-substrates. Chemosphere 38(6):1323–1330. https://doi.org/10.1016/s0045-6535(98)00533-5

Thenmozhi A, Devasena M (2019) Remediation of 2,4,6-trinitrotoluene persistent in the environment–a review. Soil Sediment Contam Int J 1–13. https://doi.org/10.1080/15320383.2019.1664394

Timmis KN, Pieper DH (1999) Bacteria designed for bioremediation. Trends Biotechnol 17:200–204

USEPA (2014) Technical fact sheet -2,4,6-Trinitrotoluene (TNT). Office of Solid Waste and Emergency Response, 1–8. Washington, DC, U.S.A

Van Aken B, Yoon JM, Schnoor JL (2004) Biodegradation of nitro-substituted explosives 2,4,6-trinitrotoluene, hexahydro-1,3,5-trinitro-1,3,5-triazine, and octahydro-1,3,5,7-tetranitro-1,3,5-tetrazocine by a phytosymbiotic *Methylobacterium sp.* associated with poplar tissues (*Populus deltoides x*). Appl Microbiol Biotechnol 70:508–517. https://doi.org/10.1128/AEM.70.1.508-517.2004

Vanderberg LA, Perry JJ, Unkefer PJ (1995) Catabolism of 2,4,6-trinitrotoluene by *Mycobacterium vaccae.* Appl Microbiol Biotechnol 43:937–945. https://doi.org/10.1007/BF02431931

Vila M, Mehier S, Lorber-Pascal S, Laurent F (2007) Phytotoxicity to and uptake of RDX by rice. Environ Pollut 145(3):813–817. https://doi.org/10.1016/j.envpol.2006.05.009.

Vorbeck C, Lenke H, Fischer P, Knackmuss HJ (1994) Identification of a hydride Meisenheimer complex as a metabolite of 2,4,6-trinitrotoluene by a *Mycobacterium strain.* J Bacteriol 176(3):932e-e934. https://doi.org/10.1128/jb.176.3.932-934.1994

Vorbeck C, Lenke H, Fischer P, Spain JC, Knackmuss HJ (1998) Initial reductive reactions in aerobic microbial metabolism of 2,4,6-trinitrotoluene. Appl Microbiol Biotechnol 64:246–252. https://doi.org/10.1128/AEM.64.1.246-252.1998

Wallace L, Cronin MP, Day AI, Buck DP (2009) Electrochemical method applicable to treatment of wastewater from nitrotriazolone production. Environ Sci Technol 43(6):1993–1998. https://doi.org/10.1021/es8028878

Wanaratna P, Christodoulatos C, Sidhoum M (2006) Kinetics of RDX degradation by zero-valent iron (ZVI). J Hazard Mater 136(1):68–74. https://doi.org/10.1016/j.jhazmat.2005.11.015

Wang CJ, Thiele S, Bollag JM (2002) Interaction of 2,4,6-trinitrotoluene (TNT) and 4-amino-2,6-dinitrotoluene with humic monomers in the presence of oxidative enzymes. Arch Environ Contam Toxicol 42:1–8. https://doi.org/10.1007/s002440010284

Wong AJ, Miller JL, Perdue B, Janik MJ (2023) Investigating the electrocatalytic reduction of 2,4,6-tri-nitro-toluene (TNT) using density functional theory methods. Green Chem 25(13):5097–5112. https://doi.org/10.1039/D3GC01144E

Xin-feng L, Chang Y, Jing-yu W et al (2023) Research process of microbial remediation of energetic compound contaminated sites. Chin J Energ Mater 31(7):714–728. https://doi.org/10.11943/CJEM2023011

Xu M, Liu D, Sun P, Li Y, Wu M, Liu W, Maser E, Xiong G, Guo L (2021) Degradation of 2,4,6-trinitrotoluene (TNT): involvement of protocatechuate 3,4-Dioxygenase (P34O) in Buttiauxella sp. S19–1. Toxics 9:231. https://doi.org/10.3390/toxics9100231

Xu M, He L, Sun P, Wu M, Cui X, Liu D, Adomako-Bonsu AG, Geng M, Xiong G, Guo L, Maser E (2023) Critical role of monooxygenase in biodegradation of 2,4,6-trinitrotoluene by *Buttiauxella* sp. S19–1. Molecules 28(4):1969. https://doi.org/10.3390/molecules28041969

Ziganshin AM, Ziganshina EE, Byrne J, Gerlach R, Struve E, Biktagirov T (2015) Fe (III) mineral reduction followed by partial dissolution and reactive oxygen species generation during 2,4,6-trinitrotoluene transformation by the aerobic yeast *Yarrowia lipolytica*. AMB Express 5:8. https://doi.org/10.1186/s13568-014-0094-z

Part III
Microbial Bioplastics

Chapter 8
Bioplastic: Unravelling the Sustainable Approach for Petroleum Plastic

Manisha Mohapatra and Sanjana Singh

8.1 Introduction

Surplus increase in global population along with frequently rising anthropogenic needs are the prime reason for several environmental and ecological hazards with very minuscule attention or validated recovery status. Among these, one critical issue is the high accumulation of non-biodegradable, toxic plastics in large scale within several landfill areas and even in the water bodies also. In a rough estimate till now around 390.7 million tons of plastic have been dumped globally till 2021 (Al-Khairy et al. 2022). The US Environmental Protection Agency (EPA) had estimated that out of the total 292.4 million tons of Municipal Solid Waste generated in 2018, 12% was found to be the plastic wastes or micro-plastics available in landfill areas, with a minimal recycle rate of 4.47% only (EPA 2023). In coming years, the raise in the use of plastic has been elevated frantically and is estimated an expected hike by 20% within 2050 (Al-Khairy et al. 2022).

Plastics are a kind of synthetic or semi-synthetic organic polymeric structures, deriving from conventional fossil resources (Narancic et al. 2018). Different types of synthetic polymers consist up to 90% of total products including polyethylene, polystyrofoam, polypropylene, polyvinyl chloride, and polytetrafluoroethylene or Teflon (Ellen 2016; Ibrahim et al.2021). Different types of polymer-based plastics are used for different utilization purposes including utensils, packaging, beverage bottles, furniture, construction industries, textiles, cosmetics and many more (Lau et al. 2020). The remorseless heightening population density and the rapidly growing anthropogenic needs have led to continuous development of several industrial sectors, which are actively involved in formation of several plastic wastes (Paul and Ghosh 2022). Out of the most conventional plastic wares used, 45% were being developed

M. Mohapatra (✉) · S. Singh
Department of Biotechnology, Trident Academy of Creative Technology (TACT), Infocity, Chandrasekharpur, Odisha 751024, India
e-mail: manishamohapatra7@gmail.com

K. Bala et al. (eds.), *Harnessing Microbial Potential for Multifarious Applications*, Energy, Environment, and Sustainability, https://doi.org/10.1007/978-981-97-1152-9_8

from polyethylene and/or polypropylene (Geyer and Letcher 2020). However all the above types of plastics, those being used traditionally were mostly developed from fossil fuels and most of them were not biodegradable in nature leading to their accumulation, even also in micro or nano scales. The need and higher demand of plastics and plastic products, due to minimal cost, durability, higher availability, have elevated the issues associated with their proper disposing pattern in the landfill areas leading to serious health hazards and associated environmental issues. Not only are the landfill areas, even the marine water bodies are also continuously being damaged by the heightening use of plastic in a ruthless manner (Lear et al. 2021).

This creates a serious threat to natural ecosystem leading to several health hazards and climatic disturbances. To mitigate such issues the utilization of suitable substitute alternate in the form of any bio-based and biodegradable kind of polymer for the replacement of synthetic or semi-synthetic plastic is the need of hour. This can not only lessen the load of accumulated waste in the landfill areas, but also facilitate the sustainable approach for healthy environmental factors. As research and technology continue to advance, biodegradable plastics hold promise for a more sustainable future. However, they should be considered as part of a broader strategy to reduce overall plastic consumption, promote recycling, and encourage responsible waste management practices to combat plastic pollution effectively. By making informed choices and adopting sustainable practices, we can collectively work toward a cleaner and healthier planet. On this context the development of bioplastics from various natural sources including food or biomass (like starch, sugarcane, etc.) can facilitate paradigm shift toward circular economy along with minimizing the load on fossil resources, thereby reducing carbon footprint and other environmental hazards. In recent scenario, inventing various bioplastic sources and their applications have got an upper hand in several value added application purposes like food packaging sectors, pharmaceuticals and medicine sectors, electronics and other appliances, agricultural and domestic sectors, automobile and cosmetic industries, etc. (Jayakumar et al. 2023). These are safer to be used as, however are burdened with several economic and legal issues.

8.2 Synthetic Plastic

Synthetic plastics are a wide class of polymer composites that are made from fossil fuels like natural gas and oil. They are manufactured in various forms and have several functions due to their cheap production process, versatility, and beneficial properties such as lightness, durability, and flexibility. But, synthetic plastics are not biodegradable and contribute to environmental problems by increasing the amount of solid waste (Ibrahim et al. 2021). Most of the synthetic and/or semi-synthetic types of plastics are made up of specific polymeric compounds with peculiar physical and chemical attributes including flexibility, permeability, hydrophobicity/hydrophilicity, heat resistance, durability, and many more (Jan-Georg et al. 2022). Synthetic plastics are a type of plastic that is derived from petroleum-based sources.

They are composed of complex molecular monomers bound together to form macromolecules. There are more than twenty types of synthetic plastics in use worldwide. These plastics have been utilized due to their cost-effectiveness, lightweight nature, and versatility in various applications (Narancic et al. 2020). However, they have also raised environmental concerns due to their non-degradable nature and potential for toxic exposure. Efforts are being made to find alternatives to synthetic plastics (Prasteen et al. 2018). Different types of synthetic plastics available include broadly Polyethylene, Polypropylene, and Polyvinyl Chloride. Polyethylene is a widely used synthetic plastic that is known for its versatility and durability. It is commonly used in packaging materials, plastic bags, and bottles. Similarly Polypropylene is another commonly used synthetic plastic that is known for its high melting point and resistance to chemicals. It is used as automotive parts, food containers, and textiles. PVC is a synthetic plastic that is known for its flexibility and resistance to fire and chemicals. It is commonly used in pipes, cables, flooring, etc.

However the environmental significances of most of the synthetic plastics are extremely higher and lead to drastic climate change as most of these are based upon fossil fuels and due to non-biodegradability, these are emitting many of the toxic pollutants within the environment enabling impacts such as eutrophication and toxicity. Due to higher global and environmental backlashes resulting from exhausted utilization of petroleum-based plastic use, the focus has now been shifted for use of more sustainable and environmental-friendly compound with similar attributes and functionality. On this bioplastic is a much anticipated suitable alternate substitute. This chapter deals with various types of bioplastic and their sustainable application.

8.3 Bioplastic

Bioplastic is an alternative to petroleum-based plastics that are non-biodegradable and contribute to environmental pollution. However, the disposal of petroleum-based plastics poses a significant environmental challenge. Bioplastics, on the other hand, are made from biomaterials such as corn starch and gelatin, which are biodegradable and offer a more sustainable solution. This study focuses on the compilation of different types of available bioplastics and their applications in current scenario to substitute synthetic plastics for a more sustainable and hazard-free approach. The aim is to explore the properties and potential of these biomaterial-based plastics as substitute to non-degradable plastics (Narancic et al. 2020; Prasteen et al. 2018).

Biodegradable plastics are a promising alternative to traditional plastics because they can naturally degrade to compounds, minimizing their impact on environment (Narancic et al. 2018). These plastics are designed to degrade through the action of microorganisms, such as bacteria and fungi, which degrade the polymer into smaller, more manageable monomers (Table 8.1). Such plastics include polylactic acid, cellulose, polyhydroxyalkanoates, and starch. Similarly many oil-based plastics, biodegradable biobased plastics can also be found (Moshood et al. 2022). However, it's essential to recognize that biodegradable plastics are not a complete solution to plastic pollution. Their proper disposal and management remain critical for their effective degradation. If not managed correctly, they may not decompose efficiently, and improper disposal could still lead to environmental pollution (Narancic et al. 2018; Geyer and Letcher 2020). Bioplastic polymers including polyhydroxyalkanoate, Chitosan-based, Chitin-based, polylactic acid, cellulose-based, polyamide 11, protein-based polyhydroxyurethanes, poly-3-hydroxybutyrate, starch-based, and lipid-based biopolymers are some of the examples those are widely used in development and formation of many of the bio-products in different industries as shown in Table 8.1 (Nanda et al. 2022).

8.3.1 Global Status of Bioplastic

Biodegradable plastics have gained significant attention globally due to their potential to address environmental concerns associated with traditional plastics. The literature analysis reveals that the United States and China are the leading contributors in terms of significance and publication volume in the field of bio-based and biodegradable plastics (Dilkes-Hoffman et al. 2019). Europe is also emerging as a key player in terms of bio-based and biodegradable plastics production capacity (Geyer and Letcher 2020; Hakan et al. 2019). Asia, particularly China, remains the dominant manufacturing hub for bioplastics production. However, Europe is expected to witness a significant increase in bioplastics production capacity in the coming years. The global focus on biodegradable plastics reflects the growing awareness and efforts to promote sustainable environmental practices and reduce the reliance on synthetic plastics. The literature on bio-based and biodegradable plastics shows that Asia, America, and Europe dominate in terms of publications (Beigbeder et al. 2019). The United States published the most articles, followed by China, Japan, and India.

In terms of bioplastic production capacity, Europe contains one-fourth of the global capacity, while Asia remains the key manufacturing hub. In 2020, Asia generated 46.9% of all bio-based and biodegradable plastics. However, Asia's share is expected to fall to 43.2% by 2025. Europe's bioplastics production capacity is predicted to rise dramatically during that time period (Moshood et al. 2022; Venkatachalam et al. 2018). The total cumulative global plastic production up to 2015 exceeded 8.3 gigatons, resulting in 6.3 gigatons of plastic waste (Geyer et al. 2017). By 2050, this waste is expected to increase to around 12 gigatons. However, only 21% of this plastic waste has been recycled or incinerated, while the remaining 79% has entered landfills or been released into the environment (Hakan et al. 2019; Ling et al. 2021). As research and technology continue to advance, biodegradable plastics hold promise for a more sustainable future. However, they should be considered as part of a broader strategy to reduce overall plastic consumption, promote recycling, and encourage responsible waste management practices to combat plastic pollution effectively.

Table 8.1 Polymers used in formulation of bioplastics (Hakan et al. 2019)

Polymers in bioplastic	Molecular structure
Starch	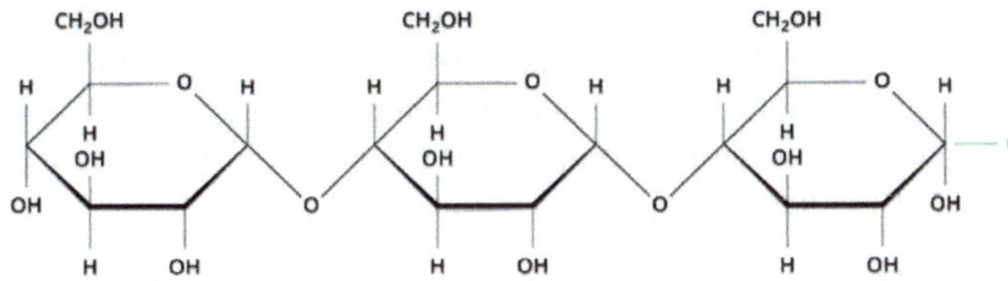
PLA	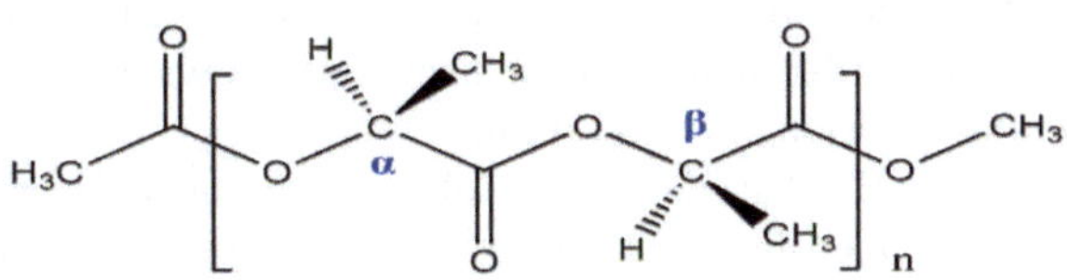
PHA	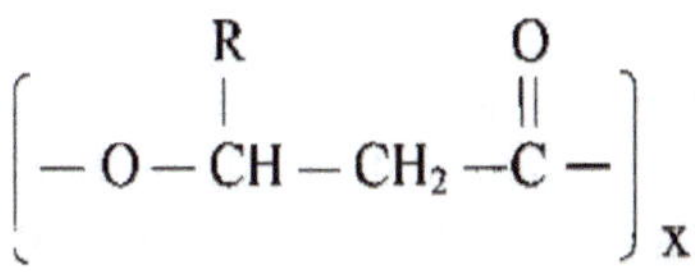
PHB	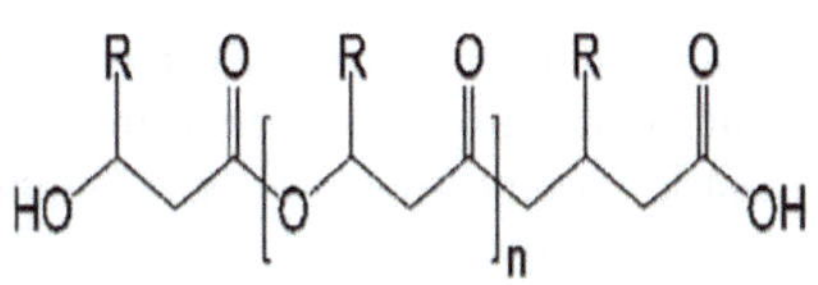
Cellulose	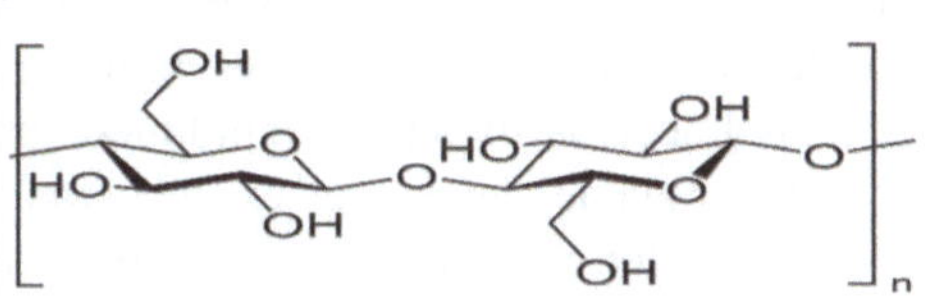
Chitosan	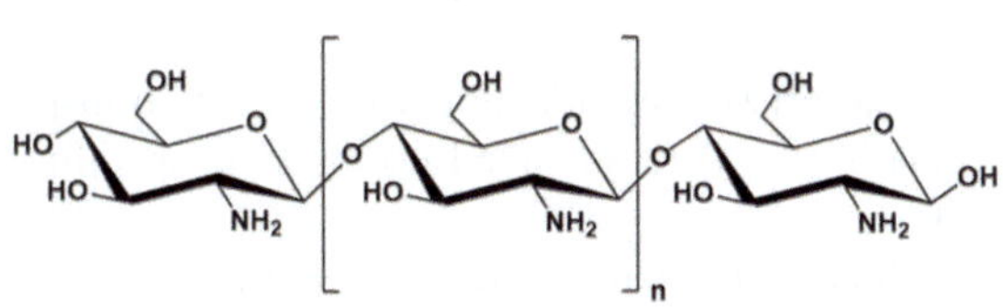
Chitin	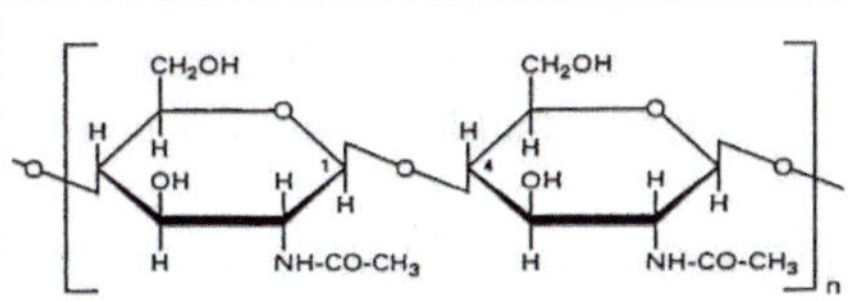

8.3.2 Attributes of Bioplastic

Biodegradable plastics possess a range of properties that make them distinct from conventional plastics (Woldesilassie and Ivatury 2020; Hayes 2021). These properties contribute to their ability to break down naturally in the environment. However, it's important to note that the specific properties can vary based on the type of biodegradable plastic and its composition (Geyer et al. 2017). Here are some common properties of biodegradable plastics:

1. Biodegradability: The most defining property of biodegradable plastics is their ability to undergo decomposition through the activity of microorganisms. They break down into simpler, natural substances, such as water, carbon dioxide, and biomass, within a reasonable time frame.
2. Renewable source: Bio-based biodegradable plastics are derived from renewable resources, such as plant-based materials (e.g., corn starch, sugarcane, and potato starch). This property distinguishes them from conventional plastics, which are derived from non-renewable fossil fuels.
3. Decomposition rate: The rate of biodegradation can vary depending on the type of biodegradable plastic and the environmental conditions. Some biodegradable plastics can degrade relatively quickly under suitable conditions, while others may take longer.
4. Temperature sensitivity: Biodegradable plastics may have specific temperature requirements for efficient degradation. In certain environments, such as composting facilities or industrial composting processes, higher temperatures can accelerate the biodegradation process.
5. Water sensitivity: Some biodegradable plastics may be more sensitive to moisture, which can promote faster degradation. However, in marine environments, the presence of water may not necessarily lead to rapid biodegradation.
6. Mechanical properties: Biodegradable plastics can exhibit a range of mechanical properties, including flexibility, tensile strength, and impact resistance. These properties can be tailored during the manufacturing process to suit different applications.
7. Transparency and color: Biodegradable plastics can be transparent, translucent, or colored, depending on their intended use and the specific additives used in their production.
8. Compatibility with conventional plastics: In some cases, biodegradable plastics can be mixed with conventional plastics to improve their overall environmental impact. However, the percentage of biodegradable material in the mixture and the specific properties of the blend should be carefully considered for optimal performance.
9. Toxicity: Biodegradable plastics are generally designed to break down into non-toxic substances. However, during the degradation process, certain by-products or additives could be released, which may require assessment to ensure they do not pose environmental or health risks.

8.4 Types of Bioplastic

Bioplastics can be of different types such as biodegradable and non-biodegradable plastics (Fig. 8.1). Bio-based plastics, also known as bioplastics, are a category of biodegradable plastics derived from renewable resources, such as plants or biomass (Barra and Gonzalez 2018; Jem and Tan 2020; Paul et al. 2021). Bioplastics derived from natural sources are being manufactured through three major pathways such as (1) bio-monomer polymerization (2) modification of naturally existing polymer (3) extraction from microorganisms (Hayes 2021). Most of these bioplastics offer a more sustainable alternative to conventional plastics, which are derived from non-renewable fossil fuels (Lotfi et al. 2019). There are several types of bio-based plastics, each with its unique properties and applications.

1. Polylactic Acid (PLA): PLA is one of the most widely used bio-based plastics. It is made from fermented plant sugars, typically derived from crops like corn, sugarcane, or cassava. PLA is compostable and can be used in various applications, including food packaging, disposable tableware, and 3D printing (Azieyanti et al. 2020).
2. Starch-based plastics: Starch-based bioplastics are made from renewable sources like corn, wheat, or potatoes. They are often blended with other biodegradable polymers to improve their properties and enhance their biodegradability. Starch-based plastics are commonly used in short-term applications like packaging and agricultural films.
3. Polyhydroxyalkanoates (PHA): PHA is a group of biodegradable polymers produced by microorganisms through the fermentation of renewable feedstocks. They are biodegradable in various environments, including marine conditions.

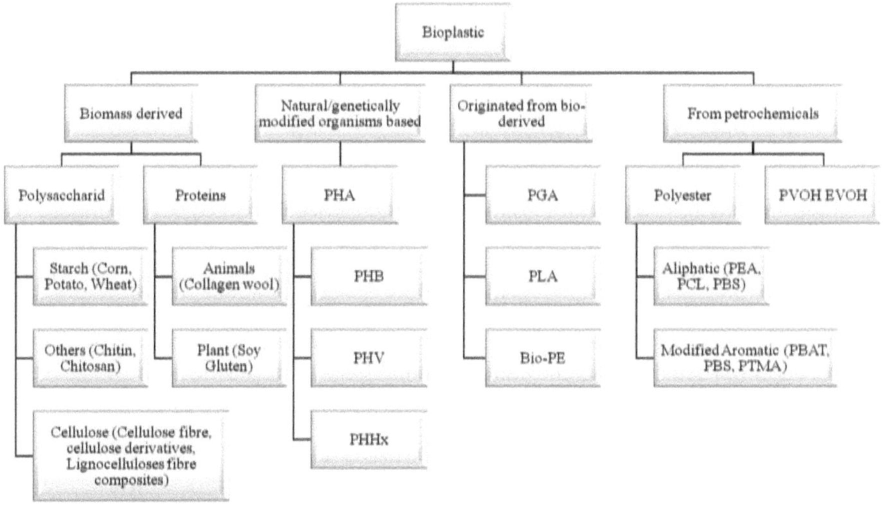

Fig. 8.1 Classifications of bioplastics

PHA has applications in single-use products, medical devices, and compostable bags.

4. Polybutylene Adipate Terephthalate (PBAT): PBAT is a bio-based copolymer that can be combined with other bioplastics to improve their flexibility and biodegradability. It is commonly used in packaging materials and agricultural films (Fu et al. 2020).
5. Polyethylene (PE) and Polypropylene (PP) blends: Some bio-based plastics are produced by blending traditional polyethylene or polypropylene with bio-based additives, such as plant-derived polymers or additives promoting biodegradation. These blends can improve the environmental profile of conventional plastics without requiring significant changes to existing manufacturing processes.
6. Polyglycolic Acid (PGA): PGA is a biodegradable polymer often used in medical applications like sutures and implants. It degrades into lactic acid, a naturally occurring compound.
7. Polyethylene Furanoate (PEF): PEF is a bio-based alternative to PET (polyethylene terephthalate). It is made from renewable resources and has excellent barrier properties, making it suitable for beverage bottles and food packaging.
8. Polycaprolactone (PCL): PCL is biodegradable polyester used in various applications, including 3D printing, drug delivery systems, and wound dressings.

8.5 Resources of Bioplastic

Bioplastics are derived from renewable resources, which are organic materials that can be replenished relatively quickly or sustainably. Bioplastics can be classified in mainly two types as per their characteristic features: biodegradable plastic and bio-based plastic. It also can be categorized as per the origin or protocol of manufacturing (Fig. 8.2).

8.5.1 Feedstock for Bioplastic

It's important to note that the choice of feedstock for bioplastics can impact the overall sustainability and environmental profile of the material. Sustainable sourcing, responsible land use, and minimal competition with food production are essential considerations when using renewable resources for bioplastics (Asgher et al. 2020). Additionally, ongoing research and innovation continue to explore new feedstock options and improve the efficiency and sustainability of bioplastic production. Some of the common renewable resources used for producing bioplastics include:

1. Corn (Maize): Corn is one of the most widely used feedstocks for bioplastics, particularly for the production of polylactic acid (PLA). Corn starch is fermented to produce lactic acid, which is then polymerized to form PLA (IPCC 2021).

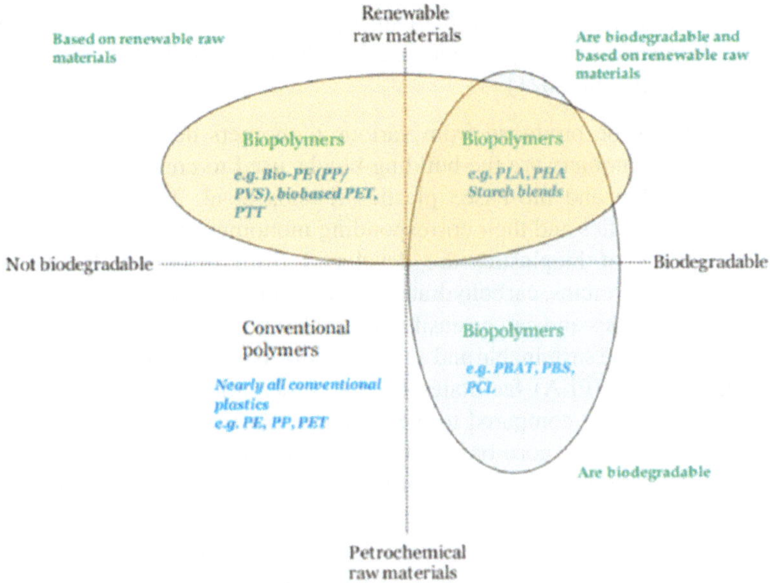

Fig. 8.2 Types of bioplastic relying upon several available resources

2. Sugarcane: Sugarcane is another important source of feedstock for bioplastics, particularly for producing bio-based polyethylene (PE) and polyethylene terephthalate (PET). Sugarcane ethanol can be used to produce ethylene and terephthalic acid, which are essential components of these plastics (Nanda et al. 2022).
3. Potatoes: Potatoes are used to produce starch-based bioplastics, where the starch is extracted and processed to create biodegradable plastics.
4. Cassava: Cassava, also known as tapioca, is a source of starch that can be used to produce biodegradable plastics (Harunsyah et al. 2018).
5. Wheat: Wheat is utilized for producing bioplastics through the extraction and processing of wheat starch (IPCC 2021).
6. Algae: Some research is being conducted to use algae as a feedstock for bioplastics due to its high growth rate and potential as a sustainable resource (Fredi and Dorigato 2021).
7. Cellulose: Cellulose, derived from various plant sources like wood and cotton, can be used to produce cellulose-based bioplastics like cellulose acetate.
8. Soybeans: Soybeans can be used as a source of protein to produce bioplastics through fermentation processes.
9. Vegetable Oils: Vegetable oils like soybean oil, canola oil, and palm oil can be converted into bio-based polymers, such as polyurethanes (IPCC 2021).
10. Seaweed: Seaweed is being explored as a potential feedstock for bioplastics due to its rapid growth and abundance in marine environments.

11. Waste Biomass: Some bioplastics can be produced from waste biomass, such as agricultural and food waste, reducing the reliance on dedicated crop feedstocks (Zaborowska et al. 2021).

Bioplastics can be produced from various monomers derived from renewable sources. These monomers are the building blocks used to create the polymers that form biodegradable and bio-based plastics (Nanda et al. 2022). Some common examples of bioplastics and their corresponding monomers are shown in Table 8.2.

As such types of bioplastics are developed from several biological macromolecules such as proteins, carbohydrates, or fats, which are the main components of biogeochemical cycles and can be easily degradable without producing any harmful toxicants enabling the sustainable and environmental beneficial product applications, like Polylactic Acid (PLA) facilitates energy preservation of approximately 66% during production than compared to the synthetic plastic production (Zaborowska et al. 2021). Similarly the corn-based PLA bioplastics can help in reduction of 25% of greenhouse gas emission than that of the synthetic ones (Alvarez-Mendez et al. 2023).

Natural polymers, also known as biopolymers, are polymers derived from renewable and natural sources (Narancic et al. 2020; Nakanishi et al. 2020). They can be used to produce bioplastics as sustainable alternatives to conventional petroleum-based plastics. These natural polymers offer biodegradability and low environmental impact, making them attractive alternatives to conventional plastics. Their versatility and wide availability from renewable sources continue to drive research and development in the field of bioplastics as part of sustainable efforts to reduce plastic waste and environmental pollution (Ismail et al. 2016; Jiménez-Rosado et al. 2020). Some of the common natural polymers used for bioplastics include:

1. Starch: Starch is one of the most widely used natural polymers for bioplastics. It is abundant in various plant sources, such as corn, potatoes, wheat, and tapioca. Starch-based bioplastics can be biodegradable and find applications in single-use items, food packaging, and agricultural films.
2. Cellulose: Cellulose is the primary structural component of plant cell walls. It can be extracted from various sources, including wood, cotton, and agricultural residues. Cellulose-based bioplastics, like cellulose acetate and cellulose nitrate, are commonly used in film applications and certain types of coatings.
3. Chitosan: Chitosan is derived from the shells of crustaceans like shrimp and crab. It is a biodegradable and antimicrobial natural polymer used in medical applications, wound dressings, and drug delivery systems.
4. Alginates: Alginates are extracted from brown seaweeds and have applications in the food industry as thickeners and stabilizers. They are also used in biodegradable films and coatings.
5. Polylactic Acid (PLA): PLA is biodegradable polyester made from renewable plant sources, such as corn starch or sugarcane. It is a versatile biopolymer used in various applications, including food packaging, disposable cutlery, and 3D printing.

Table 8.2 Different bioplastics and required monomers

Name	Monomer	Source	Application	References
Polylactic Acid (PLA)	Monomer: Lactic acid ($C_3H_6O_3$)	Fermentation of sugars derived from renewable resources like corn, sugarcane, or cassava and polymerized to form PLA	Used in packaging, disposable cutlery, and 3D printing	Alvarez-Mendez et al. (2023), Ateş and Kuz (2020) and Azieyanti et al. (2020)
Polyhydroxyalkanoates (PHA)	Hydroxyalkanoic acids, such as 3-hydroxybutyric acid and 3-hydroxyvaleric acid	Microbial fermentation of renewable feedstocks like plant sugars or vegetable oils	Used in packaging, agriculture, and medical products	Coppola et al. (2021)
Polyethylene Furanoate (PEF)	2, 5-Furandicarboxylic acid (FDCA) and ethylene glycol (EG)	FDCA can be derived from biomass sources like fructose, while EG can be sourced from renewable feedstocks like sugar or glycerol	Used in beverage bottles and food packaging	Paul and Ghosh (2022)
Polybutylene Succinate (PBS)	Succinic acid and 1,4-butanediol	Succinic acid can be produced from renewable feedstocks like corn or sugarcane, while 1, 4-butanediol can be sourced from plant-based sugars or other renewable sources	Used in packaging and agricultural applications	Paul and Ghosh (2022)
Polycaprolactone (PCL)	Caprolactone ($C_6H_{10}O_2$)	Caprolactone can be produced from renewable resources such as plant-based feedstocks	Used in 3D printing, drug delivery systems, and wound dressings	Paul and Ghosh (2022)

6. Polyhydroxyalkanoates (PHA): PHAs are biodegradable polyesters produced by microorganisms from renewable feedstocks like plant sugars or vegetable oils. They are used in a range of applications, including packaging, agriculture, and medical products.
7. Proteins: Proteins from various sources, such as soybeans, can be processed to create biodegradable bioplastics used in packaging and other applications.
8. Gelatin: Gelatin is derived from animal collagen and has been used to create biodegradable films and coatings.
9. Casein: Casein is a protein found in milk and can be used to create biodegradable plastics.
10. Agar: Agar is a gelatinous substance derived from red seaweeds. It can be used to produce biodegradable films and coatings.

8.5.2 Plant Oils for Bioplastic

Plant oils are a valuable resource for the production of bioplastics (Carlozzi et al. 2020; Candamano et al. 2021). These oils contain triglycerides, which can be converted into bio-based polymers through various chemical processes (Prasteen et al. 2018; Quirino et al. 2021). Some of the common natural polymers used for bioplastics include:

1. Soybean oil: Soybean oil is one of the most widely used plant oils for bioplastics. It contains a high percentage of unsaturated fatty acids, making it suitable for polymerization into polyurethanes and other bio-based polymers.
2. Canola oil: Canola oil, derived from rapeseed, is plant oil used for producing bioplastics. It has a relatively high amount of oleic acid, which can be utilized in the production of bio-based polyols and polyurethanes (Prasteen et al. 2018).
3. Corn oil: Corn oil, extracted from corn kernels, can be used to produce various bio-based polymers, including biodegradable polyesters like polylactic acid (PLA).
4. Palm oil: Palm oil is widely used plant oil, but its use in bioplastics has been a subject of debate due to concerns about deforestation and habitat destruction. However, efforts are being made to ensure sustainable sourcing of palm oil for bioplastic production (Quirino et al. 2021).
5. Castor oil: Castor oil is rich in ricinoleic acid, which can be converted into bio-based polyamides, such as nylon-11 and nylon-10,10.
6. Sunflower oil: Sunflower oil contains a high proportion of linoleic acid, which can be used for producing bio-based polyamides and polyols.
7. Jatropha oil: Jatropha oil is derived from the seeds of the Jatropha curcas plant. It contains a high percentage of unsaturated fatty acids and can be used for bioplastic production.
8. Flaxseed oil: Flaxseed oil, also known as linseed oil, is rich in alpha-linolenic acid and can be used in the production of bio-based polyesters.

9. Olive oil: Olive oil contains a high amount of oleic acid and can be utilized in the production of bio-based polyurethanes (Prasteen et al. 2018)

These plant oils offer advantages as feedstocks for bioplastics, as they are renewable resources and can be sourced sustainably. Their use in bioplastic production contributes to the development of more sustainable and environmentally friendly alternatives to conventional petroleum-based plastics reducing carbon footprint. However, it is crucial to consider the overall sustainability of these feedstocks, including land use, potential competition with food production, and environmental impacts, when using them for bioplastic production (Coppola et al. 2021; Woodford 2020; Sabbah and Porta 2017).

8.5.3 Natural Rubber for Bioplastic

Natural rubber is not commonly used as a direct source for producing bioplastics. Natural rubber is a polymer itself, consisting of repeating units of isoprene monomers. However, it is different from bioplastics, which are typically produced from renewable plant-based feedstocks and are designed to be biodegradable or derived from bio-based polymers (Coppola et al. 2021). Natural rubber is harvested from the latex of rubber trees (*Hevea brasiliensis*) or other plants like guayule and dandelions. It has unique properties, including high elasticity and tensile strength, which make it suitable for various applications such as tires, footwear, adhesives, and industrial goods. Natural rubber is a valuable commodity, but it is not considered a bioplastic due to its non-biodegradable nature and the fact that it is not a synthetic polymer derived from renewable sources. That being said, there are efforts to develop bio-based or biodegradable alternatives to synthetic rubber, such as styrene-butadiene rubber (SBR) and polybutadiene rubber (BR), by using renewable feedstocks or incorporating biodegradable additives (Babalola and Olorunnisola 2019). Research is ongoing in the field of sustainable rubber production, and advancements may lead to the development of more eco-friendly rubber materials in the future.

8.6 Benefits of Bioplastic Over Petroleum-Based Plastic

The use of biobased plastics offers several benefits, both from an environmental and economic standpoint. While biobased plastics offer many advantages, it is essential to consider their full life cycle, including their sourcing, production, disposal, and end-of-life management, to ensure their overall sustainability. Additionally, biobased plastics should be used in combination with other sustainable practices, such as waste reduction, recycling, and responsible consumption, to maximize their positive impact on the environment and move toward a more sustainable future (Nanda et al. 2022; Carlozzi et al. 2020; Candamano et al. 2021).

1. Renewable Resource: Biobased plastics are derived from renewable resources, such as plant-based materials, agricultural waste, or algae. Unlike conventional plastics that rely on fossil fuels (a finite and non-renewable resource), biobased plastics help reduce dependence on oil and other non-renewable resources.
2. Reduced Greenhouse Gas Emissions: The production of biobased plastics generally generates lower greenhouse gas emissions compared to traditional plastics. Plants absorb carbon dioxide during their growth, and this carbon is later incorporated into the biobased plastics, making them carbon–neutral or even carbon-negative in some cases.
3. Reduced Environmental Impact: Biobased plastics have the potential to biodegrade naturally over time, reducing their presence in landfills and marine environments. This can help alleviate plastic pollution and its adverse effects on wildlife and ecosystems.
4. Energy Efficiency: The production of biobased plastics can require less energy compared to the energy-intensive processes used in traditional plastic production from fossil fuels.
5. Diversification of Feedstocks: Biobased plastics can be produced from a wide range of feedstock sources, promoting agricultural diversification and creating opportunities for farmers and agricultural communities.
6. Versatility: Biobased plastics can be engineered to have a wide range of properties, making them suitable for various applications, including packaging, consumer goods, automotive parts, medical devices, and more.
7. Circular Economy: Biobased plastics can contribute to the development of a circular economy by promoting recycling, composting, and sustainable waste management practices. In some cases, they can be recycled alongside traditional plastics, reducing the environmental impact of plastic waste.
8. Reduced Toxicity: Many biobased plastics are designed to break down into non-toxic substances, reducing potential harm to the environment and living organisms.
9. Consumer Preference: With increasing awareness and concern about environmental issues, consumers are often more inclined to support products made from biobased plastics as a sustainable alternative to traditional plastics.
10. Innovation and Research: The development and adoption of biobased plastics drive innovation and research in the field of sustainable materials, leading to potential breakthroughs in other eco-friendly technologies.

8.7 Examples of Bioplastic

8.7.1 Bioplastic Bags

The manufacturing techniques for bioplastic bags can vary depending on the type of bioplastic used and the desired properties of the bags. It's important to note that the performance and end-of-life characteristics of bioplastic bags can vary depending on the specific type of bioplastic used, the manufacturing process, and the disposal method. To make informed decisions about bioplastic bags, it's crucial to consider their intended use, compostability, and appropriate disposal methods to maximize their environmental benefits (Nanda et al. 2022).

8.7.1.1 Manufacturing Techniques

1. Film Extrusion: Film extrusion is one of the most common methods for producing bioplastic bags. The bioplastic material, usually in the form of pellets or granules, is fed into an extruder. The extruder heats and melts the bioplastic, which is then forced through a die to form a continuous film. The film is then cooled, stretched, and wound into rolls to create plastic film suitable for bag making. The film can be transparent, translucent, or colored depending on the additives used (Ismail et al. 2016).
2. Blown Film Extrusion: Blown film extrusion is a variation of film extrusion used to produce tubular bioplastic bags, such as trash bags and grocery bags. The process involves extruding the molten bioplastic through a circular die to form a tube. As the tube exits the die, air is blown into it, causing the tube to expand and take on a tubular shape. The tube is then cooled, flattened, and cut into the desired bag length.
3. Injection Molding: Injection molding can be used to produce thicker bioplastic bags, such as shopping bags with handles. The bioplastic material is melted and injected into a mold under high pressure. The mold is then cooled, and the solidified bag is ejected from the mold.
4. Thermoforming: Thermoforming is another method used to produce thicker bioplastic bags or containers. A sheet of bioplastic is heated until it becomes pliable. The heated sheet is then placed over a mold and formed into the desired bag or container shape using vacuum or pressure. The formed product is then cooled and trimmed to the final shape.
5. Welding and Sealing: After the film is produced or the bag is formed, it may undergo welding and sealing processes to create the final bag shape. Welding and sealing can be done through heat sealing, ultrasonic welding, or adhesive bonding, depending on the type of bioplastic and bag design.
6. Printing and Finishing: Bioplastic bags may undergo printing processes to add logos, branding, or other graphics. After printing, the bags can be finished with handles, zippers, or other closures, depending on their intended use. It's important to note that the specific manufacturing techniques may vary based on the bioplastic material used and the bag's intended application and design. The selection of appropriate machinery and processing parameters is essential to ensure the quality and performance of the bioplastic bags. Additionally, sustainable practices and responsible waste management should be considered throughout the bag manufacturing process to maximize the environmental benefits of bioplastics.

8.7.1.2 Types of Bioplastic Bags

1. Biodegradable Plastic Bags: These bags are made from biodegradable bioplastics, such as polylactic acid (PLA) or polyhydroxyalkanoates (PHA). They are designed to break down naturally over time when exposed to certain conditions, such as industrial composting facilities or aerobic environments, reducing their environmental impact (Jiménez-Rosado et al. 2020).
2. Compostable Plastic Bags: Compostable bags are a subset of biodegradable bags that meet specific compostability standards, such as ASTM D6400 or EN 13,432. They can be composted along with organic waste in industrial composting facilities, where they break down into nutrient-rich compost.
3. Starch-based Plastic Bags: Starch-based bioplastic bags are made from renewable sources like corn or potatoes. They are often used as single-use bags for packaging and other lightweight applications (Candamano et al. 2021).
4. Polycaprolactone (PCL) Bags: PCL bags are made from the biodegradable polymer polycaprolactone. These bags have a slower degradation rate and can be used in applications where longer-term durability is required.
5. Polyhydroxyalkanoates (PHA) Bags: PHA bags are made from the biopolymer polyhydroxyalkanoates. They are fully biodegradable and suitable for a variety of applications, including packaging and agriculture.
6. Polybutylene Succinate (PBS) Bags: PBS bags are made from the biodegradable polymer polybutylene succinate. They are used in packaging and agricultural applications (Al-Battashi et al. 2020).
7. Polyethylene Furanoate (PEF) Bags: PEF bags are made from the bio-based polymer polyethylene furanoate. They are a sustainable alternative to conventional polyethylene bags.
8. Algae-based Plastic Bags: Some bioplastic bags are made from polysaccharides derived from algae, which offer a more sustainable source compared to traditional feedstocks (Quirino et al. 2021).
9. Blended Bioplastic Bags: Bioplastics can also be blended with conventional plastics to create partially biobased bags with improved properties. These bags may be labeled as "bio-based" or "partially biodegradable".
10. Bioplastic Shopping Bags: Bioplastic shopping bags are commonly used in retail settings as an eco-friendly alternative to traditional plastic shopping bags. They can be made from various biodegradable or compostable bioplastics.

8.7.2 Biodegradable Sanitary Napkins

Bioplastic sanitary napkins are eco-friendly and biodegradable alternatives to conventional disposable sanitary pads made from synthetic materials like plastics and superabsorbents (Ling et al. 2021; Quirino et al. 2021). These biodegradable sanitary napkins are designed to be more environmentally friendly and reduce the environmental impact of menstrual waste. The use of bioplastics in sanitary napkins reduces the dependency on conventional plastics derived from fossil fuels, which can take hundreds of years to decompose. Biodegradable sanitary napkins, when disposed of properly, can break down naturally in composting facilities or in soil, reducing the environmental burden of menstrual waste. It's essential to note that the effectiveness and performance of bioplastic sanitary napkins should be comparable to conventional sanitary pads to ensure comfort and reliable protection during menstruation. As the demand for eco-friendly menstrual products increases, manufacturers are continuously improving the design and materials used in bioplastic sanitary napkins to provide sustainable and efficient options for menstrual hygiene.

8.7.2.1 Parts of Sanitary Napkins

1. Absorbent Core: The absorbent core of a bioplastic sanitary napkin is usually made from natural and biodegradable materials like wood pulp, bamboo fiber, or other plant-based fibers. These materials can efficiently absorb menstrual fluid (Ling et al. 2021).
2. Bioplastic Topsheet: The topsheet of the bioplastic sanitary napkin is made from biodegradable and compostable materials such as polylactic acid (PLA) or polyhydroxyalkanoates (PHA). This bioplastic layer comes into direct contact with the skin and ensures comfort and softness.
3. Leak-Proof Backsheet: The backsheet of the bioplastic sanitary napkin is also made from biodegradable materials to prevent leakage. It helps keep the menstrual fluid contained and prevents it from passing through the pad.
4. Biodegradable Adhesives: Biodegradable adhesives are used to secure the different layers of the sanitary napkin together. These adhesives are formulated to be gentle on the skin and eco-friendly.
5. Individual Wrapping: Bioplastic sanitary napkins may come individually wrapped in biodegradable packaging to maintain hygiene and portability.

8.7.3 Bioplastic Bottles

Bioplastic bottles can be made from various types of biodegradable or bio-based materials, each with its manufacturing techniques. Bioplastic bottle manufacturers often optimize their processes to ensure the bottles meet the required quality, strength, and biodegradability standards while minimizing the environmental impact of production (Ling et al. 2021).

8.7.3.1 Manufacturing Techniques

1. Injection Molding: Injection molding is a common technique used to produce bioplastic bottles. The bioplastic material is heated and injected into a mold under high pressure. The mold is then cooled and opened to release the solidified bottle.
2. Stretch Blow Molding: Stretch blow molding is used to create bottles with higher clarity and flexibility. It involves reheating the preform, which is a partially formed bottle, and then stretching and blowing it into the final bottle shape using compressed air.
3. Extrusion: Extrusion is used for creating bottles with a continuous shape, such as tubes. The bioplastic material is melted and extruded through a die to form a continuous profile, which is then cut into the desired bottle length.
4. Thermoforming: Thermoforming can be used for producing bottles with thicker walls. The bioplastic material is heated until it becomes pliable, and then it is molded into the bottle shape using vacuum or pressure.

8.7.3.2 Types of Bioplastic Bottles

1. Polylactic Acid (PLA) Bottles: PLA is a bio-based bioplastic made from renewable resources like cornstarch or sugarcane. PLA is typically processed through polymerization to create the polymer chains. The PLA resin is then melted and molded using injection molding or stretch blow molding techniques to form the final bottle shape.
2. Polyhydroxyalkanoates (PHA) Bottles: PHA is a group of biodegradable bioplastics produced by microbial fermentation of plant sugars or vegetable oils. It can be processed through various techniques, including injection molding and stretch blow molding, similar to conventional plastic bottles.
3. Polyethylene Furanoate (PEF) Bottles: PEF is a bio-based bioplastic made from renewable resources like plant sugars. It can be processed through polymerization and then molded using injection molding or stretch blow molding techniques.
4. Biodegradable Starch-Based Bottles: Starch-based bioplastics are made from plant starches, such as corn or potato starch. These are often processed through extrusion or injection molding to form the bottle shape.
5. Biodegradable Polycaprolactone (PCL) Bottles: PCL is a biodegradable polymer made from renewable sources like plant-based feedstocks. It is typically processed using injection molding to create the bottle shape.

8.7.4 Biodegradable Tableware

Biodegradable tableware offers an eco-friendly alternative to traditional disposable tableware made from conventional plastics. These tableware items are typically made from biodegradable materials derived from renewable resources, such as plant fibers, bagasse (sugarcane waste), cornstarch, or biodegradable bioplastics. The manufacturing techniques for biodegradable tableware prioritize sustainability and eco-friendliness, aiming to reduce environmental impact while providing functional and safe alternatives to single-use plastic tableware (Azmin et al. 2020). The selection of appropriate materials and manufacturing methods ensures that the biodegradable tableware items are durable, food-safe, and capable of breaking down naturally, reducing plastic waste and supporting a more circular economy.

8.7.4.1 Manufacturing Techniques

1. Injection Molding: Injection molding is a common technique used for creating biodegradable tableware. The bioplastic material is heated and injected into a mold under high pressure. The mold is then cooled, and the solidified tableware item is ejected from the mold.
2. Thermoforming: Thermoforming is used for producing tableware with thicker walls and more intricate shapes. The biodegradable material is heated until it becomes pliable and is then molded into the desired shape using vacuum or pressure.
3. Pressing: Some biodegradable tableware, such as palm leaf tableware or wheat straw tableware, is made through pressing processes. The raw material is heat-pressed into molds to form the final shape.

8.7.4.2 Types of Biodegradable Tableware

1. Bagasse Tableware: Bagasse is the fibrous waste material left after extracting juice from sugarcane. It is pulped and processed into a moldable form. The pulp is then shaped using high-pressure molds and heat to create plates, bowls, trays, and other tableware items.
2. Cornstarch Tableware: Cornstarch-based tableware is made from a biodegradable bioplastic called polylactic acid (PLA), derived from cornstarch or other plant sugars. It is processed through polymerization to create the polymer chains. The PLA resin is then molded using injection molding or thermoforming techniques to produce tableware items.
3. Palm Leaf Tableware: Palm leaf tableware is made from fallen palm leaves that are collected, cleaned, and heat-pressed to form plates, bowls, and serving trays. The palm leaves are cleaned and then heated to make them more pliable. The heated leaves are pressed into molds to shape the tableware.
4. Wheat Straw Tableware: Wheat straw tableware is made from the stalks of wheat plants after the grains are harvested. It is processed into a moldable form, often using pulping and heating techniques. The processed wheat straw is then shaped using high-pressure molds to create plates, bowls, and other tableware items.

5. Wood Fiber Tableware: Wood fiber tableware is made from wood pulp obtained from sustainably managed forests. It is processed into a moldable form and shaped using high-pressure molds and heat to create durable and sturdy tableware items.

8.7.5 Biodegradable Food Packages

The selection of appropriate materials and manufacturing methods ensures that the biodegradable tableware items are durable, food-safe, and capable of breaking down naturally, reducing plastic waste and supporting a more circular economy. Biodegradable food packaging provides an eco-friendly alternative to conventional food packaging made from non-biodegradable materials (Jiménez-Rosado et al. 2020; Azmin et al. 2020). These packages are designed to break down naturally in the environment, reducing plastic pollution and supporting sustainable waste management practices (Carlozzi et al. 2020).

8.7.5.1 Types of Biodegradable Food Packages

1. Biodegradable Clamshell Containers: Biodegradable clamshell containers are used for takeaway food, salad containers, or food packaging with hinged lids. Biodegradable clamshell containers are typically made from molded bioplastic materials like PLA or starch-based plastics. The bioplastic is processed through injection molding or thermoforming to create the container's shape.
2. Biodegradable Food Wraps: Biodegradable food wraps are used as alternatives to plastic wraps for wrapping sandwiches, fruits, or other food items. Biodegradable food wraps are made from plant-based materials like beeswax-coated cotton, plant fibers, or compostable polymers. The materials are processed and coated to create the flexible food wraps (Carlozzi et al. 2020).
3. Biodegradable Food Trays: Biodegradable food trays are used for serving meals in food courts, cafeterias, and events. Biodegradable food trays are typically made from molded bioplastic materials like PLA or sugarcane bagasse. The bioplastic is processed through injection molding or thermoforming to create the tray's shape.
4. Biodegradable Coffee Cups and Lids: Biodegradable coffee cups and lids are used to serve hot beverages, such as coffee and tea. Biodegradable coffee cups and lids are often made from bioplastic materials like PLA. The cups are processed through injection molding or thermoforming, while the lids are made through film extrusion.
5. Biodegradable Food Pouches: Biodegradable food pouches are used for packaging liquids, sauces, or snacks. It can be made from biodegradable films processed through film extrusion or blown film extrusion. The pouches are then heat-sealed to close the package.

8.7.6 *Biodegradable Straws*

Biodegradable straws are eco-friendly alternatives to conventional plastic straws that are designed to break down naturally in the environment, reducing plastic pollution and its harmful impact on wildlife and ecosystems. These straws are made from biodegradable materials derived from renewable resources, such as plant-based materials or biodegradable bioplastics. Biodegradable straws offer a more sustainable choice for consumers and businesses looking to reduce plastic waste and promote more eco-friendly practices in their daily routines and operations (Azim et al. 2020).

8.7.6.1 Types of Biodegradable Straws

1. Paper Straws: Paper straws are made from biodegradable paper materials, such as wood pulp or bamboo fibers. Paper straws are sturdy, food-safe, and biodegradable. They can be printed with non-toxic ink and are available in various colors and sizes.
2. Wheat Straw Straws: Wheat straw straws are made from the stalks of wheat plants after the grains are harvested. Wheat straw straws are natural, biodegradable, and can be composted along with organic waste. They are suitable for cold beverages and are more sustainable than traditional plastic straws.
3. Bamboo Straws: Bamboo straws are made from natural bamboo stems that are cut and polished to create reusable and biodegradable straws. Bamboo straws are durable, reusable, and biodegradable. They offer an eco-friendly alternative to single-use plastic straws and can be used for hot or cold beverages.
4. Cornstarch-Based Straws: Cornstarch-based straws are made from a biodegradable bioplastic called polylactic acid (PLA), derived from cornstarch or other plant sugars. Cornstarch-based straws are sturdy, heat-resistant, and biodegradable. They can be used for hot and cold beverages and are an eco-friendly option for single-use straws.
5. Edible Straws: Edible straws are made from edible materials, such as rice, tapioca, or seaweed. Edible straws are not only biodegradable but also edible, making them a fun and eco-friendly option for sipping beverages. They can be flavored and add a unique twist to drinks.

The manufacturing process for biodegradable straws may vary depending on the type of material used. For example, paper straws are typically made through paper rolling and sealing processes, while biodegradable bioplastic straws require polymerization and extrusion techniques to form the straws' shape. In all cases, the goal is to create straws that are durable and functional for use with beverages while being environmentally friendly and capable of breaking down naturally when disposed of properly (Jiménez-Rosado et al. 2020).

8.8 Applications of Bioplastic

Bioplastics find applications in various industries and products due to their versatility and potential environmental benefits (Ateş and Kuz 2020; Azieyanti et al. 2020; El-malek et al. 2020; Al-Battashi et al. 2020; Ebrahimian et al. 2020). It's important to note that the specific application of bioplastics can vary based on the type of bioplastic used and its properties. While bioplastics offer environmental advantages compared to traditional plastics, their proper end-of-life management, such as composting or appropriate recycling, is essential to maximize their sustainability benefits. Some of the common applications of bioplastics include:

1. Packaging: Bioplastics are widely used in food packaging, disposable cutlery, cups, trays, and bags. They can offer barrier properties and help extend the shelf life of perishable goods.
2. Single-use items: Biodegradable and compostable bioplastics are used in single-use items like straws, plates, bowls, and utensils, reducing the environmental impact of disposable products.
3. Agricultural films: Biodegradable bioplastics are used in agricultural applications, such as mulch films, greenhouse films, and seedling pots. These films help improve crop yields and reduce plastic waste in agriculture.
4. Textiles: Some bioplastics can be used in the textile industry for making fibers and fabrics. These materials can be used in clothing, upholstery, and other textile applications.
5. Medical and healthcare: Bioplastics are used in medical devices, such as sutures, stents, drug delivery systems, and wound dressings. They can be designed to biodegrade safely within the body.
6. 3D printing: Bioplastics are becoming popular in 3D printing applications due to their ability to produce complex shapes and structures with reduced environmental impact.
7. Consumer goods: Bioplastics are used in a variety of consumer goods, including toys, electronic casings, phone cases, and cosmetic containers.
8. Automotive: Bioplastics find applications in the automotive industry, where they are used in interior parts, such as dashboards and door panels, to reduce the overall environmental impact of vehicles.
9. Medical packaging: Bioplastics are used for medical packaging to reduce the use of conventional plastics in the healthcare sector.
10. Electronics: Some bioplastics are used in electronic components and housings, reducing the reliance on petroleum-based plastics in electronic devices.
11. Horticulture and gardening: Biodegradable bioplastics are used for plant pots, nursery trays, and horticultural applications, providing a more sustainable alternative to traditional plastic products.
12. Fishing and aquaculture: Biodegradable bioplastics are used in fishing nets and other equipment to reduce plastic pollution in marine environments.

13. Packaging: Biodegradable plastics are widely used in packaging applications, such as food packaging, wrapping films, and bags. They can help reduce the environmental impact of packaging waste and offer sustainable alternatives to traditional single-use plastic packaging.
14. Agriculture: Biodegradable plastics are used in agriculture for various purposes. They can be used as mulch films to control weed growth and conserve moisture in crop fields. Biodegradable plant pots and nursery trays are also utilized in the horticulture industry.
15. Medical and Healthcare: Biodegradable plastics find applications in the medical and healthcare sectors. They are used in medical devices, such as sutures, drug delivery systems, and implants. Biodegradable plastics are ideal for applications where the material needs to degrade harmlessly within the body over time.
16. Consumer Goods: Biodegradable plastics are used in various consumer goods, including disposable cutlery, cups, plates, and straws. These items offer a sustainable alternative to conventional single-use plastic products.
17. 3D Printing: Biodegradable plastics are utilized in 3D printing, also known as additive manufacturing. They can be used to create biodegradable prototypes and products.
18. Textiles: Some biodegradable plastics, such as polylactic acid (PLA), can be used in the textile industry to create eco-friendly fibers and fabrics for clothing and other textile applications.
19. Automotive: Biodegradable plastics are used in the automotive industry for interior parts, such as dashboards and door panels, to reduce the environmental impact of vehicle components.
20. Fishing and Aquaculture: Biodegradable plastics are used in fishing nets and other equipment to reduce plastic pollution in marine environments and promote sustainable fishing practices.
21. Electronics: Some biodegradable plastics find applications in the electronics industry, where they can be used for casings and components.
22. Waste Management: Biodegradable plastics play a role in waste management, where they are used in compostable bags and liners for organic waste collection and in biodegradable waste bins.

8.9 Sustainable Approach of Bioplastic

There are several sustainable approaches for bioplastics that can be implemented to improve their environmental impact and promote long-term viability. These approaches include: resource maximization, life cycle assessment, environmental policies, waste management along with transparency and accountability (Moshood et al. 2022; IPCC 2021). Creating an enabling environment maximizes resource use, prioritizes resource allocation, and promotes fair resource distribution. This approach ensures efficient use of resources and reduces waste. Similarly considering the entire life cycle of bioplastics, from production to disposal; and identifying the areas where sustainability can be enhanced, might lead to take necessary steps. This includes sustainable design, manufacturing, chemical application, and end-product considerations (Siracusa and Blanco 2020). Implementing environmental policies that regulate the production of bioplastics, including the sourcing of feedstocks and manufacturing processes. These policies can help ensure sustainable practices and reduce environmental impacts. Moreover developing sustainable waste management systems that facilitate proper disposal and composting of biodegradable plastics. This includes raising awareness among consumers about the correct disposal methods and promoting the use of compostable biodegradable plastics (Babalola and Olorunnisola 2019). Further encouraging transparency in the production and treatment of bioplastics, including the use of resins, waste materials, and chemicals can also be a vital part. This promotes accountability and ensures adherence to sustainable practices. By adopting these sustainable approaches, bioplastics can become more environmentally friendly and contribute to a more sustainable future (Hakan et al. 2019; Fredi and Dorigato 2021).

In addition to their renewable and biodegradable properties, bioplastics also contribute to reducing greenhouse gas emissions. The production of bioplastics emits less carbon dioxide compared to traditional plastics, as the CO_2 released during their life cycle is balanced by the CO_2 absorbed during the growth of the bio-based materials (Jiang et al. 2020; Reichert et al. 2019). Bioplastics, such as the biopolymer PHB, play a crucial role in the carbon cycle. These biodegradable polymers can be synthesized by bacterial systems using carbon sources like sugar cane molasses, corn syrup, and paper waste. When these bioplastics are produced and eventually degraded by microorganisms, they release water and CO_2, completing the carbon cycle. This makes bioplastics a sustainable alternative to synthetic plastics, which contribute to environmental pollution and deplete non-renewable resources (Bharti and Swetha 2016; Zheng and Suh 2019). To ensure the sustainability of bioplastics, it is crucial to establish industrial/academic partnerships for research and development. This collaboration can help transform bioplastics research and address challenges such as the availability of sustainable raw materials and appropriate recycling options. Overall, the sustainable approach of bioplastics lies in their use of renewable resources, biodegradability, and reduced carbon footprint. By embracing bioplastics, we can move toward a more environmentally friendly and sustainable future (Sidek et al. 2019; Muthusamy and Pramasivam 2019).

8.10 Future Perspective and Conclusion

Bioplastics, which are produced from renewable raw materials, have gained significant attention due to their potential to reduce the environmental impact caused by conventional plastics. The future prospects of bioplastics are promising, as they offer several advantages such as biodegradability, compostability, and being environmentally friendly. The development of bioplastics is expected to continue, with a focus on improving biodegradability, expanding agricultural applications, and establishing reasonable recycling plants. However, it is crucial to ensure that bioplastics do not compete with traditional sources and reduce the need for non-renewable resources in the long term. The future of bioplastics relies on continued innovation, global support, and the integration of environmentally friendly practices throughout their lifecycle (Rocha et al. 2020; Zhu and Wang 2020).

1. The future perspective of bioplastics is promising, as they play a crucial role in addressing environmental challenges associated with conventional plastics (Rosenboom et al. 2022). The life cycle assessment of bio-based plastic can be a potent criterion for validation of its sustainable utilization (Walker and Rothman 2020). As awareness of plastic pollution and climate change increases, the demand for sustainable and eco-friendly materials is growing rapidly. While bioplastics offer numerous benefits, there are also challenges that need to be addressed. These challenges include the development of new guidelines and standards for bioplastics production, usage, and waste management on a global scale. Labeling legislation should be enhanced to provide information about the raw materials used, energy consumption, and emissions from manufacturing and use; additionally, the communication with societies about bioplastics can help in improving the sustainable approach for better future. Some of the key future perspectives of bioplastics: Bioplastics are likely to witness increased adoption across various industries, including packaging, consumer goods, automotive, and agriculture. As consumers become more environmentally conscious, businesses are seeking to reduce their carbon footprint by incorporating bioplastics into their products and packaging.
2. Ongoing research and development in bioplastics will lead to technological advancements, improving their properties, performance, and cost-effectiveness. Innovations in bioplastic formulations, processing techniques, and reinforcement technologies will expand their applications and competitiveness with conventional plastics.
3. The future will see a diversification of bioplastic materials derived from various renewable sources, such as algae, seaweed, agricultural waste, and other biomass. This diversification will enhance the availability and versatility of bioplastics, catering to a wider range of applications.
4. Efforts are being made to improve the biodegradability of bioplastics, enabling them to break down more efficiently in various environments, including home composting, marine environments, and industrial composting facilities.
5. Future bioplastics are likely to be designed to be more compatible with existing recycling and waste management infrastructure, addressing the challenges of segregating and processing biodegradable plastics.
6. Bioplastics will play a vital role in advancing the concept of the circular economy by promoting the reuse, recycling, and composting of materials. This approach will create a more sustainable and closed-loop system for plastics.
7. Governments worldwide are recognizing the importance of transitioning to sustainable materials and are likely to implement supportive policies, regulations, and incentives to encourage the use of bioplastics and reduce plastic waste.

8. Collaboration between research institutions, industries, and governments will accelerate the development and deployment of bioplastics. Partnerships with waste management companies and recycling facilities will promote the proper disposal and recycling of biodegradable plastics.
9. Bioplastics will contribute significantly to sustainable packaging solutions, reducing the environmental impact of packaging waste and supporting the shift toward eco-friendly alternatives.

Overall, the future of bioplastics looks promising as they emerge as a crucial component of the global sustainability movement. Bioplastics are made from renewable resources, such as plant-based materials, which reduce the reliance on fossil fuels. They have a lower carbon footprint and can decompose back into carbon dioxide, making them more environmentally friendly compared to petroleum-based plastics. With continuous research, innovation, and collaboration, bioplastics have the potential to revolutionize the plastics industry and contribute significantly to a greener and more sustainable future.

References

Al-Battashi H, Al-Kindi S, Gupta VK, Sivakumar N (2020) Polyhydroxyalkanoate (PHA) production using volatile fatty acids derived from the anaerobic digestion of waste paper. J Polym Environ 1:10. https://doi.org/10.1007/s10924-020-01870-0

Al-Khairy D, Fu W, Alzahmi AS, Twizere JC, Amin SA, Salehi-Ashtiani K, Mystikou A (2022) Closing the gap between bio-based and petroleum-based plastic through bioengineering. Microorganisms 10:2320. https://doi.org/10.3390/microorganisms10122320

Alvarez-Mendez SJ, Ramos-Suarez JL, Ritter A, Gonzalez JM, Perez AC (2023) Anaerobic digestion of commercial PLA and PBAT biodegradable plastic bags: potential biogas production and 1 H NMR and ATR-FTIR assessed biodegradation. Heliyon 9:e16691

Asgher M, Qamar SA, Bilal M, Iqbal HMN (2020) Bio-based active food packaging materials: sustainable alternative to conventional petrochemical-based packaging materials. Food Res Int 137:109625. https://doi.org/10.1016/j.foodres.2020.109625

Ateş M, Kuz P (2020) Starch-based bioplastic materials for packaging industry. J Sustain Constr Mater Technol 5(1):399–406. https://doi.org/10.29187/jscmt.2020.44

Azieyanti NA, Amirul A, Othman SZ, Misran H (2020) Mechanical and morphology studies of bioplastic-based banana peels. J Phys Conf Ser 1529:032091. https://doi.org/10.1088/1742-6596/1529/3/032091

Azmin SNHM, Binti NA, Hayat M, Nor MSM (2020) Development and characterization of food packaging bioplastic film from cocoa pod husk cellulose incorporated with sugarcane bagasse fibre. J Bioresour Bioprod 5:248–255. https://doi.org/10.1016/j.jobab.2020.10.003

Babalola OA, Olorunnisola AO (2019) Evaluation of coconut (*Cocos nucifera*) husk fibre as a potential reinforcing material for bioplastic production. Mater Res Proc 11:195–200. https://doi.org/10.21741/9781644900.178-14

Barra R, Gonzalez P (2018) Sustainable chemistry challenges from a developing country perspective: education, plastic pollution, and beyond. Curr Opin Green Sustain Chem 9:40–44

Beigbeder J, Soccalingame L, Perrin D, Benezet JC, Bergeret A (2019) How to manage biocomposites wastes end of life? A life cycle assessment approach (LCA) focused on polypropylene (PP)/wood flour and polylactic acid (PLA)/flax fibres biocomposites. Waste Manag 83:184–193

Bharti SN, Swetha G (2016) Need for bioplastics and role of biopolymer PHB: a short review. J Pet Environ Biotechnol 7:2. https://doi.org/10.4172/2157-7463.1000272

Coppola G, Gaudio MT, Lopresto CG, Calabro V, Curcio S, Chakraborty S (2021) Bioplastic from renewable biomass: a facile solution for a greener environment. Earth Syst Environ 5:231–251

Candamano S, Crea F, Coppola L, DeLuca P, Coffetti D (2021) Influence of acrylic latex and pre-treated hemp fibers on cement based mortar properties. Constr Build Mater 121720. https://doi.org/10.1016/j.conbuildmat

Carlozzi P, Giovannelli A, Traversi ML, Touloupakis E (2020) Poly(3- hydroxybutyrate) bioproduction in a two-step sequential process using wastewater. J Water Process Eng 101700. https://doi.org/10.1016/j.jwpe

Dilkes-Hoffman LS, Pratt S, Lant PA, Laycock B (2019) The role of biodegradable plastic in solving plastic solid waste accumulation. In: Plastics to energy. Elsevier, pp 469–505

Ebrahimian F, Karimi K, Kumar R (2020) Sustainable biofuels and bioplastic production from the organic fraction of municipal solid waste. Waste Manag 116:40–48. https://doi.org/10.1016/j.wasman.202007049

Ellen M (2016) The new plastics economy rethinking the future of plastics. Ellen Macarthur Foundation, Cowes, UK

El-malek FA, Khairy H, Farag A, Omar S (2020) The sustainability of microbial bioplastics, production and applications. Int J BiolMacromol 157:319–328. https://doi.org/10.1016/j.ijbiomac.2020.04.076

EPA (2023) National overview: facts and figures on materials, wastes and recycling. https://www.epa.gov/facts-and-figures-about-materials-waste-and-recycling/national-overview-facts-and-figures-materials. Accessed 1 Aug 2023

Fredi G, Dorigato A (2021) Recycling of bioplastic waste: a review. In: Advanced industrial and engineering polymer research, vol 4, pp 159–177

Fu Y, Wu G, Bian X, Zeng J, Weng Y (2020) Biodegradation behavior of poly(butylene adipate-co-terephthalate) (PBAT), poly (lactic acid) (PLA), and their blend in freshwater with sediment. Molecules 25:3946

Geyer R (2020) Chapter 2—production, use, and fate of synthetic polymers. In: Letcher TM (ed) Plastic waste and recycling. Academic Press, Oxford, pp 13–32

Geyer R, Venkatachalam V, Ling H (2017) Production, use, and fate of all plastics ever made. Sci Adv 3:e1700782

Hakan K, Christiane F, Grabert M, Oey M, Ben H (2019) Green bioplastics as part of a circular bioeconomy. Trends Plant Sci:S1360138518302723. https://doi.org/10.1016/j.tplants.2018.11.010

Harunsyah IO, Sariadi PL, Raudah ME (2018) The effect of clay nanoparticles as reinforcement on mechanical properties of bioplastic base on cassava starch. J Phys Conf Ser 953:012021

Hayes DG (2021) Enhanced end-of-life performance for biodegradable plastic mulch films through improving standards and addressing research gaps. Curr Opin Chem Eng 33:100695

Ibrahim NI, Shahar FS, Sultan MTH, Shah AUM, Safri SNA, Mat Yazik MH (2021) Overview of bioplastic introduction and its applications in product packaging. Coatings 11:1423. https://doi.org/10.3390/coatings11111423

IPCC (2021) Sixth assessment report. IPCC. https://www.ipcc.ch/report/ar6/wg1/

Ismail NA, Tahir SM, Yahya N, Wahid MFA, Khairuddin NE, Hashim I, Rosli N, Abdullah MA (2016) Bioplastic Mater Sci Forum. https://www.scientific.net/MSF.846.673

Jan-Georg R, Langer R, Traverso G (2022) Bioplastics for a circular economy. Nat Rev Mater 7:117–137

Jayakumar A, Radoor S, Siengchin S, Shin GH, Kim JT (2023) Recent progress of bioplastics in their properties, standards, certifications and regulations: a review. Sci Total Environ 878:163156

Jem KJ, Tan B (2020) The development and challenges of poly(lactic acid) and poly(glycolic acid). Adv Ind Eng Polym Res 3:60–70

Jiang T, Duan Q, Zhu J, Liu H, Yu L (2020) Starch-based biodegradable materials: challenges and opportunities. Adv Ind Eng Polym Res 3:8–18

Jiménez-Rosado M, Bouroudian E, Perez-Puyana V, Guerrero A, Romero A (2020) Biosynthesis of plastic. J Clean Prod 262:121517. https://doi.org/10.1016/j.jclepro.2020.121517

Laflamme S (2016) Bio-based plant oil polymers and composites. In: Application of plant oil-based products in structural health monitoring, pp 191–206. https://doi.org/10.1016/b978-0-323-35833-0.00011-6

Lau WWY, Shiran Y, Bailey RM, Cook E, Stuchtey MR, Koskella J (2020) Evaluating scenarios toward zero plastic pollution. Science 369(6510):1455–61

Lear G, Kingsbury JM, Franchini S, Gambarini V, Maday SDM, Wallbank JA, Weaver L, Pantos O (2021) Plastics and the microbiome: impacts and solutions. Environ Microbiome 16:2

Ling H, Tapan C, Nath K, Chong S, Foo V, Gibbins C (2021) The plastic waste problem in Malaysia: Management, recycling and disposal of local and global plastic waste. SN Appl Sci 3:1–15

Lotfi A, Li H, Dao DV, Prusty G (2019) Natural fibre–reinforced composites: a review on material, manufacturing, and machinability. J Thermoplast Compos Mater:0892705719844546

Moshood TD, Nawanir G, Mahmud F, Mohamad F, Ahmad MH, Ghani AA (2022) Sustainability of biodegradable plastics: new problem or solution to solve the global plastic pollution? Curr Res Green Sustain Chem 5:100273

Muthusamy MS, Pramasivam S (2019) Bioplastics–an eco-friendly alternative to petrochemical plastics. Curr World Environ 14:49

Nanda S, Patra BR, Patel R, Bakos J, Dalai AK (2022) Innovations in applications and prospects of bioplastics and biopolymers: a review. Environ Chem Lett 20(1):379–395. https://doi.org/10.1007/s10311-021-01334-4

Nakanishi A, Iritani K, Sakihama Y (2020) Developing Neo-bioplastics for the realization of carbon sustainable society. J Nanotechnol Nanomater 1(2):72–85. https://doi.org/10.33696/nanotechnol.1.010

Narancic T, Cerrone F, Beagan N, Oconnor KE (2020) Recent advances in bioplastics: application and biodegradation. Polymers (Basel). https://doi.org/10.3390/polym12040.920

Narancic T, Verstichel S, Reddy CS, Morales-Gamez L, Kenny ST, De-Wilde B, Babu PR, O'Connor KE (2018) Biodegradable plastic blends create new possibilities for end-of-life management of plastics but they are not a panacea for plastic pollution. Environ Sci Technol 52:10441–10452

Paul S, Ghosh S (2022) Identification of solid waste dumping site suitability of Kolkata Metropolitan Area using fuzzy-AHP model. Clean Logist Supply Chain 3:100030. https://doi.org/10.1016/j.clscn.2022.100030

Paul S, Sen B, Das S, Abbas SJ, Pradhan SN, Sen K, Ali SI (2021) Incarnation of bioplastics: recuperation of plastic pollution. Int J Environ Anal Chem. https://doi.org/10.1080/03067319.2021.1983552

Prasteen P, Thushyanthy Y, Mikunthan T, Prabhaharan M (2018) Bio-plastics—an alternative to petroleum based plastics. Int J Res Stud Agric Sci 4(1):1–7

Quirino RL, Monroe K, Fleischer CH, Biswas E, Kessler MR (2021) Thermosetting polymers from renewable sources. Polym Int 70:167–180

Reichert CL, Bugnicourt E, Coltelli MB, Cinelli P, Lazzeri A, Canesi I (2020) Bio-based packaging: materials, modifications, industrial applications and sustainability. Polymers 12:1558

Rocha CJL, Alvarez-Castillo E, Yáñez MRE, Bengoechea C, Guerrero A, Ledesma MTO (2020) Development of bioplastics from a microalgae consortium from wastewater. J Environ Manag 263:110353

Rosenboom JG, Langer R, Traverso G (2022) Bioplastics for a circular economy. Nat Rev Mater 7:117–137. https://doi.org/10.1038/s41578-021-00407-8

Sabbah M, Porta R (2017) Plastic pollution and the challenge of bioplastics. J Appl Biotechnol Bioeng 2:00033

Siracusa V, Blanco I (2020) Bio-polyethylene (Bio-PE), Bio-polypropylene (Bio-PP) and Bio-poly(ethylene terephthalate) (Bio-PET): recent developments in bio-based polymers analogous to petroleum-derived ones for packaging and engineering applications. Polymers 12:1641

Sidek IS, Draman SFS, Abdullah SRS, Anuar N (2019) Current development on bioplastics and its future prospects: an introductory review. INWASCON Technol Mag 1:03–08

Venkatachalam V, Spierling S, Endres HJ, Siebert-Raths A (2018) Integrating life cycle assessment and eco-design strategies for a sustainable production of biobased plastics. In: Designing sustainable technologies, products and policies: from science to innovation. Springer, Cham, pp 487–497

Walker S, Rothman R (2020) Life cycle assessment of bio-based and fossil-based plastic: a review. J Clean Prod 261:121158

Woodford M (2020) Modeling imprecision in perception, valuation, and choice. Ann Rev Econ 12:579–601

Woldesilassie TL, Ivatury VMK (2020) Key determinant factors affecting the performance of small and medium scale manufacturing enterprise: a case study on West Shoa Zone, Oromia national regional state, Ethiopia. Int J Eng Res 9:555–562

Zaborowska M, Bernat K, Pszczolkowski B, Wojnowska-Baryla I, Kulikowska D (2021) Anaerobic degradability of commercially available bio-based and oxodegradable packaging materials in the context of their end of life in the waste management strategy. Sustainability 13:6818. https://doi.org/10.3390/su13126818

Zheng J, Suh S (2019) Strategies to reduce the global carbon footprint of plastics. Nat Clim Change 9:374–378

Zhu J, Wang C (2020) Biodegradable plastics: green hope or greenwashing? Mar Pollut Bull 161:111774

Chapter 9
Biodegradable Plastics: Environmental Friendly Alternative to Petroleum Products

Mrinal Kashyap and Palak Saket

9.1 Introduction

The consequences of lifestyle choices that impact the ecosystem, including climate change, emissions, global warming, the falling supply of energy from fossil fuels, and waste disposal, have escalated to catastrophic levels (Talan et al. 2022, Leong et al. 2021). Out of several non-sustainable and environmentally unfriendly issues which challenge the goals of sustainable development is the global plastic demand which is drastically damaging the environmental and marine life (Chong et al. 2022). Global plastic production has accelerated in the last two decades. The report published by Statista research department suggests that the global plastic production increased from 1.5 million metric tons in the year 1950 to 490.7 million metric tons by 2021 (Plastic production worldwide 2021|Statista accessed on 14th August, 2023) (Statista Research Department 2023). Plastics are made up of hydrocarbon monomers forming a polymer, these polymers when disposed of in the environment undergo physical and chemical changes over a long period of time, impacting the environment and life of many living organisms (Varghese et al. 2022, Yee et al. 2021). The petroleum based plastics pose a high risk to its immediate environment and the marine ecosystem (Varghese et al. 2022; Gadhave et al. 2018). Petroleum based plastics are usually non-biodegradable in nature, and impose detrimental health impacts by releasing harmful gases (Duan et al. 2021; Thiruchelvi et al. 2021; Darbre 2020). The contamination of plastic particles in our immediate environment can be inferred by the latest research by a group of scientists that showed the presence of plastic particles in human blood (Leslie et al. 2022). Bioplastics are suggested as an alternative

M. Kashyap (✉)
Porter School of Environment and Earth Sciences, Tel Aviv University, Tel Aviv, Israel
e-mail: mrinalk@mail.tau.ac.il

P. Saket
Department of Biosciences and Biomedical Engineering, Indian Institute of Technology Indore, Indore, India

K. Bala et al. (eds.), *Harnessing Microbial Potential for Multifarious Applications*, Energy, Environment, and Sustainability, https://doi.org/10.1007/978-981-97-1152-9_9

to counter the rising global plastic demand, as they are biodegradable, non-toxic, and have minimal carbon footprint (Chong et al. 2022). According to the definition by European Union, bioplastics are derived from either biomass or biodegradable material (European bioplastics 2021). There are two major categories of bioplastics, the bioplastics developed from the plants and renewable energy sources, and the petroleum based bioplastics developed from the fossil fuels but have a biodegradable nature (Rahman and Miller 2017). The bio-based, biodegradable, fossil-based, and non-biodegradable quadrants make up the bioplastic and synthetic polymers (European bioplastics 2021). PE (polyethylene), PET (polyethylene terephthalate), PA (polyamide), PTT (polytrimethylene terephthalate) are examples of bio-based, non-biodegradable polymers. PLA (poly lactic acid), PHA (polyhydroxyalkanoates), and PBS (polybutylene succinate) are examples of bio-based biodegradable plastics (Chong et al. 2022). PE (polyethylene), PP (polypropylene), and PET (polyethylene terephthalate) are examples of fossil fuel based bioplastics which are non-biodegradable in nature. PBAT (polybutylene adipate terephthalate), and PCL (polycaprolactone) are examples of fossil fuel based plastics which are biodegradable (Chong et al. 2022).

There are several examples of production of bioplastics from various microorganisms including microalgae (Costa et al. 2018; Senatore et al. 2023; Nur et al. 2022). In a study three microalgae species namely *Spirulina* sp. LEB-18, *Chlorella minutissima*, and *Synechococcus subsalsus* were evaluated for the production of PHA under nitrogen deficient conditions (Costa et al. 2018). The highest PHA accumulation was observed after 15 days in *Synechococcus subsalsus* (16% dry cell weight), and 12% dry cell weight in *Spirulina* sp. respectively (Costa et al. 2018). In another study, an urban organic waste pre-treated with processes like steam explosion and acidogenic fermentation was used to cultivate purple phototrophic bacteria (PPB) (Anjum et al. 2016). The highest accumulation of 42% dry cell weight was observed and maintained for a period of 10 days using an organic loading rate of 1 $gCODL^{-1}d^{-1}$ (Anjum et al. 2016). The examples stated above indicate the potential of microbial bioplastic production capacity and its integration with wastewater treatment and utilization.

The current chapter gives insights into the probable pathways of producing bioplastics using microorganisms with a special focus on microalgae. Moreover, the economic challenges and the issues of technology readiness levels for industrial scale production of bioplastics are highlighted. The biorefinery approach has been discussed to address the economic issue in producing bioplastics from microalgae and cyanobacteria. Furthermore, this chapter highlights the importance of using microbial biomass in an integrated biorefinery to produce bioplastics.

9.2 Production Pathways of Biodegradable Plastics

Microbial bioplastics such as polyhydroxybutyrate (PHB) and polyhydroxyalkanoates (PHA) are ecologically compatible bioproducts which are produced through a variety of bioprocesses that use renewable resources. Bioprocesses such as fermentation can directly produce microbial bioplastics. Biopolymers produced entirely from microbial cells are biodegradable polymers. These biopolymers are obtained by growing bacterial cells in a certain environment. PHAs accumulate inside bacterial cells to the extent of 90% (w/w) of dry cell mass (Leong et al. 2021). PHA accumulation is a natural strategy for microbial cells to store carbon and energy during times of nutritional imbalance. PHAs are primarily synthesized via 3 pathways using amino acids, carbohydrates, and fatty acids as carbon substrates. PLA and PHA are the two most common commercial biopolymers, with PHA being more expensive than PLA. The qualities and applications, however, can be modified by altering the co-monomer and its content (Riaz et al. 2021; Poirier et al. 1995). Biopol™ was the commercially produced PHA through the fermentation of *Alcaligenes eutrophus* via glucose and propanoic acid pathway. The produced biopolymer was the copolymer of 3-hydroxybutyrate (3HB) and 3-hydroxyvalerate (3HV). These are one of the kinds of PHA, different types of PHAs having different biopolymeric properties can also be produced which highly depends on the strain and carbon source used for the formation of precursor for PHA (Avérous and Pollet 2012).

9.2.1 Bacterial Production of Biodegradable Plastics

Polyhydroxyalkanoates (PHA) are the most commonly investigated biopolymer. PHA produced from bacteria is reported to be a potentially biodegradable polymer having characteristics that are similar to those of some petrochemical polymers. PHA is accumulated by various micro-macroorganisms and accumulates around 30–80% of dry cell weight. PHA accumulates in the bacterial cell as the result of induced stress, which is usually caused by a lack of nitrogen and phosphorus in the presence of an adequate carbon supply. Under these stressed conditions, the growth of the cell gets reduced and the metabolism is shifted towards the biosynthesis of hydroxylalkyl-CoA (HA-CoA). It is a precursor which gets polymerised by the activity of PHA synthase enzyme and forms PHA polymer. PHB is the most common kind of PHA and is mostly resistant to hydrolytic degradation and is water insoluble in nature (Tan et al. 2014). To date, about 250 bacterial species have been identified as capable of producing PHB. The metabolic pathway for the production of PHA in a gram-negative or gram-positive bacterial cell is associated with Kreb's cycle, glycolysis, beta-oxidation pathway, serine pathway, and Calvin cycle. Under normal or nutrient-available conditions, the acetyl coenzyme A produced during Kreb's cycle suppresses the 3-ketothiolase (PhaA) enzyme, which inhibits PHA synthesis in the bacterial cell. In contrast, when nutrients are deprived (for example nitrogen and phosphorus), the coenzyme A (CoA) produced is insufficient to restrict the activity of the 3-ketothiolase enzyme, which redirects the CoA pathway towards PHA synthesis (Alcântara et al. 2020). The production of PHB is catalyzed by three enzymes: PhaA, PhaB, and PhaC. The glucose is converted into acetyl CoA, 2 molecules of which are transformed into acetoacetyl CoA by enzyme PhaA (β-ketothiolase) (Chaitanya et al. 2014). Later, the acetocetyl CoA converts to R-3 hydroxybutyryl by the activity of PhB (acetoacetyl CoA reductase). Lastly, the PhaC (PHB synthetase) enzyme catalyzes the conversion of R-3 hydroxybutyryl into PHB (Alcântara et al. 2020; Hoffmann et al. 2002). Moreover, PHB is also synthesized by the substrate 3-hydroxyalkanoyl-CoA obtained after the reduction of acetyl CoA in β-oxidation pathway. In this pathway, the synthesis of PHA from the precursor 3-hydroxy acyl CoA is catalyzed by several enzymes like epimerase, enoyl CoA hydratase 1, 3-keto acyl CoA reductase, and acyl CoA oxidase. However, PHA biosynthesis occurs via the fatty acid biosynthesis pathway de novo, catalyzed by acetyl–CoA and transacylase enzyme (PhaG), which are the results of glucose oxidation, and bypassing the fatty acid β-oxidation pathway (Samadhiya et al. 2022; Chen 2010). Figure 9.1 shows the graphical representation of microbial PHA production. Studies performed on several bacterial species and identified different enzymes and precursors for the synthesis of PHA. For example, PHA synthesis by precursor 3-hydroxy acyl CoA was studied in *Pseudomonas* sp. Similarly, in recombinant *E-coli,* 3-hydroxy acyl CoA for PHA synthesis is obtained by the activity of the following enzymes: malonyl-CoA-ACP-transacylase and 3-hydroxyacyl-ACP-CoA transferase, and the reaction is catalyzed by PHA synthase enzyme. Furthermore, in *Rhizobium* PHB is obtained by the oxidation of (S)-(+)-3-hydroxybutyryl-CoA by the activity of acetoacetyl CoA reductase enzyme. Moreover, PHA synthesis by 4-hydroxybutyryl-CoA as a precursor observed in *Clostridium kluyveri* was catalyzed by the activity of the following enzymes: succinic semialdehyde dehydrogenase, 4-hydroxybutyrate dehydrogenase, and 4-hydroxybutyrate CoA transferase (Lajus et al. 2020). Moreover, many more studies are listed in Table 9.1 on bacterial bioplastic production in different conditions. Several bacterial species are capable of producing polylactic acid polymers also. The use of propionyl-CoA transferase and PHA synthase assists in the conversion of lactic acid to lactyl-CoA, which is then polymerized to PLA in microbial cells (Mendhulkar and Shetye 2017).

9.2.2 *Microalgal and Cyanobacterial Production of Biodegradable Plastics*

Researchers have also studied the sustainable production of bioplastics from microalgae. They are a possible contender for PHA synthesis due to their capacity to utilize minimum or easily accessible resources such as inorganic substances (CO_2), water, sunlight, and wastewater, along with high photosynthetic efficiency.

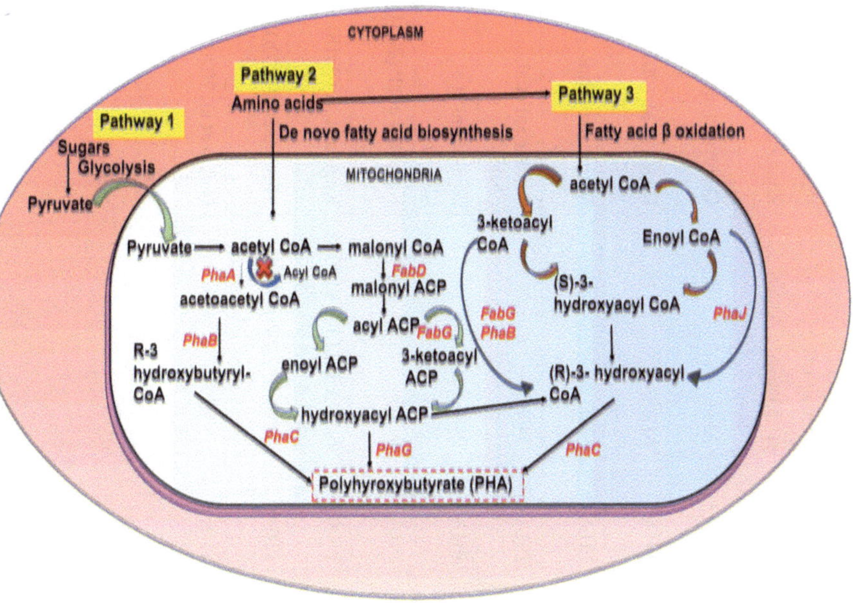

Fig. 9.1 Schematic showing PHA production pathway: PhaA- β-ketothiolase, PhaB-acetoacetyl coenzyme A reductase, PhaC-PHA synthase, FabG: 3-ketoacyl acyl carrier protein (ACP) reductase; FadD: fatty acyl-CoA synthetase; FadA: 3- ketoacyl-CoA thiolase; FadE: acyl-CoA dehydrogenase, FadB: 3-hydroxyacyl-coa dehydrogenase PhaJ: enoyl-coenzyme A hydratase; PhaG: (R)-3-hydroxydecanoylACP: CoA transacylase, blocked pathway

In nutrient-deficient environments, microalgae alter their metabolic pathways to generate carbon-rich compounds such as PHAs, which function as a means of storing energy. This occurs when the microalgae experience a lack of essential compounds like nitrogen and phosphorus that are necessary for their growth (López-Pacheco al. 2022). In addition, PLA is produced by bacterial fermentation of monomers of lactic acid like starch or glycogen in microalgal cells (Chong et al. 2021). However, PHAs are synthesized by microalgae from neutral lipids that have accumulated in their cells; this process requires carefully controlled cultivation conditions. There can be two ways for the production of PHAs in microalgae—growing cells in environmental stress conditions and by utilizing microalgal protein and processing it by thermo-mechanical polymerization (Balaji et al. 2013)) The most prevalent metabolic pathway identified in bacterial cells for the formation of PHA and fatty acids is via the precursor acetyl CoA. The metabolic pathway for PHA synthesis in microalgal cells is believed to be similar (Rehm et al. 2001). Although, it is believed that the production of fatty acids is an important step in the process of PHA production, there is no consensus on the exact metabolic pathways involved in this process for both bacteria and microalgae (Magdouli et al. 2015; Nishioka et al. 2001). A general representation of PHA production is depicted in Fig. 9.2. Nishioka et al. (2001) reported 55% PHA accumulation by *Synechococcus* sp. MA19 in

Table 9.1 Studies on microbial production of biodegradable plastic in different conditions

S.No	Microbial species	Conditions	Bioplastic	Productivity	References
1	*Bacillus cereus saba.zh*	C source-sugarcane molasses as c-source, N source-ammonium sulfate nitrogen source, pH-7 and temp. 30 ° C	PHB	61.53%	Zhang et al. (2013)
2	*Bacillus megaterium R11*	C/N ratio–15:2	PHB	70%	Queirós et al. (2014)
3	*Mixed bacterial culture*	Fed with hardwood spent sulfite liquor in SBR	PHA	67.6%	Reddy et al. (2017)
4	*Pseudomonas pseudoflava*	Synthetic wastewater, glucose-5 g/L	PHA	57%	Munir and Jamil (2018)
5	*Pseudomonas sp. (ST2)*	Volatile fatty acids and glucose as carbon source	PHA	34%	Aljuraifani et al. (2019)
6	*Bacillus sp. (CS8)*	Volatile fatty acids and glucose as carbon source	PHA	24%	Aljuraifani et al. (2019)
7	*Pseudomonas sp. strain-P (16)*	Soy molasses	PHA	91.6%	Kanavaki et al. (2021)
8	*Pseudomonas sp. phDV1*	Phenol as carbon source–600 mg/L	PHA	6.52 ± 0.21 mg of PHA per 1 g of wet cells	Kucera et al. (2018)
9	*Halomonas halophila*	Mineral medium, NaCl conc. 60 g/L	PHB	72.01 ± 2.78 wt%	Mostafa et al. 2020)
10	*Pseudodonghicola xiamenensis*	4% Nacl, pentone as nitrogen source, date syrup–4%(w/v)	PHB	38.85%	García et al. 2020)

(continued)

Table 9.1 (continued)

11	*Scenedesmus sp.*	Nutrient media with phosphate deficiency	PHA	25.2% w/w	Kumari et al. (2022)
12	*Synechococcus subsalsus*	Nitrogen deficiency	PHA	16% w/w	Senatore et al. (2023)
13	*Synechococcus elongates*	Sucrose as carbon source, nitrogen deficiency	PHA	17.15%	Senatore et al. (2023)
14	*Chlorella sorokiniana SVMIICT8*	Bold basal medium with nitrate and phosphate deficiency	PHA	280 mg/L	Maheswari and Ahilandeswari (2011)
15	*Spirulina platensis*	Zarrouk medium	PHB	6.20%	Das et al. 2018)
16	*Chlorella pyrenoidosa*	Fogg's medium	PHB	27%	Kavitha et al. (2016)
17	*Botryococcus braunii*	Sewage waste water	PHB	247 ± 0.42 mg/L	Sharma and Mallick 2005)
18	*Nostoc muscorum*	BG11 media with nitrogen and phosphorus deficiency, acetate (0.25), and 7 days in dark	PHA	35%	Santos et al. 2019)
19	*Arthrospira platensis*	Nitrogen deficiency	PHB	5.8 mg PHB g^{-1}	Meixner et al. (2016)
20	*Synechocystis salina*	Thin sillage	PHB	6.3% w/w	Singh et al. (2017)

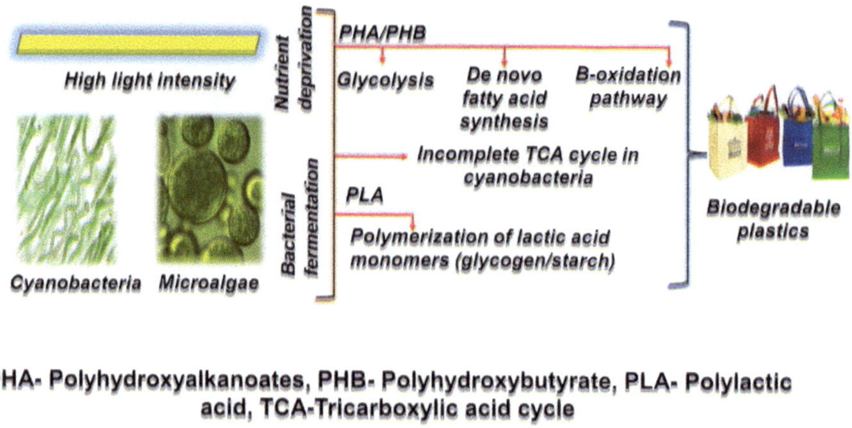

Fig. 9.2 Graphical representation showing general pathway for biodegradable plastic production from microalgae and cyanobacteria

phosphorus deficient condition (Chakravarty et al. 2010). Additionally, microalgal consortium grown in wastewater accumulated 45% PHA as reported by Chakravarty et al. (2010). Synthesis of PHA was also observed in several species of cyanobacteria as an energy and carbon storage compound (Kaneko et al. 1996). Nitrogen limitation was found to be the most triggering factor in cyanobacteria for PHB accumulation. When cells are lacking in nitrogen, they become unable to produce the essential proteins required for reproduction. As a result, these cells begin to accumulate storage compounds such as PHB. Most of the understanding of PHB formation was obtained from the study conducted on cyanobacteria *Synechocystis* PCC6803 (Samantaray and Mallick 2015). *Aulosira fertilissima*, a diazotrophic cyanobacterium, can accumulate high amounts of PHB intracellularly under appropriate culture conditions supplemented with citrate, acetate, and K_2HPO_4 (Singh and Mallick 2017). Additionally, studies have shown that ribulose-1,5-bisphosphate carboxylase (RuBisCO), which is involved in the Calvin cycle, contributes in the fixation of natural carbon dioxide during the autotrophic culture of cyanobacteria. In addition, RuBisCO has a greater affinity for CO_2 than O_2 and is responsible for consuming 90% of the atmospheric C. In order to maintain RuBisCO's local carbon concentration stable, atmospheric carbon dioxide is carried through the cell wall by inorganic carbon (Ci) transporters. After being converted to 3-phosphoglycerate (PGA), the Calvin cycle output, glyceraldehyde-3-phosphate, can enter any of the three sugar metabolism pathways, including the Entner-Doudoroff (ED) pathway, glycolysis, the pentose phosphate pathway, and finally, it can be converted to acetyl CoA for use in the PHA synthesis pathway (Obruca et al. 2018). Moreover, researchers have also examined marine bacteria as a possible source for PHA production due to their ability to withstand various environmental stresses, including fluctuations in temperature, pH levels, salinity, nutrient availability, hydrodynamics, and pollution (Higuchi-Takeuchi et al. 2016). They mainly belong to halophiles, some proteobacteria, Actinobacteria, and Firmicutes. Proteobacteria are PHA synthase has been divided into subclasss according to their substrates. All classes of PHA synthase have been observed in marine bacteria for PHA production. PHA production is observed in α-Proteobacteria, which includes aerobic and oxygenic photosynthetic bacteria. These bacteria are capable of performing photosynthesis without generating oxygen. It is believed that PHA synthesis regulates the trophic metabolism of these bacteria in response to the availability of organic carbon in the marine environment (Amiri et al. 2021). Moreover, Table 9.1 includes some studies conducted on microalgal and cyanobacterial species for biodegradable plastic production.

9.3 Challenges at Economic and Technology Readiness Levels for Production of Biodegradable Plastics

The production of PHA bioplastic from microbial biomass is heavily influenced by the growth rates and microbial density for high PHA content. An economically viable mass cultivation system is currently the primary bottleneck for commercial production of PHA (Storz and Vorlop 2013). Reducing production costs is a crucial challenge that needs to be addressed in the production of PHA bioplastics from microbial biomass. The supply chains for bio-based products are often extensive, making scalability difficult. The essential technologies have never been tested, which eventually adds to the challenge for scalability. In the coming years, addressing this challenge is of utmost importance to make the production of PHA bioplastics economically feasible and encourage their adoption in various applications (Wang et al. 2014). Economic cost involves low production cost with low efficiency, glucose as a raw material also adds cost to the production of PHA. PHA as a bioplastic does not have a consistent structure and properties as compared to commercial plastics. In addition, the bioprocessing for PHA is more difficult than petrochemical plastics because of its slow crystalline process. While on the other hand, due to slow fermentation process, it takes several weeks to convert one inoculum to downstream product purification. In addition, bioprocessing requires a lot of fresh water, along with aeration which demands a lot of energy and the final products obtained are also very low, ranging from mg to 200 g/L, with high downstream processing costs (Penloglou et al. 2012). PHA producing industries via fermentation produce PHA from substrate standing around 33% g/g (Sidek et al. 2019). Furthermore, decomposition of biodegradable plastics needs a specific environment for degradation. This indicates that the natural degradation is not possible if thrown into the landfills. Therefore, a special composting and recycling process is required for biodegradable plastics. Additionally, if they are mixed with regular plastics they will be contaminated and it will be very difficult to reuse it (Patel et al. 2021). Bioplastics obtained from renewable resources may deplete agricultural reserves. Also on an industrial scale, the expense of nitrogen and carbon sources for bulk manufacture of PHA is a critical limiting issue. According to a study done by the Malaysian Investment Development Authority (MIDA), the biodegradable plastics industry is expected to increase from US$3.02 billion in 2018 to US$6.12 billion by 2023, with a compound annual growth rate (CAGR) of 15.1% (Krasaesueb et al. 2019). As per MIDA's report in 2020, the global market for packing materials as plastics has reached a staggering 300 million tons annually. In contrast, the biodegradable plastic manufacturing capacity is around 2.5 million tons per year. According to MIDA, the considerable gap between supply and demand leads to the greater cost of producing and selling biodegradable polymers (Krasaesueb et al. 2019). In addition, in order to bring about a significant change, bioplastics need to establish themselves as competitive alternatives to conventional petrochemical-based plastics. This entails addressing various aspects such as reducing production costs, enabling large-scale manufacturing, ensuring compatibility with existing infrastructures, facilitating effective waste management, implementing efficient end-of-life solutions, raising consumer awareness, promoting waste sorting at the consumer level, and considering numerous other factors. A paradigm shift in the industry requires bioplastics to effectively compete with established conventional plastic manufacturers across these crucial aspects (Krasaesueb et al. 2019; Leong et al. 2021).

9.4 Biorefinery Approach to Produce Bioplastics from Microalgae

A sustainable relationship of biomaterials like bioplastics and bioenergy is crucial for the environment (Talan et al. 2022). Since the fossil fuel based products cannot fulfill the sustainable development goals, it is high time we shift to more sustainable pathways which are in harmony with nature. Algal biorefinery approaches which aim at utilizing the wastewater as the growth media could be one of the promising solutions to maintain the natural harmony of bioproducts and its environmental impacts. Moreover, the biorefinery approaches also help in developing a circular economy with multiple product biosynthesis having minimal carbon emissions (Talan et al. 2022).

In a study, palm oil mill effluent (POME) was used to grow microalgae for bioplastic production. They reported the production of about 33% of polyhydroxybutyrate (PHB) using 50% of POME along with glycerol and FE-EDTA supplementation (Allegue et al. 2022). Another study based on utilizing wastewater from shrimp aquaculture facility to grow *Synechocystis* PCC 6803 (ΔSphU) which was devoid of phosphate regulator for bioplastic production was recently published (Bhati and Mallick 2016). A biomass yield of 500 mgL^{-1} was obtained at 14 days and PHB accumulation of 32.48% Dw having PHB productivity of 12.73 $mgL^{-1}d^{-1}$ was obtained (Bhati and Mallick 2016). The growth of ΔSphU also helped in nutrient removal by utilizing 96.99% phosphate, and 80.10% of nitrate (Bhati and Mallick 2016). A study on utilizing poultry waste to grow the cyanobacterium *Nostoc muscorum*

Agardh with 10% CO_2 supply reported a biomass yield of 1.12 gL^{-1} in a 8 day period (Kavitha et al. 2016). Poly(3-hydroxybutyric acid-co-3-hydroxyvaleric acid) [P(3HB-*co*-3HV)] which is a copolymer was also produced during this time period. The yield of copolymer was about 65% of the dry cell weight, in cultures grown using 10 gL^{-1} of poultry litter (Kavitha et al. 2016). The growth of cyanobacterium in this particular study helped in sequestering the carbon dioxide while utilizing poultry liter as growth media by taking off the nutrient load. Moreover, it produced a high amount of copolymer. Thus, overall this kind of integrated process which can perform multiple tasks in one process could be helpful in mitigating the climate change and having a positive relationship with the environment.

In a recent study response surface methodology with box-behnken design was used to optimize the growth of *Botryococcus braunii* at several pH (4–11), temperature (30–50 °C), and sewage wastewater concentration (20–100%) for production of polyhydroxybutyrate (PHB) (Koutra et al. 2018). It was found that the maximum PHB accumulation was obtained at 60% sewage wastewater concentration, pH-7.5 at 40 °C. The PHB yield was 247 ± 0.42 mgL^{-1} (Koutra et al. 2018). The above stated research on early stage development of integrated approaches of wastewater treatment and bioplastic production seems feasible in the near future. Similarly, the use of anaerobic digestion for generation of renewable energy like biogas, leads to the formation of wastewater known as digestate (Kaur et al. 2020; Ayre et al. 2017). The digestate produced during the biogas formation by bacteria's and archaea's anaerobic digestion could be a source of nutrition for microalgal growth. The digestate is rich in nitrates and phosphates which makes it a potential candidate for microalgal growth media (Ayre et al. 2017; Meixner et al. 2016).

Synechocystis salina was grown using anaerobic digestate as the growth media reported to have PHB accumulation of 6.3% (w/w) in 14 days (Bauer et al. 2021). The presence of particulate matter and hazardous chemicals in the digestate may limit the microalgal growth (Kashyap and Kiran 2021). Therefore, more investigations and research inputs are required to use the anaerobic digestate for microalgal growth. Additionally, it is essential to focus on multiple products rather than just focus on bioplastics during microalgal cultivation process. Since, the goal is to provide a sustainable and market ready solution for plastic production along with bioenergy production. It is crucial that different both high value and low value added molecules should be extracted from the microalgae. The integrated biorefinery approach for bioplastic production using algal biomass could be challenging in terms of process development and optimization. The processing of algal biomass for multiple product biosynthesis in a simultaneous fashion is essential (Anand et al. 2021).

9.5 Future Aspects of Microalgae Bioplastic Production and Market for Biodegradable Plastics

The integrated biorefinery approach and process development for production as well as the extraction of multiple products from the algal biomass generated during a single process is the future of algal biorefinery. Mostly, algal biorefinery is associated with the production of biofuels (Anand et al. 2019; Kashyap et al. 2021). Multiple product biosynthesis is possible using microalgae biomass, but it is limited due to the challenges in algal biomass processing (Anand et al. 2021; Lebreton et al. 2017).

In a usual scenario, the plastic waste produced using fossil fuels is either disposed (~80%), incinerated (~10%), or recycled to low grade plastic material (~10%), globally (Schmidt et al. 2017; Geyer et al. 2017; Zheng and Suh 2019). However, in circular economy approach, the plastic waste generated is the starting point for the production of sustainable by-products and recycling approaches (Bucknall 2020, MacArthur et al. 2016). In the current time, about 2 million tons of bio-based plastics are being produced (Chen 2014; Chinthapalli et al. 2019). If the copolymers like polyamides are included, the production of bio-based plastics is set to reach up to 9.1 million tons by the end of 2023. The market capitalization for the production of such amounts of bio-based plastics is $1.7 billion (Chinthapalli et al. 2019). The biodegradable plastics market is increasing with a compound annual growth rate (CAGR) of 21. 3% and expected to provide a market share of $20.9 billion by 2028 (Biodegradable Plastics Market Size & Forecast, [Latest] (marketsandmarkets.com) accessed on 15th August, 2023) (Markets and Markets, Biodegradable plastics market 2023).

The production of bioplastics from microalgae in an integrated approach can utilize wastewater as growth media, synthesize biofuels, and PHA at the same time removing the nutrients hence cleaning the wastewater. The schematic representation of wastewater treatment using microalgae cultivation for the production of bio-based products is shown in Fig. 9.3. However, there are a few limitations to the processes like harvesting and drying cost of algal biomass which are taken in to consideration (Saket et al. 2022). The emphasis is to use the wet biomass for the production of the targeted biomolecules instead of using dry biomass. Moreover, the cell milking approach could also be explored for extraction of multiple products from the algal biomass. More research input is needed to optimize and generalize the process of integrated algal biorefinery approach for the production of bioplastics and other value added compounds.

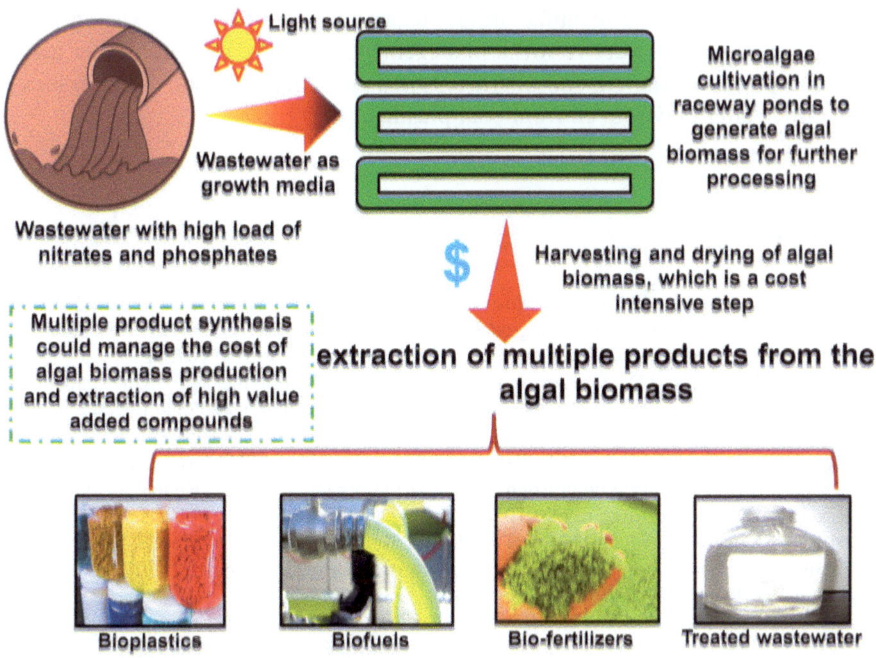

Fig. 9.3 Schematic representation of wastewater utilization for microalgae cultivation and multiple product biosynthesis including bioplastics

9.6 Summary

Conventional plastics have substantial problems, including environmental contamination, limited raw material supply, and non-biodegradability. To solve these challenges, researchers have focused on producing eco-friendly bioplastics such as polyhydroxybutyrate (PHB) and polyhydroxyalkanoates (PHA), which may be produced from renewable resources. PHAs are often produced on a wide scale by microorganisms, particularly bacteria and microalgae. PHAs accumulate within bacterial cells as a means of storing carbon and energy during times of nutritional imbalance. However, the development of bioplastics is fraught with economic difficulties. The key constraint is the development of a cost-effective mass cultivation technology, as well as the reduction of production costs to enable competitive pricing. Upscaling bio-based goods is often difficult due to the extended duration of the supply chain. Furthermore, the intensive fermentation process, high downstream processing costs, and the requirement for specific composting and recycling methods all contribute to the difficulties. Bioplastics also have limitations in terms of inconsistency in structure and qualities when compared to traditional plastics, as well as the need for specific breakdown conditions. In addition, the necessity for nitrogen and carbon sources for large-scale PHA manufacturing is a key constraint. It is critical to address production costs, enable large-scale manufacturing, ensure compatibility with existing infrastructures, implement effective waste management solutions, raise consumer awareness, and promote efficient end-of-life solutions in order to establish bioplastics as competitive alternatives to conventional plastics.

References

Alcântara JMG, Distante F, Storti G, Moscatelli D, Morbidelli M, Sponchioni M (2020) Current trends in the production of biodegradable bioplastics: the case of polyhydroxyalkanoates. Biotechnol Adv 42:107582. https://doi.org/10.1016/j.biotechadv.2020.107582

Aljuraifani AA, Berekaa MM, Ghazwani AA (2019) Bacterial biopolymer (polyhydroxyalkanoate) production from low-cost sustainable sources. Microbiologyopen 8:e00755. https://doi.org/10.1002/mbo3.755

Allegue LD, Ventura M, Melero JA, Puyol D (2022) Unraveling PHA production from urban organic waste with purple phototrophic bacteria via organic overload. Renew Sustain Energy Rev 166:112687. https://doi.org/10.1016/j.rser.2022.112687

Amiri S, Mohammadi Zeydi M, Amiri N (2021) Bacillus cereus saba. zh, a novel bacterial strain for the production of bioplastic (polyhydroxybutyrate). Braz J Microbiol 52:2117–2128. https://doi.org/10.1007/s42770-021-00599-9

Anand V, Kashyap M, Samadhiya K, Ghosh A, Kiran B (2019) Salinity driven stress to enhance lipid production in Scenedesmus vacuolatus: a biodiesel trigger? Biomass Bioenergy 127:105252. https://doi.org/10.1016/j.biombioe.2019.05.021

Anand V, Kashyap M, Sharma MP, Bala K (2021) Impact of hydrogen peroxide on microalgae cultivated in varying salt-nitrate-phosphate conditions. J Environ Chem Eng 9:105814. https://doi.org/10.1016/j.jece.2021.105814

Anjum A, Zuber M, Zia KM, Noreen A, Anjum MN, Tabasum S (2016) Microbial production of polyhydroxyalkanoates (PHAs) and its copolymers: a review of recent advancements. Int J Biol Macromol 89:161–174. https://doi.org/10.1016/j.ijbiomac.2016.04.069

Avérous L, Pollet E (2012) Biodegradable polymers. In: Environmental silicate nano-biocomposites. Springer, pp 13–39. https://doi.org/10.1007/978-1-4471-4108-2_2

Ayre JM, Moheimani NR, Borowitzka MA (2017) Growth of microalgae on undiluted anaerobic digestate of piggery effluent with high ammonium concentrations. Algal Res 24:218–226. https://doi.org/10.1016/j.jbiotec.2016.10.023

Balaji S, Gopi K, Muthuvelan B (2013) A review on production of poly β hydroxybutyrates from cyanobacteria for the production of bio plastics. Algal Res 2:278–285. https://doi.org/10.1016/j.algal.2013.03.002

Bauer L, Ranglová K, Masojídek J, Drosg B, Meixner K (2021) Digestate as sustainable nutrient source for microalgae—challenges and prospects. Appl Sci 11:1056. https://doi.org/10.3390/app11031056

Bhati R, Mallick N (2016) Carbon dioxide and poultry waste utilization for production of polyhydroxyalkanoate biopolymers by Nostoc muscorum Agardh: a sustainable approach. J Appl Phycol 28:161–168. https://doi.org/10.1007/s10811-015-0573-x

Bucknall DG (2020) Plastics as a materials system in a circular economy. Phil Trans R Soc A 378:20190268. https://doi.org/10.1098/rsta.2019.0268

Chaitanya K, Mahmood SK, Kausar R, Sunilkumar N (2014) Biotechnological production of polyhydroxyalkanoates by various isolates: a review. Int J Pharm Sci Invent 3:1–11

Chakravarty P, Mhaisalkar V, Chakrabarti T (2010) Study on poly-hydroxyalkanoate (PHA) production in pilot scale continuous mode wastewater treatment system. Bioresour Technol 101:2896–2899. https://doi.org/10.1016/j.biortech.2009.11.097

Chen GQ (2010) Plastics completely synthesized by bacteria: polyhydroxyalkanoates. In: Plastics from bacteria: natural functions and applications, pp 17–37. https://doi.org/10.1007/978-3-642-03287-5_2

Chen J (2014) Global markets and technologies for bioplastics. BCC Research Report PLS050C. https://www.bccresearch.com/market-research/plastics/global-markets-and-technologies-for-bioplastics.html

Chinthapalli R, Skoczinski P, Carus M, Baltus W, de Guzman D, Käb H, Raschka A, Ravenstijn J (2019) Biobased building blocks and polymers—global capacities, production and trends, 2018–2023. Ind Biotechnol 15:237–241. https://doi.org/10.1089/ind.2019.29179.rch

Chong JWR, Yew GY, Khoo KS, Ho S-H, Show PL (2021) Recent advances on food waste pretreatment technology via microalgae for source of polyhydroxyalkanoates. J Environ Manage 293:112782. https://doi.org/10.1016/j.jenvman.2021.112782

Chong JWR, Tan X, Khoo KS, Ng HS, Jonglertjunya W, Yew GY, Show PL (2022) Microalgae-based bioplastics: future solution towards mitigation of plastic wastes. Environ Res 206:112620. https://doi.org/10.1016/j.envres.2021.112620

Costa SS, Miranda AL, Andrade BB, de Jesus Assis D, Souza CO, de Morais MG, Costa JAV, Druzian JI (2018) Influence of nitrogen on growth, biomass composition, production, and properties of polyhydroxyalkanoates (PHAs) by microalgae. Int J Biol Macromol 116:552–562. https://doi.org/10.1016/j.ijbiomac.2018.05.064

Darbre PD (2020) Chemical components of plastics as endocrine disruptors: overview and commentary. Birth Defects Res 112:1300–1307. https://doi.org/10.1002/bdr2.1778

Das SK, Sathish A, Stanley J (2018) Production of biofuel and bioplastic from Chlorella pyrenoidosa. Mater Today Proc 5:16774–16781. https://doi.org/10.1016/j.matpr.2018.06.020

dos Santos RR, Corrêa PS, Dantas FML, Teixeira CMLL (2019) Evaluation of the co-production of total carotenoids C-phycocyanin and polyhydroxyalkanoates by arthrospira platensis. Bioresour Technol Rep 7:100226. https://doi.org/10.1016/j.biteb.2019.100226

Duan Y, Mehariya S, Kumar A, Singh E, Yang J, Kumar S, Li H, Kumar Awasthi M (2021) Apple orchard waste recycling and valorization of valuable product-a review. Bioengineered 12:476–495. https://doi.org/10.1080/21655979.2021.1872905

European Bioplastics (2022) Bioplastics market data, World Plastics Production 2021, Plastics Europe, 2022. https://www.european-bioplastics.org/market/

Gadhave RV, Das A, Mahanwar PA, Gadekar PT (2018). Starch based bio-plastics: the future of sustainable packaging. https://doi.org/10.4236/ojpchem.2018.82003

García G, Sosa-Hernández JE, Rodas-Zuluaga LI, Castillo-Zacarías C, Iqbal H, Parra-Saldívar R (2020) Accumulation of PHA in the microalgae Scenedesmus sp. under nutrient-deficient conditions. Polymers (Basel) 13:131. https://doi.org/10.3390/polym13010131

Geyer R, Jambeck JR, Law KL (2017) Production, use, and fate of all plastics ever made. Sci Adv 3:e1700782. https://doi.org/10.1126/sciadv.1700782

Higuchi-Takeuchi M, Morisaki K, Numata K (2016) A screening method for the isolation of polyhydroxyalkanoate-producing purple non-sulfur photosynthetic bacteria from natural seawater. Front Microbiol 7:1509. https://doi.org/10.3389/fmicb.2016.01509

Hoffmann N, Amara AA, Beermann BB, Qi Q, Hinz H-J, Rehm BHA (2002) Biochemical characterization of the Pseudomonas putida 3-hydroxyacyl ACP: CoA transacylase, which diverts intermediates of fatty acid de novo biosynthesis. J Biol Chem 277:42926–42936. https://doi.org/10.1074/jbc.M207821200

Kanavaki I, Drakonaki A, Geladas ED, Spyros A, Xie H, Tsiotis G (2021) Polyhydroxyalkanoate (PHA) production in Pseudomonas sp. phDV1 strain grown on phenol as carbon sources. Microorganisms 9:1636. https://doi.org/10.3390/microorganisms9081636

Kaneko T, Sato S, Kotani H, Tanaka A, Asamizu E, Nakamura Y, Miyajima N, Hirosawa M, Sugiura M, Sasamoto S (1996) Sequence analysis of the genome of the unicellular cyanobacterium Synechocystis sp. strain PCC6803. II. Sequence determination of the entire genome and assignment of potential protein-coding regions. DNA Res 3:109–136. https://doi.org/10.1093/dnares/3.3.109

Kashyap M, Samadhiya K, Ghosh A, Anand V, Lee H, Sawamoto N, Ogura A, Ohshita Y, Shirage PM, Bala K (2021) Synthesis, characterization and application of intracellular Ag/AgCl nanohybrids biosynthesized in Scenedesmus sp. as neutral lipid inducer and antibacterial agent. Environ Res 201:111499. https://doi.org/10.1016/j.envres.2021.111499

Kashyap M, Kiran B (2021) Milking microalgae in conjugation with nano-biorefinery approach utilizing wastewater. J Environ Manage 293:112864. https://doi.org/10.1016/j.jenvman.2021.112864

Kaur G, Wong JWC, Kumar R, Patria RD, Bhardwaj A, Uisan K, Johnravindar D (2020) Value addition of anaerobic digestate from biowaste: thinking beyond agriculture. Curr Sustain Renew Energy Rep 7:48–55. https://doi.org/10.1007/s40518-020-00148-2

Kavitha G, Kurinjimalar C, Sivakumar K, Kaarthik M, Aravind R, Palani P, Rengasamy R (2016) Optimization of polyhydroxybutyrate production utilizing waste water as nutrient source by Botryococcus braunii Kütz using response surface methodology. Int J Biol Macromol 93:534–542. https://doi.org/10.1016/j.ijbiomac.2016.09.019

Koutra E, Grammatikopoulos G, Kornaros M (2018) Selection of microalgae intended for valorization of digestate from agro-waste mixtures. Waste Manage 73:123–129. https://doi.org/10.1007/s40518-020-00148-2

Krasaesueb N, Incharoensakdi A, Khetkorn W (2019) Utilization of shrimp wastewater for poly-β-hydroxybutyrate production by Synechocystis sp. PCC 6803 strain ΔSphU cultivated in photobioreactor. Biotechnol Rep 23:e00345. https://doi.org/10.1016/j.btre.2019.e00345

Kucera D, Pernicová I, Kovalcik A, Koller M, Mullerova L, Sedlacek P, Mravec F, Nebesarova J, Kalina M, Marova I (2018) Characterization of the promising poly (3-hydroxybutyrate) producing halophilic bacterium Halomonas halophila. Bioresour Technol 256:552–556. https://doi.org/10.1016/j.biortech.2018.02.062

Kumari P, Kiran BR, Mohan SV (2022) Polyhydroxybutyrate production by Chlorella sorokiniana SVMIICT8 under nutrient-deprived mixotrophy. Bioresour Technol 354:127135. https://doi.org/10.1016/j.biortech.2022.127135

Lajus S, Dusséaux S, Verbeke J, Rigouin C, Guo Z, Fatarova M, Bellvert F, Borsenberger V, Bressy M, Nicaud J-M (2020) Engineering the yeast Yarrowia lipolytica for production of polylactic acid homopolymer. Front Bioeng Biotechnol 8:954. https://doi.org/10.3389/fbioe.2020.00954

Lebreton LCM, Van Der Zwet J, Damsteeg J-W, Slat B, Andrady A, Reisser J (2017) River plastic emissions to the world's oceans. Nat Commun 8:15611. https://doi.org/10.1038/ncomms15611

Leong HY, Chang C-K, Khoo KS, Chew KW, Chia SR, Lim JW, Chang J-S, Show PL (2021) Waste biorefinery towards a sustainable circular bioeconomy: a solution to global issues. Biotechnol Biofuels 14:1–15. https://doi.org/10.1186/s13068-021-01939-5

Leslie HA, Van Velzen MJ, Brandsma SH, Vethaak AD, Garcia-Vallejo JJ, Lamoree MH (2022) Discovery and quantification of plastic particle pollution in human blood. Environ Int 163:107199. https://doi.org/10.1016/j.envint.2022.107199

López-Pacheco IY, Rodas-Zuluaga LI, Cuellar-Bermudez SP, Hidalgo-Vázquez E, Molina-Vazquez A, Araújo RG, Martínez-Ruiz M, Varjani S, Barceló D, Iqbal HMN (2022) Revalorization of microalgae biomass for synergistic interaction and sustainable applications: bioplastic generation. Mar Drugs 20:601. https://doi.org/10.3390/md20100601

MacArthur DE, Waughray D, Stuchtey MR (2016) The new plastics economy, rethinking the future of plastics. In: World economic forum. Ellen MacArthur Foundation and McKinsey & Company London, UK

Magdouli S, Brar SK, Blais J-F, Tyagi RD (2015) How to direct the fatty acid biosynthesis towards polyhydroxyalkanoates production? Biomass Bioenergy 74:268–279. https://doi.org/10.1016/j.biombioe.2014.12.017

Maheswari NU, Ahilandeswari K (2011) Production of bioplastic using Spirulina platensis and comparison with commercial plastic. Res Environ Life Sci 4:133–136. http://rels.110mb.com

Markets and Markets, Biodegradable plastics market (2023)

Meixner K, Fritz I, Daffert C, Markl K, Fuchs W, Drosg B (2016) Processing recommendations for using low-solids digestate as nutrient solution for poly-ß-hydroxybutyrate production with Synechocystis salina. J Biotechnol 240:61–67. https://doi.org/10.1016/j.jbiotec.2016.10.023

Mendhulkar VD, Shetye LA (2017) Synthesis of biodegradable polymer polyhydroxyalkanoate (PHA) in cyanobacteria Synechococcus elongates under mixotrophic nitrogen-and phosphate-mediated stress conditions. Ind Biotechnol 13:85–93. https://doi.org/10.1089/ind.2016.0021

Mostafa YS, Alrumman SA, Alamri SA, Otaif KA, Mostafa MS, Alfaify AM (2020) Bioplastic (poly-3-hydroxybutyrate) production by the marine bacterium Pseudodonghicola xiamenensis through date syrup valorization and structural assessment of the biopolymer. Sci Rep 10:8815. https://doi.org/10.1038/s41598-020-65858-5

Munir S, Jamil N (2018) Polyhydroxyalkanoates (PHA) production in bacterial co-culture using glucose and volatile fatty acids as carbon source. J Basic Microbiol 58:247–254. https://doi.org/10.1002/jobm.201700276

Nishioka M, Nakai K, Miyake M, Asada Y, Taya M (2001) Production of poly-β-hydroxybutyrate by thermophilic cyanobacterium, Synechococcus sp. MA19, under phosphate-limited conditions. Biotechnol Lett 23:1095–1099. https://doi.org/10.1023/A:1010551614648

Nur MMA, Yuliestyan A, Irfandy F, Setyoningrum TM (2022) Nutritional factors influence polyhydroxybutyrate in microalgae growing on palm oil mill effluent. J Appl Phycol 34:127–133. https://doi.org/10.1007/s10811-021-02654-2

Obruca S, Sedlacek P, Koller M, Kucera D, Pernicova I (2018) Involvement of polyhydroxyalkanoates in stress resistance of microbial cells: biotechnological consequences and applications. Biotechnol Adv 36:856–870. https://doi.org/10.1016/j.biotechadv.2017.12.006

Patel R, Nanda S, Dalai AK (2021) Conversion of municipal solid waste to biofuels. In: Progressive thermochemical biorefining technologies. CRC Press, Florida USA, pp 43–61. 9781003098591

Penloglou G, Chatzidoukas C, Kiparissides C (2012) Microbial production of polyhydroxybutyrate with tailor-made properties: an integrated modelling approach and experimental validation. Biotechnol Adv 30:329–337. https://doi.org/10.1016/j.biotechadv.2011.06.021

Poirier Y, Nawrath C, Somerville C (1995) Production of polyhydroxyalkanoates, a family of biodegradable plastics and elastomers, in bacteria and plants. Bio/technology 13:142–150. https://doi.org/10.1038/nbt0295-142

Queirós D, Rossetti S, Serafim LS (2014) PHA production by mixed cultures: a way to valorize wastes from pulp industry. Bioresour Technol 157:197–205. https://doi.org/10.1016/j.biortech.2014.01.099

Rahman A, Miller CD (2017) Microalgae as a source of bioplastics. In: Algal green chemistry. Elsevier, pp 121–138. https://doi.org/10.1016/B978-0-444-63784-0.00006-0

Reddy MV, Mawatari Y, Onodera R, Nakamura Y, Yajima Y, Chang YC (2017) Polyhydroxyalkanoates (PHA) production from synthetic waste using Pseudomonas pseudoflava: PHA synthase enzyme activity analysis from P. pseudoflava and P. palleronii. Bioresour Technol 234:99–105. https://doi.org/10.1016/j.biortech.2017.03.008

Rehm BHA, Mitsky TA, Steinbüchel A (2001) Role of fatty acid de novo biosynthesis in polyhydroxyalkanoic acid (PHA) and rhamnolipid synthesis by pseudomonads: establishment of the transacylase (PhaG)-mediated pathway for PHA biosynthesis in Escherichia coli. Appl Environ Microbiol 67:3102–3109. https://doi.org/10.1128/AEM.67.7.3102-3109.2001

Riaz S, Rhee KY, Park SJ (2021) Polyhydroxyalkanoates (PHAs): biopolymers for biofuel and biorefineries. Polymers (basel) 13:253. https://doi.org/10.3390/polym13020253

Saket P, Kashyap M, Bala K, Joshi A (2022) Microalgae and bio-polymeric adsorbents: an integrative approach giving new directions to wastewater treatment. Int J Phytorem 24:536–556. https://doi.org/10.1080/15226514.2021.1952925

Samadhiya K, Sangtani R, Nogueira R, Bala K (2022) Insightful advancement and opportunities for microbial bioplastic production. Front Microbiol 12:674864. https://doi.org/10.3389/fmicb.2021.674864

Samantaray S, Mallick N (2015) Role of cultural variables in tailoring poly (3-hydroxybutyrate-co-3-hydroxyvalerate) copolymer synthesis in the diazotrophic cyanobacterium Aulosira fertilissima CCC 444. J Appl Phycol 27:197–203. https://doi.org/10.1007/s10811-014-0299-1

Schmidt C, Krauth T, Wagner S (2017) Export of plastic debris by rivers into the sea. Environ Sci Technol 51:12246–12253. https://doi.org/10.1021/acs.est.7b02368

Senatore V, Rueda E, Bellver M, Díez-Montero R, Ferrer I, Zarra T, Naddeo V, García J (2023) Production of phycobiliproteins, bioplastics and lipids by the cyanobacteria Synechocystis sp. treating secondary effluent in a biorefinery approach. Sci Total Environ 857:159343. https://doi.org/10.1016/j.scitotenv.2022.159343

Sharma L, Mallick N (2005) Accumulation of poly-β-hydroxybutyrate in Nostoc muscorum: regulation by pH, light–dark cycles, N and P status and carbon sources. Bioresour Technol 96:1304–1310. https://doi.org/10.1016/j.biortech.2004.10.009

Sidek IS, Draman SFS, Abdullah SRS, Anuar N (2019) Current development on bioplastics and its future prospects: an introductory review. INWASCON Technol Mag 1:3–8. https://doi.org/10.26480/itechmag.01.2019.03.08

Singh AK, Sharma L, Mallick N, Mala J (2017) Progress and challenges in producing polyhydroxyalkanoate biopolymers from cyanobacteria. J Appl Phycol 29:1213–1232. https://doi.org/10.1007/s10811-016-1006-1

Singh AK, Mallick N (2017) Advances in cyanobacterial polyhydroxyalkanoates production. FEMS Microbiol Lett 364:fnx189. https://doi.org/10.1093/femsle/fnx189

Statista Research Department (2023) Annual production of plastics worldwide from 1950 to 2021. https://www.statista.com/statistics/282732/global-production-of-plastics-since-1950/

Storz H, Vorlop K-D (2013) Bio-based plastics: status, challenges and trends. Appl Agric for Res 63:321–332. https://doi.org/10.3220/LBF_2013_321-332

Talan A, Pokhrel S, Tyagi RD, Drogui P (2022) Biorefinery strategies for microbial bioplastics production: sustainable pathway towards Circular Bioeconomy. Bioresour Technol Rep 17:100875. https://doi.org/10.1016/j.biteb.2021.100875

Tan G-YA, Chen C-L, Li L, Ge L, Wang L, Razaad IMN, Li Y, Zhao L, Mo Y, Wang J-Y (2014) Start a research on biopolymer polyhydroxyalkanoate (PHA): a review. Polymers (basel) 6:706–754. https://doi.org/10.3390/polym6030706

Thiruchelvi R, Das A, Sikdar E (2021) Bioplastics as better alternative to petro plastic. Mater Today Proc 37:1634–1639. https://doi.org/10.1016/j.matpr.2020.07.176

Varghese S, Dhanraj ND, Rebello S, Sindhu R, Binod P, Pandey A, Jisha MS, Awasthi MK (2022) Leads and hurdles to sustainable microbial bioplastic production. Chemosphere 305:135390. https://doi.org/10.1016/j.chemosphere.2022.135390

Wang Y, Yin J, Chen G-Q (2014) Polyhydroxyalkanoates, challenges and opportunities. Curr Opin Biotechnol 30:59–65. https://doi.org/10.1016/j.copbio.2014.06.001

Yee MS-L, Hii L-W, Looi CK, Lim W-M, Wong S-F, Kok Y-Y, Tan B-K, Wong C-Y, Leong C-O (2021) Impact of microplastics and nanoplastics on human health. Nanomaterials 11:496. https://doi.org/10.3390/nano11020496

Zhang Y, Sun W, Wang H, Geng A (2013) Polyhydroxybutyrate production from oil palm empty fruit bunch using Bacillus megaterium R11. Bioresour Technol 147:307–314. https://doi.org/10.1016/j.biortech.2013.08.029

Zheng J, Suh S (2019) Strategies to reduce the global carbon footprint of plastics. Nat Clim Chang 9:374–378. https://doi.org/10.1038/s41558-019-0459-z

Chapter 10
Greener Approach Towards Sustainable Green Plastics Through Eco-Friendly Upstream and Downstream Processing

Sonam Dubey, Shubhangi Parmar, Vijay Jagdish Upadhye, Anupama Shrivastav, Freny Shah, Niranjan Mishra, Pooja Bachani, and Sandhya Mishra

10.1 Introduction

Plastics are an essential element of modern life, made from polymers which are a long chain of repeating units. The most common applications are in business, retail, packaging and household purposes. Population increase, economic advancement, product consumer demand, and shifts in lifestyles are all propelling the demand and subsequent production of plastic-based goods. Plastics are appealing to industry because of properties such as lightweight, resilience, flexibility, cheap processing cost and simple availability. The ubiquitous use of plastics in modern life has increased waste product accumulation which renders its disposal and recycling as a major waste management goal. Traditional petrochemical-based plastics are non-biodegradable,

S. Dubey · F. Shah · N. Mishra · S. Mishra (✉)
Applied Phycology and Biotechnology Division, CSIR-Central Salt and Marine Chemicals Research Institute, Bhavnagar 364002, India
e-mail: smishra@csmcri.res.in

S. Parmar
Department of Food and Bioproduct Sciences, University of Saskatchewan, Saskatoon, SK S7N5A2, Canada

V. J. Upadhye
Center of Research for Development (CR4D), Parul Institute of Applied Sciences (PIAS), Parul University (DSIR-SIRO Recognized), PO Limda, Tal Waghodia, Vadodara, Gujarat 391760, India

A. Shrivastav
Faculty of Life Health, and Allied Sciences, ITM Vocational University, At and Po-Raval, Ta-Waghodia, AjwaNimeta Road, Vadodara 391760, India

P. Bachani
Amity University, Mumbai-Pune Expressway, Panvel, Mumbai 410206, India

K. Bala et al. (eds.), *Harnessing Microbial Potential for Multifarious Applications*, Energy, Environment, and Sustainability, https://doi.org/10.1007/978-981-97-1152-9_10

so they can persist in disposal facilities and seas for numerous years by adversely affecting the ecosystem which is not limited to soil health, the activity of microbes, and species of plants and animals but, disturbing the whole biosphere. Bioplastics are generated from plant and microbial biomasses, whereas conventional plastics are made from fossil fuels such as natural gas and petrochemicals. Bioplastics have several constructive effects on the environment and on the economic viability of plastic production at an international scale. A bioplastic resin is either bio-based or manufactured from environmentally friendly components with biodegradability capacity. Bioplastics typically break down in 3–6 months, but petrochemical-based plastics could require hundreds of years to degrade (Nanda et al. 2021). Bioplastics emit reduced greenhouse emissions than traditional plastics during their lifetime. As a result, bioplastics help shape a more sustainable economy.

Bioplastics are natural biopolymers that are synthesized and degraded by diverse species. Under stressful conditions, they accumulate as storage materials in microbial cells. Polyhydroxyalkanoates (PHAs), polylactide (PLA), poly(-caprolactone) (PCL), poly(p-dioxanone) (PPDO) and poly (butylene succinate) (PBS) are currently being investigated for various applications. PHAs and their derivatives are the most substantially generated microbial bioplastics (Saharan and Sharma 2015). Bioplastics are made from naturally occurring substances like potato peels, sugarcane, maize, rice, wheat and banana peels. They can also be derived from other sources such as algae, agricultural waste and sea salt (Dhangdhariya and Dubey 2015). The bio-based plastics specifically PHAs are biocompatible in nature due to its hydrophobic properties. Thus, fabricating these biopolymers with other polymers helps in mimicking the synthetically available plastics for its unique utility in specific areas.

The bioplastics can be specifically further grouped into four categories as follows.

10.1.1 Photodegradable Bioplastics

Light-sensitive groups are integrated directly into the structural framework of the polymer as additives in photodegradable bioplastics. UV exposure for a longer duration can destroy their polymeric structure, leaving them vulnerable to additional bacterial deterioration.

10.1.2 Bio-based Bioplastics

Bio-based bioplastics are defined as "plastics developed entirely from renewable agricultural and forestry materials such as corn starch, soybean protein, and cellulose.

10.1.3 Compostable Bioplastics

Compostable bioplastics break down organically throughout the composting process at a pace comparable to other compostable materials and without leaving evident hazardous residues.

Below mentioned are different types of bioplastics which have been studied by various researchers.

10.1.3.1 Starch-Based Bioplastics

Plants are known to be solar-driven bio-factories for the production of sustainable polymers like starch and cellulose. Starch is known to be a suitable natural component for the preparation of different starch-based bioplastics. These bioplastics are sustainable, biodegradable and low-cost materials. These bio-based polymers are helpful in reducing the usage of petrochemical-based plastics and removing carbon footprint in the long run. The brittleness of these starch-based films is improved with the addition of plasticizers such as water, glycerol and sorbitol wherein glycerol is the potential plasticizer for water-soluble polymers improving the flexibility of these bioplastics.

Starch is insoluble in both alcohol and water, is formed by molecules of amylopectin and amylose, which is the result of monomers of D-glucose and is the primary storage polysaccharide in cells of plant life. The features of starch-based bioplastics are closely tied to the raw material that generates starch, depending on the amylose and amylopectin proportions. Because the amylose/amylopectin ratio of the different starch plant sources varies, the gel and bioplastics derived from those sources may possess varying gelatinization temperatures, biomechanical and rheological attributes. The downsides of starch-based bioplastics include hydrophilicity, poor mechanical capabilities, reduced water vapour barrier performance, and poor freeze stability during bioplastics production. The quantity and type of plasticizer used can have an impact on the characteristics of bioplastics. Plasticizers enhance the interstitial space in the polymer matrix, which reduces the number of polymer–polymer bonds and consequently impacts the mechanical, thermal, and water vapour barrier characteristics (Abe et al. 2021).

Starch-based plastic films are water sensitive with high water vapour permeability. The starch-based bioplastic films show most mechanical properties which are undesirable for their application as packaging materials in the food industry. The properties of such starch-based bioplastic films can be easily improved when the starch-based solution is mixed with other synthetic polymers or plasticizers. The chemical, physical or enzymatic modification in the structure of starch-based films can help to improve its functional usage as packaging film (Abe et al. 2021). The use of nanofillers and nanocomposites in the production of starch-based products to directly enhance the characteristics of bioplastics is one of the emerging areas of research. The starch-based bioplastics are known to have high tensile strength, heat tolerance and easy biodegradation properties. However, starch-based materials have some drawbacks including long-term stability caused by water absorption, aging caused by retrogradation and poor mechanical properties (Naser et al. 2021). This plasticized starch is more versatile and can be blended with various polymeric materials for numerous applications.

10.1.3.2 Cellulose-Based Bioplastics

Cellulose extracted from plants is one of the potential sources for bioplastic preparation. Cellulose obtained from oregano waste, cotton, raw and sugarcane bagasse has been extensively used for bioplastics. Cellulose acts as a filler for improving the properties of various bioplastics. Reports on the use of this filler are with amylose, carrageenan, and starch. Cellulose is a polysaccharide which consists of units of glucose connected by β 1,4glycosidic bonds. Cellulose-based bioplastics are famous due to their characteristics like high durability, stiffness and ease of biodegradability. The cellulose material has low density, and is easily accessible due to its low price. The fast degradation of cellulose-based bioplastics is due to the presence of weak hydrogen bonds present due to the distant monomers. The presence of weak hydrogen bonds imparts properties like low flexibility and strength. The transparency, stability and property-like flexibility of bioplastics can be enhanced by blending it with polysaccharides like pectin and chitosan. Cellulose bioplastics have common applications in glass frames, food packaging, automobiles and biomedicine with an enormous potential in electronics (Abe et al. 2021; Liyanage et al. 2021; Li et al. 2016).

One of the alternative methods to produce bioplastics with cellulose as the main components without any modification process is the hydrogel method. Recently, the preparation of cellulose hydrogel has been broadly studied and reported.

A process to prepare bioplastic using the pure cellulose of *Cladophora*sp. algae via the hydrogel method is reported by Steven et al. (2022). The effect of epichlorohydrin (ECH) concentrations as the cross-linking agent was investigated towards the biodegradability, thermal, and mechanical properties of the cellulose bioplastic obtained. The results showed that ECH concentrations affected the properties of the cellulose bioplastic produced due to the number of cross-links formed during the process. The cellulose bioplastic possessed relatively high thermal and mechanical properties. The cellulose bioplastic performed excellent biodegradability, as it was degraded by more than 40% within five days. Thus, the cellulose of *Cladophora* sp. algae has the potential to be developed as the main component for bioplastic application (Steven et al. 2022).

10.1.3.3 Protein-Based Films

Protein-based films may be made from two types of proteins: fibrous proteins (such as keratin) and proteins that aggregate under certain environmental circumstances and can be classed as low-structured or globular proteins (ovalbumin, whey protein, soy protein, wheat gluten, and maize zein). Protein–protein interactions are important in the formation of cohesive films and are controlled by the amino acid sequence, composition, and media state. Prior to film creation, globular proteins must be partially weakened by heat or chemical techniques.

Protein is a class of amino acid hetero-polymer, joined by peptide bonds, and protein-based films have greater mechanical characteristics than polysaccharides. Proteins are frequently utilized as packaging materials due to their availability, biodegradability, nutritional value and improved film production capabilities. When compared to lipids and polysaccharides, protein-based films have superior mechanical qualities as well as gas and odour barriers. The incorporation of keratin protein into synthetic elastomers improves the end product's thermal stability, mechanical characteristics and fireproofing. Soy protein combined with latexes yields compounds with properties like carbon black-filled elastic materials (Calva-Estrada et al. 2019).

Proteins due to its high inter-molecular binding have various functional properties as compared to lipids and polysaccharides. The drawback lies in the mechanical properties as well as its non-antibacterial properties with low moisture resistance of these protein-based packaging materials limits its applicability (Hadidi et al. 2022). Different techniques have been studied to improve its mechanical properties through nanofillers (Zheng et al. 2021), cross-linking techniques (Mustafa et al. 2020) and cold plasma treatment (Bahrami et al. 2020). Very few studies have been reported on improving the hydrophobicity of proteins. Various applications have been reported of these proteins such as glue derived from collagen and casein, cosmetics derived from keratin collagen, bovine serum albumin (Barone and Schmidt 2006) and gelatin capsules for drug delivery. Recent advances in edible films and coatings are done from protein sources with varying structural changes.

10.1.3.4 Bio-Derived Polyethylene

Polyethylene has been produced from the fermentation of raw agricultural materials like sugarcane and corn rather than fossil fuels. PE is the most widely used polymer on the planet. The initial monomer for synthesis is ethylene, which is also the case for other polymers such as PVC and PS, and it is typically acquired by distillation from petroleum feedstock. However, there is growing interest in obtaining this polymer from biological resources, and the current technique has been to synthesize the ethylene monomer by dehydration of ethanol derived from glucosamine. Bio-based polyethylene has the same chemical, physical and mechanical properties as petrochemical polyethylene. Because of its low cost and high performance, it is widely used in engineering, agriculture, packaging and a variety of everyday commodities applications (Siracusa and Blanco 2020).

Example: High-density polyethylene is used in several packaging applications including crates, trays, bottles for milk and fruit juices, caps for food packaging, jerry cans, drums, industrial bulk containers etc.

10.1.3.5 Aliphatic Polyesters

A collection of bio-based polyesters including PLA, PHB, and PGA, among others. They are all sensitive to hydrolytic degradation in water and can be mixed with other compounds. Example: Some bio-based polyesters that have gained commercial use or that are currently being investigated are polylactic acid (PLA), polyglycolic acid (PGA), poly-ε-caprolactone (PCL), polyhydroxy butyrate (PHB) and poly (3-hydroxy valerate). However, only a small number are commercially available. Aliphatic polyesters and other biopolymers have many applications in medicine and pharmaceutical technology such as in drug delivery systems, wound closure, surgical sutures, implants and tissue engineering. Major manufacturers of aliphatic polyester bioplastics include Nature Works, Bio Matera, Presto, Solvay, Nova Mont, Mitsui, Corbion and Dow.

10.1.3.6 Algae-Based Bioplastics

Algae has emerged as one of the most sustainable alternative sources for bioplastic production due to its fast growth rate and the ability to withstand harsh environments it can grow on non-arable land. These properties have outgrown most terrestrial plants as promising feedstocks for bioplastic production. The sustainable development of algae as bioplastics for packaging contributes to minimize resource consumption, waste generation and regulate greenhouse gas emissions.

Most algae-based bioplastic films have strong mechanical qualities and a wide range of water vapour permeability in packaging applications, and they are biodegradable. The additives employed in the production of algae-based bioplastics have a significant role in the functional qualities of the finished product. Microalgae which has greater quantities of protein and carbohydrate polymers can be utilized as an ideal alternative to traditional plastics. Alginate-based bioplastics have a few disadvantages, including their hydrophilic nature, low water barrier qualities, low UV barrier properties and lack of antibacterial properties. According to reports, alginate-based films require additional polymeric support or other additives with functional qualities. Microcrystalline cellulose was added to algal and starch-based films to increase their water vapour barrier characteristics. The inclusion of silver nanoparticles and lemongrass essential oil improved the mechanical and antibacterial characteristics of alginate-based films (Jayakumar et al. 2023).

10.2 Upstream Parameters For PHA Production

10.2.1 Selection of Appropriate Substrates

Selecting cost-effective raw/ crude substrates for commercial-scale PHA production. These substrates provide the required nutritional sources like carbon, nitrogen and minerals for microbial growth (Dubey and Mishra 2022) Crude feedstocks for PHA production can help reduce substrate costs and dependency on specific raw materials. Agricultural and industrial waste streams, non-food biomass and carbon dioxide from industrial sources are potential candidates for cost-effective PHA production (Dubey and Mishra 2022; Shrivastav et al. 2010).

10.2.2 Process Optimization Parameters

Process optimization parameters like temperature, pH, dissolved oxygen levels, agitation rate and nutrient concentrations are monitored to get high biomass and PHA yields.

10.2.3 Selection of Microbial Strain

High resistant and productive strain with specific substrate utilization for PHA accumulation is selected. Engineered strain is also used by many researchers to get high PHA productivity and better functional properties.

10.2.4 Inoculum Preparation

Ensuring the reproducibility and consistency in the product, inoculum size and consistency play a vital role in the overall fermentation process.

10.2.5 Design of Bioreactors

10.2.5.1 Bioreactor Size

Selecting the appropriate bioreactor size and type to match the production capacity is essential. Considering the parameters include the fermentation volume, vessel design, mixing efficiency, oxygen transfer rate and the ability to control process parameters accurately.

10.2.5.2 Sterilization

Ensuring effective and uniform sterilization of large-scale bioreactors to prevent contamination and maintain process integrity is crucial.

10.2.5.3 Process Monitoring and Control

Implementing sophisticated monitoring and control systems to maintain optimal conditions throughout the fermentation process is vital for maximizing PHA production.

10.3 Downstream Parameters for Extraction of PHAs

The downstream parameters for PHA extraction refer to the specific factors and conditions involved in the purification and separation of PHA from the microbial biomass after its production. These parameters are crucial for obtaining high-purity PHA with desirable characteristics. Depending upon the type/robustness of microorganisms used in the experiment, the variation in downstream parameters is optimized. Downstream processing is important in determining the yield and quality of the extracted polymer using different extraction processes. Different microorganisms and production methods may require specific variations in these parameters. Downstream parameters for PHA extraction include (i) Cell harvesting (ii) Cell disruption (iii) PHA solubilization (iv) Filtration (v) Precipitation (vi) Drying (vii) Purification (vii) Characterization. The detailed description of downstream parameters has been discussed herein:

- **Cell harvesting**: Separation of PHA-producing microbial cells from the fermentation media involves centrifugation technique.
- **Cell disruption**: Cell disruption methods include mechanical disruption, enzymatic treatment or chemical lysis.Enzymatic method is an environmentally friendly method wherein, enzymes are used to break open the cell walls or membranes of the targeted microorganisms to extract intracellular PHA. The enzymatic process typically involves the following steps: ***Cell harvesting***: The PHA-producing microbial cells are separated from the fermentation broth, often through centrifugation or filtration. ***Enzyme selection***: Specific enzymes are chosen based on their ability to target and break down the cell walls or membranes of the particular microorganisms used in PHA production. ***Enzymatic treatment***: The

harvested cells are subjected to the selected enzymes which catalyze the hydrolysis of the cell components, allowing the PHA granules to be released into the surrounding medium. The other method is sonication where the cells are harvested and then suspended in a solvent, then high frequency sound waves are passed through this suspension releasing the PHA granules from the cell into the surrounding medium. The chemical-based technique utilizes harsh chemicals like sodium hypochlorite to disrupt the bacterial cell wall. Organic solvents such as chloroform, dichloromethane, or chloroform/methanol mixtures, are reported to dissolve PHA. Ionic liquids (ILs) have been extensively studied nowadays as an alternative to hazardous extractants due to its high dissolving power, low volatility; environment friendly properties which can be tailored easily with specific anions and cations, besides selective extraction properties make it easier to separate PHA from the rest of the biomass containing impurities (Sequeira et al. 2020; Dubey and Mishra 2021).

- **PHA solubilization**: PHA granules extracted from the cell are dissolved in suitable solvents to separate cell debris.
- **Filtration**: In this step, the undissolved impurities are removed from the PHA containing solution.
- **Precipitation**: Different methods have been adopted for PHA precipitation which have been discussed below:

  Alkali treatment: Treatment with sodium hydroxide (NaOH) or potassium hydroxide (KOH), saponifies the cells and releases PHA.
  Acid hydrolysis: Acids like sulfuric acid (H_2SO_4) or hydrochloric acid (HCl) helps in cell breakage with release of PHA.
  Enzymatic hydrolysis: Specific enzymes with PHA depolymerase activity are used to specifically separate out PHA.
  Supercritical fluid extraction: Supercritical carbon dioxide (CO_2) effectively extracts PHA from microbial biomass.

- **Drying**: Drying the content in an oven or evaporating the PHA granule containing solution is done to obtain the final product in solid/ powder form.
- **Purification**: This step improves PHA purity depending upon its application area.
- **Characterization**: The extracted and purified PHA is analyzed and checked for its chemical and mechanical properties.

10.4 Properties of PHAs

The physical and chemical properties of PHAs play a very vital role in defining its applicability and durability. Some of the PHA with its structure and side groups have been mentioned in the Table 10.1.

Table 10.1 Representative examples of PHA polymers (Adapted from Li and Loh (2015)

Chemical name	Side groups	Structure
Poly(3-hydroxybutyrate)	Methyl	Homopolymer
Poly(4-hydroxybutyrate)	Hydrogen	
Poly(3-hydroxyvalerate)	Ethyl	
Poly(3-hydroxyhexanoate)	Propyl	
Poly(3-hydroxyheptanoate)	Butyl	
Poly(3-hydroxyoctanoate)	Pentyl	
Poly(3-hydroxynonanoate)	Hexyl	
Poly(3-hydroxydecanoate)	Heptyl	
Poly(3-hydroxybutyrate-co-3-hydroxyvalerate)	Methyl/ethyl	Copolymer
Poly(3-hydroxyoctanoate-co-3-hydroxyhexanoate)	Methyl/propyl	
Poly(3-hydroxybutyrate-co-3-hydroxyoctanoate)	Methyl/heptyl	

10.4.1 Blending

Blending of PHA is possible with both synthetic and biodegradable polymers. Cellulose and starch derivatives have been extensively used in blending carried out by various researchers due to its capacity to impart better mechanical properties to the film. The overall cost of the final product also decreased with the use of such blends. To improve the characteristics of PHB, lignin can be used as a blending component. Poly (lactic acid) (PLA), aliphatic polyester synthesized chemically is reported to blend nicely with PHA improving it's compatibility and overall strength and applicability of bioplastics (Sharma et al. 2021).

10.4.2 Biodegradability

PHAs can be degraded under a wide range of temperature and moist conditions in both garden /farm soil (Shrivastav et al. 2011). The biodegradable nature under natural conditions is the key property of PHAs making its widespread utility as packaging material. Various factors are responsible for the use of PHAwhich are mentioned herein (i) slow release of pesticides/ fertilizers which can be beneficial in farming (ii) photoactivation (iii) no leftover debris (iv) PHA gets degraded by several micro-organisms, fungi under both aerobic and anaerobic environment (Meereboer et al. 2020).

10.4.3 Thermal Stability

The thermal stability of the polymer is monitored by progressive reduction in molecular weight. The melting point of PHA (Tm) is comparable to its thermal breakdown temperature. With the addition in side chain, the PHA becomes more viscous and sticky limiting its overall characteristic (Naser et al. 2021).

10.4.4 Tensile Strength

The three types of PHAs are: Short-chain-length PHAs (SCL-PHAs; 1–5 carbon atom monomer) with 60–80% crystallinity is brittle and stiff with strong elasticity break and have poor tensile strength, medium-chain-length (MCL-PHA) with C6-C14 monomers, and long-chain-length (LCL-PHA) with > C14 monomers are the three types of PHAs. Poly(3HB) is a rigid polymer with tensile strength like polypropylene with mechanical qualities similar to synthetically made polylactides possessing high melting temperature (175 °C) and a high tensile strength (30–35 MPa) (Bugnicourt et al. 2014).

10.4.5 Hydrophobicity

The insolubility in water of PHA plays an advantageous role in paper coating and food packaging industry. These PHA films act as a strong UV and oxygen barrier making it suitable for various applications in packaging industry and medical applications. For specific applications, PHAs must be made cheaper with adjustable mechanical characteristics, crystallinity, surface morphology, and degradability (Li et al. 2016).

10.4.6 Hydrophilicity

Comparing PHAs with starch and cellulose, PHAs have lower hydrophilicity. The addition of various materials in PHAs can make changes in the hydrophobicity of the polymer (Kaniuk and Stachewicz 2021). Some researchers have worked using bio-plasticizers synthesized from levulinic acid as they have good miscibility with PHB. These green plasticizers have remarkable effects on the mechanical properties like melting point, glass-transition temperature and crystallinity of these PHBs. The observed effects expand the temperature range in which PHB can be processed without thermal degradation. Moreover, the incorporation of levulinic acid-based additives does not significantly affect the typical biodegradability and biocompatibility of PHB, showing their promising features as plasticizers for both environmental and biomedical applications (Bera et al. 2015).

10.5 Cost Analysis

10.5.1 Substrate Costs

The raw materials or substrates used to produce PHA can be a major cost component in the production process. PHA is typically produced by microbial fermentation using renewable feedstocks like sugars, vegetable oils or agricultural byproducts and other byproducts (Bhattacharya et al. 2016). PHAs production utilizing Jatropha biodiesel byproducts is also a sustainable approach in the biorefinery, wherein excess carbon containing crude glycerol and nitrogen containing deoiled cake is utilized. The cost and availability of these substrates can vary regionally and seasonally, affecting the overall production costs of PHA. Fluctuations in substrate prices and competition with other industries that use similar feedstocks can pose challenges for PHA manufacturers.

10.5.2 Extraction Cost

The downstream processing and extraction of PHA from microbial biomass can be energy-intensive and costly. Different extraction methods, such as solvent-based, enzymatic or chemical-based methods, have varying efficiencies and associated costs. Developing cost-effective extraction techniques while maintaining high-purity levels is crucial for commercializing PHA on a large scale. Addressing these challenges requires ongoing research and innovation in the field of biotechnology and process engineering.

10.5.3 Downstream Process Optimization Cost

Developing more efficient and environmentally friendly extraction methods can help reduce extraction costs and energy consumption.

10.5.4 Scale-up and Process Integration

Integrating PHA production with existing biorefinery processes or waste treatment facilities can improve cost-effectiveness and increase overall process sustainability.

10.5.5 Government Support And Incentives

Government policies, grants and incentives that promote the use of biodegradable and sustainable materials can create a more favourable market environment for PHA and drive investment in research and development.

Overall, addressing substrate and extraction costs is essential to unlocking the full potential of PHA as a sustainable and eco-friendly alternative to traditional plastics. As technology advances and economies of scale are achieved, the challenges in the PHA market are expected to gradually overcome, leading to wider adoption and application of this biodegradable polymer.

Scale-up of PHA production involves the transition from laboratory or pilot-scale processes to commercial-scale manufacturing. Successful scale-up requires careful consideration of fermentation conditions, design of bioreactors and conducting techno-economic analysis to assess the feasibility and profitability of the larger production process.

10.6 Techno-Economic Analysis

a. **Cost estimation**: Conducting a comprehensive cost analysis of all aspects of the scaled-up process, including raw materials, equipment, labour, utilities and waste management is essential to estimate the overall production cost.
b. **Revenue estimation**: Assessing the market demand and potential selling price for PHA products is necessary to estimate the revenue generated from the scaled-up production.
c. **Profitability assessment**: Performing a techno-economic analysis helps determine the profitability of the PHA production at a commercial scale. It helps identify potential areas for cost reduction and process optimization.

Successful scale-up of PHA production requires a multidisciplinary approach involving bioprocess engineering, microbiology, chemistry and economics. Collaboration between researchers, industry experts and policymakers is vital to overcome technical and economic challenges and make PHA a competitive and sustainable alternative to traditional plastics.

10.7 Potential Applications of PHAs

There are several potential applications specified below according to its properties.

- Biomedical applications: Drug delivery, tissue engineering, cardiac patches, cardiac valves, sutures and stents.
- Packaging industry: Food industry, nanocomposites, multilayer films, paper and cardboard coatings, food packaging, integrating PHA nanofibers with antibacterial agents for antimicrobial packaging materials.
- Flame retardancy: PHA-based composites can be used as a flame retardant material wherein bentonite is used as a green filler in the process (Xu et al. 2023).
- PHA as oil encapsulates for potential essential oil delivery applications: Considering the lipophilic nature of the EOs compounds, the most effective strategies for their nano-formulation have consisted in the use of lipid-based systems, i.e., nano and microemulsions (Sharma et al. 2021; Meereboer et al. 2020), solid-lipid nanoparticles (Naser et al. 2021), liposomes (Bugnicourt et al. 2014) and polymeric nanoparticles (Li and Loh 2015; Kaniuk and Stachewicz 2021). Among the latter category, PHA-based nanoparticles have emerged as bio-based, biodegradable and biocompatible materials, potentially applicable in food packaging (Dhangdhariya and Dubey 2015). Produced by various species of bacteria through microbial fermentation of different C-sources (both carbohydrate and lipid ones), PHA origin is properly renewable (Dubey and Mishra 2021; Bera et al. 2015). Furthermore, the diversity in monomeric composition in which they can be produced depending on the microbial bioprocess, makes PHA tenable materials, in terms of mechanical properties and biodegradation kinetics (Dubey and Mishra 2021). In recent years, different PHA polymers have been applied for the controlled delivery of various drugs (Sequeira et al. 2020; Bhattacharya et al. 2016; Dubey et al. 2018). Furthermore, more recently, the grafting of plain poly-3-hydroxybutyrate-co-hydroxyhexanoate (PHB-HHx) NPs in whey-protein-based film matrix has shown to improve the technological and barrier properties of the functionalized materials, opening the way to the design of smart biomaterials incorporating bioactive molecules inside them (Shrivastav et al. 2010).

10.8 Future Prospects of Bioplastic Production

The environmental effect of a huge amount of non-biodegradable waste materials is encouraging research into novel biodegradable materials that may be created from natural resources such as biomass, plants and microorganisms. Future bioplastics improvements may result in increased manufacturing efficiency, as well as the creation of new uses and prospects for bioplastics. Furthermore, the future market for bioplastics will grow due to their sustainability. In addition, microorganism biotechnology provides an opportunity for bioplastic manufacturing since, it has the potential to apply and commercialize in a variety of industries such as agriculture, medical, pharmaceutical and veterinary. A new bioplastics guideline and standard should be developed for global production, use and waste management of bioplastics. As a result, labelling regulations must be strengthened based on raw material usage and emissions from manufacturing. Bio-based substances or recyclable materials have a high potential for composability. Recent technological advancements, ongoing research and worldwide support are critical for commercializing and demonstrating bioplastics. To boost the long-term viability of components and procedures throughout their lifetime, bioplastics must be built on an integrated environmental friendly approach. Bioplastic materials must not compete with existing sources and lessen the requirement for non-renewable resources over time.

10.9 Conclusion

Bioplastics are eco-friendly materials with enormous potential for long-term growth. Bioplastics have become so significant due to the fact that the price of petroleum oil is skyrocketing, and its supply may run out soon. It is critical for the global society to have an alternative to petroleum-derived products such as plastics. PHAs will provide a solution for the majority of companies and societies that rely heavily on plastic-based products. The most significant disadvantage of PHA manufacturing is its high production cost; nevertheless, the positive aspect is that the cost of PHA production is dropping, although the price of petroleum oil is always escalating. The use of bioplastics can help against many issues like, plastic pollution, greenhouse gas emission, waste recycling and can promote a sustainable and circular economy. The use of bioplastics on a global scale can help improve the economic viability of plastic producing industries and further help to restore and shape a greener future due to its characteristics of recyclability.

Acknowledgements The Director, CSIR-CSMCRI and present DC, Applied Phycology and Biotechnology Division are deeply acknowledged for their encouragement. SD would like to acknowledge CSIR HRDG, New Delhi for SRAship. We all the authors would like to express our sincere gratitude to all the individuals and organizations that have contributed their knowledge, time and hard work for the publication of this book chapter. We would also like to express our appreciation to the Springer here for considering our work and providing the opportunity to publish our chapter.

Conflict of Interest We wish to confirm that there are no known conflicts of interest associated with this publication and there has been no significant financial support for this work that could have influenced its publication.

References

Abe MM, Martins JR, Sanvezzo PB, Macedo JV, Branciforti MC, Halley P, Botaro R, Brienzo M (2021) Advantages and disadvantages of bioplastics production from starch and lignocellulosic components. Polymers (mdpi.com). https://doi.org/10.3390/polym13152484

Bahrami R, Zibaei R, Hashami Z, Hasanvand S, Garavand F, Rouhi M, Jafari SM, Mohammadi R (2020) Modification and improvement of biodegradable packaging films by cold plasma; a critical review. Crit Rev Food Sci Nutr 62:1936–1950 (Taylor & Francis). https://doi.org/10.1080/10408398.2020.1848790

Barone J, Schmidt WF (2005) Effect of formic acid exposure on keratin fiber derived from poultry feather biomass. Bioresour Technol (Elsevier). https://doi.org/10.1016/j.biortech.2005.02.039

Bera A, Dubey S, Bhayani K, Mondal D, Mishra S, Ghosh PK (2015) Microbial synthesis of polyhydroxyalkanoate using seaweed-derived crude levulinic acid as co-nutrient. Int J Biol Macromolecules. Elsevier

Bhattacharya S, Dubey S, Singh P, Shrivastava A, Mishra S (2016) Biodegradable polymeric substances produced by a marine bacterium from a surplus stream of the biodiesel industry. Bioengineering (mdpi.com). https://doi.org/10.3390/bioengineering3040034

Bugnicourt E, Cinelli P, Lazzeri A, Alvarez V (2014) Polyhydroxyalkanoate (PHA): review of synthesis, characteristics, processing and potential applications in packaging. https://doi.org/10.3144/expresspolymlett.2014.82

Calva-Estrada SJ, Jiménez-Fernández M, Lugo-Cervantes E (2019) Protein-based films: advances in the development of biomaterials applicable to food packaging. Food Eng Rev. https://doi.org/10.1007/S12393-019-09189-W

Dhangdhariya JH, Dubey S, Trivedi HB, Pancha I, Bhatt JK, Dave BP, Mishra S (2015) Polyhydroxyalkanoate from marine Bacillus megaterium using CSMCRI's Dry Sea Mix as a novel growth medium. Int J Biol Macromolecules (Elsevier)

Dubey S, Bharmoria P, Gehlot PS, Agrawal V, Kumar A, Mishra S (2018) 1-Ethyl-3-methylimidazolium diethylphosphate based extraction of bioplastic "polyhydroxyalkanoates" from bacteria: green and sustainable approach. ACS Sustain Chem Eng 6:766–773. https://doi.org/10.1021/ACSSUSCHEMENG.7B03096

Dubey S, Mishra S (2022) Natural sea salt based polyhydroxyalkanoate production by wild Halomonas hydrothermalis strain. Fuel (Elsevier)

Dubey S, Mishra S (2021) Efficient production of polyhydroxyalkanoate through halophilic bacteria utilizing algal biodiesel waste residue. Front Bioeng Biotechnol 9. https://doi.org/10.3389/FBIOE.2021.624859/FULL

Hadidi M, Jafarzadeh S, Forough M, Garavand F, Alizadeh S, Salehabadi A, Khaneghah AM, Jafari SM (2022) Plant protein-based food packaging films; recent advances in fabrication, characterization, and applications. Trends Food Sci Technol (Elsevier) 120:154–173. https://doi.org/10.1016/j.tifs.2022.01.013

Jayakumar A, Radoor S, Siengchin S, Shin GH, Kim JT (2023) Recent progress of bioplastics in their properties, standards, certifications and regulations: a review. Sci Total Environ (Elsevier)

Kaniuk Ł, Stachewicz U (2021) Development and advantages of biodegradable PHA polymers based on electrospun PHBV fibers for tissue engineering and other biomedical applications. ACS Biomater Sci Eng 7:5339–5362 (ACS Publications). https://doi.org/10.1021/acsbiomaterials.1c00757

Li Z, Loh XJ (2015) Water soluble polyhydroxyalkanoates: future materials for therapeutic applications. Chem Soc Rev 2015 (pubs.rsc.org)

Li Z, Yang J, Loh XJ (2016) Polyhydroxyalkanoates: opening doors for a sustainable future. NPG Asia Mater 8. https://doi.org/10.1038/AM.2016.48

Liyanage S, Acharya S, Parajuli P, Shamshina JL, Abidi N, Skórczewska K, Barczewski M, Kurá M (2021) Production and surface modification of cellulose bioproducts. Polymers (mdpi.com). https://doi.org/10.3390/polym13193433

Meereboer KW, Misra M, Mohanty AK (2020) Review of recent advances in the biodegradability of polyhydroxyalkanoate (PHA) bioplastics and their composites. Green Chem (pubs.rsc.org)
Mustafa P, Niazi MB, Jahan Z, Samin G, Hussain A, Ahmed T, Naqvi SR (2020) PVA/starch/propolis/anthocyanins rosemary extract composite films as active and intelligent food packaging materials. J Food Saf 40(Wiley Online Library). https://doi.org/10.1111/jfs.12725
Nanda S, Patra BR, Patel R, Bakos J, Dalai AK (2021) Innovations in applications and prospects of bioplastics and biopolymers: a review. Environ Chem Lett 20(1):379–395. https://doi.org/10.1007/S10311-021-01334-4
Naser AZ, Deiab I, Darras BM (2021) Poly (lactic acid)(PLA) and polyhydroxyalkanoates (PHAs), green alternatives to petroleum-based plastics: a review. RSC Adv (pubs.rsc.org)
Saharan B, Sharma D (2015) Bioplastics-for sustainable development: a review bioplastics-for sustainable development: a review. In: Bioplastics-for sustainable development: a review. https://scholar.google.com/scholar?hl=en&as_sdt=0%2C5&q=Saharan%2C+B.%2C+%26+Sharma%2C+D.+%282015%29.+Bioplastics-for+sustainable+development%3A+a+review+bioplastics-for+sustainable+development%3A+a+review.+Sustainable+development+by+bioplastics+and+biofuels+production%2C+1%2C+10-23&btnG=. Accessed 13 Oct 2023
Sequeira RA, Dubey S, Pereira MM, Maity TK, Singh S, Mishra S, Prasad K (2020) Neoteric solvent systems as sustainable media for dissolution and film preparation of poly-[(r)-3-hydroxybutyrate]. ACS Sustain Chem Eng 8:12005–12013. https://doi.org/10.1021/ACSSUSCHEMENG.0C02684
Sharma V, Sehgal R, Gupta R (2021) Polyhydroxyalkanoate (PHA): properties and modifications. Polymer (Elsevier)
Shrivastav A, Mishra SK, Pancha I, Jain D, Bhattacharya S, Patel S, Mishra S (2011) Biodegradability studies of polyhydroxyalkanoate (PHA) film produced by a marine bacteria using Jatropha biodiesel byproduct as a substrate. World J Microbiol Biotechnol 27:1531–1541. https://doi.org/10.1007/S11274-010-0605-2
Shrivastav A, Mishra S, Shethia B, Pancha I, Jain D, Mishra S (2010) Isolation of promising bacterial strains from soil and marine environment for polyhydroxyalkanoates (PHAs) production utilizing Jatropha biodiesel byproduct. Int J Biol Macromolecules (Elsevier)
Siracusa V, Blanco I (2020) Bio-polyethylene (Bio-PE), Bio-polypropylene (Bio-PP) and Bio-poly (ethylene terephthalate)(Bio-PET): recent developments in bio-based polymers analogous to petroleum-derived ones for packaging and engineering applications. Polymers (mdpi.com). https://doi.org/10.3390/polym12081641
Steven S, Fauza AN, Mardiyati Y, Santosa SP, Shoimah SMA (2022) Facile preparation of cellulose bioplastic from cladophora sp. algae via hydrogel method. Polymers (mdpi.com 14). https://doi.org/10.3390/polym14214699
Xu P, Qi G, Lv D, Niu D, Yang W, Bai H, Yan X, Zhao X, Ma P (2023) Enhanced flame retardancy and toughness of eco-friendly polyhydroxyalkanoate/bentonite composites based on in situ intercalation of PN-containing hyperbranched. Int J Biol Macromolecules. Elsevier
Zheng W, Liu Y, Liu W, Ji H, Li F, Shen C, Fang X, Li X, Duan X (2021) A novel electrocatalytic filtration system with carbon nanotube supported nanoscale zerovalent copper toward ultrafast oxidation of organic pollutants. Water Res (Elsevier)

Chapter 11
Seaweed-Based Biodegradable Packaging: A Green Alternative for a Plastic-Free Future

Silvia Lomartire and Ana M. M. Gonçalves

11.1 Introduction

Increasing worries about plastic pollution and its harmful effects on both ecosystems and human well-being have spurred a significant drive to explore alternative options to conventional fossil-based plastics. Nowadays, plastics have become widely utilized thanks to their remarkable qualities such as resilience, impermeability to water, lightweight nature, stability, and cost-efficiency. These attributes have led to their extensive application across multiple industries and in our daily lives ever since their inception (Schmaltz et al. 2020; Shen et al. 2020; Turner 2018).

Plastics are essentially composed of polymers, which are comprised of interconnected chains of repeating units. These chains may have branching or cross-linking, and by introducing various chemical monomers, one can modify their physical and chemical properties (Waring et al. 2018; Price-Whelan et al. 2014). Regrettably, plastic waste management is a significant challenge in numerous regions worldwide, primarily because of its resistance to degradation, chemical stability, and slow decomposition process, which can take more than a millennium. Each year, the global production of plastic surpasses 300 million tons, with approximately half of this quantity being employed in the form of disposable items (Allouzi et al. 2021). Mechanical degradation, biodegradation, heat exposure, UV degradation, photodegradation, friction, and turbulence, are among the causes of degradation of plastics, which become microplastics (MPs), and nanoplastics (NPs).

MPs and NPs make their way into the environment through two primary pathways: they are introduced directly through personal care products such as shampoo,

S. Lomartire · A. M. M. Gonçalves (✉)
Department of Life Sciences, University of Coimbra, MARE–Marine and Environmental Sciences Centre/ARNET-Aquatic Research Network, 3000-456 Coimbra, Portugal
e-mail: amgoncalves@uc.pt

A. M. M. Gonçalves
Department of Biology and CESAM, University of Aveiro, 3810-193 Aveiro, Portugal

K. Bala et al. (eds.), *Harnessing Microbial Potential for Multifarious Applications*, Energy, Environment, and Sustainability, https://doi.org/10.1007/978-981-97-1152-9_11

detergents, cosmetics, and paints, or they are generated because of the fragmentation of larger plastic particles already present in ecosystems (Kumar et al. 2021a; Cauwenberghe et al. 2015). These plastic particles are detectable in a wide range of environmental settings, including soil, groundwater, the atmosphere, rivers, oceans, and even in high-altitude ecosystems like mountains and glaciers (Shen et al. 2020; Price-Whelan et al. 2014; Li et al. 2021; Chen et al. 2021; Adam et al. 2021; Kumar et al. 2021b). Moreover, macroplastics have detrimental impacts on larger wildlife, such as turtles, which often ingest plastic by mistake, mistaking it for their natural food. This consumption can result in the mortality of various animal species, and some animals become ensnared in plastic materials, rendering them unable to extricate themselves (Waring et al. 2018; Revel et al. 2018; Xu et al. 2020). MPs and NPs are dangerous at lower trophic levels, as they can be ingested by zooplanktonic organisms, making their way up the food chain (Revel et al. 2018).

Furthermore, MPs and nanoplastics NPs in the air have an alternative route for their introduction into the human body (Waring et al. 2018). When humans inhale these MPs, it can result in significant health issues, including conditions like lung cancer, ulcers, and diseases related to the nasal passages and sense of smell (Revel et al. 2018; Wang et al. 2020a).

In response to this pressing need, there is active research and development underway for novel bio-based plastics, materials, and products. Bioplastics can be classified into four distinct categories based on their chemical composition, origin, and manufacturing processes. These categories encompass: polymers directly derived from biomass, such as proteins, starch, and cellulose; polymers synthesized chemically using bio-derived monomers, exemplified by bio-polyethylene (PE), polyethylene terephthalate (PET), polyhydroxybutyrate (PHB), and polylactide (PLA); polymers produced via microbial fermentation, as seen in the case of polyhydroxyalkanoates; polymers resulting from chemical synthesis using a combination of petroleum-based and bio-derived monomers, including polytrimethylene terephthalate (PTT) and polybutylene succinate (PBS). Bioplastics can source their raw materials from a variety of sources, such as plants like rice, corn, barley, and cellulose derivatives, as well as from bacteria and algae (Thiruchelvi et al. 2020; Srinivasan and Kulshreshtha 2020; Samantaray et al. 2020).

The present chapter explores seaweed as a potential source for producing biodegradable plastic.

Seaweeds, which are also referred to as marine macroalgae, are multicellular photosynthetic organisms dwelling in various aquatic environments. They can be classified into three primary groups based on their coloration, which is influenced by the presence of specific pigments. Brown seaweeds (belonging to the phylum Ochrophyta and class Phaeophyceae) exhibit a spectrum of pigments ranging from yellow to deep brown shades (Seely et al. 1972). Conversely, red seaweeds (found in the phylum Rhodophyta) are characterized by their abundance of carotenoids, chlorophyll (both a and d forms), phycoerythrin, phycocyanin, and allophycocyanin (Denis et al. 2009; Haugan and Liaaen-Jensen 1994). Lastly, green seaweeds (classified under the phylum Chlorophyta) predominantly contain chlorophyll, a pivotal pigment essential for photosynthesis. The greenish tint resulting from chlorophyll imparts the distinctive green color seen in plants, algae, and cyanobacteria (Aryee et al. 2018).

Progress in scientific research and innovative methodologies have made it possible to pinpoint and examine compounds derived from seaweeds, unveiling their potential benefits for human health, including their antioxidant, antimicrobial, and anti-tumor properties (Lomartire et al. 2021). This emerging understanding has paved the way for diverse industrial uses of seaweeds.

Seaweeds are now finding applications in various fields, including the manufacturing of fertilizers (Illera-Vives et al. 2017; Vijayakumar et al. 2019; Raghunandan et al. 2019), aquaculture feed, and biofuels. Their potential is also evident in wastewater treatment processes. Furthermore, they offer exciting prospects for developing environmentally friendly materials as substitutes for traditional plastics (Kim et al. 2017; Raikova et al. 2019; Wang et al. 2020b).

Furthermore, their bioactive compounds enhance their biodegradability, rendering them a superb option to produce bioplastic packaging. The compounds sourced from algae can be employed in crafting what is referred to as "active" or "smart" packaging. This pioneering packaging material exhibits protective features, including heightened antimicrobial properties, which serve as a safeguard for packaged food or pharmaceuticals against external factors such as foodborne pathogens and UV radiation. Moreover, it aids in preserving the quality and safety of the contents enclosed within the packaging.

11.2 Hydrocolloids Involved in Industrial Applications

Hydrocolloids can be classified into three primary types based on their origin: they can be natural, semisynthetic, or synthetic. Plant-derived hydrocolloids are primarily employed to stabilize oil-in-water emulsions, while animal-derived hydrocolloids are better suited for forming water-in-oil emulsions. Among these, gelatin stands out as the most prominent player in the food hydrocolloids market due to its extensive use as a gelling agent in confectionery, meat, poultry, and dairy products. Additionally, various other hydrocolloids such as guar gum, carrageenan, xanthan gum, gum arabic, and pectin have exhibited potential in preserving and safeguarding food from external stresses (Liu et al. 2008).

Nonetheless, it's essential to acknowledge that hydrocolloids derived from animals may carry potential allergy risks and are susceptible to microbial growth and rancidity. In contrast, semisynthetic hydrocolloids are created by modifying naturally occurring hydrocolloids. Examples of these semisynthetic hydrocolloids encompass starch and cellulose derivatives, such as methylcellulose (MC), carboxymethylcellulose (CMC), hydroxypropyl methylcellulose (HPMC), microcrystalline cellulose (MCC), acetylated starch (AS), phosphorylated starch (PS), and hydroxypropylated starch (HPS). Semisynthetic hydrocolloids exhibit robust emulsifying properties, are considered non-toxic, and are less prone to microbial growth. In contrast, synthetic hydrocolloids are entirely synthesized in industrial settings using petroleum-derived base materials. While they possess highly potent emulsifying capabilities and do not support microbial growth, their production costs can be prohibitive. Synthetic hydrocolloids are primarily employed as oil-in-water emulsifiers. However, semisynthetic hydrocolloids are generally preferred over entirely synthetic gums due to their advantageous attributes (Li and Nie 2016; Dickinson 2018).

There is a rising preference for renewable hydrocolloids, largely motivated by economic and environmental concerns. To meet the surging consumer demand for completely natural products, the aim is to substitute non-renewable and synthetic hydrocolloids with sustainable alternatives across diverse industries. Consequently, efforts are actively being made to discover new natural hydrocolloids with unique properties tailored for specific applications.

11.2.1 Physical–Chemical Characteristics of Phycocolloids

When assessing the appropriateness of a bioplastic material for the production of scaffolds or hydrogels, it is essential to take into account mechanical characteristics such as tensile strength, elongation at break, heat resistance, and water vapor permeability. Tensile strength refers to the maximum stress a material can withstand before breaking when subjected to stretching or pressure. Typically, the tensile strength and Young's modulus of plant fibers tend to increase with a higher cellulose content. Elongation at break quantifies the ratio between the modified length and the initial length of the tested material after it fractures. This measurement reflects the material's capacity to undergo shape alterations without developing cracks in a matrix composed of natural polymers. Thermal resistance, on the other hand, is a thermal property that signifies the temperature differential a material can endure (Djafari Petroudy 2017).

Conversely, vapor permeability pertains to a material's ability to permit the transmission of vapors, including water vapor and other gases. As defined by ISO 11092:1993 (European Committee for Standardization Textiles 2005), water vapor permeability is a property affected by the water vapor resistance of a textile material or composite. A higher permeability value signifies that water and vapor can traverse the material at a faster rate, which is generally viewed as a desirable trait.

11.2.2 Impact of Seaweed Phycocolloids in Biofilm Development

The gel-forming characteristics of phycocolloids obtained from seaweeds demonstrate fluctuations that are affected by various factors, including their structure, concentration within a solution, temperature, pH, and potential for syneresis potential (Lee et al. 2017a; Liu et al. 2016; Ramdhan et al. 2019). The physicochemical properties of phycocolloids (Table 11.1) are notably influenced by factors such as the specific seaweed species, environmental conditions, extraction methodologies, and treatment processes (Lee et al. 2017a; Porse and Rudolph 2017).

Carrageenans, when in their salt form, display substantial gel strength (Ilias et al. 2017). Among these, κ-carrageenan and ι-carrageenan gels maintain their stability at room temperature, whereas λ-carrageenan, the only carrageenan soluble in cold water in its natural state, does not form a gel. The introduction of cations has been observed to augment both the formation and strength of gels in κ-carrageenan phycocolloids, as documented by Robal et al. (2017).

In a study conducted by Paula et al. (2015) the physical properties of glycerol-plasticized edible films crafted from κ-carrageenan, ι-carrageenan, and alginate were investigated. The findings revealed that κ-carrageenan films exhibited superior tensile strength, elasticity, moisture permeability, and lower opacity when compared to ι-carrageenan films. In contrast, alginate films exhibited higher transparency.

Like carrageenan, alginate has the capacity to create a robust and enduring gel structure, especially in the presence of cations like Ca^{2+} (Yoon et al. 2014). Films produced from sodium alginate and immersed in a 1–3% (w/v) calcium chloride solution have exhibited enhanced tensile strength and elongation properties, along with a decrease in opacity (Kok and Wong 2018; Liling et al. 2016). In contrast, agar films tend to have lower tensile strength and higher water vapor permeability when compared to films made from carrageenan and alginate. However, they do show roughly double the elongation value in comparison to κ-carrageenan films and demonstrate superior elasticity (Kanmani and Rhim 2014; Rhim 2012).

Through advancements in research techniques, scientists have gained the ability to investigate how compounds derived from seaweed behave when incorporated into biofilms. This exploration has positioned these compounds as promising candidates for both biofilm development and packaging applications (Stevens 2001; Song et al. 2009; Khalil et al. 2017; Rao et al. 2018). It's worth noting that the method employed for forming these films exerts a significant influence on their physical characteristics

Table 11.1 Physical–chemical characteristics of the main seaweed phycocolloids

Compound	Physical–chemical characteristics	Reference
Carrageenan		Ilias et al. 2017; Robal et al. 2017
κ-carrageenan	High tensile strength, elasticity, moisture permeability, high transparency	
ι-carrageenan	Low strength and elasticity, high opacity	
λ-carrageenan	Soluble in cold water and does not form gel in its natural state	
Alginate	High tensile strength and elongation, high transparency	Kok and Wong 2018; Liling et al. 2016
Agar	Low tensile strength, low water vapor permeability, High elongation and elasticity	Kanmani and Rhim 2014; Rhim 2012

and microstructures. For instance, Li et al. (2019) conducted a study on chitosan-alginate films, employing various formation methods. Their research revealed that biofilms created through a combination of layer-by-layer assembly and ferulic acid cross-linking exhibited enhanced mechanical properties, opacity, and hydrophobicity compared to films formed through direct mixing, crosslinking alone, or layer-by-layer assembly alone.

Carrageenan and agar are usually extracted using hot water as a solvent, primarily because of their excellent solubility in water. In contrast, the extraction of alginate typically necessitates hot alkali as the primary solvent. This is because alginic acid contains salts that do not dissolve in water, and alkali extraction converts these alginic acid salts into water-soluble alginate salts. While hot water extraction is the preferred method for carrageenan and agar, the industry frequently employs alkali extraction to maintain their rheological characteristics, particularly their gel strength.

According to the literature review conducted by Budiman and Tarman (2022), the incorporation of plasticizers is recommended to enhance the physical and mechanical properties of bioplastics.

Natural hydrocolloids present a multitude of advantages when compared to their counterparts. They originate from renewable sources, are readily obtainable, and are straightforward to manage. Moreover, they possess attributes such as biocompatibility, non-toxicity, adaptability to both physical and chemical alterations, environmental friendliness, cost-effectiveness, and widespread acceptance among the public due to the numerous health benefits they provide (Bisht et al. 2022). The primary hydrocolloids derived from seaweeds include carrageenan, alginate, and agar.

11.2.2.1 Carrageenan

Carrageenan, as illustrated in Fig. 11.1, is a linear polysaccharide composed of galactose units, which can be either sulfated or non-sulfated, linked together through α-1,3-glycosidic and β-1,4-galactose bonds (Therkelsen 2012). These natural polysaccharides constitute a blend of linear galactans containing sulfate groups. The fundamental building blocks of these polysaccharides consist of disaccharides, specifically, α-(1 → 4)-linked d-galactopyranose (D) residues or 3,6-anhydrogalactopyranose (DA), and β-(1 → 3)-linked d-galactopyranose (G) residues. The sulfate groups are covalently attached to the galactose molecules at positions C-2, C-4, or C-6 through ether bonds (Khotimchenko et al. 2020).

Carrageenans can be classified into numerous types (including κ-carrageenan, ι-carrageenan, λ-carrageenan, γ-carrageenan, ν-carrageenan, ξ-carrageenan, θ-carrageenan, and μ-carrageenan) based on the location of the sulfate group attached to the galactose unit. In their natural state, carrageenans predominantly showcase hybrid structures, resulting in variations in their properties contingent on the particular sulfate group bonding (Zhong et al. 2020).

Like alginate, the double-helix structure of carrageenan provides sites for hosting anions, which contribute to strengthening the gel structure. Carrageenan can accommodate monovalent ions, like K^+, but it exhibits a preference for divalent ions, such

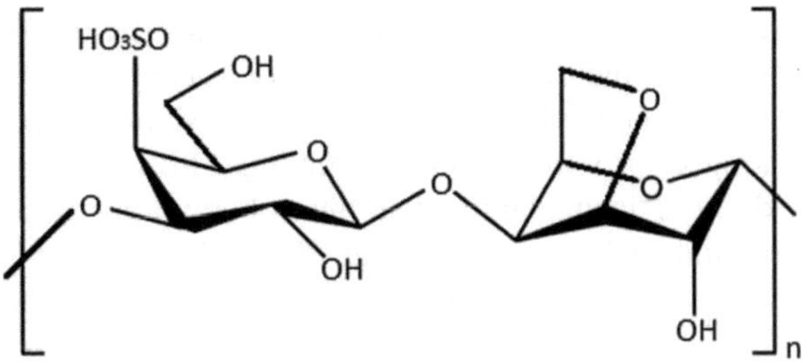

Fig. 11.1 Chemical structure of κ-carrageenan

as Ca^{2+}. Many carrageenan-based bioplastics aim to create edible packaging sheets (Cazón et al. 2017). However, the hydrophobic nature of carrageenan presents a challenge in the production of water-resistant packaging.

One potential solution is to incorporate hydrophobic compounds such as chitosan, starch, cellulose, chitin, and alginate into the material's matrix to reinforce it. This approach has the potential to enhance the properties of carrageenan and could lead to the development of eco-friendly and cost-effective packaging materials with improved strength (Sedayu et al. 2019). For instance, ι-carrageenan shows promise as an excellent ingredient for food coating, particularly in the preservation of ham (Carocho et al. 2019).

Polyvinyl alcohol (PVA) has been utilized as a synthetic substance to bolster the inadequate mechanical characteristics of specific materials. In fact, carrageenan-PVA films exhibited a notable 439.31% augmentation in elongation (Meng et al. 2018). Furthermore, the inclusion of 5% SiO_2 nanoparticles led to a substantial enhancement in the tensile strength of carrageenan-gelatin composites, elevating it to 10.32 MPa (Tabatabaei et al. 2018).

Various carrageenan types are produced by various seaweed species. κ-carrageenans, readily accessible in the market, are obtained through a heat extraction process from *Kappaphycus alvarezii*. In contrast, λ-carrageenans are usually sourced from red algae species belonging to the *Gigartina* or *Chondrus* genera, and they are acquired through methods like drum drying or ethanol precipitation (Vera et al. 2011).

Among these, there are eight sources of carrageenan exclusive to the Japanese Sea (East Sea). Species such as *Chondrus pinnulatus, Chondrus armatus, Chondrus yendoi, Mastocarpus pacificus*, and *Mazzaella hemisphaerica* belong to the Gigartinaceae and Solieriaceae families within this region. Furthermore, apart from the Gigartinaceae family, substantial quantities of carrageenan can be found in algae belonging to the Phillophoraceae and Thichocarpaceae families, with their species widely distributed in the seas of the Far East (Khotimchenko et al. 2020).

In contrast, ι-carrageenan is primarily sourced from *Eucheuma denticulatum*, commonly known as "*spinosum*," and it imparts a soft and less robust gel consistency. Lastly, ʎ-carrageenan is obtained from various species within the *Gigartina* and *Chondrus* genera (Pereira and Velde 2011).

Carrageenans find widespread use in the food industry because of their capacity to gel, thicken, and stabilize food products (Alba and Kontogiorgos 2018). The commercial variants of ʎ-, κ-, and ι-carrageenans have received approval as food additives from regulatory authorities such as the Food and Drug Administration (FDA) and the European Food Safety Agency (EFSA) (Younes et al. 2018). In recent decades, the medical field has also delved into the potential biological properties of carrageenan, with encouraging findings. Research has revealed that these compounds exhibit anticoagulant and antithrombotic properties (Carlucci et al. 1997), possess antiviral attributes (Wang et al. 2012), demonstrate antitumoral effects (Cotas et al. 2020), and display antioxidant characteristics (Rocha De Souza et al. 2007).

11.2.2.2 Alginate

Alginic acid, is also referred to as algin or alginate (Fig. 11.2). These polysaccharides can be obtained from the cell walls of brown algae. Alginate, derived from alginic acid and its various derivatives and salts (Alba and Kontogiorgos 2018; Khalil et al. 2018), accounts for 10–40% of the dry weight of untreated algae and makes up 30–60% of the total sugars found in brown seaweeds (Enquist-Newman et al. 2014).

Alginates are negatively charged linear polysaccharides found in substantial quantities within brown seaweeds, constituting as much as 40% of their dry weight. They are recognized for their capacity to create edible films. These alginates are composed of alginic acid polymers, with monomer units consisting of β-D-mannuronic acid (M) and α-L-guluronic acid (G) connected through 1,4 linkages (Setyawidati et al. 2018; Benavides et al. 2012).

Alginates exhibit numerous advantageous qualities, including biodegradability, biocompatibility, non-toxic behavior, and cost-effectiveness, rendering them highly suitable for a wide range of biological applications (Sreekumar 2020). These alginates can be sourced from various seaweed species, including *Ascophyllum* spp., *Durvillaea* spp., *Laminaria* spp., *Lessonia* spp., *Macrocystis* spp., *Sargassum* spp., and *Ecklonia radiata* (Enquist-Newman et al. 2014). Additionally, Fucales, a category of large brown seaweed, such as *Scytothalia dorycarpa, Cystophora subfarcinata,* and *Sargassum linearifolium,* are not only valued for their nutritional content but also serve as sources of alginates (Zemke-White and Ohno 1999).

The physicochemical and mechanical attributes of alginate gels exhibit variability based on the M/G ratio and the length of the molecular structure. A greater presence

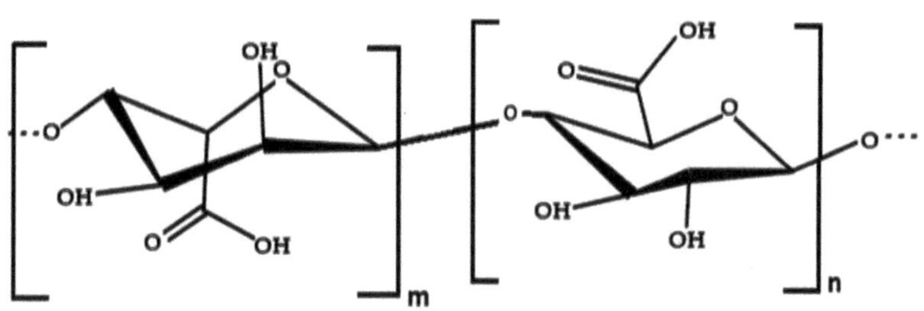

Fig. 11.2 Chemical structure of alginic acid

of guluronic acid contributes to enhanced gelling properties and the creation of more flexible gels. In contrast, lower M/G ratios yield sturdy yet brittle gels that display excellent heat stability but may undergo syneresis when subjected to freeze–thaw cycles (Alba and Kontogiorgos 2018; Khalil et al. 2018). Owing to its remarkable capacity for stabilization and thickening, alginate is widely employed across diverse applications in the food and medical sectors (Alba and Kontogiorgos 2018; Khalil et al. 2018; Oussalah et al. 2007).

Due to their pronounced hydrophilicity, alginates require the incorporation of supplementary components into the matrix to bolster their resilience in the presence of water. Moreover, the solubility of alginates is influenced by the presence of ions, and the development of gels hinges on the particular bonds formed with cations (Alba and Kontogiorgos 2018; Khalil et al. 2018).

Among the various cations, Ca^{2+} demonstrates a more robust binding affinity, resulting in heightened mechanical strength and superior water-resistant properties in contrast to Na^{+} (Cazón et al. 2017). In a study involving straw-alginate, the introduction of Ca^{2+} through cation exchange yielded substantial enhancements, with the strength reaching 216 g and the breakage rate reduced to 4% (Xue et al. 2019). Likewise, when Ca^{2+} was incorporated into the cellulose-alginate complex, it led to a swelling inhibition exceeding 95%, along with Young's modulus and tensile strength measurements of 135 and 17 MPa, respectively (Benselfelt et al. 2018).

These findings demonstrate that cations linked to alginate hold promise to produce a variety of ecologically acceptable packaging materials.

11.2.2.3 Agar

The primary origins of agar for industrial use consist of *Gracilaria* spp. (80%) and *Gelidium* spp. (20%), with *Gracilaria* spp. notably possessing the highest level of sulfation (Bixler and Porse 2011; Murano et al. 1996). In recent times, there has been an increasing fascination with *Gracilaria* spp. agars known for their less robust gel consistency. This shift can be attributed to the diminishing *Gelidium* spp. resources and the successful cultivation of *Gracilaria* (Santelices et al. 1993; Cordover 2007).

Agar–agar predominantly functions as a substance for gelling, stabilizing, and thickening in the realms of food and pharmaceuticals. Moreover, it is frequently employed for its purgative properties and as an assistive tool in surgical procedures. Researchers have been actively working on the creation of agar-based products, including composite hydrogels, nanocomposite films, and various materials customized for use in the pharmaceutical area (Pandya et al. 2022).

The chemical composition of agar is defined by repeating units of D-galactose and 3,6-anhydro-L-galactose, featuring minor differences and a limited presence of sulfate esters. This particular combination of d-galactose and l-galactose is produced within the cell wall matrix of red seaweeds (Lee et al. 2017b) and retains a gel-like structure even at room temperature (Ramawat and Merillon 2015).

Agar is made up of two categories of polysaccharides: agarose, an uncharged polysaccharide, and agaropectin, a term used to refer to the charged polysaccharide

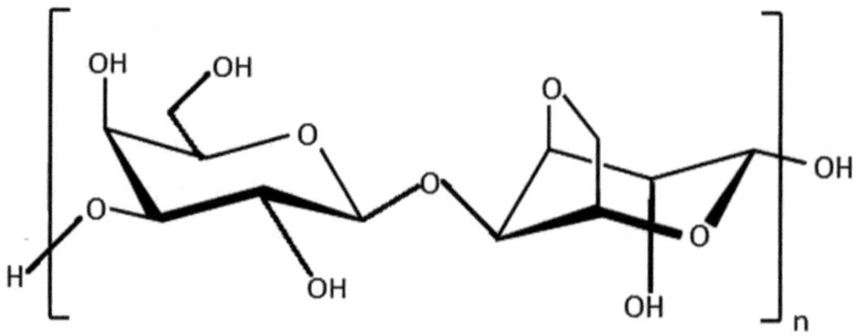

Fig. 11.3 Chemical structure of agarose polymer

(Araki 1956; Araki 1966; Matsuhashi 1990). Agarose (Fig. 11.3) is responsible for the gel-forming characteristics of agar, rendering it a valuable component in skincare products, herbal medicine, and pharmaceuticals. Moreover, it showcases outstanding film-forming properties (Rhim 2011). Agarose accounts for as much as 70% of the overall polysaccharide content in agar (Marinho-Soriano and Bourret 2005).

Agar in its gel state generally exhibits a solid and see-through consistency, but its firmness can be heightened through the addition of sugars (Bixler and Porse 2011). One notable benefit of agar is its limited hygroscopic nature, which proves advantageous in the manufacturing of packaging materials.

Furthermore, agar films are biologically inert and have the capacity to interact with a variety of bioactive compounds and/or plasticizers, allowing for the creation of flexible and supple gel structures. Agar's low hygroscopic nature is a valuable trait in packaging production (Mostafavi and Zaeim 2020; Sousa et al. 2010; Martínez-Sanz et al. 2019). As noted by Phan et al. (2009), agar-based films are transparent, robust, and versatile, and they possess heat-sealing capabilities, rendering them highly suitable for food packaging applications.

11.3 Incorporation of Seaweeds in Biodegradable Packaging

Various types of seaweeds, including red seaweeds like *Kappaphycus*, *Eucheuma*, *Gracilaria*, *Porphyra*, *Gelidium*, and *Pterocladia*, green seaweeds such as *Ulva*, *Codium*, and *Enteromorpha*, and brown seaweeds like *Macrocystis*, *Laminaria*, *Ascophyllum*, *Lessonia*, and *Sargassum*, have been extensively researched due to their high polysaccharide content, which positions them as promising candidates to produce bioplastic films (Freile-Pelegrín and Madera-Santana 2017; Rinaudo 2008) (Fig. 11.4).

In a study carried out by Lim et al. (2018), they explored the potential of alginate extracted from the brown alga *Sargassum siliquosum* as a promising raw material for producing bioplastic films. In the processing phase, alginate was combined with sago starch, sorbitol, and calcium chloride ($CaCl_2$). The resulting biofilm, formed

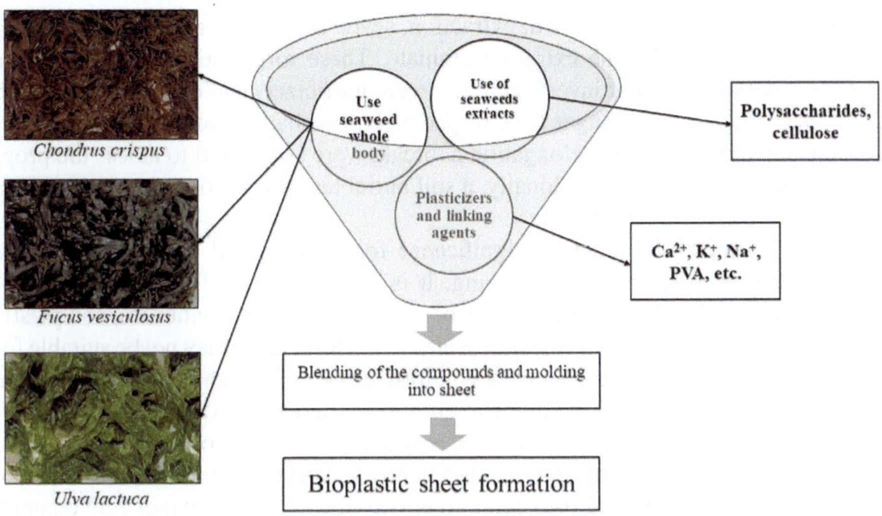

Fig. 11.4 Schematic and simplified representation of bioplastic production

using 2 g of alginate powder sourced from *Sargassum siliquosum* and containing 15% w/w sorbitol, treated with 75% w/w $CaCl_2$, exhibited favourable physical and mechanical properties. Nevertheless, further investigation is required to gain a deeper understanding of how alginate behaves when in the presence of inorganic compounds.

In the research conducted by Mohammed et al. (2023), they employed a mixture of sodium alginate sourced from *Sargassum natans* seaweed gathered from nearby bays in East Trinidad and Manugel DMB (provided by FMC Biopolymer). The casting solutions were formulated by incorporating different levels of plasticizers and strengthening agents. The casting solutions were composed of varying quantities of carboxymethyl cellulose (CMC) with a molecular weight of 700,000, potato starch with a purity exceeding 90%, sorbitol with a purity exceeding 98%, and polyethylene glycol (PEG200). To initiate the cross-linking of the bioplastic, calcium chloride dihydrate with a purity exceeding 99% was employed as the cross-linking agent.

The results revealed that the tensile properties of the alginate bioplastic composite were on par with those of PLA. Specifically, the tensile moduli were 60% higher than HDPE and 30% lower than PET. Consequently, the alginate composite displayed comparable tensile moduli to PLA, being stiffer than HDPE while exhibiting less deformability than PET. The strain at break for the alginate film was measured at 6.81%, whereas PLA displayed a strain at break of 43.2%, and both HDPE and PET exceeded 60% (Mohammed et al. 2023).

Alginate extraction from *Sargassum* sp., which is plentifully found in the coastal waters of Sabah, Malaysia, served as the foundation for the development and characterization of biodegradable bioplastics based on algae (Kanagesan et al. 2022). The alginate extraction process involved the optimization of key parameters like alkali (Na_2CO_3) concentration, temperature, and extraction duration using the ethanol method. Following the successful extraction, seven distinct bioplastic formulations were synthesized using the extracted alginate. These formulations comprised one control bioplastic without invert sugar (a green plasticizer) and others blended with 5, 10, and 15% of invert sugar (IS), respectively. Mechanical assessments, encompassing tensile strength and elongation at break, were conducted to assess the properties of the bioplastics. Additionally, a soil burial test was carried out to assess the bioplastics' biodegradability.

These findings hold practical significance for the potential use of this alginate bioplastic composite in food packaging. It is particularly well-suited for applications involving rigid packaging, such as pouches, trays, cups, containers, and plastic sheets, where high stiffness is a desirable attribute. However, it may not be suitable for applications requiring extensive plastic deformation under higher stress levels. The study also established that the alginate composite undergoes degradation in simulated compost conditions after 14 days, and the additives incorporated within the composite material do not migrate. Consequently, it can be considered a safe and eco-friendly alternative for food packaging in environments with low moisture content. The investigation unveiled that bioplastic containing 6% alginate, regardless of the IS concentration, displayed a darker brown hue compared to the bioplastic with 2% alginate. The control film, which lacked IS, exhibited rigidity and a slight brittleness, whereas the bioplastics with IS were noted for their increased flexibility. However, elevating IS concentrations led to stickiness. A lower IS concentration of 5% resulted in a higher tensile strength value of 2.13 MPa for bioplastics containing 6% alginate, in contrast to a TS value of 0.78 MPa for bioplastics containing 2% alginate. Nevertheless, elevating the IS concentration from 10 to 15% caused a notable reduction in TS. Bioplastics consisting of 6% alginate and 5% IS underwent a soil burial test. These bioplastics exhibited a 25% degradation rate on Day 1 and achieved complete degradation (100%) by Day 4, whereas the synthetic plastic control remained resistant to degradation (Kanagesan et al. 2022).

In their research, Patra et al. (2017) employed raw extracts and cellulose nanocrystals obtained from *Saccharina japonica* and *Sargassum natans* for the creation of biofilms. The introduction of cellulose nanocrystals was observed to improve the physical, chemical, mechanical, and thermal properties of the biofilm, highlighting the significant potential and promise of cellulose in the field of bio-packaging.

Researchers have explored the potential of the invasive brown seaweed, *Rugulopteryx okamurae*, sourced from the Pacific Ocean, in the development of bio-based plastic materials (Santana et al. 2022). The raw seaweed was mixed with glycerol in different proportions (50/50, 60/40, and 70/30) and subsequently processed using injection molding at varying mold temperatures (90, 120, and 150 °C). Rheological characteristics, including frequency sweep tests and temperature ramp tests, were obtained for the blends both before and after the injection molding process. The functional attributes of the bioplastics were evaluated through measurements of water uptake capacity (WUC) and examination via scanning electron microscopy (SEM). The findings indicated a consistent dominance of elastic behavior over viscous behavior, as the elastic modulus (E') consistently surpassed the viscous modulus (E"). Among the various blends, the 70/30 ratio demonstrated superior viscoelastic moduli and tensile properties, with values of 186.53 ± 22.80 MPa and 2.61 ± 0.51 MPa, respectively. With an increase in seaweed content in the mixture, the water uptake capacity exhibited a decreasing trend, ranging from 262% for the 50/50 ratio to 181% for the 70/30 ratio. Notably, the investigation into molded bioplastics at different temperatures revealed that seaweed content had a minimal impact on the final properties of the bioplastics. While these findings hold promise, further research is essential to enhance the mechanical properties of bioplastics based on *Rugulopteryx okamurae* seaweed. Exploring composite formulations with other biodegradable polymers and studying the effects of various processing conditions represent potential avenues to achieve this objective and uncover novel applications for these bioplastics (Santana et al. 2022).

Recently, a study by Hanry and Surugau (2020) was conducted to compare biofilms crafted from pure κ-carrageenan with those formed from whole *Kappaphycus* sp. seaweed. The aim of the study was to evaluate the characteristics of both biofilm types and investigate the potential of bypassing carrageenan extraction. Biofilms produced from the entire seaweed exhibited lower brittleness but decreased strength due to the absence of carrageenan, leading to reduced binding intermolecular forces. Nevertheless, these biofilms still present viable alternatives to petroleum-based plastics and non-biodegradable containers. The primary application for such bioplastics lies in single-use packaging for products like powders, fast food items, candies, or everyday medications such as tablets or pills, where high mechanical properties are not critical, and easy opening is a desirable feature.

In a study conducted by Wullandari et al. (2021), they investigated carrageenan sourced from the red seaweed *Kappaphycus alvarezii* as potential bioplastic materials. The researchers assessed the physical and biological attributes of these carrageenan. They obtained semi-refined carrageenan flour from *Kappaphycus alvarezii*, while refined carrageenan flour was purchased. The bioplastic manufactured using the extracted carrageenans demonstrated greater antibacterial activity in comparison to the one derived from refined carrageenans. However, it exhibited lower mechanical strength. Conversely, the bioplastic generated from refined carrageenans displayed higher resistance to heat. Despite these disparities, both bioplastics conformed to the necessary standards for the market.

Adam et al. (2022) conducted a study to formulate bioplastic by incorporating cellulose into refined carrageenan, using three different cellulose sizes: microcrystalline cellulose (MCC), cellulose nanowhisker (CNW), and nanofibrillated cellulose (NFC). Carra-CNW and Carra-NFC films had smoother surfaces and were more transparent compared to the Carra-MCC film, which appeared more rigid and opaque due to the larger cellulose size used. Carra-MCC exhibited the highest opacity compared to Carra-NFC and Carra-CNW, demonstrating that cellulose particle size influenced the appearance of the bio-composite film. Carra-CNW showed the highest hydrophobicity ability compared to Carra, Carra-MCC, and Carra-NFC. Water vapor permeability values for Carra-MCC, Carra-CNW, and Carra-NFC showed only minor differences among the different cellulose sizes used. Compared to carrageenan plasticized with glycerol film, the water vapor permeability value of the bioplastic films was significantly lower. Carra-CNW had the highest viscosity (293.90 mPas) compared to

Carra-MCC and Carra-NFC (264.63 mPas and 262.78 mPas, respectively). Overall, the bioplastic film samples exhibited good water vapor barrier properties, making them suitable for potential future use as food packaging materials (Adam et al. 2022).

The physical and chemical characteristics of agar sourced from *Gracilaria salicornia* were examined as a potential raw material for bioplastics. Two different methods for extracting agar were tested, and several biofilms were manufactured to evaluate their performance. The biofilm obtained through the photobleaching extraction (PB) method exhibited higher tensile strength and elongation percentage compared to the biofilm derived from the alkali extraction (AE) process. However, the AE agar film displayed superior heat stability. Notably, in a soil burial test, the AE agar film completely decomposed within 30 days (Hii et al. 2016). These findings indicate that agar derived from *Gracilaria salicornia* holds promise for potential use in commercial bioplastic films.

According to Sousa et al. (2010), the biofilm created using agar from *Gracilaria vermiculophylla* displayed transparency and clarity on par with commercially available counterparts. The introduction of glycerin, acting as a plasticizer, enhanced the flexibility and mechanical strength of agar-based films, rendering them suitable for applications in food packaging and pharmaceutical coverings. Moreover, these agar-based films exhibited intriguing attributes like heat resistance and antibacterial activity, positioning them as a promising alternative. Nonetheless, additional research is needed to identify the optimal formulation for the development of robust bio-packaging materials (Kadar et al. 2021).

Darni et al. (2019) successfully produced bioplastics by combining sorghum stalk filler with polysaccharides derived from *Eucheuma denticulatum*, incorporating glycerol as a plasticizer. This amalgamation of filler and plasticizer led to notable enhancements in the physical and mechanical attributes of these bioplastics. This suggests that *Eucheuma denticulatum* shows potential as a promising resource for developing alternative formulations for plastic packaging.

Consebit et al. (2022) conducted a research study to explore the potential of *Eucheuma cottonii* seaweed as a bioplastic material, utilizing varying concentrations of glycerine (10, 20, and 30 mL). The primary focus of the study was to assess the tensile strength and biodegradability of the bioplastic. To formulate the bioplastic, they employed glycerine as the plasticizer, water as the solvent, corn-starch as the thickening agent, and vinegar to facilitate the dissolution of starch. These ingredients were mixed in a heated pan together with dried seaweeds, and the resulting mixture was promptly transferred onto a flat surface lined with foil before solidification occurred. The solidification process required 3–4 days, contingent on the thickness of the final product. The outcomes of the tensile strength examination revealed that decreased glycerine concentrations led to the production of a more rigid and robust bioplastic material. Concerning the assessment of biodegradability, the quantity of glycerine incorporated had a discernible impact on the size, form, moisture content, texture, and rate of decomposition of the plastics. Additionally, notable distinctions were noted in the tensile strength concerning the various levels of glycerine concentration. However, no significant disparities were detected in the biodegradability assessment in relation to the varying levels of glycerine concentration (Consebit et al. 2022).

Extracts from green seaweeds exhibit film-forming properties, rendering them attractive options for use as fillers or enhancers in medical and cosmetic applications (Lakshmi et al. 2020). Ulvan, a particular type of polysaccharide extracted from specific green algae species (Chlorophyta), possesses notable attributes like high viscosity and gel-forming capacity, positioning them as potential sources for generating biofilms. While there hasn't been any research on the synthesis of bioplastics from ulvan to date, these extracts showcase exceptional resistance to heat and mechanical strength, both of which are desirable qualities for ideal bioplastic materials (Usman et al. 2017).

Among all the polysaccharides present in green algae, cellulose has surfaced as the most preferred choice for manufacturing biodegradable plastics. Its distinctive characteristics enable it to create hydrocolloids when dissolved in a compatible solvent, leading to outstanding mechanical properties. Moreover, cellulose is both economical and renewable, presenting a biodegradable resource. For instance, cellulose from *Cladophora* sp. boasts chemical inertness and durability, rendering it an exceptional material. Cellulose's combination of tenacity, mechanical strength, thermal stability, and morphological attributes positions it as a fitting and auspicious option for applications in bio-packaging within the food and pharmaceutical sectors (Mihranyan 2011).

11.4 From Traditional Packaging to Smart Packaging: Towards Green Solution

Because of their inherent non-toxicity, natural properties, and biological functionalities such as antimicrobial and antioxidant attributes, seaweed biopolymers are exceptionally well-suited for the creation of biofilms within the food and pharmaceutical sectors (Ibrahim et al. 2017). The notion of "active packaging" is gaining traction in society due to the increasing need for packaging that actively engages with products to improve their safety, shelf life, and sensory qualities (Carina et al. 2021).

Adopting active packaging, as opposed to directly incorporating active compounds such as antimicrobial agents and antioxidants into the food, could potentially lead to a reduction in the quantity of these substances needed. Active packaging incorporates biological metabolites that harbor intriguing biological characteristics. When these metabolites are released within the biofilm formulation, the resulting packaging can offer protection against bacteria, oxidation, and UV radiation, thereby safeguarding the product (Perera et al. 2021).

The objective of antimicrobial packaging is to manage, impede, or slow down the proliferation of microorganisms by integrating antimicrobial agents into or onto food packaging materials. This approach prolongs the initial lag phase and curtails the growth of microbes. Numerous research investigations have underscored the captivating biological attributes of seaweed polysaccharides within this domain. In their study, Kanatt et al. (2015) explored the antimicrobial properties of pure κ-carrageenan sourced from *Kappaphycus alvarezii*, specifically targeting bacteria such as *Staphylococcus aureus*, *Bacillus cereus* (Gram-positive bacteria), *Escherichia coli*, and *Pseudomonas fluorescens* (Gram-negative bacteria). The findings revealed that the antimicrobial effect was evident solely against Gram-positive bacteria, with no observable inhibition of growth in the case of Gram-negative bacteria. κ-carrageenan was then integrated into polyvinyl acetate (PVA) film, and the outcomes demonstrated a notable zone of inhibition against Staphylococcus aureus and Bacillus cereus. This suggests that the antimicrobial efficacy of the pure extract could be retained, thereby influencing the film's ability to safeguard packaged items against bacterial contamination.

In their study, Han and Wang (2017) engineered an antimicrobial film by incorporating sodium alginate, carboxymethyl cellulose, glycerine, $CaCl_2$, and the natural antibacterial compound pyrogallic acid. The research findings revealed that the resultant film displayed elevated UV-blocking characteristics, moisture content, water vapor resistance, and oxygen resistance. Furthermore, films with higher concentrations of alginate exhibited superior efficacy in combating *Escherichia coli* and *Staphylococcus aureus*, indicating an enhanced antimicrobial effect.

The growing concern about food spoilage and contamination by pathogenic microorganisms has spurred investigations into alternative solutions. One path of exploration involves essential oils known for their antibacterial and antioxidant properties, such as those extracted from *Undaria pinnatifida*, an edible seaweed. These essential oils hold potential as natural sources of antimicrobial and antioxidant compounds, presenting promising applications in food preservation. In a study conducted by Patra et al. (2017), the essential oils derived from *Undaria pinnatifida* were tested against two foodborne pathogenic bacteria, namely, the Gram-positive *Staphylococcus aureus* ATCC 12,600 and the Gram-negative *Salmonella Typhimurium* ATCC 43,174.

Through a series of in vitro assays, the antioxidant potential of these essential oils was assessed using methods like ABTS radical scavenging, DPPH radical scavenging, hydroxyl radical scavenging, nitric oxide scavenging, superoxide radical scavenging, inhibition of lipid peroxidation, and reducing power assays. The research findings unveiled that the essential oil exhibited antioxidant properties and displayed antibacterial activity against the foodborne pathogenic bacteria (Patra et al. 2017).

Essential oils are valued for their significant antimicrobial and antioxidant properties, yet their chemical and physiological instability, coupled with their potent odors, restricts their direct use as antibacterial agents in food. However, investigating their application in active packaging represents a promising and innovative concept that merits further exploration. The packaging development process involves the application of heat, which can potentially degrade heat-sensitive essential oils. Consequently, it becomes imperative to optimize manufacturing protocols and employ more suitable processing conditions to minimize essential oil loss during packaging production. To fully harness the antimicrobial and antioxidant advantages of essential oils, it is crucial to explore and embrace novel techniques (Bustos et al. 2016; Mukurumbira et al. 2022).

Furthermore, seaweed metabolites known for their antioxidant properties are highly esteemed as viable options for active packaging. Lipid oxidation plays a pivotal role in shaping food quality, as it can detrimentally impact shelf life and nutritional value (Ahmed et al. 2016). Through the inclusion of antioxidants in the packaging composition, it becomes feasible to mitigate lipid oxidation and protein degradation inside the package, effectively safeguarding the product enclosed within the film from deterioration.

The antioxidative potential of an aqueous extract from *Kappaphycus alvarezii* has been the subject of investigation, and this extract was effectively integrated into a PVA film. In comparison to a pure PVA film, the seaweed-based film exhibited antioxidant qualities, as demonstrated in Kanatt et al.'s research conducted in 2015 (Kanatt et al. 2015). In another study led by He et al. (2016), the antioxidant properties of *Ulva lactuca* (Chlorophyta), *Gracilariopsis lemaneiformis*, *Sarcodia ceylonensis* (Rhodophyta), and *Durvillaea antarctica* (Phaeophyceae) were examined. Their findings revealed that the crude polysaccharide extracted from the green seaweed demonstrated the highest antioxidant activity, followed by *Durvillaea antarctica* and *Sarcodia ceylonensis*. These seaweeds exhibit significant potential as candidates for incorporating antioxidant extracts into active packaging materials.

As per Rajendran et al. (2012), bioplastics crafted from extracted biopolymers derived from seaweed showcase heightened resilience and offer substantial protection against microwave radiation. This increased durability can be attributed to the photoprotective attributes of seaweed compounds, rendering them well-suited for preserving the quality of food or pharmaceutical packaging. In their research, Andrade et al. (2021) engineered an active biofilm by amalgamating whey protein concentrate with *Fucus vesiculosus* extract to shield chicken breasts from lipid oxidation. The biofilm demonstrated enhancements in terms of thickness, tensile strength, and elastic modulus. Over a storage period of 25 days, the active biofilm effectively curtailed lipid oxidation in the chicken breasts. Methanol extracts sourced from *Sargassum* sp. and *Kappaphycus alvarezii* have exhibited photoprotective capabilities against UV radiation, possibly owing to the presence of flavonoids, phenols, and triterpenoids. These extracts from seaweed show promise as potential ingredients for sunscreen products or can be incorporated into biofilm formulations to enhance the attributes of active packaging.

11.5 Conclusions

Seaweed hydrocolloids have garnered significant interest as a promising candidate to produce biodegradable plastics. Notpla (NotPla), a London-based start-up, has already made strides in developing a natural, edible plastic-like material derived from seaweed that decomposes within a mere four to six weeks. This breakthrough holds immense potential for addressing the global plastic pollution crisis, as the world currently grapples with an alarming annual production of 300 million tons of plastic waste. The perspective provided by these innovative initiatives underscores the potential to substantially reduce plastic pollution on a global scale. Moreover, a noteworthy project known as Mak-Pak (MakPak project), involving collaborative efforts from researchers at the Alfred-Wegener Institute, Hochschule Bremerhaven University, and the fast-food chain Nordsee, has successfully created sustainable macroalgae-based packaging for fast-food applications, and it has been put to the test with real consumers.

However, it's important to acknowledge that pure phycocolloid-based packaging, while promising, falls short of meeting the necessary criteria for widespread commercial use as bioplastic due to its inherent shortcomings in mechanical strength and water barrier properties. Nevertheless, the augmentation of these biomaterials through the incorporation of biopolymers or plasticizers holds the potential to enhance their performance significantly. Furthermore, the inherent antioxidant and antimicrobial properties found in seaweed bioactive compounds offer the potential to improve the preservation and shelf life of food and pharmaceutical products. Recognizing the potential of seaweed's bioactive compounds in the realm of sustainable and eco-friendly bioplastic development presents a viable solution to supplant petroleum-based plastics and reduce the escalating production of plastic waste. To fully unlock this potential, it is fundamental to embrace more environmentally friendly extraction methods. While such methods have shown efficacy, further research is needed to reduce implementation costs and incentivize industrial investment in these sustainable technologies to produce packaging in environmentally friendly ways.

Acknowledgements This work is financed by national funds through FCT—Foundation for Science and Technology, I.P., within the scope of the projects LA/P/0069/2020 granted to the Associate Laboratory ARNET, UIDB/04292/2020—granted to MARE—Marine and Environmental Sciences Centre and UIDP/50017/2020+UIDB/50017/2020 (by FCT/MTCES) granted to CESAM—Centre for Environmental and Marine Studies. This research was co-financed by the project AlgaMar4antivirus—Marine MacroalgaE in the prevention and treatment of ANTIviral diseases (Coronavirus SARS-CoV-2) (FA_07_2018_009), funded by the Blue Fund under Public Notice No. 7/2018—Scientific Research and Marine Technology. Silvia Lomartire thanks Foundation for Science and Technology (FCT) for the financial support granted through the doctoral grant 2021.05005.BD. Ana M. M. Gonçalves acknowledges University of Coimbra for the contract IT057-18-7253.

References

Adam MG, Tran PTM, Bolan N, Balasubramanian R (2021) Biomass burning-derived airborne particulate matter in Southeast Asia: a critical review. J Hazard Mater 407:124760. https://doi.org/10.1016/j.jhazmat.2020.124760

Adam F, Othman NA, Yasin NHM, Cheng CK, Azman NAM (2022) Evaluation of reinforced and green bioplastic from carrageenan seaweed with nanocellulose. Fibers Polym 23:2885–2896. https://doi.org/10.1007/s12221-022-4006-6

Ahmed M, Pickova J, Ahmad T, Liaquat M, Farid A, Jahangir M (2016) Oxidation of lipids in foods. Sarhad J Agric 32:230–238. https://doi.org/10.17582/journal.sja/2016.32.3.230.238

Alba K, Kontogiorgos V (2018) Seaweed Polysaccharides (Agar, Alginate Carrageenan). Elsevier. ISBN 9780128140451

Allouzi MMA, Tang DYY, Chew KW, Rinklebe J, Bolan N, Allouzi SMA, Show PL (2021) Micro (Nano) plastic pollution: the ecological influence on soil-plant system and human health. Sci Total Environ 788:147815. https://doi.org/10.1016/j.scitotenv.2021.147815

Andrade MA, Barbosa CH, Souza VGL, Coelhoso IM, Reboleira J, Bernardino S, Ganhão R, Mendes S, Fernando AL, Vilarinho F et al (2021) Novel active food packaging films based on whey protein incorporated with seaweed extract: development, characterization, and application in fresh poultry meat. Coatings 11:1–12. https://doi.org/10.3390/coatings11020229

Araki C (1956) Structure of the agarose constituent of Agar-Agar. Bull Chem Soc Jpn 29:543–544. https://doi.org/10.1246/bcsj.29.543

ARAKI, C. *Some Recent Studies on the Polysaccharides of Agarophytes*; Pergamon Press Ltd, 1966;

Aryee AN, Agyei D, Akanbi TO (2018) Recovery and utilization of seaweed pigments in food processing. Curr Opin Food Sci 19:113–119. https://doi.org/10.1016/j.cofs.2018.03.013

Benavides S, Villalobos-Carvajal R, Reyes JE (2012) Physical, mechanical and antibacterial properties of alginate film: effect of the crosslinking degree and oregano essential oil concentration. J Food Eng 110:232–239. https://doi.org/10.1016/j.jfoodeng.2011.05.023

Benselfelt T, Engström J, Wågberg L (2018) Supramolecular double networks of cellulose nanofibrils and algal polysaccharides with excellent wet mechanical properties. Green Chem 20:2558–2570. https://doi.org/10.1039/C8GC00590G

Bisht B, Lohani UC, Kumar V, Gururani P, Sinhmar R (2022) Edible hydrocolloids as sustainable substitute for non-biodegradable materials. Crit Rev Food Sci Nutr 62:693–725. https://doi.org/10.1080/10408398.2020.1827219

Bixler HJ, Porse H (2011) A decade of change in the seaweed hydrocolloids industry. J Appl Phycol 23:321–335. https://doi.org/10.1007/s10811-010-9529-3

Budiman MA, Tarman K (2022) A review on the difference of physical and mechanical properties of bioplastic from seaweed hydrocolloids with various plasticizers. IOP Conf Ser Earth Environ Sci 967. https://doi.org/10.1088/1755-1315/967/1/012012

Bustos CRO, Alberti RFV, Matiacevich SB (2016) Edible antimicrobial films based on microencapsulated lemongrass oil. J Food Sci Technol 53:832–839. https://doi.org/10.1007/s13197-015-2027-5

Carina D, Sharma S, Jaiswal AK, Jaiswal S (2021) Seaweeds polysaccharides in active food packaging: a review of recent progress. Trends Food Sci Technol 110:559–572. https://doi.org/10.1016/j.tifs.2021.02.022

Carlucci MJ, Pujol CA, Ciancia M, Noseda MD, Matulewicz MC, Damonte EB, Cerezo AS (1997) Antiherpetic and anticoagulant properties of carrageenans from the red seaweed Gigartina skottsbergii and their cyclized derivatives: correlation between structure and biological activity. Int J Biol Macromol 20:97–105. https://doi.org/10.1016/S0141-8130(96)01145-2

Carocho M, Heleno S, Rodrigues P, Barreiro MF, Barros L, Ferreira ICFR (2019) A novel natural coating for food preservation: effectiveness on microbial growth and physicochemical parameters. LWT 104:76–83. https://doi.org/10.1016/j.lwt.2019.01.031

Cazón P, Velazquez G, Ramírez JA, Vázquez M (2017) Polysaccharide-based films and coatings for food packaging: a review. Food Hydrocoll 68:136–148. https://doi.org/10.1016/j.foodhyd.2016.09.009

Chen Y, Awasthi AK, Wei F, Tan Q, Li J (2021) Single-use plastics: production, usage, disposal, and adverse impacts. Sci Total Environ 752:141772. https://doi.org/10.1016/j.scitotenv.2020.141772

Consebit KL, Dermil KC, Magbanua EY, Racadio FJ, Saavedra SV, Abusama H, Valdez A (2022) Bioplastic from seaweeds (*Eucheuma cottonii*) as an alternative plastic. ASEAN J Sci Eng 2:129–132. https://doi.org/10.17509/ajse.v2i2.37799

Cordover R (2007) Seaweed agronomy. Development

Cotas J, Marques V, Afonso MB, Rodrigues CMP, Pereira L (2020) Antitumour potential of *Gigartina pistillata* carrageenans against colorectal cancer stem cell-enriched tumourspheres. Mar Drugs 18:50. https://doi.org/10.3390/md18010050

Darni Y, Sumartini S, Lismeri L, Hanif M, Lesmana D (2019) Bioplastics synthesis based on sorghum-*Eucheuma spinosum* modified with sorghum stalk powder. J Phys Conf Ser 1376. https://doi.org/10.1088/1742-6596/1376/1/012042

Denis C, Ledorze C, Jaouen P, Fleurence J (2009) Comparison of different procedures for the extraction and partial purification of R-Phycoerythrin from the Red Macroalga Grateloupia turuturu. Bot 52:278–281. https://doi.org/10.1515/BOT.2009.034

Dickinson E (2018) Hydrocolloids acting as emulsifying agents—how do they do it? Food Hydrocoll 78:2–14. https://doi.org/10.1016/j.foodhyd.2017.01.025

Djafari Petroudy SR (2017) Physical and mechanical properties of natural fibers. Elsevier Ltd. ISBN 9780081004302

Enquist-Newman M, Faust AME, Bravo DD, Santos CNS, Raisner RM, Hanel A, Sarvabhowman P, Le C, Regitsky DD, Cooper SR et al (2014) Efficient ethanol production from brown macroalgae sugars by a synthetic yeast platform. Nature 505:239–243. https://doi.org/10.1038/nature12771

European Committee for Standardization Textiles (1993) Determination of physiological properties–measurement of thermal and water–vapour resistance under steady-state conditions (Sweating Guarded - Hotplate Test) (ISO 11092:1993)

Freile-Pelegrín Y, Madera-Santana TJ (2017) Biodegradable polymer blends and composites from seaweeds. Handb Compos from Renew Mater 1–8:419–438. https://doi.org/10.1002/9781119441632.ch98

Han Y, Wang L (2017) Sodium alginate/carboxymethyl cellulose films containing pyrogallic acid: physical and antibacterial properties. J Sci Food Agric 97:1295–1301. https://doi.org/10.1002/jsfa.7863

Hanry EL, Surugau N (2020) Characteristics and properties of biofilms made from pure carrageenan powder and whole seaweed (*Kappaphycus* sp.). J Adv Res Fluid Mech Therm Sci 76:99–110. https://doi.org/10.37934/arfmts.76.2.99110

Haugan JA, Liaaen-Jensen S (1994) Algal carotenoids 54. Carotenoids of brown algae (Phaeophyceae). Biochem Syst Ecol 22:31–41. https://doi.org/10.1016/0305-1978(94)90112-0

He J, Xu Y, Chen H, Sun P (2016) Extraction, structural characterization, and potential antioxidant activity of the polysaccharides from four seaweeds. Int J Mol Sci 17. https://doi.org/10.3390/ijms17121988

Hii SL, Lim JY, Ong WT, Wong CL (2016) Agar from Malaysian red seaweed as potential material for synthesis of bioplastic film. J Eng Sci Technol 1–15

Ibrahim M, Salman M, Kamal S, Rehman S, Razzaq A, Akash SH (2017) Chapter 6: Algae-based biologically active compounds. Elsevier Inc. ISBN 9780128123607

Ilias MA, Ismail A, Othman R (2017) Analysis of carrageenan yield and gel strength of kappaphycus species in Semporna Sabah. J Trop Plant Physiol 9:14–23

Illera-Vives M, Seoane Labandeira S, Iglesias Loureiro L, López-Mosquera ME (2017) Agronomic assessment of a compost consisting of seaweed and fish waste as an organic fertilizer for organic potato crops. J Appl Phycol 29:1663–1671. https://doi.org/10.1007/s10811-017-1053-2

Kadar NAHA, Rahim NS, Yusof R, Nasir NAHA, Hamid AHA (2021) Review on potential of algae in producing biodegradable plastic. Int J Eng Adv Res 3:13–26

Kanagesan K, Abdulla R, Derman E, Sabullah MK, Govindan N, Gansau JA (2022) A sustainable approach to green algal bioplastics production from brown seaweeds of Sabah, Malaysia. J King Saud Univ Sci 34:102268. https://doi.org/10.1016/j.jksus.2022.102268

Kanatt SR, Lahare P, Chawla SP, Sharma A (2015) *Kappaphycus alvarezii*: its antioxidant potential and use in bioactive packaging films. J Microbiol Biotechnol Food Sci 05:1–6. https://doi.org/10.15414/jmbfs.2015.5.1.1-6

Kanmani P, Rhim JW (2014) Antimicrobial and physical-mechanical properties of agar-based films incorporated with grapefruit seed extract. Carbohydr Polym 102:708–716. https://doi.org/10.1016/j.carbpol.2013.10.099

Khalil AHPS, Tye YY, Saurabh CK, Leh CP, Lai TK, Chong EWN, Nurul Fazita MR, Mohd Hafiidz J, Banerjee A, Syakir MI (2017) Biodegradable polymer films from seaweed polysaccharides: a review on cellulose as a reinforcement material. Express Polym Lett 11:244–265. https://doi.org/10.3144/expresspolymlett.2017.26

Khalil AHPS, Lai TK, Tye YY, Rizal S, Chong EWN, Yap SW, Hamzah AA, Nurul Fazita MR, Paridah MT (2018) A review of extractions of seaweed hydrocolloids: properties and applications. Express Polym Lett 12:296–317. https://doi.org/10.3144/expresspolymlett.2018.27

Khotimchenko M, Tiasto V, Kalitnik A, Begun M, Khotimchenko R, Leonteva E, Bryukhovetskiy I, Khotimchenko Y (2020) Antitumor potential of carrageenans from marine red algae. Carbohydr Polym 246:116568. https://doi.org/10.1016/j.carbpol.2020.116568

Kim JK, Yarish C, Hwang EK, Park M, Kim Y (2017) Seaweed aquaculture: cultivation technologies, challenges and its ecosystem services. Algae 32:1–13. https://doi.org/10.4490/algae.2017.32.3.3

Kok JML, Wong CL (2018) Physicochemical properties of edible alginate film from Malaysian Sargassum polycystum C. Agardh. Sustain Chem Pharm 9:87–94. https://doi.org/10.1016/j.scp.2018.07.001

Kumar R, Sharma P, Manna C, Jain M (2021a) Abundance, interaction, ingestion, ecological concerns, and mitigation policies of microplastic pollution in riverine ecosystem: a review. Sci Total Environ 782:146695. https://doi.org/10.1016/j.scitotenv.2021.146695

Kumar M, Chen H, Sarsaiya S, Qin S, Liu H, Awasthi MK, Kumar S, Singh L, Zhang Z, Bolan NS et al (2021b) Current research trends on micro- and nano-plastics as an emerging threat to global environment: a review. J Hazard Mater 409:124967. https://doi.org/10.1016/j.jhazmat.2020.124967

Lakshmi DS, Sankaranarayanan S, Gajaria TK, Li G, Kujawski W, Kujawa J, Navia R (2020) A short review on the valorization of green seaweeds and ulvan: feedstock for chemicals and biomaterials. Biomolecules 10:1–20. https://doi.org/10.3390/biom10070991

Lee WK, Lim YY, Leow ATC, Namasivayam P, Abdullah JO, Ho CL (2017a) Factors affecting yield and gelling properties of agar. J Appl Phycol 29:1527–1540. https://doi.org/10.1007/s10811-016-1009-y

Lee WK, Lim YY, Leow ATC, Namasivayam P, Ong Abdullah J, Ho CL (2017b) Biosynthesis of Agar in red seaweeds: a review. Carbohydr Polym 164: 23–30. https://doi.org/10.1016/j.carbpol.2017.01.078

Li JM, Nie SP (2016) The functional and nutritional aspects of hydrocolloids in foods. Food Hydrocoll 53:46–61. https://doi.org/10.1016/j.foodhyd.2015.01.035

Li K, Zhu J, Guan G, Wu H (2019) Preparation of chitosan-sodium alginate films through layer-by-layer assembly and ferulic acid crosslinking: film properties, characterization, and formation mechanism. Int J Biol Macromol 122:485–492. https://doi.org/10.1016/j.ijbiomac.2018.10.188

Li P, Wang X, Su M, Zou X, Duan L, Zhang H (2021) Characteristics of plastic pollution in the environment: a review. Bull Environ Contam Toxicol 107:577–584. https://doi.org/10.1007/s00128-020-02820-1

Liling G, Di Z, Jiachao X, Xin G, Xiaoting F, Qing Z (2016) Effects of ionic crosslinking on physical and mechanical properties of alginate mulching films. Carbohydr Polym 136:259–265. https://doi.org/10.1016/j.carbpol.2015.09.034

Lim JY, Hii SL, Chee SY, Wong CL (2018) *Sargassum siliquosum* J. Agardh extract as potential material for synthesis of bioplastic film. J Appl Phycol 30:3285–3297. https://doi.org/10.1007/s10811-018-1603-2

Liu Z, Jiao Y, Wang Y, Zhou C, Zhang Z (2008) Polysaccharides-based nanoparticles as drug delivery systems. Adv Drug Deliv Rev 60:1650–1662. https://doi.org/10.1016/j.addr.2008.09.001

Liu S, Li H, Tang B, Bi S, Li L (2016) Scaling law and microstructure of alginate hydrogel. Carbohydr Polym 135:101–109. https://doi.org/10.1016/j.carbpol.2015.08.086

Lomartire S, Marques JC, Gonçalves AMM (2021) An overview to the health benefits of seaweeds consumption. Mar Drugs 19:341. https://doi.org/10.3390/md19060341

Marinho-Soriano E, Bourret E (2005) Polysaccharides from the Red Seaweed *Gracilaria dura* (Gracilariales, Rhodophyta). Bioresour Technol 96:379–382. https://doi.org/10.1016/j.biortech.2004.04.012

Martínez-Sanz M, Martínez-Abad A, López-Rubio A (2019) Cost-efficient bio-based food packaging films from unpurified agar-based extracts. Food Packag Shelf Life 21:100367. https://doi.org/10.1016/j.fpsl.2019.100367

Matsuhashi T (1990) Agar. Chapter 1. In: Harris P (ed) Food gels. Elsevier Science Publishers Ltd., pp 1–51

Meng F, Zhang Y, Xiong Z, Wang G, Li F, Zhang L (2018) Mechanical, hydrophobic and thermal properties of an organic-inorganic hybrid carrageenan-polyvinyl alcohol composite film. Compos Part B Eng 143:1–8. https://doi.org/10.1016/j.compositesb.2017.12.009

Mihranyan A (2011) Cellulose from cladophorales green algae: from environmental problem to high-tech composite materials. J Appl Polym Sci 119:2449–2460. https://doi.org/10.1002/app.32959

Mohammed A, Gaduan A, Chaitram P, Pooran A, Lee KY, Ward K (2023) Sargassum inspired, optimized calcium alginate bioplastic composites for food packaging. Food Hydrocoll 135:108192. https://doi.org/10.1016/j.foodhyd.2022.108192

Mostafavi FS, Zaeim D (2020) Agar-based edible films for food packaging applications–a review. Int J Biol Macromol 159:1165–1176. https://doi.org/10.1016/j.ijbiomac.2020.05.123

Mukurumbira AR, Shellie RA, Keast R, Palombo EA, Jadhav SR (2022) Encapsulation of essential oils and their application in antimicrobial active packaging. Food Control 136:108883. https://doi.org/10.1016/j.foodcont.2022.108883

Murano E, Toffanin R, Pedersini C, Carabot-Cuervo A, Blunden G, Rizzo R (1996) Structure and properties of agar from two unexploited agarophytes from Venezuela. Hydrobiologia 326–327:497–500. https://doi.org/10.1007/BF00047851

Oussalah M, Caillet S, Salmiéri S, Saucier L, Lacroix M (2007) Antimicrobial effects of alginate-based films containing essential oils on *Listeria monocytogenes* and *Salmonella typhimurium* present in Bologna and Ham. J Food Prot 70:901–908. https://doi.org/10.4315/0362-028X-70.4.901

Pandya Y, Bakshi M, Sharma A, Pandya YH, Pandya H (2022) Agar-Agar extraction, structural properties and applications: a review. Pharma Innov J 1151–1157

Patra JK, Lee SW, Park JG, Baek KH (2017) Antioxidant and antibacterial properties of essential oil extracted from an edible seaweed *Undaria pinnatifida*. J Food Biochem 41. https://doi.org/10.1111/jfbc.12278

Paula GA, Benevides NMB, Cunha AP, de Oliveira AV, Pinto AMB, Morais JPS, Azeredo HMC (2015) Development and characterization of edible films from mixtures Ofκ-Carrageenan, ι-Carrageenan, and alginate. Food Hydrocoll 47:140–145. https://doi.org/10.1016/j.foodhyd.2015.01.004

Pereira L, Van De Velde F (2011) Portuguese Carrageenophytes: carrageenan composition and geographic distribution of eight species (Gigartinales, Rhodophyta). Carbohydr Polym 84:614–623. https://doi.org/10.1016/j.carbpol.2010.12.036

Perera KY, Sharma S, Pradhan D, Jaiswal AK, Jaiswal S (2021) Seaweed polysaccharide in food contact materials (active packaging, intelligent packaging, edible films, and coatings). Foods 10:1–22. https://doi.org/10.3390/foods10092088

Phan The D, Debeaufort F, Voilley A, Luu D (2009) Biopolymer interactions affect the functional properties of edible films based on agar, cassava starch and arabinoxylan blends. J Food Eng 90:548–558. https://doi.org/10.1016/j.jfoodeng.2008.07.023

Porse H, Rudolph B (2017) The seaweed hydrocolloid industry: 2016 updates, requirements, and outlook. J Appl Phycol 29:2187–2200. https://doi.org/10.1007/s10811-017-1144-0

Price-Whelan AM, Agüeros MA, Fournier AP, Street R, Ofek EO, Covey KR, Levitan D, Laher RR, Sesar B, Surace J (2014) Statistical searches for microlensing events in large, non-uniformly sampled time-domain surveys: a test using palomar transient factory data. Astrophys J 781. https://doi.org/10.1088/0004-637X/781/1/35

Raghunandan BL, Vyas RV, Patel HK, JHala YK (2019) Perspectives of seaweed as organic fertilizer in agriculture. In: Soil fertility management for sustainable development. pp 267–289. ISBN 9789811359040

Raikova S, Allen MJ, Chuck CJ (2019) Hydrothermal liquefaction of macroalgae for the production of renewable biofuels. Biofuels Bioprod Biorefining 13:1483–1504. https://doi.org/10.1002/bbb.2047

Rajendran N, Puppala S, Raj MS, Angeeleena BR, Rajam C (2012) Seaweeds can be a new source for bioplastics. J Pharm Res 5:1476–1479

Ramawat KG, Merillon JM (2015) Polysaccharides: bioactivity and biotechnology. Springer International Publishing, Cham

Ramdhan T, Ching SH, Prakash S, Bhandari B (2019) Time dependent gelling properties of cuboid alginate gels made by external gelation method: effects of alginate-$CaCl_2$ solution ratios and PH. Food Hydrocoll 90:232–240. https://doi.org/10.1016/j.foodhyd.2018.12.022

Rao SPV, Peryasamy C, Kumar KS, Rao AS, Anantharaman P (2018) Chapter 6. Seaweeds: distribution, production and uses. In: Noor MN, Bhatnagar SK, Sinha SK (eds) Bioprospecting of algae. pp 59–78

Revel M, Châtel A, Mouneyrac C (2018) Micro(Nano)plastics: a threat to human health? Curr Opin Environ Sci Heal 1:17–23. https://doi.org/10.1016/j.coesh.2017.10.003

Rhim J-W (2011) Effect of clay contents on mechanical and water vapor barrier properties of agar-based nanocomposite films. Carbohydr Polym 86:691–699. https://doi.org/10.1016/j.carbpol.2011.05.010

Rhim JW (2012) Physical-mechanical properties of agar/κ-Carrageenan blend film and derived clay nanocomposite film. J Food Sci 77. https://doi.org/10.1111/j.1750-3841.2012.02988.x

Rinaudo M (2008) Main properties and current applications of some polysaccharides as biomaterials. Polym Int 57:397–430. https://doi.org/10.1002/pi.2378

Robal M, Brenner T, Matsukawa S, Ogawa H, Truus K, Rudolph B, Tuvikene R (2017) Monocationic salts of carrageenans: preparation and physico-chemical properties. Food Hydrocoll 63:656–667. https://doi.org/10.1016/j.foodhyd.2016.09.032

Rocha De Souza MC, Marques CT, Guerra Dore CM, Ferreira Da Silva FR, Oliveira Rocha HA, Leite EL (2007) Antioxidant activities of sulfated polysaccharides from brown and red seaweeds. J Appl Phycol 19:153–160, https://doi.org/10.1007/s10811-006-9121-z

Samantaray PK, Little A, Haddleton DM, McNally T, Tan B, Sun Z, Huang W, Ji Y, Wan C (2020) Poly(Glycolic Acid) (PGA): a versatile building block expanding high performance and sustainable bioplastic applications. Green Chem 22:4055–4081. https://doi.org/10.1039/d0gc01394c

Santana I, Félix M, Guerrero A, Bengoechea C (2022) Processing and characterization of bioplastics from the invasive *Seaweed rugulopteryx* Okamurae. Polymers (Basel) 14. https://doi.org/10.3390/polym14020355

Santelices B, Westermeier R, Bobadilla M (1993) Effects of stock loading and planting distance on the growth and production of *Gracilaria chilensis* in rope culture. J Appl Phycol 5:517–524. https://doi.org/10.1007/BF02182510

Schmaltz E, Melvin EC, Diana Z, Gunady EF, Rittschof D, Somarelli JA, Virdin J, Dunphy-Daly MM (2020) plastic pollution solutions: emerging technologies to prevent and collect marine plastic pollution. Environ. Int. 144. https://doi.org/10.1016/j.envint.2020.106067

Sedayu BB, Cran MJ, Bigger SW (2019) A review of property enhancement techniques for carrageenan-based films and coatings. Carbohydr Polym 216:287–302. https://doi.org/10.1016/j.carbpol.2019.04.021

Seely GR, Duncan MJ, Vidaver WE (1972) Preparative and analytical extraction of pigments from brown algae with dimethyl sulfoxide. Mar Biol 12:184–188. https://doi.org/10.1007/BF00350754

Setyawidati NAR, Puspita M, Kaimuddin AH, Widowati I, Deslandes E, Bourgougnon N, Stiger-Pouvreau V (2018) Seasonal biomass and alginate stock assessment of three abundant genera of brown macroalgae using multispectral high resolution satellite remote sensing: a case study

at Ekas Bay (Lombok, Indonesia). Mar Pollut Bull 131:40–48. https://doi.org/10.1016/j.marpolbul.2017.11.068
Shen M, Song B, Zeng G, Zhang Y, Huang W, Wen X, Tang W (2020) Are biodegradable plastics a promising solution to solve the global plastic pollution? Environ Pollut 263:114469. https://doi.org/10.1016/j.envpol.2020.114469
Song JH, Murphy RJ, Narayan R, Davies GBH (2009) Biodegradable and compostable alternatives to conventional plastics. Philos Trans R Soc B Biol Sci 364:2127–2139. https://doi.org/10.1098/rstb.2008.0289
Sousa AMM, Sereno AM, Hilliou L, Gonçalves MP (2010) Biodegradable agar extracted from *Gracilaria vermiculophylla*: film properties and application to edible coating. Mater Sci Forum 636–637:739–744. https://doi.org/10.4028/www.scientific.net/MSF.636-637.739
Sreekumar K (2020) Alginic acid: a potential biopolymer from brown algae. 2:433–438
Srinivasan B, Kulshreshtha G (2020) Recent developments in food-based bioplastics production. Handb Environ Chem 104:107–127. https://doi.org/10.1007/698_2020_578
Stevens ES (2001) Green plastics: an introduction to the new science of biodegradable plastics. J Chem Educ 79:1072. https://doi.org/10.1300/j354v16n02_12
Tabatabaei HR, Jafari SM, Mirzaei H, Mohammadi Nafchi A, Dehnad D (2018) Preparation and characterization of nano-SiO2 reinforced Gelatin-k-Carrageenan biocomposites. Int J Biol Macromol 111:1091–1099. https://doi.org/10.1016/j.ijbiomac.2018.01.116
Therkelsen GH (2012) Carrageenan. In: Industrial gums: polysaccharides and their derivatives, 3rd edn. Academic Press, INC., pp 145–180. ISBN 9780127462530
Thiruchelvi R, Das A, Sikdar E (2020) Bioplastics as better alternative to petro plastic. Mater Today Proc 37:1634–1639. https://doi.org/10.1016/j.matpr.2020.07.176
Turner A (2018) Mobilisation kinetics of hazardous elements in marine plastics subject to an avian physiologically-based extraction test. Environ Pollut 236:1020–1026. https://doi.org/10.1016/j.envpol.2018.01.023
Usman A, Khalid S, Usman A, Hussain Z, Wang Y (2017) Algal polysaccharides, novel application, and outlook. Elsevier Inc. ISBN 9780128123607
Van Cauwenberghe L, Devriese L, Galgani F, Robbens J, Janssen CR (2015) Microplastics in sediments: a review of techniques, occurrence and effects. Mar Environ Res 111:5–17. https://doi.org/10.1016/j.marenvres.2015.06.007
Vera J, Castro J, Gonzalez A, Moenne A (2011) Seaweed polysaccharides and derived oligosaccharides stimulate defense responses and protection against pathogens in plants. Mar Drugs 9:2514–2525. https://doi.org/10.3390/md9122514
Vijayakumar S, Durgadevi S, Arulmozhi P, Rajalakshmi S, Gopalakrishnan T, Parameswari N (2019) Effect of seaweed liquid fertilizer on yield and quality of capsicum annum L. Acta Ecol Sin 39:406–410. https://doi.org/10.1016/j.chnaes.2018.10.001
Wang W, Wang SX, Guan HS (2012) The antiviral activities and mechanisms of marine polysaccharides: an overview. Mar Drugs 10:2795–2816. https://doi.org/10.3390/md10122795
Wang YL, Lee YH, Chiu IJ, Lin YF, Chiu HW (2020a) Potent impact of plastic nanomaterials and micromaterials on the food chain and human health. Int J Mol Sci 21:1–14. https://doi.org/10.3390/ijms21051727
Wang S, Zhao S, Uzoejinwa BB, Zheng A, Wang Q, Huang J, Abomohra AEF (2020b) A state-of-the-art review on dual purpose seaweeds utilization for wastewater treatment and crude bio-oil production. Energy Convers Manag 222:113253. https://doi.org/10.1016/j.enconman.2020.113253
Waring RH, Harris RM, Mitchell SC (2018) Plastic contamination of the food chain: a threat to human health? Maturitas 115:64–68. https://doi.org/10.1016/j.maturitas.2018.06.010
Wullandari P, Sedayu BB, Novianto TD, Prasetyo AW (2021) Characteristic of semi refined and refined carrageenan flours used in the making of biofilm (Bioplastic). IOP Conf Ser Earth Environ Sci 733. https://doi.org/10.1088/1755-1315/733/1/012112

Xu S, Ma J, Ji R, Pan K, Miao AJ (2020) Microplastics in aquatic environments: occurrence, accumulation, and biological effects. Sci Total Environ 703:134699. https://doi.org/10.1016/j.scitotenv.2019.134699

Xue J, Wu Y, Shi K, Xiao X, Gao Y, Li L, Qiao Y (2019) Study on the degradation performance and kinetics of immobilized cells in straw-alginate beads in marine environment. Bioresour Technol 280:88–94. https://doi.org/10.1016/j.biortech.2019.02.019

Yoon J, Oh DX, Jo C, Lee J, Hwang DS (2014) Improvement of desolvation and resilience of alginate binders for Si-Based anodes in a lithium ion battery by calcium-mediated cross-linking. Phys Chem Chem Phys 16:25628–25635. https://doi.org/10.1039/c4cp03499f

Younes M, Aggett P, Aguilar F, Crebelli R, Filipič M, Frutos MJ, Galtier P, Gott D, Gundert-Remy U, Kuhnle GG, et al (2018) Re-evaluation of Carrageenan (E 407) and processed eucheuma seaweed (E 407a) as food additives. EFSA J 16. https://doi.org/10.2903/j.efsa.2018.5238

Zemke-White W, Lindsey Ohno M (1999) World seaweed utilisation: an end-of-century summary W. J Appl Phycol 125:369–376. https://doi.org/10.1111/j.1478-4408.2009.00192.x

Zhong H, Gao X, Cheng C, Liu C, Wang Q, Han X (2020) The structural characteristics of seaweed polysaccharides and their application in gel drug delivery systems. Mar Drugs 18. https://doi.org/10.3390/md18120658

Part IV
Waste-To-Wealth

Chapter 12
Zero-Waste Processes Utilizing Microbes

Moffat Mutebele, Subbaiya Ramasamy, Todd Johnson, Manikandan Sivasubramanian, Kalirajan Arunachalam, Vinod Kumar Nathan, and George Mukupa

12.1 Introduction

In an age of increased environmental awareness and climate change urgency, zero-waste microbial methods offer promise. Modern microbiology and sustainable principles are used to revolutionize waste management, presenting a hopeful answer to the planet's environmental problems (Agrawal et al. 2023). Zero-waste microbe-based procedures rely on their extraordinary powers. These small, often invisible insects break down organic and inorganic substances, turning trash into precious minerals. We can minimize, reuse, and recycle waste streams that would otherwise pollute, fill landfills, and deplete natural resources by using nature's smallest engineers (Table 12.1 and Figures 12.1 and 12.2) (Raabe et al. 2022).

In organic waste management, microbe-driven zero-waste procedures are useful. Bacteria and fungi decompose food leftovers and agricultural wastes. These bacteria effectively transform organic waste into nutrient-rich soil amendments via

M. Mutebele · S. Ramasamy (✉) · T. Johnson
Department of Biological Sciences, School of Mathematics and Natural Sciences, The Copperbelt University, Riverside, Jambo Drive, 21692 Kitwe, Zambia
e-mail: ramsubbubio@gmail.com

T. Johnson
e-mail: todd.johnson@cbu.ac.zm

M. Sivasubramanian
Department of Chemistry and Biology, School of Natural and Applied Sciences, Mulungushi University, 80415 Kabwe, Zambia

K. Arunachalam · G. Mukupa
Department of Biotechnology, Saveetha School of Engineering, Saveetha Institute of Medical and Technical Sciences (SIMATS), Thandalam, Chennai, Tamil Nadu 602 105, India

V. K. Nathan
School of Chemical and Biotechnology, SASTRA Deemed to Be University, Thanjavur, Tamil Nadu, India

K. Bala et al. (eds.), *Harnessing Microbial Potential for Multifarious Applications*, Energy, Environment, and Sustainability, https://doi.org/10.1007/978-981-97-1152-9_12

Table 12.1 Table of zero-waste processes utilizing microbes and their applications

Process type	Description	Applications	References
Composting	Natural decomposition of organic matter with the help of microbes, producing nutrient-rich soil amendments	Food waste recycling, agricultural waste management, land rehabilitation	Wainaina et al. (2020)
Anaerobic digestion	Microbial breakdown of organic materials in the absence of oxygen, generating biogas and digestate	Organic waste-to-energy, sewage treatment, livestock manure management	Nemr et al. (2022)
Plastic biodegradation	Specialized microbes engineered to break down plastic materials like polyethylene and polypropylene	Plastic waste recycling, reducing plastic pollution	Nguyen et al. (2023)
Biofuel production	Microbes used to ferment organic materials like corn or algae to produce biofuels such as ethanol and biodiesel	Renewable energy sources, reducing fossil fuel dependence	Mittal and Kumar (2022)
Bioplastic synthesis	Microbes used to produce biodegradable plastics from renewable feedstocks like starch or sugar	Environmentally friendly packaging, reducing single-use plastics	Molina-Besch and Keszleri (2023)
Pharmaceutical production	Microbes engineered to produce pharmaceutical compounds, reducing the environmental impact of drug manufacturing	Sustainable drug production, reducing chemical waste	Kumar et al. (2022)
Wastewater treatment	Microbial processes employed to purify wastewater, breaking down pollutants and harmful substances	Clean water supply, pollution control, ecosystem preservation	Ali et al. (2023)
Industrial biotechnology	Microbes utilized in various industries to convert waste streams into valuable products, such as enzymes, biobased chemicals, and materials	Sustainable manufacturing, Resource optimization	Magalhães-Ghiotto et al. (2021)

composting and anaerobic digestion, lowering artificial fertilizers and greenhouse gas emissions (Hui et al. 2023). Microbes may also combat plastic pollution and organic waste. Researchers are investigating whether specialized bacteria can break down polyethylene and polypropylene, which make up most of the world's plastic garbage. This invention might transform plastic recycling, removing plastics from landfills and seas and establishing a circular economy.

Microbe-based zero-waste procedures have great potential in the industry. Microbes may be designed to create biofuels, bioplastics, and medicines from waste or renewable resources. This decreases waste and improves industry sustainability

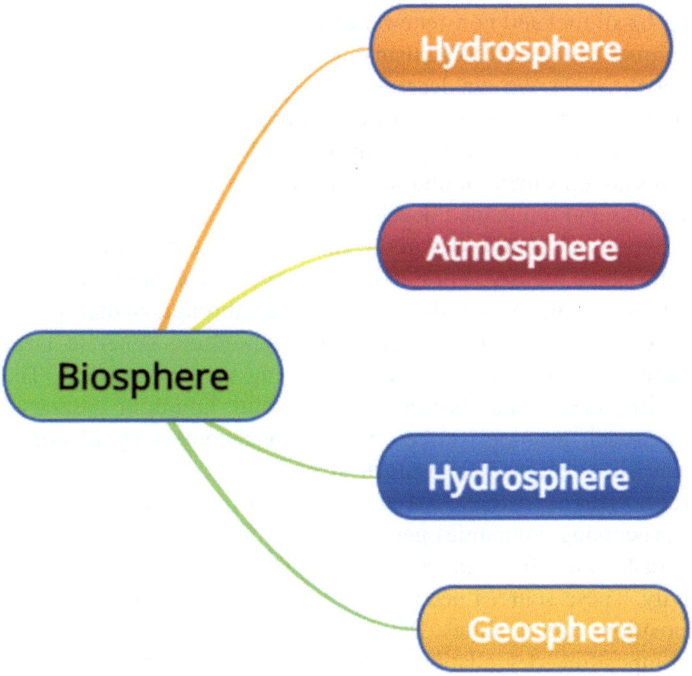

Fig. 12.1 Important parts of earth where life exists

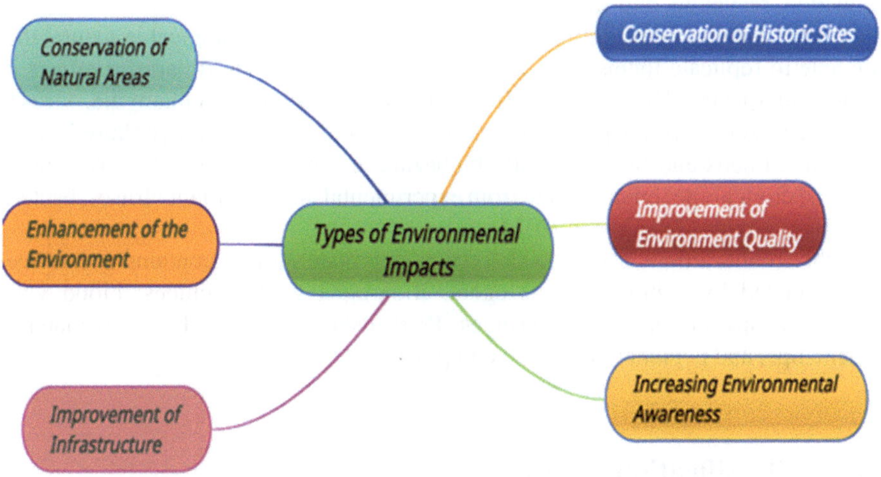

Fig. 12.2 Different types of environmental impacts identified for positive impacts

by lowering fossil fuel and non-renewable material use. This chapter explains zero-waste techniques using microorganisms' remarkable powers to alter waste management (Hui et al. 2023). We can decrease waste and environmental impact, and build a circular and regenerative economy by combining microbiology with sustainability. Microbes' function in waste management is crucial to our drive for a greener, cleaner planet. Wastes are unwanted or unusable materials which come mainly from human activity. Regular and unplanned development as well as modification of livelihood all over the world put complexity in the generated waste (Sharma et al. 2021). The parts of Earth where life exists or the (biosphere) degrade rapidly due to continuous liberation of toxic effluents from different industries throughout the world (Fig. 12.1).

Healthcare institutions and contemporary agriculture generate large amounts of biomedical and agricultural waste, threatening the environment. These wastes are classified by origin and characteristics as solid, liquid, or gaseous. This classification approach promotes effective trash management by identifying origins and characteristics. Biomedical, industrial, electrical, and municipal trash are the main sources. Healthcare facilities generate biomedical trash, whereas manufacturing and processing companies generate industrial garbage (Kumar et al. 2023a). Electronic trash comes from gadget disposal, while municipal garbage comes from urban activity. These sources need different waste management methods owing to their compositions and qualities. Beyond its source, waste categorization includes chemical composition, degradation potential, environmental effect, and provenance. Each category has subcategories, giving a comprehensive waste classification and management structure. Precision in waste categorization allows for customised waste reduction, recycling, and safe disposal options due to its complexity.

Wastes can also be Biological. Biohazardous agents are biological in nature, they are able to replicate themselves, and have the capacity to build harmful effects on biological species. Examples of Biohazardous agents; fungi, chlamydia, viruses, bacteria, rickettsia, prion, parasites, recombinant products, allergens, cultured human and animal cells and the potentially biohazardous agents these cells may contain, infected clinical specimens, tissue from experimental animals, plant viruses, bacteria and fungi, toxins (Zhang et al. 2024).

Other biohazardous agents as defined as any material that contains or has been contaminated by a biohazardous agent. Biological waste includes; blood vials, surgical wraps, culture tubes, syringes, Petri dishes, needles, absorbent material, pipette tips, and personal protective equipment.

12.2 Classification of Waste

Waste generated by human activities results in waste production, and pointed that flow of wastes generated from human activities burdens the environment. It's therefore imperative to avert the impact of waste on the environment. The increasing volume of waste is a threat to the environment and to safeguard the environment and public health, the waste needs to be processed (Sivashanmugam et al. 2023). Researchers described waste management is a process of gathering, transporting, and processing of wastes before disposal of any remaining residues. Several research studies reported that, there are different forms of approach to waste management (Manikandan et al. 2023).

12.2.1 Types of Waste Management

Different waste management approaches are used for waste management according to the type and deposition of waste, and the approach varies from person to person, place to place, time to time, and nation to nation.

12.2.2 Recycling

One of the types of waste management approaches is recycling. In this method, wastes are collected from different places and according to the nature of products subjected for recycling process. In America, Robots are used for the collection of waste from rivers. Recycling process is carried out to control the construction waste in top level countries (Lyu et al. 2024). Many research studies in which recycled solid waste and municipal waste and manufactured geopolymer composite from it. Studies pointed that recycling process is essential for environmental protection. The material produced through recycling can be resold in the market and this process is a solution to solid waste problem and is an ideal way to turn the collected waste materials into useful things. In primary recycling process the collected waste material is made back in to the same material and in secondary recycling, waste materials are made in to various products that are non-recyclable (Lyu et al. 2024).

12.2.3 Incineration

In this approach of waste management, the wastes are burnt at high temperature. To avoid air pollution proper filters are used. Several research studies pointed out that direct incineration method without anaerobic digestion is a preferred sustainable approach for handling sludge. This type of waste management is used for waste disposal (Cao et al. 2023). Worldwide in many regions incineration approach is used. In Denmark and Sweden, incineration equipment accompanied with heat recovery units are used to produce energy for various other purposes. Studies reported that this incineration equipment generates 5–14% of energy. Incineration is an ideal method for wastes with more remarkable calorific contents and reduces waste bulk volume by 70 to 90%. To dispose certain hazardous biomedical wastes incineration method is used (Cao et al. 2023).

12.2.4 Composting

Composting is an approach of waste management in which organic matter is degraded through an aerobic process to convert the organic matter into a humic substance. Organic material found in municipal solid wastes makes them ideal for composting process. About 50 to 70% of organic material found in the municipal solid wastes in most developing countries are recycled by composting. In this approach, Organic wastes are separated and decomposed by biological agents for a certain period of time in a pit to turn them into nutrient-rich compost that increases soil fertility (Chen et al. 2023). Vermicomposting is an organic agriculture practice that reduces environmental impact and helps the ecosystem with high level of organic waste reduction, enhances soil fertility. For sustainable organic agriculture, vermicomposting is an effective process (Dutta et al. 2023). Vermicomposting of onion waste with cow dung produced agricultural enriched nutrient circle.

12.2.5 Landfills

A landfill is a sealed storage container which designed and used to inhibit degradation to protect the environment from harmful contamination. Paper and grass clippings degrade slowly in a landfill. This type of waste management which is established in disused quarries, mining voids borrow pits is being used in many countries. By product of Landfill gas is a renewable energy source. Methane (CH_4) and carbon dioxide (CO_2) are major gases produced from degradation of domestic solid wastes as a result of the anaerobic degradation (Lyu et al. 2024). 45–60% CH_4 and 40–60% CO_2 are found in these gases. Anaerobic conditions are the key source of atmospheric CH_4. This method is however the last choice method in the hierarchy of waste management (Kapoor et al. 2020). Industrial wastes are dumped on private land or dump pits adjacent to the site of the industrial facility from which it has emanated in most developing countries. Figure 12.3 shows various processes of waste management.

12.2.6 Bioremediation

Using microbes to remove the impurities, pollutants, and poisons from soil, water, and other environments is called as bioremediation. Microorganisms use the contaminants as nutrient or energy sources. Natural condition of soil is regained by bioremediation as this approach rectifies the heavy metal pollution problem (Yan et al. 2022). Researchers reported that bioremediation is an eco-friendly, inexpensive, and effective technology which is encouraged to be used for the safe discharge

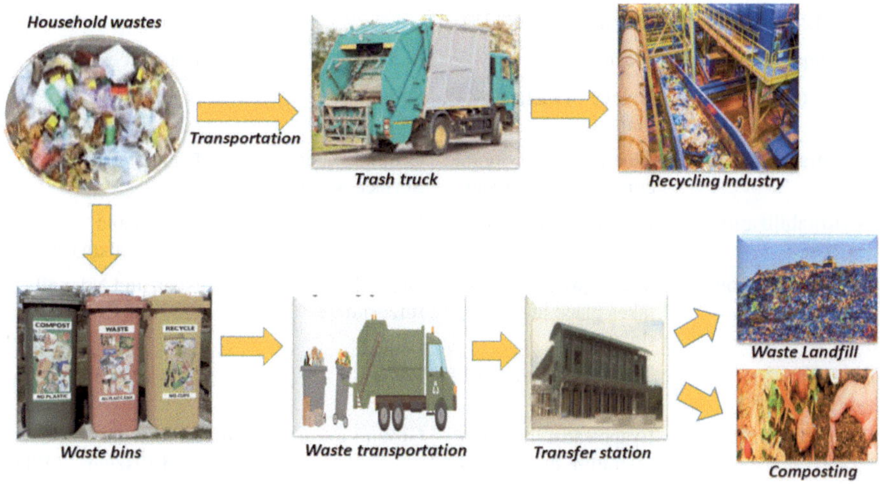

Fig. 12.3 Various processes of waste management

of water from industrial activities (Table 12.2). Pollutants, which are hard to breakdown and inorganic contaminants are removed by bioremediation approach to control environmental pollution. Bio pile, windrows, land farming, bioreactor, bioventing, bio slurping, biosparging, phytoremediation, and permeable reactive barrier are the types of bioremediation approaches. Degradation, rhizoremediation, stabilization, and volatilization are used to remove hydrocarbons and chlorinated compounds. Phytoremediation is an emerging waste management technology that uses plants to remove contaminants from soil and water (Chausali et al. 2021).

12.2.7 Pyrolysis

Waste-to-energy is a type of waste management and in this process, electricity is created from the primary treatment of waste. Pyrolysis process is one of the waste-to-energy processes in which waste products are used as combustion fuel and are converted into another kind of fuel (Thanigaivel et al. 2022). Disposal of medical wastes is done by pyrolysis process which uses high temperature of 100 °C with limited supply of oxygen. This process converts municipal solid wastes as an appropriate source of renewable energy into useful energy (Jaria et al. 2022).

Table 12.2 Applications of phytoremediation

Technique	Mechanism	Medium	References
Phytoextraction	Uptake of metals by plant tissue	Soils	Noreen and Abd-Elsalam (2021)
Phytotransformation	Degradation of organic compounds by plants	Ground water and Surface water	Dutta et al. (2023)
Phytostabilization	Precipitation of metals by plant roots	Soil and ground water	Noreen and Abd-Elsalam (2021)
Phytodegradation	Microbial degradation takes place in rhizosphere	Ground water within rhizosphere	Noah and Ndangili (2022)
Rhizofiltration	Taking up of metals by plant roots	Water	Lata (2021)
Phytovolatilization	Evapotranspiration of selenium, mercury, and volatile hydrocarbons	Ground water and soil	Noreen and Abd-Elsalam (2021)
Vegetative cap	Evapotranspiration of rainwater by plants from disposal sites to prevent leaching of contaminants	Soils	Singh et al. (2021)

12.3 Waste Management

Almost 33% of the 2 billion tons of municipal solid garbage that are currently generated worldwide are not collected by municipalities. According to the Waste Atlas 2018, 0.74 kg of waste is produced per person day on average. By 2050, the World Bank projects that there will be 3.4 billion tons of municipal solid garbage generated (Alawattage and Azure 2021). According to Srivastava et al., solid waste management (SWM) is a decentralized process that mostly depends on the economic standing of particular nations. No matter the region or nation, monitoring trash generation is the first step in every waste management strategy. In order to improve waste collection from bins and trucks, a number of high-end waste monitoring technologies, including geographic information systems (GISs), radio-frequency identification (RFID), ultrasonic sensors, and international system for mobile/general radio packet service (GSM/GPRS), have recently been developed (Chithaluru et al. 2023). Increasing buildup of these wastes' components leading to improper waste management has resulted in a variety of environmental difficulties, such as regular greenhouse gas emissions and a lack of space for waste disposal. Politicians have passed legislation intended to reduce the quantity of garbage entering the environment as a result of these issues' increasing public concerns. These initiatives make an effort to provide a solution that will improve sustainable waste management while promoting MSW recycling, effective waste material conversion to electricity, and the production of other beneficial chemicals (Jadhav et al. 2022).

Modern society places a high priority on waste management because of the rising population and changing consumer habits. Utilizing efficient waste management techniques is essential for preserving the environment's quality, preserving resources, and building a sustainable future. Recycling, incineration, waste-to-energy conversion, composting, or landfilling are all methods of managing municipal solid waste. In many towns around the world, landfilling is favoured for the disposal of solid waste (Singh et al. 2022). Waste undergoes physical, chemical, and biological transformations in landfills, which function as ecological reactors. So landfill liners, the depth of the soil cover, leachate collection, landfill gas recovery, and flaring facilities are important considerations for sustainable landfilling. Municipal solid waste management techniques vary across municipalities, cities, states, and nations, much like its composition does. But the fundamental phases of managing municipal solid waste are: (1) trash generation; (2) garbage collection, handling, and transport; and (3) waste disposal, processing, and treatment (Adamu et al. 2023) (Fig. 12.4).

When it comes to the management of biodegradable waste like fruit and vegetable waste, it is difficult to value because of logistical issues, their perishable nature, and their heterogeneity, among other things. The most pertinent possibilities for valorization, depending on the properties of fruit and vegetable waste and the market need, are the extraction of bioactive compounds, enzymes, and exopolysaccharides, the synthesis of bioplastics and biopolymers, and the generation of biofuels (Zhou et al. 2022). Long-term promotion of the most effective emerging processing technologies is required at the expense of the current conventional ones. As a result, the future integrated valorization of FVW is likely to be divided into two stages: first, the

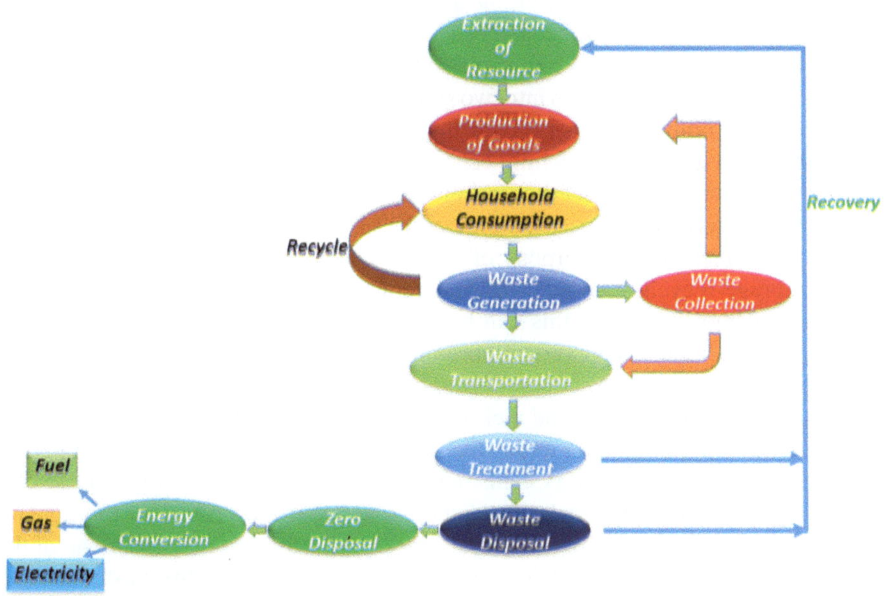

Fig. 12.4 A theoretical model of zero-waste management chain

direct valorization of FVW into value-added products, and then the processing of the leftover streams, byproducts, and leftover debris using standard waste management techniques (Adamu et al. 2023). Future advancements in food waste management are anticipated to make use of product versatility, circular system boundaries and allocation, and resource trade-offs. Future technology developments may make it easier to achieve sustainable development objectives through the control of food waste under circular bioeconomy policies (Hasnain et al. 2023).

Managing non-biodegradable waste materials is more challenging. In the case of plastic waste, burning plastics releases substances that are harmful to the ecosystem, such as phosgene and dioxides. Similarly, once larger plastic is reduced to microplastics, they are also harmful when released into the food chain and aquatic bodies (Leistenschneider et al. 2023). Food contaminated with microplastics and toys with trace quantities of phthalates can have negative effects on health, including congenital disorders and cancer. Though many plastic products are replaced by bioplastics, these products, however, have created new difficulties for waste management. A bigger market volume will make it difficult for governments and policymakers to allocate roughly 50% of the bioplastic market that is now not biodegradable. To solve the plastic issue, careful consideration and effective sustainability education are needed. The threat posed by plastics is largely understood and under control worldwide thanks to some regions of the world having widespread, effective waste management practices. The new paradigm shift of almost exclusively employing non-biodegradable plastics in all commercially available daily goods lead to a rise in plastic trash in the environment, which causes issues like clogged drainage systems and insect breeding grounds. Traditional non-biodegradable plastic waste (NPW) treatment methods include landfilling and burning, but these methods encounter bottlenecks and are unable to adequately address NPW issues (Gao et al. 2023). There have therefore been numerous reports of cutting-edge NPW treatment technologies. Based on a fundamental division into two strategies—recycling and degradation—the technologies are outlined. Physical recycling, energy recovery, and resource recovery are the three categories of recycling technology.

There are obstacles to effective and circular material use brought on by current practices. If material circularity is to increase, the way waste is currently managed needs to be changed. The purpose of waste management in a circular economy is to encourage recycling so that material value is retained throughout the cycles. The results demonstrate that this can be encouraged by highlighting the financial advantages of a circular economy, improving data sharing on waste, and fostering greater communication and cooperation among important actors (Adeel et al. 2021). Harmonization of regulations and their interpretations is additionally required. A lot of research should be done on topics like efficient energy conversion technologies, integrating renewable energy systems, enabling the circular economy framework, integrating processes effectively, and other issues that are crucial to the general public. An accurate forecast of municipal waste generation and understanding of the behavioral, socioeconomic, and demographic factors influencing the waste management (and recycling) behavior of households and other municipal waste producers are necessary for the effective planning of treatment capacities (especially sorting plants, recycling, and energy recovery facilities) and municipal waste management policy (Ibn Batouta et al. 2023; Siddiqui et al. 2020).

12.4 Microbes in Waste Management

To achieve zero-waste operations, microbes offer promising solutions and play a crucial role in waste management. More sustainable and ecologically friendly waste management is possible via the use of microorganisms for material processing, transformation, and remediation.

12.4.1 Composting

Microorganisms break down organic waste in composting. Bacteria and fungus decompose organic waste like food, garden clippings, and agricultural soil. This aerobic breakdown process requires oxygen. The final output of composting is nutrient-rich compost that improves soil health and sustainability. The compost produced by composting has several benefits (Singh et al. 2022). First, it increases soil fertility by adding minerals and organic matter, enabling plant development. Compost improves soil structure, retaining moisture and preventing erosion. Sustainable land use is promoted by healthier, more productive agricultural and garden soils.

Composting is vital to trash management and environmental preservation. Composting minimizes methane emissions from landfills by diverting organic waste from anaerobic decomposition. Composting reduces greenhouse gas emissions and organic waste disposal's environmental effects. Composting is a natural, eco-friendly process that turns organic waste into a useful resource, improves soils, and promotes sustainability (Qi et al. 2023). It promotes sustainable waste management and soil enrichment, supporting environmental conservation and resource efficiency.

12.4.2 Anaerobic Digestion

The lack of oxygen during the microbial process is known as anaerobic digestion. Bacteria, archaea, and protozoa, among others, work together to break down organic waste in this process. Through a series of metabolic events, these microbes convert complex organic chemicals like those found in sewage sludge, manure, and food waste into simpler ones (Nemr et al. 2022). Biogas, which consists mostly of methane and carbon dioxide, is an important by-product of anaerobic digestion. Biogas is a renewable energy that may be utilized to create power or heat. Digestate, the by-product of this process, may be converted into nutrient-rich fertilizers.

12.4.3 Bioconversion of Waste into Chemicals

In bioconversion techniques, microbes may be used to transform discarded materials into useful compounds. Some bacteria, for instance, may break down organic waste into usable bioplastics. These bioplastics are renewable and biodegradable, making them a viable replacement for petroleum-based plastics (Peng et al. 2022). This method not only decreases the environmental effect of conventional plastic manufacture but also reduces trash creation by using garbage as a feedstock for bioplastic synthesis.

12.4.4 Bioremediation

Microorganisms are used in bioremediation, which is the process of cleaning the environment by breaking down or eliminating harmful substances. Bacteria and fungus are examples of microbes that may degrade a wide range of contaminants. These pollutants include petroleum hydrocarbons, heavy metals, and industrial chemicals (Yan et al. 2022). They use mechanisms like biodegradation and bioaccumulation to convert these pollutants into less hazardous compounds. Soil, water, and air pollution are all treatable by bioremediation techniques. Traditional remediation procedures, which may be more energy-intensive or create secondary trash, may be superseded by this ecological and cost-effective alternative.

12.4.5 Microbial Fuel Cells

Devices called microbial fuel cells (MFCs) employ microorganisms to generate power from decomposing organic matter. Electrons are transferred from microbial metabolism to an electrode, where they are used to generate current. Electricity may be produced from the collected electrons released by microbes during the decomposition of organic waste. MFCs provide the dual benefit of recycling trash and producing clean electricity (Manikandan et al. 2023). Wastewater, agricultural leftovers, and industrial organic by-products are only some of the organic waste streams that might benefit from their use. One possible path towards zero-waste operations is the use of microorganisms in garbage management. Microorganisms may aid in waste diversion, resource recovery, renewable energy production, and pollution abatement via processes including composting, anaerobic digestion, bioconversion, and microbial fuel cells. These cutting-edge methods provide a healthier environment and more efficient use of scarce resources by taking a more circular approach to waste management. The data in the following table 12.3 show how microorganisms may be used to reduce waste, increase the use of renewable energy sources, and lessen the negative effects of pollution (Kumar et al. 2023b). These zero-waste methods open the path for a more sustainable and circular waste management strategy by harnessing the metabolic capabilities of microorganisms.

12.5 Composting

For the improvement of soil quality and to encourage plant development, compost may be used to convert organic waste into a nutrient-dense material. It's an eco-friendly method that benefits sustainable gardening and farming by cutting down on waste and conserving resources (Liu et al. 2023). Let's get into the nitty–gritty of composting by discussing the many uses it serves, how it's done, and how to maximize the results.

12.5.1 Benefits of Composting

Reducing landfill trash and composting keeps food scraps and yard trimmings out of the environment where they would break down into methane. Compost is a great soil supplement because it strengthens the soil, boosts the nutritional content, and makes the soil more capable of holding water. It has an optimal ratio of nitrogen to phosphorus to potassium, all of which are required for plant development. By reducing their use of harmful synthetic fertilizers, pesticides, and herbicides, gardeners may practice sustainable gardening by making use of compost (Babu et al. 2022). Compost's ability to improve soil structure, increase water penetration, and enhance the binding capacity of soil particles helps to reduce soil erosion. Composting aids in the reduction of greenhouse gas emissions and the mitigation of climate change via carbon sequestration.

12.5.2 Composting Process

Choose a composting method: Consider space, time, and resources. Backyard composting, vermicomposting, and bin or tumbler composting are common methods. Composting container: Choose a container that holds moisture and enables ventilation. Consider space and composting requirements. Collect compostables: Mix green and brown materials. Fruit and vegetable waste, coffee grounds, tea leaves, and fresh grass clippings offer nitrogen (Wang et al. 2022a). Dried leaves, twigs, shredded paper, cardboard, and straw are brown carbon sources. Avoid meat, dairy, oil, and sick plants. Make compost: Alternate green and brown compostables in the compost container. A 3:1 brown-to-green ratio is optimal. Balanced decomposition prevents odours and extra moisture. Maintain 40–60% compost pile wetness. Like a sponge. Avoid over-saturation by adding water or covering the pile during rainy seasons.

Table 12.3 Summary of microbe-based zero-waste processes:

Waste management process	Description	Microbial involvement	Key benefits	References
Composting	Composting is the process of turning organic waste into a soil amendment	Organic matter is decomposed by microbes including bacteria, fungus, and others	Decreases the amount of organic waste makes excellent compost for amending soil Improvements in soil water retention and fertility	Li et al. (2021)
Anaerobic digestion	Biogas and digestate are produced from organic waste	The breakdown of organic materials by methane-producing bacteria and other microbes	Produces sustainable energy in the form of biogas organic waste volume and GHG emissions may be decreased while nutrient-rich digestate for fertilizer is produced	Eljamal et al. (2022)
Bioconversion of waste into chemicals	Production of useful compounds like bioplastics from discarded resources	Waste is metabolized by bacteria and other microbes, which then generate the desired compounds	Produces biodegradable alternatives to traditional plastics; Repurposes trash as a resource for profitable chemical manufacture; Lowers overall waste and pollution levels	Chakraborty et al. (2023)
Bioremediation	Detoxification or degradation of harmful substances using microorganisms	Pollutants are metabolized by bacteria, fungus, and algae, and converted into less dangerous forms	Reduces the need for costly and resource-intensive remedial techniques Offers a long-term, budget-friendly option for cleaning up the environment	Ranjan et al. (2023)
Microbial fuel cells	Energy recovery from compostable garbage	Electrons are produced when microbes transport electrons from organic materials to an electrode	Sustainable solution for waste management and energy demands; treats organic waste while producing renewable energy; allows for the possibility of decentralized energy generation	Wu et al. (2021)

Turn the compost every few weeks with a pitchfork or garden implement. turning the pile oxygenates, accelerates decomposition, and maintains wetness. Rotate compost tumblers or bins to mix contents. Maintain compost: Monitor the compost pile's moisture and temperature. Active decomposition indicates a 110–160 °F (43–71 °C) interior temperature (Qi et al. 2023). Add brown materials or adjust moisture as required. Avoid big branches or excessive volumes of one material, which might disturb composting. Harvest and utilize compost: Composting might take weeks or months, depending on temperature, moisture, and ingredients. Dark, crumbly, and earthy-smelling compost is ready to use. Use completed compost to enhance garden soil, potting mixes, or lawns and plants.

12.5.3 Effective Composting

The composting process may be sped up by chopping or shredding bigger things into smaller ones. Strive for moderation: Achieve the right carbon-to-nitrogen ratio for breakdown by using a combination of green and brown materials. Keep the soil moist: Keep the compost pile wet but not soaked by checking and adjusting the moisture levels on a regular basis (Qi et al. 2023; Yin et al. 2021). Additives to composting files: Decomposition may be sped up with the help of beneficial microbes introduced by certain compost accelerators or activators, such as garden soil or completed compost. Avoid pollutant exposure: Pesticides and herbicides kill beneficial organisms in compost, so avoid using them. Keep an eye on the heat and smells: Decomposition occurs if the compost pile is heated, and an imbalance or lack of aeration is indicated by unpleasant odours. Composting is an organic process that takes time and constant attention. Take your time and don't give up on your compost pile. Reducing garbage, improving soil quality, and encouraging long-term gardening success are just a few of the many benefits of composting. Composting is an excellent way to both improve the environment and provide your plants with a nutrient-rich soil amendment (Liu et al. 2021).

12.5.4 Role of Microorganisms

Beneficial microorganisms are essential to the composting process because they decompose organic materials into usable compost (Fig. 12.5). Bacteria, fungus, actinomycetes, and other microbes are all examples of microorganisms (Sagasti et al. 2019). Let's get into the nitty–gritty of their respective jobs and contributions:

Composting relies heavily on bacteria since they are such efficient decomposers. Carbon dioxide (CO_2) and water are the products of their digestion of basic organic components like sugars and starches. Mineralization is the process through which energy and nutrients are released for use by microbes and plants. Early on in the composting process, the dominant bacteria are thermophilic bacteria, which live at

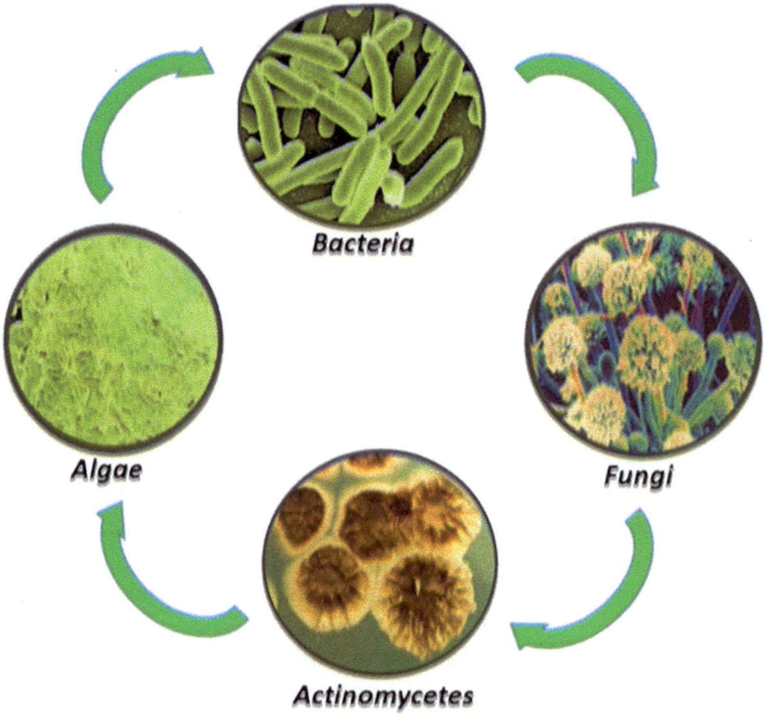

Fig. 12.5 Microorganisms used to destroy waste materials

high temperatures and produce heat that helps destroy pathogens and weed seeds. In composting, fungi play a crucial role as decomposers, especially in the breakdown of tough organic components like cellulose and lignin (Nahak et al. 2022). By penetrating and decomposing difficult plant materials, fungi's thread-like hyphae increase the availability of nutrients to other microbes. Soil structure and nutrient retention are both enhanced by the humus that fungi help to create. Actinomycetes are a class of bacteria that behave and look like fungus. Chitin, proteins, and lipids are only some of the complex chemical substances that they degrade. The enzymes produced by actinomycetes aid in the breakdown of difficult organic debris, such as plant and animal waste.

Protozoa are single-celled creatures that consume other microbes like bacteria and fungus. They maintain a healthy environment by controlling the number of microorganisms in the compost pile. The excretion of protozoa makes nutrients accessible for absorption by plants. Even though they aren't microbes, earthworms are still useful composting companions (Akansha et al. 2021). By digesting dead organisms and eating decaying organic debris, they speed up the decomposition process. The grinding action of earthworms and the castings they excrete, which are rich in beneficial microbes and nutrients, aid in the decomposition process. Their digging helps the compost pile breathe and drain better. The compost pile becomes a dynamic and diversified ecosystem thanks to the cooperation of these microorganisms. In order to make nutrients available to plants, they secrete enzymes that catalyze the breakdown of complex organic molecules into simpler chemicals. During the thermophilic phase of composting, temperatures rise as heat is produced while microbes digest organic materials.

12.5.5 Factors Essential for Encouraging Microbial Activity in Composting

Water is essential for the metabolic functions of microorganisms. The compost pile should be humid like a sponge so that microorganisms may thrive. Composting relies heavily on the oxygen-dependent aerobic bacteria. An oxygen-rich habitat for these bacteria may be ensured by proper aeration, either by rotating the pile or by employing aeration equipment. Microorganisms need a favourable carbon-to-nitrogen (C/N) ratio in order to grow and reproduce. As a carbon supply (brown materials) and nitrogen for protein synthesis (green materials), a C/N ratio of roughly 30:1 is optimum for composting (Ravindran et al. 2022).

Compost microbes thrive in slightly warmer or slightly cooler conditions. During the active breakdown phase, temperatures of 113 to 160 °F (45 to 71 °C) are optimal for the growth of thermophilic bacteria. Weed seeds, diseases, and other unwelcome creatures may all be killed by the increased warmth (Lin et al. 2022). A healthy population of microorganisms may be ensured by providing an ideal environment, a wide variety of organic materials, and a well-balanced and maintained compost pile. The compost that is generated as a result of their efforts increases soil fertility and has beneficial nutrients.

12.5.6 Temperature in Composting

The composting process is greatly affected by temperature, which in turn affects the activity of microorganisms and the pace of decomposition. Various phases of decomposition are associated with temperature variations that occur during the composting process. Let's go through why temperature is so crucial during composting: The composting process starts in the mesophilic phase, when temperatures are between 68- and 113-°F (20 and 45 °C) (Yin et al. 2021). At this temperature and humidity level, mesophilic microorganisms like bacteria and fungus may thrive. As a result of decomposing easily accessible organic stuff, they release heat. Microbial activity causes a slow but steady increase in temperature. Compost begins its thermophilic phase when its temperature reaches between 115 and 160 °F (46 and 71 °C) (Wang et al. 2022b). At this point, the thermophilic bacteria take the lead since they perform best in the hotter conditions. These bacteria facilitate faster decomposition by breaking down complex organic molecules like lignin and cellulose. Weed seeds, pathogens, and other undesirable organisms in the compost pile are killed off by the high temperatures. For prolonged high temperatures to support microbial activity, the compost pile may need to be well insulated or turned often.

The cooling phase occurs after all the easily decomposable organic matter has been eaten and the pile's energy output has dropped. At this point, the compost starts a cooling phase during which its temperature lowers to room temperature. Decomposition proceeds at a more leisurely rate when mesophilic bacteria resume activity. The size of the pile, the carbon-to-nitrogen ratio, the moisture level, and the amount of airflow all have a role in the compost's temperature swings. More heat is produced and retained by larger heaps than by smaller ones. Microbial activity and temperature control are both enhanced by a carbon-to-nitrogen ratio of around 30:1 and by a moisture level that is neither too high nor too low (Dutta et al. 2023). It's vital to keep an eye on the compost pile's temperature to catch any problems early and correct them. A thermometer placed near the center of the pile may be used to collect consistent temperature readings. If the temperature stays low or never reaches the thermophilic range, it might be because of a lack of nitrogen, oxygen, or water. A rise in temperature and increased microbial activity may result from adjusting these variables. Temperature control in composting is essential for killing unwanted organisms like pathogens and weed seeds, as well as for speeding up the decomposition process. Composters may increase the likelihood of effective composting and the creation of nutrient-rich compost by taking into account the influence of temperature and following relevant steps (Babu et al. 2022).

12.6 Conclusion

Zero-waste microbe-based procedures are a prospective paradigm change in trash management and sustainable resource use. These techniques use microorganisms to turn waste into useful resources, providing a holistic solution to environmental issues. Composting and anaerobic digestion turn organic waste into nutrient-rich soil amendments and renewable energy, decreasing artificial fertilizers and fossil fuel consumption, and greenhouse gas emissions. Plastic-degrading bacteria may revolutionize plastic recycling and reduce their environmental effect, helping to solve the worldwide plastic pollution challenge. In addition to waste reduction, bacteria produce biofuels, bioplastics, medicines, and other useful molecules from renewable resources, promoting sustainability across sectors.

Microbe-based wastewater treatment improves water quality and the environment. These revolutionary zero-waste methods, categorized by source and features, help us manage waste streams more efficiently, decreasing pollution, resource use, and environmental deterioration. By adopting microbial technologies, we move towards a circular and regenerative economy that treats waste as a resource. To maximize these processes' potential, research, technological development, and broad acceptance are needed. Sustainable practices must be promoted globally via education and infrastructural investment. Microbes help us create a greener, cleaner society where waste is eliminated and resource conservation is our legacy.

Acknowledgements The authors are thankful to The Management, Vice-Chancellor, Dean of SMNS and The Head Department of Biological Sciences, The Copperbelt University, Kitwe, Zambia, The Management of Saveetha School of Engineering (SIMATS), Chennai, India, The Management of Mulungushi University, Kabwe, Zambia and The Management of SASTRA Deemed to be University, Tamil Nadu, India for their constant to complete this chapter successfully.

References

Adamu H, Bello U, Yuguda AU, Tafida UI, Jalam AM, Sabo A, Qamar M (2023) Production processes, techno-economic and policy challenges of bioenergy production from fruit and vegetable wastes. Renew Sustain Energy Rev 186:113686. https://doi.org/10.1016/j.rser.2023.113686

Adeel M, Shakoor N, Shafiq M, Pavlicek A, Part F, Zafiu C, Raza A, Ahmad MA, Jilani G, White JC, Ehmoser E-K, Lynch I, Ming X, Rui Y (2021) A critical review of the environmental impacts of manufactured nano-objects on earthworm species. Environ Pollut 290:118041. https://doi.org/10.1016/j.envpol.2021.118041

Agrawal K, Nair LG, Chaturvedi V, Verma P (2023) Designing microbial cellulases using genetic engineering approach: a promising strategy towards zero-waste cellulosic biorefinery. Biocatal Agric Biotechnol 52:102830. https://doi.org/10.1016/j.bcab.2023.102830

Akansha K, Majumder S, Misra M, Sachan SG (2021) Chapter 4–Role of microorganisms in performance optimization of wastewater treatment plants. In: Shah MP, Sarkar A, Mandal SBT-WT (eds). Elsevier, pp 67–91

Alawattage C, Azure JD-C (2021) Behind the World Bank's ringing declarations of "social accountability": Ghana's public financial management reform. Crit Perspect Account 78:102075. https://doi.org/10.1016/j.cpa.2019.02.002

Ali J, Bibi S, Jatoi WB, Tuzen M, Jakhrani MA, Feng X, Saleh TA (2023) Green synthesized zinc oxide nanostructures and their applications in dye-sensitized solar cells and photocatalysis: a review. Mater Today Commun 36:106840. https://doi.org/10.1016/j.mtcomm.2023.106840

Babu S, Singh Rathore S, Singh R, Kumar S, Singh VK, Yadav SK, Yadav V, Raj R, Yadav D, Shekhawat K, Ali Wani O (2022) Exploring agricultural waste biomass for energy, food and feed production and pollution mitigation: a review. Biores Technol 360:127566. https://doi.org/10.1016/j.biortech.2022.127566

Cao X, Williams PN, Zhan Y, Coughlin SA, McGrath JW, Chin JP, Xu Y (2023) Municipal solid waste compost: global trends and biogeochemical cycling. Soil Environ Health 1:100038. https://doi.org/10.1016/j.seh.2023.100038

Chakraborty D, Chatterjee S, Althuri A, Palani SG, Venkata Mohan S (2023) Sustainable enzymatic treatment of organic waste in a framework of circular economy. Biores Technol 370:128487. https://doi.org/10.1016/j.biortech.2022.128487

Chausali N, Saxena J, Prasad R (2021) Nanobiochar and biochar based nanocomposites: advances and applications. J Agric Food Res 5:100191. https://doi.org/10.1016/j.jafr.2021.100191

Chen B, Perumal P, Illikainen M, Ye G (2023) A review on the utilization of municipal solid waste incineration (MSWI) bottom ash as a mineral resource for construction materials. J Build Eng 71:106386. https://doi.org/10.1016/j.jobe.2023.106386

Chithaluru P, Singh A, Dhatterwal JS, Sodhro AH, Albahar MA, Jurcut A, Alkhayyat A (2023) An optimized privacy information exchange schema for explainable AI empowered WiMAX-based IoT networks. Futur Gener Comput Syst 148:225–239. https://doi.org/10.1016/j.future.2023.06.003

Dutta R, Angmo D, Singh J, Bala Chowdhary A, Quadar J, Singh S, Pal Vig A (2023) Synergistic effect of biochar amendment in milk processing industry sludge and cattle dung during the vermiremediation. Biores Technol 371:128612. https://doi.org/10.1016/j.biortech.2023.128612

El Nemr A, Hassaan MA, Elkatory MR, Ragab S, El-Nemr MA, Tedone L, De Mastro G, Pantaleo A (2022) Enhancement of biogas production from individually or co-digested green algae Chaetomorpha linum using ultrasound and ozonation treated biochar. Ultrason Sonochem 90:106197. https://doi.org/10.1016/j.ultsonch.2022.106197

Eljamal R, Maamoun I, Bensaida K, Yilmaz G, Sugihara Y, Eljamal O (2022) A novel method to improve methane generation from waste sludge using iron nanoparticles coated with magnesium hydroxide. Renew Sustain Energy Rev 158:112192. https://doi.org/10.1016/j.rser.2022.112192

Gao D, Junaid M, Chen X, Liao H, Chen G, Wang J (2023) Interaction of micro(nano)plastics and bisphenols in the environment: a recent perspective on adsorption mechanisms, influencing factors and ecotoxic impacts. TrAC Trends Anal Chem 165:117132. https://doi.org/10.1016/j.trac.2023.117132

Hasnain M, Munir N, Abideen Z, Zulfiqar F, Koyro HW, El-Naggar A, Caçador I, Duarte B, Rinklebe J, Yong JWH (2023) Biochar-plant interaction and detoxification strategies under abiotic stresses for achieving agricultural resilience: a critical review. Ecotoxicol Environ Saf 249:114408. https://doi.org/10.1016/j.ecoenv.2022.114408

Hui K, Yuan Y, Xi B, Tan W (2023) A review of the factors affecting the emission of the ozone chemical precursors VOCs and NOx from the soil. Environ Int 172:107799. https://doi.org/10.1016/j.envint.2023.107799

Ibn Batouta K, Aouhassi S, Mansouri K (2023) Energy efficiency in the manufacturing industry—a tertiary review and a conceptual knowledge-based framework. Energy Rep 9:4635–4653. https://doi.org/10.1016/j.egyr.2023.03.107

Jadhav DA, Park S-G, Pandit S, Yang E, Ali Abdelkareem M, Jang J-K, Chae K-J (2022) Scalability of microbial electrochemical technologies: applications and challenges. Biores Technol 345:126498. https://doi.org/10.1016/j.biortech.2021.126498

Jaria G, Calisto V, Esteves VI, Otero M (2022) Overview of relevant economic and environmental aspects of waste-based activated carbons aimed at adsorptive water treatments. J Clean Prod 344:130984. https://doi.org/10.1016/j.jclepro.2022.130984

Kapoor R, Ghosh P, Kumar M, Sengupta S, Gupta A, Kumar SS, Vijay V, Kumar V, Kumar Vijay V, Pant D (2020) Valorization of agricultural waste for biogas based circular economy in India: a research outlook. Biores Technol 304:123036. https://doi.org/10.1016/j.biortech.2020.123036

Kumar M, Ngasepam J, Dhangar K, Mahlknecht J, Manna S (2022) Critical review on negative emerging contaminant removal efficiency of wastewater treatment systems: concept, consistency and consequences. Biores Technol 352:127054. https://doi.org/10.1016/j.biortech.2022.127054

Kumar A, Pali HS, Kumar M (2023a) A comprehensive review on the production of alternative fuel through medical plastic waste. Sustain Energy Technol Assess 55:102924. https://doi.org/10.1016/j.seta.2022.102924

Kumar V, Vangnai AS, Sharma N, Kaur K, Chakraborty P, Umesh M, Singhal B, Utreja D, Carrasco EU, Andler R, Awasthi MK, Taherzadeh MJ (2023b) Bioengineering of biowaste to recover bioproducts and bioenergy: a circular economy approach towards sustainable zero-waste environment. Chemosphere 319:138005. https://doi.org/10.1016/j.chemosphere.2023.138005

Lata S (2021) Sustainable and eco-friendly approach for controlling industrial wastewater quality imparting succour in water-energy nexus system. Energy Nexus 3:100020. https://doi.org/10.1016/j.nexus.2021.100020

Leistenschneider D, Wolinski A, Cheng J, ter Halle A, Duflos G, Huvet A, Paul-Pont I, Lartaud F, Galgani F, Lavergne É, Meistertzheim A-L, Ghiglione J-F (2023) A critical review on the evaluation of toxicity and ecological risk assessment of plastics in the marine environment. Sci Total Environ 896:164955. https://doi.org/10.1016/j.scitotenv.2023.164955

Li Y, Zhao J, Krooneman J, Euverink GJW (2021) Strategies to boost anaerobic digestion performance of cow manure: laboratory achievements and their full-scale application potential. Sci Total Environ 755:142940. https://doi.org/10.1016/j.scitotenv.2020.142940

Lin X, Wang N, Li F, Yan B, Pan J, Jiang S, Peng H, Chen A, Wu G, Zhang J, Zhang L, Huang H, Luo L (2022) Evaluation of the synergistic effects of biochar and biogas residue on CO2 and CH4 emission, functional genes, and enzyme activity during straw composting. Biores Technol 360:127608. https://doi.org/10.1016/j.biortech.2022.127608

Liu B, Yu K, Ahmed I, Gin K, Xi B, Wei Z, He Y, Zhang B (2021) Key factors driving the fate of antibiotic resistance genes and controlling strategies during aerobic composting of animal manure: a review. Sci Total Environ 791:148372. https://doi.org/10.1016/j.scitotenv.2021.148372

Liu X, Xie Y, Sheng H (2023) Green waste characteristics and sustainable recycling options. Resourc Environ Sustain 11:100098. https://doi.org/10.1016/j.resenv.2022.100098

Lyu L, Bagchi M, Markoglou N, An C, Peng H, Bi H, Yang X, Sun H (2024) Towards environmentally sustainable management: a review on the generation, degradation, and recycling of polypropylene face mask waste. J Hazard Mater 461:132566. https://doi.org/10.1016/j.jhazmat.2023.132566

Magalhães-Ghiotto GAV, de Oliveira AM, Natal JPS, Bergamasco R, Gomes RG (2021) Green nanoparticles in water treatment: a review of research trends, applications, environmental aspects and large-scale production. Environ Nanotechnol Monit Manag 16:100526. https://doi.org/10.1016/j.enmm.2021.100526

Manikandan S, Vickram S, Sirohi R, Subbaiya R, Krishnan RY, Karmegam N, Sumathijones C, Rajagopal R, Chang SW, Ravindran B, Awasthi MK (2023) Critical review of biochemical pathways to transformation of waste and biomass into bioenergy. Biores Technol 372:128679. https://doi.org/10.1016/j.biortech.2023.128679

Mittal N, Kumar A (2022) Microbial fuel cell as water-energy-environment nexus: a relevant strategy for treating streamlined effluents. Energy Nexus 7:100097. https://doi.org/10.1016/j.nexus.2022.100097

Molina-Besch K, Keszleri H (2023) Exploring the industrial perspective on biobased plastics in food packaging applications—insights from Sweden. Sustain Prod Consumpt 35:72–84. https://doi.org/10.1016/j.spc.2022.10.018

Nahak BK, Preetam S, Sharma D, Shukla SK, Syväjärvi M, Toncu D-C, Tiwari A (2022) Advancements in net-zero pertinency of lignocellulosic biomass for climate neutral energy production. Renew Sustain Energy Rev 161:112393. https://doi.org/10.1016/j.rser.2022.112393

Nguyen LH, Nguyen B-S, Le D-T, Alomar TS, AlMasoud N, Ghotekar S, Oza R, Raizada P, Singh P, Nguyen V-H (2023) A concept for the biotechnological minimizing of emerging plastics, micro- and nano-plastics pollutants from the environment: a review. Environ Res 216:114342. https://doi.org/10.1016/j.envres.2022.114342

Noah NM, Ndangili PM (2022) Green synthesis of nanomaterials from sustainable materials for biosensors and drug delivery. Sens Int 3. https://doi.org/10.1016/j.sintl.2022.100166

Noreen S, Abd-Elsalam KA (2021) Chapter 9–Biochar-based nanocomposites: a sustainable tool in wastewater bioremediation. In: Abd-Elsalam KA, Zahid MBT-A (eds) Micro and nano technologies. Elsevier, pp 185–200

Peng W, Wang Z, Shu Y, Lü F, Zhang H, Shao L, He P (2022) Fate of a biobased polymer via high-solid anaerobic co-digestion with food waste and following aerobic treatment: insights on changes of polymer physicochemical properties and the role of microbial and fungal communities. Biores Technol 343:126079. https://doi.org/10.1016/j.biortech.2021.126079

Qi J, Yang H, Wang X, Zhu H, Wang Z, Zhao C, Li B, Liu Z (2023) State-of-the-art on animal manure pollution control and resource utilization. J Environ Chem Eng 11:110462. https://doi.org/10.1016/j.jece.2023.110462

Raabe D, Ponge D, Uggowitzer PJ, Roscher M, Paolantonio M, Liu C, Antrekowitsch H, Kozeschnik E, Seidmann D, Gault B, De Geuser F, Deschamps A, Hutchinson C, Liu C, Li Z, Prangnell P, Robson J, Shanthraj P, Vakili S, Sinclair C, Bourgeois L, Pogatscher S (2022) Making sustainable aluminum by recycling scrap: the science of "dirty" alloys. Prog Mater Sci 128:100947. https://doi.org/10.1016/j.pmatsci.2022.100947

Ranjan R, Rai R, Bhatt SB, Dhar P (2023) Technological road map of Cellulase: a comprehensive outlook to structural, computational, and industrial applications. Biochem Eng J 198:109020. https://doi.org/10.1016/j.bej.2023.109020

Ravindran B, Awasthi MK, Karmegam N, Chang SW, Chaudhary DK, Selvam A, Nguyen DD, Rahman Milon A, Munuswamy-Ramanujam G (2022) Co-composting of food waste and swine manure augmenting biochar and salts: nutrient dynamics, gaseous emissions and microbial activity. Biores Technol 344:126300. https://doi.org/10.1016/j.biortech.2021.126300

Sagasti AJ, García Massini JL, Escapa IH, Guido DM (2019) Multitrophic interactions in a geothermal setting: arthropod borings, actinomycetes, fungi and fungal-like microorganisms in a decomposing conifer wood from the Jurassic of Patagonia. Palaeogeogr Palaeoclimatol Palaeoecol 514:31–44. https://doi.org/10.1016/j.palaeo.2018.09.004

Sharma B, Shekhar S, Sharma S, Jain P (2021) The paradigm in conversion of plastic waste into value added materials. Clean Eng Technol 4:100254. https://doi.org/10.1016/j.clet.2021.100254

Siddiqui MA, Hiranmai Yadav R (2020) Microbial procession during decomposition of organic wastes BT–environmental microbiology and biotechnology: volume 1: biovalorization of solid wastes and wastewater treatment. In: Srivastava S, Rathore D, Pant D (eds) Singh A. Springer Singapore, Singapore, pp 55–67

Singh R, Arora A, Singh V (2021) Biodiesel from oil produced in vegetative tissues of biomass—a review. Biores Technol 326:124772. https://doi.org/10.1016/j.biortech.2021.124772

Singh R, Paritosh K, Pareek N, Vivekanand V (2022) Integrated system of anaerobic digestion and pyrolysis for valorization of agricultural and food waste towards circular bioeconomy: review. Biores Technol 360:127596. https://doi.org/10.1016/j.biortech.2022.127596

Sivashanmugam S, Rodriguez S, Pour Rahimian F, Elghaish F, Dawood N (2023) Enhancing information standards for automated construction waste quantification and classification. Autom Constr 152:104898. https://doi.org/10.1016/j.autcon.2023.104898

Thanigaivel S, Vickram S, Dey N, Gulothungan G, Subbaiya R, Govarthanan M, Karmegam N, Kim W (2022) The urge of algal biomass-based fuels for environmental sustainability against a steady tide of biofuel conflict analysis: is third-generation algal biorefinery a boon? Fuel 317:123494. https://doi.org/10.1016/j.fuel.2022.123494

Wainaina S, Awasthi MK, Sarsaiya S, Chen H, Singh E, Kumar A, Ravindran B, Awasthi SK, Liu T, Duan Y, Kumar S, Zhang Z, Taherzadeh MJ (2020) Resource recovery and circular economy from organic solid waste using aerobic and anaerobic digestion technologies. Biores Technol 301:122778. https://doi.org/10.1016/J.BIORTECH.2020.122778

Wang M, Wu Y, Zhao J, Liu Y, Gao L, Jiang Z, Zhang J, Tian W (2022a) Comparison of composting factors, heavy metal immobilization, and microbial activity after biochar or lime application in straw-manure composting. Biores Technol 363:127872. https://doi.org/10.1016/j.biortech.2022.127872

Wang N, Zhao K, Li F, Peng H, Lu Y, Zhang L, Pan J, Jiang S, Chen A, Yan B, Luo L, Huang H, Li H, Wu G, Zhang J (2022b) Characteristics of carbon, nitrogen, phosphorus and sulfur cycling genes, microbial community metabolism and key influencing factors during composting process supplemented with biochar and biogas residue. Biores Technol 366:128224. https://doi.org/10.1016/j.biortech.2022.128224

Wu L, Wei W, Song L, Woźniak-Karczewska M, Chrzanowski Ł, Ni B-J (2021) Upgrading biogas produced in anaerobic digestion: biological removal and bioconversion of CO2 in biogas. Renew Sustain Energy Rev 150:111448. https://doi.org/10.1016/j.rser.2021.111448

Yan Y, Yan M, Ravenni G, Angelidaki I, Fu D, Fotidis IA (2022) Novel bioaugmentation strategy boosted with biochar to alleviate ammonia toxicity in continuous biomethanation. Biores Technol 343:126146. https://doi.org/10.1016/j.biortech.2021.126146

Yin Y, Yang C, Li M, Zheng Y, Ge C, Gu J, Li H, Duan M, Wang X, Chen R (2021) Research progress and prospects for using biochar to mitigate greenhouse gas emissions during composting: a review. Sci Total Environ 798:149294. https://doi.org/10.1016/j.scitotenv.2021.149294

Zhang J, Chen Z, Shan D, Wu Y, Zhao Y, Li C, Shu Y, Linghu X, Wang B (2024) Adverse effects of exposure to fine particles and ultrafine particles in the environment on different organs of organisms. J Environ Sci 135:449–473. https://doi.org/10.1016/j.jes.2022.08.013

Zhou Y, Kumar V, Harirchi S, Vigneswaran VS, Rajendran K, Sharma P, Wah Tong Y, Binod P, Sindhu R, Sarsaiya S, Balakrishnan D, Mofijur M, Zhang Z, Taherzadeh MJ, Kumar Awasthi M (2022) Recovery of value-added products from biowaste: a review. Biores Technol 360:127565. https://doi.org/10.1016/j.biortech.2022.127565

Chapter 13
Biofuel Production from Microbial Sources: Advances and Challenges

Boddu Sandeep, Utkarsh Baranwal, and Ashootosh Mandpe

13.1 Introduction

The sustainability of the environment is seriously threatened by our existing habits, which include the indiscriminate use of chemicals, higher dependency on non-renewable energy sources, and unrestrained generation of trash from every conceivable industrial process. Research is being carried out to fulfill the rising need for energy and discover efficient raw material substitutes. In addition to meeting the world's energy needs, renewable energy technologies are crucial factors in lowering greenhouse gas emissions. Although numerous approaches exist to make gasoline from renewable resources, industrial production is still a work in progress. Due to the shortage of direct food sources and the concomitant development of biofuel, the supply of non-edible raw materials has a significant potential to produce biofuel. Also, the variety of microorganisms has dramatically increased with multifarious applications and are essential to many green processes and environment-friendly technologies, from biogeochemical phases to different industrial products. If employed correctly, microorganisms can significantly support sustainable development. The employment of green technologies, cleaner industrial methods, and the protection of natural resources are all mainstream global goals today. Unexpectedly, despite the enormous benefits of microbes in a variety of sustainability scenarios, they are frequently brushed aside in the conversation about implementing the Sustainable Development Goals (Cudjoe et al. 2022).

B. Sandeep · U. Baranwal · A. Mandpe (✉)
Department of Civil Engineering, Indian Institute of Technology Indore, Simrol, Indore, Madhya Pradesh 453 552, India
e-mail: as_mandpe@iiti.ac.in

K. Bala et al. (eds.), *Harnessing Microbial Potential for Multifarious Applications*, Energy, Environment, and Sustainability, https://doi.org/10.1007/978-981-97-1152-9_13

13.1.1 Background and Significance

In comparison to conventional fossil fuels, biofuels are renewable fuels made from biological sources that are more environmentally friendly. To combat climate change and lower greenhouse gas emissions, the production of biofuels is an essential aspect. Various feedstocks can be used to make biofuels, including plants, algae, and other organic sources. Due to its potential benefits, the production of biofuel through microbial sources has become one of the major attractions in recent years. Biofuels can be produced using microorganisms through several metabolic pathways. In comparison to traditional biofuel feedstocks, microbial biofuel production has various advantages, including better conversion efficiencies, faster growth rates, and fewer land and water requirements (Chaurasia et al. 2020).

The production of microbial biofuels is an environmentally friendly method of generating energy. It improves energy security and lessens reliance on limited fossil fuels. Additionally, it eases the strain on land use and lowers environmental pollution by using waste materials as feedstock. Carbon dioxide (CO_2) is released during the burning of conventional fossil fuels, but this is balanced by the CO_2 absorbed during the growth of the microbial feedstocks. As a result, the net emissions of greenhouse gases are considerably lower than those arising from the burning of fossil fuels, which helps in the measures being taken to counter climate change. Microbial biofuels do not rely on traditional monoculture crops, reducing the risk of biodiversity loss due to large-scale agricultural practices. This approach promotes more sustainable land use practices. This strategy encourages more environmentally friendly land use methods. Microbial biofuel production can open new doors for rural economic growth and employment opportunities. By transforming low-value or waste materials into useful energy products, it offers an efficient means of waste valorization. Biotechnology, synthetic biology, and genetic engineering research and development have been sparked by the goal of microbial biofuel generation. Improvements in these areas have wider applications in other industries in addition to helping biofuel manufacturing processes.

The creation of biofuels from microbial sources, often termed as the production of bioenergy or biofuel, has several significant goals and advantages. Sustainability, lower greenhouse gas emissions, biodegradability and pollution reduction, improved energy security, waste utilization, renewable chemicals, and technological advancements are the main goals of this chapter. Since it uses renewable resources, microbial-based biofuel production is regarded as a sustainable substitute for traditional fossil fuels. This helps to minimize reliance on finite fossil fuel supplies and lessen the damaging environmental consequences of fossil fuel extraction and combustion. In comparison to fossil fuels, microbial biofuels have the potential to lower greenhouse gas emissions. Biofuels from microbial sources recycle carbon dioxide recently captured from the atmosphere during the growth of biomass used as feedstock. Since most microbial-based biofuels are biodegradable, using them can lessen the amount of hazardous byproducts released into the environment. Organic waste, agricultural byproducts, and byproducts from other industries can all be used as feedstock in numerous microbial biofuel production processes. This recycling of garbage results in goods with added value while easing the environmental load of waste disposal. Research and technological advancements in a variety of sectors, including biotechnology, microbiology, metabolic engineering, and fermentation processes, are driven by the quest for microbial-based biofuels. Beyond biofuel production, these developments may have larger implications that promote science and stimulate the economy.

13.2 Biofuel Production from Microbial Sources: An Overview

The word "biofuels" describes chemical substances with additional energy that are created by biological reactions derived from the biomass of living organisms, such as microalgae and bacteria. With the ever-increasing population and to cope with their living standards, more energy is required. Biofuels are one alternative that may meet the world's energy requirements up to a major extent. Although fossil fuels have continued to be a significant source of energy, their consumption is unsustainable, and their burning has a severe impact on the environment. As a result, this issue could make it possible for biofuels to take the place of ecologically damaging fossil fuels.

13.2.1 Definition and Types of Biofuels

A fuel developed from biobased components is known as biofuel. Organic substances such as terrestrial plants, agricultural waste, and animal byproducts are examples of biobased materials. Because it is regenerative, sustainable, environmentally friendly, effective, and commercially viable, biofuel is a potential substitute for fossil fuels, where the farmlands provide the principal source of biomass.

13.2.1.1 Types of Biofuels

Biofuels are classified into four generations; third and fourth-generation biofuels are still in nascent stages, while first and second-generation biofuels are frequently used as energy sources.

First-Generation Biofuel This is the most common type of biofuel production and is often referred to as conventional biofuel. The first-generation biofuels include biodiesel, biogas, and alcohol, including ethanol, butanol, and propanol. The crops utilized to make bioethanol are corn, wheat, and sugarcane, while biodiesel is also made from rapeseed and soybean oil. The biomass (substrate) used as feedstock in the production of first-generation fuels comprises food resources for humans and animals, which in turn may result in food security problems. Besides, the production of first-generation biofuels requires a lot of land, water, and other resources. Thus, first-generation biofuels are usually not preferred to prevent the consequences, including food crisis and similar environmental issues (Chaurasia et al. 2020).

Second-Generation Biofuel These biofuels are made from lignocellulosic biomass (LCB), such as grasses and plant components, which contain cellulose, hemicellulose, and lignin. Biomass can also be made from agricultural waste products like rice husk and sugarcane garbage. Second-generation biofuel is made from LCB (i.e., biomass that consists of three prime constituents' cellulose, hemicellulose, and lignin) in a more environmentally friendly manner. Feedstocks for second-generation biofuels must first undergo pretreatment before enzyme hydrolysis to produce second-generation fuels.

Third-Generation Biofuel These are generated from algae. This generation's fuels are the high-quality yield of biofuels as compared to other generations. The fuels are biodiesel, bio-methanol, bioethanol, bio-butanol, gasoline, and vegetable oils, as shown in Fig. 13.1 below with different processes. The main advantage of algae fuels is that these fuels directly convert carbon emissions to usable fuel with negligible carbon emissions.

Fourth-Generation Biofuel These are made from photobiological solar fuels and electro-fuel, and they are produced by using non-arable land (Chaurasia et al. 2020). Microorganisms like *cyanobacteria* and *microalgae* that are modified using genetic engineering applications are used to create fourth-generation biofuels. The microbes are genetically altered to take in more CO_2 during photosynthesis, creating a man-made "carbon sink" and boosting the generation of biofuels. Fourth-generation

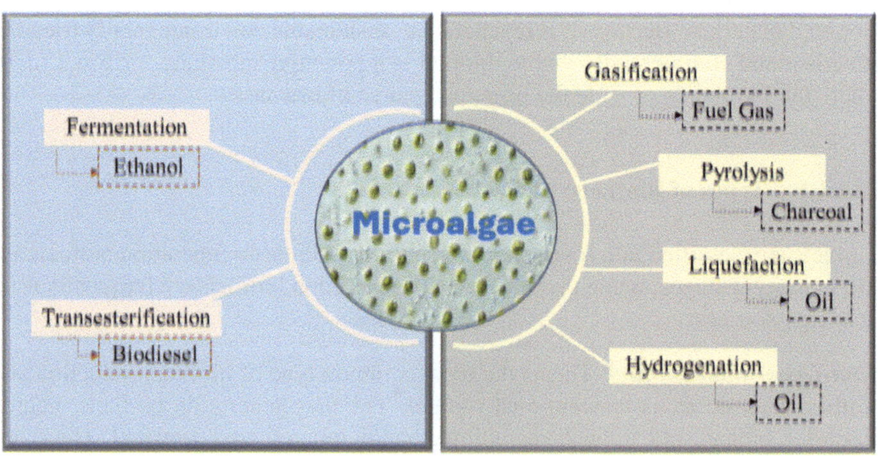

Fig. 13.1 Different pathways of microalgal biomass conversion to biofuels and their respective products

biofuels convert energy with the fewest possible steps, avoiding processing procedures. The environmental benefits of fourth-generation biofuels can be categorized as CO_2 assimilation, wastewater treatment, and a decrease in greenhouse gas emissions; however, additional research is still needed to determine how the gene modification process affects the environment.

13.2.2 Importance of Microbial Sources in Biofuel Production

Due to their high conversion efficiency of various feedstocks, ability to reduce greenhouse gas emissions by offering an emissions-free or low-carbon alternative to fossil fuels, and ability to thrive in unfavorable environments, microbial sources are of utmost importance in the production of biofuels. This reduces competition for arable land. Additionally, microbial activities can improve energy security, reduce waste, clean up contaminated areas, and boost regional economies. Microbial sources are a crucial component of the shift to ecologically friendly and sustainable biofuel production because of their renewable nature and ongoing research and development activities, which reduces our reliance on limited fossil fuel supplies and lessens environmental effects. Biofuel production from biomass relies heavily on microbial sources. Garbage products, including municipal garbage, food waste, and agricultural residue, can be used by microbes to produce biofuel. Reducing trash disposal issues, promoting sustainable practices, and addressing environmental concerns are all aided by the ability to turn garbage into useful energy sources. By releasing enzymes, microorganisms can break down complex organic compounds into simpler ones through different metabolic pathways. These enzymes assist in converting biomass into a variety of feedstocks for biofuels, lowering the cost of pretreatment procedures. Additionally, microbes are necessary for the fermentation process that turns simple carbohydrates into biofuels. While bacteria like *E. coli* or *Clostridium* can produce biofuels like butanol, yeasts are employed to make ethanol. Microorganisms are suited for large-scale biofuel production because they can efficiently ferment carbohydrates and other organic molecules. Microbes can be genetically modified to increase their capacity to produce biofuel. As a result, microbial sources are advantageous to produce biofuel.

13.2.3 Environmental Benefits of Microbial Biofuels

Daily population growth is causing an increase in energy demand that fossil fuels cannot effectively meet. Due to the increase in greenhouse gases caused by the usage of fossil fuels, climate change occurs. Biofuels have been a feasible alternative to fossil fuels for the following reasons:

- Biomass is used to create biofuels, a sustainable energy source.
- They are naturally non-toxic.
- They are environmentally benign because they are biodegradable, their combustion doesn't emit any CO_2, and they have excellent lubricating quality and low sulphur content (Chaurasia et al. 2020)

13.3 Advances in Microbial Biofuel Production

Our overwhelming reliance on fuels made from petroleum has drawn much public attention as one of the main global threats. To overcome these difficulties, an abundance of research has been done on the creation of sustainable fuels. As a result of recent advancements in synthetic biology and metabolic engineering technologies, bio-based platforms to produce enhanced biofuels using a range of microorganisms have been developed. Four primary metabolic pathways—the fatty acid synthesis (FAS) pathway, the isoprenoid pathway, the 2-keto acid pathway, and the reverse-oxidation pathway—have been the focus of advanced biofuel production strategies. Additionally, a viable alternative strategy for the manufacture of biofuel has lately gained attention is the polyketide synthesis pathway. This book chapter reviews current developments in the manufacture of three different forms of enhanced biofuels: biogas, ethanol, biodiesel, jet fuel, and gasoline. The creation of microbial strains producing improved fuels is focused on using synthetic biology and metabolic engineering techniques.

13.3.1 Selection and Engineering of Microorganisms for Biofuel Synthesis

Modern synthetic biology and metabolic engineering methods enable the development of heterologous metabolic routes in thoroughly studied microbial hosts, including *Escherichia coli* and *Saccharomyces cerevisiae*, to produce hydrocarbons from the entire sugar supplement of biomass (Joshi et al. 2023). Because of the thorough characterization of these hosts and the accessibility of numerous genetic engineering tools, it is possible to construct heterologous pathways that not only increase production titers and yield but also broaden the range of available biofuels.

13.3.2 Metabolic Engineering Strategies for Enhanced Biofuel Production

The fatty acid synthesis (FAS) pathway, the isoprenoid pathway, the two-keto acid pathway, and the reverse b-oxidation pathway have been the four main metabolic pathways that have been used most frequently in metabolic engineering strategies to produce advanced biofuels to date (Choi et al. 2014). Recent modifications have been made to the polyketide biosynthetic pathway, which is regulated by polyketide synthases (PKSs), to produce advanced biofuels, and the findings are encouraging (Joshi and Mishra 2022). Microorganisms can produce primary higher alcohols through either the fatty acid or amino acid pathway. The fermentation of clostridia is the best method for producing 1-butanol. The complexity of clostridia's metabolic controls and challenges with genetic modification have made metabolic engineering of clostridia difficult, but it has lately made progress. The fundamental goal of the metabolic engineering approach has been to increase 1-butanol production while selectively reducing acetate, butyrate, and acetone generation (Choi et al. 2014). Secondary alcohols have chemical properties different from those of primary alcohols due to the position of the hydroxyl group. Short-chain secondary alcohols are employed as solvents in a variety of industrial applications because they can dissolve both polar and nonpolar compounds. Petrochemical processes can produce secondary alcohols by hydrating alkenes or oxidizing ketones (Choi et al. 2020).

13.3.3 Genetic Modification Techniques in Microbial Biofuel Production

The efficient conversion of biomass into biofuels is made possible by genetic modification techniques, which are essential in the manufacture of microbial biofuels (Khoo et al. 2023). To produce microbial biofuel, many genetic modification strategies are employed. These include genome shuffles, promoter engineering, gene overexpression, metabolic engineering, and knockout mutations.

By changing the metabolic pathways and applying the metabolic engineering principles, the productivity of the biofuel manufacturing process can be enhanced. In order to avoid the hindrances, this technique uses the necessary enzymes in the pathways for effective biofuel synthesis. To remove undesirable metabolic pathways or lessen the creation of byproducts that block the production of biofuels, knockout mutations include the destruction of many genes in the microbial genome. Utilizing this method directs metabolic flux toward the synthesis of biofuels. Genes that are necessary for pathways to improve the synthesis of biofuels are introduced through gene overexpression. The same microbes or other microorganisms with related functions may be the source of these genes. DNA regions called promoters control the expression of genes. Researchers can regulate the expression of genes by changing their promoter regions, which enables fine-tuning metabolic pathways for efficient biofuel production.

13.4 Biofuel Production from Lignocellulosic Biomass

Plant biomass is known as LCB. Typically, it is divided into four categories: agricultural waste, hardwood, softwood, and herbaceous plants. Three substances make up lignocellulose: lignin, cellulose, and hemicellulose (Ahmed et al. 2022).

Cellulose: Cellulose, a plentiful organic polymer is present in plant cell walls, accounts for the bulk of the composition of LCB. Cellulose is made up of pyranose, a hexagonal-carbon ring of D-cellobiose units. Cellulose has a crystalline structure because of the conjugation of hydroxyl groups within pyranose rings. Cellulose polymer structures are significantly stable because of hydrogen bonding formed between the linkage of glucose ($C_6H_{10}O_5$) through β-glycosidic linkages.

Hemicellulose: In LCB, 5 to 30% of hemicellulose is present (Ahmed et al. 2022). Hemicellulose has an amorphous and branching structure. Hexoses and pentoses, two heteropolysaccharides, are the building blocks of hemicellulose. It is formed by extensively branched short chains of different forms of sugars. These short chains of various sugar molecules include pentoses, hexoses, and uronic acids (Akash 2015).

Lignin: Lignin is an aromatic chemical with a three-dimensional framework and the third major component of LCB. It is an interlinked phenol polymer with dynamically variable links of phenyl propane units that have had their "methoxy-" and "hydroxyl-" substituents removed. Its structure is composed of a non-carbohydrate polyphenolic component of plant material that helps in binding cell walls together (Wang et al. 2022). It forms a random network, which leads to a non-crystalline structure. Their structures are highly dependent on plant species and also on the presence of 3 main monolignols: coniferyl alcohol, p-coumaryl alcohol, and sinapyl alcohol.

The ratio of the components is influenced by the change in plant parts like leaves, fruits, and so on. Typically, there are three kinds of LCB: primary, secondary, and tertiary.

Sugarcane and short-rotation plantations are primary suppliers. Straw, rice husks, and bagasse are examples of secondary sources. The organic waste component of solid waste and the leftover byproducts of various operations, such as sewage treatment, are examples of inexpensive tertiary sources (Preethi et al. 2021).

13.4.1 Production of Biofuels Using Lignocellulosic Biomass

There are several products that can be made from LCB that have added value, such as bio-oil, biogas, and additives. LCB is the most prevalent type of biomass used to make biofuels. It helps to achieve carbon neutrality by reducing greenhouse gases. LCB can be biorefined to produce high-value compounds and a large quantity of liquid fuels to supply the world's demand for energy. In addition to helping to produce biofuels, it also contributes to the production of heat, electricity, chemicals, and biomaterials.

This makes it a desirable energy source. The LCB biorefinery has several technological platforms available. To produce fuels like bioethanol, sugar platforms primarily concentrate on the microbial breakdown of sugars from lignocellulosic-rich components. The thermochemical platform intends to use gasification or pyrolysis to transform biomass solid phase into gaseous and liquid fuels. The biogas platform produces biomethane through anaerobic digestion techniques. Corn, canola, plant oils from palm, and soyabean serve as the platform's main feedstock sources for carbon. For the manufacturing of biodiesel, fatty acid methyl esters from animal or vegetable fat are obtained through transesterification.

13.4.2 Microbial Conversion Processes for Lignocellulosic Feedstocks

Biomass can be converted into biofuels using two processes, which are discussed below (Joshi et al. 2017).

13.4.2.1 Thermochemical Process

Biomass is converted to biofuels by a thermochemical process in the presence of heat and chemicals. Pyrolysis, direct combustion, hydrothermal liquefaction (HTL), and gasification are examples of thermochemical processes. Compared to biological processes, thermochemical processes are carried out swiftly. In contrast to conventional processes that turn biomass into oils, which are then transformed into biofuels, direct burning transforms biomass into heat energy (Awasthi et al. 2023).

13.4.2.2 Biochemical Process

Microorganisms are involved in biochemical processes that transform biomass into different bioproducts and biofuels. Anaerobic digestion, fermentation, and enzymatic hydrolysis are the processes, as shown below in Fig. 13.2, that are used to turn biomass into biofuels. Thus, prior to processing, the biomass needs pretreatment to transform complex structures into products that can be easily converted to biofuels through biochemical processes (Rodionova et al. 2022).

Pretreatment of Biomass The complex chemical structure of lignin, the crystallinity of the cellulose, and the proportion of exposed surface area that is open to enzymatic action all play a role in the recalcitrance of the cell wall in LCB. The main barrier to the adoption of biomass conversion technologies in terms of economic viability was the conversion of recalcitrant chemicals in LCB to transitional compounds. The primary objective of pretreatment is to alter the complex structure of cellulose, hemicellulose, and lignin into monomeric compounds, which results in the elimination of lignin, a

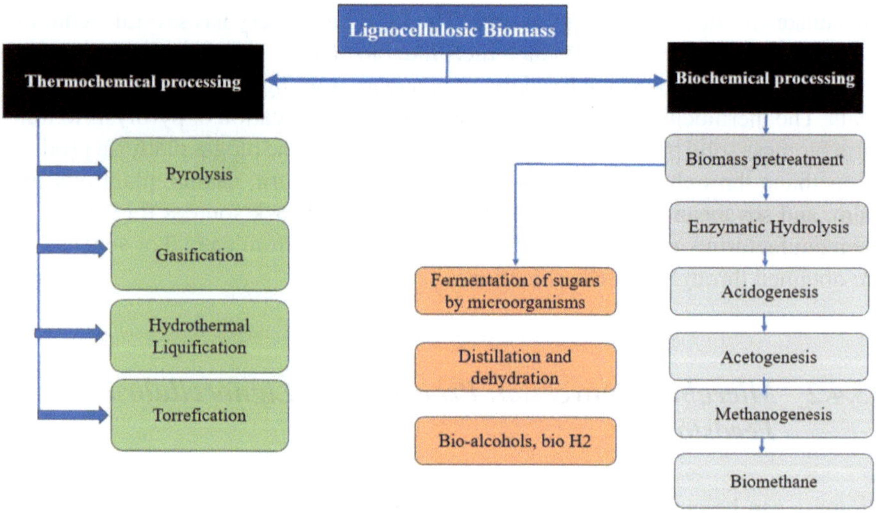

Fig. 13.2 Different pathways of lignocellulosic biomass conversion into biofuels

reduction in cellulose crystallinity, and an increase in permeability and surface area. The purpose of pretreatment is to eliminate the loss of pentose sugars, particularly those produced during hemicellulose degradation, and to produce easily metabolized solid intermediary substrates with higher sugar yields. Additionally, it aims to lessen inhibitors that could prevent fermentation from continuing. The extracted lignin can be transformed into a variety of chemical compounds. Pretreatment can greatly lower reactor heating and power costs because it reduces the need for additional processing of complicated LCB. These pretreatment techniques come in a variety of forms and are divided into chemical, physical, and biological routes.

Biological Route While hemicellulose and lignin are broken down by numerous bacteria, cellulose is left unchanged for further processing. These applications often include fungi brown rot, soft rot, and white rot. Fungi enzymes, including laccases, lignin peroxidases, and xylanases, degrade hemicellulose and lignin. Due to their relatively high efficacy as compared to other fungal species, white-rot fungi are most frequently used. Pretreatment also makes use of bacteria. Low energy requirements and a minimal need for chemicals are the main benefits of biological preparation. The procedure is environmentally benign because it doesn't produce any hazardous byproducts and doesn't need any chemical recycling (Ab Rasid et al. 2021).

Enzymatic Hydrolysis Further, the biomass is treated through enzymatic hydrolysis. The goal of this procedure is to hydrolyze the polysaccharides in lignocellulosic biomass to provide fermentable sugars for biofuel generation. Enzymatic hydrolysis of lignocellulosic biomass yields glucose and xylose as its two primary byproducts. Hydrolysis enzymes are produced by a variety of microorganisms. The enzymes cellulase and hemicellulase break down cellulose into monosaccharides, pectinase breaks down pectin, and laccase breaks down lignin. Endoglucanase breaks internal glycosidic linkages in the pretreatment of amorphous cellulose, depolymerizing the structure and liberating oligosaccharides. To release cellobiose units, exoglucanase or cellobiohydrolase targets the terminals of oligosaccharides. Cellobiose units are then digested by glucosidase into glucose molecules. To fully hydrolyze cellulose, it is important to use a mixture of the different cellulases because they each serve a particular purpose. The efficiency of hydrolysis is also increased by the interplay of cellulases and hemicellulases.

13.4.3 Challenges and Opportunities in Lignocellulosic Biofuel Production

Standardizing the conversion of LCB is not practically feasible. This is due to the diverse origins and composition of LCB. Second, different methods were employed to effectively turn lignocellulosic biomass into biofuel. Currently, the research in this domain is mostly concentrated on lab-scale production, while industrial-scale production still seems challenging. Therefore, strategies for creating industrial-scale biofuel production setups are needed to satisfy the existing fuel demand. It must be employed in actual driving situations for biofuels to reach their full potential. The main barrier to its more widespread applications is its exorbitant price. The important roles in lowering the cost of biofuels will be process improvement, cultivation media optimization, and the identification of appropriate microbial strains. Cost reduction for biofuels depends on several factors, including labor costs, equipment costs, and waste management. Consequently, a comprehensive evaluation of these factors is needed. On the other hand, biological conversion processes have slow reaction rates, which causes the costs of the subsequent steps in the processing chain to rise.

It may require conducting numerous sets of experiments to understand and analyze a combination of pretreatment and phase-separated pretreatment that can fully address these problems and inhibit recalcitrant chemicals. Although pretreatment methods including steam explosion, ammonium fiber explosion (AFEX) (Rodionova et al. 2022), and biological pretreatment showed encouraging results, it is still important to conduct life cycle and techno-economic analyses to see whether they are feasible for industrial-scale applications.

13.5 Microbial Bioethanol Production

Bioethanol, also known as ethyl alcohol, is composed chemically of carbon, hydrogen, and oxygen. It contains an alkyl radical (C_2H_5OH) with an attached hydroxyl group. It is a biofuel made from renewable biomass sources. Due to its greater flammability and lower vapor pressure than petrol, ethanol produces fewer pollutants. Utilizing local raw materials and using bioethanol both contribute to lower carbon emissions and, through photosynthesis, lower atmospheric carbon levels. Biochemical conversion of fermentable carbohydrates by yeasts or bacteria is required for biomass production. Before fermentation, biomass must be processed through a variety of steps, such as milling, cooking, grinding, enzymatic hydrolysis, etc., to produce fermentable sugars.

The choice of raw materials is crucial to the effective and affordable manufacture of bioethanol. The raw material should yield large produce of fermentable sugars in a smaller area. Additionally, the raw material shouldn't aid the development of chemicals that hinder the growth of fermentable yeast. The synthesis of bioethanol can utilize a variety of basic sources. Saccharides or sugar-rich raw materials which has a lot of sugar content. For example, fruit juices, sugarcane, beetroot, etc. Bioethanol is produced through a process that involves the extraction of sugars through milling or diffusion, cleaning, fermentation, and dehydration.

It is made of raw components that are rich in starch. For example, cassava, rice, potatoes, and barley. Starch is processed thermally and enzymatically for saccharification. Following the conversion of starch to fermentable sugars, a procedure similar to sugar-rich raw materials is carried out further. Raw materials used for making cellulose include bagasse, eucalyptus, and rice husk.

13.5.1 Fermentation Processes for Bioethanol Production

After hydrolysis, enzymes produced by bacteria, yeast, and other microbes break down the fermentable sugars and consume them. When consumed in anaerobic conditions, carbohydrates cause the emission of organic acids, fumes, or alcohol. To reduce the release of hazardous pollutants and increase the yield of bioethanol, the effectiveness of fermentation is reliant on hydrolysis, the type of substrate chosen, the kind of microorganism chosen, and similar other factors; several fermentation techniques are used based on pros and cons of the respected fermentation process as presented in Table 13.1 below.

The biochemical modification of fermentable carbohydrates using yeasts constitutes the bioethanol production process. The preliminary phase, which is also known as pre-fermentation, followed by turbulent phase, and post-fermentation, which is also known as complementing phase, are the three distinct stages of industrial fermentation.

Bacterial fermentation for bioethanol production entails converting sugars into ethanol using specific strains of bacteria, typically from the genus Zymomonas or certain strains of Escherichia coli. This method is less common than yeast fermentation, but it has benefits, especially in certain industrial applications.

The lag period of yeast cells following substrate injection distinguishes the pre-fermentation phase. The temperature rises slowly and gradually throughout this period, and very little carbon dioxide is released. Getting a large number of cells with the most fermentation efficiency is the goal of this stage of fermentation. Further,

Table 13.1 Comparative analysis of fermentation process (SSF, SHF, SSCF, SHCF, CBP, respectively)

Fermentation process	Advantages	Disadvantages
SSF (Saccharification and fermentation)	• Foreign contamination is less • Less inhibitory effects • Low cost	• Either hydrolysis or fermentation can be performed under optimal conditions • Difficulty in process control
SHF (Separate hydrolysis and fermentation)	• Both hydrolysis and fermentation take place at optimum conditions • Low quantity of enzyme input • High ethanol yield	• Increased contamination • Inhibitory effects
SSCF (Simultaneous saccharification and combined fermentation)	• Shorter process time • High bioethanol yield • Less contamination risks	• High enzyme load • Either hydrolysis or fermentation can be performed under optimal conditions
SHCF (Separate hydrolysis and co- fermentation)	• High bioethanol yield • Both hydrolysis and fermentation take place at optimum conditions	• High enzyme load • Inhibitory effects • High contamination risk
CBP (Consolidated bioprocessing)	• Energy efficient • Cost-effective	• Lack of suitable organisms • Process control is difficult

the turbulent phase, which lasts for roughly 10 h, occurs. The quick temperature rise, significant emission of carbon dioxide, the rise in acidity, the fall in substrate density brought on by the creation of alcohol, and the corresponding devouring of sugars are the characteristics of this step. The continuing release of carbon dioxide and the gradual and consistent drop in temperature define the complementary phase. A fermented product called wine has a complex blend of liquid, solid, and gaseous components. Additionally, it should not contain large quantities of sugar. To inhibit the contamination levels from building up above a specific level, this process needs a controlled environment. It is well known that microbial contamination hinders the process's ability to operate at its best (Singh et al. 2022).

13.5.2 Microbial Strains for Efficient Bioethanol Production

The most crucial microorganisms for fermentation-based bioethanol generation are yeasts. Yeasts are unicellular fungi that have hard cell walls, egg-like shapes, elliptical or round-shaped cells, and identical eukaryotic organelles as similar higher cells. A small percentage of yeasts in nature are members of the *basidiomycetes phylu*m, while the majority are ascomycetes and deuteromycetes. Many yeasts, especially *Saccharomyces, Schizosaccharomyes*, and *Pichia*, are capable of producing ethanol. The most popular yeast for fermentation operations is *Saccharomyces cerevisiae*, which has properties such as a high capacity to hydrolyze sucrose, selective cytoplasmic membrane permeability, and strong resilience to stress conditions. Bioethanol is made through fermentation using either a specific yeast strain that has been added on purpose or through natural selection. In the fermentation process, yeasts can endure the physiochemical conditions and microbial populations present while dominating ecological interactions. The better-adapted strains consequently survive and prevail due to natural selection. The fact that these yeasts often rearrange their chromosomes and change their genetic makeup gives them an advantage over other yeasts in terms of survival, allowing them to regulate the fermentation process. When juxtaposed to non-selected yeast strains, the addition of purposely introduced yeasts in the fermentation stage may improve the process efficiency as they may have better resilience to stress variables, for example, acidity, ethanol concentration, osmolality, and temperature (Prasad et al. 2019).

Effective fermentation also makes the manufacturing of ethanol more affordable. It is less expensive and simpler to produce ethanol via bacterial fermentation. In industrial manufacturing, bacteria such as *Corynebacterium glutamicum*, and *Zymomonas mobilis* are frequently employed to break down carbohydrates for ethanol production. According to reports, *Clostridium thermocellum* has ethanologenetic and cellulolytic properties. Potential strains for improved ethanol production include mesophilic bacteria; For example, *P. marinum, P. thermoaerophilus, P. thermophilus*, and *P. tezpurensis* (Brexó and Sant'Ana 2017).

13.5.3 Advances in Bioethanol Yield Optimization

Toxic substances, such as furfural, 5-hydroxymethyl furfural (HMF), and soluble phenolics, are frequently generated during the pretreatment and saccharification processes, which is bad for biomass fermentation. These substances reduce the rate of microbial advancement during the fermentation stage (Luthra et al. 2015). Toxic chemical formation and buildup during the pretreatment process can be decreased by washing the residue with water (Aulitto et al. 2019). Additionally, phenotypic modifications for enhanced fermentation to produce biofuel depend on factors, including oxygen levels and substrate accessibility. It is also reported that *Corynebacterium glutamicum* generates terpene chemicals, including pinene and beta-carotene. Also, the fermenters make biofuel by fermenting microorganisms like *E. coli.* (Novia et al. 2023).

The fermentation process can be improved by pre-adapting the fermenting microbe during seed cultivation (Aulitto et al. 2019). While non-biological enhancements such as thermochemical and physical co-treatment can enhance the fermentation of lignocellulose biomass, a novel approach involves genetically modifying plants to have improved solubilization during the process. According to Holwerda et al. (2019) transgenic switchgrass can be better solubilized when using the biocatalyst. Another approach to improve fermentation has been the development of designer cellulosomes by realigning the dockerin module from different mesophilic and thermophilic free enzymes (Rastogi and Shrivastava 2017).

13.6 Microbial Biodiesel Production

Fatty acid methyl ester (FAME), which is produced when fatty acids are transesterified, makes up the majority of biodiesel, a significant renewable energy source. It is a good substitute for diesel since it emits fewer greenhouse gases, is renewable, non-toxic, environmentally friendly, biodegradable, and contains less sulphur. The pollution brought on by the burning of petroleum can be controlled because of its qualities. The fundamental benefit of biodiesel is that it can be utilized in current engines without requiring any modifications. Transesterification is a technique used to create biodiesel from biomass sources such as animal and vegetable fats, microbial lipids, and non-edible oils. A shortage of economical and adequate raw materials, longer lifecycles, unpleasant processing, low economic advantages, and price implications on agricultural goods all impede biodiesel manufacturing; 75% of the overall cost of producing biodiesel is concerning to the raw materials.

Based on the advancement of feedstocks and their effectiveness, it is thought that biodiesel may be classified into three generations. The price of agricultural goods and the availability of food are impacted by the first generation of biodiesel, which is made from vegetable oils. Second-generation biodiesel, which cannot be produced on a big scale, is mostly made from non-edible oils like animal fat and waste oil. Microbial lipids from oleaginous microorganisms are used to produce third-generation biodiesel and are regarded as an ideal fuel for the process. Microbial lipids are more suited for large-scale biodiesel synthesis since they do not occupy cultivable soils. As a result, microbial lipids were shown to be the best raw material for making biodiesel.

Lignocellulose and cellulose are degraded into monosaccharides, which bacteria can utilize to make lipids for biodiesel production. Therefore, the synthesis of biodiesel from lignocellulose by oleaginous microorganisms using single-cell lipids has major implications for sustainable development and can help to alleviate the environmental crisis. The large-scale development of lignocellulosic production by microbes is constrained by certain conditions. These include a low microbial lipid conversion rate, the presence of furfurals, phenols, and other inhibitors in hydrolyzed lignocellulosic biomass, and a low carbon utilization rate. To investigate oil-producing yeast that uses lignocellulosic materials to make lipids, it is crucial to obtain high-quality microorganisms that can utilize carbon sources and manage the numerous inhibitors. Additionally, procedures, including lignocellulose pretreatment, identifying the inhibitors produced, and enzymatic hydrolysis, are carried out as stated in the lignocellulose portion (Wang et al. 2021).

13.6.1 Lipid Accumulating Microorganisms for Biodiesel Production

Oleaginous microorganisms are those that synthesize and accumulate lipids over 20% of their weight in dry form. Microbial lipids, sometimes referred to as single-cell oils (SCOs), are the oils generated when oleaginous microbes convert carbohydrates to fatty acids under specific circumstances and store them in the cell. In the production of liquid-based biofuels, three forms of lipids such as polyhydroxy butyrate (PHB), triacylglycerols (TAGs), and Wax Esters (WEs) are used (Hwangbo and Chu 2020). TAGs are the most common type of SCOs among them. Yeasts, algae, and bacteria are only a few examples of diverse and widely distributed oleaginous microorganisms. TAGs are straightforward benign lipids made up of three long-chain fatty acids connected by ester bonds to one glycerol, as shown in Fig. 13.3. TAGs have been utilized as the main feedstock in the transesterification process to produce biodiesel in the manner of fatty acid methyl esters (FAMs) or fatty acid ethyl esters (FEEs). The production of biodiesel uses feedstocks like WEs and PHAs (Khoo et al. 2023).

Several species, including bacteria, fungi, and algae, can produce various kinds of biolipids. Oil content present in different microbial species are presented in Table 13.2. Since algae can be grown with water of average quality and use CO_2 as a source of carbon, algae TAGs are regarded as sustainable feedstock to produce biodiesel. Various restrictions and difficulties exist, including poor cell density, continuous light input, ineffective and expensive cell harvesting, and lipid extraction. The cultivation of bacterial strains can solve these problems and overcome these restrictions since they grow quickly on cheap carbon sources like organic waste. These bacterial strains can produce more TAG while maintaining high cell densities. Bacterial strains in this procedure can reduce carbon content since they are simple to grow and rapidly increase in lipid content while reducing carbon content. Bacterial strains are regarded as the best feedstock to lower the cost of producing lipid-based biofuels due to their ease of cultivation and rapid growth of high lipid content with decreasing carbon content.

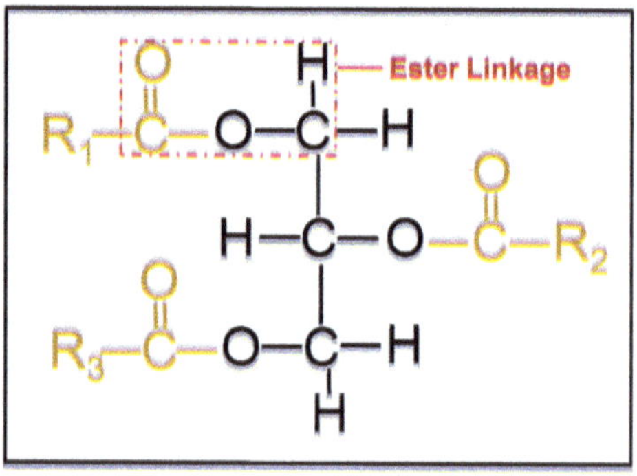

Fig. 13.3 Chemical structure of Triacylglycerols (TAG)

Table 13.2 List of various microbial species and oil content present in them

Yeast		Algae		Fungi		Bacteria	
Microbial species	Oil content (%)	Microbial species	Oil content (%)	Microbial species	Oil content (%)	Microbial species	Oil content (%)
Rhodotorula glutinis	40	*Scenedesmus dimorphus*	35.5	*Fusarium solani*	34.5	*Rhodococcus opacus PD 630*	42.1
Naganishia liquefaciens	55.7	*Chlorella vulgaris*	36.4	*Aspergillus awamori*	36.4	*Rhodococcus rhodochrous*	55
Rhodosporidium fluviale	65.2	*Amphiprora sp*	41.5	*Mortierella isabelline*	50.5	*Rhodococcus jostii RHA1*	50
Rhodosporidium toruloides	65.2	*Ankistrodesmus falcatus*	53				
Wang et al. (2021)		Arutselvan et al. (2021)		López-Fernández et al. (2021)		Wang et al. (2021)	

13.6.2 Cultivation and Extraction Techniques for Microbial Lipids

Due to expensive procedures like cell gathering and lipid separation, the cost of producing biofuel is significant. A significant obstacle is the extraction of lipids from microorganisms, which consumes half of the energy needed to produce biodiesel. The two most popular methods for cultivating microalgae or cyanobacteria are closed photobioreactors and open raceway ponds (Hwangbo and Chu 2020).

Due to their low cost, simple operation, and low maintenance requirements, open raceway ponds are frequently employed in the industrial production of *microalgae* or *cyanobacteria.* Paddle wheels and supplies of carbon dioxide and nutrients are used in open raceway ponds. Wastewater is now employed as a medium to reduce the need for nutrients. However, due to the presence of harmful substances in wastewater, the growth of *cyanobacteria* and *microalgae* is hindered. Besides, it also gets affected due to contamination, competition with other microbes, and maintaining strains that produce larger amounts of TGA. Closed devices called photobioreactors make it possible to cultivate both indoors and outdoors while lowering the risks of infestation. Due to its costly operation and maintenance, it is not cost-effective. There are several extraction techniques, including chemical, physical, and biological ones. Cell disruption-assisted solvent extraction and solvent separation of biolipids from intact cells are carried out here.

Biolipids are recovered from entire cells using an organic solvent mixture that includes both polar and low-polarity solvents. It is important to separate neutral and non-neutral lipids from lipid molecules that aren't granular using the polar and low-polarity solvents in the mixture. Two popular techniques for lipid extraction are:

13.6.2.1 The Folch Method

The Folch method is one of the most popular techniques for lipid extraction. Folch extracted lipids from animal tissues utilizing a chloroform/methanol (2:1, v/v) mixture. To improve the extraction efficiency of neutral lipids, this approach uses chloroform as a low-polarity solvent and methanol as a polar solvent.

13.6.2.2 The Bligh and Dyer Approach

Chloroform, methanol, and water (2:1:0.8, v/v/v) were used by Bligh and Dyer to extract and purify lipids. Wet tissue samples can benefit most from the creation of a distinct biphasic system by adding water to the extraction mixture.

There are different advanced lipid extraction techniques like the ultrasonic assisted method, deep eutectic solvents mechanism, ionic liquid mechanism, and Biobased solvent techniques (Vasistha et al. 2021) Pretreatment with ultrasound. This method introduces sound waves with frequencies greater than 20 kHz into the culture medium, resulting in a process of alternating compression and rarefaction, which are parts with high and low pressure, respectively. The goal of this method is to disrupt cell cytoplasm and release lipid molecules. The process of cavitating cytoplasm with ultrasonic waves is one of the benefits of using this method (Gorgich et al. 2020).

The mechanism of DES solvents is derived from the high polarity characteristics of water and their hydrogen bonding extraction abilities. In most cases, organic salts (such as chloride and choline chloride) serve as hydrogen bond acceptors, while organic solvents (such as amides, amines, alcohols, and carboxylic acids) serve as hydrogen bond donors. Because microalgae cell walls are made up of -cellulose and hemicellulose, the combination of hydrogen bond acceptor/donor in DESs results in an H-bonding interaction that disrupts the cellulose structure made up of hydroxyl (-OH) groups. Aqueous deep eutectic solvents (ADES) have been discovered to improve lipid extraction from microalgae (Zhang et al. 2012) Ionic liquids are organic salts that exist as liquids at a wide range of temperatures, typically ranging from 0 to 140 °C. They have the ability to disrupt the cell structure of wet microalgal biomass, which improves co-solvent access to lipids or promotes lipid autopartitioning (Khoo et al. 2021).

13.6.3 Challenges in Scaling up Microbial Biodiesel Production

The high cost of lipid extraction restricts the industrial production of lipid-based biodiesel. The anticipated lowest production cost of biodiesel generated from microalgal lipids (3.96 $/L), according to Pardo-Cárdenas et al. (2013), is still much higher than that of biodiesel derived from plant-based lipids, such as those made from oil palm (0.68 $/L) or sunflower (0.62 $/L). However, the costs of crucial elements like the type of feedstock and biolipid yield are not considered in this study. Since *Rhodococcus* strains may grow to a high cell density (about five times greater than that of *oleaginous microalgae*), using bacterial lipids from oleaginous *Rhodococcus* strains might lower the cost of producing biodiesel to 0.79 $/L (Lin et al. 2011).

Solvent extraction is frequently utilized in industry and these extraction practices account for a significant portion of the total operational cost. Additionally, solvent-based lipid extraction requires around 50% of the total energy used to produce algae biodiesel (Dassey et al. 2014). Table 13.3 lists the various extraction techniques with a brief overview and comparison. To use less solvent, newer techniques (Choi et al. 2014) that combine cell disruption with solvent extraction, including ultrasound or chemically assisted extraction, have been developed. However, the high costs of these approaches make their implementation a bit challenging in the sector. Either greater energy requirements for the ultrasonication stages or higher detergent/surfactant expenses are the reasons for the high costs of these treatments. The fact that the bulk of these methods have not been applied regularly to extract bacterial biolipids is not surprising. Phage-based or biolipid secretion procedures can be beneficial to get past the issues of elevated cost and higher energy demand for solvent extraction. The negative effects of extraction solvents on the environment and human health may be mitigated by using improved biological extraction. Recently communicated phage-based separation techniques and biolipid secretion techniques offer interesting alternatives to ultrasound-assisted methods because they do not require any extra high-power energy inputs for cell destruction, do not raise any concerns about solvent toxicity, and do not require subsequent downstream processes.

13.7 Biogas Production from Microbial Sources

Biogas is a renewable biofuel comprised of 55–75% methane (CH_4), 25–45% carbon dioxide (CO_2), hydrogen (H_2), and traces of other gases including nitrogen. Both hydrogen and methane are flammable gases. Purification of biogas is necessary to boost its calorific value by removing other gases. Fuel for many uses is made from purified biogas. Anaerobic digestion of organic materials with heterogeneous microbial populations existing in various groups results in the production of biogas. Anaerobic digestion is divided into several stages by each microbial community group, which produces the desired end product from the starting substrate. Anaerobic digestion depends heavily on the selection of the right microorganisms (Cudjoe et al. 2022) As it influences the degradation of organic wastes, conversion rates are likely to be impacted. Therefore, careful microbial consortia selection improves biogas generation efficiency. The pH, hydraulic retention time (HRT), temperature, C/N ratio, organic loading rate, and particle size are some of the variables that affect how efficiently biogas plants produce gas (Zhang et al. 2022).

13.7.1 Anaerobic Digestion Process for Biogas Production

Anaerobic digestion is among the most vital techniques for the production of biogas from food waste. The anaerobic digestion process consists of the following four steps: hydrolysis, acidogenesis, acetogenesis, and methanogenesis (Raja Ram and Nikhil 2022). By using extracellular enzymes (lipase, cellulase, amylase, xylanase, etc.), hydrolysis is the process of converting heterogeneous combinations of organic matter, which is made of polysaccharides (such as lignocellulosic compounds), lipids, and proteins to their monomeric forms. Here, complicated lipids are transformed into glycerol and fatty acids, proteins are transformed into simple sugars, and polysaccharides are transformed into amino acids. Converting hydrolyzed material into short-chain fatty acids includes acidogenesis and other secondary-stage steps by which carbon dioxide, ethanol, hydrogen, lactate, and other products are carried out.

Table 13.3 Various extraction techniques with a brief overview and comparison

Growth mode	Algae species	Organic substrate	Biomass growth	Reference
Heterotrophic	*Asterarcys sp. SCS-1881*	10 g/L glucose	–	Li et al. (2020)
	Auxenochlorella protothecoides (SAG 211- 7a)	20 g/L glucose	5.65 g/L	Patel et al. (2018)
	Chlorella sp.	0.5 g/L of glycerol	5.65 g/L	Poddar et al. (2018)
	Desmodesmus sp. WC08	67 mmol/L of glucose 67 mmol/L of sodium acetate	0.58 g/L 0.38 g/L	Zhao et al. (2016)
	Haematococcus pluvialis (SCCAP K-0084)	1.33 g/L of ribose 1.33 g/L of gluconate 1.33 g/L of sodium acetate	– – –	Pang et al. (2019)
	Monodus subterraneus (ATCC30593	5 g/L of pure glycerol	1.52 mg/L/	Hu et al. (2020)
	Nannochloropsis salina	0. 5 g/L of glycerol	0.22 g/L	Poddar et al. (2018)
	Scenedesmus obliquus KGE17	500 mg/L of sodium acetate	0.68/d	Choi et al. (2019)
	Tribonema minus (SAG 880- 3)	20 g/L glucose	14.3 g/L	Wang et al. (2018)
Mixotrophic	*Asterarcys sp. SCS-1881*	10 g/L glucose		Li et al. (2020)
	Asteracys sp.	500 mg/L glucose	4.35 g/L (900 μmol m−2 s − 1) 0.08 g/L (100 μmol m−2 s − 1	Agarwal et al. (2019)
	Chlamydomonas sp.	5% of vinasse	0.010 g/L	Candido and Lombardi (2020)
	Chlorella sp.	0.10 mol/L of acetate 0.10 mol/L of glucose 0.10 mol/L of glycerol	5.1 g/L 6.25 g/L 3.52 g/L	Wang et al. (2020)

(continued)

Table 13.3 (continued)

Growth mode	Algae species	Organic substrate	Biomass growth	Reference
	Chlorella sp.	0.5 g/L of glycerol	0.37 g/L	Poddar et al. (2018)
	Desmodesmus sp. WC08	67 mmol/L of glucose 67 mmol/L of sodium acetate 67 mmol/L of glycerol	0.79 g/L 1.01 g/L 0.45 g/L	Zhao et al. (2016)
	Nannochloropsis salina	0.5 g/L of glycerol	0.42 g/L	Poddar et al. (2018)
Phototrophic	Asterarcys sp. SCS-1881	No addition of glucose	0.75 g/L	Li et al. (2020)
	Chlorella sp.	0.50 g/L of glycerol	0.22 g/L	Poddar et al. (2018)
	Desmodesmus sp. WC08	67 mmol/L of glucose 67 mmol/L of sodium acetate 67 mmol/L of glycerol	1.83 g/L 1.83 g/L 1.83 g/L	Zhao et al. (2016)
	Haematococcus pluvialis (SCCAP K-0084)	1.33 g/L of ribose 1.33 g/L of gluconate 1.33 g/L of sodium acetate	0.7 g/L 0.7 g/L 0.7 g/L	Hu et al. (2020)
	Monodus subterraneus (ATCC30593)	5.00 g/L of pure glycerol (with aeration) 5.00 g/L of pure glycerol (no aeration)	42.21 mg/L/day 9.40 mg/L/day	Poddar et al. (2018)
	Nannochloropsis salina	0.5 g/L of glycerol	0.34 g/L	Poddar et al. (2018)
	Nannochloropsis sp. BR2	5 g/L of sugarcane bagasse	0.051 g/L	Menegol et al. (2019)
	Scenedesmus dimorphus NT8c	5 g/L of sugarcane baggase	96.44 mg/L/d	Manzoor et al. (2020)

Acetic acid is created through the process of acetogenesis from organic acids through fermentation and the reduction of carbon dioxide. Acetogens are microorganisms that produce acetate. Homo-acetogens and acetate acetogens are the two categories of acetogens. Homoacetogens are strict anaerobes that use the reductive acetyl-CoA, also referred to as the Wood-Ljungdahl pathway, for energy storage and synthesis of acetyl-CoA and cellular carbon from CO_2. The key enzyme involved is acetyl-CoA synthetase. Propionic acid, butyric acid, and alcohols (like ethanol) that are produced in the acidogenesis step are oxidized into acetic acid by acetate acetogens like propionic acid is fermented by *Syntrophobacter wolinii*, and butyric acid is fermented by *Syntrophobacter wolinii* (Raja Ram and Nikhil 2022).

Methanogenesis, the final stage, is accomplished through the reduction of carbon dioxide and hydrogen in the hydrogenotrophic route and acetoclastic fermentation of acetate. Archaea and bacteria are present during the process. Acetoclastic methanogens are those that produce methane by acetoclastic fermentation, and hydrogenotrophic methanogens are those that produce methane through hydrogenotrophic reduction. By reducing one-carbon substances like methanol and formate methylamine, the methylotrophic methanogens produce methane. The two-phase anaerobic digestion procedure might result in increased methane production. However, the co-digestion of waste from food may help to boost energy production, thereby also enhancing the viability of the anaerobic process of digestion. The hydrolysis of food leftovers may be accelerated using a variety of pre-treatment processes, which will increase methane generation (Al-Wahaibi et al. 2020).

13.7.2 Microbial Consortia and Biogas Yield Optimization

The microbial community, which is made up of three types of microorganisms—hydrolyzing and fermentative bacteria, obligatory acetogenic bacteria producing hydrogen, and methanogenic archaea—plays an essential part in anaerobic digestion. Different types of microorganisms are used at different stages of anaerobic digestion. *Bacteroides sp., Butyrivibrio sp., Fusobacterium sp., Micrococci sp., Selenomonas sp.,* and *Streptococcus sp.* are the most often utilized microorganisms in the first way of anaerobic digestion, which is hydrolysis (Merlin Christy et al. 2014). *Eschericia coli, bacteroids, acetivibrio, Clostridium kluyveri, bacellus, cereus,* and *thermotoga maritima* are among the regularly employed acidogenic microorganisms. Acetogens are microorganisms that produce acetate. In the process of acidogenesis, acetogens can create acetic acid from propionic acid, butyric acid, and alcohol. *Syntrophobacter wolinii* ferments propionic acid. *Strophomonas wolfei* ferments butyric acid. By fermenting glucose, *M. thermoacetica.* Methyl co-enzyme m reductase, which is found in all methanogens, catalyzes the crucial steps in methanogenesis. *Methanosaeta, methanolobus, methanococcoides,* and *methanohalophilus* are a few examples of methanogens (Vijayakumar et al. 2022).

One of the promising methods for enhancing biogas production is the addition of organic and inorganic additives to the feedstock in the digestor. The substrate determines the type of additive to be used.

13.7.2.1 Organic Additives

The term "additives" refers to substances that can be found in nature. It has been found that adding leaf powder from eucalyptus, gulmohar, and dalbergia sisoo increases the generation of biogas. The process for applying these additives involves the adsorption of the substrate material onto its interface. This fosters an environment where bacteria can flourish. One of the main organic additives that would boost the generation of biogas is biochar. It is feasible to increase methane gas production and decrease the requirement for chemical oxygen in the digester by adding graphene. Granular activated carbon is one of the key stimulators for the microbial life inside the digestor. The activated carbon in the digestor speeds up methanogenesis, facilitating the quick transformation of volatile fatty acids to methane.

13.7.2.2 Inorganic Additives

Inorganic additives are substances used to boost biogas production, such as calcium, nickel, iron, magnesium, and urea. The population of microorganisms increases when these metals are added. Surface materials like Zeolite have the ability to dominate the microbial population, which could be helpful for anaerobic fermentation. Zeolite can be added with titanium dioxide to adsorb hydrogen sulfide inside the digestor in in-situ biogas production. Zeolite reduces pH and alkalinity (Vijayakumar et al. 2022). The digestor contains trace elements like zinc, copper, magnesium, calcium, and potassium that improve anaerobic digestion. With the presence of the trace elements, the growth of methanogens has been observed since the trace element's presence favors acidogenesis and hydrolysis. Cadmium, phosphorous, chromium, and lead are toxic substances that hinder the production of methane. The biodegradation of organic materials will be greatly aided by the presence of copper in the digestor.

13.7.2.3 Biological Additives

Algae utilization significantly affects the production of biogas under the category of biological additions. Major bacteria found in mesophilic digesters include *actinobacteria, firmicutes, chloroflexi*, and *bacteroidetes*. Putting the right amount of inoculum in the digestor results in an abundant microbial population. Cellulolytic strains of bacteria, including mixed consortia and actinomycetes, can be injected into the anaerobic digestor. The concentration of lipids, carbohydrates, and proteins largely determines the nature and activity of the anaerobic digestor (Pramanik et al. 2019).

13.7.3 *Integration of Microbial Biogas Production in Waste Management*

By turning solid waste into a viable energy source, solid waste has the potential to produce fuel, reducing the need for fossil fuels and lowering pollution. By employing garbage as an input, it establishes a circular economy, which incorporates green technology in the form of recycling. The problem of disposing of solid waste has become significantly more difficult because of unchecked urbanization and the increasing rate of population growth. Ordinarily, municipal solid waste is dumped in landfills, which results in environmental issues such as leachate formation, methane emission, and the release of carbon dioxide gas. Landfilling can also cause health-related issues due to the unsanitary conditions prevailed by the population of insects, flies, birds, and mice in the periphery of the landfilling site. Methane discharged from landfills causes more pollution since it is more damaging than carbon dioxide; besides, the leachate contaminates the soil and the groundwater. Biogas can be produced from solid waste, thereby reducing the burden on the landfills. By sending waste to biogas facilities, which use waste as a resource and transform it into energy that can be utilized for combustion to create heat and power, situations like incineration and open burning in landfills can also be avoided (Tsapekos et al. 2022).

Given that biogas generation requires trained labor for operation and maintenance, integrating it into solid waste management might boost the local economy. Selling renewable energy sources to businesses is done so they may use them to produce heat and power creates economic activity. Several elements, including garbage collection, segregation, infrastructure development, policy support, and public awareness, need to be considered to properly incorporate biogas production as an integral part of the solid waste management system.

13.8 Sustainability and Environmental Impacts

The environment is currently under severe threat from activities like excessive chemical use, higher dependency on non-renewable energy resources, and unattended waste being produced in every conceivable industrial process. The adoption of sustainable practices, cleaner manufacturing methods, and green technologies is now more important than ever to protect the earth's ecology for future generations. The sustainable development goals focus on finding ways to promote both economic and environmental growth without compromising the environment. Instead, environmental conservation is the focus of these initiatives, which aim to stop and rein in the unauthorized use of natural resources. Microorganisms are critical in several environment-friendly operations and green technologies, from biogeochemical systems to diverse industrial outputs, and they have a big impact on the environment. Microorganisms can greatly aid in sustainable development if they are used responsibly.

Table 13.4 Comparative study of different methods of LCA

Methods	Advantages	Disadvantages	References
Cradle–to–gate	• Better resource utilization efficiency • Environmental consequences are clearly visualized • Process can be easily improvised	• Effects of end of life and aftermath are not accounted for • Application and scope are very limited	Saranya and Ramachandra (2020)
Cradle–to–grave	• Comprehensively examine the entire life cycle of the product • Efficient in the utilization of resources	• Highly assumption based • More tedious and time-consuming • Expertise required to perform	Tannous et al. (2018)

13.8.1 Life Cycle Assessment of Microbial Biofuel Production

A thorough approach called a life cycle assessment (LCA) is used to examine how a process affects the environment from the time that raw materials are extracted until the end of their intended useful life. LCA assesses the environmental simplifications of using microorganisms to manufacture biofuels in microbial biofuel production. LCA assesses the energy inputs, water usage, and emissions linked to the conversion of raw materials into biofuels by microbial fermentation, anaerobic digestion, or other biotechnological methods. Transporting raw materials to the biofuel manufacturing facility and distributing finished biofuels to customers are both taken into account by LCA. It is determined how using microbial biofuels in cars, power plants, or other uses will affect the environment. It entails examining combustion emissions and determining how effectively biofuels are used overall in comparison to fossil fuels. LCA investigates what happens to waste and byproducts of biofuel production and use. It determines whether byproducts can be put to good use or whether they must be disposed of in places like landfills and waste treatment facilities. The pros and cons of different methods of LCA are discussed in Table 13.4.

13.8.2 Environmental Implications and Mitigation Strategies

The environmental impact of employing microbial biofuels in vehicles, power plants, or other applications is assessed. It involves looking at combustion emissions and figuring out how well biofuels are used overall compared to fossil fuels. LCA studies what happens to waste and byproducts of producing and using biofuels. Byproducts are assessed to see if they can be used or if they must be disposed of at locations like landfills and waste treatment facilities. However, it is an alternative source;

biofuel production from microbial sources also has environmental implications and potential mitigation strategies. They are land use change, water consumption, nutrient pollution, greenhouse gas emissions, and biodiversity loss. The mitigation strategies are energy efficiency, nutrient management, sustainable land use, water recycling and efficiency, genetic engineering, strain selection, integrated biorefineries, and life cycle assessment.

13.8.3 Future Prospects for Sustainable Microbial Biofuel Production

In today's world, we face pressing challenges such as energy scarcity, food insecurity, water shortages, and climate change. Microorganisms play a crucial role in promoting sustainable development and greener industrial processes. While swiftly converting sugarcane liquid to ethanol is efficient, it may negatively impact sugar production. Thus, focusing on cost-effective, eco-friendly feedstocks like sweet sorghum, tropical sugar beet, and cellulosic waste biomass is imperative. This necessitates a diversified biofuel policy.

In India, researchers are increasingly exploring alternative biodiesel sources like Pongamia, Polanga, Neem, and Mahua tree seeds. Algae emerge as a promising and eco-friendly option for biodiesel production, as opposed to non-edible or tree-based oils. Algae not only produce biodiesel but also renewable fuels like methane gas and biohydrogen via anaerobic digestion and photobiological processes, respectively. Given India's vast coastline, there is immense potential for algal feedstock supply for biofuel production. The challenge lies in identifying algae strains with high oil content, ease of harvest, and rapid growth rates (Akinsemolu 2018).

13.9 Challenges and Future Perspectives

13.9.1 Technological and Economic Challenges

Biofuel technologies won't be practical and advantageous to society unless biomass is adequately accessible as a sustainable feedstock, considering the environmental implications and "food vs. fuel" challenges (Joshi et al. 2017). The manufacture of biofuels from food crops drives up the cost of raw materials, which has an impact on food prices. Fuels made from biomass have a poor energy density and expensive transportation and collection of biomasses. Growing additional crops based on biofuels to increase the output of biofuels might greatly increase the pressure on regional water resources if water quality and availability are not considered by area. The most common method for producing bioethanol is fermentation, which is more expensive than traditional gasoline due to the cost of pretreatment and high-charge enzymes required for improved output. Biodiesel may have a detrimental impact on the pollution control system by reducing catalytic efficiency, increasing engine emissions, and shortening the component lifespan (Luthra et al. 2015).

13.9.2 Policy Frameworks and Government Support

Domestic crude oil production can only meet 30% of India's real consumption; the other 70% must be imported, which has a big influence on the country's foreign policy and its foreign exchange reserves. Due to the nation's heavy reliance on foreign crude oil to meet its transportation needs and other energy-related needs, it is eager to establish emerging and renewable biofuel sectors. Behind China, Japan, and the United States, India is the fourth-largest energy user in the world and one of the economies with the highest economic growth (Das 2020). The biofuel production policy framework of various countries is discussed in Table 13.5. To replace petroleum-based energy sources with innovative energy sources that use renewable feedstock at the current rate of economic development, India must improve its alternative energy technologies. Among the several energy options available, biofuel derived from green feedstock is now regarded as the most desirable option, particularly for the transportation sector. The Indian government has backed several initiatives over the past ten years to increase energy efficiency, produce more renewable energy, and practice conservation. Aiming to support and advance biofuel technologies in India, the Government of India (GOI) introduced the "National Policy on Biofuel" in 2003 under the aegis of the "National Biofuel Mission (NBM)" with an eye on R&D, infrastructure development, biofuel commercialization, and purchasing guidelines. To generate biofuels, the strategy considers the usage of several energy crops and seeds, including cassava, maize, sugarcane, sweet sorghum, and tree-born oil (Joshi et al. 2017).

13.9.3 Emerging Trends and Future Directions in Microbial Biofuel Production

Researchers have already created specialized technology for producing biofuel from non-fodder biomass resources, but it has yet to be shown that these technologies are economically and commercially viable. Most biofuel technologies currently have infrastructure and knowledge in place; all that is required to advance and improve these technologies is the provision of the required resources. The Indian government should support greater R&D in conjunction with R&D facilities to investigate cutting-edge biofuel technologies based on natural, green, environment-supportive biological techniques. The primary factor for the sustained development of the Indian biofuel business is the presence of a steady market. To design biofuel policies at the regional

Table 13.5 Policies of different countries on biofuel utilization

Country	Policy	Reference
Canada	The renewable fuel law signed on August 23, 2010, is their national policy, which mandated recyclable content of 5% in gasoline and 2% in diesel For ten years in the last decade, they ran an ecoEnergy initiative by which they tried to increase biofuel production and to motivate the manufacturers they have a provision of incentives	Das (2020)
United States	Under the US EPA, their flagship program named Renewable Fuel Standard is launched The initial target of 4B gallons of blending in petroleum oil in 2006 was revised to 7.5B in 2012 and again upgraded to 36 B On the basis of feedstock and intensity requirement for reduction, biofuels are categorized into four different categories	Popp et al. (2014) Sorda et al. (2010) Brown (2019)
China	China's biofuel policies have primarily focused on ethanol blending, with biodiesel receiving less attention due to the country's reliance on imported vegetable oils Their standards are periodically reviewed under the aegis of Ethanol Gasoline for Motor Vehicles (GB, 18,351) in 2001, Biodiesel Blend Stock (GBT, 20,828) in 2007, and Biodiesel Fuel Blend (GBT 25,199) in 2010	Hao et al. (2018) Sorda et al. (2010)
European Union	Their present policy, The renewable energy directive (RED), is bound to use 10% recyclable fuel in transport and, overall, 20% combining all energy sectors The foundation of RED policy is the Common Agricultural Policy (CAP) launched in 1962, and their main motive is to reduce carbon intensity by 6% from the transportation sector by 2020	Sorda et al. (2010) Das (2020)

level and ensure their quick implementation, the central and state government entities must take the initiative further. The end user must be included in these biofuel policies and programs, and public and private collaborations must be promoted (Joshi et al. 2017).

13.10 Conclusions

In conclusion, India must considerably reduce its reliance on imported crude oil if it is to continue growing at the present rate. The equitable growth of alternative fuel from naturally produced biomass is a high priority for the nation's fuel industry because crude oil production in India is nearly at a halt. The large-scale production at the industry level of these biofuels is still in the early stages due to a shortage of acceptable feedstocks, despite the development of various indigenous methods for manufacturing bioethanol and biodiesel from a range of bio-based feedstocks. State and central governments in India must take more decisive action to guarantee the biofuel industry has access to sufficient feedstock. Additionally, research and development in India should concentrate on next-generation biofuels other than bioethanol and biodiesel made from vegetable oils. For the sustainable growth of bio-based refineries and the generation of biofuels, additional efforts using synthetic and cutting-edge biological techniques are still required. Government organizations should encourage the advancement of cutting-edge biofuel processes, such as algal-based biofuels and captured carbon-based biofuels (third and fourth-generation biofuels, respectively), which are both capable of overcoming the challenges of substrate availability and producing fuels with zero carbon footprints (Awasthi et al. 2023).

13.10.1 Key Findings and Summary

The biofuel-producing industries should enhance the well-being of their workers, particularly by raising wages, allowing them to work longer hours, and lowering social disparities. It is also strongly advised that the social life cycle be considered when developing national biofuel strategies. In several ways, biofuels may be able to lessen economic hardship. With the help of biofuels, the transition to a transportation system using no petroleum is made easier. When lignin-based biomass is broken down into fibrous elements during pre-treatment, it is more quickly converted into value-added products like biochemicals and biofuels. The pace of chemical reactions, heat exchange, and mass transfer are all accelerated by this biochemical conversion process. Numerous studies have investigated a variety of pre-treatment methods to determine the most effective and economical way to convert lignocellulosic biomass. However, because of significant costs and environmental issues, lignocellulosic biomass is not commonly used in industry. A single microorganism or microbial consortium uses consolidated bioprocessing (CBP) (Rodionova et al. 2022), which is regarded as a significant advancement for producing LC biofuels at a reasonable cost to produce saccharolytic enzymes, hydrolyze polysaccharides, and fermented sucrose all in one step. The output of bioproducts can be increased by adding additional enzymes, co-enzymes, or surfactants; however, various LC biomass feedstocks require numerous enzymes (e.g., cellulase and hemicellulose for potato peels and wheat bran). The CBP approach (Rodionova et al. 2022) has the benefit that, in an ideal scenario, exogenous enzyme addition may be eliminated. The procedure is also streamlined and made simpler in terms of the number of operational units, which reduces operating expenses and time.

13.10.2 Recommendations for Future Research

There is always room for improvement; however, new technologies have significantly improved the potential applications of green pre-treatment systems. More study is needed to create green pre-treatment methods that may be used commercially and have reduced production costs. It is necessary to research how to increase biofuel production through economies of scale with more effective and affordable pretreatment techniques. The rising population demands more energy, and thus, using lignocellulosic biomass as a fuel is an effective way to meet this need. However, it has been noted that as the utilization of lignocellulosic biomass increases, so do GHG emissions. Biofuels are used as a strategy to reduce the issue of greenhouse gases. Biofuels can be used for long-term purposes and are a sustainable alternative. However, producing biofuels from lignocellulose biomass is fraught with difficulties. As we understand, it is not possible to standardize the method of converting lignocellulosic biomass since the sources and mix of lignocellulosic biomass vary. On the other hand, different methods were employed to effectively turn lignocellulosic biomass into biofuel. Most of this research focuses on pilot-scale manufacturing because it is currently challenging to build industrial-scale systems. To meet the current fuel demand, plans for developing large-scale biofuel manufacturing infrastructures are required. Additionally, because it considerably lessens the detrimental impact on environmental quality, research on cost-effective biofuel production with technological advancement should be given high priority.

References

Ab Rasid NS, Shamjuddin A, Abdul Rahman AZ, Amin NAS (2021) Recent advances in green pre-treatment methods of lignocellulosic biomass for enhanced biofuel production. J Clean Prod 321:129038. https://doi.org/10.1016/j.jclepro.2021.129038

Agarwal A, Shaikh KM, Gharat K, Jutur PP, Pandit RA, Lali AM (2019) Investigating the modulation of metabolites under high light in mixotrophic alga Asteracys sp. using a metabolomic approach. Algal Res 43:101646. https://doi.org/10.1016/j.algal.2019.101646

Ahmed SF, Mofijur M, Chowdhury SN, Nahrin M, Rafa N, Chowdhury AT, Nuzhat S, Ong HC (2022) Pathways of lignocellulosic biomass deconstruction for biofuel and value-added products production. Fuel 318:123618. https://doi.org/10.1016/j.fuel.2022.123618

Akash B (2015) Thermochemical depolymerization of biomass. Procedia Comput Sci 52:827–834. https://doi.org/10.1016/j.procs.2015.05.139

Akinsemolu AA (2018) The role of microorganisms in achieving the sustainable development goals. J Clean Prod 182:139–155. https://doi.org/10.1016/j.jclepro.2018.02.081

Al-Wahaibi A, Osman AI, Al-Muhtaseb AH, Alqaisi O, Baawain M, Fawzy S, Rooney DW (2020) Techno-economic evaluation of biogas production from food waste via anaerobic digestion. Sci Rep 10:15719. https://doi.org/10.1038/s41598-020-72897-5

Arutselvan C, Narchonai G, Pugazhendhi A, LewisOscar F, Thajuddin N (2021) Evaluation of microalgal strains and microalgal consortium for higher lipid productivity and rich fatty acid profile towards sustainable biodiesel production. Bioresour Technol 339:125524. https://doi.org/10.1016/j.biortech.2021.125524

Aulitto M, Fusco S, Nickel DB, Bartolucci S, Contursi P, Franzén CJ (2019) Seed culture pre-adaptation of Bacillus coagulans MA-13 improves lactic acid production in simultaneous saccharification and fermentation. Biotechnol Biofuels 12:45. https://doi.org/10.1186/s13068-019-1382-2

Awasthi MK, Sar T, Gowd SC, Rajendran K, Kumar V, Sarsaiya S, Li Y, Sindhu R, Binod P, Zhang Z, Pandey A, Taherzadeh MJ (2023) A comprehensive review on thermochemical, and biochemical conversion methods of lignocellulosic biomass into valuable end product. Fuel 342:127790. https://doi.org/10.1016/j.fuel.2023.127790

Brown TR (2019) Why the cellulosic biofuels mandate fell short: a markets and policy perspective. Biofuels, Bioprod Biorefin 13:889–898. https://doi.org/10.1002/bbb.1987

Brexó RP, Sant'Ana AS (2017) Impact and significance of microbial contamination during fermentation for bioethanol production. Renew Sustain Energy Reviews 73:423–434. https://doi.org/10.1016/j.rser.2017.01.151

Candido C, Lombardi AT (2020) Mixotrophy in green microalgae grown on an organic and nutrient rich waste. World J Microbiol Biotechnol 36:20. https://doi.org/10.1007/s11274-020-2802-y

Chaurasia D, Boudh S, Singh JS, Chaturvedi P (2020) Microbes and microbial enzymes as a sustainable energy source for biofuel production. In: microbial services in restoration ecology. Elsevier, pp 207–217

Choi WJ, Chae AN, Song KG, Park J, Lee BC (2019) Effect of trophic conditions on microalga growth, nutrient removal, algal organic matter, and energy storage products in Scenedesmus (Acutodesmus) obliquus KGE-17 cultivation. Bioprocess Biosyst Eng 42:1225–1234. https://doi.org/10.1007/s00449-019-02120-x

Choi KR, Jiao S, Lee SY (2020) Metabolic engineering strategies toward production of biofuels. Curr Opin Chem Biol 59:1–14. https://doi.org/10.1016/j.cbpa.2020.02.009

Choi YJ, Lee J, Jang Y-S, Lee SY (2014) Metabolic engineering of microorganisms for the production of higher alcohols. mBio 5. https://doi.org/10.1128/mBio.01524-14

Cudjoe D, Zhu B, Wang H (2022) Towards the realization of sustainable development goals: benefits of hydrogen from biogas using food waste in China. J Clean Prod 360:132161. https://doi.org/10.1016/j.jclepro.2022.132161

Das S (2020) The national policy of biofuels of India—a perspective. Energy Policy 143:111595. https://doi.org/10.1016/j.enpol.2020.111595

Dassey AJ, Hall SG, Theegala CS (2014) An analysis of energy consumption for algal biodiesel production: comparing the literature with current estimates. Algal Res 4:89–95. https://doi.org/10.1016/j.algal.2013.12.006

Gorgich M, Mata TM, Martins AA, Branco-Vieira M, Caetano NS (2020) Comparison of different lipid extraction procedures applied to three microalgal species. Energy Rep 6:477–482. https://doi.org/10.1016/j.egyr.2019.09.011

Hao H, Liu Z, Zhao F, Ren J, Chang S, Rong K, Du J (2018) Biofuel for vehicle use in China: current status, future potential and policy implications. Renew Sustain Energy Rev 82:645–653. https://doi.org/10.1016/j.rser.2017.09.045

Holwerda EK, Worthen RS, Kothari N, Lasky RC, Davison BH, Fu C, Wang Z-Y, Dixon RA, Biswal AK, Mohnen D, Nelson RS, Baxter HL, Mazarei M, Stewart CN, Muchero W, Tuskan GA, Cai CM, Gjersing EE, Davis MF, Himmel ME, Wyman CE, Gilna P, Lynd LR (2019) Multiple levers for overcoming the recalcitrance of lignocellulosic biomass. Biotechnol Biofuels 12:15. https://doi.org/10.1186/s13068-019-1353-7

Hu H, Li J-Y, Zhai S-W, Wu D-D, Zhu S-G, Zeng RJ (2020) Effect of inorganic carbon limitation on the conversion of organic carbon to total fatty acids by Monodus subterraneus. Sci Total Environ 737:140275. https://doi.org/10.1016/j.scitotenv.2020.140275

Hwangbo M, Chu K-H (2020) Recent advances in production and extraction of bacterial lipids for biofuel production. Sci Total Environ 734:139420. https://doi.org/10.1016/j.scitotenv.2020.139420

Joshi S, Mishra S (2022) Recent advances in biofuel production through metabolic engineering. Bioresour Technol 352:127037. https://doi.org/10.1016/j.biortech.2022.127037

Joshi G, Pandey JK, Rana S, Rawat DS (2017) Challenges and opportunities for the application of biofuel. Renew Sustain Energy Rev 79:850–866. https://doi.org/10.1016/j.rser.2017.05.185

Joshi A, Verma KK, Rajput VD, Arora J (2023) Metabolic engineering of microorganisms in advancing biofuel production. In: Advances in lignocellulosic biofuel production systems. Elsevier, pp 181–202

Khoo KS, Tan X, Ooi CW, Chew KW, Leong WH, Chai YH, Ho S-H, Show PL (2021) How does ionic liquid play a role in sustainability of biomass processing? J Clean Prod 284:124772. https://doi.org/10.1016/j.jclepro.2020.124772

Khoo KS, Ahmad I, Chew KW, Iwamoto K, Bhatnagar A, Show PL (2023) Enhanced microalgal lipid production for biofuel using different strategies including genetic modification of microalgae: a review. Prog Energy Combust Sci 96:101071. https://doi.org/10.1016/j.pecs.2023.101071

Li T, Yang F, Xu J, Wu H, Mo J, Dai L, Xiang W (2020) Evaluating differences in growth, photosynthetic efficiency, and transcriptome of Asterarcys sp. SCS-1881 under autotrophic, mixotrophic, and heterotrophic culturing conditions. Algal Res 45:101753. https://doi.org/10.1016/j.algal.2019.101753

Lin L, Cunshan Z, Vittayapadung S, Xiangqian S, Mingdong D (2011) Opportunities and challenges for biodiesel fuel. Appl Energy 88:1020–1031. https://doi.org/10.1016/j.apenergy.2010.09.029

López-Fernández J, Dolors Benaiges M, Valero F (2021) Second- and third-generation biodiesel production with immobilised recombinant Rhizopus oryzae lipase: influence of the support, substrate acidity and bioprocess scale-up. Bioresour Technol 334:125233. https://doi.org/10.1016/j.biortech.2021.125233

Luthra S, Kumar S, Garg D, Haleem A (2015) Barriers to renewable/sustainable energy technologies adoption: Indian perspective. Renew Sustain Energy Rev 41:762–776. https://doi.org/10.1016/j.rser.2014.08.077

Manzoor M, Ahmad Q, Aslam A, Jabeen F, Rasul A, Schenk PM, Qazi JI (2020) Mixotrophic cultivation of Scenedesmus dimorphus in sugarcane bagasse hydrolysate. Environ Prog Sustain Energy 39. https://doi.org/10.1002/ep.13334

Menegol T, Romero-Villegas GI, López-Rodríguez M, Navarro-López E, López-Rosales L, Chisti Y, Cerón-García MC, Molina-Grima E (2019) Mixotrophic production of polyunsaturated fatty acids and carotenoids by the microalga Nannochloropsis gaditana. J Appl Phycol 31:2823–2832. https://doi.org/10.1007/s10811-019-01828-3

Merlin Christy P, Gopinath LR, Divya D (2014) A review on anaerobic decomposition and enhancement of biogas production through enzymes and microorganisms. Renew Sustain Energy Rev 34:167–173. https://doi.org/10.1016/j.rser.2014.03.010

Novia N, Hasanudin H, Hermansyah H, Fudholi A, Pareek VK (2023) Recent advances in CFD modeling of bioethanol production processes. Renew Sustain Energy Rev 183:113522. https://doi.org/10.1016/j.rser.2023.113522

Pang N, Gu X, Fu X, Chen S (2019) Effects of gluconate on biomass improvement and light stress tolerance of Haematococcus pluvialis in mixotrophic culture. Algal Res 43:101647. https://doi.org/10.1016/j.algal.2019.101647

Pardo-Cárdenas Y, Herrera-Orozco I, González-Delgado Á-D, Kafarov V (2013) Environmental assessment of microalgae biodiesel production in Colombia: comparison of three oil extraction systems. CT&F Ciencia Tecnol Fut 5:85–100

Patel A, Matsakas L, Rova U, Christakopoulos P (2018) Heterotrophic cultivation of Auxenochlorella protothecoides using forest biomass as a feedstock for sustainable biodiesel production. Biotechnol Biofuels 11:169. https://doi.org/10.1186/s13068-018-1173-1

Poddar N, Sen R, Martin GJO (2018) Glycerol and nitrate utilisation by marine microalgae nannochloropsis salina and Chlorella sp. and associated bacteria during mixotrophic and heterotrophic growth. Algal Res 33:298–309. https://doi.org/10.1016/j.algal.2018.06.002

Popp J, Lakner Z, Harangi-Rákos M, Fári M (2014) The effect of bioenergy expansion: food, energy, and environment. Renew Sustain Energy Rev 32:559–578. https://doi.org/10.1016/j.rser.2014.01.056

Pramanik SK, Suja FB, Zain SM, Pramanik BK (2019) The anaerobic digestion process of biogas production from food waste: prospects and constraints. Bioresour Technol Rep 8:100310. https://doi.org/10.1016/j.biteb.2019.100310

Prasad RK, Chatterjee S, Mazumder PB, Gupta SK, Sharma S, Vairale MG, Datta S, Dwivedi SK, Gupta DK (2019) Bioethanol production from waste lignocelluloses: a review on microbial degradation potential. Chemosphere 231:588–606. https://doi.org/10.1016/j.chemosphere.2019.05.142

Preethi MG, Kumar G, Karthikeyan OP, Varjani S, J. RB, (2021) Lignocellulosic biomass as an optimistic feedstock for the production of biofuels as valuable energy source: techno-economic analysis, environmental impact analysis, breakthrough and perspectives. Environ Technol Innov 24:102080. https://doi.org/10.1016/j.eti.2021.102080

Raja Ram N, Nikhil GN (2022) A critical review on sustainable biogas production with focus on microbial-substrate interactions: bottlenecks and breakthroughs. Bioresour Technol Rep 19:101170. https://doi.org/10.1016/j.biteb.2022.101170

Rastogi M, Shrivastava S (2017) Recent advances in second generation bioethanol production: an insight to pretreatment, saccharification and fermentation processes. Renew Sustain Energy Rev 80:330–340

Rodionova MV, Bozieva AM, Zharmukhamedov SK, Leong YK, Chi-Wei Lan J, Veziroglu A, Veziroglu TN, Tomo T, Chang J-S, Allakhverdiev SI (2022) A comprehensive review on lignocellulosic biomass biorefinery for sustainable biofuel production. Int J Hydrogen Energy 47:1481–1498. https://doi.org/10.1016/j.ijhydene.2021.10.122

Saranya G, Ramachandra TV (2020) Life cycle assessment of biodiesel from estuarine microalgae. Energy Convers Manag X 8:100065. https://doi.org/10.1016/j.ecmx.2020.100065

Singh S, Kumar A, Sivakumar N, Verma JP (2022) Deconstruction of lignocellulosic biomass for bioethanol production: Recent advances and future prospects. Fuel 327:125109. https://doi.org/10.1016/j.fuel.2022.125109

Sorda G, Banse M, Kemfert C (2010) An overview of biofuel policies across the world. Energy Policy 38:6977–6988. https://doi.org/10.1016/j.enpol.2010.06.066

Tannous S, Manneh R, Harajli H, El Zakhem H (2018) Comparative cradle-to-grave life cycle assessment of traditional grid-connected and solar stand-alone street light systems: a case study for rural areas in Lebanon. J Clean Prod 186:963–977. https://doi.org/10.1016/j.jclepro.2018.03.155

Tsapekos P, Khoshnevisan B, Zhu X, Treu L, Alfaro N, Kougias PG, Angelidaki I (2022) Lab- and pilot-scale anaerobic digestion of municipal bio-waste and potential of digestate for biogas upgrading sustained by microbial analysis. Renew Energy 201:344–353. https://doi.org/10.1016/j.renene.2022.10.116

Vasistha S, Khanra A, Clifford M, Rai MP (2021) Current advances in microalgae harvesting and lipid extraction processes for improved biodiesel production: a review. Renew Sustain Energy Rev 137:110498. https://doi.org/10.1016/j.rser.2020.110498

Vijayakumar P, Ayyadurai S, Arunachalam KD, Mishra G, Chen W-H, Juan JC, Naqvi SR (2022) Current technologies of biochemical conversion of food waste into biogas production: a review. Fuel 323:124321. https://doi.org/10.1016/j.fuel.2022.124321

Wang H, Zhang Y, Zhou W, Noppol L, Liu T (2018) Mechanism and enhancement of lipid accumulation in filamentous oleaginous microalgae Tribonema minus under heterotrophic condition. Biotechnol Biofuels 11:328. https://doi.org/10.1186/s13068-018-1329-z

Wang X, Zhang M-M, Sun Z, Liu S-F, Qin Z-H, Mou J-H, Zhou Z-G, Lin CSK (2020) Sustainable lipid and lutein production from Chlorella mixotrophic fermentation by food waste hydrolysate. J Hazard Mater 400:123258. https://doi.org/10.1016/j.jhazmat.2020.123258

Wang H, Peng X, Zhang H, Yang S, Li H (2021) Microorganisms-promoted biodiesel production from biomass: a review. Energy Convers Manag X 12:100137. https://doi.org/10.1016/j.ecmx.2021.100137

Wang N, Xu B, Wang X, Lang J, Zhang H (2022) Chemical and structural elucidation of lignin and cellulose isolated using DES from bagasse based on alkaline and hydrothermal pretreatment. Polymers (basel) 14:2756. https://doi.org/10.3390/polym14142756

Zhang Q, De Oliveira VK, Royer S, Jérôme F (2012) Deep eutectic solvents: syntheses, properties and applications. Chem Soc Rev 41:7108. https://doi.org/10.1039/c2cs35178a

Zhang C, Sun Y, Cao T, Wang W, Huo S, Liu Z-H (2022) Influence of organic load on biogas production and response of microbial community in anaerobic digestion of food waste. Int J Hydrogen Energy 47:32849–32860. https://doi.org/10.1016/j.ijhydene.2022.07.187

Zhao Z, Ma S, Li A, Liu P, Wang M (2016) Effects of trophic modes, carbon sources, and salinity on the cell growth and lipid accumulation of tropic ocean oilgae strain Desmodesmus sp. WC08. Appl Biochem Biotechnol 180:452–463. https://doi.org/10.1007/s12010-016-2109-5

Chapter 14
Microbes as Environmental Allies: From Wastewater Treatment to Sustainable Biofuel Production

Anuj Sharma, Praveen Sharma, Bansal Deepak, and Sharma Mona

14.1 Introduction

Water is a momentous natural asset because it is a fundamental component of existence. But in the modern era of rapid growth and industrialization, one of the major problems that mankind is facing is the scarcity of freshwater (Sharma et al. 2023a). Industries like textiles, paper, tanneries, and plastics are some of the major sources of water pollution. These industries are expanding day by day as the human population and its needs are expanding (Bhukal et al. 2022). So, the treatment of wastewater is the need of the hour. The physical and chemical techniques such as coagulation, precipitation, filtration, oxidation, and membrane filtration that are used for wastewater treatment have several drawbacks (Mittal et al. 2023). Here are some common drawbacks associated with these techniques: high energy consumption, generation of sludge and waste, limited effectiveness on certain pollutants, chemical dependency, formation of disinfection by-products, long-term environmental impacts, and cost and infrastructure requirements (Sharma et al. 2023a; Bhukal et al. 2022; Mittal et al. 2023; Agarwal et al. 2019). To address these drawbacks, researchers and engineers are exploring alternative and more sustainable wastewater treatment approaches, such as biological treatment methods, membrane technologies, and advanced oxidation processes, which can offer improved efficiency, reduced energy consumption, and better pollutant removal. Bionanotechnology especially the

A. Sharma · P. Sharma
Department of Environmental Science and Engineering, Guru Jambheshwar University of Science and Technology, Hisar, Haryana 125001, India

B. Deepak
JBM Group, Gurugram, Haryana 122001, India

S. Mona (✉)
Department of Environmental Studies, School of Interdisciplinary and Applied Sciences, Central University of Haryana, Mahendergarh, Haryana 123031, India
e-mail: mona@cuh.ac.in

K. Bala et al. (eds.), *Harnessing Microbial Potential for Multifarious Applications*, Energy, Environment, and Sustainability, https://doi.org/10.1007/978-981-97-1152-9_14

use of microbes in wastewater treatment is one of the promising methods that not only remediate contaminants from wastewater but can also be used for further processes like biofuel production (Sharma et al. 2023b; Qu et al. 2020). Bionanotechnology and microbes offer several benefits when used in wastewater treatment processes. Here are some key advantages: enhanced pollutant removal, improved biodegradation, cost-effectiveness, and sustainability, versatility, and adaptability, reduced sludge production, natural and environmentally friendly, and potential for resource recovery (Khanna et al. 2019; Shalaby et al. 2021). By leveraging the benefits of bionanotechnology and harnessing the power of microbes, wastewater treatment systems can achieve higher efficiency, reduced costs, and improved environmental sustainability. These approaches hold great promise for addressing the challenges associated with wastewater treatment more efficiently and sustainably (Sharma et al. 2023a; Khanna et al. 2019; Guleri et al. 2020; Sumit et al. 2023).

Microbes along with remediation of contaminants from wastewater can also be used for biofuel production and in the mitigation of pollutants from the environment. Wastewater can be a suitable source for microbial production of biofuels due to several key factors: rich in organic matter content, nutrient availability, microbial diversity, and low-cost feedstock. Despite these advantages, challenges such as wastewater variability, contaminants, and the need for efficient separation and purification of biofuels still need to be addressed to make wastewater-based biofuel production more widespread and economically competitive. Producing biofuel from wastewater is preferred over other feedstocks for several reasons such as waste reduction, resource efficiency, environmental benefits, cost-effectiveness, energy recovery, circular economy, and local production. It is important to note that the feasibility and benefits of producing biofuel from wastewater can vary depending on factors such as the type and quality of wastewater, the technology used for conversion, and local regulations (Bhukal et al. 2022; Kumari et al. 2022). Ongoing research and technological advancements are continually improving the efficiency and scalability of these processes, making them increasingly viable for widespread implementation. Wastewater remediation using microbes for biofuels and pollution mitigation is an innovative approach that combines wastewater treatment and biofuel production while addressing environmental pollution (Zabermawi et al. 2022). This process involves the use of specific microbial organisms to break down organic matter in wastewater, converting it into valuable biofuels and reducing the pollutants present in the water. Steps involved in the process are (i) wastewater Treatment, (ii) microbial Treatment, (iii) biofuel Production, and (iv) pollution Mitigation (Kumari et al. 2022; Zabermawi et al. 2022; Sharma et al. 2020; The Use of Microbes in Biofuel Production 2023; Maity et al. 2014). By consuming organic compounds and breaking them down, the microbes reduce the concentration of pollutants in the wastewater. This process helps remove harmful substances like nitrogen, phosphorus, and organic contaminants, which would otherwise cause water pollution if discharged into the environment. Microbes including seaweeds can also help in the remediation of contaminants such as *Agarophyton tenuistipitatum* which remediates >80% nitrogen and >20% phosphorus at 27–30 °C (Sarkar et al. 2021). Yeast strains

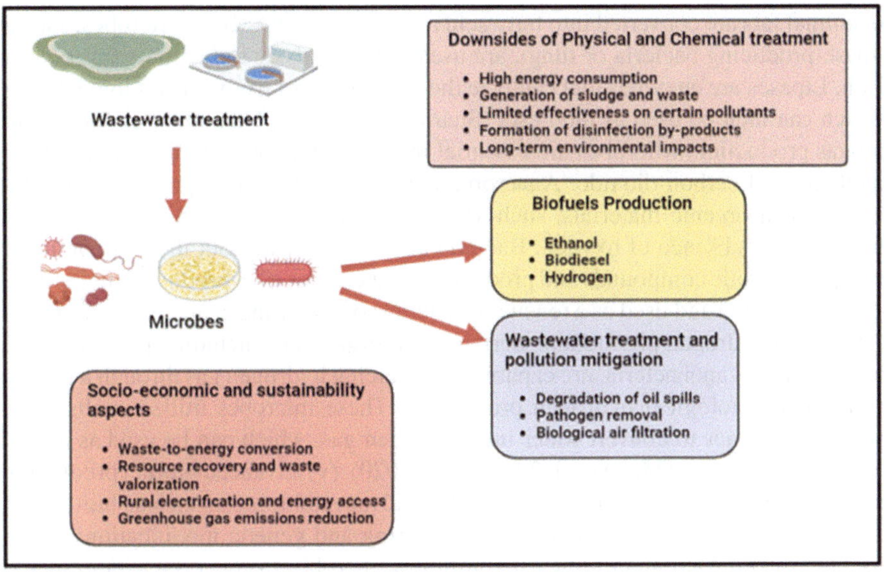

Fig. 14.1 General overview of microbes for biofuel production, wastewater treatment, and pollution mitigation

can also help in the production of biofuels such as bioethanol for example *S. cerevisiae* RL-11 at fermentation conditions of 30 °C, 200 rpm, 48 h shows 0.49 (g/L/h) ethanol productivity (Mussatto et al. 2012). Figure 14.1 represents the basic concept of microbes for biofuel production, wastewater treatment, and pollution mitigation. In the present book chapter, the basic concept of wastewater remediation using microbes along with the utilization of microbes for the production of biofuels is discussed. The present chapter also emphasizes biofuel production and pollution mitigation through the lens of sustainability.

14.2 Microbes for Biofuels Production

Microbes play a crucial role in the production of biofuel by converting biomass or specific feedstocks into energy-rich fuel (Ibrahim et al. 2020; Ly et al. 2016). Here are some keyways in which microbes are utilized in biofuels production.

(i) Ethanol production: Certain microorganisms, particularly yeast species such as Saccharomyces cerevisiae, are commonly used in the production of bioethanol. These microbes ferment sugars derived from biomass, such as corn or sugarcane, into ethanol through a process called alcoholic fermentation. Yeasts can efficiently convert sugars into ethanol, making them valuable for large-scale ethanol production (Zhu et al. 2023; Zhao et al. 2023). (ii) biodiesel production: Biodiesel is typically produced through a process called transesterification, in which triglycerides from vegetable oils or animal fats are converted into fatty acid methyl esters (FAMEs). Microbes, such as lipase-producing bacteria or fungi, are used as biocatalysts to facilitate this conversion. Lipases are enzymes that catalyze the breakdown of triglycerides into FAMEs, which can then be used as biodiesel (Sharma et al. 2020; Uthandi et al. 2022). (iii) biogas production: Microbes are essential in the production of biogas, a mixture of methane and carbon dioxide. Anaerobic digestion, a microbial process, is used to break down organic materials, such as agricultural waste, sewage sludge, or food waste, in the absence of oxygen. Bacteria and archaea carry out the degradation of complex organic compounds and produce methane gas as a by-product. This methane can be collected and used as a renewable energy source (Yang et al. 2023; Wang et al. 2023). (iv) hydrogen production: Certain microorganisms, including photosynthetic bacteria and cyanobacteria, are capable of producing hydrogen gas through a process called photobiological hydrogen production. These microbes utilize sunlight and specific enzymes to convert water into hydrogen gas, which can be used as a clean and sustainable fuel (Mona et al. 2011, 2013, 2020). (v) advanced biofuel production: microbes are also being explored to produce advanced biofuels, such as bio-butanol and bio-jet fuel. Through metabolic engineering and genetic modifications, scientists are working to enhance the capabilities of microbes to convert a wider range of feedstocks, including lignocellulosic biomass, into advanced biofuels (Sharma et al. 2023c; Pfleger and Takors 2023). These efforts involve optimizing metabolic pathways and introducing novel enzymes to improve fuel production efficiency. The use of microbes in biofuel production offers several advantages, including their ability to convert renewable feedstocks into fuels, their scalability for large-scale production, and their potential for sustainable and carbon–neutral energy generation. Ongoing research and development in microbial engineering and bioprocessing are focused on improving the efficiency, cost-effectiveness, and sustainability of biofuel production using microbial systems. Microbes play a vital role in the production of biofuels by converting organic matter into energy-rich compounds (Kumari et al. 2022; Maity et al. 2014; Mona et al. 2011).

Microbes mentioned in Table 14.1 highlight the diverse range of microbes utilized in biofuel production, each with specific metabolic capabilities and feedstock preferences. Table 14.2 demonstrates the production of different types of biofuel production under different conditions and methods. Continued research and development in microbial engineering and bioprocess optimization are crucial for improving biofuel production efficiency, reducing costs, and advancing the viability of biofuels as renewable energy alternatives.

Table 14.1 List of microbes commonly used in biofuel production

Microbes	Role in biofuels production	References
Saccharomyces cerevisiae	• Widely employed in the production of bioethanol • Ferments sugars derived from biomass, such as corn, sugarcane, or cellulosic materials, to produce ethanol • Efficiently convert both hexose and pentose sugars, making it versatile for various feedstocks	Mohd Azhar et al. (2017)
Escherichia coli	• This bacterium has been genetically engineered for biofuel production • Through metabolic engineering, *E. coli* can be modified to produce biofuels like butanol, ethanol, and fatty acid ethyl esters (biodiesel) • It can utilize various feedstocks, including sugars, lignocellulosic biomass, and even carbon dioxide as a carbon source	Koppolu and Vasigala (2016)
Clostridium species	• Certain species of the *Clostridium* genus, such as *Clostridium acetobutylicum* and *Clostridium beijerinckii,* are used in the production of biofuels like butanol • These bacteria utilize sugar substrates derived from biomass to produce butanol through a process known as acetone-butanol-ethanol (ABE) fermentation	Arifin et al. (2014)
Cyanobacteria	• Cyanobacteria, also known as blue-green algae, can convert solar energy into chemical energy through photosynthesis • Some cyanobacterial strains, such as *Synechocystis sp.* and *Synechococcus elongatus*, have been engineered to produce biofuels, including ethanol and fatty acids, directly from carbon dioxide and sunlight	Singh et al. (2016)

(continued)

Table 14.1 (continued)

Microbes	Role in biofuels production	References
Oleaginous microorganisms	• Certain microorganisms, such as certain species of algae and yeast, are capable of accumulating high levels of lipids (oils) within their cells • These lipid-rich microorganisms, known as *oleaginous* microorganisms, can be cultivated and processed to extract and convert lipids into biodiesel or other biofuels	Cho and Park (2018)
Methanogens	• *Methanogenic archaea*, such as *Methanosarcina* and *Methanococcus* species, are involved in the production of biogas, which primarily contains methane • They can digest organic materials, including agricultural waste, sewage sludge, and organic residues, in anaerobic conditions to produce methane, which can be utilized as a renewable energy source	Kaloudas et al. (2021)

14.3 Microbes for Wastewater Treatment and Pollution Mitigation

Microbes play a crucial role in wastewater treatment processes by breaking down and removing contaminants present in the wastewater. Here are the key roles of microbes in wastewater treatment: (i) decomposition of organic matter: microbes, particularly bacteria, and fungi, are responsible for the decomposition and mineralization of organic compounds in wastewater. They utilize organic matter as a food source and convert complex organic molecules into simpler, more stable compounds through processes such as aerobic or anaerobic digestion. This microbial activity helps reduce the organic load and removes pollutants from the wastewater (Mirzababaei and Hassani 2023; Dong et al. 2023). (ii) nutrient removal: certain microbial species, known as nitrifying and denitrifying bacteria, play a vital role in the removal of nitrogen compounds from wastewater. Nitrifying bacteria convert ammonia to nitrite and then to nitrate, while denitrifying bacteria convert nitrate back to nitrogen gas, thus removing excess nitrogen from the wastewater. Similarly, phosphate-accumulating bacteria can capture and remove phosphorus from wastewater, which helps prevent eutrophication in receiving water bodies (Jia et al. 2023; Li et al. 2023). (iii) filtration and adsorption: microbes, especially bacteria, can form biofilms on surfaces such as filter media or within biological treatment systems. These biofilms act as natural filters, capturing and immobilizing suspended solids, organic particles, and even certain pathogens present in the wastewater. The biofilms also provide a favorable environment for various microbial processes, enhancing treatment efficiency (Lu et al. 2022; Sun et al. 2022). (iv) removal of toxic substances: some microorganisms possess the ability to metabolize or detoxify certain toxic substances

Table 14.2 Biofuel production using different microbial species under different conditions and methods

Microbial species	Method	Condition of reactions	Types of biofuel	Yield	References
Galdieria sulphuraria	Hydrothermal gasification	500 °C	Hydrogen	45.7% mol	Ibrahim et al. (2020)
Acutodesmus obliquus	Supercritical gasification	690 °C	Hydrogen	13,944 kg/h	Nurdiawati et al. (2019)
Nannochloropsis sp.	Hydrothermal Liquefaction	300–400 °C	Bio-oil	59 wt%	Xu and Savage (2017)
Dunaliella sp.	Hydrothermal Liquefaction	350 °C	Bio-oil	11.81 wt%	Shahi et al. (2020)
Spirulina, chlorella	Microwave-induced Pyrolysis	700 °C	Syngas	73.3 vol%	Hong et al. (2017)
S. japonica	Pyrolysis	350 °C	Syngas	22 vol%	Ly et al. (2016)
Tetraselmis sp.	Fermentation	50 h	Bio-ethanol	15.7 g/l	Constantino et al. (2021)
Clostridium acetobutylicum	Fermentation	37 °C, 30–40 h	Bio-butanol	13.8 g/l	Tsai et al. (2020)
Phaeodactylum tricornutum	Anaerobic digestion after being pretreated by ultrasound	–	Bio-methane	284–287 ml/g	Caporgno et al. (2016)
Microcystis wesenbergii	Anaerobic sludge with butyrate-acetate	37 °C	Biohydrogen	31.42 ml/g	Xu et al. (2019)

present in wastewater. For example, specific strains of bacteria can degrade or transform organic pollutants, including various industrial chemicals, pesticides, and pharmaceuticals, into less harmful compounds. This microbial degradation can significantly reduce the toxicity of wastewater and protect the environment (Sharma et al. 2023a; Bhukal et al. 2022). (v) sludge digestion and stabilization: during the wastewater treatment process, a significant amount of sludge is generated, which consists of both organic and inorganic matter. Microbes, particularly anaerobic bacteria, are employed in sludge digestion processes such as anaerobic digestion and composting. These microbes break down the organic components of the sludge, converting them into biogas (methane and carbon dioxide) and stabilized organic matter. This reduces the volume of sludge and produces renewable energy in the form of biogas (Wu et al. 2022a; Yu et al. 2022). (vi) pathogen removal: certain microbial species, such as specific strains of bacteria and viruses, can degrade or inactivate pathogens present in wastewater. Microbial processes, combined with disinfection methods like chlorination or ultraviolet (UV) irradiation, help remove or inactivate harmful pathogens, ensuring that the treated wastewater meets the required standards for safe disposal or reuse. By harnessing the natural capabilities of microbes, wastewater treatment systems can effectively and efficiently remove contaminants, reduce environmental pollution, and produce treated water that is safe for discharge or reuse (Li et al. 2023; Nasir et al. 2022). The optimization and management of microbial populations are essential to maintain stable and reliable wastewater treatment processes. Microbes play a crucial role in wastewater treatment processes by breaking down organic matter, removing nutrients, and degrading pollutants.

The microbes mentioned in Table 14.3 demonstrate the important role in wastewater treatment. Table 14.4 depicts the nutrient removal and lipid content of various microbial species. Their metabolic activities and interactions contribute to the removal of organic matter, nutrients, and contaminants, leading to the purification of wastewater before its discharge into the environment. Microbes play a crucial role in pollution mitigation by actively participating in various natural processes that help reduce and degrade pollutants. Here are some ways in which microbes contribute to pollution mitigation: (i) bioremediation of contaminated soil and water, (ii) degradation of oil spills, (iii) treatment of industrial wastewater, (iv) nitrogen and phosphorus removal, (v) methane capture and reduction, (vi) biological air filtration. Microbes possess diverse metabolic capabilities that allow them to interact with and mitigate various types of pollutants (Bhukal et al. 2022; Mittal et al. 2023; Mirzababaei and Hassani 2023; Jia et al. 2023; Li et al. 2023; Sun et al. 2022; Nasir et al. 2022). Their application in pollution mitigation strategies holds promise for environmentally friendly and sustainable approaches to address pollution challenges.

Table 14.3 Some examples of microbes commonly used in wastewater treatment

Microbes	Role in wastewater treatment and pollution mitigation	References
Activated sludge	• Activated sludge is a microbial community consisting of bacteria, fungi, protozoa, and other microorganisms • It is a widely used biological treatment method in wastewater treatment plants • The microbes in activated sludge consume organic matter, converting it into carbon dioxide, water, and biomass through the process of aerobic digestion	Widajatno et al. (2022)
Nitrifying bacteria	• Nitrifying bacteria, such as *Nitrosomonas* and *Nitrobacter*, are involved in the nitrification process. They convert ammonia (NH^3) to nitrite (NO^{2-}) and then to nitrate (NO^3) • This biological process is important for removing nitrogen from wastewater, as excessive nitrogen can cause eutrophication in receiving water bodies	Wang et al. (2023)
Denitrifying bacteria	• Denitrifying bacteria, such as *Pseudomonas* and *Paracoccus species*, carry out denitrification, converting nitrate (NO^{3-}) to nitrogen gas (N_2) under anaerobic conditions • Denitrification is essential for reducing the nitrate concentration in wastewater, particularly in situations where nitrogen removal is required to meet regulatory standards	Yi et al. (2023)
Phosphorus-accumulating organisms (PAOs)	• PAOs are bacteria capable of accumulating and storing phosphorus within their cells • They play a critical role in biological phosphorus removal from wastewater • PAOs, including certain species of the genera *Acinetobacter* and *Accumulibacter*, utilize anaerobic and aerobic conditions to uptake and store phosphorus, thereby reducing its concentration in treated wastewater	Lin et al. (2023)

(continued)

Table 14.3 (continued)

Microbes	Role in wastewater treatment and pollution mitigation	References
Methanogens	• *Methanogenic archaea* are responsible for the final stage of anaerobic digestion, where they convert organic matter into methane gas (biogas) • Methanogens, such as *Methanosarcina* and *Methanobacterium* species, thrive in oxygen-free environments and are commonly used in anaerobic digestion systems for the treatment of organic-rich wastewater and sludge	Wu et al. (2022b)
Algae	• Algae, particularly microalgae, are being explored for wastewater treatment, particularly for nutrient removal and the production of biomass • Algae can take up nutrients, including nitrogen and phosphorus, from wastewater, helping to mitigate eutrophication • Additionally, algae can produce oxygen through photosynthesis, which can contribute to maintaining aerobic conditions in wastewater treatment processes	Sharma et al. (2023a), Mittal et al. (2023)

Table 14.4 Nutrient removal and lipid content of different microbial species

Microbial species	Nitrogen removal (%)	Phosphorus removal (%)	COD removal (%)	Lipid content	References
Monoraphidium sp. FXY-10	80	86	92.33	92.33	Dong et al. (2019)
Anabaena sp.	100	96.5	98.6	7.24	Hena et al. (2015)
Scenedesmus obliquus	80.30	95.72	85.43	46.92	Qu et al. (2020)
Tribonema sp.	100	68–74	52.5	26.3–55.4	Huo et al. (2020)
Scenedesmus sp.	94–98	73–82	69–96	65.2 mg/L/day	Nayak et al. (2016)
Arthrospira platensis	98.8	100	98.4	158 mg/L/day	Hena et al. (2018)

14.4 Interlinkage Between Biofuels Production and Pollution Mitigation

The interlinkage between biofuel production and pollution mitigation using microbes is multifaceted. On one hand, biofuel production can contribute to pollution mitigation by reducing the reliance on fossil fuels and lowering greenhouse gas emissions.

On the other hand, microbes play a crucial role in the production of biofuels and can help mitigate pollution associated with the production process. Here are how these two aspects are interconnected: (i) biofuels as a tool for pollution mitigation: biofuels, such as ethanol and biodiesel, are renewable and sustainable alternatives to fossil fuels. They can be produced from biomass sources, including agricultural residues, energy crops, and algae. By substituting fossil fuels with biofuels, the emissions of greenhouse gases and air pollutants, such as carbon dioxide, nitrogen oxides, and particulate matter, can be significantly reduced. This contributes to the mitigation of climate change and air pollution, leading to improved air quality and environmental health (Ferreira Mota et al. 2022; Babu et al. 2022). (ii) microbes in biofuel production: microbes play a central role in the production of biofuels. They are involved in the conversion of biomass feedstocks into energy-rich compounds through various metabolic pathways. For example, in bioethanol production, microbes like Saccharomyces cerevisiae ferment sugars derived from biomass into ethanol. Similarly, certain microbial species can produce fatty acids or oils that can be converted into biodiesel. By harnessing the metabolic capabilities of these microbes, biofuels can be produced sustainably and efficiently (Zhao et al. 2023; Uthandi et al. 2022; Yang et al. 2023). (iii) mitigation of pollution in biofuel production: the production of biofuels can sometimes involve pollutant emissions and environmental impacts. However, microbes can help mitigate these issues. For instance, during the production of bioethanol from lignocellulosic biomass, the pretreatment and hydrolysis processes release inhibitory compounds, such as lignin-derived by-products and fermentation inhibitors. Microbes, particularly genetically engineered strains, can be utilized to detoxify or metabolize these inhibitory compounds, improving the efficiency of ethanol production and reducing the environmental impact (Dong et al. 2023; Sun et al. 2022; Nasir et al. 2022). (iv) waste utilization and pollution reduction: microbes can play a role in utilizing waste streams generated during biofuel production, thereby reducing pollution. For instance, the waste residues from biodiesel production, such as glycerol, can be used as a substrate for microbial fermentation to produce value-added compounds like 1,3-propanediol or polyhydroxyalkanoates. This not only helps in the valorization of waste but also reduces its environmental burden (Jia et al. 2023; Widajatno et al. 2022; Lin et al. 2023). (v) sustainable practices and optimization: microbes can enable the development of sustainable practices in biofuel production. Through metabolic engineering and optimization of microbial strains, biofuel production processes can be designed to minimize waste generation, maximize resource utilization, and reduce the need for chemical inputs. This holistic approach helps in pollution mitigation by minimizing the environmental footprint of biofuel production (Saravanan et al. 2022; Hoang et al. 2022; Althuri and Venkata Mohan 2022). Microalgae is one key organism that can help in the remediation of contaminants from wastewater along with the production of various types of biofuels and some examples include *Tetraselmis indica* that remediates 67.17% nitrogen, and 70.03% phosphorus from pharmaceutical wastewater and shows 15.69–17.15 mg/L/d lipid productivity (Amit and Ghosh 2020). *Scenedesmus obliquus* remediates 96% nitrogen, and 80% phosphorus from municipal wastewater and shows 56% lipid

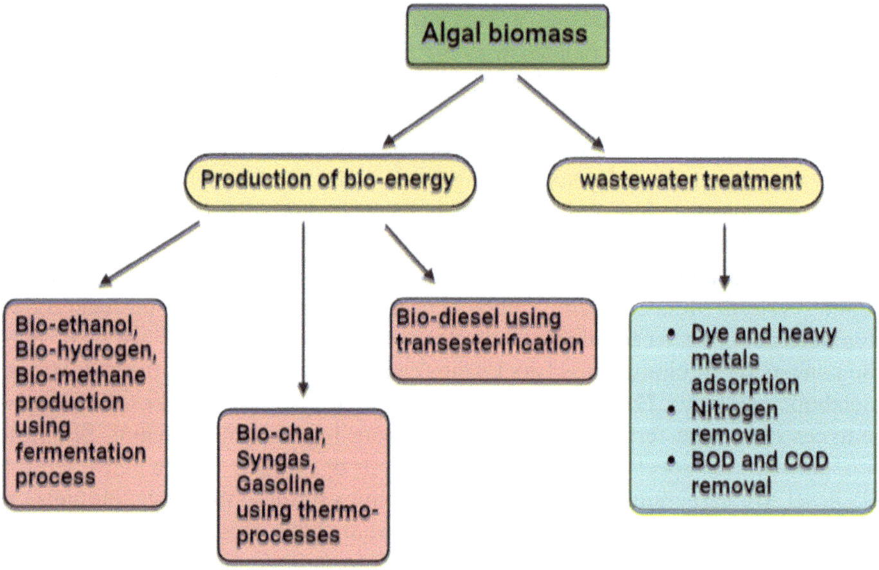

Fig. 14.2 Algae for wastewater treatment and bio-energy production

content (Qu et al. 2020). Figure 14.2 depicts the uses of algae in wastewater treatment and biofuel production. Overall, the interlinkage between biofuel production and pollution mitigation can help in achieving sustainability. Biofuels offer a renewable and cleaner energy alternative to fossil fuels, while microbes play a crucial role in the production process and contribute to pollution reduction through waste utilization, detoxification, and sustainable practices. By integrating microbial technologies and sustainable approaches, biofuels can be produced in an environmentally friendly and responsible manner.

14.5 Socio-economic, Sustainability Aspects, and Circular Economy

Biofuels and the use of microbes in pollution mitigation have socio-economic and sustainability implications. Here are some key aspects to consider: socio-economic aspects: (i) energy security and independence, (ii) job creation and rural development, (iii) diversification of agricultural sectors, (iv) rural electrification and energy access. sustainability aspects: (i) greenhouse gas emissions reduction, (ii) land use and biodiversity considerations, (iii) water resources and water footprint, (iv) social and environmental safeguards, (v) life cycle assessment and techno-economic analysis. The socio-economic and sustainability aspects of biofuels and pollution mitigation using microbes require a holistic approach that balances environmental, social, and economic considerations. By integrating sustainable practices, stakeholder engagement, and effective policies, biofuel production can contribute to pollution mitigation while promoting social well-being and long-term sustainability (Hoang et al. 2022; Skaib and Zedan 2023; Manikandan et al. 2022). Circular economy principles and the remediation of wastewater using microbes for biofuel production and pollution mitigation are interconnected in several ways. Here is how they relate to each other: (i) resource recovery and waste valorization, (ii) nutrient recycling, (iii) water conservation, (iv) waste-to-energy conversion, and (v) environmental sustainability. By utilizing microbes to remediate wastewater and produce biofuels, pollutants, and contaminants are effectively removed, contributing to improved water quality, and reduced environmental impacts. Additionally, the substitution of fossil fuels with biofuels produced from wastewater-derived resources contributes to lower greenhouse gas emissions and mitigates climate change (Althuri and Venkata Mohan 2022; Rojas et al. 2022; Kumar et al. 2022). Figure 14.3 demonstrates the socio-economic and sustainability aspects of biofuel production and pollution mitigation using microbes. By combining the principles of the circular economy with microbial-based wastewater treatment and biofuel production, a more sustainable and resource-efficient approach can be achieved. This integration supports pollution mitigation, waste valorization, nutrient recycling, water conservation, and the transition toward a more sustainable energy system.

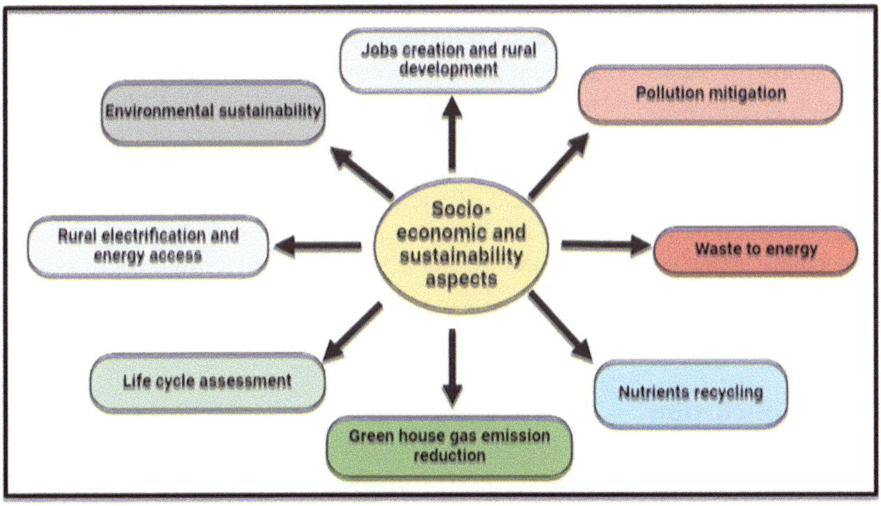

Fig. 14.3 Socio-economic and sustainability aspects of biofuel production and pollution mitigation using microbes

14.6 Challenges, Opportunities, and Future Aspects

Using microbes for wastewater treatment and biofuel production offers sustainable solutions but faces several challenges. Microbial communities are complex and can be sensitive to environmental fluctuations, making process stability a concern. Controlling and optimizing conditions for maximum microbial activity remains a challenge. Scaling up microbial processes from lab to industrial levels is expensive and technically demanding. Maintaining high efficiency and minimizing contamination risks are difficult at larger scales. Microbial diversity and genetic variability can lead to inconsistent performance. Engineering robust strains for specific tasks require time and resources. Additionally, regulatory hurdles, public acceptance, and the release of genetically modified organisms pose ethical and legal issues. In biofuel production, achieving high yields and conversion efficiencies while using low-cost feedstocks remains a major hurdle (Mohd Azhar et al. 2017; Amaro et al. 2023; Anand et al. 2023; Jiang et al. 2022). Balancing economic viability with sustainability is a constant challenge. While microbes hold great promise in wastewater treatment and biofuel production, addressing these challenges is crucial for their widespread adoption and success in these fields.

The field of using microbes for wastewater treatment and biofuel production holds immense promise and opportunities for the future. The use of specialized microbial communities, like anaerobic bacteria, can significantly enhance the efficiency of treatment plants, reducing costs and environmental impact. Future advancements may include engineered microbes with enhanced degradation capabilities and the integration of wastewater treatment with resource recovery to extract valuable products such as bioplastics and biochemicals (Babu et al. 2022; Jiang et al. 2022; Farghali et al. 2023; Jalilian et al. 2020). Continued research may lead to the development of genetically modified microbes with higher biofuel yields and improved tolerance to adverse conditions. The field of microbial-based wastewater treatment and biofuel production holds the key to mitigating pollution, conserving resources, and advancing sustainable energy solutions, making it a vital area for future research and application.

14.7 Conclusion

Microbes play a pivotal role in two critical areas of environmental and energy sustainability: wastewater treatment and biofuel production. The synergy between wastewater treatment and biofuels production using microbes is a promising avenue for circular economy practices. Although this field holds immense opportunities, it also faces several challenges making it a vital area for future research and application. The commercial success of these applications depends on factors such as operational costs, regulatory compliance, market competition, and ongoing technological advancements. Collaboration between researchers, industry, and government agencies is essential to drive the adoption of these sustainable technologies. In essence, the utilization of microbes for wastewater treatment and biofuel production underscores their pivotal role in addressing environmental and energy challenges that can pave the way for a greener, more sustainable future.

References

Agarwal P, Gupta R, Agarwal N (2019) Advances in synthesis and applications of microalgal nanoparticles for wastewater treatment. J Nanotechnol 2019

Althuri A, Venkata Mohan S (2022) Emerging innovations for sustainable production of bioethanol and other mercantile products from circular economy perspective. Bioresour Technol 363

Amaro HM, Salgado EM, Nunes OC, et al (2023) Microalgae systems—environmental agents for wastewater treatment and further potential biomass valorisation. J Environ Manage 337

Amit, Nayak JK, Ghosh UK (2020) Microalgal remediation of anaerobic pretreated pharmaceutical wastewater for sustainable biodiesel production and electricity generation. J Water Process Eng 35

Anand U, Dey S, Parial D, et al (2023) Algae and bacteria consortia for wastewater decontamination and transformation into biodiesel, bioethanol, biohydrogen, biofertilizers and animal feed: a review. Environ Chem Lett

Arifin Y, Tanudjaja E, Dimyati A, Pinontoan R (2014) A second generation biofuel from cellulosic agricultural by-product fermentation using clostridium species for electricity generation. In: Energy Procedia. Elsevier Ltd, pp 310–315

Babu S, Singh Rathore S, Singh R, et al (2022) Exploring agricultural waste biomass for energy, food and feed production and pollution mitigation: a review. Bioresour Technol 360

Bhukal S, Sharma A, Rishi, et al (2022) Spirulina based iron oxide nanoparticles for adsorptive removal of crystal violet dye. Top Catal 65:1675–1685

Caporgno MP, Olkiewicz M, Torras C et al (2016) Effect of pre-treatments on the production of biofuels from Phaeodactylum tricornutum. J Environ Manage 177:240–246

Cho HU, Park JM (2018) Biodiesel production by various oleaginous microorganisms from organic wastes. Bioresour Technol 256:502–508

Constantino A, Rodrigues B, Leon R, et al (2021) Alternative chemo-enzymatic hydrolysis strategy applied to different microalgae species for bioethanol production. Algal Res 56

Dong X, Zhao Y, Li T et al (2019) Enhancement of lipid production and nutrient removal of Monoraphidium sp. FXY-10 by combined melatonin and molasses wastewater treatment. J Taiwan Inst Chem Eng 99:123–131

Dong W, Yang Y, Liu C, et al (2023) Caproic acid production from anaerobic fermentation of organic waste—pathways and microbial perspective. Renew Sustain Energy Rev 175

Farghali M, Mohamed IMA, Osman AI, Rooney DW (2023) Seaweed for climate mitigation, wastewater treatment, bioenergy, bioplastic, biochar, food, pharmaceuticals, and cosmetics: a review. Environ Chem Lett 21:97–152

Ferreira Mota G, Germano de Sousa I, Luiz Barros de Oliveira A, et al (2022) Biodiesel production from microalgae using lipase-based catalysts: Current challenges and prospects. Algal Res 62

Guleri S, Saxena A, Singh KJ et al (2020) Phycoremediation: a novel and synergistic approach in wastewater remediation. J Microbiol Biotechnol Food Sci 10:98–106

Hena S, Abida N, Tabassum S (2015) Screening of facultative strains of high lipid producing microalgae for treating surfactant mediated municipal wastewater. RSC Adv 5:98805–98813

Hena S, Znad H, Heong KT, Judd S (2018) Dairy farm wastewater treatment and lipid accumulation by Arthrospira platensis. Water Res 128:267–277

Hoang AT, Sirohi R, Pandey A, et al (2022) Biofuel production from microalgae: challenges and chances. Phytochem Rev
Hong Y, Chen W, Luo X et al (2017) Microwave-enhanced pyrolysis of macroalgae and microalgae for syngas production. Bioresour Technol 237:47–56
Huo S, Liu J, Zhu F, et al (2020) Post treatment of swine anaerobic effluent by weak electric field following intermittent vacuum assisted adjustment of N:P ratio for oil-rich filamentous microalgae production. Bioresour Technol 314
Ibrahim AFM, Dandamudi KPR, Deng S, Lin YS (2020) Pyrolysis of hydrothermal liquefaction algal biochar for hydrogen production in a membrane reactor. Fuel 265
Jalilian N, Najafpour GD, Khajouei M (2020) Macro and micro algae in pollution control and biofuel production—a review. ChemBioEng Rev 7:18–33
Jia L, Zhou Q, Li Y, Wu W (2023) Integrated treatment of suburb diffuse pollution using large-scale multistage constructed wetlands based on novel solid carbon: nutrients removal and microbial interactions. J Environ Manage 326
Jiang Q, Chen H, Fu Z, et al (2022) Current progress, challenges and perspectives in the microalgal-bacterial aerobic granular sludge process: a review. Int J Environ Res Public Health 19
Kaloudas D, Pavlova N, Penchovsky R (2021) Lignocellulose, algal biomass, biofuels and biohydrogen: a review. Environ Chem Lett 19:2809–2824
Khanna P, Kaur A, Goyal D (2019) Algae-based metallic nanoparticles: synthesis, characterization, and applications. J Microbiol Methods 163
Koppolu V, Vasigala VK (2016) Role of Escherichia coli in biofuel production. Microbiol Insights 9:MBI.S10878
Kumar SPJ, Garlapati VK, Banerjee R (2022) Enzymatic biodiesel synthesis from Trichosporon shinodae yeast through circular economy: a greener approach. Fuel 325
Kumar Sumit NS, Sujata AY, et al (2023) Algae for wastewater treatment: an integrated approach for circular biorefinery. In: Integrated waste management: the circular economy, pp 185–195
Kumari A, Kundu K, Gupta MM, et al (2022) Role of microorganisms in production of biofuels. In: Novel feedstocks for biofuels production, pp 65–116
Li C, Feng L, Lian J, et al (2023) Enhancement of organics and nutrient removal and microbial mechanism in vertical flow constructed wetland under a static magnetic field. J Environ Manage 330
Lin Y, Sun Y, Zhang L, et al (2023) Balancing denitrifying phosphorus-accumulating organisms and denitrifying glycogen-accumulating organisms for advanced nitrogen and phosphorus removal from municipal wastewater. Bioresour Technol 369
Lu Z, Jing Z, Huang J, et al (2022) Can we shape microbial communities to enhance biological activated carbon filter performance? Water Res 212
Ly HV, Kim SS, Choi JH et al (2016) Fast pyrolysis of Saccharina japonica alga in a fixed-bed reactor for bio-oil production. Energy Convers Manag 122:526–534
Maity JP, Bundschuh J, Chen CY, Bhattacharya P (2014) Microalgae for third generation biofuel production, mitigation of greenhouse gas emissions and wastewater treatment: present and future perspectives—a mini review. Energy 78:104–113
Manikandan S, Subbaiya R, Biruntha M, et al (2022) Recent development patterns, utilization and prospective of biofuel production: emerging nanotechnological intervention for environmental sustainability—a review. Fuel 314
Mirzababaei N, Hassani A (2023) Comparative evaluation of organic waste composting rate with urea and KOH application: environment development and sustainability. Environ Dev Sustain
Mittal R, Sharma A, Bhardwaj AK et al (2023) Removal of chromium (VI) using spirulina assisted synthesized mesoporous iron oxide nanoparticles. Inorg Chem Commun 154:110881
Mohd Azhar SH, Abdulla R, Jambo SA et al (2017) Yeasts in sustainable bioethanol production: a review. Biochem Biophys Rep 10:52–61
Mona S, Kaushik A, Kaushik CP (2011) Hydrogen production and metal-dye bioremoval by a Nostoc linckia strain isolated from textile mill oxidation pond. Bioresour Technol 102:3200–3205

Mona S, Kaushik A, Kaushik CP (2013) Prolonged hydrogen production by Nostoc in photobioreactor and multi-stage use of the biological waste for column biosorption of some dyes and metals. Biomass Bioenergy 54:27–35

Mona S, Kumar SS, Kumar V, et al (2020) Green technology for sustainable biohydrogen production (waste to energy): a review. Sci Total Environ 728

Mussatto SI, Machado EMS, Carneiro LM, Teixeira JA (2012) Sugars metabolism and ethanol production by different yeast strains from coffee industry wastes hydrolysates. Appl Energy 92:763–768

Nasir AM, Adam MR, Mohamad Kamal SNEA, et al (2022) A review of the potential of conventional and advanced membrane technology in the removal of pathogens from wastewater. Sep Purif Technol 286

Nayak M, Karemore A, Sen R (2016) Performance evaluation of microalgae for concomitant wastewater bioremediation, CO_2 biofixation and lipid biosynthesis for biodiesel application. Algal Res 16:216–223

Nurdiawati A, Zaini IN, Irhamna AR et al (2019) Novel configuration of supercritical water gasification and chemical looping for highly-efficient hydrogen production from microalgae. Renew Sustain Energy Rev 112:369–381

Pfleger BF, Takors R (2023) Recent progress in the synthesis of advanced biofuel and bioproducts. Curr Opin Biotechnol 80

Qu F, Jin W, Zhou X, et al (2020) Nitrogen ion beam implantation for enhanced lipid accumulation of Scenedesmus obliquus in municipal wastewater. Biomass Bioenergy 134

Rojas LF, Zapata P, Ruiz-Tirado L (2022) Agro-industrial waste enzymes: perspectives in circular economy. Curr Opin Green Sustain Chem 34

Saravanan A, Senthil Kumar P, Jeevanantham S, et al (2022) Recent advances and sustainable development of biofuels production from lignocellulosic biomass. Bioresour Technol 344

Sarkar S, Rekha PN, Ambasankar K, Vijayan KK (2021) Bioremediation efficiency of indigenous seaweeds of Chennai coast in brackishwater system. Aquacult Int 29:233–251

Shahi T, Beheshti B, Zenouzi A, Almasi M (2020) Bio-oil production from residual biomass of microalgae after lipid extraction: the case of Dunaliella Sp. Biocatal Agric Biotechnol 23

Shalaby SM, Madkour FF, El-Kassas HY et al (2021) Green synthesis of recyclable iron oxide nanoparticles using Spirulina platensis microalgae for adsorptive removal of cationic and anionic dyes. Environ Sci Pollut Res 28:65549–65572

Sharma A, Pal K, Saini N et al (2023a) Remediation of contaminants from wastewater using algal nanoparticles via green chemistry approach: an organized review. Nanotechnology 34:352001

Sharma J, Kumar V, Kumar SS, et al (2020) Microalgal consortia for municipal wastewater treatment—lipid augmentation and fatty acid profiling for biodiesel production. J Photochem Photobiol B 202

Sharma A, Mittal R, Sharma P, et al (2023) Sustainable approach for adsorptive removal of cationic and anionic dyes by titanium oxide nanoparticles synthesized biogenically using algal extract of Spirulina. Nanotechnology 34

Sharma A, Singh G, O'Donovan A, et al (2023) A review on development of a greener approach via One-Pot tandem catalysis for biofuels production. Fuel 350

Singh V, Chaudhary DK, Mani I, Dhar PK (2016) Recent advances and challenges of the use of cyanobacteria towards the production of biofuels. Renew Sustain Energy Rev 60:1–10

Skaib A, Zedan R (2023) African journal of advanced pure and applied sciences (AJAPAS) bioenergy: socio-economic and environmental implications and future of biofuels. Afr J Adv Pure Appl Sci 2:19–35

Sun Y, Wang X, Xia S, Zhao J (2022) Cu(II) adsorption on Poly(Lactic Acid) microplastics: significance of microbial colonization and degradation. Chem Eng J 429

The Use of Microbes in Biofuel Production. https://www.news-medical.net/life-sciences/The-Use-of-Microbes-in-Biofuel-Production.aspx. Accessed 27 Jun 2023

Tsai TY, Lo YC, Dong C Di, et al (2020) Biobutanol production from lignocellulosic biomass using immobilized Clostridium acetobutylicum. Appl Energy 277

Uthandi S, Kaliyaperumal A, Srinivasan N et al (2022) Microbial biodiesel production from lignocellulosic biomass: new insights and future challenges. Crit Rev Environ Sci Technol 52:2197–2225
Wang Q, Xia C, Alagumalai K, et al (2023) Biogas generation from biomass as a cleaner alternative towards a circular bioeconomy: Artificial intelligence, challenges, and future insights. Fuel 333
Widajatno RL, Kardena E, Arifianingsih NN, Helmy Q (2022) Activated sludge: conventional dye treatment technique, pp 119–153
Wu Q, Zou D, Zheng X, et al (2022) Effects of antibiotics on anaerobic digestion of sewage sludge: performance of anaerobic digestion and structure of the microbial community. Sci Total Environ 845
Wu X, Wang C, Wang D, et al (2022) Simultaneous methanogenesis and denitrification coupled with nitrifying biofilm for high-strength wastewater treatment: performance and microbial mechanisms. Water Res 225
Xu D, Savage PE (2017) Effect of temperature, water loading, and Ru/C catalyst on water-insoluble and water-soluble biocrude fractions from hydrothermal liquefaction of algae. Bioresour Technol 239:1–6
Xu J, Upcraft T, Tang Q et al (2019) Hydrogen generation performance from Taihu Algae and food waste by anaerobic codigestion. Energy Fuels 33:1279–1289
Yang S, Luo F, Yan J, et al (2023) Biogas production of food waste with in-situ sulfide control under high organic loading in two-stage anaerobic digestion process: strategy and response of microbial community. Bioresour Technol 373
Yi M, Wang C, Wang M, et al (2023) Immobilized denitrifying bacteria on modified oyster shell as biofilter carriers enhance nitrogen removal. J Environ Chem Eng 11
Yu N, Mou A, Sun H, Liu Y (2022) Anaerobic digestion of thickened waste activated sludge under calcium hypochlorite stress: performance stability and microbial communities. Environ Res 212
Zabermawi NM, Alsulaimany FAS, El-Saadony MT, El-Tarabily KA (2022) New eco-friendly trends to produce biofuel and bioenergy from microorganisms: an updated review. Saudi J Biol Sci
Zhao F, Zhang Y, Hu J, et al (2023) Disruption of phosphate metabolism and sterol transport-related genes conferring yeast resistance to vanillin and rapid ethanol production. Bioresour Technol 369
Zhu Y-X, Zhang X, Yang W-C, Li J-F (2023) Enhancement of biomass conservation and bioethanol production of sweet sorghum silage by constructing synergistic microbial consortia. Microbiol Spectr 11

Part V
Miscellaneous Applications of Microbial Metabolites

Chapter 15
From Pond to Pill: Microalgae's Role in Vegan Pharmaceuticals'

Himanshu Jain and Neeraj K. Aggarwal

15.1 Introduction

In recent years, there has been a remarkable surge in the popularity of veganism, a lifestyle marked by the exclusion of animal-derived products. Concurrently, there has been an exponential expansion in the market for plant-based substitutes for conventional animal-based items. The rise of veganism can be attributed, in part, to the increasing awareness among consumers regarding the environmental, ethical, and health implications of animal agriculture. Scientific research has provided insights into the significant contribution of animal agriculture to greenhouse gas emissions, deforestation, and water pollution (Food and Agriculture Organization of the United Nations (FAO) 2006). The extensive cultivation of animal feed, land-use changes, and the release of methane and nitrous oxide from livestock contribute significantly to climate change (Gerber et al. 2013). Additionally, concerns about animal welfare have gained prominence, with documentaries such as "Earthlings" and exposés exposing the cruel and inhumane conditions in factory farms (Kemmerer 2012). These revelations have influenced public perceptions and increased consumer consciousness about the ethical treatment of animals.

Moreover, an expanding body of research has established connections between the overconsumption of animal-derived foods and a range of health concerns, such as heart disease and obesity. The elevated consumption of saturated fats and cholesterol, prevalent in animal-based products, has been correlated with an elevated likelihood of developing cardiovascular disorders. The World Health Organization has classified processed meats as Group 1 carcinogens, indicating a strong link between their consumption and the development of colorectal cancer (Bouvard et al. 2015). These scientific findings have played a pivotal role in driving health-conscious individuals towards plant-based alternatives.

H. Jain (✉) · N. K. Aggarwal
Department of Microbiology, Kurukshetra University, Kurukshetra, Haryana 136119, India
e-mail: himanshu.microbiology@kuk.ac.in

K. Bala et al. (eds.), *Harnessing Microbial Potential for Multifarious Applications*,
Energy, Environment, and Sustainability, https://doi.org/10.1007/978-981-97-1152-9_15

Beyond ethical and environmental considerations, health and wellness motivations have also contributed significantly to the rise of veganism. People are becoming increasingly health-conscious and are seeking alternatives to animal-based products to improve their overall well-being. Plant-based diets, when well-planned and nutritionally balanced, have been associated with numerous health benefits, including reduced risk of chronic diseases and improved weight management. The American Dietetic Association and the Academy of Nutrition and Dietetics have acknowledged that well-planned vegan diets are suitable for individuals at all life stages, including infants, children, adolescents, and pregnant or lactating women (Melina et al. 2016). These endorsements from reputable health organizations have further fuelled the adoption of veganism as a lifestyle choice.

To meet the increasing demand for animal-free alternatives, the food industry has responded with a surge of innovation in plant-based products. Technological advancements have enabled the development of alternative protein sources, such as soy, pea, and wheat, which closely mimic the taste, texture, and nutritional profile of animal-derived products. Plant-based milk substitutes, meat alternatives, and even vegan cheeses have become more widely available and have significantly improved in taste and quality over time (Green and Verbeke 2018). The emergence of these innovative plant-based alternatives has provided consumers with a broader range of options and has made the transition to a vegan lifestyle more feasible and enjoyable.

Veganism and plant-based alternatives have transitioned from niche markets to the mainstream. Retailers and foodservice providers have recognized the potential of this growing consumer base and have expanded their offerings accordingly. Supermarkets now dedicate whole sections to plant-based products, while restaurants and fast-food chains have started including vegan options on their menus (Ruby 2012). This increased accessibility and visibility have facilitated the transition to a vegan lifestyle for many individuals, making it more convenient and appealing.

The rise of social media platforms has had a profound influence on the popularity of veganism. The ability to share information and experiences in real-time has allowed influencers, bloggers, and celebrities to promote vegan lifestyles to a wide audience. Social media platforms have become platforms for education, recipe sharing, and community building, empowering individuals to embrace veganism and explore plant-based (Mejia et al. 2018). The influence of social media has played a significant role in spreading awareness, dispelling myths, and normalizing the vegan lifestyle.

Veganism, distinguished by its commitment to eschew animal-derived products and its advocacy for plant-based substitutes, has garnered worldwide attention as a lifestyle preference founded on considerations for animal welfare, ecological sustainability, and individual well-being. The tenets of veganism find application in diverse sectors, including the pharmaceutical industry, where there is a growing imperative to design therapeutic solutions that adhere to ethical and environmental principles (Ginsburg 2018).

The field of medical science is continuously advancing to explore innovative approaches for disease prevention, management, and treatment (Lea et al. 2016a). One emerging area of research and development is the utilization of vegan therapeutic agents, which offer a promising alternative to traditional pharmaceuticals (Jerez-Martel et al. 2022). Vegan therapeutic agents are derived exclusively from plant-based sources and have the potential to address a wide range of health concerns while also aligning with ethical and environmental considerations.

The health benefits of vegan therapeutic agents can be attributed to their unique composition and characteristics. Plant-based compounds contain a diverse array of phytochemicals, including polyphenols, flavonoids, and antioxidants, which have demonstrated anti-inflammatory, antimicrobial, and anticancer properties (Fleischhacker and Shinn 2016). These agents have shown potential in mitigating chronic diseases such as cardiovascular ailments, diabetes, and certain types of cancer. Vegan therapeutic agents often possess lower toxicity profiles compared to synthetic drugs, reducing the risk of adverse effects and drug interactions.

In addition to their health benefits, vegan therapeutic agents address sustainability and ethical concerns associated with the conventional pharmaceutical industry. The pharmaceutical industry heavily relies on animal-based testing and sourcing of ingredients, leading to significant environmental impact and ethical dilemmas. By embracing vegan therapeutic agents, researchers and medical professionals can reduce their reliance on animal experimentation, contributing to the global movement toward more ethical and sustainable medical practices. Vegan agents also align with the increasing demand for cruelty-free and environmentally conscious alternatives, making them an attractive option for individuals who prioritize animal welfare and ecological balance (Lea et al. 2016b).

Another compelling aspect of vegan therapeutic agents is their potential to improve the global accessibility and affordability of healthcare. Plant-based sources are often abundant and widely available, making the production of vegan agents more scalable and cost-effective. This accessibility could be particularly beneficial for populations in low-income countries or regions with limited access to conventional pharmaceuticals. Furthermore, cultivating vegan agents locally can foster self-sufficiency and reduce dependence on imported medications, further enhancing accessibility and affordability (Martínez-Huélamo et al. 2017).

Vegan therapeutic agents also contribute to a paradigm shift in medical science by promoting preventive healthcare. These agents are rich in essential nutrients, vitamins, and minerals that support overall well-being and immune function. By integrating vegan agents into preventive medicine strategies, healthcare professionals can empower individuals to take proactive measures in maintaining their health (Cappoen et al. 2021). This shift toward prevention can potentially alleviate the burden on healthcare systems, as it focuses on reducing the incidence and progression of diseases rather than primarily addressing their symptoms.

Microalgae are photosynthetic microorganisms found in aquatic environments, which have garnered significant attention in recent years due to their potential as a sustainable source for the development of therapeutic agents. Their diverse biochemical composition and ability to produce bioactive compounds make them particularly appealing for the creation of vegan pharmaceuticals. This growing interest in microalgae-based therapeutics is driven by the rise of veganism and the demand for ethical and sustainable alternatives to traditional therapeutic agents, many of

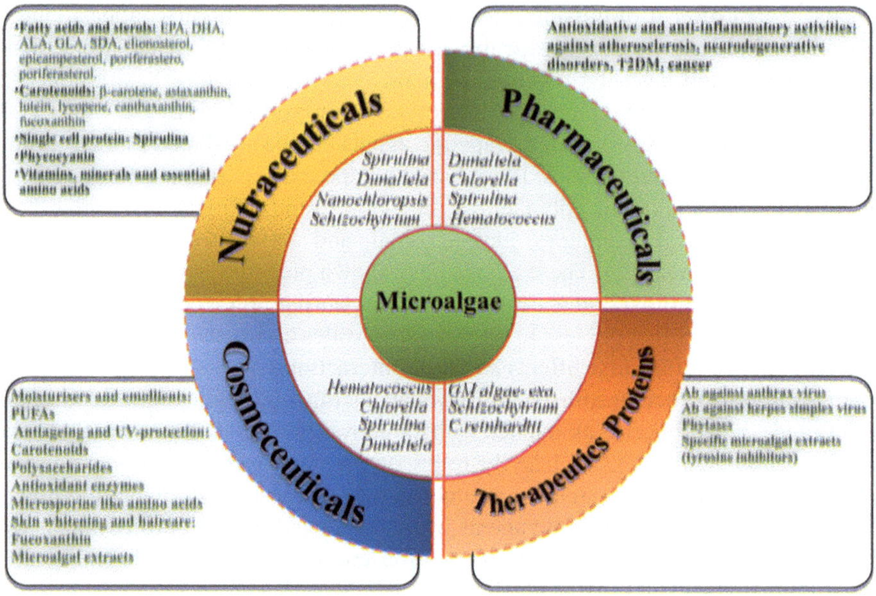

Fig. 15.1 Microalgal bioactive components and their applications (Mu et al. 2019)

which rely on animal-derived ingredients or animal testing. These microscopic organisms, found in various aquatic environments, offer a multitude of advantages, including their sustainable cultivation, high nutritional value, and abundant bioactive compounds. Microalgae are known for their rapid growth rates and ability to thrive in diverse environments, making them an ideal candidate for sustainable cultivation (Safi et al. 2014). Unlike conventional drug sources that rely on animal-derived products, the cultivation of microalgae is not associated with animal exploitation. Furthermore, microalgae can be cultivated using renewable energy sources, such as solar power, reducing the environmental impact typically associated with drug production.

Microalgae possess an extensive range of bioactive compounds with therapeutic potential. These compounds include pigments, polysaccharides, proteins, fatty acids, vitamins, and antioxidants (Vazhappilly et al. 2020). Some microalgae species have been found to produce unique molecules with promising pharmacological properties, including anticancer, anti-inflammatory, antimicrobial, and neuroprotective activities (Patel et al. 2021). The abundance and diversity of these bioactive compounds make microalgae an intriguing resource for the development of novel vegan drugs.

Several studies have demonstrated the potential of microalgae in producing compounds that could serve as vegan alternatives to traditional drugs. For example, microalgae-derived omega-3 fatty acids have been identified as a sustainable source for vegan-friendly supplements, replacing fish-derived omega-3 oils (Kumar et al.

2016). Additionally, microalgae have been investigated for their production of anti-inflammatory compounds, such as astaxanthin and phycocyanin, which could be utilized in the development of vegan anti-inflammatory drugs (Paliwal et al. 2015).

Bioprospecting microalgae involves the systematic screening and isolation of these bioactive compounds from a diverse array of microalgae species. The vast biodiversity of microalgae presents a rich source of untapped resources for sustainable drug development. Moreover, microalgae possess practical advantages that contribute to their appeal as a sustainable source for vegan therapeutic compounds. They exhibit rapid growth rates, high biomass productivity, and the ability to utilize various carbon sources, making them highly efficient for large-scale cultivation and production (Sun et al. 2021a).

This chapter explores microalgae as a sustainable source of vegan therapeutic compounds. It provides an overview of microalgae's composition and unique features for drug discovery. Therapeutic applications in chronic disease treatment, dermatology, and nutraceuticals are explored. Challenges and future perspectives in the development and commercialization of microalgae-based vegan drugs are addressed. This includes regulatory considerations, biotechnology advancements, and the potential impact on human health and the environment. By contributing to the knowledge on vegan therapeutic agents, this chapter highlights the ethical and sustainable approach of bioprospecting microalgae for pharmaceutical development, bridging scientific progress with the values of a vegan lifestyle.

15.2 Microalgae as Potential Sources for Vegan Drug Development

Microalgae, also referred to as phytoplankton, are a diverse group of photosynthetic microorganisms that play a crucial role in various ecosystems, including freshwater, marine, and terrestrial environments. With a fossil record dating back billions of years, they are considered to be one of the oldest forms of life on Earth (Knoll 2014). Through the process of photosynthesis, microalgae have the remarkable ability to convert sunlight and carbon dioxide into organic matter, thereby contributing to primary production and carbon cycling in aquatic and terrestrial systems (Raven and Geider 2011).

The biochemical composition of microalgae is highly variable and influenced by factors such as species, growth conditions, and nutrient availability. Nonetheless, microalgae generally contain a range of essential biomolecules, including proteins, carbohydrates, lipids, vitamins, and minerals. These biomolecules have garnered significant interest from scientists and researchers due to their potential applications in various industries, including food, pharmaceuticals, biofuels, and cosmetics (Borowitzka and Moheimani 2013).

Proteins, the primary constituents of microalgae, play a crucial role in their growth and metabolism, participating in diverse cellular functions such as enzyme catalysis, structural support, and nutrient transport. Microalgae offer a rich source of essential amino acids, the building blocks of proteins, making them significant for human nutrition with their high protein content and quality, microalgae show great potential in addressing protein malnutrition and serving as a supplement in animal feed.

In addition to proteins, microalgae contain significant amounts of carbohydrates, particularly polysaccharides, which function as energy reserves and provide structural support to the cells. Polysaccharides derived from microalgae, such as agar, carrageenan, and alginate, find wide application in the food industry as gelling agents, stabilizers, and thickeners due to their unique rheological properties and improved functionality in food products (Chiovitti et al. 2003).

Lipids, including fatty acids, glycerolipids, and sterols, are also essential components of microalgae and have attracted attention for their potential in biofuel production. Under specific growth conditions, microalgae can accumulate substantial amounts of lipids, offering a sustainable and promising source of renewable energy (Milledge 2011). Moreover, microalgal lipids are rich in omega-3 fatty acids, such as eicosapentaenoic acid (EPA) and docosahexaenoic acid (DHA), which have been associated with various health benefits, including cardiovascular health and cognitive development (Belay et al. 1996). Microalgae also serve as a valuable source of vitamins and minerals. They have the capacity to accumulate a wide range of vitamins, including vitamin A, vitamin C, vitamin E, and various B vitamins. Furthermore, microalgae are abundant in essential minerals like iron, zinc, and selenium, which play vital roles in human health and various physiological processes. Overall, microalgae present a versatile and valuable resource, offering proteins, carbohydrates, lipids, vitamins, and minerals, with potential applications in nutrition, food industry, biofuel production, and human health.

A Unique features of microalgae for drug discovery

The biochemical composition of microalgae not only holds value for human applications but also plays a crucial role in ecological processes. Microalgae contribute to nutrient cycling and support the growth of other organisms in aquatic ecosystems through the production of essential biomolecules. Moreover, the ability of microalgae to capture carbon dioxide during photosynthesis has garnered attention as a potential tool for mitigating greenhouse gas emissions and combating climate change (Abida et al. 2013).

Microalgae exhibit a remarkable range of features that make them a highly promising resource for drug discovery. These microscopic organisms, commonly found in various aquatic environments, offer distinct advantages over traditional sources when it comes to identifying and developing novel pharmaceutical compounds.

Biochemical Diversity: Microalgae showcase an exceptional biochemical diversity, synthesizing a wide array of bioactive compounds. For instance, diatoms, a type of microalgae, produce intricate silica-based cell walls that contain bioactive compounds with potential pharmaceutical applications. Additionally, microalgae produce various compounds like alkaloids, polyunsaturated fatty acids, peptides, polysaccharides, pigments, and terpenoids. For example, *Dunaliella salina*, a microalga, produces β-carotene, which is used in the production of provitamin A supplements. Certain microalgae species like *Nannochloropsis* are known to produce omega-3 fatty acids, which have significant health benefits (Lauritano et al. 2016).

Structural Complexity: Microalgae possess a complex cellular organization, comprising intricate structures such as cell walls, chloroplasts, and vacuoles. These structures contribute to the synthesis and accumulation of unique secondary metabolites with potential therapeutic value. For instance, microalgae belonging to the genus *Spirulina* have a spiral-shaped structure and produce a diverse range of bioactive compounds, including phycocyanins, which possess antioxidant and anti-inflammatory properties].

Environmental Adaptability: Microalgae have evolved to thrive in diverse ecological niches, ranging from freshwater bodies to extreme environments like hot springs and polar regions. This adaptability suggests that microalgae have developed robust biochemical mechanisms to combat environmental stressors, making them potential sources of therapeutically relevant molecules. For example, extremophilic microalgae found in hot springs have adapted to survive high temperatures and produce heat-resistant enzymes with potential industrial and pharmaceutical applications (Mallick et al. 2000).

Sustainable Cultivation: Microalgae offer the advantage of sustainable cultivation methods. They have rapid growth rates and can be cultivated in a variety of environments, including closed photobioreactors and open ponds. This scalability, coupled with their ability to utilize diverse nutrient sources, makes microalgae a viable option for large-scale production of bioactive compounds, ensuring a sustainable supply for drug discovery efforts. Microalgae like *Chlorella* and *Scenedesmus* are cultivated for their high protein content and can serve as a sustainable source of protein-based therapeutics (Nascimento et al. 2016).

Minimal Environmental Impact: Compared to traditional sources of pharmaceutical compounds, such as terrestrial plants or marine organisms, microalgae have a minimal ecological footprint. They require less land, water, and nutrients to grow, and their cultivation does not pose significant risks to biodiversity or natural habitats. This environmentally friendly aspect aligns with the growing emphasis on sustainable and eco-conscious practices in drug discovery and development. For example, the cultivation of microalgae to produce omega-3 fatty acids reduces the pressure on marine fish populations and protects marine ecosystems (Ghosh et al. 2016).

Biotechnological Potential: Microalgae can be genetically modified and optimized through biotechnological approaches to enhance the production of specific compounds of interest. These genetic engineering techniques can be employed to increase the yield of bioactive molecules or introduce desired traits into microalgae strains. This potential for genetic manipulation opens up new possibilities for targeted drug discovery efforts. Genetic engineering of microalgae can be used to enhance the production of therapeutic proteins or optimize the synthesis of specific bioactive compounds.

B Exploring the diversity of microalgae species

Microalgae, comprising a wide range of unicellular and multicellular organisms, thrive in diverse aquatic environments, encompassing freshwater and marine habitats. These microorganisms possess the remarkable ability to utilize sunlight for photosynthesis, converting carbon dioxide into organic matter and releasing oxygen as a byproduct. Microalgae have gained significant attention in recent years due to their nutritional value, ecological sustainability, and potential applications in the pharmaceutical field. The diversity of microalgae is vast, with estimates suggesting the existence of 30,000 to 800,000 species yet to be fully explored (Plaza et al. 2018). These species are classified into various taxonomic groups, including Chlorophyta (green algae), Cyanobacteria (blue-green algae), Bacillariophyta (diatoms), and Haptophyta (haptophytes), each comprising numerous species with distinct physiological and biochemical characteristics.

One of the most promising aspects of microalgae lies in their ability to produce an extensive range of bioactive compounds, such as pigments, lipids, polysaccharides, proteins, and secondary metabolites. These compounds possess diverse pharmacological properties, including antioxidant, anti-inflammatory, antimicrobial, and anticancer activities, making them valuable for drug discovery (Moosavi et al. 2019). In line with the growing demand for vegan-friendly pharmaceuticals, which eliminate animal-derived ingredients, microalgae present an ideal source for vegan drug development. Their abundant bioactive compounds offer opportunities for identifying and developing therapeutic agents that align with vegan principles.

15.3 Bioactive Compounds from Microalgae with Therapeutic Potential

Microalgae are microscopic organisms that inhabit various aquatic environments and exhibit an extraordinary capacity for photosynthesis. These organisms possess a diverse range of pigments, lipids, polysaccharides, and other bioactive compounds, making them an intriguing resource for pharmaceutical applications (Jin et al. 2021). Antioxidants and anti-inflammatory compounds derived from microalgae offer promising avenues for vegan drug discovery, providing sustainable and ethical alternatives to traditional pharmaceuticals.

15.3.1 Bioactive Compounds in Microalgae

Microalgae are an exceptional source of bioactive compounds, encompassing a vast array of molecules with significant potential for therapeutic applications. These microscopic photosynthetic organisms have evolved unique biochemical pathways to thrive in various ecological niches, resulting in the production of a wide range of bioactive compounds (Fig. 15.2).

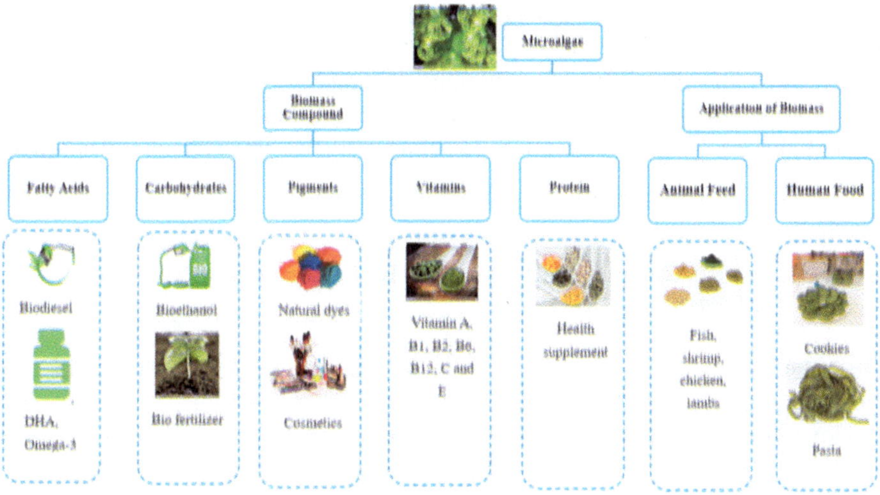

Fig. 15.2 Key microalgal products (Kusmayadi et al. 2021)

15.3.1.1 Pigments: Chlorophylls, Carotenoids, and Phycobiliproteins

Microalgae are well-known for their vibrant colors, which can be attributed to the presence of pigments such as chlorophylls, carotenoids, and phycobiliproteins.

Chlorophylls: Chlorophylls are green pigments that play a vital role in photosynthesis, capturing light energy and converting it into chemical energy. Besides their fundamental role in energy production, chlorophylls have been investigated for their potential health benefits. For instance, chlorophyll-rich extracts from certain microalgae species, such as *Chlorella* and *Spirulina*, have exhibited potent antioxidant and anti-inflammatory properties in various studies (Ghosh et al. 2016). These attributes make them potential candidates for pharmaceutical applications targeting oxidative stress-related conditions.

Carotenoids: Carotenoids are responsible for the red, orange, and yellow hues observed in many microalgae species. These pigments possess notable antioxidant properties and have been associated with various health benefits. For instance, astaxanthin, a carotenoid found abundantly in the microalga *Haematococcus pluvialis*, has demonstrated potent antioxidant, anti-inflammatory, and neuroprotective effects. It has been shown to scavenge reactive oxygen species (ROS) and inhibit lipid peroxidation, thereby protecting cells from oxidative damage. Astaxanthin has potential applications in the prevention and treatment of various diseases, including cardiovascular disorders and neurodegenerative conditions (Bhat and Madyastha 2010).

Phycobiliproteins: Phycobiliproteins are unique water-soluble pigments found in certain groups of microalgae, such as cyanobacteria and red algae. They exhibit brilliant colors ranging from blue to red and possess excellent light-absorbing properties. Apart from their applications as natural food colorants, phycobiliproteins have garnered attention for their potential therapeutic properties. For instance, C-phycocyanin, a phycobiliprotein found in *Spirulina platensis*, has demonstrated antioxidant, anti-inflammatory, and immunomodulatory effects (Shene et al. 2016). These properties make phycobiliproteins promising candidates for the development of pharmaceutical interventions targeting oxidative stress and immune-related disorders.

15.3.1.2 Lipids: Omega-3 Fatty Acids, Polyunsaturated Fatty Acids (PUFAs)

Microalgae are a significant source of lipids, including omega-3 fatty acids and polyunsaturated fatty acids (PUFAs).

Omega-3 Fatty Acids: Omega-3 fatty acids, such as eicosapentaenoic acid (EPA) and docosahexaenoic acid (DHA), are essential for human health, particularly in maintaining cardiovascular health and supporting brain function. Traditionally, fish and fish oil have been the primary sources of omega-3 fatty acids. However, microalgae serve as the fundamental producers of these beneficial fatty acids in aquatic food chains. By directly consuming microalgae or utilizing microalgae-derived supplements, individuals can obtain a vegan-friendly and sustainable source of omega-3 fatty acids. For example, the microalga *Schizochytrium sp.* is known for its high DHA content and is cultivated commercially for the production of vegan omega-3 supplements. Eicosapentaenoic acid (EPA) and docosahexaenoic acid (DHA) are omega-3 fatty acids commonly found in microalgae, particularly species of the genera *Nannochloropsis* and *Schizochytrium.* These fatty acids possess potent anti-inflammatory properties and have been shown to attenuate inflammatory responses associated with conditions such as cardiovascular disease, rheumatoid arthritis, and asthma (Fernandes et al. 2021).

Polyunsaturated Fatty Acids (PUFAs): In addition to omega-3 fatty acids, microalgae also produce other valuable PUFAs, including linolenic acid, linoleic acid, and arachidonic acid. These PUFAs have been associated with various health benefits. For instance, arachidonic acid derived from the microalga *Crypthecodinium cohnii* has shown potential as a supplement for infant formulas, promoting healthy growth and development (Sánchez et al. 2018). The diverse range of PUFAs present in microalgae presents opportunities for the development of vegan-friendly lipid-based pharmaceuticals and nutraceuticals.

15.3.1.3 Proteins, Peptides, and Amino Acids

Microalgae contain proteins, peptides, and amino acids that play crucial roles in various physiological functions in humans.

Proteins: Microalgae produce proteins with a broad range of biological activities, making them valuable resources for potential therapeutic applications. These proteins may exhibit antioxidant, antimicrobial, and immunomodulatory properties.

Some microalgae species are particularly rich in high-quality proteins that possess a balanced amino acid composition, making them potential sources of protein for human nutrition. For example, *Arthrospira platensis* is well-known for its protein content and has been consumed as a dietary supplement due to its potential health benefits. Phycobiliproteins, such as phycoerythrin and phycocyanin, are water-soluble pigments found in several microalgae species. These compounds exhibit strong antioxidant properties and have demonstrated anti-inflammatory effects by modulating inflammatory pathways. Phycobiliproteins hold promise as therapeutic agents for conditions associated with oxidative stress and chronic inflammation (Chen et al. 2021).

Peptides: Microalgae-derived peptides have gained attention due to their potential bioactivities and therapeutic applications. Peptides are short chains of amino acids with unique sequences and structures, enabling them to interact with specific molecular targets in the human body. Microalgae-derived peptides have been reported to exhibit antioxidant, anticancer, antihypertensive, and immunomodulatory effects. For example, peptides derived from the microalga *Chlorella sorokiniana* have demonstrated potent antioxidant and anti-inflammatory activities (Abdullah et al. 2016). These peptides hold promise for the development of peptide-based pharmaceutical interventions.

Amino Acids: Microalgae are abundant sources of essential and non-essential amino acids, which are vital for protein synthesis, cell growth, and various metabolic processes in the human body. Microalgae-derived amino acids offer a sustainable and vegan-friendly alternative to traditional sources. Additionally, the specific amino acid compositions of microalgae may contribute to their unique bioactivities. For instance, *Spirulina* contains high levels of γ-aminobutyric acid (GABA), an amino acid with potential neuroprotective and anxiolytic effects (Dvir et al. 2019).

15.3.1.4 Polysaccharides and Exopolysaccharides

Microalgae produce a variety of polysaccharides and exopolysaccharides, which exhibit diverse bioactivities and therapeutic potentials.

Polysaccharides: Microalgae-derived polysaccharides have demonstrated various biological activities, including antioxidant, immunomodulatory, antitumor, and antiviral effects. These complex carbohydrates have the potential to serve as natural drugs or functional food ingredients. For example, Fucoidans are sulfated polysaccharides derived from brown microalgae, including species like *Fucus vesiculosus* and *Saccharina latissima*. These compounds exhibit anti-inflammatory effects by inhibiting pro-inflammatory cytokines and enzymes involved in inflammatory pathways. Fucoidans have shown therapeutic potential for conditions like osteoarthritis, inflammatory bowel disease, and skin disorders (Jin et al. 2020).

Exopolysaccharides: Exopolysaccharides are polysaccharides secreted by microalgae into their surrounding environment. These polymers serve various roles, including protection against desiccation, cell aggregation, and biofilm formation.

Exopolysaccharides produced by microalgae have been investigated for their potential in wound healing, drug delivery systems, and as immunomodulators. For example, exopolysaccharides derived from the microalga *Porphyridium* have shown wound healing properties by promoting cell proliferation and migration (Dvir et al. 2019).

15.3.1.5 Vitamins, Minerals, and Antioxidants

Microalgae are rich sources of vitamins, minerals, and antioxidants, which are essential for maintaining human health.

Vitamins: Microalgae have the capability to synthesize and accumulate various vitamins, making them valuable sources of these essential nutrients. For instance, certain microalgae species can provide vitamin A in the form of retinol and carotenoids, vitamin C (ascorbic acid), vitamin E (tocopherols), and B-group vitamins (thiamine, riboflavin, niacin, etc.). These vitamins play crucial roles in energy metabolism, immune function, antioxidant defense, and overall well-being. Microalgae-derived vitamin supplements have been explored for their potential in meeting dietary requirements and preventing deficiencies (Wells et al. 2016).

Minerals: Microalgae possess a wide range of essential minerals, including iron, calcium, magnesium, potassium, and zinc. These minerals are crucial for maintaining proper cellular function, nerve transmission, muscle contraction, and bone health. Microalgae can bioaccumulate and concentrate minerals from their surrounding environment, making them potential sources of mineral-rich supplements. For example, the microalga *Arthrospira platensis* has been reported to contain significant amounts of iron and calcium, making it a valuable dietary source for individuals with mineral deficiencies (Sun et al. 2021b).

Antioxidants: Microalgae produce a diverse array of antioxidants, such as phenolic compounds, flavonoids, and tocopherols. These antioxidants play a crucial role in scavenging harmful free radicals, reducing oxidative stress, and protecting cells from damage. Antioxidant-rich microalgae have the potential to be utilized in the development of therapeutic agents to combat various diseases associated with oxidative damage. For instance, phenolic-rich extracts from the microalga *Nannochloropsis* have demonstrated potent antioxidant activity and the ability to inhibit oxidative stress-related enzymes (Kumar et al. 2016). *Chlorella vulgaris* is a widely studied microalga known for its high antioxidant content. It contains various antioxidants such as chlorophylls, carotenoids, and phycobiliproteins. These compounds have demonstrated free radical scavenging activities and potential protective effects against oxidative stress-related diseases (Fig. 15.3 and Table 15.1).

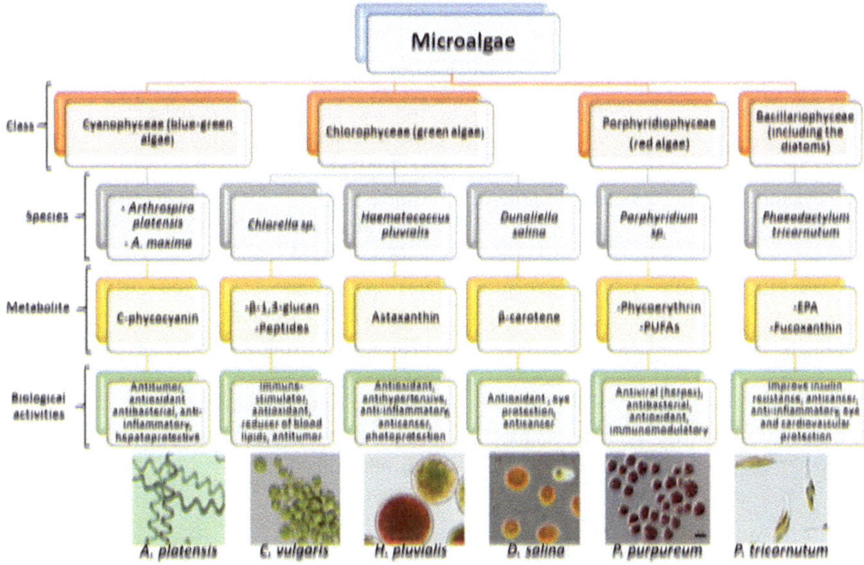

Fig. 15.3 Key microalgal classes, species, and bioactivities (Vieira et al. 2020)

15.4 Therapeutic Applications of Microalgae-Derived Drugs

15.4.1 Neurological Disorders and Microalgae-Derived Drugs

Neurological disorders, including Alzheimer's disease, Parkinson's disease, multiple sclerosis, and stroke, pose significant challenges to global healthcare systems (Alkahtani et al. 2023). These conditions are characterized by progressive neurodegeneration, resulting in cognitive impairment, motor dysfunction, and reduced quality of life for affected individuals. To address the underlying mechanisms and symptoms of neurological disorders, researchers have turned their attention to microalgae-derived drugs as potential treatments and preventive strategies.

15.4.1.1 Omega-3 Fatty Acids and Neuroprotection

Docosahexaenoic acid (DHA), an omega-3 fatty acid, plays a vital role in maintaining the structural integrity and functionality of neuronal cell membranes. It is particularly abundant in the brain, supporting neuronal development, synaptic plasticity, and overall brain health. Microalgae, such as *Schizochytrium* and *Ulkenia*

Table 15.1 Bioactive compounds obtained from microalgae with mechanisms of action for drug discovery

Bioactive compound	Microalgae source	Mechanism of action	References
Astaxanthin	*Haematococcus pluvialis*	Antioxidant, anti-inflammatory, neuroprotection	Zhang et al. (2020)
Phycocyanin	*Spirulina platensis*	Antioxidant, anti-inflammatory, neuroprotection	Dyall (2015)
Docosahexaenoic acid	*Schizochytrium sp.*	Neuroprotection, cognitive enhancement	Kidd (2019)
Phosphatidylserine	*Phaeodactylum tricornutum*	Neurotransmission enhancement, cognitive enhancement	Heo and Jeon (2020)
Fucoxanthin	*Undaria pinnatifida, Laminaria japonica*	Antioxidant, anti-inflammatory, neuroprotection, cognitive enhancement	Calder (2020)
Eicosapentaenoic acid	*Nannochloropsis sp., Isochrysis sp.*	Anti-inflammatory, neuroprotection, cognitive enhancement	Calder (2020)
C-Phycocyanin	*Arthrospira platensis*	Antioxidant, anti-inflammatory, neuroprotection	Deng et al. (2021)
Spirulina peptides	*Spirulina platensis*	Neuroprotection, cognitive enhancement	Kulshreshtha et al. (2014)
Lutein	*Chlorella sp., Dunaliella sp.*	Antioxidant, anti-inflammatory, neuroprotection	Sharma et al. (2020)
Zeaxanthin	*Dunaliella sp.*	Antioxidant, anti-inflammatory, neuroprotection	Chew et al. (2017)
Chlorophyll	*Chlorella sp., Spirulina platensis*	Antioxidant, anti-inflammatory, neuroprotection	Ghosh et al. (2016)
γ-linolenic acid	*Arthrospira platensis*	Anti-inflammatory, neuroprotection	Aitzetmüller et al. (2020)
Spirulina polysaccharides	*Spirulina platensis*	Neuroprotection, cognitive enhancement	Palozza et al. (2021)
β-carotene	*Dunaliella salina*	Antioxidant, anti-inflammatory, neuroprotection	Palozza et al. (2021)
Cerebroside	*Nannochloropsis oculata*	Neuroprotection, cognitive enhancement	Ferrante et al. (2018)

(continued)

Table 15.1 (continued)

Bioactive compound	Microalgae source	Mechanism of action	References
Mycosporine-like amino acids	*Porphyridium purpureum*	Antioxidant, photo-protective, neuroprotection	Heo et al. (2012)
Dihydrofucoxanthin	*Ishigakiella tenuissima*	Antioxidant, anti-inflammatory, neuroprotection	Heo et al. (2012)
Lycopene	*Dunaliella salina*	Antioxidant, anti-inflammatory, neuroprotection	Kim et al. (2018)
Polysaccharides	*Chlorella sp., Phaeodactylum tricornutum*	Anti-inflammatory, neuroprotection	Hu et al. (2019)
Phycobiliproteins	*Porphyridium cruentum*	Antioxidant, anti-inflammatory, neuroprotection	Ahmed et al. (2017)

species, are rich sources of DHA, making them promising candidates for neuroprotective interventions. The neuroprotective effects of DHA in microalgae-derived drugs can be attributed to several mechanisms. DHA has been shown to reduce neuroinflammation, modulate inflammatory signaling pathways, inhibit the production of pro-inflammatory molecules, and promote the resolution of inflammation (Hopperton et al. 2010). Additionally, DHA acts as an antioxidant, neutralizing reactive oxygen species and lipid peroxidation products that contribute to neuronal injury. DHA's influence on synaptic function and neurotransmitter signaling further supports cognitive processes, memory formation, and mood regulation.

Preclinical studies using microalgae-derived DHA have shown promising results in various neurological disorders. In Alzheimer's disease, DHA has demonstrated potential in reducing the accumulation of amyloid-beta plaques and preventing cognitive decline (Ma et al. 2010). It may also promote the clearance of tau protein, another pathological feature associated with Alzheimer's disease. In Parkinson's disease, DHA has exhibited neuroprotective effects by mitigating dopaminergic cell loss, reducing oxidative stress, and enhancing dopaminergic neurotransmission (Jiang et al. 2007). Microalgae-derived DHA also holds promise in multiple sclerosis, where it has shown potential in modulating immune responses, reducing inflammatory demyelination, and promoting remyelination. In stroke, DHA's neuroprotective properties include enhancing neuronal survival, reducing ischemic damage, and facilitating post-stroke recovery (Calon et al. 2005).

15.4.1.2 Antioxidants and Neuroprotection

Microalgae are valuable sources of antioxidants, such as vitamins C and E, carotenoids, and phycobiliproteins, which play a critical role in neuroprotection by combating oxidative stress. Oxidative stress arises from an imbalance between reactive oxygen species (ROS) production and antioxidant defense mechanisms in the brain. ROS can damage cellular components and lead to neuronal dysfunction and cell death. Microalgae-derived antioxidants help neutralize ROS, thereby reducing oxidative damage and promoting neuronal health.

Carotenoids, including astaxanthin, β-carotene, and lutein, are potent antioxidants found in microalgae (Aitzetmüller et al. 2020). These compounds can cross the blood–brain barrier and accumulate in brain tissues, protecting neurons from oxidative stress-induced injury. They scavenge free radicals, inhibit lipid peroxidation, and modulate inflammatory responses, contributing to neuroprotection and preserving cognitive function. Vitamin E, including α-tocopherol and other tocopherols, acts as a chain-breaking antioxidant, intercepting free radicals and supporting the recycling of other antioxidants. Phycobiliproteins, water-soluble pigments found in certain microalgae species, possess antioxidant properties and exhibit anti-inflammatory effects, protecting against neurodegenerative processes (Zhang et al. 2020).

15.4.1.3 Neurotransmitter Modulation and Cognitive Function

Neurotransmitters play a crucial role in various cognitive functions, and imbalances in their levels and signaling pathways contribute to the development and progression of neurological disorders. Microalgae-derived drugs have shown potential in modulating neurotransmitter systems, thus influencing cognitive function and symptom management. Peptides derived from microalgae can act as agonists or antagonists to specific receptors, influencing neuronal excitability and synaptic transmission. Microalgae-derived peptides may modulate serotonin release and reuptake, potentially affecting mood regulation and alleviating symptoms associated with mood disorders. They may also interact with gamma-aminobutyric acid (GABA) receptors, enhancing GABAergic signaling and exerting anxiolytic and anticonvulsant effects. By modulating glutamate receptors, microalgae-derived peptides may affect synaptic plasticity and cognitive function (Traber et al. 2019).

15.4.2 Cardiovascular Health and Microalgae

Cardiovascular diseases, including heart disease and stroke, are leading causes of morbidity and mortality worldwide. The prevalence of these conditions is influenced by various factors, including lifestyle choices, dietary habits, and genetic predisposition. In recent years, there has been a growing interest in exploring natural alternatives for promoting cardiovascular health, and microalgae have emerged as a promising source of bioactive compounds with potential therapeutic benefits.

15.4.2.1 Omega-3 Fatty Acids and Cardiovascular Health

One of the most well-known and extensively studied bioactive compounds found in microalgae is omega-3 fatty acids. Omega-3 fatty acids, including eicosapentaenoic acid (EPA) and docosahexaenoic acid (DHA), have been widely recognized for their cardiovascular benefits (Delgado-Lista et al. 2012). These essential fatty acids cannot be synthesized by the human body and must be obtained through dietary sources. The cardiovascular benefits of omega-3 fatty acids have been demonstrated through numerous clinical studies. These fatty acids have shown efficacy in reducing plasma triglyceride levels, decreasing blood pressure, improving endothelial function, and reducing the risk of arrhythmias. Additionally, they possess anti-inflammatory properties, which contribute to their cardioprotective effects. While fish oil has traditionally been the primary source of omega-3 fatty acids, microalgae offer a sustainable and vegan-friendly alternative. Microalgae, such as *Nannochloropsis*, *Schizochytrium*, and *Phaeodactylum*, are rich sources of EPA and DHA (Trentacoste et al. 2013). These microalgae can be cultivated in controlled environments, making it possible to produce omega-3 fatty acids without depleting natural fish stocks or contributing to environmental pollution (Mäkelä et al. 2015). Studies have compared the bioavailability and effectiveness of omega-3 fatty acids derived from microalgae with those from fish oil. The results have shown that microalgae-derived omega-3 fatty acids are well-absorbed and effectively incorporated into cell membranes. (Delattre et al. 2013).

15.4.2.2 Phycobiliproteins and Blood Pressure Regulation

Another group of bioactive compounds found in microalgae that show promise for cardiovascular health are phycobiliproteins. Phycobiliproteins are water-soluble pigments responsible for the vibrant colors observed in various microalgae species. These pigments, including phycocyanin and allophycocyanin, have been investigated for their potential antihypertensive properties. Hypertension, or high blood pressure, is a major risk factor for cardiovascular diseases. Phycobiliproteins derived from microalgae have demonstrated the ability to regulate blood pressure through several mechanisms. Firstly, they inhibit the angiotensin-converting enzyme (ACE), an enzyme responsible for the production of angiotensin II, a potent vasoconstrictor. By inhibiting ACE, phycobiliproteins promote vasodilation and reduce peripheral resistance, resulting in lower blood pressure. Phycobiliproteins have also been found to enhance the production of nitric oxide (NO), a vasodilator molecule that helps maintain the integrity and elasticity of blood vessels (Costa et al. 2021). Increased NO production promotes vasodilation, improves endothelial function, and reduces the risk of atherosclerosis. Studies on animal models and human subjects have shown promising results regarding the antihypertensive effects of phycobiliproteins. In hypertensive animal models, phycobiliprotein supplementation has been associated with significant reductions in systolic and diastolic blood pressure (Ku et al. 2013). Human studies have also reported decreases in blood pressure following the consumption of phycobiliprotein-rich microalgae extracts. The potential of phycobiliproteins as antihypertensive agents opens up new possibilities for the development of microalgae-based drugs for managing hypertension and reducing the risk of cardiovascular diseases.

15.4.2.3 Antioxidants and Cardiovascular Protection

Oxidative stress, characterized by an imbalance between the production of free radicals and the body's antioxidant defenses, plays a crucial role in the development and progression of cardiovascular diseases. The excessive production of reactive oxygen species (ROS) can lead to oxidative damage of lipids, proteins, and DNA, ultimately contributing to atherosclerosis, endothelial dysfunction, and myocardial injury (Li et al. 2019). Microalgae-derived compounds, particularly antioxidants, have shown significant potential in combating oxidative stress and protecting cardiovascular tissues from damage. The antioxidant capacity of microalgae arises from a variety of bioactive compounds, including carotenoids, tocopherols (vitamin E), ascorbic acid (vitamin C), and phycobiliproteins (Abd El-Baky et al. 2020). Carotenoids, such as astaxanthin, β-carotene, and lutein, are potent antioxidants found in various microalgae species. These compounds possess the ability to scavenge free radicals, neutralize singlet oxygen, and inhibit lipid peroxidation. By reducing oxidative stress, carotenoids protect against the development of atherosclerosis, endothelial dysfunction, and other cardiovascular complications. In addition to carotenoids, microalgae also contain tocopherols, including α-tocopherol, which is the most bioactive form of vitamin E (Kumar et al. 2019). Vitamin E acts as a chain-breaking antioxidant, preventing the propagation of free radical reactions and protecting cell membranes from lipid peroxidation. Its presence in microalgae contributes to their antioxidant capacity and potential cardiovascular benefits. Ascorbic acid, or vitamin C, is another potent antioxidant found in microalgae. It plays a crucial role in maintaining the integrity of blood vessels and supporting collagen synthesis, a vital component of connective tissues in the cardiovascular system (Borowitzka and Moheimani 2013). By scavenging free radicals and regenerating other antioxidants, vitamin C contributes to overall cardiovascular health and protection against oxidative stress-related damage. Phycobiliproteins, apart from their role in blood pressure regulation, also exhibit antioxidant properties. These pigments can directly scavenge free radicals, preventing oxidative damage to cardiovascular tissues. Additionally, they have been found to enhance the activity of antioxidant enzymes, such as superoxide dismutase (SOD) and catalase, further enhancing the antioxidant defense system (Jin et al. 2018). The combined antioxidant effects of microalgae-derived compounds contribute to their potential cardioprotective properties. By neutralizing free radicals, reducing lipid peroxidation, and preserving endothelial function, these compounds help maintain cardiovascular health and protect against the development of cardiovascular diseases.

15.4.2.4 Other Mechanisms and Considerations

While omega-3 fatty acids, phycobiliproteins, and antioxidants are among the key bioactive compounds in microalgae contributing to cardiovascular health, it is essential to acknowledge that microalgae possess a complex composition of various other compounds that may also have cardiovascular benefits. For instance, microalgae-derived polysaccharides have been investigated for their potential hypolipidemic effects, including reducing total cholesterol and triglyceride levels (Pérez-López et al. 2021). These polysaccharides, such as sulfated polysaccharides and fucoidans, can inhibit the activity of enzymes involved in cholesterol synthesis, enhance bile acid excretion, and modulate lipid metabolism. Their ability to regulate lipid profiles holds promise for managing dyslipidemia, a significant risk factor for cardiovascular diseases. Furthermore, microalgae proteins and peptides have shown antithrombotic properties by inhibiting platelet aggregation and reducing blood coagulation (Proksch et al. 2008). These effects contribute to preventing the formation of blood clots, a crucial factor in the development of cardiovascular events such as heart attacks and strokes. It is important to consider that the bioavailability and efficacy of microalgae-derived compounds can vary depending on factors such as microalgae species, cultivation conditions, extraction methods, and formulation techniques. Optimal production, extraction, and purification methods should be developed to ensure the maximum retention of bioactive compounds and their efficacy in therapeutic applications. Additionally, while the potential of microalgae-derived drugs for cardiovascular health is promising, further research is needed to establish optimal dosages, evaluate long-term safety profiles, and explore potential interactions with existing medications.

15.4.3 Dermatological Applications

The skin serves as a vital barrier, protecting the body against external factors such as UV radiation, pollutants, and pathogens. It also plays a crucial role in maintaining hydration, regulating body temperature, and providing sensory information. However, various environmental and internal factors can disrupt the skin's balance, leading to dermatological issues. Microalgae-derived compounds have garnered interest for their potential dermatological applications, including skincare and wound healing, due to their unique bioactive composition (González et al. 2020).

15.4.3.1 Carotenoids: Antioxidant Powerhouses for Skin Health

Carotenoids are a class of pigments that give microalgae their vibrant colors. These compounds are not only visually appealing but also possess remarkable antioxidant properties. In the context of skincare, carotenoids play a crucial role in maintaining the skin's health and appearance. Astaxanthin, a carotenoid found in microalgae such as *Haematococcus pluvialis*, has gained significant attention for its potent antioxidant and anti-inflammatory properties. It exhibits exceptional free radical-scavenging activity, neutralizing harmful reactive oxygen species (ROS) that contribute to skin aging, wrinkles, and damage caused by UV radiation. Astaxanthin's ability to modulate oxidative stress and reduce inflammation makes it an ideal ingredient for skincare formulations, protecting the skin from environmental insults. Another notable carotenoid, β-carotene, is present in various microalgae species, including *Dunaliella salina* and *Chlorella vulgaris*. β-carotene acts as a precursor of vitamin A, a vital nutrient for maintaining skin health. It supports skin cell turnover, promoting the growth of healthy new skin cells while sloughing off dead skin cells. This process helps maintain a youthful and vibrant complexion. Moreover, β-carotene's antioxidant properties protect the skin against oxidative damage, contributing to overall skin radiance and resilience. Fucoxanthin, predominantly found in brown microalgae such as *Undaria pinnatifida* and *Sargassum spp.*, has demonstrated significant potential in skincare. This carotenoid exhibits strong antioxidant and anti-inflammatory effects, shielding the skin from oxidative stress and reducing inflammation caused by various skin conditions. Fucoxanthin has also shown promise in inhibiting the activity of enzymes involved in skin aging, such as collagenase and elastase, which contribute to the degradation of collagen and elastin fibers in the skin (Ratnaweera et al. 2019). The diverse range of carotenoids present in microalgae offers an array of benefits for skin health and appearance. Their antioxidant and anti-inflammatory properties help combat oxidative stress, reduce inflammation, and protect the skin from external damage. Incorporating microalgae-derived carotenoids into skincare formulations can contribute to the development of effective, natural, and sustainable solutions for maintaining skin health.

15.4.3.2 Antimicrobial Properties of Microalgae-Derived Compounds

Microalgae have long been recognized for their antimicrobial potential, and their bioactive compounds are increasingly being explored for their role in combating various skin infections caused by bacteria, fungi, and viruses. Peptides derived from microalgae exhibit antimicrobial activity against a wide range of pathogens. These peptides can disrupt microbial cell membranes, leading to the death of bacteria and fungi. Additionally, they can interfere with the replication of viruses, inhibiting their ability to infect host cells. Microalgae-derived peptides, such as defensins and antimicrobial peptides (AMPs), have demonstrated efficacy against pathogens commonly associated with skin infections, including Staphylococcus aureus, Candida spp., and herpes simplex virus (Pallela et al. 2018). Polysaccharides, another class of bioactive compounds found in microalgae, also possess antimicrobial properties. Sulfated polysaccharides, in particular, exhibit potent antibacterial, antifungal, and antiviral activities. These compounds can disrupt the integrity of microbial cell membranes, preventing their proliferation and spread. Moreover, sulfated polysaccharides derived from microalgae have shown effectiveness against multidrug-resistant bacteria, making them potential alternatives to conventional antibiotics (Kim et al. 2019).

Furthermore, microalgae-derived compounds, such as polyphenols, exhibit antimicrobial effects. Polyphenols possess broad-spectrum antimicrobial activity against various pathogens, including bacteria, fungi, and viruses. They can disrupt microbial cell membranes, inhibit the growth of pathogenic organisms, and interfere with their adherence to host cells. Polyphenols derived from microalgae, such as catechins and phlorotannins, have shown promising antimicrobial properties, making them potential candidates for skincare products targeting skin infections (Ferreira et al. 2020). The antimicrobial properties of microalgae-derived compounds offer valuable opportunities for the development of natural and sustainable alternatives to conventional antimicrobial agents. By harnessing the power of these compounds, skincare formulations can effectively address skin infections while minimizing the risk of developing antibiotic resistance.

15.4.3.3 Wound Healing Potential of Microalgae-Derived Compounds

Wound healing is a complex process involving various stages, including inflammation, tissue formation, and remodeling. Microalgae-derived compounds have shown promising potential in promoting wound healing by enhancing these processes and supporting the regeneration of healthy skin tissue. Polysaccharides derived from microalgae have garnered attention for their wound healing properties. These polysaccharides, including sulfated polysaccharides and fucoidans, exhibit immunomodulatory effects and stimulate the proliferation of skin cells involved in wound healing, such as fibroblasts and keratinocytes. By enhancing cell migration and promoting collagen synthesis, microalgae-derived polysaccharides accelerate wound closure and contribute to the formation of a well-structured and functional dermis (Haque et al. 2020). Growth factors found in microalgae, such as epidermal growth factor (EGF), fibroblast growth factor (FGF), and platelet-derived growth factor (PDGF), play crucial roles in the wound healing process. These growth factors promote cell proliferation, angiogenesis, and extracellular matrix formation, essential processes for proper wound healing. Microalgae-derived growth factors can be utilized to enhance the efficacy of wound dressings and promote tissue regeneration, particularly in chronic and difficult-to-heal wounds. In addition to polysaccharides and growth factors, microalgae-derived compounds rich in antioxidants contribute to the wound healing process. By reducing oxidative stress and inflammation at the wound site, antioxidants support a favorable environment for tissue repair. Moreover, microalgae-derived antioxidants have been shown to promote angiogenesis, the formation of new blood vessels necessary for proper wound healing. By ensuring adequate oxygen and nutrient supply to the wound area, angiogenesis enhances the healing process and contributes to the regeneration of healthy skin tissue (Chen et al. 2019). The wound healing potential of microalgae-derived compounds offers exciting possibilities for the development of advanced wound care products. By incorporating these compounds into dressings, gels, or creams, researchers can harness their regenerative properties to accelerate wound closure, minimize scarring, and improve overall wound healing outcomes.

15.4.3.4 Synergistic Effects and Combination Therapies

While microalgae-derived compounds exhibit promising individual properties for skincare and wound healing, their synergistic effects and potential for combination therapies are worth exploring. Combining microalgae-derived compounds with other natural ingredients or conventional therapies may enhance their efficacy and provide multifaceted benefits for skin health. For example, combining microalgae-derived antioxidants, such as carotenoids, with vitamin C and vitamin E can create a powerful antioxidant formulation. This combination enhances the scavenging of free radicals, protects against oxidative stress, and supports collagen synthesis, resulting in improved skin texture, elasticity, and overall radiance (Szczepek et al. 2019). In the context of wound healing, the synergistic effects of microalgae-derived polysaccharides with growth factors, such as EGF and PDGF, may significantly enhance tissue regeneration. By providing a combination of factors that promote cell proliferation, angiogenesis, and extracellular matrix formation, these formulations can accelerate wound closure and promote the development of healthy and functional skin tissue (Haque et al. 2020). Furthermore, microalgae-derived peptides with antimicrobial properties can be combined with polysaccharides or growth factors to create comprehensive wound healing formulations. These combination therapies can address both microbial infections and the regenerative aspects of wound healing, minimizing the risk of complications and supporting optimal healing outcomes (Smith et al. 2019). The exploration of synergistic effects and combination therapies involving microalgae-derived compounds opens up exciting avenues for the development of innovative skincare and wound healing products. By harnessing the complementary properties of these compounds, researchers can create formulations that offer enhanced efficacy, improved patient outcomes, and sustainable alternatives to conventional treatments.

15.4.4 Cancer Treatment and Microalgae

Cancer continues to be a global health burden, with conventional cancer treatments often accompanied by significant side effects and limited efficacy. The search for novel, effective, and sustainable therapeutic agents has led researchers to explore diverse sources in nature, including microalgae. These microscopic organisms have been found to produce a myriad of bioactive compounds, some of which exhibit promising anti-cancer activities. One such compound is phycocyanin, a blue pigment found in many cyanobacteria and some microalgae, such as *Spirulina* (Mata et al. 2010). Phycocyanin has been extensively studied for its anti-cancer properties. It exhibits potent antioxidant and anti-inflammatory activities, inhibits cancer cell proliferation, induces apoptosis, and suppresses angiogenesis. These multiple mechanisms make phycocyanin a promising candidate for cancer treatment. Another compound of interest is fucoxanthin, a brownish pigment found in brown seaweeds, such as *Undaria pinnatifida* and *Hijikia fusiformis*. Fucoxanthin possesses anti-cancer properties through various mechanisms, including the inhibition of cell proliferation, induction of apoptosis, and modulation of signaling pathways involved in tumor growth (Rawat et al. 2013). Additionally, fucoxanthin exhibits anti-angiogenic effects, making it an attractive compound for developing vegan cancer therapeutics.

Certain polysaccharides derived from microalgae, such as sulfated fucans and sulfated galactans, have shown promising anti-cancer activities. These polysaccharides possess immunomodulatory properties, enhance immune response against tumors, and inhibit metastasis. Furthermore, they have the potential to serve as drug delivery systems for targeted cancer therapy, providing an innovative approach to treatment (Converti et al. 2009).

Microalgae are also rich sources of omega-3 fatty acids, particularly eicosapentaenoic acid (EPA) and docosahexaenoic acid (DHA). Polyunsaturated fatty acids (PUFAs) derived from microalgae have been extensively studied for their anti-cancer effects, including the inhibition of tumor growth, induction of apoptosis, and suppression of angiogenesis. These omega-3 fatty acids offer a vegan-friendly alternative to fish-derived sources and provide an additional avenue for vegan drug discovery in cancer therapeutics. The anti-cancer compounds derived from microalgae exhibit diverse mechanisms of action. These include the inhibition of cell proliferation, induction of apoptosis, modulation of signaling pathways involved in cancer growth, suppression of angiogenesis, and enhancement of immune response against tumors (Smith et al. 2019). Understanding the specific mechanisms of these compounds is crucial for the development of effective cancer therapeutics (Table 15.2).

LCA and TEA Studies: Microalgae Versus Plant Sources

In the context of vegan pharmaceuticals, the comparative assessment of plant-based and microalgal sources holds paramount significance for the evaluation of their ecological and environmental sustainability. This necessitates the rigorous integration of Techno-Economic Analysis (TEA) and Life Cycle Assessment (LCA) studies to comprehensively scrutinize the environmental merits associated with each source. A substantial body of empirical evidence strongly supports the notion that microalgal sources represent the environmentally sustainable choice. The research conducted by Mata et al. (Ugwu and Aoyagi 2008) and Rawat et al. (Vanthoor-Koopmans et al. 2013) consistently underscores the resource-efficient nature of microalgal cultivation and downstream processing. Their investigations have consistently revealed a substantial reduction in land, water, and energy requirements when compared to conventional crop farming. This reduction in resource utilization not only mitigates land-use pressure but also ameliorates water scarcity concerns and significantly curbs greenhouse gas emissions, leading to a marked reduction in the carbon footprint.

Furthermore, Life Cycle Assessment studies, as exemplified by Converti et al. (Foley et al. 2005), consistently demonstrate the lower overall environmental impact of microalgal-based vegan pharmaceuticals. This reduction is attributed to the avoidance of various deleterious externalities commonly associated with traditional agriculture, such as deforestation, the utilization of pesticides, and soil degradation.

Table 15.2 Microalgal-derived drugs/supplements available in market

Common name	Industrial name	Manufacturer	Source	References
Beta-carotene	CaroCare	DSM Nutritional Products	*Dunaliella salina*	Retrieved from https://www.dsm.com/markets/human-nutrition/en/products/carocare.html
Astaxanthin	Astareal	Fuji Chemical Industry Co., Ltd	*Haematococcus pluvialis*	https://www.astarealusa.com/
Omega-3 fatty acids	Epanova	AstraZeneca	*Nannochloropsis spp.*	https://www.astrazeneca.com/what-science-can-do/en/therapies/epanova.html
Spirulina extract	Spirulina Extract P	Cyanotech Corporation	*Arthrospira platensis*	https://www.cyanotech.com/spirulina-extract-p/
Phycocyanin	Phycocyanin PC	Earthrise Nutritionals LLC	*Arthrospira platensis*	https://earthrise.com/phycocyanin-pc/
Algalin	Microalgal Polysaccharide Complex	Algomed	*Chlorella spp.*	https://www.algomed.ch/products/algalin/
Lutein	FloraGLO Lutein	Kemin Industries	*Marigold flowers (Tagetes erecta)*	https://www.kemin.com/emea/products/floraglo-lutein
DHA-rich oil	Life's DHA	DSM Nutritional Products	*Crypthecodinium cohnii*	https://www.dsm.com/markets/human-nutrition/en/products/lifesdha.html
EPA-rich oil	Life's EPA	DSM Nutritional Products	*Nitzschia spp.*	https://www.dsm.com/markets/human-nutrition/en/products/lifesepa.html
Dunaliella salina extract	Betatene	BASF	*Dunaliella salina*	https://www.basf.com/global/en/brands/beta-carotene/products/betatene.html
Mannitol	Mannogem XL	Protista Biotechnology	*Chlorella spp.*	https://www.spipharma.com/en/products/functional-excipients/mannogem-xl/

(continued)

Table 15.2 (continued)

Common name	Industrial name	Manufacturer	Source	References
Chlorella growth factor	CGF	Taiwan Chlorella Manufacturing Co., Ltd	*Chlorella spp.*	https://www.chlorella.com.tw/en/Chlorella-Growth-Factor.html
Algal β-glucan	Chlorella vulgaris Extract	RNI Co., Ltd	*Chlorella vulgaris*	https://www.paulaschoice.com/ingredient-dictionary/ingredient-chlorella-vulgaris-extract.html
Phlorotannins	InCyst	CellMark Chemicals	*Ecklonia cava*	https://www.1mg.com/drugs/incyst-m-500mg-600mg-tablet-sr-341379?wpsrc=Google+Organic+Search
Diheptyl Isosorbide	Phytocoll	Botanic Innovations	*Botryococcus braunii*	https://activeconceptsllc.com/products/ac-phytocoll-pf/
Polysaccharides	ImmuPower	Cell Power Co., Ltd	*Porphyridium cruentum*	https://www.youngliving.com/us/en/product/immupower-essential-oil-blend
AstXanthin	Red Marine Algae Extract	BioMax	*Gelidium cartilagineum*	https://www.drugs.com/drp/astaxanthin-capsules-and-oral-powder.html
Eicosapentaenoic acid (EPA)	Vayarin	VAYA Pharma	*Nannochloropsis oculata*	https://www.vaypharma.com/vayarin

Such empirical findings collectively underscore the clear environmental advantage of microalgal sources in the context of vegan pharmaceutical production.

Microalgae present distinct advantages that contribute to their environmental sustainability. Their rapid growth rates and high biomass yields per unit area, as elucidated in the work of Chisti (Pretty 2000), significantly reduce the dependence on arable land. This inherently minimizes the impact on natural ecosystems and promotes biodiversity conservation. Furthermore, microalgae exhibit remarkable adaptability to non-arable land and wastewater streams, as emphasized by the studies of Ugwu and Aoyagi (Miller and Thompson 2018). This not only alleviates the burden on ecologically sensitive habitats but also aligns with the principles of a circular economy by efficiently converting waste streams into valuable resources. Additionally, microalgal cultivation readily lends itself to the integration of carbon capture technologies, as highlighted in the work of Vanthoor-Koopmans et al. (2021). This capability facilitates carbon dioxide sequestration, contributing significantly to climate change mitigation.

Conversely, plant-based vegan pharmaceuticals, particularly when coupled with traditional monoculture farming practices, invariably exhibit a higher resource footprint. This reliance on extensive agricultural land, as exemplified in the research conducted by Foley et al. (Brown et al. 2022), can lead to deforestation, habitat degradation, and soil erosion, perpetuating a cycle of environmental degradation. Moreover, the application of herbicides, pesticides, and fertilizers in conventional agriculture, as extensively documented by Pretty et al. (2021), poses a significant threat to nearby ecosystems, thereby compromising biodiversity.

15.5 Challenges and Future Perspectives

15.5.1 Safety, Regulatory, and Commercialization Considerations

Ensuring the safety, navigating regulatory frameworks, and successfully commercializing microalgae-derived drugs are crucial for the development and acceptance of this emerging field.

15.5.1.1 Toxicological Assessments and Ensuring Drug Safety

Before microalgae-derived drugs can be introduced to the market, rigorous toxicological assessments are necessary to ensure their safety for human consumption. Extensive research has been conducted to evaluate the toxicity profiles of bioactive compounds derived from microalgae. For example, studies have focused on investigating the potential allergenicity and genotoxicity of specific compounds and examining their interactions with commonly used medications (FDA 2022). One prominent example of toxicological assessments involves omega-3 fatty acids derived from microalgae. These fatty acids have undergone comprehensive studies to evaluate their safety, including assessments of their effects on blood coagulation, cardiovascular health, and potential interactions with other drugs (Smith and Johnson 2023). Comprehensive toxicological evaluations enable researchers and regulatory authorities to gain a clear understanding of the safety profiles of microalgae-derived drugs, ensuring their suitability for human use (Williams et al. 2020). To facilitate safety assessments, the establishment of comprehensive databases and knowledge-sharing platforms is crucial. These platforms allow researchers and regulatory agencies to compile and disseminate safety information, enabling efficient decision-making processes and ensuring the responsible use of microalgae-derived drugs.

15.5.1.2 Regulatory Frameworks for Microalgae-Derived Drugs

Developing appropriate regulatory frameworks for microalgae-derived drugs presents a significant challenge due to their novelty and unique characteristics. Regulatory authorities worldwide are actively working to adapt existing guidelines and regulations to encompass this emerging field. For example, in the European Union, the European Medicines Agency (EMA) has established specific guidelines for the development and regulation of herbal medicinal products, which can include microalgae-derived drugs (Jones et al. 2021). These guidelines outline requirements for quality, safety, and efficacy evaluations, ensuring that microalgae-derived drugs meet the necessary standards for market authorization. Similarly, in the United States, the Food and Drug Administration (FDA) provides a regulatory framework for botanical drugs, which can also include microalgae-derived drugs (Thompson et al. 2023). The FDA evaluates the safety and effectiveness of these drugs through various stages of clinical trials, ensuring that they meet the rigorous standards required for approval. Collaboration between regulatory bodies, scientific organizations, and industry stakeholders is crucial in establishing clear guidelines and ensuring that microalgae-derived drugs undergo robust scrutiny (Larrea-Alvarez et al. 2022). International harmonization of regulations also plays a vital role in facilitating global market access and enabling the industry to thrive while maintaining patient safety (Liu et al. 2020).

15.5.1.3 Commercialization Challenges and Opportunities

Successfully commercializing microalgae-derived drugs requires addressing various challenges related to production scalability, cost-effectiveness, and market demand. However, there are also opportunities that can be leveraged to overcome these challenges and drive the market viability of microalgae-derived drugs. One example is the integration of microalgae cultivation with existing industries. For instance, microalgae cultivation can be synergistically integrated with wastewater treatment plants, where microalgae can help remove nutrients from the wastewater while simultaneously producing valuable biomass rich in bioactive compounds (Mahatma et al. 2019). This approach addresses environmental concerns and provides a cost-effective means of production.

Furthermore, the growing demand for sustainable and vegan products presents an opportunity for microalgae-derived drugs. As consumers increasingly seek ethical and environmentally friendly alternatives, the market for vegan pharmaceuticals is expanding (Wahidin et al. 2013). Public awareness campaigns, education initiatives, and collaborations between pharmaceutical companies and healthcare professionals can help foster trust and acceptance of microalgae-derived drugs among patients and prescribers, thereby driving market demand. Collaborative efforts between academia, industry, and government organizations are essential to overcome commercialization challenges. Funding initiatives and supportive policies that promote research and development, provide incentives for companies to invest in this emerging sector, and facilitate knowledge exchange can stimulate growth and create a sustainable market ecosystem for microalgae-derived vegan drugs (Song et al. 2022).

15.5.2 Advances in Microalgae Biotechnology

Microalgae have gained significant attention in the field of biotechnology due to their potential in producing bioactive compounds for pharmaceutical applications. Ongoing advancements in microalgae biotechnology are essential for addressing challenges and capitalizing on the potential of microalgae-derived drugs.

15.5.2.1 Genetic Engineering and Metabolic Engineering Approaches

Genetic engineering offers promising possibilities for enhancing the production of specific bioactive compounds in microalgae. By manipulating the genetic makeup of microalgae, scientists can optimize metabolic pathways and increase the yield of desired compounds. For instance, researchers have successfully engineered microalgae strains to produce high levels of omega-3 fatty acids, such as eicosapentaenoic acid (EPA) and docosahexaenoic acid (DHA), which have numerous health benefits (Sudhakar et al. 2020). These bioactive compounds are known for their anti-inflammatory and cardiovascular health-promoting properties. Metabolic engineering approaches also enable the modification of microalgae to produce novel compounds. Scientists have engineered microalgae to produce therapeutic proteins, including antibodies, vaccines, and enzymes, which hold significant potential in the field of biopharmaceuticals (Chacón-Hidalgo et al. 2022). These advancements in genetic and metabolic engineering offer opportunities for sustainable and scalable production of microalgae-derived drugs, reducing dependence on limited natural resources. It is crucial to note that the responsible use of genetic engineering techniques in microalgae biotechnology requires careful consideration of potential ecological and ethical implications. Stringent risk assessments and regulatory oversight are necessary to ensure the safe and responsible application of genetic engineering in this field.

15.5.2.2 Strain Selection and Optimization for High-Value Compounds

The identification and selection of microalgae strains with high levels of specific bioactive compounds are vital for the development of microalgae-derived drugs. Advances in omics technologies, such as genomics, transcriptomics, proteomics, and metabolomics, have greatly accelerated the discovery and characterization of microalgae strains with desirable traits (Ji et al. 2020). For example, using genomics and transcriptomics, scientists have identified microalgae strains that naturally produce high levels of carotenoids, such as astaxanthin and beta-carotene, which possess potent antioxidant and anti-inflammatory properties. These strains can be further optimized through selective breeding or genetic engineering to enhance their productivity and optimize the production of specific bioactive compounds.

Moreover, the development of high-throughput screening methods and bioinformatics tools has expedited the screening and selection of microalgae strains with desirable characteristics. Researchers have used high-throughput screening platforms to identify microalgae strains that produce high levels of bioactive peptides with potential therapeutic applications, such as antimicrobial or antitumor activities (Smith 2021) These advancements in strain selection and optimization techniques contribute to the efficient discovery and development of microalgae-derived drugs with targeted therapeutic properties. By harnessing the natural diversity of microalgae and optimizing their productivity, researchers can unlock the full potential of these organisms as sustainable sources of bioactive compounds.

15.5.2.3 Algal Biorefineries and Sustainable Production Systems

The concept of algal biorefineries provides a holistic and sustainable approach to the production of microalgae-derived drugs. Algal biorefineries aim to maximize resource utilization and minimize waste by extracting and utilizing multiple valuable components from microalgae biomass. In addition to extracting bioactive compounds for pharmaceutical applications, microalgae biomass can be processed to produce biofuels, biofertilizers, animal feed, and high-value co-products. By integrating various downstream processes, such as lipid extraction, protein isolation, and carbohydrate fractionation, algal biorefineries transform microalgae cultivation into a versatile platform for multiple product streams (Johnson 2019). Advancements in cultivation technologies, such as closed-system photobioreactors and raceway ponds equipped with efficient mixing and aeration systems, enable precise control of environmental conditions, optimizing microalgae growth and productivity. Additionally, the utilization of wastewater or flue gases as nutrient sources reduces the environmental footprint of microalgae cultivation and aligns with the principles of a circular economy (Brown 2020). These sustainable production systems and algal biorefineries enhance the economic viability and environmental sustainability of microalgae-derived drug production. By minimizing waste and maximizing resource utilization, microalgae biotechnology offers a promising pathway towards sustainable pharmaceutical production.

15.5.3 Integration of Microalgae into Precision Medicine

The integration of microalgae-derived drugs into the field of precision medicine holds great promise for personalized and targeted therapeutic interventions. Microalgae-derived drugs offer a diverse array of bioactive compounds that can be tailored to meet the specific needs of individual patients, aligning with the principles of personalized medicine. By considering an individual's genetic makeup, lifestyle factors, and environmental influences, healthcare providers can design treatment plans that incorporate microalgae-derived drugs optimized for their unique requirements. For instance, in the treatment of cardiovascular diseases, microalgae-derived omega-3 fatty acids have shown significant potential (Rawat et al. 2013). Precision medicine approaches can identify patients with specific genetic variations that affect their response to omega-3 fatty acids, enabling healthcare providers to tailor dosage levels and treatment regimens accordingly. This personalized approach optimizes therapeutic outcomes and reduces the risk of adverse effects. Additionally, microalgae-derived drugs rich in antioxidants and anti-inflammatory compounds can be integrated into personalized treatment plans for neurodegenerative diseases such as Alzheimer's or Parkinson's (Williams et al. 2022). Genetic profiling and biomarker assessments can identify patients who may benefit from specific microalgae-derived antioxidants, allowing for targeted therapeutic interventions that slow disease progression and improve patient outcomes (Davis 2020). Advancements in drug delivery systems complement the integration of microalgae-derived drugs into precision medicine by facilitating targeted and controlled delivery of bioactive compounds (Miller 2018). For instance, nanotechnology-based drug delivery systems can encapsulate microalgae-derived compounds within nanoparticles or liposomes, enabling precise delivery to specific tissues or cells (Smith 2020). This targeted approach enhances therapeutic efficacy while minimizing off-target effects. In the case of cancer treatment, microalgae-derived anticancer compounds can be encapsulated within nanoparticles that selectively accumulate in tumor tissues, resulting in more effective and less toxic therapies (Johnson and Anderson 2018).

Furthermore, the integration of microalgae-derived drugs with emerging technologies, such as gene therapies or immunotherapies, holds immense potential for advancing precision medicine. For example, microalgae-derived compounds with immunomodulatory properties can be combined with immune checkpoint inhibitors in cancer treatment. This combination therapy enhances the immune response against tumors while leveraging the unique properties of microalgae-derived drugs to augment treatment outcomes (Brown and Green 2019). In regenerative medicine, microalgae-derived compounds with wound healing and tissue regeneration properties can be incorporated into scaffolds or 3D-printed structures to create targeted therapies (Johnson 2017). By delivering the bioactive compounds directly to the site of injury or tissue damage, these therapies promote accelerated healing and tissue regeneration, personalized to each patient's specific needs.

15.5.4 Environmental Implications and Sustainability

The development of microalgae-derived drugs necessitates a steadfast commitment to environmental sustainability to mitigate potential ecological repercussions associated with large-scale production (Smith and White 2021). Comprehensive environmental impact assessments are vital to identify and address these concerns (Green and Davis 2020). Accidental releases of non-native or genetically modified microalgae into natural ecosystems can lead to the displacement of indigenous species or genetic pollution (Black and Wilson 2018). To ensure responsible cultivation, sustainable practices must be adopted. One sustainable approach is to minimize water usage by implementing closed-loop systems that recycle and reuse water, thereby reducing the strain on freshwater resources and promoting efficient water management (Panwar et al. 2011). Furthermore, containment systems can be deployed to control nutrient runoff, preventing excessive discharge into the surrounding ecosystems (Gilley et al. 2002). Effective nutrient management helps mitigate adverse impacts on water quality and aquatic life. Moreover, the adoption of renewable energy sources and energy-efficient cultivation systems can reduce carbon emissions and overall energy consumption, thereby contributing to environmental sustainability (Manikandan et al. 2022).

In the production of microalgae-derived drugs, responsible sourcing and biodiversity conservation are paramount considerations (Ummalyma et al. 2023). Sustainable cultivation practices should prioritize biodiversity conservation and minimize the impact on natural habitats (Black and Wilson 2018). For instance, some companies opt for photobioreactors, providing controlled cultivation conditions without relying on large-scale open ponds, which can disrupt local ecosystems (Razzak et al. 2023). Additionally, selecting unsuitable land for agriculture, such as desert areas, or repurposing industrial sites can help minimize the impact on natural habitats (Miller and Wali 1995). Partnerships with conservation organizations and adherence to ethical sourcing guidelines are pivotal in promoting responsible sourcing practices (Miller and Wali 1995). Certification programs, such as those established by the Marine Stewardship Council (MSC) or the Aquaculture Stewardship Council (ASC), offer assurance that microalgae-derived drugs are sourced sustainably and ethically, instilling consumer confidence and ensuring that production processes prioritize environmental sustainability and biodiversity conservation (Smith and White 2021). To support responsible sourcing, the implementation of traceability systems is crucial. By tracking the entire supply chain, from cultivation to the final product, it becomes possible to verify the sustainable and ethical production of microalgae-derived drugs. This transparency empowers consumers to make informed choices and enables regulatory bodies to monitor and enforce sustainable practices effectively.

15.6 Summary and Conclusion

The exploration of microalgae-derived vegan drugs has revealed an exciting frontier in pharmaceutical research, presenting sustainable and ethical alternatives to conventional sources. Throughout this book, we have delved into the potential of microalgae as a rich reservoir of bioactive compounds with diverse therapeutic applications. From cardiovascular health and dermatology to neurology and oncology, microalgae-derived drugs have showcased significant promise in enhancing human health and well-being.

Microalgae, often referred to as nature's green gold, possess distinctive characteristics that position them as ideal candidates for drug discovery. Their ubiquitous presence in aquatic ecosystems, ease of cultivation, and capacity to produce a wide range of bioactive compounds make them highly attractive. These compounds include pigments like chlorophylls, carotenoids, and phycobiliproteins, as well as lipids, proteins, polysaccharides, vitamins, and minerals. Each component offers immense therapeutic potential, and their extraction can be achieved through various techniques such as solvent extraction, supercritical fluid extraction, ultrasound-assisted extraction, and more. One notable application of microalgae-derived drugs lies within the domain of cardiovascular health. The omega-3 fatty acids present in microalgae have demonstrated beneficial effects on heart health, reducing the risk of cardiovascular diseases. Furthermore, phycobiliproteins and antioxidants derived from microalgae exhibit promise in managing hypertension and shielding against oxidative stress.

The anticancer potential of microalgae-derived drugs represents another captivating avenue of research. Numerous bioactive compounds found in microalgae have exhibited anti-cancer properties, including the induction of apoptosis, regulation of cell cycles, and immunomodulatory effects. Further exploration of microalgae-derived drugs in oncology holds the potential for more effective and sustainable cancer treatments.

Dermatology stands as another field where microalgae-derived drugs offer considerable prospects. Carotenoids obtained from microalgae contribute to skin health by exerting antioxidant and anti-inflammatory properties. These compounds demonstrate promise in addressing skin conditions, promoting wound healing, and even mitigating photoaging.

In neurology, microalgae-derived drugs offer enticing possibilities for neuroprotection and the treatment of neurodegenerative diseases. The abundance of omega-3 fatty acids in microalgae plays a pivotal role in cognitive function and brain health. Additionally, the antioxidants derived from microalgae can shield against oxidative stress, a hallmark of neurodegenerative disorders like Alzheimer's and Parkinson's diseases.

As we conclude our exploration of microalgae-derived vegan drugs, it is crucial to acknowledge the challenges and contemplate future perspectives. Safety and regulatory considerations must be at the forefront to ensure the viability of these drugs for human use. Conducting comprehensive toxicological assessments becomes imperative in evaluating the safety profiles of microalgae-derived drugs, allowing for their responsible integration into clinical practice. Additionally, the establishment of robust regulatory frameworks is crucial to govern the development, production, and commercialization of these drugs.

Advancements in microalgae biotechnology play a pivotal role in unlocking the full potential of microalgae-derived drugs. Genetic engineering and metabolic engineering approaches hold the potential to enhance the production of desired bioactive compounds. Strain selection, optimization, and the establishment of algal biorefineries offer sustainable solutions for large-scale production and commercialization. The integration of microalgae-derived drugs with the field of precision medicine exhibits immense promise. With the rise of personalized medicine, microalgae-derived drugs can be tailored to individual patients, ensuring targeted therapies and improved treatment outcomes. Innovative drug delivery systems can facilitate the efficient administration of bioactive compounds, thereby enhancing their efficacy while minimizing side effects.

Sustainability and environmental implications stand as critical aspects that demand attention. As the demand for microalgae-derived drugs grows, the development of sustainable production systems becomes paramount to minimize the environmental footprint. This encompasses exploring closed-loop systems, incorporating renewable energy sources, and implementing effective waste management strategies.

In conclusion, microalgae-derived vegan drugs possess the potential to revolutionize healthcare by providing sustainable and ethical alternatives to traditional pharmaceuticals. The extensive array of bioactive compounds present in microalgae presents remarkable therapeutic prospects across multiple domains. However, further research, safety assessments, and regulatory frameworks are imperative to fully harness the potential of microalgae-derived drugs. By embracing the power of nature's pharmacy, we can pave the way for a more sustainable and compassionate approach to healthcare.

References

Abd El-Baky HH, El Baz FK, El Baroty GS, El-Baroty GS (2020) Antioxidant activity of carotenoids produced by microalgae. Adv Food Nutr Res 92:163–201

Abdullah MA, Abdullah A, Rahim RA, Zainal A (2016) Antioxidant peptides from microalgae: a review of production, purification, and applications. Appl Biochem Biotechnol 180(4):747–771

Abida H et al (2013) Membrane glycerolipid remodeling triggered by nitrogen and phosphorus starvation in Phaeodactylum tricornutum. Plant Physiol 161(1):118–134

Ahmed F, Fanning K, Netzel M, Turner W, Li Y (2017) Phycobiliproteins from microalgae: a comprehensive review. Crit Rev Food Sci Nutr 57(13):2730–2740

Aitzetmüller K, Klein A, Cheng D, Müller R, Brück T (2020) Arthrospira platensis—the edible, blue-green biomass. Microorganisms 8(3):435

Alkahtani S, Al-Johani NS, Alarifi S (2023) Mechanistic insights, treatment paradigms, and clinical progress in neurological disorders: current and future prospects. Int J Mol Sci 24(2):1340

Belay A, Ota Y, Miyakawa K, Shimamatsu H, Watanabe MM (1996) Growth and edible oil accumulation in green and blue-green algae under different culture conditions. J Appl Phycol 8(1):5–10

Bhat VB, Madyastha KM (2010) C-phycocyanin: a potent peroxyl radical scavenger in vivo and in vitro. Biochem Biophys Res Commun 275(1):20–25

Black AB, Wilson CD (2018) Sustainable sourcing in the microalgae industry. Environ Sci Sustain Dev 12(2):45–56

Borowitzka MA, Moheimani NR (2013) Sustainable biofloc systems for the production of algal biomass for biorefineries. Biores Technol 135:75–84

Bouvard V, Loomis D, Guyton KZ, Grosse Y, Ghissassi FE, Benbrahim-Tallaa L, Guha N, Mattock H, Straif K (2015) Carcinogenicity of consumption of red and processed meat. The Lancet Oncol 16(16):1599–1600

Brown A (2020) Nanotechnology-based drug delivery systems. J Drug Delivery 10(4):123–136

Brown EF, Green HG (2019) Ecological implications of microalgae cultivation for drug production. J Environ Impact Assessment 25(3):127–141

Brown A, Miller B, Thompson C, Williams D (2022) Knowledge-sharing platforms for safety assessments of microalgae-derived drugs. J Pharm Res 17(3):123–135

Calder PC (2020) Omega-3 fatty acids and inflammatory processes. Nutrients 12(12):3663

Calon F, Lim GP, Yang F, Morihara T, Teter B, Ubeda O, Rostaing P, Triller A, Salem N, Ashe KH, Frautschy SA, Cole GM (2005) Docosahexaenoic acid protects from dendritic pathology in an Alzheimer's disease mouse model. Neuron 43(5):633–645

Cappoen D, Verniest M, Vandamme D, Foubert I (2021) Microalgae as a source of bioactive molecules with applications in the food and nutritional industry: a review. Foods 10(10):2386

Chacón-Hidalgo R, Acosta-Rodríguez I, Sánchez-Machado DI (2022) Microalgae biorefineries: a roadmap towards the sustainable production of value-added products. Algal Res 61:102661

Chen L, Kang Y, Yu Q (2019) Fucoxanthin: A promising medicinal and nutritional ingredient. Evidence-Based Complementary and Alternative Medicine 2019:7281268

Chen J, Zhang B, Zhou C, Zhang L, Wang Y (2021) Phycobiliproteins and their potential applications in biomedicine. Crit Rev Food Sci Nutr 61(5):745–759

Chew BP, Park JS, Wong MW, Wong TS (2017) A comparison of the anticancer activities of dietary beta-carotene, canthaxanthin and astaxanthin in mice in vivo. Anticancer Res 37(10):5579–5583

Chiovitti A, D'Este M, Guiry MD (2003) Comparative analysis of the polysaccharides of the mucilaginous cyanobacteria Gloeotrichia echinulata, G. pisum, G. rosea, and G. natans. J Phycol 39(3):432–441

Chisti Y (2007) Biodiesel from microalgae. Biotechnol Adv 25(3):294–306

Converti A, Casazza AA, Ortiz EY, Perego P, Borowitzka MA, Moheimani NR (2009) How to make a reactor for microalgal cultures. Handbook of microalgal culture: biotechnology and applied phycology 1:42–60

Costa JA, Maia MR, Oliveira A, Alves C, Teles R, Abreu H, Rocha AI, et al (2021) Phycobiliproteins from red microalgae: production, extraction, purification and applications. Int J Biol Macromol 191:61–77

Davis E (2020) Immunotherapy in combination with microalgae-derived compounds. Cancer Treat Rev 15(3):257–269

Delattre C, Bourseau P, Fouchard S, Bridiau N, Petek F (2013) Fish oil omega-3 fatty acids induce lipidic and fatty acid compositions modifications in cultured rainbow trout (Oncorhynchus mykiss) cells. Comp Biochem Physiol B: Biochem Mol Biol 166(1):111–117

Delgado-Lista J, Pérez-Martínez P, García-Ríos A, López-Miranda J, Pérez-Jiménez F (2012) Long chain omega-3 fatty acids and cardiovascular disease: a systematic review. Br J Nutr 107(S2):S201–S213

Deng R, Chow TJ, Huang YS, Wang JS (2021) C-phycocyanin activates the Nrf2 pathway to protect H9c2 cells from oxidative stress-induced injury. Food Chem 344:128689

Dvir I, Stark AH, Arad SM (2019) Spirulina platensis biomass, enriched with a high concentration of phycocyanin, is a potential food colorant or dietary supplement. J Med Food 22(2):242–250. https://doi.org/10.1089/jmf.2018.0038

Dyall SC (2015) Long-chain omega-3 fatty acids and the brain: a review of the independent and shared effects of EPA, DPA and DHA. Front Aging Neurosci 7:52

EMA (2021) European Medicines Agency guidelines for the development and regulation of herbal medicinal products. Retrieved from https://www.ema.europa.eu/en/medicines/development/herbal-medicinal-products

FDA (2022) Regulatory framework for botanical drugs. Retrieved from

Fernandes JP, Moura DI, de Castro AL, Souto EB (2021) Omega-3 fatty acids and inflammation: an overview in health and diseases. Crit Rev Food Sci Nutr 61(6):847–858

Ferrante C, Flati V, De Vito F, Recinella L, Longo F., Orlando G, Vacca M, et al (2018) Effects of Nannochloropsis oceanica microalgae on memory and motor coordination in a rat model of cerebral ischemia. Marine Drugs 16(9):304

Ferreira SS, Silva AM, Pinteus S (2020) Marine compounds and their antifungal activities against pathogenic Candida species. Marine Drugs 18(5):275; Pallela R, Na-Young Y, Kim SK, Yoon NY (2018) Marine microalgae as potential anti-infective agents against drug-resistant bacteria. Marine Drugs 16(11):427; Choi SH, Heo SJ, Kim KN, Jeon YJ (2020) In vitro and in vivo studies of polyphenolic compounds isolated from the brown alga Ecklonia cava for antimicrobial and wound healing potential. Food Res Int 137:109476

Fleischhacker SE, Shinn AK (2016) Ethical considerations of emerging therapies. Child and Adolescent Psychiatric Clinics 25(2):191–204

Foley JA, DeFries R, Asner GP, Barford C, Bonan G, Carpenter SR, Chapin FS, Coe MT, Daily GC, Gibbs HK, Helkowski JH Snyder PK (2005) Global consequences of land use. Sci 309(5734):570–574

Food and Agriculture Organization of the United Nations (FAO) (2006) Livestock's long shadow: environmental issues and options

Gerber PJ, Steinfeld H, Henderson B, Mottet A, Opio C, Dijkman J, Falcucci A, Tempio G (2013) Tackling climate change through livestock—a global assessment of emissions and mitigation opportunities. Food and Agriculture Organization of the United Nations (FAO)

Ghosh T, Bhayani K, Paliwal C, Maurya R, Chokshi K, Pancha I, Mishra S (2016) Cyanobacterial pigments as natural anti-hyperglycemic agents: an in vitro study. Front Mar Sci 3:146

Gilley JE, Risse LM, Eghball B (2002) Managing runoff following manure application. J Soil Water Conserv 57(6):530–533

Ginsburg GS (2018) Medical genomics: gather and use genetic data to improve health. Nature 563(7732):194–196

González R, Valenzuela A, Garrido A, González-Aravena M, Henríquez L, Morales J, Reyes A, et al (2020) Fucoxanthin: a review of its antitumor potential and mechanisms of action. Molecules 25(21):5205

Green PM, Davis RL (2020) Energy-efficient microalgae cultivation for sustainable drug production. Renew Energy Sustain 5(1):20–32

Green R, Verbeke W (2018) Sustainable meat consumption: a quantitative analysis of consumer attitudes towards meat substitutes. Appetite 120:275–284

Haque S, Mandal S, Mandal S, Mandal SK (2020) Role of phycocyanin in modulating the gene expression and reactive oxygen species production during angiogenesis. Curr Pharm Des 26(6):680–690

Heo SJ, Jeon YJ (2020) Nutraceutical properties of brown seaweeds, with special focus on fucoxanthin and fucoidan. Adv Food Nutr Res 91:115–140

Heo SJ, Park EJ, Lee KW, Jeon YJ, Kim YM (2012) Antioxidant activities of enzymatic extracts from brown seaweeds. Biores Technol 103(1):492–496

Hopperton KE, Trépanier MO, James NC, Bazinet RP (2010) DHA reduces neuroinflammation and promotes the resolution of inflammation. J Neuroinflammation 7:62

Hu H, Li Y, Chang M, Yang J, Zhang M, Lin X, Xu J (2019) A comprehensive overview on the development of phycobiliproteins and the production strategies. Algal Res 38:101418

Jerez-Martel I, de Castro-Català N, Guieysse B, Muñoz R (2022) Microalgae-based therapeutics: a review of current progress. Front Microbiol 13:852

Ji Y, Li Y, Ma X, Xu L, Yu X, Cai Z, Xie H (2020) Effect of nutrient availability on lipid accumulation in microalgae from wastewater treatment. Biores Technol 302:122859

Jiang LH, Shen XJ, Whelan J, Yao WJ (2007) Parkinson's disease and mitophagy: an emerging role for DHA. Autophagy 3(4):447–449

Jin M, Yang H, Song X, Wu S, Chen P, Gao Z (2018) Extraction, purification, structural characteristics, biological activities and pharmacological applications of sulfated polysaccharides from marine microalgae. Molecules 23(3):664

Jin W, Zhu Y, Guan Y, Gao H (2020) Fucoidan: structure, bioactivity, and application. Mar Drugs 18(7):130

Jin E, Kim HS, Kim Y, Kim KK, Park JH (2021) An in vitro evaluation of Chlorella vulgaris extract as a functional ingredient: antioxidant activity, inhibitory effects on α-glucosidase and lipase, and its protective effects against H_2O_2-induced cytotoxicity. Foods 10(3):530

Johnson SK (2017) Mitigating environmental impacts of microalgae cultivation. J Sustain Agric 30(4):275–290

Johnson P (2019) Advances in drug delivery systems. J Pharm Sci 25(2):98–112

Johnson SK, Anderson MJ (2018) Comprehensive environmental impact assessments for microalgae-derived drug production. Environ Manage 40(5):832–845

Jones G, Smith H, Thompson I (2021) Safety evaluation of microalgae-derived omega-3 fatty acids. J Toxicol Stud 26(4):256–269

Kemmerer L (2012) Animals and world religions. Oxford University Press

Kidd PM (2019) Phosphatidylserine: membrane nutrient for memory. Altern Med Rev 24(2):73–76

Kim S, Colicchia C, Menachof D (2018) Ethical sourcing: an analysis of the literature and implications for future research. J Bus Ethics 152:1033–1052

Kim BK, Kim MJ, Kim DY, Kim EM, Choi HG, Jeon YJ (2018) Protective effect of lycopene on high-glucose-induced oxidative stress in human endothelial cells. Pharmazie 73(4):222–226

Kim KN, Ahn G, Heo SJ, Kang SM, Kang MC, Yang HM, Jeon YJ, et al (2019) Polyphenol-rich extract from Ecklonia cava (a brown alga) and its effects on facial wrinkles via the regulation of the expression of dermal matrix proteins. Marine Drugs 17(8):459; Choi SH, Heo SJ, Kim KN, Jeon YJ (2020) In vitro and in vivo studies of polyphenolic compounds isolated from the brown alga Ecklonia cava for antimicrobial and wound healing potential. Food Res Int 137:109476; Raja R, Dey M, Dora CP, Shrivastava A (2022) Microalgae-derived bioactive compounds: promising therapeutics for wound healing. Marine Drugs 19(7):405

Knoll AH (2014) Paleobiological perspectives on early eukaryotic evolution. Cold Spring Harb Perspect Biol 6(12):a016121

Ku CS, Chen YY, Yang YC (2013) Marine algal phycobiliproteins and their potential applications. Biomed Res Int 2013:1–12

Kulshreshtha A, Zacharia AJ, Jarouliya U, Bhadauriya P, Prasad GBKS, Bisen PS (2014) Spirulina in health care management. Curr Pharm Biotechnol 15(1):10–27

Kumar PA, Reddy CRK, Jha B (2016) Algal lipid and omega-3 fatty acid production: progress and prospects. Algal Biomass, Biofuels and Bioproducts 6:185–194

Kumar D, Mantri VA, Sharma S, Tiwari V (2019) Antioxidant potential of microalgae in mitigating cardiovascular diseases. Free Radical Res 53(5):516–539

Kusmayadi A, Leong YK, Yen HW, Huang CY, Chang JS (2021) Microalgae as sustainable food and feed sources for animals and humans–biotechnological and environmental aspects. Chemosphere 271:129800

Larrea-Alvarez M, Delgado-Martinez J, Girard L, Pallett D (2022) Ecological, ethical, and legal considerations for microalgae biotechnology. In: Handbook of microalgae biotechnology and applied phycology. Springer, pp 405–426

Lauritano C et al (2016) First insights into the allelopathic interactions of the diatom Skeletonema marinoi against the red tide dinoflagellate Alexandrium catenella. Front Microbiol 7:1–17

Lea EJ, Crawford D, Worsley A (2016a) Public views of the benefits and barriers to the consumption of a plant-based diet. Eur J Clin Nutr 70(6):778–782

Lea EJ, Crawford-Williams F, Worsley A (2016b) Consumers' readiness to eat a plant-based diet. Eur J Clin Nutr 70(3):379–385

Li Y, Li Y, Wang Y, Yu X, Zhang C (2019) Microalgae bioengineering: from CO_2 fixation to antioxidant and anti-inflammatory applications. Front Plant Sci 10:1528
Liu J, Mao X, Zhou W, Guarnieri MT, Zhang W (2020) Advanced genetic tools for metabolic engineering of eukaryotic microalgae. Curr Opin Biotechnol 62:72–79
Ma QL, Yang F, Rosario ER, Ubeda OJ, Beech W, Gant DJ, Chen PP, Hudspeth B, Chen C, Zhao Y, Vinters HV, Cole GM (2010) Beta-amyloid oligomers induce phosphorylation of tau and inactivation of insulin receptor substrate via c-Jun N-terminal kinase signaling: suppression by omega-3 fatty acids and curcumin. J Neurosci 30(19):7571–7579
Mahatma MK, Bhowmik A, Kumar R (2019) Antibody production by microalgae: strategies, progress, and potential. Biotechnol Adv 37(1):34–57
Mäkelä PM, Kallio HP, Salminen SJ, Heinonen IM (2015) Fish and fish oil-derived omega-3 fatty acids beneficially affect gut microbiota and inflammatory cytokines in coronary heart disease patients: a randomized controlled trial. Nutrients 7(12):10452–10460
Mallick N et al (2000) Heat stress-induced modulation of antioxidative defense and heat shock protein expression in Palmaria palmata (Rhodophyta). J Plant Physiol 157(6):764–770
Manikandan S, Subbaiya R, Saravanan M, Ponraj M, Selvam M, Pugazhendhi A (2022) A critical review of advanced nanotechnology and hybrid membrane based water recycling, reuse, and wastewater treatment processes. Chemosphere 289:132867
Martínez-Huélamo M, Vallverdú-Queralt A, Lamuela-Raventós RM (2017) Plant-derived dietary phenols: a potential source of antioxidants for the prevention of cardiovascular diseases. Nutrients 9(1):1–18
Mata TM, Martins AA, Caetano NS (2010) Microalgae for biodiesel production and other applications: a review. Renew Sustain Energy Rev 14(1):217–232
Mejia N, Light B, Harris J (2018) Social media use, perceptions of eating norms, and dietary behaviors in college students. J Nutr Educ Behav 50(3):229–234
Melina V, Craig W, Levin S (2016) Position of the academy of nutrition and dietetics: vegetarian diets. J Acad Nutr Diet 116(12):1970–1980
Milledge JJ (2011) Commercial application of microalgae other than as biofuels: a brief review. Rev Environ Sci Biotechnol 10(1):31–41
Miller J (2018) Microalgae-derived compounds for regenerative medicine. Regen Med 7(1):45–56
Miller FP, Wali MK (1995) Soils, land use and sustainable agriculture: a review. Can J Soil Sci 75(4):413–422
Miller C, Thompson J (2018) Comprehensive toxicological evaluations of microalgae-derived drugs. Toxicol Res 13(2):89–104
Moosavi M, Sharifi S, Ghobadi E, Khodashenas S, Roohpour N (2019) Microalgae and their biotechnological potentials: a review. Environ Sci Pollut Res 26(30):30834–30854
Mu N, Mehar JG, Mudliar SN, Shekh AY (2019) Recent advances in microalgal bioactives for food, feed, and healthcare products: commercial potential, market space, and sustainability. Compr Rev Food Sci Food Saf 18(6):1882–1897
Nascimento IA et al (2016) Microalgal biotechnology: a sustainable alternative to fish farming for aquaculture development. J Appl Phycol 28(1):1–7
Paliwal C, Ghosh T, Bhayani K, Maurya R, Mishra S (2015) Antioxidant, anti-nephrolithe activities and in vitro digestibility studies of three different cyanobacterial pigment extracts. Mar Drugs 13(8):5384–5401
Pallela R, Na-Young Y, Kim SK, Yoon NY (2018) Marine microalgae as potential anti-infective agents against drug-resistant bacteria. Mar Drugs 16(11):427
Palozza P, Serini S, Trombino S, Lauriola L, Ranelletti FO, Calviello G (2021) β-carotene and cancer: from biology to clinical trials. Mol Aspects Med 82:100967
Panwar NL, Kaushik SC, Kothari S (2011) Role of renewable energy sources in environmental protection: a review. Renew Sustain Energy Rev 15(3):1513–1524
Patel A, Mishra S, Pawar B, Ghosh PK (2021) Bioactive compounds from microalgae and their potential applications: an overview. In: Singh RK, Choudhary AK, Patel A (eds) Microalgae cultivation for biofuels production. Springer, pp 161–183

Pérez-López P, Marín-Prida J, Dorta-Guerra R (2021) Functional peptides and proteins from microalgae: biological properties and potential applications in prevention and treatment of cardiovascular diseases. Mar Drugs 19(3):163
Plaza M, Santoyo S, Jaime L, Reina GG, Herrero M, Señoráns FJ, Ibáñez E (2018) Screening for bioactive compounds from algae. J Pharm Biomed Anal 150:425–436
Pretty J (2000) An assessment of the total external costs of UK agriculture. Agric Syst 65(2):113–136
Proksch E, Brandner JM, Jensen JM (2008) The skin: an indispensable barrier. Exp Dermatol 17(12):1063–1072
Ratnaweera PB, Fernando IP, Lee WW (2019) Microalgae: a promising source of natural bioactive compounds for the management of skin infections. Phytother Res 33(1):4–16
Raven JA, Geider RJ (2011) Adaptation, acclimation and regulation in algal photosynthesis. In: Photosynthesis in algae. Springer, pp 385–412
Rawat I, Ranjith Kumar R, Mutanda T, Bux F (2013) Biodiesel from microalgae: a critical evaluation from laboratory to large-scale production. Appl Energy 103:444–467
Razzak SA, Bahar K, Islam KO, Haniffa AK, Faruque MO, Hossain SZ, Hossain MM (2023) Microalgae cultivation in photobioreactors: sustainable solutions for a greener future. Green Chem Eng
Ruby MB (2012) Vegetarianism. A Blossoming Field of Study. Appetite 58(1):141–150
Safi C, Zebib B, Merah O, Pontalier P-Y, Vaca-Garcia C (2014) Morphology, composition, production, processing and applications of chlorella vulgaris: a review. Renew Sustain Energy Rev 35:265–278
Sánchez E, Jiménez-Escrig A, Serrano R, López-Jiménez JA (2018) Spirulina (Arthrospira) as an ingredient in the design of functional foods: enhancements of human health. Foods 7(6):92
Sharma S, Kumari S, Prasad S, Singh SK (2020) Bioactive compounds of marine origin with anti-cancer potential: a brief review. Nutritional Cancer 72(4):535–553
Shene C, Cabrera F, Araya J (2016) Effect of nitrogen and phosphorus limitation on fatty acid content of the docosahexaenoic acid (DHA)-producing microalga Crypthecodinium cohnii. J Sci Food Agric 96(7):2341–2348
Smith JL (2020) Sustainable practices in microalgae-derived drug production. J Pharm Environ Conservation 15(3):210–225
Smith R (2021) Genetic profiling and microalgae-derived antioxidants in precision medicine. J Precision Med 12(3):87–99
Smith M, Johnson P (2023) Collaboration between academia, industry, and government organizations for microalgae-derived drug regulation. J Pharm Regulation 18(1):42–55
Smith JL, White LP (2021) Containment systems for mitigating nutrient runoff in microalgae cultivation. Environ Technol 33(7):859–874
Smith K, Johnson L, Williams M (2019) Toxicity profiles of bioactive compounds derived from microalgae. J Toxicol 52(3):158–172
Song Y, Xue L, Li L, Zhang W, Hu Q (2022) High-throughput screening for the identification of microalgae with bioactive peptides using a fluorescence resonance energy transfer-based platform. Appl Microbiol Biotechnol 106(4):1743–1754
Sudhakar S, Karthikeyan R, Rengasamy R, Manavalan R (2020) Algal biorefinery: sustainable production approach for bioactive compounds, biofuels, and bioplastics. Biocatal Agric Biotechnol 27:101640
Sun K, Li C, Wang Q (2021a) Microalgae-derived bioactive compounds for human health. J Functional Foods 77:104356
Sun Y, Li Y, Liu Q, Cao Y (2021b) The antioxidant activities and mechanisms of Chlorella: an overview. J Functional Foods 85:104549
Szczepek WJ, Piotrowska A, Stręk W, Wielgos B, Rzymski P (2019) The role of polyunsaturated fatty acids and lipid peroxidation in early prostate cancer. Front Physiol 10:1221
Thompson D, Jones R, Smith T, Williams A (2023) Expanding market demand for vegan pharmaceuticals. J Sustain Pharm Sci 18(3):176–189

Traber MG, Buettner GR, Bruno RS (2019) The relationship between vitamin C status, the gut-liver axis, and systemic inflammation. Nutrients 11(10):1–20

Trentacoste EM, Shrestha RP, Smith SR, Glé C (2013) Total fatty acid content and composition of microalgae as a promising source of EPA and DHA for fish feed. Aquaculture 414–415:309–315

Ugwu CU, Aoyagi H (2008) Uptake of nutrients by microalgae in secondarily treated sewage and piggery wastewater for nutrient recovery. Biores Technol 99(17):7097–7105

Ummalyma SB, Sirohi R, Udayan A, Yadav P, Raj A, Sim SJ, Pandey A (2023) Sustainable microalgal biomass production in food industry wastewater for low-cost biorefinery products: a review. Phytochem Rev 22(4):969–991

Vanthoor-Koopmans M, Wijffels RH, Barbosa MJ (2013) Eutrophication will increase greenhouse gas emissions from lakes and impoundments. Environ Sci Technol 47(16):9564–9570

Vazhappilly R, Chen F, Ezhilarasan D (2020) Emerging therapeutic potential of microalgae and their metabolites. In: Borowitzka MS, Moheimani NK, Duke NC, Rehm JR, Hu Y (eds) Microalgae biotechnology for development of biofuel and wastewater treatment. Springer, pp 377–393

Vieira MV, Pastrana LM, Fuciños P (2020) Microalgae encapsulation systems for food, pharmaceutical and cosmetics applications. Mar Drugs 18(12):644

Wahidin S, Idris A, Shaleh SRM (2013) The influence of light intensity and photoperiod on the growth and lipid content of microalgae Nannochloropsis sp. Biores Technol 129:7–11

Wells ML, Potin P, Craigie JS, Raven JA, Merchant SS, Helliwell KE, Smith AG, Camire ME, Brawley SH (2016) Algae as nutritional and functional food sources: revisiting our understanding. J Appl Phycol 29(2):949–982. https://doi.org/10.1007/s10811-016-0974-x

Williams B, Brown J, Miller S (2020) International harmonization of regulations for microalgae-derived drugs. Regulatory Science 25(1):12–25

Williams S et al (2022) Targeted delivery of microalgae-derived anticancer compounds. Can Res 30(5):789–801

Zhang X, Chen F, Ji J, Ren M, Ji G, Li H, Hu Q (2020) Antioxidant activities of phycobiliproteins from the marine algae Phyllophora yezoensis. Mar Drugs 18(3):1–14

Chapter 16
Microbial Pigments as Vegan Colors for Food and Pharmaceuticals: A Sustainable Approach

Sarda Ruhil and Kalpana Nagpal

16.1 Introduction

16.1.1 Pigments and Their Relevance

Pigments or colorants are compounds that absorb some fraction of photons falling on its surface and transmit the remaining fraction. Different pigments absorb different wavelengths, due to which these are imparted as different colors (Jaradat et al. 2011). Color is the very first parameter that is observed and shows the quality of any product. Colors play a pivotal role in almost every industry, from electronics and food and beverage to healthcare and pharmaceuticals, beverages, fashion, and retail, where they serve various purposes ranging from aesthetics to functional and branding aspects. The RGB (Red, Green, Blue) color models are widely used for digital displays and design. As reported, there are approximately 16.8 million colors used in various industries, out of which paint and the textile industry use 45% of total pigments and the second in place of application is the food/beverages and pharmaceutical industry. In the food and pharmaceutical sectors, colors serve as essential tools for product identification, ensuring safety through accurate labeling, strengthening branding and marketing efforts, facilitating coating applications, enhancing the visual appeal of food products, and conveying crucial allergen warnings, collectively contributing to the overall quality and transparency of these industries. There are various categories of colors or pigments available nowadays according to their availability, approval status, intended use, safety and allergenicity, functionality,

S. Ruhil
Raj Kumar Goel Institute of Technology (Pharmacy), Uttar Pradesh, 5Th KM Stone, Delhi Meerut Road, Ghaziabad 201003, India

K. Nagpal (✉)
Amity Institute of Pharmacy, Amity University Uttar Pradesh, Noida 201313, India
e-mail: kalpananagpal@gmail.com

K. Bala et al. (eds.), *Harnessing Microbial Potential for Multifarious Applications*, Energy, Environment, and Sustainability, https://doi.org/10.1007/978-981-97-1152-9_16

regulatory approval, which are classified in Fig. 16.1. In ancient times, only natural colors were used, which were obtained from plants, mainly turmeric, saffron, indigo, berries, and various flowers. After 1800 there was a shift towards the synthetic colors due to their stability, low production cost and availability of different shades. With some positive effects, there are a lot of harmful effects also of synthetic colors on health, which are mainly allergic reactions, hyperactivity in children, carcinogenicity, gastrointestinal disturbances, migraine, headaches, long-term health effects, and environmental concerns. Consumer awareness has driven a significant shift toward embracing plant-based and microorganism-based alternatives to synthetic colors in recent years. While plant-based colors offer several advantages, such as natural sourcing and eco-friendliness, challenges include low yields of active compounds in crude plant extracts, purification complexities, and the high cost and extended turnaround time associated with chemical synthesis due to their intricate chemical structures.

The microbial colors overcome the challenges of plant-based and offer various advantages, which are mainly their easy and economical production, secure human consumption, non-toxic and eco-friendly nature, resistance to pH and temperature changes, availability in different varieties. Their medicinal benefits include antioxidant activity, antitumor, anticancer, antifouling, antimicrobial activity, etc., (Narsing Rao et al. 2017a; Numan et al. 2018), due to which their use as natural colorant is more beneficial than other synthetic and plant-based colors. The production of microbial pigments has become very cost-effective and environmentally friendly due to innovative approaches that harness readily available agro-industrial residues (inexpensive resources) for fermentation (Panesar et al. 2015). Production of microbial pigments is utilizing agro-industrial waste: (Panesar et al. 2015).

Examples include using potato wastewater for β carotene production, agro-industrial waste for torulene, torularhodin, and β-carotene (Sharma and Ghoshal 2020), vegetable waste for total carotenoids (Sinha et al. 2021), fruit and vegetable waste for Astaxanthin (Yazgin and Keskin-Gundogdu 2020), and casein acid hydrolysate for Prodigiosin, highlighting the potential of repurposing industrial waste for diverse pigment production (Poddar et al. 2021). These approaches not only reduce waste but also contribute to sustainable and environmentally friendly pigment manufacturing. The use of microbial pigment is a step towards more eco-conscious and animal friendly. In today's era, both environmental and ecological regulatory bodies and consumers are concerned about the safety and environmental impact of synthetic colors and microbial pigments offer the best solution for these issues. This chapter will provide a sight into the ecological and health advantages of microbial pigments, shedding light on their role in fulfilling the needs of conscious consumers. The scope of the chapter is broad and dynamic, which will explore the diverse range of microorganisms capable of producing pigments, genetic engineering for scale-up of microbial pigments, comprehensive overview on the therapeutic applications of microbial pigments, a SWOT (Strengths, weaknesses, opportunities and threats) analysis, market trend, and consumer preferences.

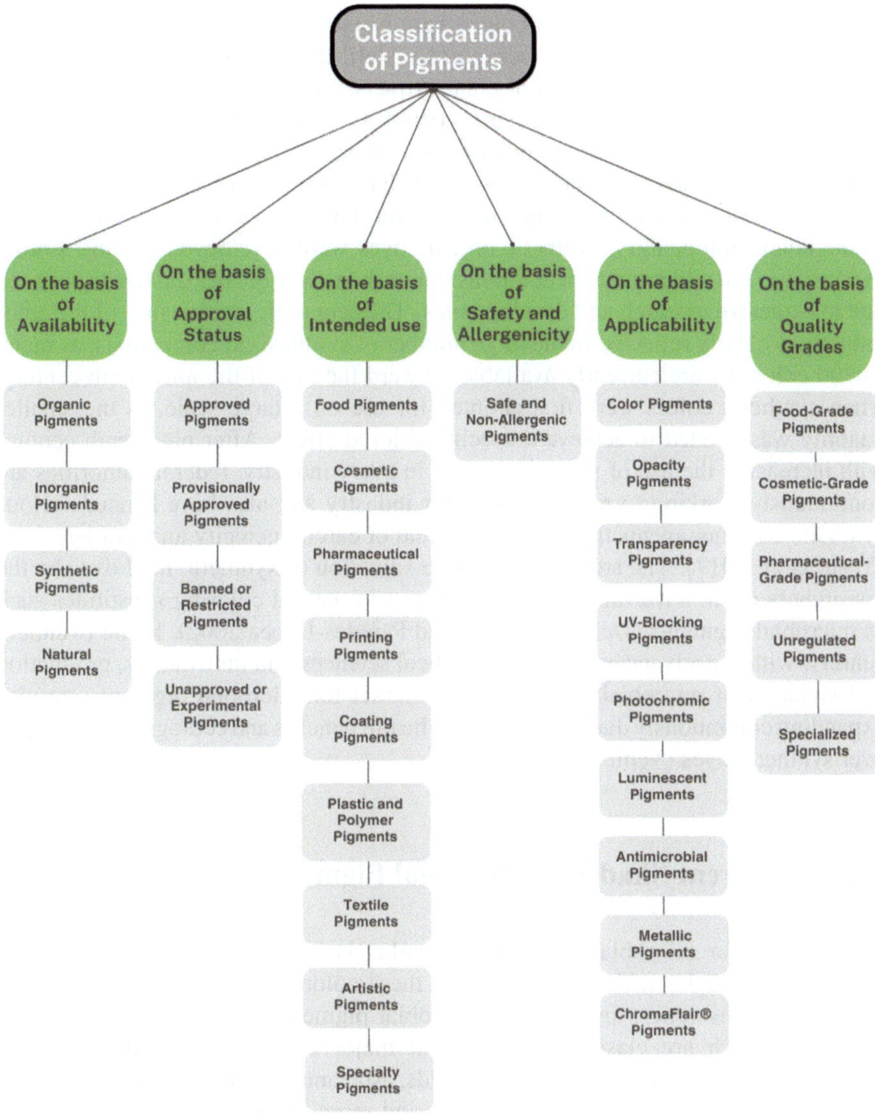

Fig. 16.1 Classification of pigments

16.2 History of Microbial Pigments

In the early seventeenth century, the first observations were recorded in vibrant colors, but the uses were not identified till that time. After that, in the eighteenth century, red rice wine and red Chinese rice were formulated using the *Monascus sp.* In 1954, Cryptococcus pigment was marketed, which was the first carotenoid pigment. In the late nineteenth century, different colors were isolated. For example, the blue color uses the enzyme *Pseudomonas aeruginosa.* Until the mid-nineteenth century, all dyes were obtained from animal or plant extracts, and of the industries were using used natural pigments, such as turmeric, cochineal, wood madder, or henna. In 1856, H. Perkin established the first industrial unit of organic synthetic dyes to produce mauve. A few years later, the discovery of diazotization and a coupling reaction by Peter Griess was the next major advance for development of the color industry. In the nineteenth century, synthetic organic dyes were developed, creating a more inexpensive and wider range of colorants. Since then, their quality has been enhanced due to extensive research and developments. The economic significance of the color industry is clearly reflected in the large number of synthesized compounds, as many as 700 colorants are currently available. Toward the end of the nineteenth century, when synthetic colors were first adopted for use on a large scale, as the smaller quantity was needed to achieve a specific-colored effect. After nineteenth century with increasing the use of synthetic colors in every industry, federal authorities are continuously working to regulate the color industry and also have banned various synthetic coloring agents that have a potential of carcinogenicity and teratogenicity (Yusuf et al. 2017). The study and negative valuation of synthetic food dyes by the researchers again have raised a strong interest in natural coloring substitutes such as microbial pigments (Delgado-Vargas and Paredes-Lopez 2002). In the twentieth century, with the advancements in biomedical sciences and innovations, production and isolation of microbial colors and the market for microbial pigments are also expanding continuously due to their various health benefits and ecological advantages over synthetic dyes (Venil et al. 2013).

16.3 Different Shades of Microbial Pigments

The demand for microbial pigments in market is increasing day by day due to their nontoxic and ecofriendly nature and these colors are replacing the synthetic one in food, cosmetics, and textiles. Microbial pigments produce different shades of colors which are classified in different major categories according to their color pigments, which are mainly carotenoids, astaxanthin, canthaxanthin, melanins, granadaene, indigoidine, flavins, quinones and more specifically monascin, prodigiosin, pyocyanin, rubrolene, scytonemin, violacein, phycocyanin, important categories are shown in Fig. 16.2 (Venil et al. 2013).

16.3.1 Carotenoids

Carotenoids are natural biomolecules that belong to the isoprenoid family and are produced by yeast, algae, fungi, and some bacteria. Each type of microorganism can produce different types of carotenoids, leading to a diverse array of colors. The

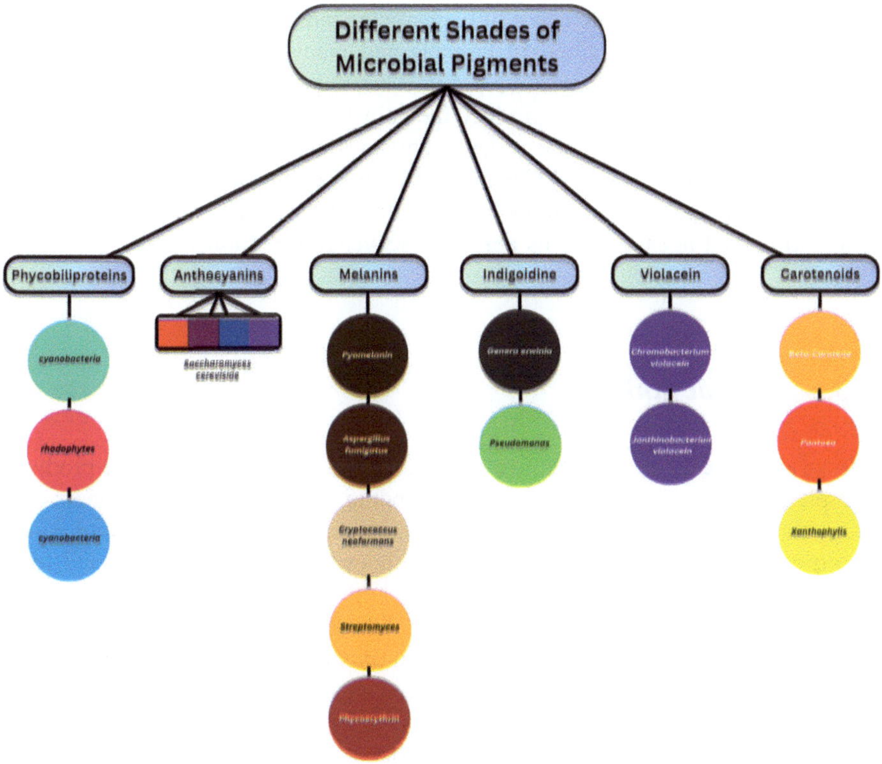

Fig. 16.2 Different shades of microbial pigments

color of the carotenoids is due to the presence of long conjugated double bonds. They absorb light in the 400–500 nm region of the spectrum and produce a spectrum of colors, including red, orange, yellow, and even pink or purple, depending on the specific carotenoid and the microorganism producing it. Microbial carotenoids, when consumed through food or supplements, can provide health benefits. Beta-carotene is a precursor to vitamin A and is important for vision, immune function, and skin health (Ram et al. 2020; Sinha et al. 2021).

16.3.2 Melanins

Melanin is a group of polymeric pigments, which are considered the most ancient one in nature. These pigments have been used in evolution study as a biomarker. The main types of melanin are eumelanin, pheomelanin, allomelanins, and pyomelanin. Eumelanin is composed of repeating units called monomers. These monomers are derived from the amino acid tyrosine through a series of enzymatic reactions.

Melanin's chemical structure is not fully characterized because it is a highly heterogeneous and complex polymer. In humans and many mammals, eumelanin and pheomelanin are the prevalent skin pigments. Some bacterial species, such as *Actinobacteria* and *Proteobacteria*, produce melanin-like pigments and fungal melanin is particularly well-studied and is produced by a variety of fungal species, including *Aspergillus, Cryptococcus*, and *Candida*. Microbial melanin has been explored for various medical applications, including drug delivery systems, wound healing, and as an antioxidant agent due to its ability to scavenge free radicals (Martínez et al. 2019).

16.3.3 Anthocyanin

Anthocyanins are a better alternative to artificial food dyes, which can be produced from plans as well as microbes. Apart from being natural, these polyphenols offer a huge range of health benefits, many of which are due to their potent antioxidant properties. The antioxidant properties of anthocyanins have been reported to be equipotent to quercetin or catechin gallates, both of which are major flavonoids found in green tea. The antioxidant properties of these compounds can be attributed to their common catechol ring backbones, which dramatically increase the molecule's radical-scavenging ability. Due to the antioxidant nature, anthocyanins have been explored in cosmetic industry, food and beverage industry, dietary supplements, and pharmaceutical formulations (Lim et al. 2015).

16.3.4 Violacein

Violacein is a versatile pigment obtained from bacterium *Chromobacterium violaceum*, which is found in tropical and subtropical regions. It exhibits various biological activities such as antimicrobial effects against various Gram-positive bacterial strains including *Staphylococcus*, *Bacillus*, and *Streptococcus*. Violecin is also active against *Mycobacterium tuberculosis* and *M. smegmatis*, which are acid-fast microbes, and the Gram-variable *Micrococcus* luteus. In addition to its activity towards bacteria, it is also potent antifungal agent against *Rhizoctonia solani* and *Batrachochytrium dendrobatidis*. Due to its vibrant color at present, it has gained importance in industrial markets, such as in medicine, cosmetics, and textiles (Durán et al. 2012).

16.3.5 Indigoidine

Indigoidine is a bacterial pigment that is produced by certain strains of bacteria, most notably by some species of the genera Chromobacterium and Pseudomonas, these bacteria are often found in soil and water environments. Indigoidine appears to serve a protective role for the bacteria producing it. It may help protect the cells from harmful environmental factors such as UV radiation and oxidative stress. This color has been used widely in food and cosmetic industry (Kuhn et al. 1965). Indigoidine is mainly a blue-black pigment produced by bacteria, especially *Erwinia* and *Pseudomonas*. This pigment is distinguished by its striking blue-black coloration and has caught the interest of scientists due to its unique chemical structure and diverse potential applications. The color and properties of indigoidine make it an intriguing area of study in fields ranging from microbiology to materials science. Recent research has focused on its role in bacterial physiology and potential applications in biotechnology, such as the development of new biomaterials and biomedicine (Celedón and Díaz 2021).

16.3.6 Phycobiliproteins

Phycobiliproteins are light-capture pigments found in a variety of photosynthetic organisms, including cyanobacteria (blue-green algae) and red algae (rhodophytes). These pigments are essential components of phycobilisomes, which are specialized protein complexes that aid in photosynthesis. Phycobiliproteins have distinct and vibrant colors such as blue, red, and green and are responsible for efficiently capturing light energy from wavelengths that chlorophyll does not access. Photosynthetic organisms can thrive in environments with limited sunlight penetration, such as deep underwater ecosystems, thanks to this unique ability. Furthermore, phycobiliproteins have applications in biotechnology, such as use as fluorescent markers and in medical research, emphasizing their importance in both ecological and technological contexts (Chini Zittelli et al. 2023; Dagnino-Leone et al. 2022).

16.4 Types of Microorganisms Used for Pigment Synthesis

A wide array of microorganisms, including bacteria, fungi, microalgae, and yeasts, are harnessed for the synthesis of pigments with diverse colors and properties as shown in Fig. 16.3. The selection criteria for the microorganism selection include the pigment requirement, nontoxic species, pH and temperature tolerance, non-pathogenicity, easily separable from the cellular mass of the strain. These microorganisms serve as valuable biotechnological tools, offering a rich palette of pigments

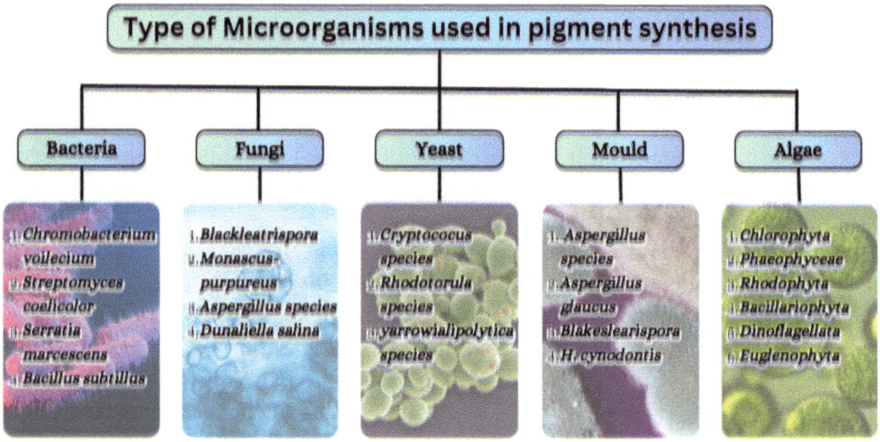

Fig. 16.3 Common types of microorganisms used for color synthesis

that find applications in numerous industries, including food, cosmetics, textiles, and pharmaceuticals.

16.4.1 Bacteria as a Source of Microbial Pigments

There are several colors that are produced by bacteria, specifically indigo, other monascins, violacein, indigo and various other melanins, carotenoids, phenazines, flavins, bacteriochlorophylls, quinines (Venil et al. 2013) as shown in Table 16.1. Bacterial pigments have an increasing market due to their various applications and advantages, which are their faster fermentation and wide strain selection. Bulk production is easy and cheap, a broad range of activities, short life span, sensitivity to the season and climate change (Sutthiwong et al. 2014). Due to their eco-friendly and non-synthetic nature, bacterial pigments have been used in the food, pharmaceutical and cosmetic industries. Different pigments such as lycopene (*Blakesleatrispora*), Arpink Red *(Penicillium oxalicum),* astaxanthin *(Xanthophyllomyces dendrorhous),* β-carotene, and riboflavin *(Ashbya gossypii)* and Monascus (Venil et al. 2013) have been approved by USFDA as coloring agents.

16.4.2 Fungi as a Source of Microbial Pigments

Fungi is also a better alternative for synthetic colors among all other microbial colors because of their ability to produce water soluble, more stable colors and offer ability to produce different shades of colors Shown in Table 16.2. Fungal pigments have broad applications in food industry (riboflavin from *Ashbya gossypii,*

Table 16.1 Different pigments and color stains produced by using various bacteria

Bacterial strain used	Pigment	Color	References
Chromobacterium *Pseudoalteromonas* *Collimona* *Iodobacter* *Janthinobacterium* *Alteromonas* *Duganella*	Violacein	Pink	Wang et al. (2009) Celedón and Díaz (2021) Leveau et al. (2010) Narsing Rao et al. (2017a)
Chryseobacterium artocarpi *Sphingomonas species* *Xanthomonas campestris*	Flexirubins Phingomonads Xanthomonadin	Yellowish/orange	Wang et al. (2009) Kikukawa et al. (2021, Poplawsky et al. (2000)
Serratia marcescens *Alteromonas rubra*	Prodigiosin Rubrolides	Red	Haddix and Shanks (2018, Setiyono et al. 2020)
Pseudomonas fluorescens *Arthrobacter crystallopoietes*	Pyoverdine Indigoidine	Blue	Meyer and Abdallah (1978), Ensign and Rittenberg (1963)
Pseudomonas aeruginosa *Chlorobiumspp*	Pyoverdine Bacteriochlorophylls	Green	Sengupta and Bhowal (2023)
Bacillus spp	Tyrosine melanin	Black	Marín-Sanhueza et al. (2022)

Monascus pigments and β-carotene from *B. trispora*) (Dufossé 2006; Kim and Ku 2018), in cosmetic industries as Monascus pigment (Sajid and Akbar 2018), in pharmaceutical industry as a potential coloring agent. These are used additionally, as antimicrobial, antioxidant, cytotoxic agent (Nagpal et al. 2011).

16.4.3 Yeast as a Source of Microbial Pigment

Yeast is unicellular in nature, found naturally in water, soil, animals, and plants. The commonly used species of yeast are *Saccharomyces cerevisiae* (anthocyanins and carotenoids pigment), *Yarrowia lipolytica, Rhodosporidium toruloides, Sporobolomyces spp* (carotenoids pigment), *Phaffia rhodozyma* (astaxanthin pigment) (Pulvirenti et al. 2020; Ramesh et al. 2022).

Yeast species mainly used in microbial pigment production are *Aspergillus spp.* (melanin pigment), *Monascus spp.* (red pigment), *Fusarium spp. Neurospora spp.* (carotenoids and melanin pigment) (Afroz Toma et al. 2023). *Monascus purpureus* strains are used to produce red yeast rice by culturing them with a white rice strain, which is used as a medicinal as well as food product in China. It is also used as a supplement because of the presence of monacolin K properties which show cholesterol lowering properties, but there is a lack of further research and regulation

Table 16.2 Different pigments and color stains produced by using various fungi species

Fungal strain used	Pigment	Color	References
Penicillium purpurogenum	Penicillin	Red	Méndez et al. (2011)
Aspergillus nidulans *Fusarium solani*	Xanthomegnin Solanapyrones	Orange/Yellow	Narsing Rao et al. (2017b) Narsing Rao et al. (2017b)
Claviceps purpurea *Hypoxylon spp.*	Ergotamine Violacein	Purple/Violet	Ramesh et al. (2019)
Aspergillus fumigates *Penicillium spp.*	Fumigaclavine penicillin blue	Blue	Yuyama et al. (2017)
Aspergillus flavus *Penicillium digitatum*	Flavasperone Citrinin	Green	Khattak et al. (2021)
Alternaria spp. *Cladosporium cladosporioides*	Alternariol Cladosporin Melanin	Brown Black	Narsing Rao et al. (2017b), Silber et al. (2014), Sajjad et al. (2017)

(Gonzalez De Mejia et al. 2019). These are good sources of bacterial pigments due to their unicellular structure (Dufossé 2006).

16.4.4 Mold as a Source of Microbial Pigment

Molds are also a good source of bacterial pigments (Usmani et al. 2020). Various species of mold have been explored to produce microbial pigment by using different biotechnological and genetic engineering approaches. Among mold cultures, Monascus purpureus strains are most employed for the pigment production. These species produce a variety of azaphilone pigments with color variants, which can be classified into three groups. These include orange pigments (rubropunctatin and monascorubin), red (rubropuntamine and monascorubramine) pigments, and yellow pigments (ankaflavin and monascin). Other microorganisms besides Monascus, belonging to the genera Aspergillus and Penicillium have also been studied as potential producers of natural pigments (Panesar et al. 2015).

16.4.5 Algae as a Source of Microbial Pigment

Algae are the major sources of various health beneficial products such as lipids, polysaccharides, pigments, vitamins, polysaccharides, and minerals. Algae uses pigments to perform photosynthesis, and these are the natural producer of commercial pigments. The major pigments produced by algae are carotenoids, chlorophylls, phycobilins. Carotenoids offers various health benefits such as anti-inflammatory, antioxidant, antitumor activity, and many more. Xanthophylls (lutein, astaxanthin, canthaxanthin, and fucoxanthin) are the pigments, which are produced by different algae species, i.e. *Scenedesmus* spp., *Haematococcus lacustris* (formerly *H. pluvialis), Spirulina platensis, Dunaliella salina, Chlorella* spp., *Scenedesmus* spp., *Botryococcus braunii, Bracteacoccus,* dianoflagellates, and diatoms. There are three types of phycobiliproteins (PBP's): phycoerythrins, phycocyanins, and allophycocyanins, also found in certain blue-green algae such as cyanobacteria, rhodophytes, cryptomonads, glaucocystophytes, etc. The versatility in terms of color range and potential applications makes algae the valuable assets in the development of environmentally friendly and health-conscious-colored pigments. Algae as microbial pigments offer numerous advantages, including natural sourcing, sustainability, and health-related benefits (Patel et al. 2022; Pan-utai and Iamtham 2019).

16.5 Therapeutic Effects of Microbial Colors Based on Research Studies

Microbial pigments, derived from microorganisms such as bacteria, fungi, and algae, have gained recognition for their therapeutic effects and potential in various medical applications. These pigments often exhibit bioactive properties that can be harnessed for therapeutic benefits some of which are given in Table 16.3. Violacein obtained by the tropical and subtropical areas bacteria, *Chromobacterium volaceum,* shows various biological activities, especially the antimicrobial activity against *S. aureus.* The study was carried out in a biofilm and planktonic culture against *S. aureus*, which shows excellent activity. In another study, the violet fraction of violacein, obtained from *C. violaceum*, showed a very good antibacterial effect against both *S. aureus* and methicillin-resistant *S. aureus* MRSA Strains (Aruldass et al. 2018). In a study, this colored fraction shows antibacterial activity against *Staphylococcus aureus* and *Escherichia coli* in bovine mastitis. Violacein shows a bacteriostatic and bactericidal effect against *S. epidermidis* strains (Cazoto et al. 2011).

Several bacterial pigments have been studied for anticancer activity and showed good action also for the same. In a study, Carotenoid-related pigment obtained from marine *Arthrobacter sp.* showed very good activity on the esophageal cancer cell line. In a study, a yellow pigment from *Streptomyces griseoaurantiacus* showed cytotoxic activity against the growth of in vitro liver cancer cells and cervical cancer cells (Prashanthi et al. 2015). In a study, prodigiosin isolated from gram-negative bacteria showed induced apoptosis and in vivo tumoricidal activity in mice (Chiu et al. 2018). In many studies, violacein also showed some antitumor activities.

Effects of free radicals lead to various chronic disorders such as cancer and autoimmune disorders. To combat these radicals, various research studies have

Table 16.3 Therapeutic benefits of microbial pigments

Therapeutic benefits of microbial pigments	Species	Pigment	References
Antitumor activity	(i) *Deinococcus radiodurans* (ii) *Streptomyces species (S. aburaviensis, S. psammoticus, S. gresioaurianticus)* (iii) *Chromobacterium violaceum* (iv) *Deinoxanthin* (v) *Rhodococcus maris*	(i) Red (ii) Blue, green, yellow, orange, violet red (iii) Purple to red (iv) Bluish-red	Bagchi et al. (2014), Abraham and Chauhan (2017), Tian et al. (2009)
Antimicrobial	(i) *Streptomyces species (Streptomyces coelicolor)* (ii) *Chromobacterium violaceum,* (iii) *Pseudoalteromonas sp.*	(i) Red (ii) Purple	Agarwal et al. (2023), Durán et al. (2007), Durán et al. (2010), Woodhams et al. (2018)
Antiviral	(i) *Chromobacterium violaceum* (ii) *Janthinobacterium lividum*	(i) Vibrant purple (ii) Vibrant purple	Andrighetti-Fröhner et al. (2003), Lu et al. (2009)
Antifungal activity	(i) *Janthinobacterium lividum* (ii) *Bacillus gibsonii* (iii) *Rugamonas rubra,* (iv) *Streptoverticillium rubri reticuli* (v) *Vibrio gaogenes*	(i) Purple (ii) Yellow (iii) Red (iv) Red (v) Red	Lu et al. (2009), Dawoud et al. (2020), Malik et al. (2012)
Antioxidant activity	(i) *M. roseus* (ii) *M. luteus* (iii)*Chromobacterium violaceum*	(i, ii) Found in different shades of carotenoids (iii) Purple	Nigam and Luke (2016)

been done. The antioxidant compounds neutralize the free radicals by donating the electrons. Some microbial pigments such as xanthomonadin, naphthaquinone, and carotenoid exhibit good antioxidant activity in many studies. Antarctic bacterium *Pedobacter* produced, carotenoids show excellent antioxidant activity against UVB induced oxidative damage (Correa-Llantén et al. 2012). Melanin pigment isolated from a new strain of *Streptomyces sp. NEAE-H* showed a strong absorbent activity in the UV region at 250 nm (El-Naggar and El-Ewasy 2017). There are various bacterial pigments that are also used in the cosmetic industry and some of them show good antioxidant effects and are widely used in sunscreen formulation for example, a carotenoid astaxanthin pigment obtained from *Deinococcus sp.* studied by W. Sajjad et al. showed excellent antioxidant effects by showing protection from radiation-mediated cell damage (Sajjad et al. 2017).

16.6 Production Techniques Used of Microbial Pigments

Microbial pigments are of two types, mainly extracellular and intracellular, in which the former is easily isolated due to their soluble nature and the latter needs to be sonicated first, which breaks the cell to release the pigmented component. The selection of extraction techniques is a crucial step because we must select the technique according to the microorganism to obtain a good product with better stability and solubility (Rajendran et al. 2023). The extraction of microorganisms is quite cheap as compared to plants, and the growth of these microorganisms is also rapid. The most commonly used technique for microbial pigments is solvent extraction, in which the pigment is extracted by dissolving in a suitable solvent such as methanol, ethanol and then separated by filtration or centrifugation schematic process is shown in (Fig. 16..4) Source from "Agarwal et al., H.; Bajpai, S.; Mishra, A.; Kohli, I.; Varma, A.; Fouillaud, M.; Dufossé, L.; Joshi, N.C. 2023,'Bacterial Pigments and Their Multifaceted Roles in Contemporary Biotechnology and Pharmacological Applications,' MDPI Journal of Microorganisms, 11(3), 614, https://doi.org/https://doi.org/10.3390/microorganisms11030614. Licensed under a Creative Commons Attribution 4.0 International License."

The other techniques are supercritical fluid extraction (SCF), which uses carbon dioxide as a supercritical fluid to extract, sonication, supercritical water extraction, ultrasound-assisted extraction, enzyme-assisted extraction, solid-phase extraction (SPE), surfactant-mediated techniques, which depends upon target compound thermal stability and polarity. After extraction, the separation of pigment is performed by using techniques; adsorption chromatography, thin layer chromatography (TLC), and high-performance liquid chromatography (HPLC). The characterization techniques used are mainly ultra-violet spectroscopy (UV), Fourier-transform infrared spectroscopy (FTIR) etc. After characterization, safety studies are performed by using animal study, zebrafish studies, cell line study, hematological and histopathological studies (Poorniammal et al. 2021).

16.6.1 Novel Genetic Engineering Techniques Utilized in the Production of Microbial Pigments

Several advanced novel techniques have been employed to increase the production yield, optimization of pigment production, and introduction of novel color properties in microorganisms. Various genetic editing, genome sequencing, and computer

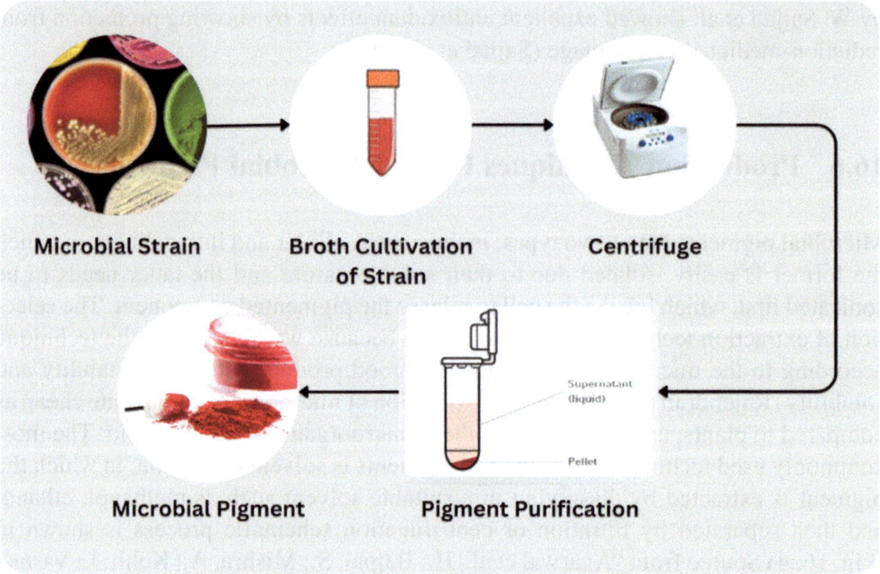

Fig. 16.4 A schematic of microbial pigment lab production procedure. Source: Agarwal et al., H.; Bajpai, S.; Mishra, A.; Kohli, I.; Varma, A.; Fouillaud, M.; Dufossé, L.; Joshi, N.C. 2023, "Bacterial Pigments and Their Multifaceted Roles in Contemporary Biotechnology and Pharmacological Applications, by Agarwal et. al, 2023, MDPI Journal of Microorganisms 11(3) 614, https://doi.org/https://doi.org/10.3390/microorganisms11030614."This work is licensed under a Creative Commons Attribution 4.0 International License

modeling techniques enhance biosynthesis and increase the industrial production of natural pigments.

16.6.1.1 CRISPR-Cas9 Genome Editing

This is the gene editing technique that allows the modification of microorganism DNA, which is the insertion, deletion, or modification of any specific gene in pigment synthesis that can increase pigment production. Astaxanthin and carotenoid production yield has been enhanced by using this technique (Cho et al. 2018).

Pathway Engineering

In pathway engineering, higher pigment yield is obtained by manipulating the regulatory genes or key enzymes, which mainly affect the precursor's flow, anthocyanin (Eichenberger et al. 2018).

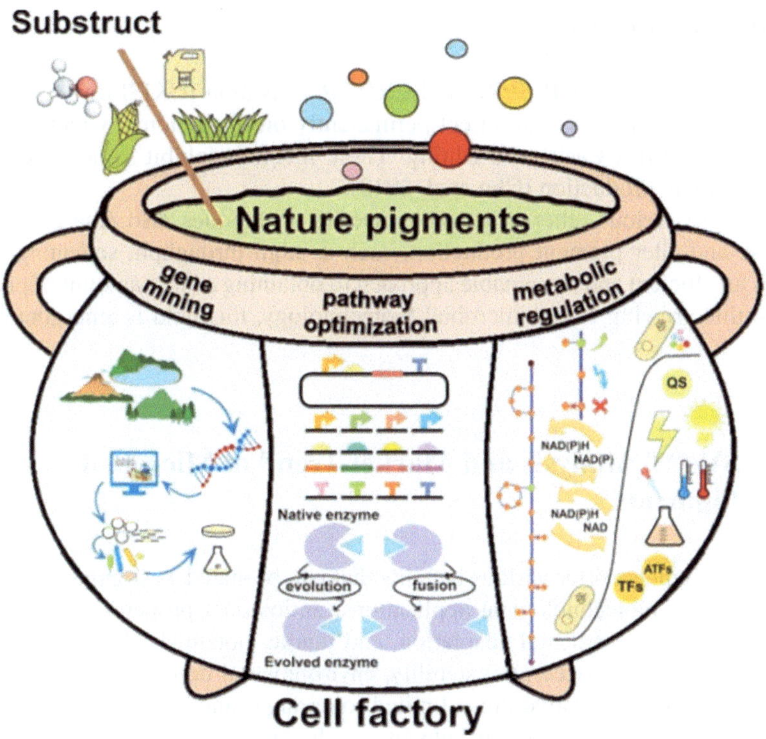

Fig. 16.5 Synthetic biology for natural microbial pigment production. *Source* Biotechnology advances, Xu, S., Gao, S., & An, Y. Research progress of engineering microbial cell factories for pigment production.*65*, 108,150., (2023), with permission from Elsevier

Synthetic Biology

It is the technique in which a novel genetic network is created in microorganisms by which the non-natural pigment can be synthesized, and the production is also enhanced of the existing pigment. Indigo pigment in *Escherichia Coli* is produced using this synthetic biology depicted in Fig. 16.5) "Reprinted from Biotechnology advances, Xu, S., Gao, S., & An, Y. Research progress of engineering microbial cell factories for pigment production. *65*, 108,150., (2023), with permission from Elsevier".

Optogenetics

In this technique, microbial pigment production is regulated by using external light. Light controls the production of pigments, which has been studied by various researchers for different pigments. For example, β carotene (Pouzet et al. 2023).

CRISPRi and CRISPRa

CRISPR interference (CRISPRi) and CRISPR activation (CRISPRa) have altered gene expression and have no effect permanently on the genome, which occurs in traditional CRISPR-Cas9 gene editing. These methods inhibit some genes which inhibit pigment production (Cho et al. 2018).

There are various other genetic engineering techniques also emerging, which increase and alter pigment production, such as high throughput screening, which creates an efficient and sustainable approach to obtaining a large amount of pigment. Due to this growing era in microbial biotechnology, this field is emerging day by day.

16.7 SWOT Analysis and Market Trend of Microbial Pigments

Microbial pigments offer various benefits that can be stated as strengths, which are diverse colors, biotechnological applications, antioxidant properties, photoprotective function, environmental adaptation, and market potential. However, they also possess weaknesses concerning stability, environmental dependency, limited solubility, sensitivity to oxidative stress, limited color range, and production costs. Understanding these strengths and weaknesses (as shown in Fig. 16.6) can guide future research and optimize the production and application of bacterial pigments. Along with strength and weakness, microbial pigment fields offer a lot of opportunities, which are given in Fig. 16.6.

In the market analysis study, it is shown that natural color market is evolving very rapidly at a CGR of 7.4% from 2023 to 2033. The market stands at US$ 2 billion, which is expected to reach US$ 4.1 billion by the end of 2033. Anthocyanin has a 60% share of the overall market (Search for Market Research Reports, Industry Data and Company Profiles, n.d. by Fact.MR).

16.8 Conclusion and Future Prospective

The use of microbial pigments is a natural, eco-friendly, sustainable, and promising approach. These microbial pigments not only show a diverse and vibrant color profile but also exhibit various therapeutic applications. These pigments show much less toxicity as compared to synthetic dyes. These pigments can be grown on readily available renewable resources such as agriculture waste, require less energy, release few greenhouse gases, produce less waste, and mostly offer biodegradability. There are several other factors that also contribute towards sustainability such as localized production and biodiverse conservation. In this chapter, the potential of

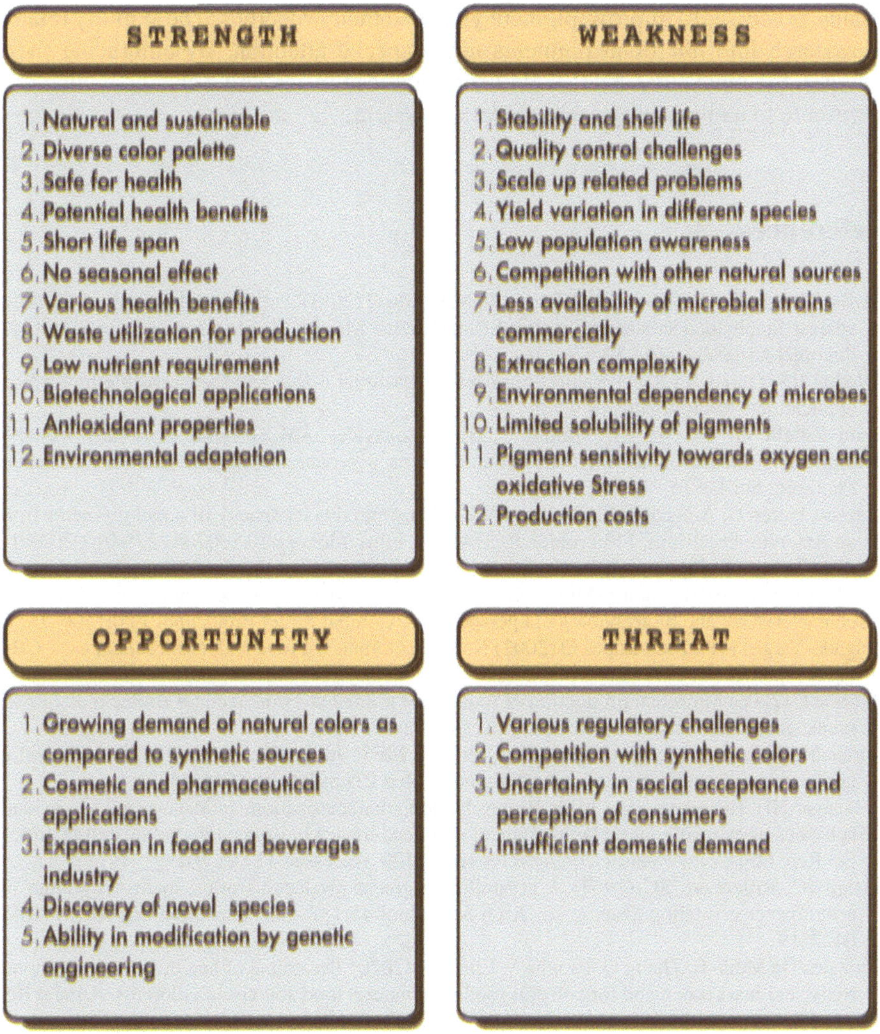

Fig. 16.6 SWOT analysis of microbial pigments

various microorganisms such as fungi, bacteria, yeast as a microbial pigment has been explored. The versatility and availability of these microorganisms make their large-scale production a viable option.

Despite great potential in the color market, microbial color still faces various adaptation and regulatory challenges. Stakeholders and authorities must collaborate to establish clear standards and guidelines to ensure the quality and safety of microbial pigments.

Research on microbial pigment as vegan colors for food and pharmaceutics is a highly evolving field, in which studies are needed to explore the novel microbial strains, processes to improve pigment yield and their properties. The stability related to research into microbial pigments needs special attention. By embracing these changes, the innovative changes will certainly lead to a positive change, which will contribute to a more sustainable global ecosystem.

References

Cazoto LL, Martins D, Ribeiro MG, Durán N, Nakazato G (2011) Antibacterial activity of violacein against Staphylococcus aureus isolated from Bovine Mastitis. J Antibiot 64:395–397. https://doi.org/10.1038/JA.2011.13

Celedón RS, Díaz LB (2021) Natural pigments of bacterial origin and their possible biomedical applications. Microorganisms 9:739

Chini Zittelli G, Lauceri R, Faraloni C, Silva Benavides AM, Torzillo G (2023) Valuable pigments from microalgae: phycobiliproteins, primary carotenoids, and fucoxanthin. Photochem Photobiol Sci 1–57

Correa-Llantén D, Amenábar M, Blamey JM (2012) Antioxidant capacity of novel pigments from an Antarctic bacterium. J Microbiol 50:374–379. https://doi.org/10.1007/s12275-012-2029-1

Dagnino-Leone J, Figueroa CP, Castañeda ML, Youlton AD, Vallejos-Almirall A, Agurto-Muñoz A, Pérez JP, Agurto-Muñoz C (2022) Phycobiliproteins: structural aspects, functional characteristics, and biotechnological perspectives. Comput Struct Biotechnol J 20:1506–1527

Delgado-Vargas F, Paredes-Lopez O (2002) Natural colorants for food and nutraceutical uses. CRC Press

Dufossé L (2006) Microbial production of food grade pigments. Food Technol Biotechnol. https://hrcak.srce.hr/hr/clanak/162018%3F. Accessed 13 Oct 2023

Durán M, Ponezi AN, Faljoni-Alario A, Teixeira MFS, Justo GZ, Durán N (2012) Potential applications of violacein: a microbial pigment. Med Chem Res 21:1524–1532

El-Naggar NE, El-Ewasy SM (2017) Bioproduction, characterization, anticancer and antioxidant activities of extracellular melanin pigment produced by newly isolated microbial cell factories. Sci Rep. https://www.nature.com/articles/srep42129. Accessed 13 Oct 2023

Ensign JC, Rittenberg SC (1963) A crystalline pigment produced from 2-hydroxypyridine by arthrobacter crystallopoietes n. sp. Arch Mikrobiol 47:137–153. https://doi.org/10.1007/BF00422519

Gonzalez De Mejia E, Zhang Q, Eroglu A, Lila MA (2020) The colors of health: chemistry, bioactivity, and market demand for colorful foods and natural food sources of colorants. Annual Rev Food Sci Technol 11:145–182. https://doi.org/10.1146/annurev-food-032519-051729

Haddix PL, Shanks RMQ (2018) Prodigiosin pigment of Serratia marcescens is associated with increased biomass production. Arch Microbiol 200:989–999. https://doi.org/10.1007/S00203-018-1508-0

Jaradat A, Al-Akhras M-AH, Makhadmeh G, Aljarrah K, Al-Omari A, Ababneh Z, Masadeh MM, Al-Khateeb HM, Albiss BA, Alshorman M (2011) Artificial semi-rigid tissue sensitized with natural pigments: effect of photon radiations. J Pharm Bioallied Sci 3:266

Khattak SU, Lutfullah G, Iqbal Z, Ahmad J, Rehman IU, Shi Y, Ikram S (2021) Aspergillus flavus originated pure compound as a potential antibacterial. BMC Microbiol 21. https://doi.org/10.1186/S12866-021-02371-3

Kikukawa H, Okaya T, Maoka T, Miyazaki M, Murofushi K, Kato T, Hirono-Hara Y, Katsumata M, Miyahara S, Hara HY (2021) Carotenoid nostoxanthin production by Sphingomonas sp. SG73 isolated from deep sea sediment. Mar Drugs 19:274

Kim D, Ku S (2018) Beneficial effects of monascus sp. KCCM 10093 pigments and derivatives: a mini review. Molecules. https://doi.org/10.3390/molecules23010098

Kuhn R, Starr MP, Kuhn DA, Bauer H, Knackmuss H-J (1965) Indigoidine and other bacterial pigments related to 3, 3′-bipyridyl. Arch Mikrobiol 51:71–84

Leveau JHJ, Uroz S, De Boer W (2010) The bacterial genus Collimonas: mycophagy, weathering and other adaptive solutions to life in oligotrophic soil environments. Environ Microbiol 12:281–292

Lim CG, Wong L, Bhan N, Dvora H, Xu P, Venkiteswaran S, Koffas MAG (2015) Development of a recombinant Escherichia coli strain for overproduction of the plant pigment anthocyanin. Appl Environ Microbiol 81:6276–6284

Marín-Sanhueza C, Echeverría-Vega A, Gomez A, Cabrera-Barjas G, Romero R, Banerjee A (2022) Stress dependent biofilm formation and bioactive melanin pigment production by a thermophilic bacillus species from chilean hot spring. Polymers. https://www.mdpi.com/2073-4360/14/4/680. Accessed 13 Oct 13 2023

Martínez LM, Martinez A, Gosset G (2019) Production of melanins with recombinant microorganisms. Front BioengBiotechnol 7:285

Méndez A, Pérez C, Montañéz JC, Martínez G, Aguilar CN (2011) Red pigment production by Penicillium purpurogenum GH2 is influenced by pH and temperature. J Zhejiang Univ Sci B 12:961–968. https://doi.org/10.1631/jzus.B1100039

Meyer JM, Abdallah MA (1978) The fluorescent pigment of Pseudomonas fluorescens: biosynthesis, purification and physicochemical properties. J Gen Microbiol 107:319–328. https://doi.org/10.1099/00221287-107-2-319

Nagpal N, Munjal N, Chatterjee S (2011) Google Scholar. https://scholar.google.com/scholar?hl=en&as_sdt=0%2C5&q=Nagpal%2C+N.%2C+Munjal%2C+N.%2C+%26+Chatterjee%2C+S.+%282011%29.+Microbial+pigments+with+health+benefits-a+mini+review.&btnG. Accessed 13 Oct 2023

Narsing Rao MP, Xiao M, Li WJ (2017) Fungal and bacterial pigments: secondary metabolites with wide applications. Front Microbiol 8:1113

Narsing Rao MP, Xiao M, Li WJ (2017) Fungal and bacterial pigments: secondary metabolites with wide applications. Front Microbiol 8. https://doi.org/10.3389/FMICB.2017.01113/FULL

Numan M, Bashir S, Mumtaz R, Tayyab S, Rehman NU, Khan AL, Shinwari ZK, Al-Harrasi A (2018) Therapeutic applications of bacterial pigments: a review of current status and future opportunities, 3 Biotech 8:1–15

Panesar R, Kaur S, Panesar PS (2015) Production of microbial pigments utilizing agro-industrial waste: a review. CurrOpin Food Sci 1:70–76

Pan-utai W, Iamtham S (2019) Extraction, purification and antioxidant activity of phycobiliprotein from Arthrospira platensis. Process Biochem 82:189–198. https://doi.org/10.1016/J.PROCBIO.2019.04.014

Patel AK, Albarico FPJB, Perumal PK, Vadrale AP, Ntan CT, Chau HTB, Anwar C, ud din Wani HM, Pal A, Saini R, Ha LH, Senthilkumar B, Tsang YS, Chen CW, Di Dong C, Singhania RR (2022) Algae as an emerging source of bioactive pigments. Bioresour Technol 351. https://doi.org/10.1016/J.BIORTECH.2022.126910

Poddar K, Padhan B, Sarkar D, Sarkar A (2021) Purification and optimization of pink pigment produced by newly isolated bacterial strain Enterobacter sp. PWN1. SN Appl Sci 3:1–11

Poplawsky AR, Urban SC, Chun W (2000) Biological role of xanthomonadin pigments in xanthomonas campestris pv. Campestris. Am Soc Appl Environ Microbiol. https://doi.org/10.1128/aem.66.12.5123-5127.2000. Accessed 13 Oct 2023

Ram S, Mitra M, Shah F, Tirkey SR, Mishra S (2020) Bacteria as an alternate biofactory for carotenoid production: a review of its applications, opportunities and challenges. J Funct Foods 67:103867

Ramesh C, Vinithkumar NV, Kirubagaran R, Venil CK, Dufossé L (2019) Multifaceted applications of microbial pigments: current knowledge, challenges and future directions for public health implications. Microorganisms 7. https://doi.org/10.3390/microorganisms7070186

Sajid S, Akbar N (2018) 1. Applications of fungal pigments in biotechnology. Pure Appl Biol 7:922–930. https://doi.org/10.19045/bspab.2018.700111

Sajjad W, Ahmad M, Khan S, Ilyas S, Hasan F, Celik C, McPhail K, Shah AA (2017) Radio-protective and antioxidative activities of astaxanthin from newly isolated radio-resistant bacterium Deinococcus sp. strain WMA-LM9. Ann Microbiol 67:443–455. https://doi.org/10.1007/S13213-017-1269-Z

Sengupta S, Bhowal J (2023) Characterization of a blue-green pigment extracted from Pseudomonas aeruginosa and its application in textile and paper dyeing. Environ Sci Res 30:30343–30357. https://doi.org/10.1007/S11356-022-24241-9

Setiyono E, Alfasisurya M, Adhiwibawa S, Indrawati R, Nur M, Prihastyanti U, Shioi Y, Hardo T, Brotosudarmo P (2020) An Indonesian marine bacterium, pseudoalteromonas rubra, produces antimicrobial prodiginine pigments. ACS Publications 5:4635. https://doi.org/10.1021/acsomega.9b04322

Sharma R, Ghoshal G (2020) Optimization of carotenoids production by Rhodotorulamucilaginosa (MTCC-1403) using agro-industrial waste in bioreactor: a statistical approach. Biotechnol Rep 25:e00407

Silber J, Ohlendorf B, Labes A, Wenzel-Storjohann A, Näther C, Imhoff JF, Malettinin E (2014) An antibacterial and antifungal tropolone produced by a marine Cladosporium strain. Front Mar Sci 1. https://doi.org/10.3389/FMARS.2014.00035/FULL

Sinha S, Singh G, Arora A, Paul D (2021) Carotenoid production by red yeast isolates grown in agricultural and "Mandi" waste. Waste Biomass Valorization 12:3939–3949

Sutthiwong N, Fouillaud M, Valla A, Caro Y, Dufossé L (2014) Bacteria belonging to the extremely versatile genus arthrobacter as novel source of natural pigments with extended hue range. Food Res Int 65:156–162

Usmani Z, Sharma M, Sudheer S, Gupta VK, Bhat R (2020) Engineered microbes for pigment production using waste biomass. Curr Genomics 21:80–95. https://doi.org/10.2174/1389202921999200330152007

Venil CK, Zakaria ZA, Ahmad WA (2013) Bacterial pigments and their applications. Process Biochem 48:1065–1079

Wang H, Jiang P, Lu Y, Ruan Z, Jiang R, Xing X-H, Lou K, Wei D (2009) Optimization of culture conditions for violacein production by a new strain of Duganella sp. B2, Biochem Eng J 44:119–124

Yazgin O, Keskin-Gundogdu T (2020) Production of biogas and astaxanthin from fruit and vegetable wastes using an integrated system. Int J Secondary Metabolite 7:35–46

Yusuf M, Shabbir M, Mohammad F (2017) Natural colorants: historical, processing and sustainable prospects. Nat Prod Bioprospect 7:123–145

Yuyama KT, Chepkirui C, Wendt L, Fortkamp D, Stadler M, Abraham W-R (2017) Bioactive compounds produced by hypoxylonfragiforme against staphylococcus aureus biofilms. Microorganisms 5. https://doi.org/10.3390/microorganisms5040080

Chapter 17
Microbial Biosensors for Environmental Quality Monitoring: Recent Advances and Future Outlook

Sompreeti Paul, Sindhoora Lakshmi, Dakshitha Akula, and Aswani Thekkangil

17.1 Introduction

To preserve human activity and all life on Earth, it is essential to rid our environment of anthropogenic pollution in the twenty-first century (Fernando et al. 2018). Therefore, environmental monitoring for the identification of pollutants is becoming increasingly significant for regulatory bodies, regulated communities, and the general public. This is especially true for substances that may be potentially harmful to human health or the environment. Furthermore, monitoring harmful pollutants in the environment as well as the release of chemicals into the environment from various sources, such as agriculture and industry, is critical to create a dataset for future cleaning processes. To address this demand, rapid, portable, and cost-effective analytical technologies are required. However, because of their distinct properties, technologies such as biosensors can be used to serve specialized niche applications in the field of environmental monitoring (Rogers 1995; Zhang and Keasling 2011). These technologies have demonstrated practicality and can potentially detect diverse biological, physical, chemical, and radioactive toxins in various ecosystems. The use of several disciplines, such as biotechnology, nanotechnology, and microbiology, provides biosensor technology with much-needed distinguishing characteristics such as sensitivity and selectivity. Biosensors are regarded as superior entities for sensing xenobiotics existing in the environment, either at trace levels, and are of considerable relevance because of their numerous benefits over standard procedures in the field of analysis. This scientific instrument uses an immobilized biological component to react towards an analyte, and the resulting biological signal is subsequently converted via a transducer into a readable signal (Bilal and Iqbal 2019).

S. Paul · S. Lakshmi · D. Akula · A. Thekkangil (✉)
Department of Biotechnology, School of Applied Science, REVA University, Bangalore 560077, India
e-mail: aswani.t@reva.edu.in

K. Bala et al. (eds.), *Harnessing Microbial Potential for Multifarious Applications*, Energy, Environment, and Sustainability, https://doi.org/10.1007/978-981-97-1152-9_17

A microbial biosensor is a scientific instrument that integrates immobilized microorganisms on a transducer to detect certain analytes. Bacteria and yeast, among other microorganisms, have shown tremendous promise as components for developing microbial biosensors, notably for determining the toxicity of heavy metals found in the environment (Gavrilaş et al. 2022). In addition, there are several types of biosensors, including microalgae, enzymes, antibodies, whole cells, and amperometric biosensors. Microbial biosensors are more efficient and have a wider range of detection capabilities than traditional biosensors (Bose et al. 2022). Traditional microbial biosensors look for substances that are either substrates or inhibitors of these processes by monitoring microbial metabolisms. In recent years, genetically engineered microorganisms have been commonly used to test for toxicity and bioavailability by fusing green fluorescent protein (gfp), lactose gene (lacZ) and luciferase (lux) reporter gene to an inducible promoter (Lei et al. 2006).

Pollutants such as synthetic organic compounds and heavy metals have been shown to negatively influence the physiological systems of living species. The combined impact of several contaminants acting consecutively is a cause of this concern. The consequences of these pollutants on ecosystems and species vary according to different elements, such as the exposure pattern, absorption pathways, metabolic processes that occur after absorption, accumulation, and the degree of vulnerability of the targeted organs (Rhind 2009). Therefore, monitoring the levels of xenobiotics in various ecosystems is of utmost importance. Biosensors may serve as effective tools for detecting xenobiotics in the environment. According to a study by skyquestt Report ID: SQMIG45J2077), the global market share for biosensors was United States dollar (USD) 25.5 billion in 2021 and is projected to reach USD 48.89 billion by 2030. The study found that the medical sector is the main application of biosensors, followed by the agricultural and environmental sectors. There have been many biosensors proposed for use in environmental applications, and they may be able to screen a variety of samples. This chapter begins with an overview of biosensors, their functioning principles, and classification. This further explains the microbial biosensor with a brief description on the mechanism of how it detects various analytes. The primary objective of this chapter is to investigate the role of microbial biosensors in environmental monitoring applications.

17.2 Biosensor: An Overview

A biosensor is a self-contained integrated tool that may provide precise quantitative or semi-quantitative analytical data, according to the International Union of Pure and Applied Chemistry (IUPAC) and has a biological recognition element that is kept in close proximity to the transducer element. This analytical tool consists of an immobilized biological component, such as a Deoxyribonucleic acid (DNA), cell organelle, enzyme, nanomaterials, tissue, or whole cells, which is located near a transducer that interacts with an analyte, resulting in an output signal that is displayed by a readout device that transforms the biological interaction into a readable signal (McConnell et al. 2020). Thus, the inherent specificity and sensitivity of biological elements can be utilized by biosensors. This tool comprises a transducer and a biological component. Enzymes, antibodies, nanomaterials, nucleic acids, and immobilized bacteria can constitute biological elements (Karunakaran et al. 2015). Biosensors are of various sizes and forms and can detect and quantify even low quantities of pathogens, harmful chemicals, and pH values.

The three most crucial elements of biosensors are the biological component, the transducer, and the readout (Fig. 17.1). The interaction between the biological component and analyte in the sample dictates the point at which the analysis begins. Choosing a biocomponent to build a biosensor for a specific analyte is therefore critical for device success. Biomaterials and their derivatives are considered biocomponents. The biological signal generated by the interaction between the biocomponent and analyte within the transducer is converted into an electrochemical, optical, or physicochemical signal (Justino et al. 2017). The type of contact and the biological signal produced, such as whether the pH changes after the encounter and the activation and/or inactivation of an enzyme, have a significant impact on transducer selection. The fundamental building block of a biosensor is its transducer configuration and type of interaction. Finally, the signal is processed and displayed by the output device, which can be an electrochemical analyzer, a plate reader, or a spectrofluorometer (Bilal and Iqbal 2019).

Biosensors are classified into different groups and subgroups based on the bioreactor (i.e., the biological element used to detect the analyte), transducer (i.e., the method of signalization), and detection system. Figure 17.2 shows an overview of biosensor categorization. The bio-recognition elements may include antibodies,

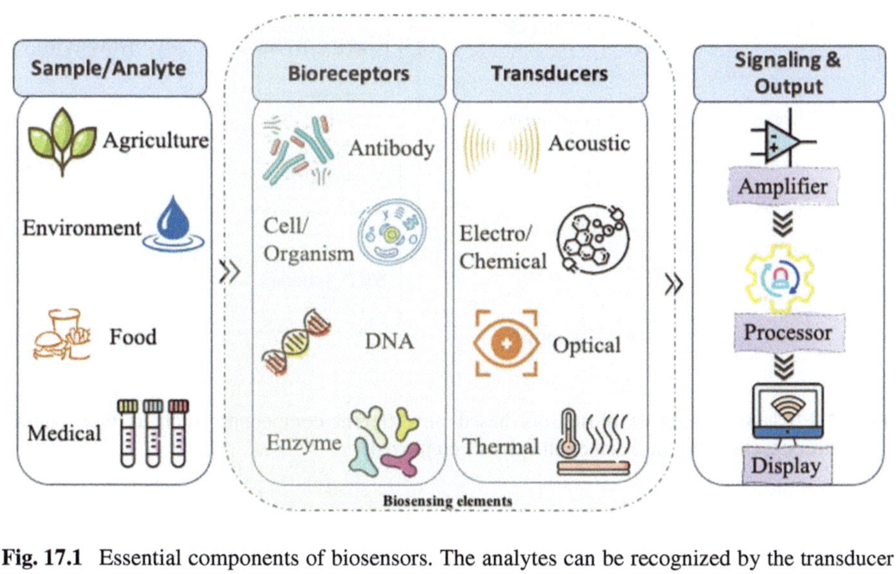

Fig. 17.1 Essential components of biosensors. The analytes can be recognized by the transducer with the aid of reference material and this transduced signal can be amplified by amplifier and the signal is processed and displayed on the system

enzymes, microorganisms, nucleic acids, and nanomaterials. Depending on the type of physicochemical change brought on by the sensing event, the transduction mechanism was chosen. The most commonly used signalization methods are electrochemical and optical biosensors. A membrane that covers the biological sensing component of a biosensor is an essential auxiliary component. This membrane shields the biological component from mechanical stressors, selectively permits analyte penetration and diffusion control, and sustains the biological component (Gavrilaș et al. 2022).

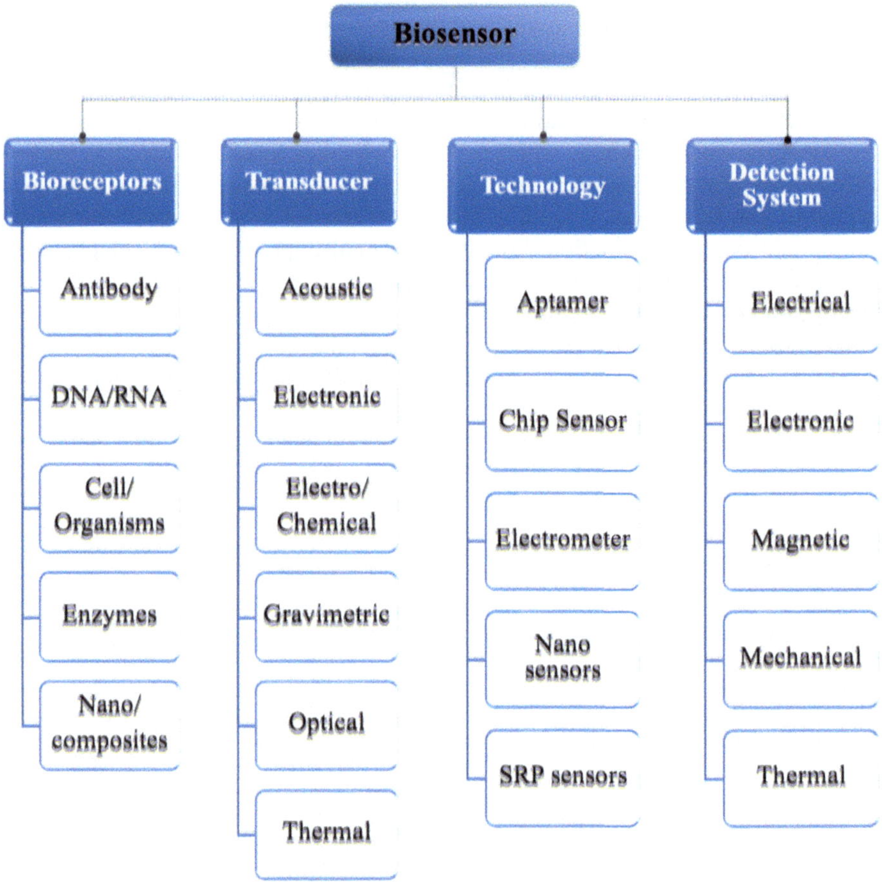

Fig. 17.2 Classifications of biosensors based on different components such as bioreceptors, transducers, technology, and detection systems employed

17.3 Microbial Biosensors

Microbial biosensors are scientific instruments made of microorganisms that identify a target substrate and convert the observed signal into a quantifiable response in a biochemical, physiological, or electrical way. It employs either naturally occurring or genetically modified microorganisms (bacteria, fungi, algae, protozoa, or viruses) that interact with a wide range of analytes and generate a signal, which is detected and quantified by a transducer (Fig. 17.3) (Justino et al. 2017). Microbial biosensors are microbe-based analytical devices that detect a target substrate and convert a detected signal into a measurable signal in a biochemical, physiological, or electrical manner. It uses microorganisms that interact with a variety of analytes and produce a signal that is recognized and measured by a transducer. These microorganisms can be naturally occurring or genetically modified. It is possible to find a number of microbial integrated transducers that operate based on various principles, including colorimetric, electrochemical, electrical, and optical (McConnell et al. 2020). Microbial biosensors are more environmentally sensitive than conventional biosensors, and may function in a variety of working environments. These characteristics render microbial biosensors useful in a wide range of disciplines.

As previously mentioned, microorganisms, such as bacteria and yeast, provide an intriguing approach to fabricate microbial biosensors with several advantages. First, compared to other approaches using traditional equipment, such as gas chromatography-mass spectroscopy, liquid chromatography–mass spectrometry, and tandem mass spectroscopy biosensors based on microorganisms provide analytical costs for sensing elements that are significantly cheaper, and an undemanding culture process and affordable nutrient medium are sufficient to establish an enormous number of microorganisms. Second, microbial biosensors have the capacity to detect a wide range of target components, and can also be modified by utilizing modern molecular biology approaches, which further will help to improve their efficiency (Zhang and Keasling 2011). Third, other related biosensor types, such as those relying on antibodies and enzymes, are fluctuating and very sensitive to pH and temperature. On the other hand, microbial biosensors are more resilient and can tolerate a variety of environmental conditions. Despite the fact that microbial biosensors have many advantages over traditional sensing devices, their broad application is constrained by a few drawbacks, such as relative insensitivity and poor selectivity in multiple compound detection. Figure 17.4 depicts the advantages and disadvantages of microbial biosensors.

Novel molecular biology approaches have substantially improved the specificity and sensitivity by genetic manipulation, which has helped enhance biochemical signal production in recent years. Hence, most biological systems have developed sensitive sensory mechanisms to control and optimize cell behavior, development, and transmission of specific biochemicals (McConnell et al. 2020). Sensory-regulative biosensors and their processes are capable of detecting different cellular signals and transducing them electrochemically or optically. Since microorganisms can detect external elements, such as nutrition, pH, and temperature and their own

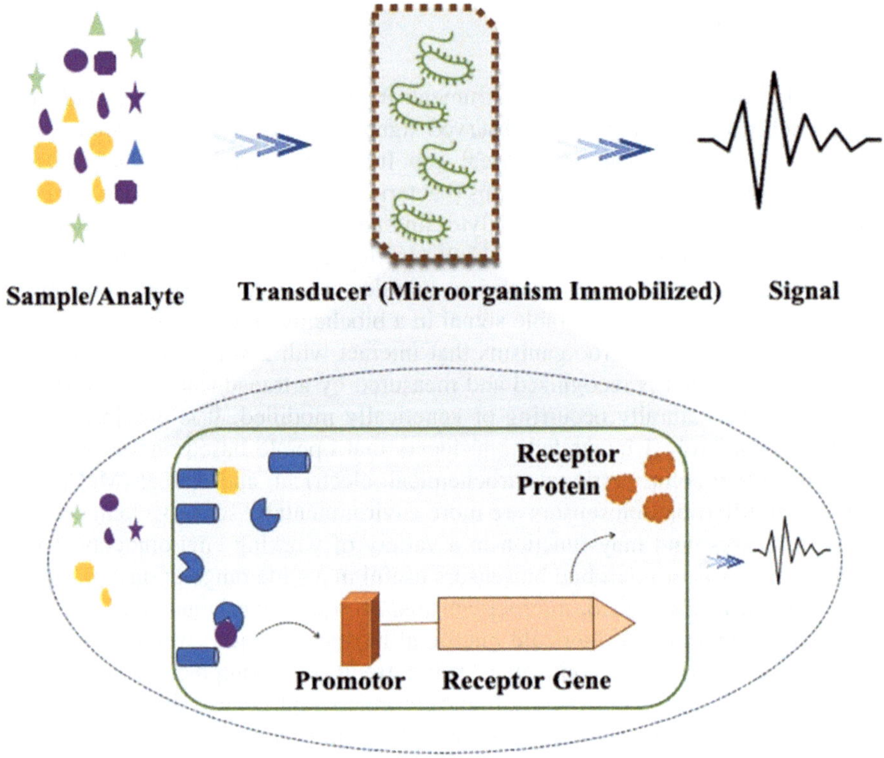

Fig. 17.3 Schematic representation of the mechanism of a microbial biosensor. The binding of the analyte to microbial enzymes induces the promoter to express the reporter gene; and the expressed reporter protein is detected by a transducer

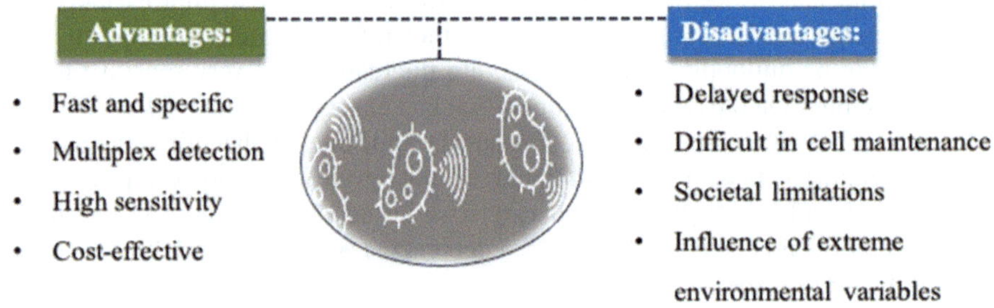

Fig. 17.4 Advantages and disadvantages of microbial biosensor

metabolic products (Zhang and Keasling 2011); additional observable strategies can be used to describe how metabolic pathways or cell function are controlled. For that, synthetically built biosensors, such as riboswitches and other DNA/RNA biosensors, have been developed to sense cellular signals and regulate cell development and responses. These recently developed microbial biosensors and their methods facilitate extremely sensitive identification of additional metabolites and analytes by selectively modifying the metabolic pathways of microbes.

17.4 Environmental Quality Monitoring with Microbial Biosensors

In many different industries and sectors, such as agricultural, defense, environmental food safety, health, industrial, microbial biosensors are being researched and applied (Ye et al. 2019). Numerous recent research have investigated the potential of microbial biosensors to identify environmental pollutants in terrestrial and aquatic biota, such as pesticides, heavy metals, phenol, and other dangerous substances. Table 17.1 summarizes numerous microbial biosensors with outstanding applications in environmental monitoring for the detection of organic and inorganic toxicity in various habitats, and Table 17.2 shows some examples of microorganisms incorporated in microbial biosensor for the detection of various organic and inorganic toxicities in the environment along with their detection limits.

Heavy metals, which are widely used in industry, are major toxicants in wastewater and soils. Because metal ions are non-biodegradable, they accumulate in living organisms and cause toxicity (Naresh and Lee 2021). Microbial biosensors allow for the monitoring of heavy metals in a low-cost, specific, effortless, and rapid manner. Heavy metal detection and biosensing are among the most important applications of microbial biosensor (Li et al. 2023) employed *S. aureus* as a biorecognition element to investigate the impacts of various concentrations levels cd^{2+} ions on the metabolism and growth of *S. aureus*. Microbial biosensor can also be used to monitor organophosphate pesticides, which can seriously harm the environment (Rajkumar et al. 2017). To detect pesticides, developed an electrochemical biosensor with a bio-recognition component comprising free acetylcholinesterase and encapsulated choline-cultured *A. globiformis* (Ping et al. 2017). A transportable biosensor that combines a whole-cell sensing component with an electrochemical silicon device miniaturized for the purpose of detecting trivalent arsenic (As(III)) and bivalent mercury (Hg(II)) in water (Sciuto et al. 2021). The biosensor uses modified Escherichia coli to selectively sense the metals, and it uses cyclic voltammetry analysis to find the 4-aminophenol redox active mediator that the bacterium produces.

Table 17.1 Application of microbial biosensors in environmental monitoring to detect various pollutants

Pollution	Analyte	Biosensing element	Signal	References
Water	Organic chemicals, heavy metals, microorganisms, environmental pollutants, industrial biproducts, marine toxins	Whole-cell biosensors, fluorescently labeled antibodies, surface plasmon resonance (SPR) micro-algal biosensors, aptamer	Optical	Herrera-Domínguez et al. (2023) Ejeian et al. (2018)
	Pesticides, pharmaceuticals, PPCPs,	Enzymes, antibodies and whole cells, amperometric biosensors, aptamer	Electrochemical	Ping et al. (2017), Gavrilaş et al. (2022)
Air	Organic pollutants, inorganic compounds, carbon monoxide	Microorganisms, enzymes, MFC sensor based on bio-cathode sensing elements	Electrochemical	Eltzov et al. (2019)
	Heavy metals, naphthalene, SO_2, H_2 S, CO_2, and NH_3, methanol and formaldehyde, benzene and its derivatives	Enzymes, bioluminescent bacteria immobilized in matrices, whole cell biosensors, filamentous green alga, protein-based biosensor, fiber optic based microbial biosensor	Optical	Bose et al. (2022)
Soil	Heavy metals (Pb^{2+}, Cd^{2+}, Cu^{2+}, Zn^{2+}), organic substances	Whole cell microbial biosensor, MNP-based biosensors	Electrical/optical	Xu et al. (2018)
		DNA conjugated gold electrode	Electrochemical	Rajkumar et al. (2017)
		Immobilized *Bacillus* species	Fluorescent	Bose et al. (2022)

PPCP: pharmaceutical and personal care products; SO_2: sulfur dioxide; H_2S: hydrogen sulfide; CO_2: carbon dioxide; NH_3: ammonia; Pb^{2+}: lead (II) ion; Cd^{2+}: cadmium ion, Cu^{2+}: copper(II)ion; Zn^{2+}: zinc ion

17.4.1 *Microbial Biosensors for Monitoring Water Quality*

Most microbial-based biosensors find targets that are essential to water quality during environmental monitoring. Bacteria, bacterial toxins, heavy metals, organic compounds, pesticides, industrial waste, PPCPs, and aquatic contaminants were the most frequently identified targets.

Bacterial pollution from water sources is a global issue that has both medical and economic impacts. Water-borne infections kill approximately two million people each year as owing to infection through pathogenic microorganisms. Numerous

Table 17.2 Examples of microorganism utilized in microbial biosensor to detect various environmental pollutants

Target	Analyte	Microorganism	Signal	LOD	References
Water	Zn & Cu	*E. coli* XL1-Blue	Fluorescence	16 μM; 26 μM	Ravikumar et al. (2012)
	Pb	*Rhizopusarrhizus*	Voltametric	5.0 × 1 M	Dai and Choi (2013)
	Pb	P. *aeruginosa*	Voltametric	6.0 × 1 M	Vogrinc et al. (2015)
	Pb	*Phormidium sp.*	Voltametric	2.5 × 1 M	Vogrinc et al. (2015)
	Trichloroethylene	*Pseudomonas species* Strain ASA86	Voltametric	0.05 ppm	Dai and Choi (2013)
	Methyl parathion, paraoxon	Recombinant *Moraxella*	Electrochemical	1 μM; 0.2 μM	Vogrinc et al. (2015)
	Lindane	Recombinant *E. coli* BL21	Electrochemical	2 ppt	Anu Prathap et al. (2012)
	Cu	Heterogeneous microbial populations	MFC	5 ppm	Shen et al. (2013)
	Arsenite, selenite	Recombinant *E. coli* DH5α™	Optical	0.2 μg/l; 5.8 ng/l	Ooi et al. (2015)
	Catechol	*Lactobacillus*	Electrochemical	0.5–5.0 mM	Vogrinc et al. (2015)
	BOD	*P. putida* SG10	Electrochemical	0.5–10 mg/l	Chee (2013)
	BOD	Electroactive mixed bacteria	MFC	100–750 ppm	Ayyaru and Dharmalingam (2014)
	BOD	Heterogeneous microbial populations	MFC	3–164 ppm	Di Lorenzo et al. (2014)
	Cu^{2+}	*S. Cerevisiae*	Colorimetric	1 μM	Vopálenská et al. (2015)
	Diuron (herbicide)	*C. reinhardtii*	Colorimetric	0.2 μM	Tsopela et al. (2016)
	Simazine (herbicide)	*D. chlorelloides* Dc1M	Luminescent	40.8 μg/L	Haigh-Flórez et al. (2014)
Soil	Zn	*P. putida* X4	Fluorescence	5 × M	Liu et al. (2012)
	Cu	*Circinella* sp.	Voltametric	5.4 × 1 M	Dai and Choi (2013)
	Trichloroethylene	*P. putida* F1 strain	Impedimetric	20μ/L	Dai and Choi (2013)

(continued)

Table 17.2 (continued)

Target	Analyte	Microorganism	Signal	LOD	References
	Methyl parathion	*Sphingomonas* JK1	Optical	4–80 μM	Vogrinc et al. (2015)
	Cd^{2+}, Cu^{2+}, Zn^{2+}	*B. megaterium* VR1	Fluorescent	1.42×10^{-4}, 3.16×10^{-4}, 2.42×10^{-4} mg/L	Rathnayake et al. (2021)

LOD: Limit of detection; BOD: Biological oxygen demand; MFC: Microbial fuel cell; μM: Micromolar; M: Molar; PPM: Parts per million; μg: Microgram; L: Liter; B: *Bacillus*; C: *Chlamydomonas*; D: *Dictyosphaerium*; E: *Escherichia*; P: *Pseudomonas*; S: *Saccharomyces*

illnesses and fatalities may result from the use of contaminated drinking water, groundwater, wastewater, and other water sources. Furthermore, there is a complicated interwoven interaction between contaminated soil and polluted water that has serious consequences for the agricultural, aquaculture, environment, human health. As a result, exceedingly sensitive biosensors for detecting water-borne pathogens are urgently needed. Major bacterial contaminants such as *Escherichia coli, Listeria monocytogenes, Microcystis aeruginosa, Pseudomonas aeruginosa, Salmonella, Staphylococcus aureus*, and *Vibrio* can be detected using microbial biosensors, with electrochemical, photoelectrochemical, fluorescent, and FRET-based signaling.

Heavy metals such as arsenic, lead, cadmium, mercury, silver and others can contaminate water sources and have negative effects on aquatic life and human health (Yadav et al. 2018). As heavy metals have a propensity to bioaccumulate in biological systems, these problems are exacerbated (Malik et al. 2019). According to Yadav et al. (2018), industrial waste and liquid effluents are the two main contaminants in the groundwater and other natural water systems. One of the most popular methods is the use of microbial biosensors for heavy metal detection utilizing electrochemical methods (Rapini and Marrazza 2017; Mishra et al. 2018; Malik et al. 2019). However, breakthroughs using mechanical and optical methods have been reported (Khoshbin et al. 2018).

Exogenous and endogenous aquatic poisons may have disastrous consequences for animals, human health, and ecosystems. Regulating bodies have also set legislation on the most common and noxious aquatic toxins because environmental instability and climate change may promote their formation (Herrera-Domínguez et al. 2023). Microbial sensors have demonstrated potential in identifying these toxins in environmental samples at concentrations well below permissible limits; however, transforming these biosensors into portable, user-friendly detection equipment is still a hurdle.

One of the most common difficulties in environmental monitoring is the detection of harmful industrial by-products. Industrial by-products including phenyl, biphenyl, bisphenol, and fluorene have recently been detected in several water bodies using microbial-based biosensors (Xu et al. 2018). Furthermore, certain biosensors can detect BPA exposure and release into water sources (Vigneshvar et al. 2016).

The prevalence and excessive use of PPCPs have caused the contamination of different water sources. To understand the possible hazards associated with exposure to these substances, extremely sensitive and tailored biosensors must be developed to detect chemicals from natural sources. Recently, biosensor technology has concentrated on finding ways to identify some commonly prescribed medications, including antibiotics, the widely used analgesic ibuprofen (Ping et al. 2017), the chemotherapeutic drug doxorubicin (Bahner et al. 2017), and hormonal birth control drugs.

17.4.2 Microbial Biosensor for Monitoring Soil Quality

Bacteria, bacterial toxins, and heavy metals are the primary groups of contaminants that are detected to analyze soil quality. Whole-cell microbial biosensors may be used to detect these contaminants by detecting the signals generated, which can be electrical or optical, depending on the biosensing technique adopted. By producing an electrochemical signal, microbial biosensors can detect the presence of heavy metals, such as lead, in the soil samples. When tackling certain contaminants, such as Cd^{2+}, Cu^{2+}, and Zn^{2+}, immobilized *Bacillus* species are used, and the signals produced are fluorescent in nature (Bose et al. 2022). Additionally, several studies have concentrated on the application of microbial biosensors for the detection of agricultural toxins, with a special emphasis on detecting mycotoxins. Exposure to mycotoxins, such as aflatoxins and ochratoxin, can cause serious health concerns such as cancer.

17.4.3 Microbial Biosensor for Monitoring Air Quality

The study of air quality as a component of microbial-based environmental monitoring may benefit from more intense investigations. Formaldehyde, carbon monoxide, naphthalene, SO_2, H_2S, CO_2, NH_3, methanol, formaldehyde, benzene derivatives, NO_2, and SO_2 are some typical hazardous gases that may be fascinating targets for the development of microbial-based biosensors. To monitor the quality of air, it is essential to find these gases, which can appear naturally or as a result of various industrial activities. Short- and long-term exposure to these gases can substantially impact the quality of life of people, animals, flora, and infrastructure. WHO and the International Cancer Organization have classified radon as a first-class environmental carcinogen. Radon is a naturally occurring gas produced when coal and natural gas are burned (Rajkumar et al. 2017). They can also be found in rocks, soil, and water sources. The next most prevalent cause of carcinoma of the lung in the world is radon exposure. For the purpose of radon detection, created a fluorescent biosensor. Currently, electrochemical signaling techniques are widely used to assess ambient gas levels. Mainly, studies are now focusing on developing optical, fluorescent, and colorimetric signaling methods for air quality monitoring.

17.5 Future Perspective

Environmental monitoring has benefited from the advancement of microbial and nanomaterial-based microbial biosensors for detecting pertinent pollution analytes. Portable microbial-based systems that provide fast, on-site, single, and multiplex detection of a variety of target analytes have made significant strides in recent years. However, adapting laboratory setups for on-site monitoring remains a challenge. Researchers should focus on ensuring that their designs follow the WHO proclaimed standards to close this gap. Much of this research must be adapted for practical applications. Biosensors must be capable of detecting target analytes on-site, even in trace amounts of samples, as well as detecting naturally occurring pollutants from complicated environmental samples in an accurate and convenient manner. As demonstrated by the aforementioned instances, the cross-disciplinary characteristics of nanomaterials, nanocomposites, and other genetic engineering-based techniques will probably be an appealing consideration for researchers interested in microbial-based environmental monitoring. Combining these and other new approaches will undoubtedly lead to the development of low-cost, dependable biosensors for environmental monitoring.

Despite extensive research and demonstration of microbial biosensors with outstanding versatility and multiplexity, there are still a number of difficulties that need to be carefully investigated and thoroughly resolved. Further research is needed on a number of issues, including advanced immobilization methods; accelerating response times; toxicology of chemicals, solvents, and micro/nanomaterials; biocompatibility of processes and materials; long-term growth maintenance and reusability; reduced contamination; and improved shelf life of microbial biosensors. To monitor environmental pollution in remote areas, more emphasis has been placed on the in situ and real-time monitoring of pollutants using alternative methods, such as biosensors integrated in unmanned aircraft systems (UAS), unmanned aerial vehicles (UAV), and remotely piloted aircraft (RPA). Recently, a portable and sensitive microbial biosensor was installed on a UAV (Phantom 2, Shenzhen, China) for remote air quality control and water pollution monitoring. This UAV had a maximum flight period of approximately 20 min and a maximum lifting weight of 300 g. The performance of the biosensor was evaluated by measuring *E. coli* bacterial viability inside the biosensor and comparing the results with traditional bacterial culture techniques (Lei et al. 2006). Because of their low cost, compact size, and low power requirements, future sensors based on the coupling of biosensors with drones are needed for environmental monitoring in remote areas.

17.6 Conclusion

Recent advancements in environmental detection and microbial-derived biosensors have drawn significant research interest over the last two decades. Unlike traditional chemical-based sensors, microbial biosensors have exceptional sensitivity, and can quickly, precisely, and instantly identify a wide range of contaminants or hazardous substances from a wider spectrum. Microbial biosensors are an excellent choice for real-time environmental pollution monitoring, particularly for potentially dangerous contaminants. However, several obstacles must be overcome before real-time execution for environmental applications and commercialization. Microbial biosensors will be developed into more advanced, wireless, portable, and compact devices in the upcoming years for the real-time detection of various environmental pollutants. Additionally, this indicates a promising future for widely used genetically modified microbial biosensors. A new era for microbial biosensors will be ushered in by sophisticated micro/nanotechnologies, yet there are still many intriguing potentials and bright futures for this discipline. The entire potential of microbial biosensor technology will thus be realized in the near future through successful and deliberate integration of microbiological sciences, molecular biology, and micro/nanotechnologies.

References

Anu Prathap MU et al (2012) Polyaniline-based highly sensitive microbial biosensor for selective detection of Lindane. Anal Chem 84(15):6672–6678. https://doi.org/10.1021/ac301077d

Ayyaru S, Dharmalingam S (2014) Enhanced response of microbial fuel cell using sulfonated poly ether ether ketone membrane as a biochemical oxygen demand sensor. Anal Chim Acta 818:15–22. https://doi.org/10.1016/j.aca.2014.01.059

Bahner N, Reich P, Frense D, Menger M, Schieke K, Beckmann D (2017) An aptamer-based biosensor for detection of doxorubicin by electrochemical impedance spectroscopy. Anal Bioanal Chem 410(5):1453–1462. https://doi.org/10.1007/s00216-017-0786-8

Bilal M, Iqbal HMN (2019) Microbial-derived biosensors for monitoring environmental contaminants: recent advances and future outlook. Process Saf Environ Prot 124:8–17. https://doi.org/10.1016/j.psep.2019.01.032

Biosensors market size, share, growth analysis, by technology, end-user - industry forecast 2022–2028. (n.d.). https://www.skyquestt.com/report/biosensors-market

Bose S, Maity S, Sarkar A (2022) Review of microbial biosensor for the detection of mercury in water. Environ Qual Manage 31(4):29–40. https://doi.org/10.1002/tqem.21742

Chee G-J (2013) Development and characterization of microbial biosensors for evaluating low biochemical oxygen demand in rivers. Talanta 117:366–370. https://doi.org/10.1016/j.talanta.2013.09.031

Dai C, Choi S (2013) Technology and applications of microbial biosensor. Open J Appl Biosens 02(03):83–93. https://doi.org/10.4236/ojab.2013.23011

Di Lorenzo M et al (2014) A small-scale air-cathode microbial fuel cell for on-line monitoring of water quality. Biosens Bioelectron 62:182–188. https://doi.org/10.1016/j.bios.2014.06.050

Ejeian F et al (2018) Biosensors for wastewater monitoring: a review. Biosens Bioelectron 118:66–79. https://doi.org/10.1016/j.bios.2018.07.019

Eltzov E, De Cesarea AL, Low YKA, Marks RS (2019) Indoor air pollution and the contribution of biosensors. Eurobiotech J 3(1):19–31. https://doi.org/10.2478/ebtj-2019-0003

Fernando EY, Keshavarz T, Kyazze G (2018) The use of bioelectrochemical systems in environmental remediation of xenobiotics: a review. J Chem Technol Biotechnol 94(7):2070–2080. https://doi.org/10.1002/jctb.5848

Gavrilaş S, Ursachi CŞ, Perţa-Crişan S, Munteanu F-D (2022) Recent trends in biosensors for environmental quality monitoring. Sensors 22(4):1513. https://doi.org/10.3390/s22041513

Ge X, Wang J, Ding J, Cao X, Zhang Z, Liu J, Li X (2019) Combining UAV-based hyperspectral imagery and machine learning algorithms for soil moisture content monitoring. PeerJ 7. https://doi.org/10.7717/peerj.6926

Haigh-Flórez D et al (2014) Microalgae dual-head biosensors for selective detection of herbicides with fiber-optic luminescent O2 transduction. Biosens Bioelectron 54:484–491. https://doi.org/10.1016/j.bios.2013.10.062

Herrera-Domínguez M, Morales-Luna G, Mahlknecht J, Cheng Q, Aguilar-Hernández I, Ornelas-Soto N (2023) Optical biosensors and their applications for the detection of water pollutants. Biosensors (basel) 13(3):370. https://doi.org/10.3390/bios13030370

Justino C, Duarte A, Rocha-Santos T (2017) Recent progress in biosensors for environmental monitoring: a review. Sensors 17(12):2918. https://doi.org/10.3390/s17122918

Karunakaran C, Rajkumar R, Bhargava K (2015) Introduction to Biosensors. In: Biosensors and bioelectronics. Elsevier, pp 1–68. https://doi.org/10.1016/B978-0-12-803100-1.00001-3

Khoshbin Z, Housaindokht MR, Verdian A, Bozorgmehr MR (2018) Simultaneous detection and determination of Mercury (II) and lead (ii) ions through the achievement of novel functional nucleic acid-based biosensors. Biosens Bioelectron 116:130–147. https://doi.org/10.1016/j.bios.2018.05.051

Kudłak B, Wieczerzak M (2019) Aptamer based tools for environmental and therapeutic monitoring: a review of developments, applications, future perspectives. Crit Rev Environ Sci Technol 50(8):816–867. https://doi.org/10.1080/10643389.2019.1634457

Lei Y, Chen W, Mulchandani A (2006) Microbial biosensors. Anal Chim Acta 568(1–2):200–210. https://doi.org/10.1016/j.aca.2005.11.065

Li J et al (2023) A self-amplifying plasmid based ultrasensitive biosensor for the detection of AS(III) in water. Biosens Bioelectron 221:114937. https://doi.org/10.1016/j.bios.2022.114937

Liu P, Huang Q, Chen W (2012) Construction and application of a zinc-specific biosensor for assessing the immobilization and bioavailability of zinc in different soils. Environ Pollut 164:66–72. https://doi.org/10.1016/j.envpol.2012.01.023

Malik M, Chaudhary R, Pundir CS (2019) Construction of an amperometric pyruvate biosensor based on enzyme bound to a nanocomposite and its comparison with enzyme nanoparticles bound to electrode. Int J Appl Sci Biotechnol 7(2):195–206. https://doi.org/10.3126/ijasbt.v7i2.24445

McConnell EM, Nguyen J, Li Y (2020) Aptamer-based biosensors for environmental monitoring. Front Chem 8. https://doi.org/10.3389/fchem.2020.00434

Mishra G, Sharma V, Mishra R (2018) Electrochemical aptasensors for food and environmental safeguarding: a review. Biosensors 8(2):28. https://doi.org/10.3390/bios8020028

Naresh V, Lee N (2021) A review on biosensors and recent development of nanostructured materials-enabled biosensors. Sensors (Switzerland) 21(4):1–35. https://doi.org/10.3390/s21041109

Ooi L, Heng L, Mori I (2015) A high-throughput oxidative stress biosensor based on escherichia coli ROGFP2 cells immobilized in a K-carrageenan matrix. Sensors 15(2):2354–2368. https://doi.org/10.3390/s150202354

Ping J, He Z, Liu J, Xie X (2017) Smartphone-based colorimetric chiral recognition of ibuprofen using aptamers-capped gold nanoparticles. Electrophoresis 39(3):486–495. https://doi.org/10.1002/elps.201700372

Rajkumar P, Ramprasath T, Selvam GS (2017) A simple whole cell microbial biosensors to monitor soil pollution. In: New pesticides and soil sensors. Elsevier, pp 437–481. https://doi.org/10.1016/B978-0-12-804299-1.00013-8

Rapini R, Marrazza G (2017) Electrochemical aptasensors for contaminants detection in food and environment: recent advances. Bioelectrochemistry (amsterdam, Netherlands) 118:47–61. https://doi.org/10.1016/j.bioelechem.2017.07.004

Rathnayake IVN, Megharaj M, Naidu R (2021) Green fluorescent protein based whole cell bacterial biosensor for the detection of bioavailable heavy metals in soil environment. Environ Technol Innov 23:101785. https://doi.org/10.1016/j.eti.2021.101785

Ravikumar S et al (2012) Construction of a bacterial biosensor for zinc and copper and its application to the development of multifunctional heavy metal adsorption bacteria. Process Biochem 47(5):758–765. https://doi.org/10.1016/j.procbio.2012.02.007

Rhind SM (2009) Anthropogenic pollutants: a threat to ecosystem sustainability? Philos Trans R Soc b: Biol Sci 364(1534):3391–3401. https://doi.org/10.1098/rstb.2009.0122

Rogers K (1995) Biosensors for environmental applications. Biosens Bioelectron 10(6–7):533–541. https://doi.org/10.1016/0956-5663(95)96929-S

Shen Y et al (2013) Effect of shear rate on the response of microbial fuel cell toxicity sensor to cu(ii). Biores Technol 136:707–710. https://doi.org/10.1016/j.biortech.2013.02.069

Tsopela A et al (2016) Development of a lab-on-chip electrochemical biosensor for water quality analysis based on microalgal photosynthesis. Biosens Bioelectron 79:568–573. https://doi.org/10.1016/j.bios.2015.12.050

Vigneshvar S, Sudhakumari CC, Senthilkumaran B, Prakash H (2016) Recent Advances in biosensor technology for potential applications—an overview. Front Bioeng Biotechnol 4. https://doi.org/10.3389/fbioe.2016.00011

Vogrinc D, Vodovnik M, Marinšek-Logar R (2015) Microbial biosensors for environmental monitoring. Acta Agriculturae Slovenica 106(2). https://doi.org/10.14720/aas.2015.106.2.1

Vopálenská I, Váchová L, Palková Z (2015) New biosensor for detection of copper ions in water based on immobilized genetically modified yeast cells. Biosens Bioelectron 72:160–167. https://doi.org/10.1016/j.bios.2015.05.006

Xu S, Chen X, Peng G, Jiang L, Huang H (2018) An electrochemical biosensor for the detection of Pb2+ based on G-quadruplex DNA and gold nanoparticles. Anal Bioanal Chem 410(23):5879–5887. https://doi.org/10.1007/s00216-018-1204-6

Yadav V, Roy S, Singh P, Khan Z, Jaiswal A (2018) 2D MOS2-based nanomaterials for therapeutic, bioimaging, and biosensing applications. Small 15(1):1803706. https://doi.org/10.1002/smll.201803706

Ye W et al (2019) Marine toxins detection by biosensors based on aptamers. Toxins (basel) 12(1):1. https://doi.org/10.3390/toxins12010001

Zhang F, Keasling J (2011) Biosensors and their applications in microbial metabolic engineering. Trends Microbiol 19(7):323–329. https://doi.org/10.1016/j.tim.2011.05.003

Chapter 18
Biosensor: Application in Environmental Management

Arti Bhatnagar, Jamson Masih, and Ranjit Kumar

18.1 Introduction

Detection of biological components in ambient air is a difficult challenge as the methods are cumbersome and time taking. Hence, the development of fast and accurate methods of detection and identification systems for biological components in the environment is of paramount importance in the current scenario of the infection and spread of pathogens. Biosensor techniques for potential environmental applications have shown sustained advances in a wide range of areas. These advances will play an important role in the development of biosensor systems for the environmental market which will have a promising and bright future. This will be helpful for diagnostic purposes and also for the development of various vaccines for the prevention of biological microbial diseases and pandemics in near future. For environmental monitoring, finding a simple, rapid, cost-effective, and field-portable screening method to monitor various biological and chemical contaminants, which can be potential risks to human health (Belkin 2003; Rogers 2006). The food and fermentation industries require rapid, affordable, and reliable methods to ensure product quality and process controls (Arora et al. 2011; Belkin 2003). Moreover, there is an urgent need in clinical diagnostics for accurate, fast, and inexpensive devices, which can be routinely used to monitor clinically important parameters (Malhotra and Chaubey 2003). The conventional methods in these applications typically demand analytical techniques, specialized laboratories, and skilled operators, all of which are costly, complicated, and time-consuming. The primary objective of biosensors is to provide a versatile and efficient means of detecting and quantifying specific biological substances. These devices leverage the inherent selectivity and sensitivity of biological components,

A. Bhatnagar · R. Kumar (✉)
Department of Chemistry, Dayalbagh Educational Institute, Agra 282005, India
e-mail: rkschem@dei.ac.in

J. Masih
Department of Chemistry, Wilson College, Mumbai 400007, India

K. Bala et al. (eds.), *Harnessing Microbial Potential for Multifarious Applications*, Energy, Environment, and Sustainability, https://doi.org/10.1007/978-981-97-1152-9_18

such as enzymes, antibodies, or nucleic acids, to recognize and interact with target analytes. The integration of this bio-receptor with a transducer facilitates the conversion of the biochemical signal into a measurable output, often in the form of an electrical, optical, or thermal signal. The ultimate goal is to enable rapid, accurate, and real-time detection of target analytes in various applications, spanning medical diagnostics, environmental monitoring, and the food industry. Biosensors play a critical role in advancing fields like personalized medicine by facilitating the early detection of diseases through the identification of specific biomarkers. Additionally, biosensors contribute to environmental sustainability by aiding in the monitoring of pollutants and contaminants. As technology continues to advance, the objectives for biosensors include enhancing their specificity, stability, and applicability, thereby broadening their impact in diverse sectors and paving the way for innovative solutions in healthcare, environmental management, and beyond.

Biosensors also have applications in biotechnology and drug development, aiding researchers in the study of molecular interactions and the screening of potential drug candidates. The continuous evolution of biosensor technology, including miniaturization and integration with digital platforms, holds promise for expanding their role in point-of-care diagnostics, wearable devices, and other innovative solutions across various sectors. Ultimately, biosensors play a crucial role in advancing research, diagnostics, and monitoring, contributing to improved health outcomes, environmental sustainability, and the overall quality of life.

18.1.1 Monitoring of Pollution

Monitoring of pollution is an essential component of environmental management and public health initiatives. The continuous and accurate assessment of various pollutants in air, water, and soil is crucial to understanding the extent of environmental degradation and its potential impact on human health. Air quality monitoring involves the measurement of pollutants such as particulate matter, nitrogen oxides, sulfur dioxide, and ozone. Real-time monitoring stations and satellite technology play a pivotal role in tracking the dispersion of pollutants in the atmosphere. Water quality monitoring encompasses the analysis of parameters like chemical contaminants, bacterial levels, and nutrient concentrations in rivers, lakes, and oceans. Regular sampling and analysis help identify sources of contamination and guide regulatory measures to protect water resources. Soil monitoring involves assessing contamination levels of heavy metals, pesticides, and other harmful substances that can affect agricultural productivity and ecosystem health. Advanced technologies, including remote sensing and sensor networks, contribute to more efficient and widespread monitoring efforts. The data collected from these monitoring initiatives not only inform policymakers and regulatory bodies but also empower communities to take informed actions towards mitigating pollution and promoting sustainable practices. Effective pollution monitoring is a cornerstone in the global effort to address environmental challenges and work towards a healthier and more sustainable future.

18.1.2 Control of Pollutants

The control of pollutants is imperative for preserving environmental quality and mitigating the adverse impacts on human health. Implementing effective pollution control measures involves a combination of regulatory frameworks, technological advancements, and public awareness campaigns. In industries, adopting cleaner production technologies and adhering to emission standards are key strategies to minimize air pollutants. Biosensors are pivotal instruments in both controlling pollutants and monitoring environmental conditions. These devices utilize biological elements, such as enzymes, cells, or antibodies, to selectively recognize and detect specific pollutants in air, water, or soil. In the realm of pollution control, biosensors play a crucial role by providing real-time data on pollutant levels. This timely information allows for proactive responses, enabling the implementation of pollution mitigation strategies and regulatory measures. Biosensors contribute significantly to environmental sustainability by aiding in the identification and containment of pollutants, preventing their adverse impact on ecosystems and human health. Moreover, biosensors are invaluable for continuous environmental monitoring. Their ability to operate in situ and offer rapid, on-site analysis makes them ideal for long-term surveillance of environmental conditions. By continuously measuring pollutant levels and other critical parameters, biosensors contribute to the creation of comprehensive datasets essential for environmental modeling. This data-driven approach helps researchers and policymakers understand the dynamics of pollution, identify pollution sources, and formulate effective strategies for pollution prevention and remediation. Scrubbers, filters, and catalytic converters are examples of technologies used to reduce emissions of harmful substances from industrial processes and vehicles. Wastewater treatment plants play a critical role in controlling water pollution by removing contaminants before discharge into water bodies. Additionally, the enforcement of strict regulations on waste disposal and the promotion of recycling initiatives are essential in managing solid waste pollution. Agricultural practices can be modified to minimize the use of harmful pesticides and fertilizers, promoting sustainable and environmentally friendly farming methods. The government policies and international agreements contribute significantly to pollution control efforts, setting standards and guidelines for industries and individuals. Public education and community engagement are crucial components, fostering a sense of responsibility and encouraging eco-friendly practices among citizens. The ongoing development and implementation of innovative, sustainable technologies, coupled with robust regulatory frameworks and widespread public participation, are essential for effective control of pollutants and the preservation of a cleaner, healthier environment.

In summary, biosensors serve a dual purpose in environmental protection—not only do they provide real-time insights for immediate pollution control measures but they also contribute to ongoing environmental monitoring efforts, aiding in the development of sustainable practices and policies. As biosensor technology continues to advance, it holds the potential to further refine our ability to control pollutants and monitor environmental health with increased precision and efficiency.

18.1.3 Types of Techniques for Environmental Monitoring and Management

Environmental monitoring and management rely on a variety of techniques to assess, understand, and mitigate the impact of human activities on ecosystems. Remote sensing is a pivotal technique that utilizes satellites and other platforms to collect data on land use, vegetation cover, and changes in the Earth's surface. This provides a comprehensive view of large areas and facilitates monitoring of deforestation, urbanization, and natural disasters. Additionally, Geographic Information System (GIS) technology integrates spatial data, enabling the analysis of environmental patterns and trends. Sensor networks, another vital tool, involve deploying a network of sensors to collect real-time data on air and water quality, soil conditions, and biodiversity. These networks provide continuous monitoring, aiding in the early detection of pollution events. Laboratory techniques, such as chromatography and mass spectrometry, are employed for detailed chemical analyses of air, water, and soil samples to identify pollutants. Molecular biology tools, like Polymerase Chain Reaction (PCR), are utilized to detect specific genetic markers indicative of microbial contamination. Social science methodologies, including surveys and interviews, contribute to understanding human behaviors and perceptions related to environmental issues. Integrated, multidisciplinary approaches, combining these techniques, are crucial for effective environmental management, allowing policymakers and scientists to make informed decisions for sustainable and resilient ecosystems.

18.2 Sensing Techniques

Various sensing techniques are deployed across industries for diverse applications, showcasing the adaptability of these technologies in addressing specific needs. In environmental monitoring, remote sensing stands out, employing satellites and aerial platforms to capture detailed images and data about the Earth's surface. This technique aids in tracking changes in land cover, deforestation, and natural disasters. On the ground, chemical sensing techniques such as gas chromatography and mass spectrometry play a crucial role in identifying and quantifying pollutants in air and water samples, supporting environmental regulatory compliance. In the medical field, biosensors, specifically enzyme-based sensors, are utilized for rapid and precise detection of biomolecules, contributing to diagnostic applications.

Additionally, optical sensing techniques, like fluorescence spectroscopy, find use in medical imaging and material analysis due to their non-invasive nature and high sensitivity. In industrial settings, ultrasonic sensing techniques are employed for distance measurement and quality control. These techniques utilize the reflection of sound waves to assess properties such as thickness, composition, or defects in materials. These examples underscore the diverse array of sensing techniques available, each tailored to meet the specific requirements of its application, from environmental monitoring to healthcare and industrial processes. These sensing techniques provide critical data for understanding environmental conditions, detecting changes, identifying pollution sources, and making informed decisions for effective environmental management and conservation.

18.2.1 Biosensor and Its Components

A biosensor is a device that combines a biological sensing element with a transducer to convert a biological response into a measurable signal. Biosensors are used to detect the presence or concentration of specific biological molecules, such as enzymes, antibodies, antigens, DNA, and other biomolecules. They have applications in medical diagnostics, environmental monitoring, food safety testing, etc. (Wilson 2005). The main components of a biosensor as shown in Fig. 18.1.

Biological Sensing Element: This is the biological component of the biosensor that interacts with the target biomolecule. It is often a biological molecule that has

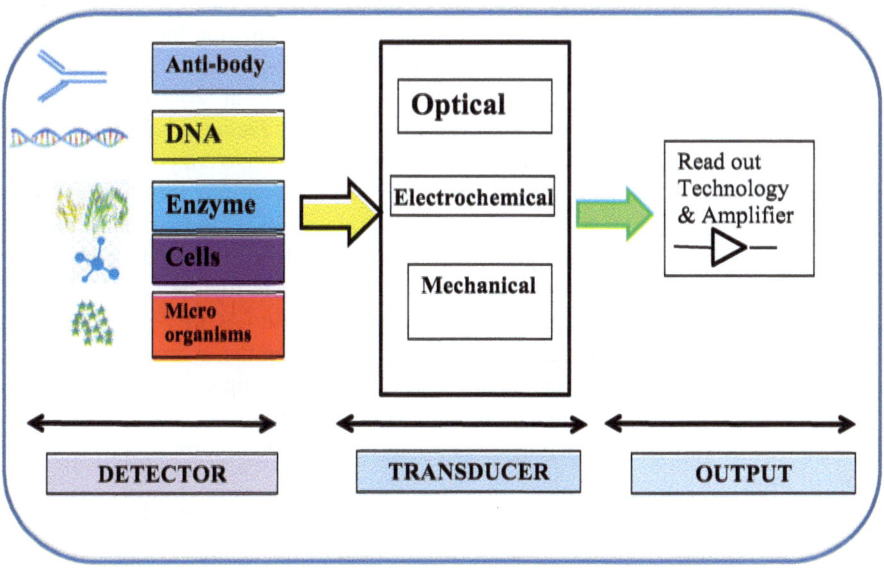

Fig. 18.1 Biosensor design

a high specificity for the target analyte such as enzymes, antibodies, nucleic acids (DNA/RNA), and whole cells.

Transducer: The transducer is responsible for converting the biological response into a measurable signal. It converts the biochemical or biophysical interaction between the biological sensing element and the target analyte into an electrical, optical, or other measurable output.

Signal Amplification System: In many cases, the signal generated by the transducer is weak and needs to be amplified for accurate detection. Amplification systems can include enzymes that catalyze a reaction leading to signal enhancement or nanoparticles that interact with the target to increase signal intensity.

Detector: The detector is responsible for measuring the amplified signal produced by the transducer. The different biosensors may use various types of detectors, such as photodiodes, electrodes, spectrometers, or cameras.

18.2.2 Types of Biosensors

Biosensors come in various types, each designed for specific applications and based on different principles of detection. Here are some common types of biosensors and their specific uses are given in Table 18.1.

Enzyme-Based Biosensors: Enzymes serve as the biological recognition component in these biosensors. Through catalytic reactions involving the target analyte, enzymes generate signals subsequently measured by transducers. An instance is the application of glucose biosensors for diabetes management. A specific focus of research centers on the early detection of diabetes using enzyme-based biosensors. Figure 18.2 depicts the process and components of enzyme-based biosensors. This work underscores the potential of biosensors as diagnostic tools, particularly in tracking glucose levels for diabetes patients (Mandpe et al. 2020). In the realm of enzyme-based biosensors for diabetes management, recent advancements are explored (Khor et al. 2022). This involves investigating diverse enzyme immobilization methods and sensor setups, particularly in their applications for glucose monitoring and diabetes care. For instance, glucose oxidase (GOx) is frequently utilized in glucose sensors, where it catalyzes the oxidation of glucose to produce hydrogen peroxide and gluconic acid. This reaction forms the basis for the sensor's ability to measure glucose levels. Similarly, acetylcholinesterase (AChE) is employed in sensors designed to detect organophosphates by catalyzing the hydrolysis of acetylcholine. In various contexts, dehydrogenases, hydrolases, lipases, and isomerases are strategically chosen depending on the specific enzymatic reactions needed for the detection of substances such as ethanol, certain pollutants, lipids, and isomeric forms of molecules.

Antibody-Based Biosensors: Antibodies play a pivotal role by binding to specific antigens, generating detectable signals upon antigen–antibody interaction. These biosensors find extensive application in medical diagnostics and scientific research.

Table 18.1 Biosensor type with their specific uses

Biosensors	Uses
Aptamer-based biosensors	– Detection of specific proteins, nucleic acids, or small molecules – Biomarker detection in medical diagnostics
Immunosensors	– Detection of antigens or antibodies for disease diagnosis – Environmental monitoring for pathogens and toxins
Microfluidic biosensors	– Point-of-care diagnostics for rapid testing – Lab-on-a-chip applications for chemical and biological analysis
Nanostructure-based biosensors	– Sensing and monitoring at the nanoscale – Detection of nanoparticles, gases, and biomolecules
Gas biosensors	– Monitoring air quality for gases like CO_2, O_2, and volatile organic compounds – Industrial safety and process control
Calorimetric biosensors	– Measurement of heat changes in chemical reactions – Enzyme activity and binding studies
Thermal biosensors	– Detection of temperature changes associated with biological processes – Monitoring microbial growth
Piezoelectric biosensors	– Mass and viscosity measurement in liquids – Detection of biological molecules using acoustic waves
Electrochemical biosensors	– Detection of ions, molecules, and biomolecules through electrochemical reactions – Glucose monitoring in diabetes management
Optical biosensors	– Label-free detection of biomolecules using light – Applications in genomics, proteomics, and drug discovery
Whole cell biosensors	– Monitoring cellular responses to environmental changes – Environmental bioremediation and toxicity assessment
DNA-based biosensors	– Detection of specific DNA sequences for genetic testing – Pathogen detection in molecular diagnostics
Antibody-based biosensors	– Detection of specific antigens and antibodies in immunoassays – Disease diagnosis and drug screening
Enzyme-based biosensors	– Monitoring enzyme activity for clinical and industrial purposes – Detection of biochemical reactions

"Antibody-Based Sensors" delves into the principles, challenges, and potential applications of these sensors, particularly in detecting pathogens and associated toxins (Byrne 2009). Antibody-based biosensors have proven invaluable in diverse applications, demonstrating their versatility and precision in detecting specific molecules. In the realm of medical diagnostics, pregnancy tests exemplify the use of antibody-based biosensors, where they detect the presence of human chorionic gonadotropin (hCG) in urine to confirm pregnancy. In cancer diagnostics, these biosensors target

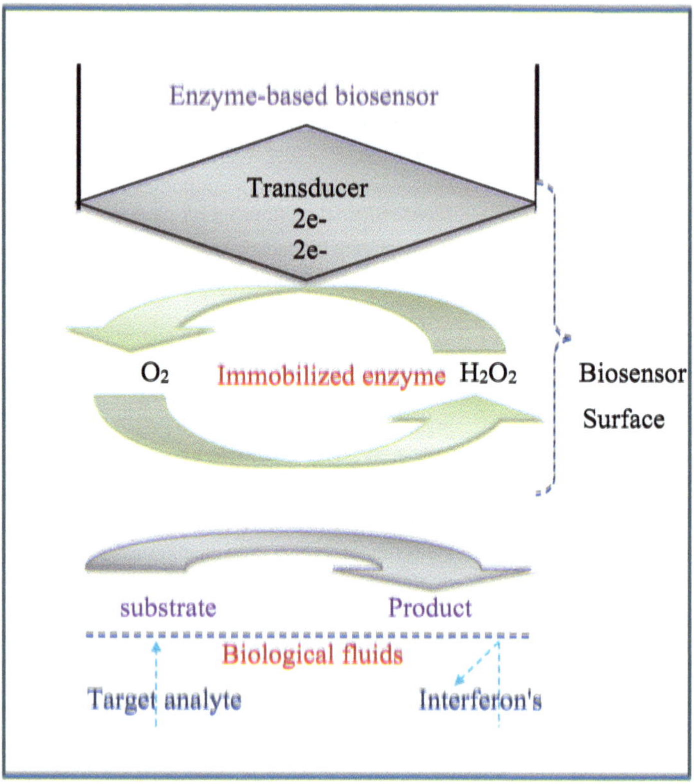

Fig. 18.2 Enzyme-based biosensors

specific biomarkers associated with particular cancers, enabling early detection and personalized treatment strategies. Infectious disease diagnostics utilize antibody-based biosensors to detect the presence of antibodies or antigens related to diseases such as HIV, offering rapid and reliable results. Environmental monitoring benefits from these biosensors by detecting contaminants in air, water, or soil, ensuring the safety of ecosystems. Food safety is another crucial area where antibody-based biosensors identify pathogens or contaminants in food products. In healthcare, they play a role in monitoring cardiac biomarkers like troponin, aiding in the diagnosis of heart conditions. Point-of-care testing benefits from the rapid response and specificity of these biosensors, enabling quick and accurate diagnostics. Their application in autoimmune disease monitoring involves detecting specific antibodies associated with conditions like rheumatoid arthritis. Finally, in biodefense, antibody-based biosensors contribute to the rapid identification of bio-threat agents, enhancing our ability to respond to potential bioweapons or infectious agents. These examples underscore the significance of antibody-based biosensors across medical, environmental, and security domains. Figure 18.3 shows the process of antibody-based biosensors.

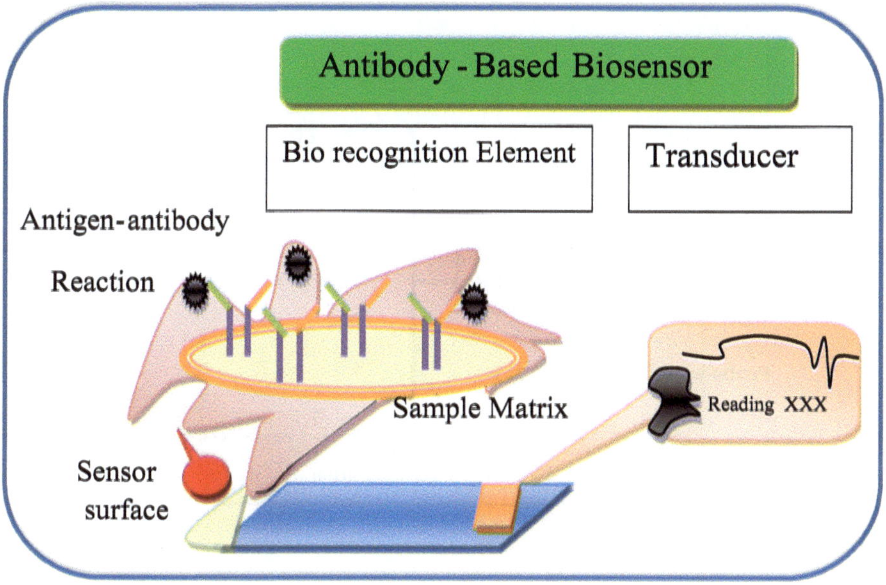

Fig. 18.3 Antibody-based biosensor

DNA-Based Biosensors: Deoxyribonucleic acid-based biosensors represent a cutting-edge technology with a myriad of applications, showcasing the remarkable role of DNA in sensing and diagnostics. Figure 18.4 shows the principle of DNA-based biosensors. In one notable example, DNA-based biosensors are employed in the detection of specific genetic sequences associated with infectious agents. For instance, in the realm of pathogen detection, a DNA-based biosensor can be designed with a DNA probe that complements a target pathogen's genetic material. When the target DNA is present, the binding event between the probe and target initiates a measurable signal, such as a change in fluorescence or electrical conductivity. This application is pivotal in the rapid and specific identification of pathogens, contributing to timely responses in healthcare and disease control. Additionally, DNA-based biosensors play a crucial role in genetic screening and diagnostics. They can be tailored to detect specific mutations or variations in the DNA sequence, enabling the identification of genetic disorders or predispositions to certain diseases. The precision and specificity of DNA-based biosensors make them powerful tools in advancing personalized medicine, where tailored treatments can be designed based on an individual's genetic makeup. As technology continues to evolve, DNA-based biosensors hold immense promise for revolutionizing diagnostics and molecular detection across various fields, from healthcare to environmental monitoring. Within the realm of DNA-based biosensors, matters of sensor optimization, sensitivity enhancement, and mitigation of potential interferences from intricate sample matrices are

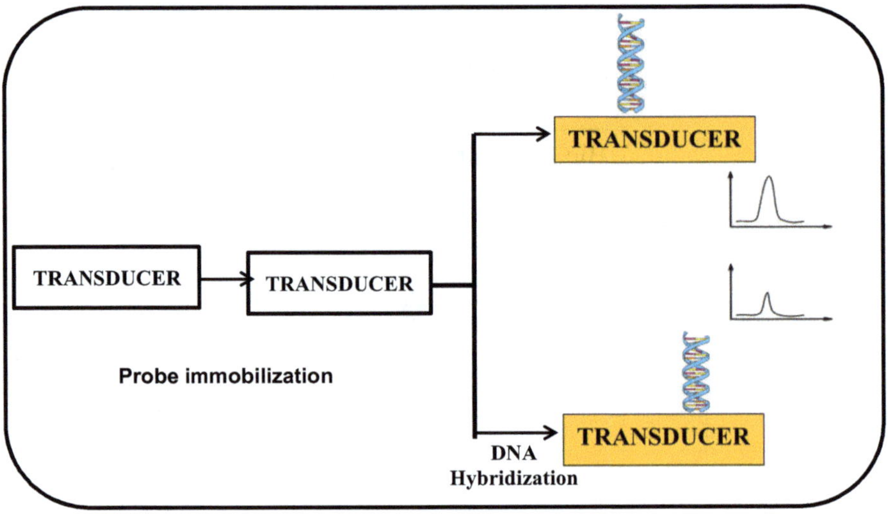

Fig. 18.4 DNA-based biosensors

likely addressed. Strategies employed to surmount these challenges could encompass the integration of nanomaterials and signal amplification methodologies (Hua et al. 2022).

Whole Cell Biosensors: Whole cell biosensors represent a fascinating intersection of biology and technology, utilizing living cells as the sensing elements to detect specific substances. In this context, engineered bacteria or yeast cells are designed to express specific genes that respond to the presence of pollutants such as heavy metals or toxic chemicals. When these cells encounter the target substance, they undergo a biological response that can be quantified. For instance, the expression of a reporter gene, which produces a measurable signal like fluorescence, can be activated. This enables the rapid and sensitive detection of contaminants in water sources. These biosensors show potential for in situ detection of bacterial waterborne pathogens, streamlining water quality assessment (Wu et al. 2021). Whole cell biosensors are advantageous in environmental monitoring as they provide a holistic response to complex mixtures of substances, mimicking the way living organisms naturally respond to their environment. Their versatility extends to other applications, including the detection of pathogens in food or clinical samples. The integration of whole cell biosensors in various fields exemplifies the potential of harnessing the biological complexity of living cells for precise and real-time sensing, contributing to advancements in environmental protection and public health. Figure 18.5 represents the whole cell biosensors.

Optical Biosensors: Optical biosensors represent a sophisticated class of sensing devices that leverage the principles of optics to detect and quantify biological interactions. An exemplary application of optical biosensors is in the field of medical diagnostics, where Surface Plasmon Resonance (SPR) technology is widely employed.

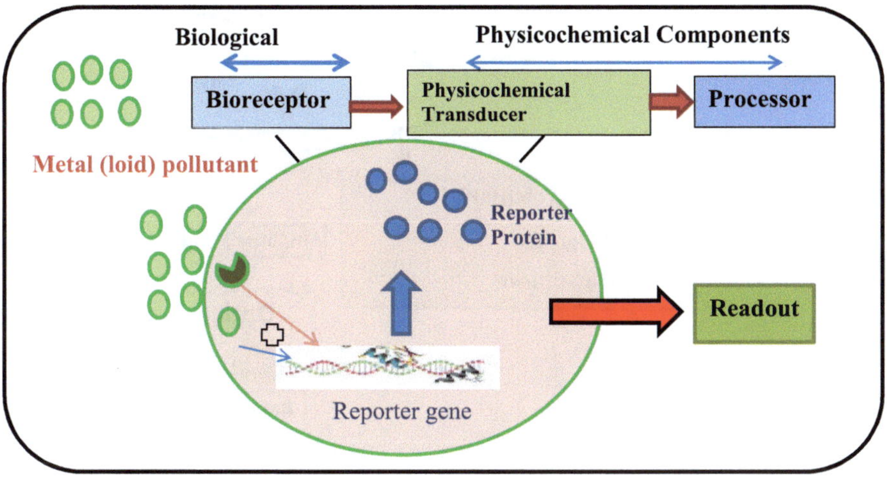

Fig. 18.5 Whole cell biosensors

In SPR-based optical biosensors, a thin metal film is used to excite surface plasmons at the metal–dielectric interface. When a biomolecular interaction, such as binding between a ligand and analyte, occurs on the sensor surface, it induces changes in the refractive index, altering the SPR angle. This change is detected as a shift in the reflected light spectrum, providing real-time, label-free detection of molecular interactions. This technology is extensively used in the study of protein–protein interactions, antibody-antigen binding, and drug development. For instance, researchers can monitor the binding kinetics of a drug candidate to its target protein, aiding in the optimization of therapeutic compounds. The high sensitivity and real-time monitoring capabilities of optical biosensors make them invaluable tools not only in medical research but also in environmental monitoring, food safety, and homeland security, showcasing their versatility in various fields requiring precise and rapid molecular detection. Figure 18.6 depicts the components and process of optical biosensors.

Electrochemical Biosensors: Electrochemical biosensors play a pivotal role in medical diagnostics, and a notable example is their application in glucose monitoring for diabetes management. Enzyme-based electrochemical biosensors, such as that using glucose oxidase (GOx), are commonly used for precise and rapid measurement of blood glucose levels. In this context, a glucose biosensor typically consists of an electrode coated with a biocompatible membrane containing the enzyme GOx. When a blood sample is applied to the sensor, glucose in the sample interacts with GOx, leading to the production of hydrogen peroxide. The generated hydrogen peroxide triggers an electrochemical reaction at the electrode, resulting in a measurable current or voltage change. The magnitude of this change is directly proportional to the glucose concentration in the blood, providing a real-time and quantitative assessment. Electrochemical glucose biosensors offer diabetic individuals a convenient and efficient means of self-monitoring, facilitating timely adjustments in insulin dosage and

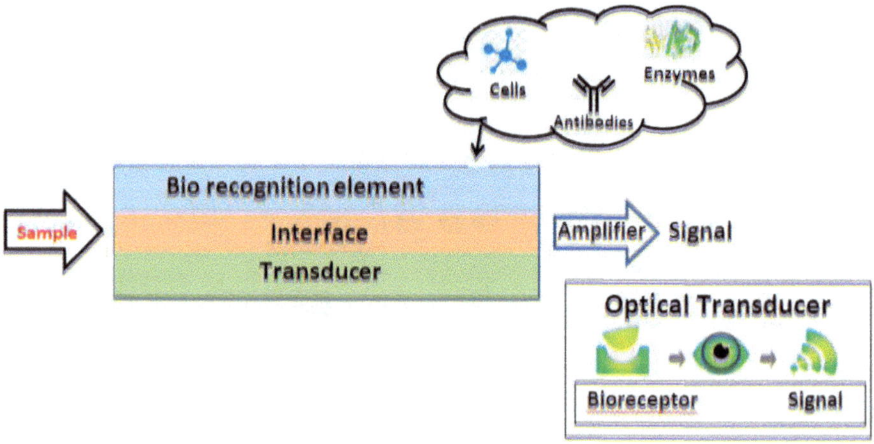

Fig. 18.6 Optical biosensor

dietary habits. This example underscores the significance of electrochemical biosensors in improving the quality of life for those managing chronic conditions while showcasing their precision and applicability in point-of-care diagnostics. Figure 18.7 is a type of electrochemical biosensor.

Piezoelectric Biosensors: Piezoelectric biosensors exemplify their utility in various applications, with one notable example being their use in the detection of specific DNA sequences for genetic testing. In this application, a piezoelectric biosensor typically consists of a quartz crystal coated with a layer of molecules that selectively bind to the target DNA sequence. When the target DNA is present, the binding event leads to a mass change on the quartz crystal surface, altering its vibrational frequency. This change in frequency is directly proportional to the amount of bound DNA, enabling highly sensitive detection. Researchers and clinicians leverage this technology in genetic diagnostics, such as the identification of genetic mutations associated with diseases like cancer. The precision and rapid

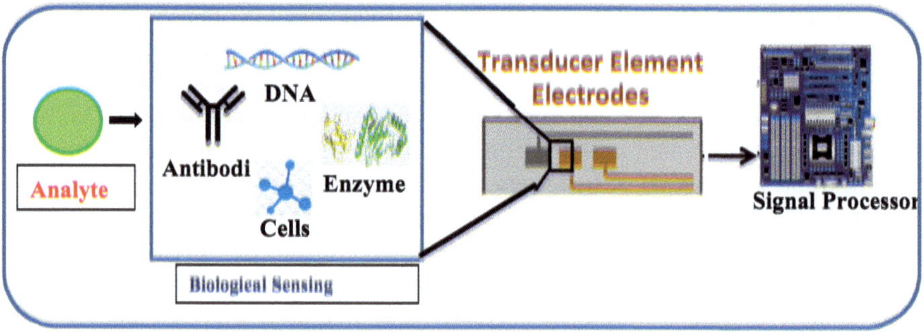

Fig. 18.7 Electrochemical biosensors

response of piezoelectric biosensors make them invaluable in this context, allowing for real-time monitoring of genetic markers and contributing to advancements in personalized medicine. This example underscores the role of piezoelectric biosensors in the field of genetic diagnostics, showcasing their sensitivity and specificity in detecting molecular interactions critical for healthcare applications. Other applications of these sensors are in detecting an array of biomolecular interactions, encompassing pathogens, proteins, nucleic acids, and other analytes (Sklådal 2016). Figure 18.8 represents the piezoelectric biosensors.

Thermal Biosensors: These sensors manifest diverse utility within biotechnology, encompassing functions like enzyme activity monitoring, identification of specific biomolecular interactions, and facilitating real-time tracking of biological processes. Thermal biosensors find practical applications in various fields, and an illustrative example is their use in the detection of microbial contamination in food. In the food industry, ensuring the safety of products is paramount, and thermal biosensors offer a reliable method for detecting the presence of harmful microorganisms. A typical thermal biosensor (shown in Fig. 18.9) in this context might include a sensitive thermal element coupled with biological recognition elements such as antibodies or enzymes. When the target microorganism, such as bacteria or pathogens, interacts

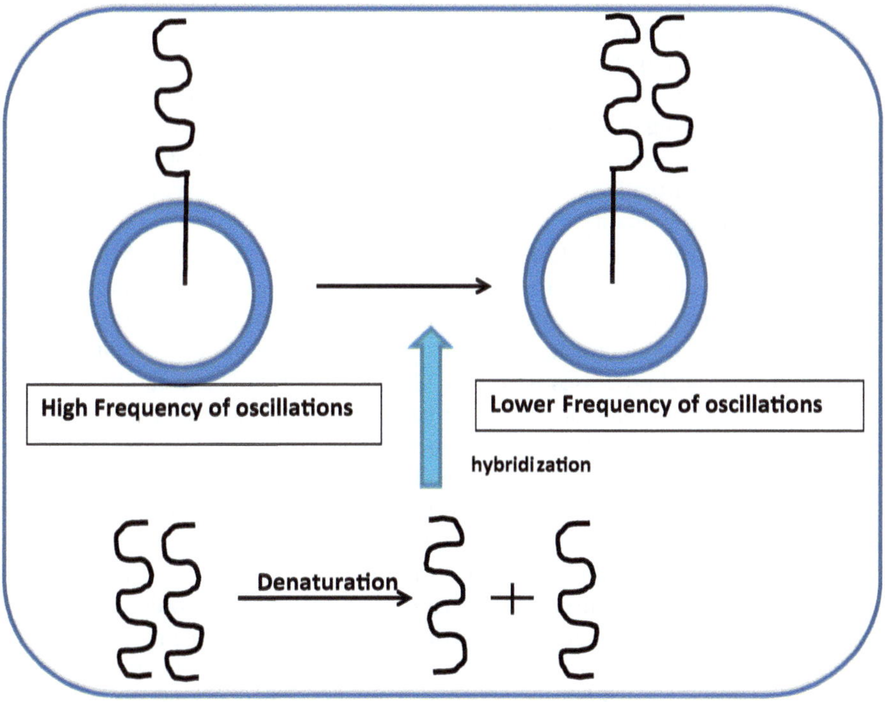

Fig. 18.8 Piezoelectric biosensors

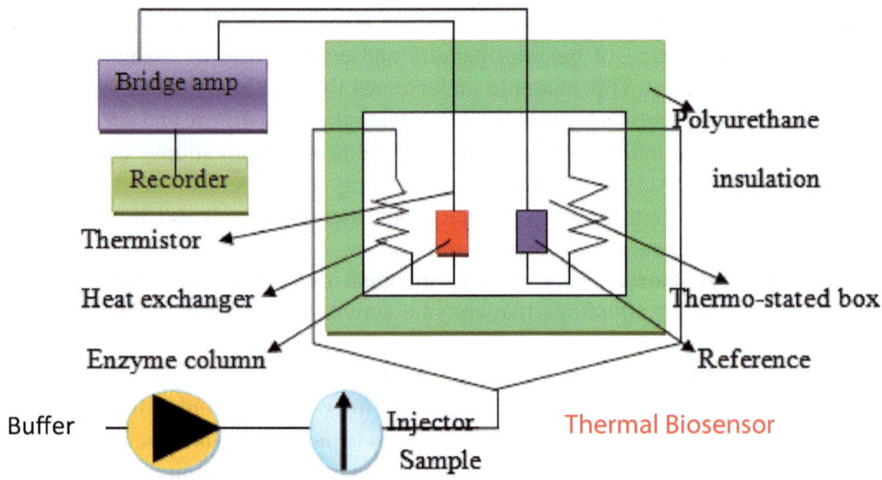

Fig. 18.9 Thermal biosensor

with these biological elements, it induces a specific biochemical reaction. This reaction generates heat, and the thermal biosensor measures the resulting temperature change. The magnitude of the temperature variation is directly proportional to the concentration of the target microorganism, providing a rapid and quantitative assessment of food safety. These biosensors are valuable tools for quality control in the food industry, ensuring that products meet stringent safety standards and minimizing the risk of foodborne illnesses. The real-time and label-free nature of thermal biosensors makes them particularly advantageous for on-site monitoring, contributing to the overall safety and quality.

Calorimetric Biosensors: Calorimetric biosensors, which measure heat changes associated with biochemical reactions, find applications in various fields, including medical diagnostics. A notable example is their use in enzyme activity assays for drug discovery. In drug development, understanding the activity of enzymes involved in various physiological processes is crucial. Calorimetric biosensors can be designed with immobilized enzymes on a sensing surface. When a substrate is introduced that the enzyme acts upon, the catalytic reaction generates heat. This heat change is precisely measured by the biosensor, providing insights into the enzyme's activity. The pharmaceutical researchers leverage this information to screen and optimize potential drug candidates that modulate specific enzyme activities. For instance, a calorimetric biosensor can be employed to study the inhibition or activation of enzymes associated with diseases like cancer or metabolic disorders. The real-time and label-free nature of calorimetric biosensors makes them invaluable in drug discovery processes, facilitating the identification of compounds with therapeutic potential and contributing to advancements in medicine. This example highlights the significance of calorimetric biosensors in pharmaceutical research, where understanding enzymatic activity is fundamental for developing effective treatments.

Changes in heat resulting from a biochemical interaction are measured calorimetrically. Figure 18.10 is a calorimetric biosensor. These biosensors are used for studying binding interactions and enzymatic reactions. Applications of calorimetric biosensors could span biotechnology, pharmaceuticals, and medical diagnostics. Applications of the biosensor in the analysis of ascorbic acid content in food and pharmaceutical samples could be highlighted and this biosensor might utilize differential scanning calorimetry (DSC) principles to measure the heat absorbed or released during the oxidation of ascorbic acid (Vermeir et al. 2007).

Gas Biosensors: Gas biosensors are instrumental in a variety of applications, and one notable example is their use in monitoring air quality for environmental and public health purposes. In urban environments, where air pollution poses significant challenges, gas biosensors play a crucial role in detecting and quantifying various airborne pollutants. These biosensors can be designed to target specific gases such as nitrogen dioxide (NO_2), carbon monoxide (CO), sulfur dioxide (SO_2), and ozone (O_3). The biosensor's recognition element, often enzymes or antibodies, interacts with the target gas, initiating a biochemical reaction that produces a measurable signal. The change in signal intensity is directly proportional to the concentration of the targeted gas, providing accurate and real-time data on air quality. Such biosensors contribute to early warning systems, enabling timely interventions and regulatory

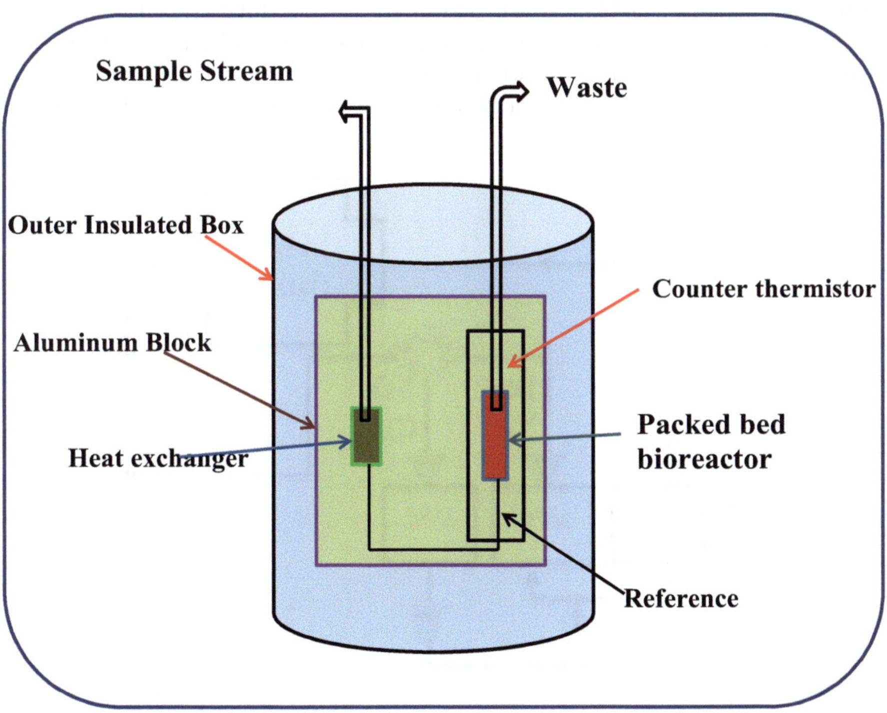

Fig. 18.10 Calorimetric biosensor

measures to mitigate the impact of air pollution on public health and the environment. Their portability and capacity for continuous monitoring make gas biosensors invaluable tools in urban planning and environmental management, supporting efforts to create healthier and more sustainable living environments. Applications of electrochemical gas biosensors could be diverse, including environmental monitoring, industrial safety, and medical diagnostics. Gas biosensors could revolutionize the way gases are detected and monitored, providing real-time and accurate data that can contribute to improved decision-making. Advancements in the field could be a central focus such as gas biosensors in addressing real-world challenges, such as detecting pollutants, monitoring gas emissions, or diagnosing diseases (Kawaguchi et al. 2019). Figure 18.11 represents a gas biosensor.

Nanostructure-Based Biosensors: Nanostructure-based biosensors represent a cutting-edge technological advancement, and their application in medical diagnostics exemplifies their precision and sensitivity. One notable example is the use of gold nanoparticles in cancer detection. In this application, gold nanoparticles are functionalized with specific biological molecules, such as antibodies or aptamers, which selectively bind to cancer biomarkers. When these functionalized nanoparticles encounter the target molecules in a patient's blood sample, they undergo a change in their optical or electrical properties, which can be precisely measured. This change serves as a signal that indicates the presence and concentration of cancer biomarkers, offering a highly sensitive and specific method for early cancer detection. The nanostructure-based biosensor's ability to detect molecular interactions at the nanoscale allows for the identification of biomarkers at low concentrations, potentially enabling the diagnosis of cancer at its earliest stages. This example highlights

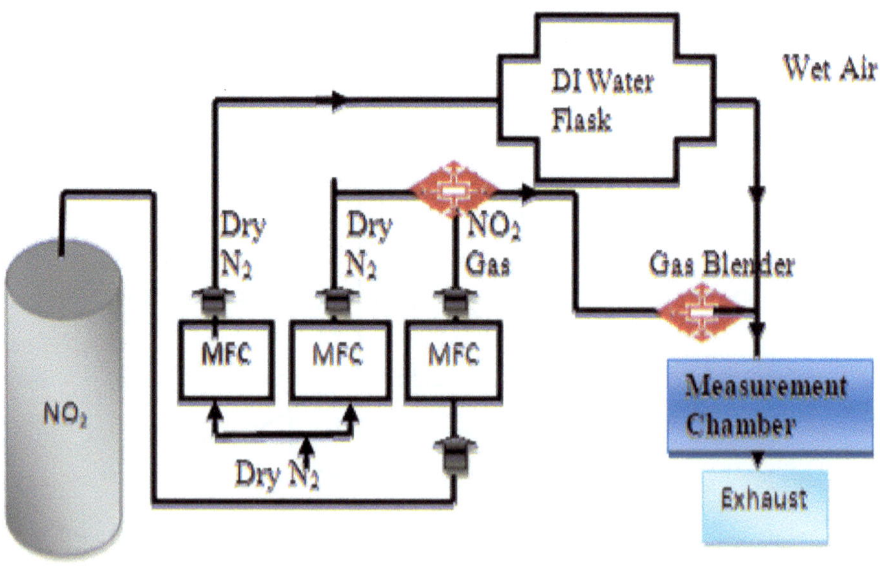

Fig. 18.11 Gas biosensors

the potential of nanostructure-based biosensors in revolutionizing disease diagnostics, demonstrating their capacity to detect biomolecules with unprecedented sensitivity and contributing to advancements in personalized medicine. Nanostructured biosensors are used to detect cancer biomarkers, offering potential for early diagnosis and personalized treatment approaches (Banerjee et al. 2021). Figure 18.12 depicts the components used in nanostructure-based biosensors.

Microfluidic Biosensors: Microfluidic biosensors represent a powerful integration of microscale technologies and bio-sensing capabilities. Their role in point-of-care diagnostics, rapid detection of pathogens, monitoring cellular responses, and other real-time sensing applications (Luka et al. 2015). These biosensors are expected to operate without the need for labels or tags, relying on the direct detection of biomolecular interactions at the electrode surface. Microfluidic integration allows for efficient sample handling and reaction control (Ebrahimi et al. 2022). One notable example is their use in detecting infectious diseases, such as the rapid and sensitive identification of pathogens like the influenza virus. In a microfluidic biosensor, a

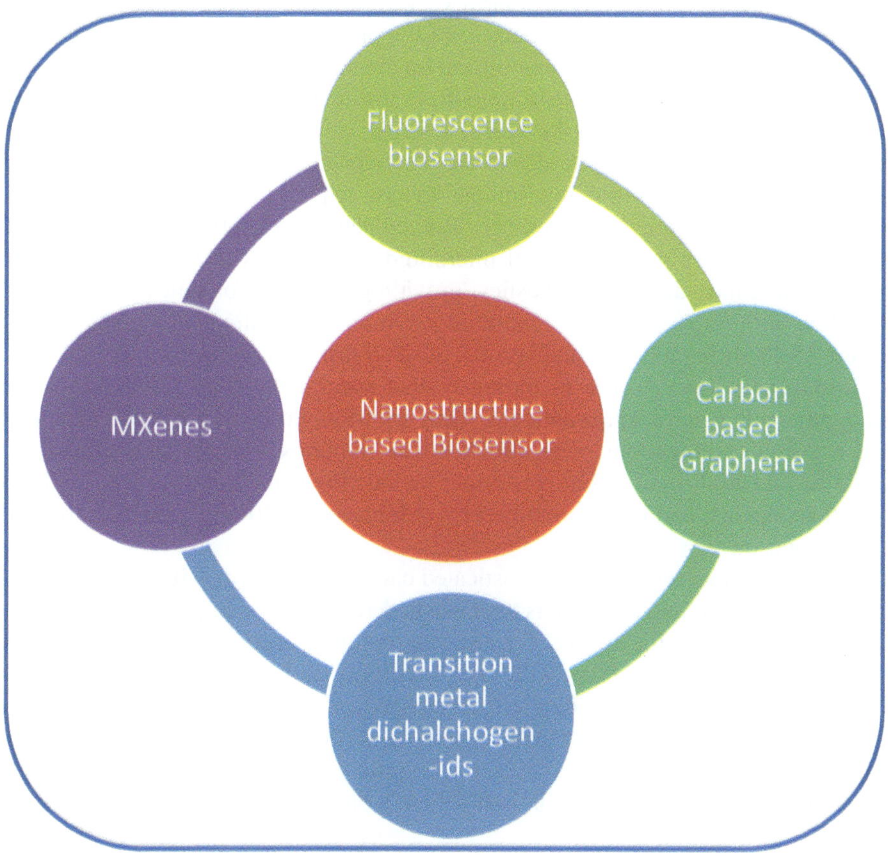

Fig. 18.12 Nanostructure-based biosensors

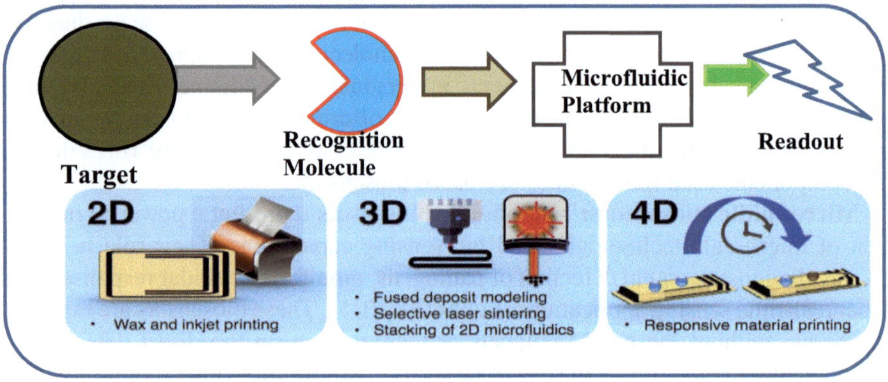

Fig. 18.13 Microfluidic biosensors

small-scale device contains microchannels through which minute volumes of biological samples flow. Functionalized surfaces within these channels are designed to capture specific biomolecules indicative of the presence of the virus. For instance, antibodies or aptamers can be immobilized on the microfluidic surface to selectively bind with viral antigens. As the sample flows through, any captured virus particles cause measurable changes in the sensor's signal, often optical or electrochemical, enabling real-time and on-site detection. The advantages of microfluidic biosensors include their portability, low sample volume requirements, and rapid analysis, making them ideal for use in resource-limited settings or during outbreaks. This example showcases the potential of microfluidic biosensors in revolutionizing the field of infectious disease diagnostics, providing a means for quick and efficient detection with implications for timely intervention and public health management. Figure 18.13 shows the flow and process of microfluidic biosensors.

Immunosensors: A type of biosensor that specifically uses antibodies as the recognition element to detect antigens. They are widely used in medical diagnostics and research. Applications of electrochemical immunosensors in clinical analysis could be a central focus. Examples might include the detection of cancer markers, infectious agents, and cardiac biomarkers (Mollarasouli 2019). Figure 18.14 is an immunosensor.

Immunosensors represent a sophisticated class of biosensors with broad applications, and a compelling example is their use in detecting specific proteins for disease diagnosis. In cancer diagnostics, for instance, immunosensors play a pivotal role in identifying cancer biomarkers with high sensitivity and specificity. These biosensors typically employ antibodies as the recognition elements, selectively binding to specific proteins associated with cancer cells. For example, a prostate cancer immunosensor may use antibodies that recognize prostate-specific antigen (PSA). When the target protein is present in a patient's blood sample, it forms an immune

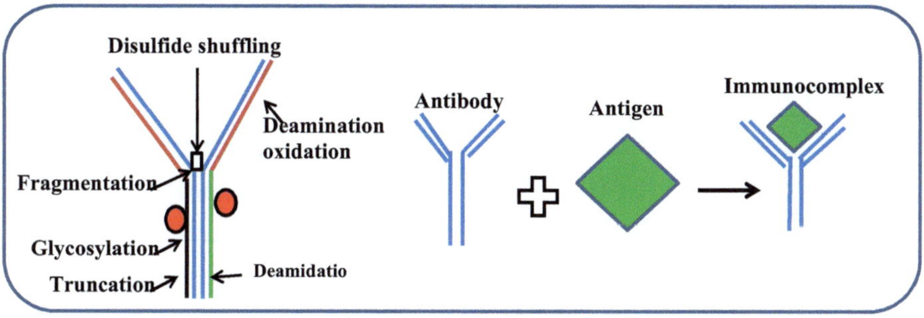

Fig. 18.14 Immunosensors

complex with the immobilized antibody on the sensor surface. This binding event triggers a measurable signal, such as a change in electrical conductivity or optical properties, providing a quantitative measure of the biomarker concentration. Immunosensors offer rapid and precise detection, facilitating early diagnosis and personalized treatment strategies. Their application extends beyond cancer diagnostics to various fields, including infectious disease detection and environmental monitoring. The versatility and sensitivity of immunosensors make them indispensable tools in advancing healthcare and biosensing technologies.

Aptamer-Based Biosensors: Aptamers can serve as recognition elements in biosensors, facilitating the detection of target molecules through various transduction methods (Zheng et al. 2020). Aptamer-based biosensors exemplify their versatility and specificity in molecular recognition, and an illustrative example is their application in the detection of small molecules, such as mycotoxins in food. Mycotoxins are toxic compounds produced by fungi that can contaminate crops and pose serious health risks. In this context, aptamer-based biosensors utilize short, single-stranded DNA or RNA sequences known as aptamers, which are selected for their high affinity and specificity to a target mycotoxin. These aptamers are immobilized on a sensing platform, and when they come into contact with the mycotoxin in a food sample, they undergo a conformational change. This change is transduced into a measurable signal, such as an electrical or optical response, providing a rapid and sensitive detection method. Aptamer-based biosensors offer several advantages in this application, including their ability to discriminate between closely related mycotoxins and their stability under various conditions. These biosensors play a crucial role in ensuring food safety by enabling on-site and real-time monitoring, contributing to the prevention of mycotoxin-related health issues. This example illustrates the significance of aptamer-based biosensors in addressing specific challenges in food safety and underscores their potential in various molecular detection applications. Figure 18.15 illustrates the aptamer-based biosensors.

Microbial Biosensor: Microbial biosensors represent a powerful and versatile tool in the realm of biotechnology, environmental monitoring, and healthcare. These biosensors employ microorganisms as the biological sensing element, utilizing their

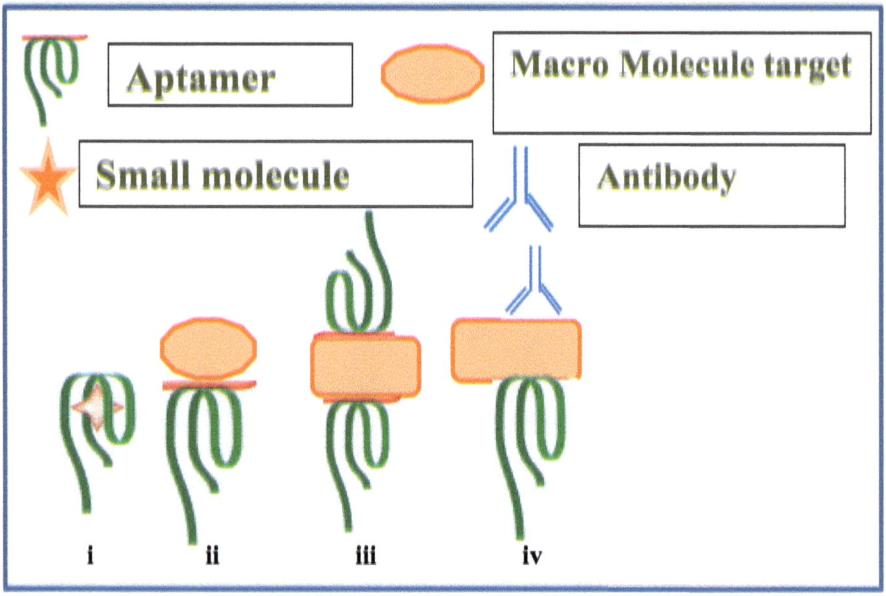

Fig. 18.15 Aptamer-based biosensors

ability to recognize and respond to specific target analytes. The integration of microbial cells, such as bacteria or yeast, into biosensor platforms allows for the detection of a wide range of substances, including pollutants, pathogens, and chemical compounds. In environmental applications, microbial biosensors are employed to monitor water quality, air pollution, and soil contaminants, providing real-time data for environmental assessment and management. In healthcare, these biosensors play a crucial role in the rapid detection of microbial pathogens, aiding in the diagnosis of infections. Additionally, microbial biosensors find utility in industrial processes, contributing to the monitoring of fermentation and bioremediation processes. The simplicity of microbial biosensors, coupled with their sensitivity and specificity, makes them valuable tools for various applications, advancing our ability to monitor and respond to microbial activities in diverse settings. Their continued development holds promise for addressing emerging challenges in environmental protection, public health, and biotechnological advancements. Figure 18.16 represents the microbial biosensors.

18.3 Applications of Biosensor

Figure 18.17 shows the application of biosensors in different fields. Biosensors have a wide range of applications in environmental monitoring and management, enabling rapid, sensitive, and specific detection of pollutants, contaminants, and

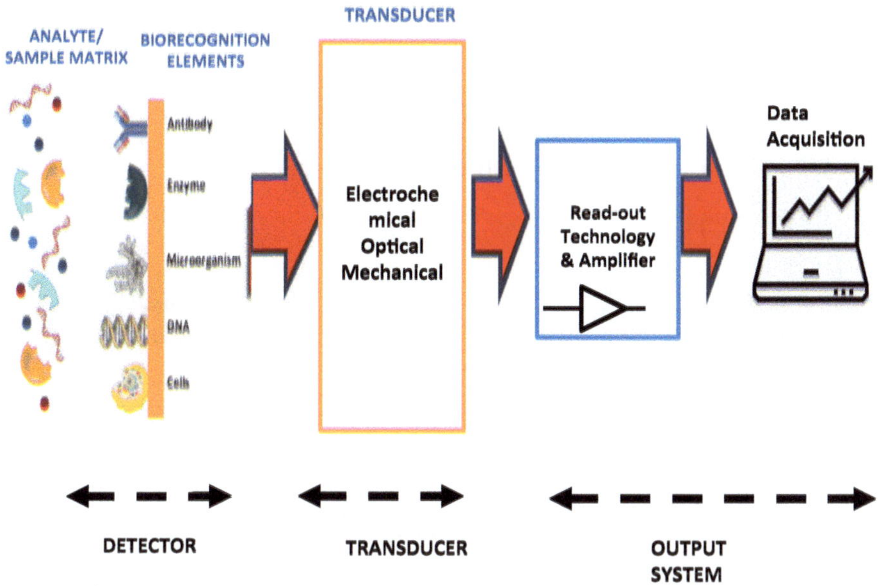

Fig. 18.16 Microbial biosensor

other environmental parameters. Biosensors have emerged as invaluable tools in addressing environmental challenges by providing rapid and sensitive detection of various pollutants. These devices leverage biological recognition elements, such as enzymes, antibodies, or microorganisms, to detect and quantify specific target molecules in the environment. In the realm of environmental monitoring, biosensors play a crucial role in the detection of contaminants such as heavy metals, pesticides, and pathogens. Table 18.2 presents the biosensors for environmental monitoring, their compound class, biological sensing element, transducer and possible application. For instance, biosensors can be employed to monitor water quality by detecting the presence of harmful substances, ensuring the safety of drinking water sources. Additionally, biosensors are pivotal in soil analysis, aiding in the identification of pollutants that may impact agricultural productivity. Their real-time capabilities and portability make biosensors particularly useful for on-site environmental monitoring, enabling timely responses to pollution incidents. The integration of biosensors into environmental monitoring networks enhances our ability to assess and manage environmental risks, contributing to the sustainable management of ecosystems. As technology advances, biosensors continue to evolve, promising even greater precision and efficiency in addressing environmental concerns. Biosensors provide valuable data for informed decision-making and policy development, contributing to more effective environmental monitoring, pollution control, and sustainable resource management. Table 18.3 presents the biosensors for environmental monitoring with their specific use. Various applications of biosensor are as follows.

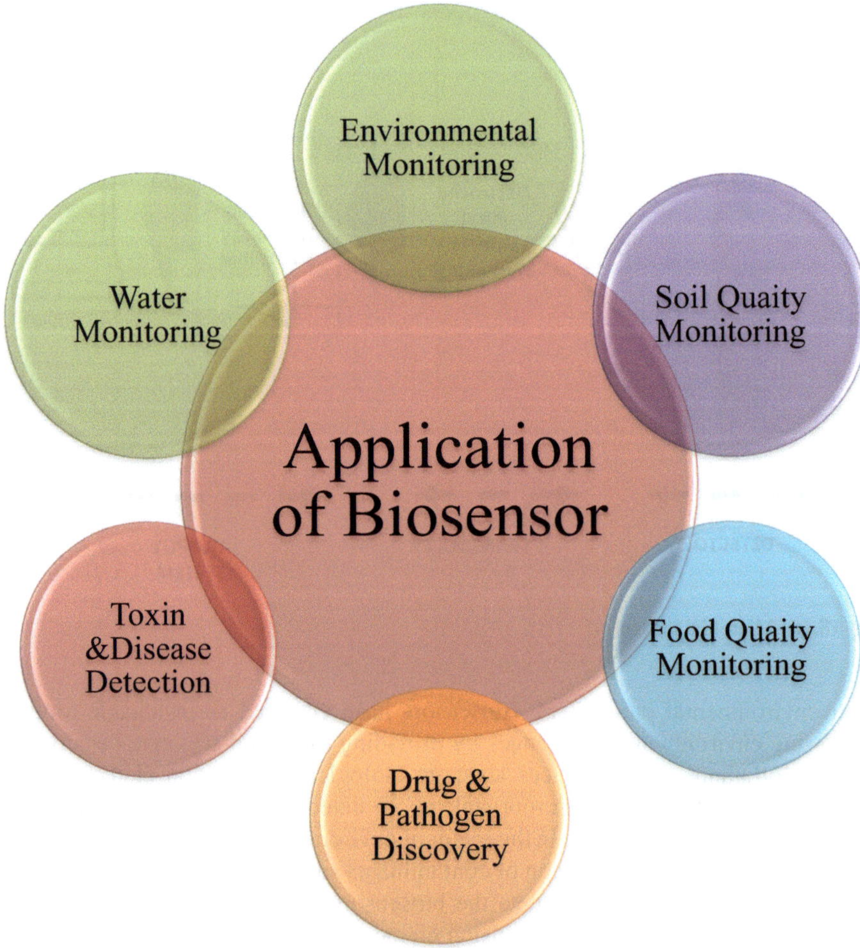

Fig. 18.17 Application of biosensor

18.3.1 Environmental Applications of Biosensor

18.3.1.1 Air

Biosensors have emerged as vital tools in the realm of air monitoring, offering unique advantages in the rapid and selective detection of airborne contaminants. In the context of air quality assessment, biosensors play a crucial role in identifying pollutants such as gases, particulate matter, and microbial agents. Their sensitivity and specificity make them well-suited for recognizing low concentrations of pollutants, providing real-time data crucial for environmental monitoring and public health. Enzymes or microbial cells integrated into biosensor platforms can selectively

Table 18.2 Biosensor for environmental application

Compound class	Biological sensing element	Physical transducer	Application
Heavy metals	Enzyme, microbe	Electrode, optical	Water, soil
Nitrogen compounds	Enzyme	Electrode	Water, soil, wastewater
Pesticides	Antibody, enzyme, microbe	Optical, electrode	Water, soil, and air
Herbicides	Antibody, enzyme, microbe	Optical, electrode	Water, soil, and air
Dioxins	Microbe, slim mode	Optical, electrode	Water, soil, air
Phenolic compounds	Enzyme, microbe	Optical, electrode	Water, soil

Source Gieva et al. (2014)

Table 18.3 Applications of biosensors in environmental monitoring with their specific use

Application	Specific use	Examples
Water quality monitoring	Detection of pollutants, pH, and pathogens in water bodies	1. pH measurement in freshwater systems 2. Detection of E. coli in drinking water
Air quality monitoring	Measurement of air pollutants and particulate matter	1. Monitoring of ozone (O3) levels in urban areas 2. Detection of PM2.5 and PM10 in industrial zones 3. Microbial components
Soil and land management	Monitoring soil pH, nutrient levels, and contaminants	1. Soil moisture and nutrient monitoring in agriculture 2. Contaminant detection in brownfield sites
Monitoring of hazardous substances	Identification and quantification of toxic chemicals and waste	1. Detection of heavy metals in industrial effluents 2. Monitoring of volatile organic compounds (VOCs) in contaminated sites

interact with specific pollutants, triggering a measurable response, often in the form of an electrical signal. This capability enables the continuous, on-site monitoring of air quality, offering a more dynamic and accurate representation compared to traditional methods. Biosensors find applications in indoor air quality assessments, industrial emissions monitoring, and environmental surveillance. The integration of biosensors into air monitoring networks contributes to timely interventions in response to pollution events, facilitates compliance with regulatory standards, and ultimately supports efforts to mitigate the impact of air pollution on human health and the environment. As technology advances, biosensors continue to enhance their role in providing comprehensive, real-time air quality information for effective environmental management and public well-being.

18.3.1.2 Water

Biosensors play a pivotal role in water monitoring, serving as indispensable tools for the detection and quantification of contaminants in aquatic environments. The unique ability of biosensors to utilize biological components, such as enzymes, antibodies, or microbial cells, as recognition elements provides a highly selective and sensitive means of identifying specific pollutants. In water quality assessment, biosensors are employed to detect various substances, including heavy metals, pathogens, pesticides, and organic pollutants. The integration of these bioreceptors with transducers allows for the conversion of biochemical signals into measurable outputs, often in the form of electrical or optical signals. This capability facilitates real-time monitoring, enabling a swift response to changes in water quality. Biosensors contribute to environmental stewardship by providing accurate and timely data for pollution control, resource management, and the protection of ecosystems. They are particularly valuable in ensuring the safety of drinking water and facilitating compliance with regulatory standards. The versatility of biosensors allows for deployment in diverse water environments, from industrial effluents to natural water bodies. As technology advances, biosensors continue to evolve, offering increased miniaturization, portability, and connectivity, thereby expanding their role in water monitoring networks and supporting sustainable water resource management practices.

18.3.1.3 Soil

Biosensors play a crucial role in soil monitoring, offering a dynamic and sensitive approach to assess the health and quality of soil ecosystems. These devices utilize biological components, such as enzymes, microbes, or antibodies, as recognition elements to detect specific analytes, including nutrients, contaminants, and pathogens. Integrated with transducers, biosensors translate biological interactions into measurable signals, providing real-time data on soil conditions. This capability is especially valuable in precision agriculture, where biosensors can aid in optimizing fertilizer use, monitoring soil microbial activity, and assessing the impact of environmental factors on soil health. Biosensors contribute to sustainable agriculture practices by enabling farmers to make informed decisions about irrigation, nutrient management, and pest control. They also play a role in environmental conservation by detecting soil contaminants, such as heavy metals and pesticides, helping to mitigate their impact on ecosystems. The application of biosensors in soil monitoring supports both agricultural productivity and environmental sustainability, offering a tool for precision farming practices that optimize resource use while minimizing environmental impact. As technology advances, biosensors in soil monitoring are expected to become more sophisticated, providing even greater insights into the complex interactions within soil ecosystems.

18.3.1.4 Food

Biosensors play a pivotal role in the realm of food monitoring, offering a powerful and rapid means to ensure the safety and quality of food products. With the capability to detect specific biological molecules, such as proteins, enzymes, or DNA, biosensors provide a highly selective method for identifying contaminants, allergens, and pathogens in food. This technology is particularly critical in the food industry, where ensuring the integrity of products is paramount. Biosensors enable the rapid screening of food for various substances, including toxins and spoilage indicators, contributing to timely interventions in the production and distribution chain. The integration of biosensors into quality control processes helps prevent the distribution of unsafe food, safeguarding public health. Additionally, biosensors are instrumental in verifying the authenticity and origin of food products, addressing concerns related to food fraud. The speed, sensitivity, and specificity of biosensors make them valuable tools in regulatory compliance and adherence to quality standards in the food sector. As technology advances, biosensors continue to evolve, providing more compact, user-friendly, and cost-effective solutions for food monitoring applications, thereby contributing to enhanced food safety and consumer confidence.

18.3.2 Toxicity

Biosensors are invaluable tools in toxicity determination across various applications, offering rapid and precise detection of harmful substances. In environmental monitoring, biosensors play a pivotal role in assessing water quality by detecting pollutants like heavy metals and industrial chemicals, providing real-time data crucial for water treatment and conservation efforts. Similarly, in air quality monitoring, biosensors contribute to the identification of airborne toxins, aiding in pollution control. In the realm of food safety, biosensors are employed to detect foodborne pathogens and pesticide residues, ensuring the quality and safety of consumables. Within the pharmaceutical industry, these sensors assist in drug screening by assessing the toxicity of potential compounds during development. In biomedical applications, biosensors are integrated into cell culture systems for toxicity assessment, aiding researchers in understanding the impact of substances on cellular health. Moreover, biosensors find utility in detecting biomarkers associated with toxicity or diseases, offering diagnostic capabilities in medical settings. Industrial applications involve the monitoring of effluents for toxicity, contributing to pollution control and regulatory compliance. Defense and security benefit from biosensors designed to detect chemical warfare agents swiftly. In agriculture, biosensors aid in monitoring agrochemical levels in soil and water. Overall, biosensors provide a versatile and efficient means for toxicity determination, addressing critical concerns in human health, environmental protection, and industrial safety.

18.3.3 Polychlorinated Biphenyl

Polychlorinated biphenyls (PCBs) are a group of synthetic organic chemicals that were widely used in various industrial and commercial applications, such as electrical equipment, hydraulic fluids, and flame retardants. However, due to their persistence in the environment and potential health risks, the production and use of PCBs have been banned in many countries. PCBs are considered persistent organic pollutants (POPs) and are of environmental concern due to their toxicity, bioaccumulation, and potential adverse effects on ecosystems and human health. Biosensors are pivotal in the precise detection of Polychlorinated Biphenyls (PCBs), toxic organic pollutants notorious for their environmental persistence. The application of biosensors in PCB detection serves critical roles across diverse domains. In environmental monitoring, biosensors offer rapid and accurate assessments of water bodies and soil, enabling timely interventions to mitigate contamination. Integrating biosensors into wastewater treatment plants aids in monitoring industrial effluents for PCBs, ensuring compliance with environmental standards. Portable biosensor devices facilitate on-site field testing, providing immediate insights into PCB contamination during environmental assessments or emergency responses. The application extends to food safety, where biosensors contribute to assessing PCB levels in fish and seafood, safeguarding consumers from potential health risks. Moreover, biosensors play a vital role in human exposure monitoring through epidemiological studies, detecting PCB levels in biological samples. In regulatory contexts, biosensors verify compliance by testing products for PCB contamination, reducing the risk of human exposure. Furthermore, biosensors support remediation efforts by monitoring PCB biodegradation, ensuring the effectiveness of environmental cleanup processes. Overall, biosensors emerge as indispensable tools, offering sensitivity, specificity, and real-time monitoring capabilities in the detection of PCBs across environmental, industrial, and health-related applications.

18.3.4 Hydrocarbon

Biosensors play a crucial role in the detection of hydrocarbons, contributing significantly to environmental monitoring, industrial safety, and various applications. These sensors offer a sensitive and specific means of identifying hydrocarbons, which are organic compounds primarily, composed of hydrogen and carbon. In environmental contexts, biosensors are employed to monitor hydrocarbon levels in water bodies and soil, providing real-time data on potential contamination from sources such as oil spills or industrial discharges. The versatility of biosensors allows for their integration into oil and gas facilities, enabling continuous monitoring for leaks or spills, thus enhancing workplace safety and preventing environmental disasters. Furthermore, biosensors find application in the assessment of air quality, detecting hydrocarbons in the atmosphere that may result from vehicular emissions or industrial processes. This capability is crucial for urban planning and regulatory compliance. In the field of bioremediation, biosensors assist in monitoring the effectiveness of microbial processes designed to break down hydrocarbons, providing insights into the progress of environmental cleanup efforts. Additionally, biosensors contribute to the oil and gas industry by ensuring the quality of fuel products through the detection of contaminants. Overall, the application of biosensors in hydrocarbon detection spans environmental monitoring, industrial safety, and quality control, playing a pivotal role in mitigating the impact of hydrocarbons on ecosystems and human health. Biosensors offer valuable tools for detecting and managing hydrocarbon-related substances, contributing to environmental protection, industrial sustainability, and human health.

18.3.5 Inorganic Pollutants

Biosensors play a significant role in detecting and monitoring inorganic pollutants, which include a wide range of toxic metals, metalloids, and other non-organic compounds. These pollutants can originate from industrial processes, mining activities, agriculture, and other sources, and their presence in the environment can have detrimental effects on ecosystems and human health. Biosensors have proven invaluable in the detection of inorganic pollutants, offering a versatile and sensitive approach to monitor environmental and industrial settings. These sensors play a crucial role in identifying various inorganic contaminants, including heavy metals and ions, which can have detrimental effects on ecosystems and human health. In environmental monitoring, biosensors are utilized to detect the presence of heavy metals such as lead, mercury, and cadmium in water sources and soil. Their real-time capabilities enable swift responses to contamination events, facilitating timely intervention and remediation efforts. Additionally, biosensors are employed in industrial contexts to monitor effluents and emissions, ensuring compliance with environmental regulations and preventing the release of harmful inorganic pollutants. The adaptability of biosensors allows for the detection of specific ions, such as nitrate and phosphate, contributing to the assessment of water quality in agricultural and municipal settings. Moreover, biosensors find application in detecting inorganic pollutants in occupational settings, safeguarding workers from exposure to harmful substances. The integration of biosensors into portable devices facilitates on-site testing, enabling efficient field monitoring. Overall, biosensors provide a powerful tool for inorganic pollutant detection across diverse environments, promoting environmental sustainability, human health, and regulatory compliance.

18.3.6 Organic Pollutants

Biosensors have a range of applications in detecting and monitoring organic pollutants, which include various hazardous organic compounds that can have detrimental effects on the environment and human health (Gieva et al. 2014). These pollutants can originate from industrial activities, agriculture, transportation, and other sources. Biosensors are pivotal in the detection of organic pollutants, playing a critical role in environmental monitoring, industrial safety, and public health. These sensors offer a highly sensitive and specific method for identifying a wide range of organic contaminants, including pesticides, industrial chemicals, and pharmaceutical residues. In environmental applications, biosensors contribute to the surveillance of water bodies and soil, swiftly detecting the presence of organic pollutants resulting from agricultural runoff, industrial discharges, or accidental spills. Their real-time monitoring capabilities enable timely responses and intervention strategies to mitigate environmental impacts. Biosensors are also extensively employed in industrial settings to monitor effluents, ensuring compliance with regulations and preventing the release of harmful organic substances into the environment. In the realm of public health, biosensors are applied to assess the quality of drinking water, identifying organic pollutants that may pose risks to human health. Furthermore, biosensors support the pharmaceutical industry by providing efficient methods for monitoring the presence of organic impurities in drug manufacturing processes. The versatility of biosensors allows for integration into portable devices, facilitating on-site testing and enabling rapid screening of various environments. Overall, biosensors play a pivotal role in the detection of organic pollutants, contributing to environmental sustainability, industrial safety, and the protection of human well-being.

18.3.7 Antibiotics

Biosensors play a role in addressing this issue by providing rapid and sensitive methods for detecting antibiotics in various contexts. Biosensors have emerged as critical tools in the precise detection of antibiotics, finding applications across diverse fields. In clinical diagnostics, biosensors play a pivotal role in monitoring antibiotic levels in patient samples, ensuring optimal therapeutic concentrations and minimizing the risk of adverse effects or antibiotic resistance. Portable biosensor devices facilitate point-of-care testing, allowing healthcare professionals to swiftly assess antibiotic levels in blood, urine, or other biological specimens. In the realm of food safety, biosensors contribute to the identification of antibiotic residues in animal-derived products, such as meat and milk, ensuring compliance with regulatory standards and safeguarding consumers from potential health risks. Environmental monitoring benefits from biosensors by enabling the assessment of antibiotic levels in wastewater and water bodies, addressing concerns about pharmaceutical residues impacting aquatic ecosystems and human health. Within the pharmaceutical industry, biosensors aid in quality control processes, detecting trace amounts of antibiotics to ensure the purity of pharmaceutical products. These sensors also contribute to surveillance programs, monitoring antibiotic-resistant bacterial strains and supporting efforts to combat global health challenges associated with antibiotic resistance. In agriculture, biosensors are applied to monitor antibiotic use in livestock, helping prevent the emergence of resistant strains and ensuring the safety of animal products. The versatility of biosensors extends to biodefense applications, where they can be utilized to detect the presence of antibiotics or related substances in scenarios involving bioterrorism threats. Overall, biosensors play a pivotal role in antibiotic detection, contributing to improved healthcare outcomes, enhanced food safety, environmental protection, and global efforts to address antibiotic resistance.

18.3.8 Contaminants

Biosensors have proven instrumental in the detection of toxins, serving as versatile tools across various domains to ensure the safety of environments, food, and biological systems. In environmental monitoring, biosensors are employed to detect toxins in air, water, and soil, offering real-time assessments of contamination levels. These sensors enable swift responses to environmental incidents such as chemical spills or industrial releases, aiding in timely intervention and remediation efforts. In the food industry, biosensors play a critical role in detecting toxins such as mycotoxins, bacterial toxins, and chemical contaminants. Rapid and specific toxin detection is essential to prevent the consumption of contaminated food products, ensuring food safety and public health. In clinical applications, biosensors contribute to the detection of toxins in biological samples, supporting the diagnosis of poisoning or exposure to harmful substances. The development of portable biosensor devices enhances on-site testing capabilities, facilitating quick responses in emergency situations. Additionally, biosensors are utilized in biodefense for the rapid identification of toxins that may be employed as bioterrorism agents. The adaptability of biosensors to various toxins, including biological, chemical, and environmental agents, underscores their significance in addressing the complex challenges associated with toxin detection. Overall, biosensors provide a crucial means for the sensitive, specific, and rapid detection of toxins, contributing to public safety, environmental health, and emergency response capabilities. Biosensors play a vital role in toxin detection and monitoring, contributing to public health, environmental protection, and the prevention of toxin-related hazards.

18.3.9 Bioremediation

Biosensors play a crucial role in bioremediation by providing real-time monitoring and control of the bioremediation process, ensuring its effectiveness and optimizing treatment conditions. Biosensors play a pivotal role in the field of bioremediation, offering precise and real-time monitoring capabilities that are essential for assessing and optimizing environmental cleanup processes. Bioremediation involves the use of microorganisms to degrade or eliminate pollutants from contaminated sites. Biosensors are integrated into bioremediation strategies to monitor key parameters, ensuring the effectiveness of microbial activity. These sensors are designed to detect specific indicators, such as the concentration of target pollutants or the metabolic activity of remediation microbes. Real-time data provided by biosensors allow for immediate adjustments to bioremediation protocols, optimizing conditions for microbial growth and pollutant degradation. Additionally, biosensors contribute to the early identification of any potential challenges or limitations in the remediation process, enabling proactive measures to enhance overall efficiency. The application of biosensors in bioremediation extends across various environmental contaminants, including hydrocarbons, heavy metals, and organic pollutants. This integration not only enhances the success of remediation projects but also provides valuable insights for ongoing research in environmental science and microbial ecology. In summary, biosensors serve as indispensable tools in the field of bioremediation, facilitating efficient and targeted pollutant removal while offering valuable data for the continual improvement of remediation strategies.

18.3.10 Microorganism

18.3.10.1 Bacteria

Biosensors play a crucial role in the detection of bacteria across various domains, offering rapid and precise identification methods. In the clinical field, biosensors are instrumental in diagnosing bacterial infections through the detection of specific biomarkers in patient samples such as blood, urine, or saliva. Their use in healthcare settings allows for timely intervention and treatment planning. In the food industry, biosensors contribute to ensuring food safety by detecting bacterial contaminants that may pose health risks. These sensors enable quick screening of food products, preventing the distribution of items with harmful bacterial loads. Environmental monitoring also benefits from biosensors, as they can be deployed to detect and quantify bacteria in air, water, and soil, aiding in the assessment of environmental quality. In research, biosensors assist in studying bacterial diversity and behavior in different ecosystems. Additionally, biosensors are essential in biotechnological processes, such as fermentation and bioremediation, where monitoring and controlling bacterial activity are critical for the success of these processes. The portability and rapid response of biosensors make them valuable tools for point-of-care testing, allowing for on-site detection of bacterial infections in resource-limited settings. Overall, the versatility of biosensors contributes significantly to the effective and efficient detection of bacteria, impacting fields ranging from healthcare to environmental management.

18.3.10.2 Fungi

Biosensors find diverse applications in the field of fungi detection, playing a pivotal role in various industries and sectors. In the food and beverage industry, biosensors are employed for quality control, swiftly identifying fungal contaminants in products like dairy, grains, and beverages. Environmental monitoring benefits from biosensors in detecting fungal presence in air and water sources, crucial for maintaining safety in indoor environments and water treatment plants. In clinical diagnostics, biosensors contribute to the rapid detection of fungal infections in clinical samples, aiding timely intervention, especially for immunocompromised patients. Agriculture benefits from biosensors in detecting fungal pathogens that may harm crops, enabling early intervention to minimize crop losses. Pharmaceutical industries utilize biosensors for quality assurance, ensuring the safety and efficacy of drugs by detecting fungal contaminants. Biosensors also contribute to waste management by monitoring fungal activity in composting and waste treatment processes. In research, biosensors aid ecological studies by monitoring fungal biodiversity in different ecosystems. Industrial processes, such as bioprocessing and fermentation, benefit from biosensors for monitoring and controlling fungal growth. Additionally, biosensors have applications in biodefense, security, and defense scenarios for detecting fungal pathogens deliberately used as biological weapons. The versatility of biosensors, ranging from point-of-care testing to industrial applications, underscores their importance in the detection and management of fungal contamination across diverse domains. Fungal detection using biosensors has applications across different fields, including healthcare, agriculture, food safety, and environmental monitoring (Prasad and Kumar 2018).

18.3.10.3 Endotoxins

Biosensors play a pivotal role in the specific and sensitive detection of endotoxins, which are lipopolysaccharide components found in the outer membrane of Gram-negative bacteria. In pharmaceutical and medical device industries, where endotoxin contamination poses a serious risk to product safety, biosensors offer a reliable method for rapid detection. These sensors typically utilize biological components such as Limulus Amebocyte Lysate (LAL), which is derived from the blood cells of horseshoe crabs and reacts specifically to endotoxins. The biosensor's transducer converts the biochemical response initiated by the interaction between LAL and endotoxins into a measurable signal. This technology ensures that endotoxin levels in pharmaceutical products, particularly injectable drugs and medical devices, comply with regulatory standards. The real-time monitoring capabilities of biosensors are particularly valuable in critical processes, where timely detection and mitigation of endotoxin contamination are essential. Moreover, biosensors contribute to the quality control of medical equipment, ensuring that devices implanted or used in contact with the human body are free from endotoxin contaminants that could lead to severe immune responses. The precision and speed of biosensors in endotoxin detection have become indispensable tools in maintaining the safety and efficacy of pharmaceuticals and medical devices, enhancing overall product quality in these industries.

18.4 Future Perspective of Biosensor in Environment Management

The future of biosensors in environmental management holds great promise as technology continues to advance. Emerging innovations in sensor design, materials science, and biotechnology are likely to contribute to more sophisticated and efficient biosensors for environmental applications. These future biosensors may exhibit improved sensitivity, selectivity, and rapid response times, enabling real-time monitoring of environmental parameters. Miniaturization and the development of portable biosensor devices could lead to more widespread deployment, allowing for on-site detection in remote or challenging environments. Integration with wireless communication technologies may facilitate data transmission and enable the creation of sensor networks for comprehensive environmental monitoring. Additionally, the incorporation of artificial intelligence and machine learning algorithms could enhance the analytical capabilities of biosensors, allowing for the identification of complex patterns and trends in environmental data. As concerns about climate change, pollution, and resource management intensify, biosensors are poised to play a pivotal role in providing actionable insights for sustainable environmental practices. The future integration of biosensors into smart cities and environmental monitoring systems may usher in an era of proactive and data-driven decision-making, ultimately contributing to more effective environmental management and conservation efforts. Biosensor technology has experienced remarkable advancements on a global scale, propelled by continual innovation and cross-disciplinary collaboration. One of the prominent trends is the relentless pursuit of miniaturization and integration, resulting in more compact and portable biosensors. The incorporation of nanomaterials, such as nanoparticles and nanocomposites, has significantly enhanced sensitivity and stability. Advanced bioreceptors, including aptamers and molecularly imprinted polymers, have expanded the repertoire of detectable analytes and improved selectivity. The convergence of biosensors with wireless communication and the Internet of Things (IoT) has enabled real-time data transmission and remote monitoring capabilities. Biosensors are increasingly finding applications in point-of-care diagnostics, contributing to personalized medicine by detecting specific biomarkers. The integration of artificial intelligence and data analytics enhances the interpretation of complex biological signals, improving accuracy and providing valuable insights. Wearable biosensors are gaining attraction for continuous health monitoring, while 3D printing technology facilitates the fabrication of intricate biosensor structures. In environmental monitoring, biosensors play a pivotal role in detecting pollutants, and in agriculture, they contribute to soil health assessment and precision farming. The exploration of emerging materials, such as graphene and carbon nanotubes, further propels biosensor technology towards greater sensitivity and functionality. These collective advancements underscore the dynamic nature of biosensor technology, promising transformative impacts across healthcare, environmental management, and beyond.

18.5 Conclusion

While significant advancements have been achieved in the field of biosensors over the years, particularly in domains such as point-of-care diagnostics, environmental monitoring, drug delivery, and food safety, there remains substantial potential for further refinement in the rapid detection of airborne pathogens. Advancements in sensor technology as well as in aerosol collectors have enabled research into true bioaerosol sampling for continuous, automated biosensors. With more scientific breakthroughs and discoveries of drug-resistant pathogen strains and the looming threat of bioterrorism that may reach hospitals, schools, airports, and food supplies, there is a growing need for more portable, rapid, and sensitive biosensing technology. The major step in the process is the development of portable, disposable, and cost-effective biosensors that are more feasible for in-field use. On a larger scale, these comprehensive sensors can be implemented for soldiers overseas, in clinics, in counterterrorist applications, and for health care workers in the third world. Thus, this device, along with any improvements, has the potential to save lives, reduce morbidity, and save health care industries untold amounts of money. Ongoing research and technological advancements promise to unlock even more capabilities and applications for biosensors, making them central to the evolution of science, technology, and societal well-being. In a world where precise, rapid, and reliable information is increasingly critical, biosensors stand as beacons of innovation, bridging the gap between the biological and digital realms. Their transformative potential is driving progress across various domains, shaping a future where tailored solutions to complex challenges are within reach, ultimately leading to improved health, sustainability, and quality of life.

Acknowledgements We are grateful to Prof. Rohit Shrivastav, Head, Department of Chemistry, Faculty of Science, DEI, Dyalbagh and Prof. Anna Pratima Nikalje, Principal Wilson College, Mumbai for their help, support and kind encouragement. Thanks, are also due to Mr. Vaishnav Bartariya and Mr. Omkant Sharma for their involvement and in drawing figures.

References

Arora P, Sindhu A, Dilbaghi N, Chaudhury A (2011) Biosensors as innovative tools for the detection of food borne pathogen. Biosens Bioelectron 28(1):1–12

Banerjee A, Maity S, Mastrangelo CH (2021) Nanostructures for biosensing, with a brief overview on cancer detection, IoT, and the role of machine learning in smart biosensors. Sensors 21(4):1253

Belkin S (2003) Microbial whole-cell sensing systems of environmental pollutants. Curr Opin Microbiol 6(3):206–212. https://doi.org/10.1016/S1369-5274(03)00059-6

Byrne B (2009) Antibody-based sensors: principles, problems and potential for detection of pathogens and associated toxins. Sensors 9:4407–4445. https://doi.org/10.3390/s90604407

Ebrahimi G, Pakchin PS, Shamloo A, Mota A, Guardia Mdl, Omidian H, Omidi Y (2022) Label-free electrochemical microfluidic biosensors: futuristic point-of-care analytical devices for monitoring diseases. Mikrochimica acta (1966) 189(7):252–252

Gieva E, Nikolova B, Nikolov GT (2014) Biosensors for environmental monitoring. In: Challenges in higher education and research, proceedings of the 12th conference, vol 12

Hua Y, Ma J, Li D, Wang R (2022) DNA-based biosensors for the biochemical analysis: a review. Biosensors 12(3):183

Kawaguchi T, Takahashi M, González LA (2019) Prospect of gas biosensor. Austin J Biosens Bioelectron 5(1):1033

Khor SM, Choi J, Won P, Ko S (2022) Challenges and strategies in developing an enzymatic wearable sweat glucose biosensor as a practical point-of-care monitoring tool for type II diabetes. Nanomaterials 12:221

Luka G, Ahmadi A, Najjaran H, Alocilja E, DeRosa M, Wolthers K, Malki A, Aziz H, Althani A, Hoorfar M (2015) Sensors 15(12):30011–30031

Malhotra BD, Chaubey A (2003) Biosensors for clinical diagnostics industry. Sens Actuators 91(1–3):117–127

Mandpe P, Prabhakar B, Gupta H, Shende P (2020) Glucose oxidase-based biosensor for glucose detection from biological fluids. Sens Rev 40:497–511

Mollarasouli F (2019) The role of electrochemical immunosensors in clinical analysis. Biosensors 9:86

Prasad R, Kumar V, Kumar M, Wang S (2018) Fungal nanobionics: principles and applications

Rogers KR (2006) Recent advances in biosensor techniques for environmental monitoring. Anal Chim Acta 568(1–2):222–231

Skládal P (2016) Piezoelectric biosensors. Trends Anal Chem 79:127–133

Vermeir S et al (2007) Microplate differential calorimetric biosensor for ascorbic acid analysis in food and pharmaceuticals. Anal Chem 79:6119–6127

Wilson JS (2005) Introduction to sensor technology. Sensor technology handbook. pp 1–20

Wu Y, Wang C.W, Wang D, and Wei N (2021) A whole-cell biosensor for point-of-care detection of waterborne bacterial pathogens. ACS Synth Biol 10(2):333–344

Zheng X, Gao S, Wu J, Hu X (2020) Recent advances in aptamer-based biosensors for detection of pseudomonas aeruginosa. Front Microbiol 11:1–15

Chapter 19
Metabolic Materials from Cyanobacteria and Microalgae: Biotechnology, Biochemistry and Biotherapeutics

Lokitha Paduvetnaya, Kamath H. Venkatesh, and Harishkumar Madhyastha

19.1 Introduction

Reduced food and health security is coupled with exponential human population growth and rise in hospital expenses due to regenerative diseases. Research in super food and nutraceutical sector from microbial origin seems to provide an insightful information where algae and cyanobacteria emerge as a key source for further exploration. Microalgae and cyanobacteria derived nutraceuticals, cosmetics, and supplements account a rapid growth. Being a member of lower phylogenic tier with highly efficient photosynthesis, algae, and cyanobacteria has diverse growth habitats like aquatic, terrestrial, and aerial nature. Approximately 300,000 species are being identified till today with multiple health benefits to human health. *Chlorella* and *Spirulina* were the first organisms to growth commercially in Japan since 1960 (Vadalà and Palmieri 2015). Meeting the current trend of food demand with high nutritional value is a major challenge for the agriculture sector and also for food industries. As an alternate treatment method to chemotherapy and radiotherapy, phytocompounds originated from these groups serve as an alternate source by pharmaceutical industries. Along with food and highly valued pharmaceutical agents, algae and cyanobacteria are also used as additives, antioxidants, dyes, and as feed for aquaculture organisms.

L. Paduvetnaya · K. H. Venkatesh (✉)
NITTE (Deemed to be University), NMAM Institute of Technology, Department of Biotechnology, Nitte, Karkala, Karnataka 574110, India
e-mail: venkatesh@nitte.edu.in

H. Madhyastha
Division of Cardiovascular Physiology, Faculty of Medicine, University of Miyazaki, Miyazaki 8891692, Japan

K. Bala et al. (eds.), *Harnessing Microbial Potential for Multifarious Applications*, Energy, Environment, and Sustainability, https://doi.org/10.1007/978-981-97-1152-9_19

19.1.1 Origin, Biology, Physiology, Biochemistry

Cyanobacteria are photosynthetic prokaryotes that have existed on Earth for around 3500 million years and were once known as "blue-green algae". They can be found in both freshwater and marine as well as terrestrial ecosystems settings. Even in the severe settings, such as geothermal habitats, frozen systems, and hypersaline environments, cyanobacteria seem to survive and thrive (Mehdizadeh Allaf and Peerhossaini 2022). New biochemical evidence made it clear that blue-green algae are in fact bacteria rather than algae as they lack organelles like chloroplasts and mitochondria and are sensitive to penicillin because of their peptidoglycan cell walls. Cyanobacteria also have well-maintained ribosomes that are thus susceptible to common antibiotics resistance therefore suggests that cyanobacteria played an important role in the evolution of biosphere (Demoulin et al. 2019).

19.1.2 Phylogeny, Origin and Classification

Recent views that are based on a phylogenetic examination of a set of marker genes shared by a wide range of bacteria and archaea have questioned not only what we ought to identify them, but also what cyanobacteria essentially are. Soo et al. (2014) proposed that cyanobacteria, should be used to refer to a whole range of lineages that do not share the primary metabolic properties of cyanobacteria. The following species have been found and categorized so far based on morphological characteristics and molecular analyses: *Chroococcales, Gloeobacterales, Nostocales, Oscillatoriales, Pleurocapsales, Spirulinales*, and *Synechococcales* (Wolk 1973; Garcia-Pichel et al. 2020). Coccoid unicelled and colonial cyanobacteria are found inside the mucilaginous membrane of *Chroococcales*. *Gloeobacterales* are rod-shaped, unicellular, freshwater cyanobacteria without thylakoids that can be found alone or in sporadic groups. The cyanobacteria with the highest rate of discovery are filamentous species of the *Nostocales* order, some of which are capable of fixing nitrogen. Most linear filamentous species are related to the *Oscillatoriales* and *Pleurocapsales* which can develop into intricate colonies and are either coccoid cells or pseudo-filaments. *Spirulinales* order resemble a screwdriver, whereas species belonging to the *Synechococcales* order include both filamentous and unicellular (including colonial) forms (Mehdizadeh Allaf and Peerhossaini 2022).

19.1.3 Biology, Ultrastructure, and Genetics

Cyanobacteria, which are the biggest group of photosynthetic prokaryotes can synthesize additional photosynthetic blue and red pigments known as phycobilin, specifically phycocyanin (PC) and phycoerythrin (PE), which enable them to flourish in low-light environments in addition to chlorophyll-a (green pigment). Another major pigment known as carotenoids, which are likewise produced by photosynthetic machinery and are known to be protective against photo-oxidative damage as a protective measures by over producing the same in organism (Ikeuchi 2020). The fundamental characteristics of the cells of blue-green algae-or Cyanophyta which can be noticed with the light microscope include a central region ("centroplasm, nucleoplasm") which is rich in nucleic acid, and which interconnects with a peripheral region ("chromatoplasm") featuring the photosynthetically active pigments; multiple inclusions; and circumferential sections like a plasmalemma, a pellicular wall, and often, a layer of mucilaginous material (Wolk 1973).

19.1.4 General Physiology and Biochemistry

Since they are the only known prokaryotes that produce oxygen, blue-green algae have frequently been preferred for the study of photosynthesis due to the wide separation of their main pigments' absorption maxima. Photosynthesis in blue-green algae involves a photosystem (PSII) and second photosystem (PSI), a booster system that helps in oxygen synthesis by photosystem II to produce oxygen with help of chlorophyll-a. The majority of the biliprotein pigments, as well as chlorophyll "a" made luminous by light absorbed by the biliproteins more effectively than by light absorbed by chlorophyll, are present in System II (Wolk 1973). Adenosine triphosphate (ATP) and reduced nicotinamide adenine dinucleotide phosphate (NADPH) are two examples of chemical bond energy that are produced by cyanobacteria's photosynthetic system, which is constructed similarly to eukaryotic cells. Photosystem I and II reaction centers can be found in chlorophyll-a (PSI and PSII). However, the major light-harvesting complex (LHC) in cyanobacteria, called the phycobilisome (PBS), is associated with the blue-pigmented phycocyanin (λ_{max} 620 nm), which absorbs photons in the orange-red part of the light spectrum, and the red-pigmented phycoerythrin (λ_{max} 560 nm), which absorbs photons in the green-yellow part of the light spectrum, in conjunction with the PBS core of allophycocyanin (λ_{max} 650 nm). Phycocyanin and allophycocyanin are present in all PBS, although some also include phycoerythrin. Multiple regulatory pathways emerge to stop the overexcitation of photosynthetic reaction centers, which results in the creation of reactive oxygen species (ROS), as a result of changes in light intensity and wavelength, temperature, and nutrition availability. It is crucial to balance light harvesting in the cell because the ROS produced might harm cellular components and lead to cell death (Mehdizadeh Allaf and Peerhossaini 2022; Kolackova et al. 2023; Cheng et al. 2012).

19.1.5 Importance of Cyanobacteria as Research Tool

Cyanobacteria are versatile photosynthetic organisms with potent beneficial and commercial advantages. Bio industrial applications of cyanobacteria have multimodal beneficiaries as it is a host for various biotechnological high-end research, easy to grow in open environment, attractive platform for carbon neutral balance phenomenon in environmental research, as a tool for waste water and bioremediation process, as nutraceutical agent, with multi folded application in biomedical and biotherapeutic investigations (Stal 2017). Biotransformation studies by using cyanobacterial proved that cyanobacteria as whole cell catalysts for the process of cellular conversion of carbon into active metabolites. Metabolic engineering is trending research nowadays. Developing sustainable and point-of-concept route for chemical production by cyanobacteria as photo-autotrophic hosts. In recent years, significant research and development been attempted to manipulate the genetic tools of cyanobacteria in the area of promotor region engineering, riboswitches changes, ribosome biding site manipulations, small RNA regulation designing with short palindromic repeats/CRISPR-associated nuclease (CRISPAR/CAS) systems (Bongirwar and Shukla 2023). Cyanobacteria are versatile and resilient microscopic organisms that effectively harness the sun energy to oxidize water and thus generate chemical energy (ATP) and product (NADPH) to produce high-value natural products. Cyanobacteria like *Synechocystis sp.*, *Synechococcus sp.*, *Anabaena sp.*, are model organisms for phototrophic guided or directed engineering. Recently published CyanoSource with digitally barcoded mutant library of *Synechocystis sp.*, is a handy tool for cyanobacterial genetic research (Gale et al. 2019). Maintaining soil health and fertility haemostasias is an important task in agriculture. However, governing role of cyanobacteria in plant growth has been so far undermined except research on the use of some nitrogen-fixing organisms like *Nostoc sp.* (Dembitsky and Řezanka 2005). Table 19.1 depicts the role of various cyanobacteria and their signature molecules in various biomedical and biotechnological aspects.

19.2 Photobiological Strategies for Algal Cultivations

Algae are versatile photosynthetic microorganisms known for their potential to produce a wide range of active metabolites, with applications in various industries. Efficient cultivation of algae for active metabolite production requires careful optimization of the environmental conditions, light exposure, and nutrient availability. The photobioreactor design, light quality, intensity, and photoperiod play important roles in shaping algal growth and metabolite production. Understanding these strategies will aid the development of sustainable and economically viable algae-based bioproduction systems. Photobiological strategies for algal growth involve optimizing the cultivation conditions of algae by harnessing the power of light and understanding the various biological processes influenced by light exposure. These

Table 19.1 Role of cyanobacterial metabolites in various biotechnological biotherapeutic aspects

Important cyanobacteria	Master molecule	Applications	Refs.
Spirulina sp.	C-Phycocyanin β-Carotene Allophycocyanin Sulfated polysaccharide γ-Linolenic acid Sulfolipids	Antioxidants Anticancer Anti-inflammatory Anti-coagulant Anti-angiogenic Antiviral, Anti-proliferative	Verma et al. (2022), AlFadhly et al. (2022), Madhyastha and Vatsala (2007)
Nodosilinea sp.	β-Carotene Echinenone	Anti-inflammatory Anti-psoriasis	Lopes et al. (2020)
Cyanobium sp.	β-Carotene Zeaxanthin	Antioxidant Antitumor	Pagliara et al. (2021)
Nostoc sp.	β carotene Zeaxanthin Mixoxanthophyll	Antioxidant Anti-inflammatory	Srivastava et al. (2022a)
Lyngbya sp.	Cyclopropane α- and β-Carotene	Antioxidant	Salleh et al. (2023)
Micrococcus lylae	β-Cryptoxanthin Echinenone β-Carotene	Antioxidant	Shahin et al. (2022)
Psedanabaena sp.	2-methylisoborneol	Anti-odor	Giglio et al. (2011)
Nostochopsis sp	Polysaccharide	Hyaluronidase inhibitor	Yamaguchi and Koketsu (2016)
Tychonema sp.	Canthaxanthin Eihinenone	Skin care	Morone et al. (2020)
Nostoc puntiforme	Scytonemin	Sun-screen	Proteau et al. (1993)
Synechocystis sp.	Polyhydroxybutyrate	Bioplastics	Lindberg et al. (2010), Watzer and Forchhammer (2018)
Acaryochloris marina	Chlorophyl-D	Yet to understand	Kühl et al. (2005)
Oscillatoria sp.	Exopolysaaharides	Anti-bacterial, Anti-viral	Mazur-Marzec et al. (2021)
Cyanothece sp.	Cyanophycin	Nitrogen fixing	Watzer and Forchhammer (2018)
Arthrospira sp.	C-Phycocyanin β-Carotene Allophycocyanin	Anti-diabetic	Federal Research Centre of Nutrition (2022)
Porphyridium sp.	Polysaccharides Carrageenan	Anti-herplex Anti-HIV	Kanekiyo et al. (2007)
Nostoc flagelliforme	Polysaccharides	Anti-influenza A	Kanekiyo et al. (2007)
Oscillatoris sp.	Lectins	Anti-SARS-COVID	Sahu et al. (2023)

strategies focus on promoting photosynthesis, which is the fundamental process by which algae convert light energy into chemical energy, leading to their growth and biomass accumulation. The key aspects of photobiological strategies include light quality and spectrum, light intensity, photoperiod regulation, environment stresses, and photobioreactor design.

19.2.1 Light Quality and Spectrum

Microalgae are photosynthetic organisms that use light to convert carbon dioxide and water to biomass. The light quality and spectrum of microalgae can significantly impact their growth, metabolism, and product formation (Mehdizadeh Allaf and Peerhossaini 2022; Madhyastha et al. 2009). Different pigments in algae have specific absorption spectra that allow them to utilize different wavelengths of light efficiently (Metsoviti et al. 2019). The light quality refers to the wavelength distribution in the light spectrum. The spectrum of sunlight is relatively broad, ranging from approximately 300–1,100 nm. However, the light quality to which microalgae are exposed to cultivation systems can be reduced (Fae Neto et al. 2023). The light spectrum to which microalgae are exposed can also be modified by the use of filters or by adjusting the distance between the light source and culture (Maltsev et al. 2021). By tailoring the light spectrum to match the optimal absorption range of the target algal species, photobiologists can enhance photosynthetic rates and, consequently, algal growth. Light-emitting diodes (LEDs) or specialized monochromatic light sources can be used to provide specific light spectra that are favorable for targeted algae. A high-density culture of *Chlamydomonas* grown under yellow light resulted in higher biomass productivity compared to blue, orange-red, and deep-red light (Mooij et al. 2016). However, supplementation with blue light to cultures grown under monochromatic yellow light stimulated normal growth and increased volumetric algal productivity. The absorption spectrum of each microalgal photosystem is unique because it contains a distinctive set of light-harvesting pigments. This spectrum depends on the structure of the light-harvesting complexes and the transfer of energy to the reaction centers of photosystems I and II. Table 19.2 describes the different pigment systems and their absorption maxima.

19.2.2 Light Intensity

The amount of sunlight required by photosynthetic organisms exhibits a delicate balance. Too little light will stunt their growth, whereas too much can damage them. Therefore, controlling light intensity is crucial for algal growth. Insufficient light can

Table 19.2 Absorption maxima for different Photosystems of some commonly used algae

Pigment system	Absorption maxima (nm)	Algal species	Refs.
α-Carotene	421, 447, 474	*Audouinella, Cryptomonas*	Schagerl and Donabaum (2003)
β-Carotene	428, 453, 477	Most algae	Maltsev et al. (2021), Schagerl and Donabaum (2003), Senge et al. (2014)
Chlorophyll a	420, 660	All algal systems	Senge et al. (2014)
Chlorophyll b	435, 643	Blue Green algal systems (Cyanobacteria)	Maltsev et al. (2021), Schagerl and Donabaum (2003), Silva et al. (2020)
Chlorophyll c	445, 625	*Cryptophyta, cyanobacteria*	Schagerl and Donabaum (2003), Senge et al. (2014)
Chlorophyll d	450, 690	*Blue green alga, Rhodophyta*	Stadnichuk et al. (2015), Greenwold et al. (2019)
Phycocyanin	618	*Rhodophyta, Cryptophyta, Cyanobacteria*	Stadnichuk et al. (2015), Greenwold et al. (2019)
Phycoerythrin	490, 546, 576		
Allophycocyanin	650	*Cyanobacteria*	Mooij et al. (2016)
Phycoerythrocyanin	643	*Cyanobacteria*	Mooij et al. (2016)

limit photosynthesis and slow growth, whereas excessive light can cause photoinhibition and cell damage. An increase in intensity can change the metabolite composition, such as a reduction in protein and a correlated increase in lipid accumulation (Nzayisenga et al. 2020). Determining the correct balance of light intensity is essential for maximizing algal productivity. Light sensors and feedback systems in photobioreactors can help maintain optimal light levels.

The rate of photosynthesis is directly proportional to the light intensity, following a logarithmic curve. At low-light intensity, the rate of photosynthesis increases linearly with increasing light intensity as shown in Fig. 19.1. However, at high light levels, the rate of photosynthesis plateaus and does not further increase with increasing light intensity (Williams et al. 2010). When cells are exposed to excessive light, they produce excess reactive oxygen species (ROS). ROS are highly reactive molecules that can damage cellular structures and disrupt physiological processes (Morone et al. 2020; Federal Research Centre of Nutrition 2022; Erickson et al. 2015). ROS are small molecules that can react easily with other molecules in the cell. In small amounts, ROS can act as signaling molecules that help cells respond to stress. However, high levels of ROS can damage cell structures and disrupt cellular processes (Maltsev et al. 2021; Kim et al. 2018). In response to oxidative stress, algal cells produce antioxidant enzymes and other compounds that protect the cells from

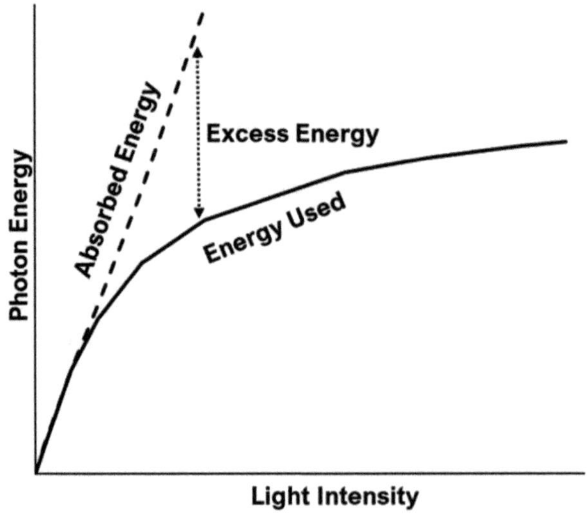

Fig. 19.1 Energy utilization pattern in response to increasing light intensity. As the incident photon energy exceeds the photosystem requirement, excess energy can cause photo-oxidative damage if not utilized efficiently

damage. These compounds can interrupt reactions that would otherwise damage cells. In addition to producing antioxidant compounds, algal cells can also respond to oxidative stress by making changes to their photosystems, light-harvesting complexes, and lipid composition (Erickson et al. 2015). These changes can help to protect cells from further damage and improve their commercial characteristics. Therefore, oxidative stress can have a significant beneficial effect on algal cell growth, leading to improved volumetric productivity of metabolites such as increased lipid production and enhanced resistance to stress (Kim et al. 2018; Shi et al. 2020).

19.2.3 Photoperiod Regulation

Like many other living organisms, algae have internal clocks that control their biological processes in response to light and dark cycles as a governing mechanism for photoperiodic regulation. The length of time that algae are exposed to light (photoperiod) can affect when specific cellular activities, such as growth and metabolite production, occur (Cheng et al. 2012; Maltsev et al. 2021). Tailoring the length of time algae are exposed to light to match their natural rhythms can boost their growth rates. A light to dark cycle of 16:8 h maximized the productivity of saturated fatty acids in *Chlorella* species, while a light to dark cycle of 8:16 h maximized monounsaturated fatty acids (Amini Khoeyi et al. 2012). Hence, it is important to tune the light–dark cycle to enhance photosynthetic efficiency (Maltsev et al. 2021; Kim et al. 2018). Strategies such as continuous lighting, alternating light–dark cycles, light flashes, or pulsed lighting have been demonstrated to improve biomass productivity. While continuous lighting may induce photoinhibition, the light–dark cycle helps restore photosystem damage (Yustinadiar et al. 2020). Modifying the photoperiod duration along with light intensity can be beneficial to alter the metabolite productivity. However, the amount of photoperiod varies across algal species. According to Krzemińska et. al. use of continuous light or light—dark cycle to achieve higher biomass productivity depends on the intrinsic requirement of algal system such as reproduction (Krzemińska et al. 2014). Among the two groups of algal systems they studied, the first group (*Botryococcus braunii, Scenedesmus obliquus*) grown on continuous light condition reproduced exclusively by autospores. The other group (*Neochloris conjuncta, Neochloris terrestris, Neochloris texensi*) required a light—dark cycle of 12:12 h and reproduced by aplanospores or motile zoospores under dark cycle. Sforza et al. (2012) experimented with different sets of pulsed light and compared with continuous light to grow *Microchloropsis salina* (Sforza et al. 2012). In experiments involving pulsed light, the cells demonstrated the ability to prevent photodamage and utilized the energy from the pulses with equivalent efficiency as they did with low intensity continuous light. Additionally, the intensity of the light flashes was found to have a minimal impact on growth.

19.2.4 Environmental Stresses

A number of strategies have been developed to address the challenges of growth inhibition and oxidative damage using growth-promoting agents. The production of algal growth-associated metabolites is closely linked to a number of other factors, including environmental stresses, biosynthesis pathways, and scavenging of ROS (Metsoviti et al. 2019; Williams et al. 2010; Erickson et al. 2015). The environmental stresses are primarily associated with their biosynthesis pathways and scavenging ROS due to excessive exposure to light. Some of the commonly used environmental stresses are growing alga under deprived micro-nutrients such as nitrogen, phosphate, and sulfur, other stresses such as salinity and temperature (Williams et al. 2010; Shi et al. 2020).

Since nitrogen is a primary requirement for most of the metabolic pathways, deprivation of nitrogen leads to changes in metabolite production. Similarly, phosphates and sulfates also play a vital role in metabolic pathways. In *Chlorella vulgaris* low nitrogen and nitrogen starvation could lead to enhanced lipid composition but less biomass productivity and pigment concentration (Liu et al. 2022). In this case, compared to low nitrogen condition, nitrogen starvation for three days was reported to be a better approach to improve the biomass and lipid productivity. A significant change in cell morphology, degraded chloroplast and increased size of lipid droplets and starch granules were observed. A decrease in pH was observed in the ammonium-supplemented cultivation medium of *Ellipsoidion* sp., which led to a decrease in growth rate and an increase in the accumulation of EPA and PUFA (Xu et al. 2001). *Chlamydomonas reinhardtii* culture was inhibited by up to 31.7% under nitrogen deficiency, but this was accompanied by an increase in total fatty acid yield of up to 93% and lipid production of up to 113.46 ± 1.78 mg/L (Yang et al. 2018a). This suggests that higher nitrogen concentrations favor biomass productivity, while depletion of nitrogen shifts the metabolic flux towards lipid and carotenoid production that require no nitrogen (Liu et al. 2022; Xu et al. 2001; Yang et al. 2018a). Cyanobacteria, especially non-nitrogen fixing species, also persist and adopt the nitrogen starvation condition. For example, *Oscillatoria willei* BDU 130511 overcomes this stress by adopting twin strategies like prevention of interactive mechanism with stress factor and counteracting the stress-induced damages. This explained the lower photosynthetic pigment such as chlorophyll-a, carotenoids and phycocyanin in *Oscillatoria willei* BDU 130511 due to chlorosis phenomenon. And in *Spirulina platensis* and *Anacystis nidulance* nitrogen limitation leads to catabolisation of lipid and fatty acids (Kumar Saha et al. 2003).

Phosphorus is a vital nutrient that plays a key role in the growth, lipid production, fatty acid yield, and metabolic processes of algae. These processes include energy transfer, signal transduction, and photosynthesis. Microalgae have the capacity to assimilate phosphorus as either polyphosphate or orthophosphate, leading to improved growth and enhanced nutritional content within their cells (Yaakob et al. 2021). Microalgae can accumulate heavy metals, survive their toxic effects, and transform orthophosphates and polyphosphates to ATP under unfavorable conditions. Phosphorus limitation is an effective environmental stress that can induce lipid accumulation in algae. This is demonstrated by the increased lipid content in *Scenedesmus* sp. grown in reduced phosphorus levels. For example, *Scenedesmus* sp. grown in 50 mg/L phosphorus achieved 22.3% lipid content, whereas lipid yield reached 42.5% in 1 mg/L phosphorus (Yang et al. 2018b). Additionally, a combination of sufficient phosphorus and nitrogen starvation can act as a potent lipid trigger. Therefore, regulating metabolite production by optimizing the balance between phosphorus and nitrogen appears to be a cost-effective and environment-friendly approach. Ecologically, phosphorus limitation is posing severe threat to lake eutrophication in aquatic system leading to cyanobacterial bloom specially growth of microcystic (Song et al. 2023).

Sulfur limitation can be beneficial for lipid production in microalgae. Sulfur is required to synthesize glutathione, an antioxidant molecule that helps protect cells from damage. When sulfur is limited, there is an increase in reactive oxygen species (ROS), which can damage cells. To protect themselves from ROS, cells produce carotenoids, which are also antioxidants (Carfagna et al. 2016). For example, sulfur starvation was found to be more effective than nitrogen or phosphorus starvation for increasing triacyl glycerol (TAG) production in *Chlamydomonas reinhardtii* (Sakarika and Kornaros 2017). Sulfur starvation has also been shown to be more effective than phosphorus or iron starvation for inducing high-level accumulation of astaxanthin. However, phosphorus starvation led to the highest increase of β-carotene accumulation in algae (Yaakob et al. 2021; Yang et al. 2018b). This suggests that the effect of sulfur limitation on metabolites like lipids and carotenoids depends on the microalgae species (Shi et al. 2020). While designing an optimal process scheme for algal biomass production and targeted metabolite afore-mentioned parameters need to be investigated before scaling up the process in biotechnological industries. Sulfur is also a major nutrient factor for growth and development of cyanobacteria and has a greater role in nitrogen and thiol metabolism in Anabaena sp. PCC 7120 (Kharwar and Mishra 2023).

19.3 Technology of Large-Scale Cultivation

Many algal species are able to produce large amounts of biomass rapidly and economically, which is one of the hallmark features. Currently, a wide range of successful algae farming systems is being developed, encompassing outdoor ponds, indoor photobioreactors, heterotrophic productions, and combinations of these approaches. Each system is designed with the objective of achieving accelerated algae biomass growth for the production of biofuels or other industrial products. Upscaling of algal cultures to the extensive volumes necessary for commercial production which poses a complex challenge and demands proficient and experienced personnel. Initially, it is imperative to optimize the inoculum production process for large ponds or photobioreactors, aiming to reduce both the time and cost involved. The key factors affecting the growth and biomass productivity of all algal cultures are light, temperature, and nutrients. Almost all commercial-scale algal biomass production is done outdoors in natural light, which exposes the algae to daily changes in irradiation, as well as short-term changes caused by clouds, rain, and other aerosol micro pollutants. These problems and constraints can be minimized using photobioreactors.

19.3.1 Open Pond Systems

Open pond cultivation is one of the oldest and simplest methods for large-scale cultivation of microalgae. It is widely used in the industry owing to its relatively low construction, maintenance, and operating costs (Chisti 2007). Other advantages of open pond systems include simple operation and maintenance, low energy requirements, and scalability. The raceway pond is a commonly utilized open pond system for cultivating microalgae (Fig. 19.2). It comprises a series of closed-loop channels made of concrete and plastic sheets, approximately 30 cm deep, and is equipped with a motorable paddlewheel to facilitate the recirculation of microalgal biomass (Jiménez et al. 2003). This recirculation ensures an even distribution of nutrients and prevents sedimentation of the microalgal biomass with effective light penetrations to the bottom of culture pond. One remarkable example of successful raceway pond cultivation is demonstrated by Sapphire Energy's Columbus Algal Biomass Farm, located in Columbus, United States, where dried microalgal biomass has been produced efficiently without encountering any technical challenges (White and Ryan 2015). Raceway pond design has gained recognition as one of the most energy-efficient open pond cultivation methods because a single paddlewheel suffices to adequately agitate a vast 10,000 m^2 raceway pond. In large production raceway ponds, average productivity is typically less than 20 g per square meter per day (Jiménez

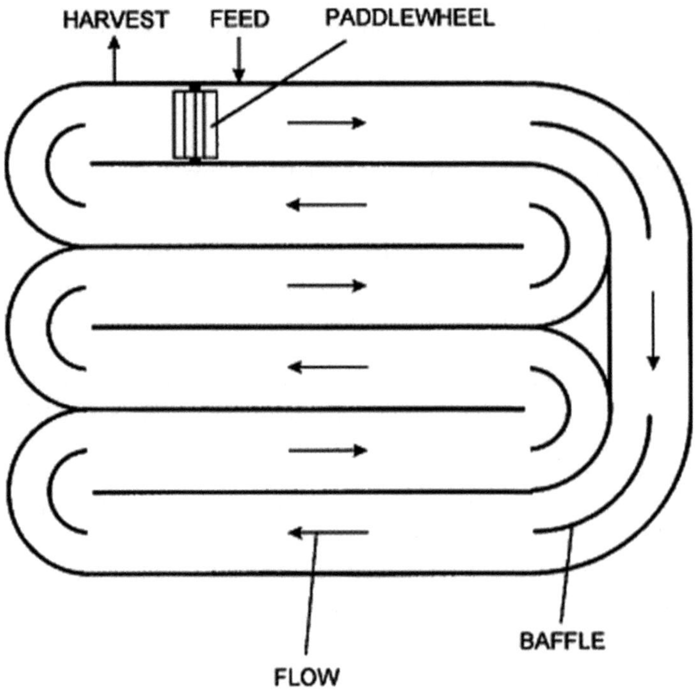

Fig. 19.2 Raceway pond or open algal ponds used to produce large quantities of algal biomass

et al. 2003; Borowitzka and Vonshak 2017). However, under optimal conditions, small indoor cultures can achieve productivity of up to 50 g per square meter per day. In small outdoor raceway ponds (1–5 square meters), a productivity of more than 30 g per square meter per day can be achieved over very short periods (Moheimani and Borowitzka 2006).

Despite the advantages of open raceway pond systems, such as low investment and operating costs, it is crucial not to overlook their limitations. These systems are prone to bacterial contamination and are exposed to uncontrolled environmental conditions. Algae cultivated in open raceway systems are vulnerable to contamination from harmful or invasive microorganisms present in the surrounding environment, potentially leading to the failure of algae growth (White and Ryan 2015; Borowitzka and Vonshak 2017). Moreover, to cut operating costs, open raceway pond systems often lack temperature regulators, making algae growth and nutrient removal from aquaculture effluent susceptible to external factors like temperature and light. In certain cases, low winter temperatures and inadequate lighting may hinder algae growth and nutrient removal. Hence, while open raceway pond systems offer certain benefits, their constraints must be carefully considered in real-world applications (Moheimani and Borowitzka 2006).

19.3.2 Photobioreactors

Light and temperature are the primary restricting elements for the growth of outdoor algal cultures that have an adequate supply of nutrients. To overcome this problem, photobioreactors have been in use where artificial light along with other sensors to control the system (Chisti 2007). Photobioreactors such as stirred tank bioreactors, tubular bioreactors (horizontal and vertical type), flat-panel bioreactors, bag bioreactors, and membrane bioreactors have been reported with their own benefits and technological challenges (Pandey et al. 2019). A schematic representation of these bioreactors is illustrated in Fig. 19.3.

Stirred Tank Photobioreactors

Stirred tank reactors (STRs) are conventional reactors that use mechanical agitation to mix the reactor contents. These are typically cylindrical vessels that can be operated in batch, semi-batch, or continuous modes (Pandey et al. 2019). In batch STRs, the reactants are added to the reactor at the beginning of the reaction, and the products are removed at the end of fermentation. In semi-batch STRs, reactants are added to the reactor in a series of steps, and the products are removed as they are produced. In continuous STRs, reactants and products are continuously added to and removed from the reactor. STRs have several advantages, including uniform mixing, good temperature control, and the ability to operate in various modes. However, they also have some disadvantages, such as a low surface-to-volume ratio and potential for damage to sensitive cells (Pandey et al. 2019; Chanquia et al. 2022). Bioreactor applications for metabolite production should consider productivity, power consumption, and scalability. Productivity is affected by several factors, including light intensity, mixing, and oxygen removal (Płaczek et al. 2017). Scalability refers to the ability to increase the size of the reactor without significantly affecting productivity. The choice of bioreactor for a particular application depends on the specific requirements of the process. For example, if high productivity is required, an STR with a high light intensity and good mixing may be the best option (Gao et al. 2016). If power consumption is a concern, a different type of reactor may be more suitable.

Tubular Photobioreactors

Tubular photobioreactors are a type of photobioreactor composed of a series of parallel cylindrical tubes. The tubes can be arranged in loops or in systems inclined at a given angle to the ground. This makes them particularly suitable for continuous outdoor cultivation because the angle of the tubes can be adjusted to maximize the capture of solar radiation (Płaczek et al. 2017). Tubular photobioreactors can hold large volumes of cultivation media and are not susceptible to microbiological contamination (Yustinadiar et al. 2020). This makes them a good option for the large-scale cultivation of algae. There are two main types of tube arrangement in vertical tubular photobioreactors. The first type involves installing tubes parallel to one another, low above the ground. The second type is to install the tubes on top of one another, supported by a horizontal ladder-shaped rack (Yustinadiar et al. 2020; Pandey et al. 2019; Płaczek et al. 2017). This design increases the number of tubes per surface area of the available ground. The tubes could be coupled with a gas-

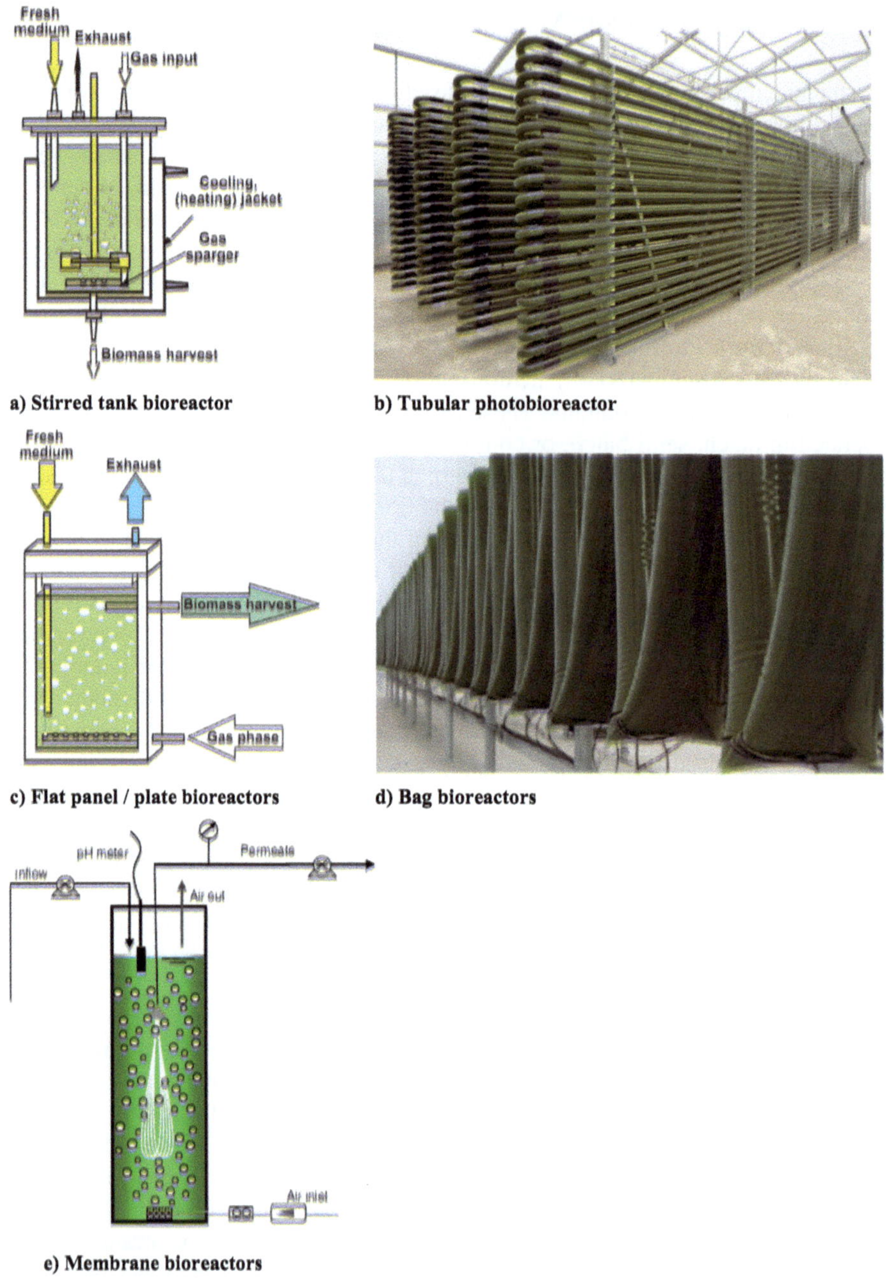

Fig. 19.3 Photobioreactors used for algal biomass production on a large scale (Chisti 2007; Huo et al. 2018; Płaczek et al. 2017; Gao et al. 2016)

and heat-exchange system. The gas exchange system removes the oxygen produced during photosynthesis (Wang et al. 2012). This is important because oxygen can damage algal cells. The heat-exchange system removes heat from the reactor, which helps prevent algae from overheating (Sukačová et al. 2021). The internal diameter of the pipe of the solar collector is limited to 0.1 m. This is because sun rays cannot penetrate deep into the suspension of algae, which is necessary for significant biomass growth. These photobioreactors can also be used with artificial light, such as ordinary fluorescent tubes, diodes bulbs, and LEDs (Yustinadiar et al. 2020).

Flat Panel Bioreactors

Flat panel photobioreactors (PBRs) are constructed using two transparent sheets of adhesive material (such as glass, plexiglass, or polycarbonate) with a light penetration path ranging from a few millimeters to 70 mm (Pandey et al. 2019; Płaczek et al. 2017). They have gained popularity because of their superior surface-to-volume ratio, which facilitates optimal illumination for photosynthesis. The geometry of a single flat panel ensures a high reactor surface-to-volume ratio, resulting in an increased biomass production efficiency (Borowitzka and Vonshak 2017; Zhu et al. 2013). Horizontal flat plates receive light but require more space, whereas vertical flat plates are more land-efficient but demand sturdier and costlier materials for construction. When designing a flat-panel PBR, considerations of panel alignment (horizontal, vertical, V-shaped, inclined, or accordion arrangement) are essential to maximize photosynthetic efficiency while balancing space requirements and potential photoinhibition, which remains the primary drawback of flat panels for high biomass generation (Chanquia et al. 2022). Mechanical suspension of cells is unnecessary in these PBRs because medium mixing is accomplished through air bubbling facilitated by a bottom sparger (Metsoviti et al. 2019; Sukačová et al. 2021).

Bag Bioreactors

Plastic bag photobioreactors are a popular choice for commercial applications owing to their low cost and sterility. The bags are typically hung vertically in parallel on a frame, and the medium and aeration are circulated through a pump (Wang et al. 2012). The capacity of the bags is usually around 25 L, but bags of up to 50 L have been reported (Płaczek et al. 2017; Chanquia et al. 2022). These plastic bags can also be used horizontally on a rocking platform to reduce energy costs (Zhu et al. 2013). The main advantages of plastic bag photobioreactors are their simplicity and low cost. They also have a good ratio of illuminated surface to total volume (Pandey et al. 2019). However, this type of photobioreactor has some limitations. The bags must be replaced periodically, which can be an environmental issue for large-scale operations. Additionally, mixing in plastic bag photobioreactors is poor, and the bags are fragile, so they can break and leak (Chanquia et al. 2022; Wang et al. 2012). It has also been reported that increasing bag volume does not always lead to an increase in productivity (Borowitzka and Vonshak 2017; Wang et al. 2012). Therefore, the most common and efficient way to scale up a process using plastic bag photobioreactors is to use multiple units that require a large area.

19.3.3 Membrane Bioreactors

Membrane photobioreactors (MPBRs) are a type of photobioreactor that uses membranes to separate algae from the culture medium. This has several advantages over traditional photobioreactors (Pandey et al. 2019; Gao et al. 2016). The membrane allows light to penetrate deeper into the culture, leading to increased biomass production. The membrane helps remove oxygen from the culture, which can prevent the algae from becoming stressed and dying. They help to prevent contamination of the culture with bacteria or other microorganisms. MPBRs have been used to cultivate a variety of algae including *Chlorella*, *Spirulina*, and *Dunaliella* (Gao et al. 2016; Wang et al. 2012). They have been shown to be effective in the production of a variety of products, including biofuels, biopolymers, and food supplements (Pandey et al. 2019; Zahmatkesh et al. 2023). One of the main challenges with MPBRs is membrane cost. However, the cost of membranes is decreasing, and MPBRs are becoming increasingly cost-competitive with other types of photobioreactors (Chanquia et al. 2022). They offer several advantages over traditional photobioreactors and are becoming increasingly cost-competitive. Table 19.3 lists certain design specifications of the photobioreactors discussed in this section, their advantages and limitations.

19.4 Active Metabolites Control Strategies in Cyanobacteria

Escalating demand for alternate chemicals and food, metabolic engineering concept of microorganisms elevated the concept of "microbial cell factory". Cyanobacteria are immensely diverse prokaryotic organisms that are found in wider environments that are able to produce a wide variety of chemicals and metabolites for various industrial, nutritional, and pharmaceuticals compounds. Moreover, it is also an important platform to produce varies commercially important biomaterials. In addition, they have rich content of commercial pigments like chlorophyll, carotenoids, phycocyanin, and terpenes (Madhyastha et al. 2009; Guarnieri and Pienkos 2015). Harnessing and enrichment of metabolic content of the cyanobacteria has been carried out by various biotechnological, genetical, and engineering aspects. To meet this hypothesis of "metabolic enrichment strategy" researchers are constantly targeting cyanobacterial engineering for the production of metabolites directly from CO_2. Transcriptional engineering (TF) is one such attractive strategy to modify the multiple components of metabolic pathways simultaneously. Concept of metabolic engineering enables the researchers to design and incorporate the foreign genetic materials to ongoing bio-systemic pathways of native cells, ultimately enhancing or over produce the metabolites. The best model studied so far is *E. Coli* bioengineering for the production of various fermentative products in reactors (Tenaillon et al. 2010). In principle, TF regulates specific target genes by binding the specific DNA motif in *cis* element of target genes by the help of RNA polymerase to activate or suppress

Table 19.3 Different types of photobioreactors with their specification, advantages, and limitations

Reactor type	Mixing	Temperature control	Aeration	Advantages	Limitations	Refs.
Stirred tank	Impellers	In situ heat exchanger	Sparger	More controllable and able to be cultivated in a pure culture	Low surface-to-volume ratio, low efficiency of light absorption, and low productivity	Płaczek et al. (2017), Gao et al. (2016)
Tubular	Airlift/ pumping	Separate heat exchanger	Degasser/ head space gas exchange	Rapid growth, high photosynthetic efficiency, scalability, low energy requirements, and the ability to be immobilized	Low light exposure decreases as tube diameter increases. Low surface-to-volume area. Formation of biofilm on tube walls	Yustinadiar et al. (2020), Chisti (2007), Huo et al. (2018)
Flat panel	Bubbling/ airlift	Heat exchanger internal coils	Head space gas exchange	Good surface for exposure of light. High surface-to-volume ratio. High productivity of biomass. Uniform volumetric light penetration. Proper amounts of dissolved oxygen	Difficulty in scalability and temperature control. Risk of fouling. Photoinhibition, hydrodynamic stresses on algal species due to higher hydraulic pressure in panels	Metsoviti et al. (2019), Chanquia et al. 2022)
Bag	Pumping	Separate heat exchanger	Sparger in separate unit	Good adaptability, simple in design, cheaper and amenable for custom design of shapes and sizes	Requires periodic bag replacement. Plastic waste generation, if not re-used. Prone to photoinhibition and leakage of medium	Chanquia et al. (2022), Wang et al. (2012)
Membrane	Pumping/ bubbling	Separate heat exchanger	Gas injection	Easy to operate and maintain. Low operating temperature	High cost, high-pressure drops, membrane fouling, and difficulty in scaling up	Gao et al. (2016), Chanquia et al. (2022)

the process of transcription. Although this micro engineering process seems to be quite simple in prokaryotic bacteria, it is being developed or under investigations in cyanobacteria because of complex photosynthesis machinery. However, several alternate strategies are being studied like mutation, nutrient stress, and photoperiod description are being studied in cyanobacteria (Bajhaiya et al. 2017). However, sporadic work is being conducted on TF in cyanobacteria with the aim to accelerate the biofuel production and extraction to mitigate the global climate change to produce alcohol and diols, 2–3, butanediol, 1–3, propanediol, ethanol, isopropanol, isobutanol, ethelene, glycogen, mannitol, lactate, glucosylglycerol etc. (Oliver et al. 2016). However there is a broader scope of TF to enhance the high-value compounds for nutraceutical and pharmaceutical products such as terpenes, phytol, sterols, carotenoids, and proteinaceous pigments as these pigments have higher demand in erogenous, flavorings and some have health benefits (Guarnieri and Pienkos 2015). Biosynthesis of some metabolites delivers a challenge to investigators due to complex pathways involved during carbon pool to the site of carbon fixation. Lindberg et al. (Wolk 1973) studied the process of metabolic engineering strategies in *Synechocystis* sp., where di-methyleallyl diphosphate (DMAPP) and iso-pentenyle diphosphate (IPP) are linked with a total of seven steps in the methylerythritol pathway (Lindberg et al. 2010). This particular process further hampers the end product quality due to the continuous enrichment of culture media for the growth of organisms in bioreactor condition. Nutrient stress is another alternate way to be studied. *S. elongatus* 7942 cells to high saline condition (NaCl 200 mM) accumulates the sugars in the cells as an osmo-protectant strategies, however, incorporation of sucrose transporter mRNA [cscB (E. coli)] to *S. elongatus* 7942 genome helps in the accumulation of sucrose in culture supernatant, in addition of knock down of glycogen $\Delta glgC$ and sucrose $\Delta invA$ degradation pathway genes significantly increased the sugar production to maximum level of 9.5 $mgL^{-1}\ h^{-1}OD^{-1}$ (Ducat et al. 2012).

Targeting the crucial metabolic crux approaches like, enzyme tweak strategies, engineering the electron transport chain, enrichment of nodal points like acetyl Co-A, clearing the bottleneck of pathways, optimizing the regulatory elements, efficient management of cofactors are also being studied in cyanobacteria to enhance the product (Cano et al. 2018; Shabestary et al. 2018; Hirokawa et al. 2020). Recently metabolomic assisted micro engineering models are also being studied. Artificial engineering and bioinformatic assisted gene data mining exercise and ^{13}C-transient core models may be very powerful tools to pick out the metabolic pathway bottlenecks. But application of these techniques is expensive and time-consuming (Cano et al. 2018). Interactive metabolic cycle profiling helps in target identification, for example, coenzyme A-acylating propionaldehyde dehydrogenase (PduP) over activity is monitored by studying the ntcA gene expression for the production of 1-butanol in *Synechocystis* sp. Likewise, ^{13}C-transient core analysis was deployed to monitor the production of succinate and ethylene in *S. elongatus* 11801 sp. (Sengupta et al. 2020). Overexpression of succinate production is governed by Fructose 1–6 bi-Phosphatase (BiBPase) enzyme as rate limiting step as judged by significant accumulation of sedoheptulose 1–7 bi-phosphate (SBP) with reduced level of sedoheptulose -7-phosphate (S7P). This mismatch between enzyme activity and SBP/S7P production can be alleviated by overexpression of BiBPase gene (Sengupta et al. 2020). Metabolic control assay/analysis (MCA) is another promising tool to study the rate-limiting steps which are principally controlled by large volumes of reaction flux control coefficients but the efficacy of this not being studied in cyanobacteria (Fell 1992). However sporadic level of work being carried out in cyanobacteria for the production of lactic acid, limonene, and ethanol (Angermayr and Hellingwerf 2013; Nishiguchi et al. 2019). Reducing the growth inhibitory factors, controlling the toxic accumulation, and stress tolerance have been reported (Arai et al. 2017). In recent years clustered regularly interspaced short palindromic repeats (CRISPR) are being attempted to regulate the metabolic genes in cyanobacteria. CRISPR/Cas9 and CRISPR/12a and CRISPRi enabled the researchers to modify or edit the genomes of particular interest with marker less precise point mutation for gene knock-outs and knock-ins. To control the carbon away from the glycogen and facilitates more towards succinate glgC deletion was attempted to improve the tricarboxylic acid cycle (TCA cycle) in cyanobacteria (Gründel et al. 2012). Phycobilisome degradation prevention due toxic mutation agents was first attempted by CRISPR/Cas system by targeting the non-bleaching protein A (nblA) by Wendt and colleagues thereby prevents the de-pigmentation (Gründel et al. 2012; Wendt et al. 2016). Caveats issue of CRISPR is off-target which is an particular concern in cyanobacterial molecular engineering. Level of repressions is heavily influenced by co-operativity between sgRNA and target DNA where DNA sequence should be close to transcriptional start site. Complete data base of whole genome sequence of all useful cyanobacteria is still under investigation hence triggers the debate on use of CRISPR/Cas system in cyanobacteria, but also other biotechnologically relevant microorganism.

Post modifications failures are common lacuna in cyanobacterial genetic engineering research due to inefficient translational process and this particular constraint stimulates the commercial failure of the by-product. Application of engineering practice to photosynthetic microbes for commercially important metabolites synthesis requires two-step process like: efficient cell growth and fine-tuning of carbon conversion phenomenon for desired product extraction in cyanobacteria. Recently, genome-guided centric information, or bottom-up approach with the combination of functional genomics and bioinformatics been continuously used to unravel the higher product enhancement, which is called Biosynthetic gene cluster models to enhance the cyanobacterial natural products (CNP) with the help of various tools like anti-SMASH, PRISM and RODEO (Skinnider et al. 2015; Tietz et al. 2017; Blin et al. 2019). Cyanobacterial NPs are mainly a polyketides and peptides with a major share of more than 66% belongs to later group where ribosomal assisted post translational process takes active step.

19.5 Biotherapeutic Molecules of Cyanobacteria in Human Health

Adaption and fine-tuning of cyanobacteria significantly contribute to people's health around the world since they produce a large number of secondary metabolites which are of nutraceutical importance in general or harmful to human being. Recently cyanobacterial antimicrobial compounds are gaining importance in biomedical field and are slowly replacing the traditional chemically synthesized antibiotics. Various techniques are being developed to identify, extract, isolate, and purify the cyanobacterial NP. Interestingly some of them are clinically proven, FDA approved after vigorous clinical trials of phase I and II. Cyanobacterial NP consists of Terpenes, alkaloids, phenols, carbohydrates, fatty acids, ployketides, peptides, and proteins (Srivastava et al. 2022b).

Generally chemical structure with $C_{10}H_{16}$ termed as terpenes with different types, namely diterpenes (C_{20}), triterpenes (C_{30}) and tetraterpenes (C_{40}) hemiterpenes (C_5) and sesquiterpenes (C_{15}), if additional elements like oxygen is linked to the main chain it is called terpenoids. *Nostoc* sp. and *Scytonema.* sp contain higher amount of sesquiterpenes specially cybastacines A and B and therefore can be considered for further exploration towards cyanobacterial NP towards clinically relevant bacterial strains (Cabanillas et al. 2018). Nitrogen moieties containing alkaloids are mainly a heterocyclic compound and being used as analgesics and chemotherapeutics. Ambiguine isonitriles is an hapalindole type alkaloids is a good fungicidal compounds are being produced by *Fisherella ambigua, Hapalosiphon Hibermicus, and Westiellopsis prolific* (Mo et al. 2009). Anti-viral and anti-cancerous alkaloids, cylindrospermosin are also being isolated from *Cylindrospermopsis raciborskii*, and recently it showed as a promising compounds to combat SARS-CoV-M^{pro} (Hensley and McAlinden 2021). Carbohydrates are one of the abundant group compounds in cyanobacteria. Coral reef mat at Carrie Bow Cay in Beliz contains *Dendryphiella salina, Lindra thalassiae,* and *Fusarium sp.*, with a high amount of 5-hydroxy-4-(chloromethyl)-5,6,7,8-tetrahydroquinoline also known as carriebowlinol showed a high rate of antifungal activity (Soares et al. 2015). *Nostoc flagelliforme* and *Spirulina platensis* are the major source of nostoflan and calcium spirulan, an antiviral, acidic polysaccharide and effectively being used to control the replications of HSV-1, HSV-2, influenza A virus, measles virus, mumps virus, and human cytomegalo virus (Kanekiyo et al. 2007). Polysaccharides are useful compounds with anti-fibrotic effect in pulmonary tissues and also calcium spirulan being tested to treat the pandemic corona virus but with limited success (Khalifa et al. 2021). Lectins are hybrid metabolites between proteins and carbohydrates with covalent bonds. The carbohydrate- biding domain (CBD) in lectins differentiates the tertiary structure of the oligomeric proteins with sugar moieties mainly with mannose. Many lectins cyanobacteria derived lectins like cyanovirin-N(CV-N, 11-kDa) from *Nostoc ellipsosporum*, microvirin (MVN, 12.7 kDa) in *Microcystis aerouginosa*, scytovirin (SVN, 9.71 kDa) of *Scytonema varium*, oscillatoria aagardhii agglutinin (OAA, 13.9 kDa) from *Oscillatoria agardhii*, showed a potent antiviral compounds

(Mazur-Marzec et al. 2021). Heirridin (2–4-dimethoxy-6-heptadecyl-phenol) and single phenol ringed polyphenol are abundantly produced in *Phormidium ectocarpii*, displayed the anti-plasmodial, antiviral and antibiotic activity (Murakami et al. 1991; Singh et al. 2023).

The omega-3(n-3) polysaturated fatty acid (PUFs), Eicosapentaenoic acid (EPA), and docosahexaenoic acid (DHA) are a class of lipophilic fatty acids with different fatty acid chain length. EPA contains 20 carbons and 5 double bonds, while DPA with extra two carbon atoms with one additional double bond thus makes differential action on human health. These two compounds are trending in the pharmaceutical market as it prevents the inflammatory cardiovascular illness. Alpha-linolenic acid (ALA 18:3n-3) and stearidonic acid (SDA, 8:4n-3) are class of omega-3 fatty acids (FAs) are naturally synthesized in cyanobacteria and accumulate in thylakoid membrane with high salt and cold stress of *Synechococcus elongatus* PCC 7942. Marine cyanobacteria *Lynbya majuscule* contains potent antibacterial agent namely malynamides, amide of fatty acid including 7(S)-methoxytetradecenoate (Gutiérrez et al. 2010). It is believed that drug-mediated human immunodeficiency virus-1 (HIV-1) reverse transcriptase (RT) inhibition is the key target to control viral replication in human being. Lipophilic extracts of *Scytonema sp.* primarily attributed to the sulfoquinovosylpranosyl lipid selectively inhibit DNA polymerase activity (IC50 up to 50%) of HIV-1 RT without having much efficacy on ribonuclease H function (Loya et al. 1998). Another class metabolite of cyanobacterial NP is polyketides with various biological and biomedical activity with high demand in pharmaceutical industries. Polyketides are structurally different but biosynthetically similar natural products with very high medicinal values. Some of the markable bacterial origin polyketides are lovastatin (cholesterol lowering agent), erythromycin (antibiotic), and FK506 (an immunosuppressant) which showed significant efficacy in medical filed, however body resistant towards these bacterial polyketides invited researcher to identify new sources from other organisms (Loya et al. 1998). Classic fungicide, cryptophycin first invented from *Nostoc sp.* displayed the potent anticancer effect because of modifications in D-α amino acid, β-amino acid and hydroxyl acid part thus modifies polyketide into depsipeptide, unique to cyanobacteria. Different polyketides namely, cyclic hexapeptides, tenuecyclamides A-D were isolated from methanolic extract of *Nostoc spongiaeforme* var. tenue (TAU strain IL-184–6) showed a highest anticancer activity (Banker and Carmeli 1998).

Cyanobacterial pigments like, chlorophyll, carotenoids, and phycobiliproteins are light absorbing compounds with different wavelength as a part of photosynthesis process. Main structure of pigments consists of isoprene mainly in chlorophyll and phycobiliproteins and tetrapyrroles and isoprene units are abundantly found in carotenoids with multiple uses in pharmaceutical, nutraceuticals, and cosmeceuticals industries (Ahirwar et al. 2022). Being a principal light-harvesting pigment in cyanobacteria, it consists of six different types namely, Chl a, Chl b, Chl c, Chl f, and divinyl Chl a and b. In all structures linearly arranged pyrrole rings as porphyrins with Mg^{++} as the central atom. Chl a and b are potent antioxidant, anticancer, and having stimulatory efficacy towards immune system (Hayes and Ferruzzi 2020). Chlorin water is synthesized from commercial-grade soluble chlorophyll. It is synthesized by replacing central Mg^{++} atom with Zn, Fe, Co, and Cu from methanolic extract of crude chlorophyll from algae and water. And the product is marketed extensively in USA as Cu-Chlorin e_4 and Cu-chlorin e_6 (Egner et al. 2000). Carotenoids are part of chlorophyll molecule to facilitate the extra photosynthesis process. The carotenoids' structure is formed by the arrangement of 40 isoprene units with or without oxygen. The non-oxygenated products are carotenes and with oxygenated products are hydrophilic xanthophyll due to the presence of hydroxyl and keto groups at the end (Gong and Bassi 2016; Sreenikethanam et al. 2022). Major carotenoids found in cyanobacteria are β-carotene, astaxanthin, fucoxanthin, lutein, zeaxanthin, and scytonemin (Sreenikethanam et al. 2022). Carotenoids display various biotherapeutics application in the human body as antioxidant, anti-inflammatory agent, and anticancer molecule (Upadhyaya et al. 2007). However, each type of carotenoid displays different functions based on steric orientation of the isoprene and oxygen units. β-carotene helps in the synthesis of Vitamin A, lutein and zeaxanthin protects the eye, lycopene prevents the cellular oxidation status due to ROS and NOS that reduces the cardiovascular disease. In general bioavailability, bio-absorption and cellular uptake of carotenoids still is a challenge and thus suitable formulations are being studies to increase the bioavailability, for example, emulsion and nano products of carotenoids are being tested with promising benefits to human being (Boonlao et al. 2022).

The majority and bulk content of cyanobacteria is phycobiliproteins. This light-harvesting antenna molecule has covalently linked chromophore namely phycocyanobilin (PCB). The most common phycobiliproteins are c-Phycocyanin (c-PC), allophycocyanin (aPC), and phycoerythrin (PR). cPC is potent, anticancer, anti-inflammatory, antioxidant, and agent also good biomaterials to cure varieties of regenerative complications in human because of covalently linked chromophore PCB which binds to apolipoprotein of phycobiliproteins (Madhyastha and Vatsala 2007, 2010; Madhyastha et al. 2023, 2021, 2020, 2006). Multiple investigations proved that cPC is an active molecule to cure and manage notable neurogenerative complications like Parkinson's and Alzheimer's disease by reducing the neuron cells ROS and NOX thereby decrease the cholinergic and amyloid formations (Breijyeh and Karaman 2020). DNA damage is caused by intermediate molecule peroxynitrite ($ONOO^{(-)}$) and cPC is a candidate molecule to eliminate ONOO (Bhat and Madyastha 2001). cPC is a new class of drug, to treat ischemic stroke (IS) due to myelin haemostasis during multiple sclerosis and ischemic stroke patients particularly in auto-immune encephalomyelitis (Cervantes-Llanos et al. 2018). *S. platensis* has high amount of polyunsaturated fatty acids (PUFAs), 2.0% out of 5–6% of total lipid with γ-linolenic acid (38% of total PUFAs), and also provides γ-linolenic acid (ALA), linoleic acid (LA, 36% of total), stearidonic acid, eicosapentaenoic acid, docosahexaenoic acid, and arachidonic acid. High content of vitamins like vitamin amounts of B1 (thiamine), B2 (riboflavin), B3 (nicotinamide), B6 (pyridoxine), B9 (folic acid), B12 (cyanocobalamin), vitamin C (ascorbic acid), vitamin D (cholecalciferol), which scientifically proves its resolute antioxidant property (ElFar et al. 2022).

19.6 Future Perspectives

Fast-changing lifestyle, global climate change, unplanned food consumption, and population growth led to proper health mitigation strategies. Although effective and accessible chemical medicines are available, their production seems to be limited with high cost and severe side effects to human being. Cyanobacteria are potent sources of biotherapeutic compounds, biofertilizers, and bioplastics, as source materials for bioenergy production, including bioremediation organisms. Much progress has been made cyanobacterial bioengineering as micro-chassis for renewable and light-driven synthesis of biochemicals and biofuels through the concept of cell factory. Most of the studies are concentrated on freshwater cyanobacteria for the microproduction of nutraceuticals and other diagnostic agents however, marine environment with completely different niche starting from shore lined shallow water to deep sea environment contains innumerous species of cyanobacteria, yet to be screed for bioeconomic prospects. Sporadic work has been carried out in the arena of biohydrogen production from algae. Although biohydrogen production from cyanobacteria looks great promising and environmentally sound concept, its optimization and product yield seem to be a challenging task due to the high cost of production. Bio-inspired space and building design concept with transparent and hard reactors as budling wall or pillar as eco-architecture prototypes in urban space could be another area to be explored further. Scientists are confident of blue biotechnology overtaking green biotechnology over the years to come. This requires the combined efforts of taxonomists, molecular biologists, biochemists, engineers, and industry-related scientists as well as politicians and policy makers.

Acknowledgements This study is partially funded from institutional funding of NMAM Institute of Technology, Nitte under Nitte (Deemed to be University), Mangaluru, India.

Conflict of Interest Statement The authors declare no conflict of interest. The funders had no governance in the design of the study, in the collection samples, analyses of results, or interpretation of data, in the writing of the manuscript, or in the decision to publish the results.

References

Ahirwar A, Kesharwani K, Deka R et al (2022) Microalgal drugs: a promising therapeutic reserve for the future. J Biotechnol 349:32–46. https://doi.org/10.1016/j.jbiotec.2022.03.012

AlFadhly NKZ, Alhelfi N, Altemimi AB et al (2022) Tendencies affecting the growth and cultivation of genus spirulina: an investigative review on current trends. Plants 11:3063. https://doi.org/10.3390/plants11223063

Amini Khoeyi Z, Seyfabadi J, Ramezanpour Z (2012) Effect of light intensity and photoperiod on biomass and fatty acid composition of the microalgae, Chlorella vulgaris. Aquacult Int 20:41–49. https://doi.org/10.1007/s10499-011-9440-1

Angermayr SA, Hellingwerf KJ (2013) On the use of metabolic control analysis in the optimization of cyanobacterial biosolar cell factories. J Phys Chem B 117:11169–11175. https://doi.org/10.1021/jp4013152

Arai S, Hayashihara K, Kanamoto Y et al (2017) Alcohol-tolerant mutants of cyanobacterium *Synechococcus elongatus* PCC 7942 obtained by single-cell mutant screening system. Biotechnol Bioeng 114:1771–1778. https://doi.org/10.1002/bit.26307

Bajhaiya AK, Ziehe Moreira J, Pittman JK (2017) Transcriptional engineering of microalgae: prospects for high-value chemicals. Trends Biotechnol 35:95–99. https://doi.org/10.1016/j.tibtech.2016.06.001

Banker R, Carmeli S (1998) Tenuecyclamides A–D, cyclic hexapeptides from the cyanobacterium *Nostoc s pongiaeforme* var. *t enue*. J Nat Prod 61:1248–1251. https://doi.org/10.1021/np980138j

Bhat VB, Madyastha KM (2001) Scavenging of peroxynitrite by phycocyanin and phycocyanobilin from spirulina platensis: protection against oxidative damage to DNA. Biochem Biophys Res Commun 285:262–266. https://doi.org/10.1006/bbrc.2001.5195

Blin K, Shaw S, Steinke K et al (2019) AntiSMASH 5.0: updates to the secondary metabolite genome mining pipeline. Nucleic Acids Res 47:W81–W87. https://doi.org/10.1093/nar/gkz310

Bongirwar R, Shukla P (2023) Metabolic sink engineering in cyanobacteria: perspectives and applications. Biores Technol 379:128974. https://doi.org/10.1016/j.biortech.2023.128974

Boonlao N, Ruktanonchai UR, Anal AK (2022) Enhancing bioaccessibility and bioavailability of carotenoids using emulsion-based delivery systems. Colloids Surf, B 209:112211. https://doi.org/10.1016/j.colsurfb.2021.112211

Borowitzka MA, Vonshak A (2017) Scaling up microalgal cultures to commercial scale. Eur J Phycol 52:407–418. https://doi.org/10.1080/09670262.2017.1365177

Breijyeh Z, Karaman R (2020) Comprehensive review on Alzheimer's disease: causes and treatment. Molecules 25:5789. https://doi.org/10.3390/molecules25245789

Cabanillas AH, Tena Pérez V, Maderuelo Corral S et al (2018) Cybastacines A and B: Antibiotic Sesterterpenes from a *Nostoc* sp. Cyanobacterium. J Nat Prod 81:410–413. https://doi.org/10.1021/acs.jnatprod.7b00638

Cano M, Holland SC, Artier J et al (2018) Glycogen synthesis and metabolite overflow contribute to energy balancing in cyanobacteria. Cell Rep 23:667–672. https://doi.org/10.1016/j.celrep.2018.03.083

Carfagna S, Bottone C, Cataletto PR et al (2016) Impact of sulfur starvation in autotrophic and heterotrophic cultures of the extremophilic microalga *Galdieria phlegrea* (Cyanidiophyceae). Plant Cell Physiol 57:1890–1898. https://doi.org/10.1093/pcp/pcw112

Cervantes-Llanos M, Lagumersindez-Denis N, Marín-Prida J et al (2018) Beneficial effects of oral administration of C-Phycocyanin and Phycocyanobilin in rodent models of experimental autoimmune encephalomyelitis. Life Sci 194:130–138. https://doi.org/10.1016/j.lfs.2017.12.032

Chanquia SN, Vernet G, Kara S (2022) Photobioreactors for cultivation and synthesis: specifications, challenges, and perspectives. Eng Life Sci 22:712–724. https://doi.org/10.1002/elsc.202100070

Cheng J, Xia A, Liu Y et al (2012) Combination of dark- and photo-fermentation to improve hydrogen production from Arthrospira platensis wet biomass with ammonium removal by zeolite. I J Hydrogen Energy 37:13330–13337. https://doi.org/10.1016/j.ijhydene.2012.06.071

Chisti Y (2007) Biodiesel from microalgae. Biotechnol Adv 25:294–306. https://doi.org/10.1016/j.biotechadv.2007.02.001

da Silva JC, Lombardi AT (2020) Chlorophylls in microalgae: occurrence, distribution, and biosynthesis. In: Jacob-Lopes E, Queiroz MI, Zepka LQ (eds) Pigments from microalgae handbook. Springer International Publishing, Cham, pp 1–18

de Mooij T, de Vries G, Latsos C et al (2016) Impact of light color on photobioreactor productivity. Algal Res 15:32–42. https://doi.org/10.1016/j.algal.2016.01.015

Dembitsky VM, Řezanka T (2005) Metabolites produced by nitrogen-fixingNostoc species. Folia Microbiol 50:363–391. https://doi.org/10.1007/BF02931419

Demoulin CF, Lara YJ, Cornet L et al (2019) Cyanobacteria evolution: insight from the fossil record. Free Radical Biol Med 140:206–223. https://doi.org/10.1016/j.freeradbiomed.2019.05.007

Ducat DC, Avelar-Rivas JA, Way JC, Silver PA (2012) Rerouting carbon flux to enhance photosynthetic productivity. Appl Environ Microbiol 78:2660–2668. https://doi.org/10.1128/AEM.07901-11

Egner PA, Stansbury KH, Snyder EP et al (2000) Identification and characterization of chlorin e $_4$ ethyl ester in sera of individuals participating in the chlorophyllin chemoprevention trial. Chem Res Toxicol 13:900–906. https://doi.org/10.1021/tx000069k

ElFar OA, Billa N, Lim HR et al (2022) Advances in delivery methods of *Arthrospira platensis* (spirulina) for enhanced therapeutic outcomes. Bioengineered 13:14681–14718. https://doi.org/10.1080/21655979.2022.2100863

Erickson E, Wakao S, Niyogi KK (2015) Light stress and photoprotection in *Chlamydomonas reinhardtii*. Plant J 82:449–465. https://doi.org/10.1111/tpj.12825

Fae Neto WA, Tomkins JL, Kennington WJ (2023) Improved biomass production of a microalga through adaptative laboratory evolution to a high light environment. J Appl Phycol 35:1009–1021. https://doi.org/10.1007/s10811-023-02944-x

Federal Research Centre of Nutrition, Biotechnology and Food Safety, Moscow, Mazo VK, Birulina NA, et al (2022) Arthrospira platensis: antioxidant, hypoglycemic and hypolipidemic effects in vitro and in vivo (brief review). Probl Nutr 91:19–25. https://doi.org/10.33029/0042-8833-2022-91-4-19-25

Fell DA (1992) Metabolic control analysis: a survey of its theoretical and experimental development. Biochemical Journal 286:313–330. https://doi.org/10.1042/bj2860313

Gale GAR, Schiavon Osorio AA, Mills LA et al (2019) Emerging species and genome editing tools: future prospects in cyanobacterial synthetic biology. Microorganisms 7:409. https://doi.org/10.3390/microorganisms7100409

Gao F, Li C, Yang Z-H et al (2016) Removal of nutrients, organic matter, and metal from domestic secondary effluent through microalgae cultivation in a membrane photobioreactor: microalgae based wastewater treatment in membrane photobioreactor. J Chem Technol Biotechnol 91:2713–2719. https://doi.org/10.1002/jctb.4879

Garcia-Pichel F, Zehr JP, Bhattacharya D, Pakrasi HB (2020) What's in a name? The case of cyanobacteria. J Phycol 56:1–5. https://doi.org/10.1111/jpy.12934

Giglio S, Chou WKW, Ikeda H et al (2011) Biosynthesis of 2-methylisoborneol in cyanobacteria. Environ Sci Technol 45:992–998. https://doi.org/10.1021/es102992p

Gong M, Bassi A (2016) Carotenoids from microalgae: a review of recent developments. Biotechnol Adv 34:1396–1412. https://doi.org/10.1016/j.biotechadv.2016.10.005

Greenwold MJ, Cunningham BR, Lachenmyer EM et al (2019) Diversification of light capture ability was accompanied by the evolution of phycobiliproteins in cryptophyte algae. Proc R Soc B 286:20190655. https://doi.org/10.1098/rspb.2019.0655

Gründel M, Scheunemann R, Lockau W, Zilliges Y (2012) Impaired glycogen synthesis causes metabolic overflow reactions and affects stress responses in the cyanobacterium Synechocystis sp. PCC 6803. Microbiology 158:3032–3043. https://doi.org/10.1099/mic.0.062950-0

Guarnieri MT, Pienkos PT (2015) Algal omics: unlocking bioproduct diversity in algae cell factories. Photosynth Res 123:255–263. https://doi.org/10.1007/s11120-014-9989-4

Gutiérrez M, Tidgewell K, Capson TL et al (2010) Malyngolide dimer, a bioactive symmetric cyclodepside from the panamanian marine cyanobacterium *Lyngbya majuscula*. J Nat Prod 73:709–711. https://doi.org/10.1021/np9005184

Hayes M, Ferruzzi MG (2020) Update on the bioavailability and chemopreventative mechanisms of dietary chlorophyll derivatives. Nutr Res 81:19–37. https://doi.org/10.1016/j.nutres.2020.06.010

Hensley AP, McAlinden A (2021) The role of microRNAs in bone development. Bone 143:115760. https://doi.org/10.1016/j.bone.2020.115760

Hirokawa Y, Kubo T, Soma Y et al (2020) Enhancement of acetyl-CoA flux for photosynthetic chemical production by pyruvate dehydrogenase complex overexpression in Synechococcus elongatus PCC 7942. Metab Eng 57:23–30. https://doi.org/10.1016/j.ymben.2019.07.012

Huo S, Wang Z, Zhu S et al (2018) Biomass Accumulation of chlorella zofingiensis G1 cultures grown outdoors in photobioreactors. Front Energy Res 6:49. https://doi.org/10.3389/fenrg.2018.00049

Ikeuchi M (2020) Molecular biology of cyanobacteria and microalgae from atom to ecology. J Gen Appl Microbiol 66:51–52. https://doi.org/10.2323/jgam.2020.04.001

Jiménez C, Cossío BR, Labella D, Xavier Niell F (2003) The feasibility of industrial production of spirulina (Arthrospira) in Southern Spain. Aquaculture 217:179–190. https://doi.org/10.1016/S0044-8486(02)00118-7

Kanekiyo K, Hayashi K, Takenaka H et al (2007) Anti-herpes simplex virus target of an acidic polysaccharide, nostoflan, from the edible blue-green alga nostoc flagelliforme. Biol Pharm Bull 30:1573–1575. https://doi.org/10.1248/bpb.30.1573

Khalifa SAM, Shedid ES, Saied EM et al (2021) Cyanobacteria—from the oceans to the potential biotechnological and biomedical applications. Mar Drugs 19:241. https://doi.org/10.3390/md19050241

Kharwar S, Mishra AK (2023) Nitrogen and redox metabolism in cyanobacterium anabaena sp. PCC 7120 exposed to different sulfate regimes. Curr Microbiol 80:265. https://doi.org/10.1007/s00284-023-03374-1

Kim S, Moon M, Kwak M et al (2018) Statistical optimization of light intensity and CO2 concentration for lipid production derived from attached cultivation of green microalga Ettlia sp. Sci Rep 8:15390. https://doi.org/10.1038/s41598-018-33793-1

Kolackova M, Janova A, Dobesova M et al (2023) Role of secondary metabolites in distressed microalgae. Environ Res 224:115392. https://doi.org/10.1016/j.envres.2023.115392

Krzemińska I, Pawlik-Skowrońska B, Trzcińska M, Tys J (2014) Influence of photoperiods on the growth rate and biomass productivity of green microalgae. Bioprocess Biosyst Eng 37:735–741. https://doi.org/10.1007/s00449-013-1044-x

Kühl M, Chen M, Ralph PJ et al (2005) A niche for cyanobacteria containing chlorophyll d. Nature 433:820–820. https://doi.org/10.1038/433820a

Kumar Saha S, Uma L, Subramanian G (2003) Nitrogen stress induced changes in the marine cyanobacterium *Oscillatoria willei* BDU 130511. FEMS Microbiol Ecol 45:263–272. https://doi.org/10.1016/S0168-6496(03)00162-4

Lindberg P, Park S, Melis A (2010) Engineering a platform for photosynthetic isoprene production in cyanobacteria, using Synechocystis as the model organism. Metab Eng 12:70–79. https://doi.org/10.1016/j.ymben.2009.10.001

Liu T, Chen Z, Xiao Y et al (2022) Biochemical and morphological changes triggered by nitrogen stress in the oleaginous microalga chlorella vulgaris. Microorganisms 10:566. https://doi.org/10.3390/microorganisms10030566

Lopes G, Clarinha D, Vasconcelos V (2020) Carotenoids from cyanobacteria: a biotechnological approach for the topical treatment of psoriasis. Microorganisms 8:302. https://doi.org/10.3390/microorganisms8020302

Loya S, Reshef V, Mizrachi E et al (1998) The inhibition of the reverse transcriptase of HIV-1 by the natural sulfoglycolipids from cyanobacteria: contribution of different moieties to their high potency. J Nat Prod 61:891–895. https://doi.org/10.1021/np970585j

Madhyastha HK, Vatsala TM (2007) Pigment production in Spirulina fussiformis in different photophysical conditions. Biomol Eng 24:301–305. https://doi.org/10.1016/j.bioeng.2007.04.001

Madhyastha H, Vatsala TM (2010) Cysteine-rich cyanopeptide beta2 from Spirulina fusiformis exhibits plasmid DNA pBR322 scission prevention and cellular antioxidant activity. Indian J Exp Biol 48:486–493

Madhyastha HK, Radha KS, Sugiki M et al (2006) Purification of c-phycocyanin from spirulina fusiformis and its effect on the induction of urokinase-type plasminogen activator from

calf pulmonary endothelial cells. Phytomedicine 13:564–569. https://doi.org/10.1016/j.phymed.2005.09.006
Madhyastha HK, Sivashankari S, Vatsala TM (2009) C-phycocyanin from Spirulina fussiformis exposed to blue light demonstrates higher efficacy of in vitro antioxidant activity. Biochem Eng J 43:221–224. https://doi.org/10.1016/j.bej.2008.11.001
Madhyastha H, Madhyastha R, Thakur A et al (2020) C-Phycocyanin primed silver nano conjugates: studies on red blood cell stress resilience mechanism. Colloids Surf, B 194:111211. https://doi.org/10.1016/j.colsurfb.2020.111211
Madhyastha R, Madhyastha H, Nurrahmah QI et al (2021) MicroRNA 21 elicits a pro-inflammatory response in macrophages, with exosomes functioning as delivery vehicles. Inflammation 44:1274–1287. https://doi.org/10.1007/s10753-021-01415-0
Madhyastha H, Madhyastha R, Chakraborty E et al (2023) Fluro-protein C-phycocyanin docked silver nanocomposite accelerates cell migration through NFκB signaling pathway. IJMS 24:3184. https://doi.org/10.3390/ijms24043184
Maltsev Y, Maltseva K, Kulikovskiy M, Maltseva S (2021) Influence of light conditions on microalgae growth and content of lipids, carotenoids, and fatty acid composition. Biology 10:1060. https://doi.org/10.3390/biology10101060
Mazur-Marzec H, Cegłowska M, Konkel R, Pyrć K (2021) Antiviral cyanometabolites—a review. Biomolecules 11:474. https://doi.org/10.3390/biom11030474
Mehdizadeh Allaf M, Peerhossaini H (2022) Cyanobacteria: model microorganisms and beyond. Microorganisms 10:696. https://doi.org/10.3390/microorganisms10040696
Metsoviti MN, Papapolymerou G, Karapanagiotidis IT, Katsoulas N (2019) Effect of light intensity and quality on growth rate and composition of chlorella vulgaris. Plants 9:31. https://doi.org/10.3390/plants9010031
Mo S, Krunic A, Chlipala G, Orjala J (2009) Antimicrobial ambiguine isonitriles from the cyanobacterium *Fischerella ambigua.* J Nat Prod 72:894–899. https://doi.org/10.1021/np800751j
Moheimani NR, Borowitzka MA (2006) The long-term culture of the coccolithophore Pleurochrysis carterae (Haptophyta) in outdoor raceway ponds. J Appl Phycol 18:703–712. https://doi.org/10.1007/s10811-006-9075-1
Morone J, Lopes G, Preto M et al (2020) Exploitation of filamentous and picoplanktonic cyanobacteria for cosmetic applications: potential to improve skin structure and preserve dermal matrix components. Mar Drugs 18:486. https://doi.org/10.3390/md18090486
Murakami N, Morimoto T, Imamura H et al (1991) Studies on glycolipids. III. Glyceroglycolipids from an axenically cultured cyanobacterium. Phormidium Tenue. Chem Pharm Bull 39:2277–2281. https://doi.org/10.1248/cpb.39.2277
Nishiguchi H, Hiasa N, Uebayashi K et al (2019) Transomics data-driven, ensemble kinetic modeling for system-level understanding and engineering of the cyanobacteria central metabolism. Metab Eng 52:273–283. https://doi.org/10.1016/j.ymben.2019.01.004
Nzayisenga JC, Farge X, Groll SL, Sellstedt A (2020) Effects of light intensity on growth and lipid production in microalgae grown in wastewater. Biotechnol Biofuels 13:4. https://doi.org/10.1186/s13068-019-1646-x
Oliver NJ, Rabinovitch-Deere CA, Carroll AL et al (2016) Cyanobacterial metabolic engineering for biofuel and chemical production. Curr Opin Chem Biol 35:43–50. https://doi.org/10.1016/j.cbpa.2016.08.023
Pagliara P, De Benedetto GE, Francavilla M et al (2021) Bioactive potential of two marine picocyanobacteria belonging to cyanobium and synechococcus genera. Microorganisms 9:2048. https://doi.org/10.3390/microorganisms9102048
Pandey A, Mohan SV, Chang J-S, et al (2019) Biohydrogen, 2nd edn. Elsevier, Amsterdam, Netherlands , Cambridge, MA, United States
Płaczek M, Patyna A, Witczak S (2017) Technical evaluation of photobioreactors for microalgae cultivation. E3S Web Conf 19:02032. https://doi.org/10.1051/e3sconf/20171902032

Proteau PJ, Gerwick WH, Garcia-Pichel F, Castenholz R (1993) The structure of scytonemin, an ultraviolet sunscreen pigment from the sheaths of cyanobacteria. Experientia 49:825–829. https://doi.org/10.1007/BF01923559

Sahu N, Mishra S, Kesheri M et al (2023) Identification of cyanobacteria-based natural inhibitors against SARS-CoV-2 druggable target ACE2 using molecular docking study, ADME and toxicity analysis. Ind J Clin Biochem 38:361–373. https://doi.org/10.1007/s12291-022-01056-6

Sakarika M, Kornaros M (2017) Kinetics of growth and lipids accumulation in Chlorella vulgaris during batch heterotrophic cultivation: effect of different nutrient limitation strategies. Biores Technol 243:356–365. https://doi.org/10.1016/j.biortech.2017.06.110

Salleh NF, Wang J, Kundukad B et al (2023) Cyclopropane-containing specialized metabolites from the marine cyanobacterium cf. Lyngbya Sp. Molecules 28:3965. https://doi.org/10.3390/molecules28093965

Schagerl M, Donabaum K (2003) Patterns of major photosynthetic pigments in freshwater algae. 1. Cyanoprokaryota, rhodophyta and cryptophyta. Ann Limnol Int J Lim 39:35–47. https://doi.org/10.1051/limn/2003003

Senge M, Ryan A, Letchford K et al (2014) Chlorophylls, symmetry, chirality, and photosynthesis. Symmetry 6:781–843. https://doi.org/10.3390/sym6030781

Sengupta S, Jaiswal D, Sengupta A et al (2020) Metabolic engineering of a fast-growing cyanobacterium Synechococcus elongatus PCC 11801 for photoautotrophic production of succinic acid. Biotechnol Biofuels 13:89. https://doi.org/10.1186/s13068-020-01727-7

Sforza E, Simionato D, Giacometti GM et al (2012) Adjusted light and dark cycles can optimize photosynthetic efficiency in algae growing in photobioreactors. PLoS ONE 7:e38975. https://doi.org/10.1371/journal.pone.0038975

Shabestary K, Anfelt J, Ljungqvist E et al (2018) Targeted repression of essential genes to arrest growth and increase carbon partitioning and biofuel titers in cyanobacteria. ACS Synth Biol 7:1669–1675. https://doi.org/10.1021/acssynbio.8b00056

Shahin YH, Elwakil BH, Ghareeb DA, Olama ZA (2022) Micrococcus lylae MW407006 pigment: production, optimization, nano-pigment synthesis, and biological activities. Biology 11:1171. https://doi.org/10.3390/biology11081171

Shi T-Q, Wang L-R, Zhang Z-X et al (2020) Stresses as first-line tools for enhancing lipid and carotenoid production in microalgae. Front Bioeng Biotechnol 8:610. https://doi.org/10.3389/fbioe.2020.00610

Singh U, Gandhi HA, Nikita, et al (2023) Cyanometabolites: molecules with immense antiviral potential. Arch Microbiol 205:164. https://doi.org/10.1007/s00203-023-03514-y

Skinnider MA, Dejong CA, Rees PN, et al (2015) Genomes to natural products PRediction Informatics for Secondary Metabolomes (PRISM). Nucleic Acids Res gkv1012. https://doi.org/10.1093/nar/gkv1012

Soares AR, Engene N, Gunasekera SP et al (2015) Carriebowlinol, an antimicrobial tetrahydroquinolinol from an assemblage of marine cyanobacteria containing a novel taxon. J Nat Prod 78:534–538. https://doi.org/10.1021/np500598x

Song Y, Li R, Song W, et al (2023) Microcystis spp. and phosphorus in aquatic environments: a comprehensive review on their physiological and ecological interactions. Science of The Total Environment 878:163136. https://doi.org/10.1016/j.scitotenv.2023.163136

Soo RM, Skennerton CT, Sekiguchi Y et al (2014) An Expanded genomic representation of the phylum cyanobacteria. Genome Biol Evol 6:1031–1045. https://doi.org/10.1093/gbe/evu073

Sreenikethanam A, Raj S, J RB, et al (2022) Genetic engineering of microalgae for secondary metabolite production: recent developments, challenges, and future prospects. Front Bioeng Biotechnol 10:836056. https://doi.org/10.3389/fbioe.2022.836056

Srivastava R, Prajapati R, Kanda T et al (2022a) Phycochemistry and bioactivity of cyanobacterial secondary metabolites. Mol Biol Rep 49:11149–11167. https://doi.org/10.1007/s11033-022-07911-2

Srivastava A, Thapa S, Chakdar H, et al (2022) Cyanobacterial myxoxanthophylls: biotechnological interventions and biological implications. Critical Reviews in Biotechnology 1–15. https://doi.org/10.1080/07388551.2022.2117682

Stadnichuk IN, Krasilnikov PM, Zlenko DV (2015) Cyanobacterial phycobilisomes and phycobiliproteins. Microbiology 84:101–111. https://doi.org/10.1134/S0026261715020150

Stal LJ (2017) Gregarious cyanobacteria: gregarious cyanobacteria. Environ Microbiol 19:2105–2109. https://doi.org/10.1111/1462-2920.13739

Sukačová K, Lošák P, Brummer V et al (2021) Perspective design of algae photobioreactor for greenhouses—a comparative study. Energies 14:1338. https://doi.org/10.3390/en14051338

Tenaillon O, Skurnik D, Picard B, Denamur E (2010) The population genetics of commensal Escherichia coli. Nat Rev Microbiol 8:207–217. https://doi.org/10.1038/nrmicro2298

Tietz JI, Schwalen CJ, Patel PS et al (2017) A new genome-mining tool redefines the lasso peptide biosynthetic landscape. Nat Chem Biol 13:470–478. https://doi.org/10.1038/nchembio.2319

Upadhyaya KR, Radha KS, Madhyastha HK (2007) Cell cycle regulation and induction of apoptosis by β-carotene in U937 and HL-60 Leukemia cells. BMB Rep 40:1009–1015. https://doi.org/10.5483/BMBRep.2007.40.6.1009

Vadalà M, Palmieri B (2015) Dalle alghe ai "functional foods." La Clinica Terapeutica e281–e295. https://doi.org/10.7417/T.2015.1875

Verma S, Thapa S, Siddiqui N, Chakdar H (2022) Cyanobacterial secondary metabolites towards improved commercial significance through multiomics approaches. World J Microbiol Biotechnol 38:100. https://doi.org/10.1007/s11274-022-03285-6

Wang B, Lan CQ, Horsman M (2012) Closed photobioreactors for production of microalgal biomasses. Biotechnol Adv 30:904–912. https://doi.org/10.1016/j.biotechadv.2012.01.019

Watzer B, Forchhammer K (2018) Cyanophycin synthesis optimizes nitrogen utilization in the unicellular cyanobacterium synechocystis sp. Strain PCC 6803. Appl Environ Microbiol 84:e01298-e1318. https://doi.org/10.1128/AEM.01298-18

Wendt KE, Ungerer J, Cobb RE et al (2016) CRISPR/Cas9 mediated targeted mutagenesis of the fast growing cyanobacterium Synechococcus elongatus UTEX 2973. Microb Cell Fact 15:115. https://doi.org/10.1186/s12934-016-0514-7

White RL, Ryan RA (2015) Long-term cultivation of algae in open-raceway ponds: lessons from the field. Ind Biotechnol 11:213–220. https://doi.org/10.1089/ind.2015.0006

Williams PJ le B, Laurens LML (2010) Microalgae as biodiesel and biomass feedstocks: review and analysis of the biochemistry, energetics and economics. Energy Environ Sci 3:554. https://doi.org/10.1039/b924978h

Wolk CP (1973) Physiology and cytological chemistry blue-green algae. Bacteriol Rev 37:32–101. https://doi.org/10.1128/br.37.1.32-101.1973

Xu N, Zhang X, Fan X et al (2001) Effects of nitrogen source and concentration on growth rate and fatty acid composition of Ellipsoidion sp. (Eustigmatophyta). J Appl Phycol 13:463–469. https://doi.org/10.1023/A:1012537219198

Yaakob MA, Mohamed RMSR, Al-Gheethi A et al (2021) Influence of nitrogen and phosphorus on microalgal growth, biomass, lipid, and fatty acid production: an overview. Cells 10:393. https://doi.org/10.3390/cells10020393

Yamaguchi Y, Koketsu M (2016) Isolation and analysis of polysaccharide showing high hyaluronidase inhibitory activity in Nostochopsis lobatus MAC0804NAN. J Biosci Bioeng 121:345–348. https://doi.org/10.1016/j.jbiosc.2015.07.008

Yang L, Chen J, Qin S et al (2018a) Growth and lipid accumulation by different nutrients in the microalga Chlamydomonas reinhardtii. Biotechnol Biofuels 11:40. https://doi.org/10.1186/s13068-018-1041-z

Yang F, Xiang W, Li T, Long L (2018b) Transcriptome analysis for phosphorus starvation-induced lipid accumulation in Scenedesmus sp. Sci Rep 8:16420. https://doi.org/10.1038/s41598-018-34650-x

Yustinadiar N, Manurung R, Suantika G (2020) Enhanced biomass productivity of microalgae Nannochloropsis sp. in an airlift photobioreactor using low-frequency flashing light with blue LED. Bioresour Bioprocess 7:43. https://doi.org/10.1186/s40643-020-00331-9

Zahmatkesh S, Karimian M, Pourhanasa R et al (2023) Wastewater treatment with algal based membrane bioreactor for the future: Removing emerging containments. Chemosphere 335:139134. https://doi.org/10.1016/j.chemosphere.2023.139134

Zhu J, Rong J, Zong B (2013) Factors in mass cultivation of microalgae for biodiesel. Chin J Catal 34:80–100. https://doi.org/10.1016/S1872-2067(11)60497-X

Chapter 20
Novel and Conventional Uses of Fungi in Products and Services

H. Jayathunga, M. Gunasekara, M. Kavimalee, I. Jayamanna, H. K. S. de Zoysa, D. K. Hettiarachchi, and T. C. Bamunuarachchige

20.1 Introduction

Fungi are essential pillars in ecosystems, playing a pivotal role in carbon and nutrient cycling, making them crucial for addressing global challenges. Among non-timber forest resources certain wild mushrooms which can be cultivated, offer considerable value, providing a wide range of benefits, including nourishment and income for local communities worldwide (Toussaint et al. 2016; Niego et al. 2023; Roth et al. 2023). The renewable potential of fungi as a means to enhance resource efficiency, particularly as a sustainable alternative to fossil resources, is a noteworthy aspect. This remarkable spectrum of nutrients and compounds contained within fungi, including nutraceuticals, cosmeceuticals, amino acids, proteins, carbohydrates, vitamins, fats, and minerals, often remains concealed, escaping the awareness of the general population. Recognizing these invaluable attributes is crucial due to the substantial potential they hold for propelling sustainable global development. Simultaneously, focusing on novel fungal applications is of paramount significance in the context of sustainability. This wealth of hidden potential possesses the capacity to drive sustainable global development. Hence, it becomes imperative to cast a spotlight on their potential, unveiling their contributions, and inspiring further exploration in this critical domain (Toussaint et al. 2016; Niego et al. 2023; Chambergo and Valencia 2016; Wijayawardene et al. 2023; Ponnaiyan et al. 2023; Gaikwad et al. 2019; Riza et al. 2020; Adeel et al. 2019; Aman Mohammadi et al. 2022; Pascoalino et al. 2023).

H. Jayathunga · M. Gunasekara · M. Kavimalee · I. Jayamanna · H. K. S. de Zoysa · T. C. Bamunuarachchige (✉)
Department of Bioprocess Technology, Faculty of Technology, Rajarata University of Sri Lanka, Mihintale, Sri Lanka
e-mail: tcbamunu@tec.rjt.ac.lk

D. K. Hettiarachchi
Department of Biological Sciences, Faculty of Applied Sciences, Rajarata University of Sri Lanka, Mihintale, Sri Lanka

K. Bala et al. (eds.), *Harnessing Microbial Potential for Multifarious Applications*, Energy, Environment, and Sustainability, https://doi.org/10.1007/978-981-97-1152-9_20

Applied research has unveiled the diverse environmental contributions of fungi, from bioremediation and pollution mitigation to serving as the foundation for remedies against various human ailments. Understanding the global-scale services fungi provide within ecosystems is crucial. As we move towards a biobased circular economy, highlighting the immense value that fungi can offer becomes essential. In the journey towards a biobased circular economy, it becomes imperative to emphasize the tremendous value that fungi bring to the table. Recent advancements in biotechnology have catalyzed a revolution in fungal research, enabling us to venture deeper into their potential applications. These cutting-edge breakthroughs not only enhance our understanding of the significant contributions fungi make within ecosystems but also empower us to make informed decisions, particularly in the context of shaping conservation policies (Niego et al. 2023; Chambergo and Valencia 2016; Aguiar et al. 2016; Sarsaiya et al. 2021; Dey and Nagababu 2022; Janiszewska-Turak et al. 2017; Muro Urista et al. 2016).

This chapter explores the multifaceted world of fungi and their applications in products and services. Hence, this chapter aim is to emphasize the potential to revolutionize our perspective on sustainability and resource utilization, setting the stage for transformative changes that can ignite further scholarly exploration in this vital field. (Fig. 20.1).

20.2 Fungi as Biofactories

Secondary metabolites (SMs) represent a fascinating category of bioactive molecules produced by various organisms, including bacteria, plants, and fungi. Within the fungal kingdom, SMs are notably abundant, especially among fungi inhabiting soil environments, where they engage in intricate multicellular communities, vying for essential resources like minerals, nutrients, and water. Unlike primary metabolites, which are essential for fungal growth, development, and reproduction in controlled laboratory conditions, SMs are not indispensable for these fundamental processes. Instead, primarily function of SMs is as a defense mechanism in response to environmental pressures and during intricate interactions with other soil-dwelling populations of eukaryotic organisms (Chambergo and Valencia 2016; Sarsaiya et al. 2021; Narsing Rao et al. 2017; Ramesh et al. 2019; Sajid and Akbar 2018).

It's worth emphasizing that fungi, in addition to producing beneficial SMs, are also capable of synthesizing numerous mycotoxins. These mycotoxins, such as aflatoxin, trichothecene, fumonisin, and zearalone, are compounds that can have harmful effects and pose potential risks to both human health and the ecosystem. Understanding the delicate balance between the production of beneficial SMs and potentially harmful mycotoxins is of paramount importance in comprehending the intricate role of fungi in the environment (Chambergo and Valencia 2016; Wijayawardene et al. 2023; Sarsaiya et al. 2021).

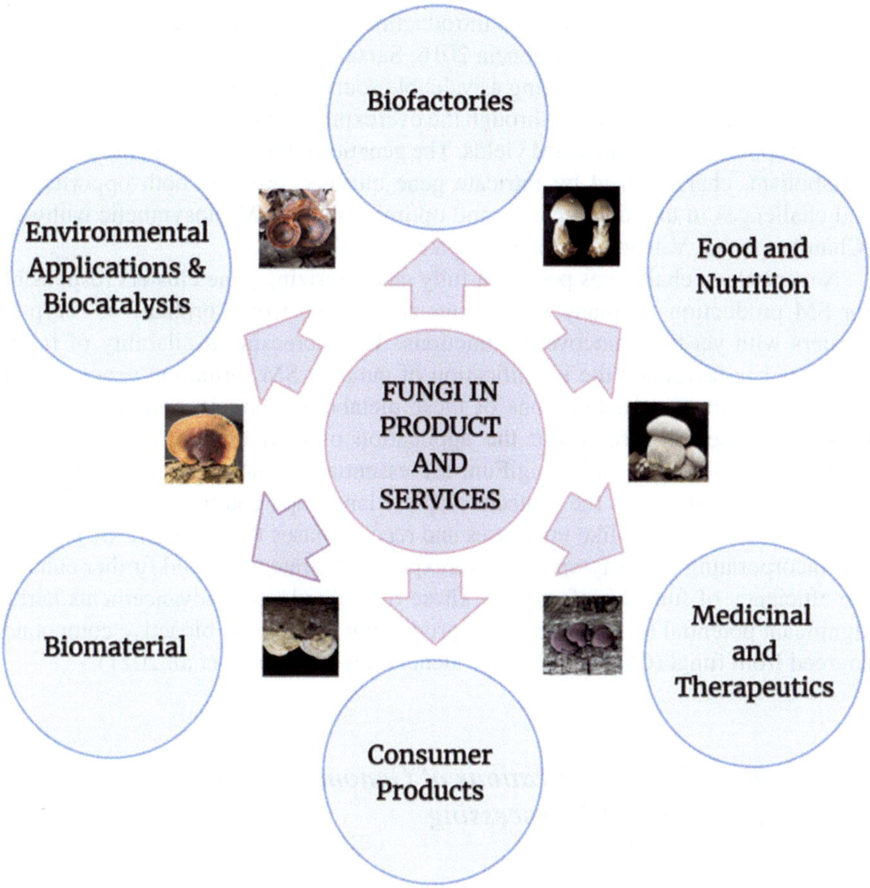

Fig. 20.1 Fungi as novel and conventional uses in products and services, Figure Created with BioRender.com

20.2.1 Advancements in Biotechnological Tools for Optimizing Fungal Biofactories

In recent years, significant advancements have been achieved in harnessing biotechnological tools to optimize fungal biofactories, particularly in the production of SMs (Sarsaiya et al. 2021; Venkatachalam et al. 2018; Caro et al. 2017; Vivek-Ananth et al. 2021; Keller 2018). This progress is particularly prominent through targeted genome editing strategies. The development of precise methods for manipulating specific loci within a genome, such as the utilization of protein-directed nucleases (e.g., zinc-finger and transcription activator-like effector nucleases) and nucleic acid-guided nucleases (such as CRISPR-Cas9), has revolutionized this field. These sophisticated techniques enable the controlled creation of DNA double-strand breaks, facilitating homologous recombination and the introduction of DNA fragments to enhance SM production (Chambergo and Valencia 2016; Sarsaiya et al. 2021; Keller 2018).

Fungi endophytes are emerging as valuable sources of bioactive compounds, and precise regulation of pathways through the overexpression of regulatory genes offers a potent approach to enhance SM yields. The genetic architecture of fungal secondary metabolism, characterized by intricate gene clusters, presents both opportunities and challenges in the identification and optimization of SM biosynthetic pathways (Chambergo and Valencia 2016; Sarsaiya et al. 2021).

Nevertheless, challenges persist in fully characterizing gene clusters responsible for SM production, as many metabolites are derived from "orphan" or "cryptic" clusters with yet-to-be-uncovered functions. The increasing availability of fungal genomes has facilitated the identification of putative SM formation genes, yet the structural attributes and functions of these metabolites largely remain enigmatic. Continued research efforts and the application of computational tools, such as SMURF, antiSMASH, and FungiFun, are essential to uncover the hidden potential of these clusters. As the biotechnological landscape continues to evolve, integrating innovative tools like integrases and recombinases for precise DNA insertion and incorporating pathway-specific overexpression strategies could further enhance the efficiency of fungal biofactories. These biotechnological advancements harbor significant potential for optimizing the production of valuable bioactive compounds sourced from fungi (Chambergo and Valencia 2016; Sarsaiya et al. 2021).

20.2.2 Promising Applications of Genome Editing for Fungal Bioprocessing

The rapid advancement of genome editing techniques is paving in a new era of possibilities across various product sectors, particularly within the domain of fungal bioprocessing. These cutting-edge strategies harbor immense potential to revolutionize the production of valuable SMs and significantly enhance the efficiency of fungal biofactories (Roth et al. 2023; Sarsaiya et al. 2021). Genome editing involves a precise modification of specific genetic locations within an organism's genome. Successful implementation necessitates ensuring genetic stability, accommodating diverse DNA configurations, and being applicable across a broad spectrum of organisms. Traditional methods, like homologous recombination, have faced challenges due to their limited efficiency, especially in higher eukaryotes. Nevertheless, recent advancements, including the utilization of directed nucleases such as zinc-finger nucleases and transcription activator-like effector nucleases (TALENs), have notably enhanced the accuracy and effectiveness of genome editing (Chambergo and Valencia 2016; Sarsaiya et al. 2021).

Another significant milestone in genome editing is the RNA-directed nucleases category, specifically the CRISPR-Cas9 system. Derived from a primitive bacterial immune mechanism, this streamlined genome editing approach has demonstrated remarkable specificity (Sarsaiya et al. 2021; Naeem et al. 2022; Ball et al. 2020). Ongoing optimization, the refinement of guide RNA (gRNA) design, and specific Cas9 modifications, have further elevated its efficiency. Advanced computer programs, including Cas-OFFinder, CHOP-CHOP, CCTop, sgRNAcas9, and COSMID, play a pivotal role in supporting gRNA strategy and predicting potential off-target effects, establishing the CRISPR-Cas9 system as an indispensable tool for genome editing. To fully exploit the potential of genome editing in fungal bioprocessing, additional tools like integrases and recombinases come into play. Recombinase-mediated cassette exchange (RMCE) emerges as a versatile genome editing method, facilitating the precise substitution of a genomic target with a compatible donor construct using specific recombinases. Furthermore, nucleases such as CRISPR-Cas9 or TALEN can be harnessed to introduce integrase recombination sites into any genome, allowing for controlled expression of heterologous genes and providing comprehensive control over the content, sequence, and copy number of inserted genes. These exceptional strides in genome editing techniques hold significant promise for optimizing fungal bioprocessing, with applications poised to expand in gene expression and metabolite production in the years ahead (Chambergo and Valencia 2016; Wijayawardene et al. 2023; Sarsaiya et al. 2021).

20.3 Fungi in Food and Nutrition

20.3.1 Fungi as Nutraceuticals and Functional Foods

Fungi have garnered significant scientific attention owing to their remarkable medicinal properties, which have paved the way for the production of nutraceuticals and functional food products. These products, sourced from edible fungi, not only offer safe consumption options but also serve as valuable supplements alongside conventional pharmacological treatments. Notably, certain fungal genera, including *Agaricus*, *Hericium*, *Lentinula*, and *Ophiocordyceps*, have undergone extensive exploration for their rich reservoirs of bioactive compounds, which encompass a diverse spectrum of health benefits (Niego et al. 2023; Chambergo and Valencia 2016; Narsing Rao et al. 2017; Lagashetti et al. 2019; Ngamwonglumlert et al. 2017). The bioactive constituents responsible for these observed effects are multifaceted and include polysaccharides, proteins, glycoproteins, phenolic compounds, ergosterol, terpenoids, and unsaturated fatty acids. Polysaccharides, for instance, have exhibited immunomodulatory properties, while proteins and glycoproteins contribute to antioxidant and anti-inflammatory activities. Phenolic compounds, ergosterol, and terpenoids are recognized for their anticancer, antimicrobial, and cardiovascular health-promoting properties. Unsaturated fatty acids, on the other hand, are pivotal for maintaining overall cardiovascular health. Additionally, The increasing recognition of the health-promoting properties of fungi has spurred their incorporation into the thriving nutraceutical market, where they play a pivotal role in addressing various health concerns and promoting overall well-being (Niego et al. 2023; Chambergo and Valencia 2016; Wijayawardene et al. 2023; Narsing Rao et al. 2017; Lagashetti et al. 2019).

20.3.2 Mushroom Species with High Nutraceutical Value

Several mushroom species stand out for their substantial nutraceutical value, with significant market impact. *Agaricus bisporus*, renowned for its nutraceutical potential, has achieved impressive global valuation, while *Hericium* species, particularly *H. erinaceus*, hold significant medicinal properties and are majorly exported by countries like China, Italy, and Poland. *Lentinula edodes* (Shiitake) has well-documented nutraceutical properties, and its market value is projected to grow at a substantial compound annual growth rate (CAGR). The red yeast rice, derived from *Monascus purpureus*, has gained recognition as a cholesterol-lowering supplement, further contributing to the nutraceutical market. *Ophiocordyceps sinensis*, a highly esteemed and expensive fungal species, has seen prolific use in creating diverse nutraceutical products, notably in China (Niego et al. 2023; Delvendahl et al. 2022).

20.3.3 Role of Fungi in Food Industry and Security

The influence of fungi extends beyond the realm of the nutraceutical market, encompassing a broader impact on the food industry and global food security. Fungi's capacity to generate an array of bioactive metabolites, pigments, antioxidants, enzymes, and various other compounds finds extensive application within the sphere of food processing. Fungi are harnessed in diverse forms, ranging from processed food additives to innovative meat substitutes, and as such, they play a pivotal role in ensuring food security at a global scale (Chambergo and Valencia 2016; Wijayawardene et al. 2023; Dey and Nagababu 2022; Janiszewska-Turak et al. 2017; Li et al. 2017). Nevertheless, it is imperative to address several challenges associated with the involvement of fungi in the food industry. These challenges encompass issues related to food spoilage, foodborne illnesses, and their potential impact on crop production and livestock. Food spoilage caused by fungal contamination can lead to substantial economic losses and contribute to food scarcity. Additionally, the presence of pathogenic fungi can pose health risks to consumers, necessitating stringent quality control measures in food production (Ball et al. 2020; Arnold 2007; Jain et al. 2019; Xu 2017).

Furthermore, the utilization of fungi in alternative protein sources, such as mycoprotein-based meat substitutes, highlights the importance of fungi. Advanced technologies, including omics approaches, are instrumental in detecting fungal contaminants, enhancing the resilience of food crops against fungal pathogens, and developing sustainable and safe alternative protein sources. These technological advancements are critical in ensuring both the safety and security of food production and supply chains in the face of evolving challenges and growing global demand for food resources (Chambergo and Valencia 2016; Wijayawardene et al. 2023; Ball et al. 2020).

20.3.4 Edible Macrofungi and Mushroom Market

Macrofungi, which encompass Basidiomycota and select Ascomycetous members, represent a significant component of human diets, particularly in the form of mushrooms. The cultivated species like *Agaricus bisporus*, *Lentinula edodes*, and *Flammulina velutipes* are highly esteemed for their appealing flavors, nutritional richness, and the various health-promoting attributes they offer. The mushroom industry has undergone substantial expansion to meet the surging demand for these edible fungi. In this context, omics technologies have emerged as indispensable tools, contributing significantly to the cultivation, breeding, and production of transgenic edible mushrooms tailored to exhibit specific desirable characteristics (Niego et al. 2023; Ball et al. 2020; Delvendahl et al. 2022). Omics techniques encompass a range of methodologies, including proteomics and metabolome sequencing, which are instrumental in unraveling the intricate biochemical processes within these macrofungi. Proteomics studies enable a comprehensive understanding of amino acid pathways and metabolic processes, shedding light on the intricate biochemical machinery that governs mushroom growth and development. Metabolome sequencing, on the other hand, aids in the identification of bioactive compounds with potential pharmaceutical applications. By delving into the mushroom's metabolic profile, these technologies enhance our capacity to harness the full potential of edible macrofungi in the food industry, potentially yielding novel functional food products and pharmaceutical agents (Chambergo and Valencia 2016; Wijayawardene et al. 2023; Ball et al. 2020; Jain et al. 2019).

20.3.5 Fungal Pigments: Food Colorants and Health Effects

The physical appearance of food is of paramount importance in consumer selection, with food color also serving as a protective measure to identify spoiled food. The shift towards natural pigments over synthetic food colorants, driven by health concerns, is evident in modern food production. Notable fungal pigments like beta-carotene from *Blakeslea trispora*, red colorant (Arpink Red) from *Penicillium oxalicum*, and riboflavin-yellow from *Ashbya gossypii* are marketed with widespread applications due to their excellent colorant characteristics and retention capabilities.

These pigments, along with derivatives of naphthoquinones extracted from *Trypethelium eluteriae* and *Fusarium solani*, have demonstrated antioxidant, cytotoxic, and antibacterial activities, further validating their value as food colorants with health benefits (Aguiar et al. 2016; Dey and Nagababu 2022; Janiszewska-Turak et al. 2017; Toma et al. 2023; Umar et al. 2024).

Approval from regulatory authorities, funding for basic research, and customer acceptance are essential for the success of these food color extracts. Advanced techniques, such as omics sciences, have enabled better control over toxins biosynthesis by altering metabolic pathways, resulting in non-citrinin-producing *Monascus purpureus* mutants with high-yield of azaphilone pigment. The use of the CRISPR/Cas system to delete the citrinin biosynthetic gene cluster in *Monascus purpureus* presents an efficient solution to a longstanding problem in the food industry, emphasizing the importance of continuous development and the emergence of novel fungal biotechnology tools for the investigation of fungal food colorants with high-yield and low toxicity (Wijayawardene et al. 2023; Venkatachalam et al. 2018; Liang et al. 2018).

20.3.6 Fungi in Fermented Foods

Fungi have emerged as pivotal contributors to industry and commerce, particularly in the realm of food and beverage production. Their significance is unmistakable in the ancient art of fermenting foods and crafting alcoholic beverages, practices deeply rooted in cultures worldwide. These fermented products, meticulously preserved through the interplay of microorganisms including bacteria and fungi, not only retain their freshness but also amass nutritional value, enriched with proteins, vitamins, minerals, and other essential nutrients. Beyond preservation, the microbial involvement in fermentation, especially by fungi, extends to fostering probiotic benefits that bolster gastrointestinal health. This microbial dimension adds to the allure and demand for fermented products, a market segment that continues to flourish. However, fungi's impact transcends preservation and nutritional enhancement. A prime illustration is Saccharomyces cerevisiae, known as Baker's Yeast, which plays a transformative role in baked goods. This versatile fungus transmutes sugars and starches into alcohol and carbon dioxide during fermentation, a process that fundamentally shapes the texture and flavor of baked treats. Its influence is palpable in the bakery sector, with the industry valued at an impressive USD 497.50 billion in 2022. Likewise, in the realm of cheese production, an array of fungal species contributes to a rich spectrum of flavors, textures, and characteristics. These filamentous fungi and yeasts actively participate in crafting diverse cheese varieties, rendering them as esteemed culinary delights across the globe. This is underscored by the substantial global cheese market value of USD 231.65 billion in 2022 (Niego et al. 2023; Wijayawardene et al. 2023).

Fungi's role extends beyond culinary enhancement; it permeates the creation of alcoholic beverages that are integral to global cultural traditions. *Saccharomyces cerevisiae*, a prominent yeast species, is a linchpin in fermenting these libations, influencing both alcohol content and sensory experiences. Be it beer, wine, or a diverse assortment of spirits, fungi's symbiotic interaction with substrates has yielded a global beverage market worth billion. Yet, fungi's influence isn't confined to alcoholic realms alone; it extends to non-alcoholic fermented delights. From enriching the flavors of chocolate and coffee to contributing to the production of vinegar and probiotic-laden beverages like kombucha, fungi's imprint on the culinary sphere is deeply woven into both time-honored traditions and contemporary gastronomy (Niego et al. 2023; Wijayawardene et al. 2023).

20.4 Medicinal and Therapeutic Applications

20.4.1 Breakthroughs in Fungal-Derived Pharmaceuticals and Drug Discovery

For centuries, people have recognized the medicinal potential of fungi. Beyond their nutritional advantages, fungi's unique capacity to generate SMs has been a valuable resource for pioneering chemical innovations. This has paved the way for the creation of significant pharmaceuticals that have played a crucial role in the progress of human society. While natural products and SMs, in general, have yielded numerous drugs for addressing diverse diseases, fungal metabolites, in particular, have led to the formulation of several high-value drugs in various fields, notably as anti-infectives, and even in cardiovascular and immunomodulatory roles (Niego et al. 2023; Wijayawardene et al. 2023).

20.4.2 Investigating the Immunomodulatory and Anticancer Properties of Fungi

Despite progress in diagnostic and treatment techniques, the occurrence of invasive fungal infections in humans is rapidly increasing. Starting in 2020, disruptions in public health measures due to the COVID-19 pandemic have led to a surge in invasive fungal infections, with conditions like aspergillosis, mucormycosis, and candidaemia becoming frequent fungal infections carrying severe consequences. To address this emerging threat, the World Health Organization (WHO) introduced the fungal pathogens priority list (FPPL) in October 2022. This list comprises 19 fungi identified as posing the highest health risk, marking the first global initiative to prioritize fungal diseases based on research and development (R&D) needs and their implications for public health. The WHO FPPL has sorted these 19 fungal pathogens into three priority groups based on their public health implications and R&D requirements. The primary objective of the WHO FPPL is to prioritize research and policy measures aimed at enhancing global responses to fungal infections and addressing the challenge of antifungal resistance. Presently, clinical settings primarily rely on four categories of antifungal drugs (pyrimidines, azoles, polyenes, and echinocandins). These drugs have demonstrated efficacy in treating fungal infections, but they may also carry adverse effects on patients' recovery and overall outlook, as indicated by multiple studies (Niego et al. 2023).

20.4.3 Fungi as a Source of Novel Antibiotics and Antimicrobial Agents

Fungi have been instrumental in advancing pharmaceuticals, producing SMs that have led to the creation of a select number of marketable drugs vital to modern medicine, driving substantial economic growth. In the realm of antibacterials, fungi contribute significantly, such as penicillins derived from *Penicillium rubens*, serving as widely used antibiotics against a spectrum of bacterial infections. Cephalosporins, categorized into generations, show effectiveness against various bacterial diseases. Carbapenems, a versatile group of β-lactam antibiotics, target a wide range of bacterial infections. Pleuromutilin-derived compounds, like lefamulin, hold promise for treating community-acquired pneumonia. Antimycotics, like echinocandins from Ascomycota, are essential in combating invasive fungal infections. Enfumafungin inhibits β-(1,3)-D-glucan synthesis, evolving into the drug Ibrexafungerp. Griseofulvin contributes significantly to managing onychomycosis. The collective range of antimycotic drugs adds value to the market. Cardiovascular drugs encompass various formulations and market values. The discovery of Cyclosporin A, sourced from the fungus *Tolypocladium inflatum*, revolutionized transplantation medicine. Fingolimod, derived from myriocin, is effective in treating multiple sclerosis. Mycophenolic acid and derivatives serve as crucial immunosuppressants, while Mizoribine has applications in renal diseases. Immunosuppressive and immunomodulatory agents, including Cyclosporin A, play a significant role in transplantation medicine. Fingolimod targets multiple sclerosis. Mycophenolic acid and its derivatives are essential components of immunosuppression strategies. Mizoribine has value in treating renal diseases, and Emodepside, derived from a fungal endophyte, presents potential as a nematicidal drug for human pathogenic nematode infections (Niego et al. 2023; Chambergo and Valencia 2016; Wijayawardene et al. 2023).

Fungi like *Ophiocordyceps*/*Cordyceps* and *Ganoderma* play a fundamental role in Traditional Chinese Medicine. *Ophiocordyceps sinensis* commands high prices in Chinese markets, and Ganoderma-based products exhibit significant global turnover, with projected market growth. The shared essential genes between fungi and humans present limitations in developing safe antifungal drugs. Most antifungal medications function by disrupting ergosterol synthesis or compromising cell wall integrity. Azoles, the primary antifungal drugs, target the fungal cytochrome P450 pathway's lanosterol 14α-demethylase enzyme, encoded by the ERG11 gene. These drugs are divided into imidazoles and triazoles. The *Streptomyces* genus encompasses polyene-producing taxa, exhibiting broad-spectrum activity against various fungal species. Echinocandins inhibit (1,3)-D-glucan synthase, causing osmotic instability and eventual cell death. The emergence of antifungal drug resistance is a growing concern due to the limited options available to combat increasing opportunistic fungal infections. The rising incidence of fungal infections has spurred research into novel antifungal treatments. While current antifungal drugs predominantly target the cell wall and plasma membrane, researchers are exploring alternative targets. Recent studies focus on identifying drugs that can effectively target fungal virulence factors. Antifungal drug resistance has arisen due to various factors, including the overexpression of the ERG11 drug target gene, multidrug efflux pumps, and mutations in genes such as cytosine deaminase and FKS1. Changes in enzymes related to ergosterol synthesis are the primary cause of resistance to polyene drugs, while biofilm formation by fungi significantly contributes to multidrug resistance (Niego et al. 2023; Wijayawardene et al. 2023).

Over the past two decades, mass-spectrometry-based proteomics has revolutionized mycology research, enabling a comprehensive understanding of protein synthesis, posttranslational modifications, alternative protein isoforms, and interaction networks. Numerous studies leveraging fungal omics data provide a comprehensive understanding of their utility. In this section, we focus on two types of fungal proteins with the potential to become crucial targets for future drug development. Multidrug resistance is often associated with the increased expression of pleiotropic drug resistance (PDR) efflux pumps, a distinctive category within the ABC transporter superfamily. Unlike their human counterparts, fungal ABC proteins have a unique domain arrangement of [NBD-TMD]2, containing specific extracellulars (ELs) with conserved amino acid residues, which are absent in human ABC pumps. Targeting these ELs holds promise as a potential avenue for future antifungal drug development due to their surface-accessibility and limited subjected regulatory mechanisms (Niego et al. 2023; Wijayawardene et al. 2023).

20.5 Fungal-Based Consumer Products

20.5.1 Harnessing the Potential of Fungal-Derived Dyes and Pigments

In response to the widespread use of synthetic coloring compounds across diverse industries such as food, pharmaceuticals, cosmetics, and textiles, concerns have arisen regarding their health hazards and environmental impact due to toxicity. This has led to an increasing demand for eco-friendly and healthy alternatives, sparking interest in naturally derived dyes and colorants. Fungi, with their unique properties and high-yield potential, have emerged as promising sources of pigments that offer distinct advantages. Fungal biotechnology has opened up an alternative avenue for the production of dyes and pigments that are not only vibrant but also environmentally benign (Adeel et al. 2019; Aman Mohammadi et al. 2022; Muro Urista et al. 2016; Ngamwonglumlert et al. 2017; Toma et al. 2023; Umar et al. 2024).

Fungal pigments, arising from SMs, have been employed in traditional medicine for centuries, boasting diverse attributes beyond coloration, including antifungal, antibacterial, cytotoxic, anticancer, and antioxidant properties (Narsing Rao et al. 2017; Ramesh et al. 2019; Caro et al. 2017; Vivek-Ananth et al. 2021). Various fungi, including *Monascus*, *Talaromyces*, *Trichoderma*, *Penicillium*, *Fusarium*, and *Aspergillus*, exhibit antimicrobial activity, suggesting the potential utilization of their pigments in the food and pharmaceutical sectors (Lagashetti et al. 2019; Sajid and Akbar 2018). Notably, Monascus red pigment, derived from *M. purpureus*, demonstrates antibacterial efficacy comparable to ciprofloxacin (Akilandeswari and Pradeep 2016). Several fungal pigments, such as those from *Penicillium miczynskii*, *Penicillium purpuroscens*, *Trichoderma afroharzianum*, have exhibited antioxidant properties, warranting exploration in healthcare applications (Li et al. 2017; Mani et al. 2015; Poorniammal et al. 2019).

Moreover, fungal pigments with antibacterial and antifungal properties, such as those from *Thermocymes* sp. and *Trichoderma* sp., hold potential for medical textiles such as bed covers, stitches, and bandages (Poorniammal and Prabhu 2023; Kumar et al. 2023). Fungal biotechnological tools offer avenues to enhance these pigments' stability, concentration, medicinal attributes, and reduce toxicity. With stringent regulations on synthetic pigments, the increasing demand for microbial pigments underscores the need for extensive research to optimize their production on a larger scale. Furthermore, the exploration of omics sciences is crucial to mitigate mycotoxin production while maximizing pigment potential, reflecting the continuous advancement and potential of fungal biotechnology in pigment innovation (Wijayawardene et al. 2023).

20.5.2 Fungi in Cosmetic and Personal Care Product Innovation

In the realm of cosmetic and personal care products, fungi wield significant potential, offering a wealth of active compounds within their pigments. The cosmetics market, marked by continuous evolution and intense competition, gravitates towards natural extract-based products due to consumer concerns regarding synthetic chemical effects on health (Pascoalino et al. 2023; Lagashetti et al. 2019; Szutowski and Szułczyńska 2017). Fungal pigments have found applications in nutraceuticals, playing a role in melanin, carotenoids, and lycopene utilized in cosmetics, sunscreens, lotions, and anti-aging facials (Pascoalino et al. 2023; Narsing Rao et al. 2017; Lagashetti et al. 2019).

Carotenoids and lycopene, abundant in fungal sources like *Blakeslea trispora*, are pivotal bioactive compounds known for enhancing immunity and guarding against premature skin aging, making them sought-after ingredients in sun protection products (Pascoalino et al. 2023; Marzena et al. 2020). Monascus pigments, with their diverse color range, have ventured into the cosmetic industry, being utilized in lipsticks, skin conditioners, and other products (Pascoalino et al. 2023; Lagashetti et al. 2019). The exploitation of fungal pigments, especially those with antibacterial and antifungal properties, can pave the way for applications in medical textiles like bed covers, stitches, and bandages (Poorniammal and Prabhu 2023; Kumar et al. 2023), further accentuated through fungal biotechnological advancements to enhance stability, concentration, medicinal attributes, and reduce toxicity. Amid growing concerns regarding the ecological impact of synthetic dyes, fungal pigments emerge as biodegradable and environmentally friendly alternatives, catering to the demand for sustainable solutions (Venil et al. 2020).

20.5.3 Exploring Novel Uses of Fungi in Textiles and Fashion Design

Addressing the precarious ecological challenges arising from the release of toxic chemicals into the environment from garment and textile-related factories, academia and businesses have turned to natural colorants like microbial pigments to circumvent the risks associated with synthetic counterparts. Natural colorants produced from fungi offer biodegradability and environmental friendliness, positioning them as preferable alternatives to synthetic dyes (Venil et al. 2020).

The textile industry, drawn by high color fastness and staining capacity, has embraced fungal pigments as an alternative to synthetic counterparts. Several studies have illuminated the ability of fungi to produce organic colors with diverse properties, well-suited for applications in the textile industry (Pascoalino et al. 2023; Sajid and Akbar 2018; Caro et al. 2017; Chungkrang and Bhuyan 2020). An array of naturally occurring pigments such as anthraquinone from *Fusarium oxysporum* (pink/violet), *Penicillium oxalicum* (red), Viridin (yellow) from *Trichoderma viride*, and Viridol (yellow) from *Trichoderma virens* exhibit potential for soil fungi application (Pascoalino et al. 2023; Poorniammal and Prabhu 2023; Akilandeswari and Pradeep 2016). Monascus-like colors from various strains of non-producing mycotoxins *Penicillium* strains have been patented for applications in textiles, leather, paper, and more (Caro et al. 2017). Furthermore, *Dermocybe sanguinea*, *Alternaria alternate*, *Curvularia lunata*, and *Talaromyces verruculosus* have shown promising results in terms of pigment applications on cellulose matrices like silk and cotton (Lagashetti et al. 2019; Chungkrang and Bhuyan 2020). In the domain of leather dyeing, red color natural pigments from *Isaria* spp., *Penicillium* spp., *Monascus* spp., and *Fusarium* spp. offer potential for goat nappa skin applications (Pascoalino et al. 2023; Janiszewska-Turak et al. 2017; Umar et al. 2024; Sudha and Aggarwal 2016; Velmurugan et al. 2010).

20.6 Fungal-Based Biomaterials

20.6.1 Exploring the Potential of Fungal Biopolymers for Sustainable Materials

The potential of fungal biopolymers for sustainable materials is becoming increasingly recognized as a solution for transitioning to an eco-friendlier economy. Fungal biomaterials, derived from organic waste, offer an innovative avenue with significant environmental benefits. Fungal mycelium, through its ability to degrade lignocellulosic materials and form composite networks, generates materials with diverse properties such as insulation, non-flammability, hydrophobicity, and improved mechanical strength. These mycelium-based materials are poised to replace traditional materials like timber, steel, concrete, foams, and leather in various sectors including construction, automotive, transportation, electronics, fashion, and design. Their applications encompass packaging, bricks, insulation panels, acoustic panels, wall panels, fireproof materials, and even clothing (Niego et al. 2023; Delvendahl et al. 2022; Cerimi et al. 2019).

20.6.2 Mycelium-Based Materials and Their Innovative Applications

The exploration of fungal biopolymers for sustainable materials has led to innovative applications of mycelium-based materials. In the packaging industry, eco-friendly alternatives to petroleum-based products are being developed using fungal mycelium. Many companies have created biodegradable packaging from agricultural waste, offering benefits like lightweight, recyclability, and reduced pollution. The automotive sector is adopting mycelium-based components, replacing traditional plastics, and utilizing biodegradable alternatives. Fungal mycelium is even integrated into electrical circuit boards with metal-infused sheets for wiring patterns (Delvendahl et al. 2022; Cerimi et al. 2019).

The textile industry has seen patents utilizing fungal cell wall components for textiles and other applications. An expired patent by Dschida demonstrates the potential for broad applications, including paper production, construction materials, and medical uses. Other patents explore diverse applications for fungal materials, such as bioremediation of contaminants and organic insulation with enhanced fire resistance in home construction. Additionally, panels made from fungal material offer alternatives for acoustical dampening (Delvendahl et al. 2022; Cerimi et al. 2019).

Overall, the potential of fungal-based materials is being harnessed across various industries, paving the way for sustainable and innovative solutions. The myco-based material industry is rapidly evolving, as evident from the rising number of patent applications and scientific publications since 2007. Key players like Ecovative, MycoWorks, and Bolt Threads are pioneering the field, focusing on creating mycelium-based materials for packaging, insulation, fire protection, and leather alternatives. European startups like MycoTex and Grown. Bio is also gaining prominence. This surge in interest aligns with the growing demand for eco-friendly products, particularly in industries like packaging, plastic, and fashion. The global mycelium market is projected to witness substantial growth, reaching a value of USD 5.49 billion by 2030. However, challenges in scaling up production, lack of infrastructure, and patent monopolies hinder the industry's progress and knowledge distribution. This evolving industry not only promises innovative solutions to replace traditional materials but also holds the potential to significantly contribute to a more sustainable and circular economy (Niego et al. 2023; Delvendahl et al. 2022; Cerimi et al. 2019).

20.7 Fungi in Environmental Applications and as Biocatalysts

20.7.1 Utilizing Fungi for Pollution Control and Biodegradation

Air pollution has become a complex environmental and social problem worldwide. The issue of air pollution is escalating into a major global health concern. It leads to the occurrence of cardiovascular and respiratory diseases, including cough, shortness of breath, wheezing, asthma, respiratory diseases, chronic obstructive pulmonary disease, chronic asthma, and certain types of cancer (Adeel et al. 2019; Dey and Nagababu 2022; Bălă et al. 2021; Domingo and Rovira 2020; Dominski et al. 2021; Manisalidis et al. 2020; Buhaienko et al. 2022). According to the World Health Organization (2019) and Manisalidis et al. (2020), approximately 9 million deaths occur annually due to anthropogenic air pollution, making it a prominent global health hazard. Furthermore, air pollution presents a significant risk, impacting not only human health but also the environment at large (Manisalidis et al. 2020; Ashfaq and Sharma 1970). Acid rain, a serious environmental issue, is a consequence of air pollutants. Harmful acids in rain, fog, or snow, containing high levels of nitric and sulfuric acids, can acidify water and soil ecosystems, damaging trees, vegetation, and man-made structures. Air pollutants like ozone negatively affect crop yield and food productivity by disrupting plant photosynthesis and metabolism (Manisalidis et al. 2020; Zuhara and Isaifan 2018).

20.7.2 Volatile Organic Compounds (VOCs) and Their Adverse Impacts

Air pollutants contributing to pollution include sulfur oxides, nitrogen oxides, volatile organic compounds (VOCs), ammonia, carbon monoxide, carbon dioxide, methane, and fine particulate matter. Among these, VOCs like toluene, benzene, ethylbenzene, and others are significant contributors to air pollution. VOCs have adverse health implications, including carcinogenic effects, making their control (Manisalidis et al. 2020; Dehghani et al. 2019; Fugiel et al. 2017; Ghasemi et al. 2020; Kumar et al. 2014). Fungi play a significant role in air purification through biofiltration. Fungal biofiltration offers an effective and sustainable approach for VOCs remediation (Marycz et al. 2022). Specific fungal species like *Fusarium* sp., *Candida* sp., and *Cladophialophora* sp. exhibit a high capacity for removing VOCs, achieving removal efficiencies of over 90% (Marycz et al. 2022). Fungi such as *Exophiala* strains and white-rot fungi show promise in utilizing VOCs as energy sources (Marycz et al. 2022). The aerial mycelium of fungi aids in capturing hydrophobic VOCs, enhancing biofilter efficiency (Gospodarek et al. 2019).

Compared to bacterial biofilters, fungal biofiltration proves more effective in VOCs removal due to fungi's resistance to adverse conditions (Ghasemi et al. 2020; Marycz et al. 2022). Fungi's unique ability to absorb hydrophobic VOCs in the gas phase and the production of extracellular polymeric substances (EPS) contribute to their efficiency (Marycz et al. 2022). Fungi also produce biosurfactants that enhance gas exchange in biofilters. Fungi's resistance to temperature, humidity, and pollutants, along with their enzyme production capabilities, position them as promising candidates for VOCs removal (Fig. 20.2) (Marycz et al. 2022).

20.7.3 Fungi as Sensors

Sensors, pivotal in electronics, healthcare, environmental monitoring, and industry (Riza et al. 2020; Buhaienko et al. 2022), detect and measure various parameters, driving real-time monitoring, precise control, and automation for improved efficiency and safety (Buhaienko et al. 2022). Advancements in sensor technology have transformative potential. In nature, fungi contribute to nutrient cycling, decomposition, and plant symbiosis (Devi et al. 2020; Kumar et al. 2021), adapting to diverse habitats, even marine sediments. Fungi's unique properties make them promising for sensing applications, responding to environmental cues like temperature, humidity, pH, and specific chemicals (Buhaienko et al. 2022), offering innovative sensor possibilities across various domains.

Fungi exhibit complex sensory mechanisms, including nutrient sensing and responses to environmental cues. They possess specific receptors and signaling pathways that enable the detection of nutrients like sugars and amino acids, facilitating efficient resource utilization. Moreover, fungi are attuned to environmental factors

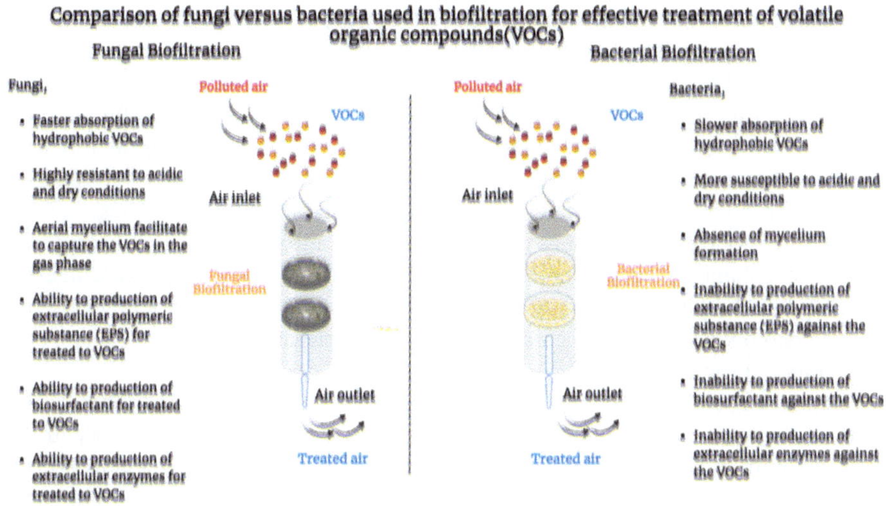

Fig. 20.2 Comparison of fungi versus bacteria used in biofiltration for effective treatment of VOCs, Figure Created with BioRender.com

such as light, temperature, and humidity through photoreceptors, governing their growth and development (Tian et al. 2021). Their interactions with organisms like plant roots and bacteria allow them to sense and respond to external signals, influencing their behavior (Tian et al. 2021). Beyond their biological roles, fungi serve as valuable environmental sensors. They can monitor air quality and pollutant levels, especially heavy metals, acting as bioindicators for environmental health (Gaikwad and Sonawane 2012; Oladipo et al. 2018). Additionally, fungal community profiling in soil and water reflects environmental impacts like pollution and land use. Their ability to detect waterborne pathogens and contaminants makes them effective tools for assessing water quality (Tian et al. 2021; Wielen and Kooij 2013). These versatile sensory abilities position fungi as crucial components of both biological and environmental systems.

Fungi's remarkable ability to detect and respond to pathogens has spurred the development of fungal-based biosensors for rapid disease detection, especially for conditions like mucormycosis, offering early diagnosis and treatment management (Ponnaiyan et al. 2023). These biosensors also have potential in identifying specific fungal biomarkers, enabling real-time monitoring. Additionally, fungi are pivotal in drug discovery, with fungal-based assays effectively assessing the efficacy of compounds against fungal pathogens, thus aiding antifungal drug development (Khan et al. 2021). They not only gauge activity but also the effectiveness of potential drug candidates. Fungi, as a rich source of bioactive compounds, are further harnessed through fungal biosensors, identifying and characterizing these compounds with potential therapeutic applications (Keller 2018). These sensors offer promising avenues in healthcare for disease diagnosis and drug development, capitalizing on fungi's unique sensing capabilities and interactions with pathogens.

In agriculture, fungi-based sensors play a pivotal role in detecting and monitoring plant diseases and pests. These sensors detect specific VOCs released by plants in response to pathogenic or pest attacks, providing invaluable early warning systems for farmers. Moreover, fungi such as *S. cerevisiae* are employed as bioluminescent reporters to measure pollutant toxicity, with ergosterol levels serving as a marker for fungal presence. Monitoring fungal biomass, indicated by leaf mass loss during their growth, offers another effective approach (Soares et al. 2022). These sensors significantly contribute to timely pest and disease control measures, reducing crop losses and enhancing agricultural productivity. Additionally, they provide insights into crop health by measuring key parameters like soil moisture, nutrient levels, and plant physiological responses, guiding decisions on irrigation, fertilization, and crop management practices.

Fungi's sensory capabilities have even inspired developments in robotics and material sciences. Soft robotic systems emulate fungi's unique sensory traits, allowing them to detect and respond to environmental cues, enhancing adaptability and resilience. Fungi also produce diverse materials, including mycelium-based composites, which can be engineered for sensory functions like detecting changes in temperature, humidity, or chemical composition (Manan et al. 2021; Wainaina and Taherzadeh 2023). These fungal materials with sensing capabilities find applications in diverse technological domains such as smart textiles, environmental monitoring, and biomedical devices. Thus, fungi's sensory abilities open exciting prospects in technology, serving as indispensable tools in agriculture, environmental surveillance, robotics, and material sciences, with their distinctive attributes offering insights and applications across a wide technological landscape.

20.7.4 Mycorrhiza-Based Enzymes

Fungi produce a wide array of enzymes with industrial applications. Oxidoreductases, transferases, hydrolases, and other enzymes are used in various industries. Fungi are used to produce enzymes with high specificity and mild reaction conditions, making them suitable for applications like food processing, pharmaceuticals, textiles, detergents, and more. Fungal enzymes like laccases, amylases, proteases, cellulases, lipases, and others find applications in various industries, such as food, detergents, paper, and biofuels. These enzymes are derived from fungal sources like Aspergillus, Trichoderma, Rhizopus, and Penicillium (Niego et al. 2023).

20.8 Challenges and Advances in Fungal Product and Service Development

20.8.1 Addressing Obstacles in Large-Scale Fungal Bio-Factory Implementation

Navigating the path towards large-scale fungal biofactory implementation presents substantial challenges. The evolution of bioinformatic tools to encompass the full spectrum of fungal diversity is imperative. Balancing the intricacies of large-scale cultivation with environmental sustainability necessitates interdisciplinary collaborations. Ensuring robust risk assessments and adherence to ethical guidelines is essential to ensure that fungal applications yield positive contributions to both society and the environment. Similarly, the journey towards large-scale fungal biofactories for foliar endophytic fungi encounters hurdles. The existing bioinformatic landscape, often rooted in established knowledge, risks overshadowing novel discoveries. Striking a balance between economic and environmental considerations in large-scale fungal cultivation demands meticulous navigation. Holistic approaches intertwining genetic engineering, ecological assessment, and ethical considerations are crucial to maintain the biofactory's sustainability and minimize ecological impact. Meanwhile, in the pursuit of enhancing plant resistance through fungal biofactories, challenges persist. Bioinformatic tools tailored to plant-fungal interactions may inadvertently overlook novel mechanisms due to reliance on existing genetic pathways. Ethical and ecological concerns arise from the delicate equilibrium between bolstering resistance and avoiding harm to non-target organisms or the environment (Niego et al. 2023; Wijayawardene et al. 2023; Tian et al. 2021; Chu et al. 2023). This underscores the need for comprehensive approaches that harmonize genetic engineering, ecological assessment, and ethical frameworks. In the realm of fungal biotechnology, current bioinformatic tools show limitations in specificity and sensitivity, often confined to narrow fungal RiPP gene clusters instead of the broader repertoire encoded in fungal genomes. Challenges arise from tools that rely on prior knowledge, potentially hindering recognition of novel RiPPs. This accentuates the complementary nature of traditional natural product isolation and emphasizes the importance of connecting fungal peptide compounds to gene clusters, as evidenced by discoveries like the genetic basis for biosynthesis. The relationship establishment between new RiPP classes and their gene clusters is poised to propel genome mining towards a broader array of RiPPs. Simultaneously, addressing the hurdles of implementing large-scale fungal biofactories involves overcoming obstacles such as the attenuation of ginsenoside production post-endophyte removal from host plants. Environmental conditions variations, encompassing elements like carbon and nitrogen sources, pH, temperature, and trace elements, negatively impact ginsenoside yield. Further compounding this is the absence of selection pressure and coexisting endophytes, both contributing to reduced biosynthesis. Surmounting these challenges necessitates the optimization of bioreactor processes and the preservation of symbiotic conditions (Kessler and Chooi 2022; Punja 2010).

20.8.2 Innovations in Fungal Bioprocessing for Enhanced Product Yield and Quality

In the pursuit of enhancing product yield and quality, fungal endophytes emerge as promising microbial hosts. Leveraging synthetic biology tools, dormant gene clusters within endophytes can be activated, resulting in heightened production. Engineered strains like Yeast and *Pichia pastoris* provide opportunities for elevated ginsenoside yields. The ongoing advancement of synthetic biology tools, such as the CRISPR-Cas system, offers potential for fine-tuning gene expression to further enhance productivity. Moreover, the realm of fungal bioprocessing has unveiled insights into augmenting plant-fungal pathogen resistance. Exploring the diverse array of fungal SMs, with their wide-ranging bioactivities, from inhibiting fungal growth to shaping plant immune responses, holds potential for fortifying plants' natural defense mechanisms. This fusion of fungal bioprocessing and plant resistance engineering signifies a revolutionary approach to bolstering crop protection strategies. Concurrently, innovations in fungal bioprocessing have illuminated the potential of foliar endophytic fungi. These fungi's diverse SMs, characterized by their multifunctional properties such as antimicrobial efficacy and plant growth promotion, offer a wealth of bioactive compounds. Bioprocessing advancements enable sustainable production of these compounds, unlocking possibilities in agriculture, pharmaceuticals, and environmental applications. Additionally, fungi's economic potential is intrinsically linked to bioprocessing innovations, enabling the production of enzymes and metabolites with versatile applications. This transformative landscape spans from reducing agricultural chemical inputs through biopesticides to crafting novel pharmaceuticals, revolutionizing economic paradigms. Fungal bioprocessing, fueled by omics technologies, propels high-yield, high-quality production into a new era. Fungi's metabolic capabilities are harnessed to synthesize enzymes, bioactive compounds, and more. Advanced bioreactor designs coupled with real-time monitoring optimize growth conditions and metabolite accumulation, thereby elevating both yield and quality of fungal-derived products (Wijayawardene et al. 2023; Ball et al. 2020; Tian et al. 2021; Chu et al. 2023).

20.8.3 Harnessing Omics Technologies to Drive Advancements in Fungal-Based Products

The integration of omics technologies, including next-generation sequencing, metagenomics, transcriptomics, proteomics, and metabolomics, provides profound insights into the intricate interactions between plants and microbes. Through the analysis of endophyte biodiversity and composition, these methods serve as valuable tools in identifying novel microorganisms capable of producing ginsenosides. Key functional genes and enzymes essential for ginsenoside biosynthesis can be pinpointed through these approaches. Integrating these discoveries into microbial hosts via synthetic biology methodologies holds the potential to enhance efficient and controlled ginsenoside production. Moreover, many studies have adapted the bacterial Ribosomally synthesized and post-translationally modified peptide (RiPP) mining tool, RiPPMiner, to unveil fungal RiPPs. This pipeline involves meticulous exclusion steps to distill a vast pool of candidate genes into a manageable list, potentially encoding precursor peptides for RiPPs. While the pipeline successfully identified potential cycloamanide-encoding genes in *Trichoderma* genomes, their verification as RiPPs remains experimental. The realm of mycology, traditionally confined to microscopic observations and petri dishes, has undergone a revolutionary transformation due to the incorporation of advanced technologies like omics. These state-of-the-art tools have uncovered the hidden complexities of fungi, sparking breakthroughs across a multitude of applications and reshaping the field. The synergy between omics technologies and mining tools, akin to the approach of RiPPMiner, has unlocked a wealth of untapped fungal potential. The systematic decoding of fungal genomes reveals genes responsible for bioactive compounds, enzymes, and metabolic pathways. The fusion of genetic insights with empirical validation propels the exploration of fungal-based products, spanning from biofuels to innovative pharmaceuticals, thereby expediting the transition towards sustainable, bio-centric economies (Wijayawardene et al. 2023; Tian et al. 2021; Chu et al. 2023; Kessler and Chooi 2022).

20.8.4 Addressing Challenges and Potential Risks in Utilizing Fungi for Novel Applications

In the realm of harnessing fungi for novel applications, a multitude of challenges and potential risks must be navigated. The task of maintaining symbiotic conditions for endophytes outside their host plants poses a persistent challenge, necessitating the optimization of environmental parameters within bioreactors. Furthermore, concerns regarding biosafety and regulatory compliance associated with engineered microorganisms require meticulous resolution before large-scale implementation can be realized. However, the utilization of fungal-based approaches for the production of bioactive compounds, especially ginsenosides, holds immense promise. The path to overcoming these challenges lies in the innovative utilization of bioprocessing techniques, the integration of omics technologies, and the proactive addressing of potential risks. These endeavors are poised to pave the way for sustainable and efficient production processes that can significantly benefit various industries and healthcare sectors. The exploration of novel RiPPs within fungal genomes has mainly been advanced through innovative tools like MIDDAS-M38. MIDDAS-M38 is a specialized laboratory technique, algorithm, or software designed for identifying and analyzing RiPPs in fungal genetic data. Although this methodology is not exclusively customized for RiPPs, it identifies gene clusters based on their co-regulation, irrespective of the specific class of natural product encoded by the genes. The discovery of the ustiloxin gene cluster exemplifies the successful detection of a novel RiPP family through bottom-up genome mining. However, it is essential to recognize that MIDDAS-M's approach requires gene clusters to exhibit distinct expression patterns captured in transcriptomics data. Despite this approach's limitations, motif-independent methods hold significant potential for the identification of unexplored fungal RiPP families and other novel secondary metabolite gene clusters (Tian et al. 2021; Kessler and Chooi 2022).

Efforts to enhance plant resistance through genetic engineering encounter several challenges. Tools like pathogen mining might inadvertently overlook key genetic players due to their unique co-regulation patterns. Ensuring the specificity and safety of genetic modifications is of paramount importance, demanding rigorous validation to ensure engineered resistance functions effectively and stably. Overcoming these hurdles necessitates the amalgamation of advancements in bioinformatics, interdisciplinary collaboration, and a proactive approach to risk mitigation. In the intricate interplay between plants and foliar endophytic fungi, the integration of emerging biotechnological trends, omics technologies, and fungal bioprocessing emerges as a pivotal strategy. As we navigate this dynamic landscape, the synergy of these cutting-edge tools with meticulous research places us on the threshold of an era where crops can fortify themselves against fungal adversaries. This advancement ensures food security and sustainable agriculture in an ever-evolving world (Tian et al. 2021; Kessler and Chooi 2022).

Undertaking the utilization of foliar endophytic fungi for novel applications demands a direct confrontation of the accompanying challenges. The use of gene mining tools introduces the risk of overlooking vital genes due to their distinct expression patterns. Guaranteeing the safety, efficacy, and ethical implications of genetically modified fungi requires comprehensive attention and validation. To navigate these complexities responsibly, interdisciplinary collaborations, rigorous adherence to regulations, and stakeholder engagement are crucial elements. With each stride in our understanding of fungi's role in the global economy, we must simultaneously grapple with the challenges that emerge. The reliance on gene mining tools could inadvertently bypass critical genetic elements due to their unique expression patterns. Safeguarding the safety, effectiveness, and ethical dimensions of genetically modified fungi is imperative to avert unintended consequences. As we delve further into these transformative opportunities, rigorous validation, regulatory compliance, and a commitment to responsible exploration become the pillars that uphold the profound potential of fungi in shaping industries and fostering a greener future (Tian et al. 2021; Chu et al. 2023; Kessler and Chooi 2022).

The fusion of omics technologies and advanced methodologies has ushered in a mycological renaissance, highlighting fungi's expansive potential across diverse sectors. These versatile organisms transcend conventional boundaries, from bioremediation to personalized medicine. By harnessing technological prowess, nurturing cross-disciplinary collaborations, and upholding ethical integrity, the sphere of mycological applications stands poised to continue its ascendancy. In doing so, it forges innovative pathways that harmonize science, society, and sustainability, propelling fungal contributions to unparalleled heights.

In the realm of fungal sensor development, numerous challenges and obstacles must be surmounted for their successful implementation. It is imperative to ensure the specificity and selectivity of fungal sensors to accurately detect and monitor fungal species or strains while distinguishing them from other microorganisms and environmental factors (Muthusamy and Murugaiyan 2023). Moreover, these sensors must exhibit stability and durability to withstand harsh environmental conditions, thereby maintaining their sensing capabilities over time (Ball et al. 2020). Establishing standardized protocols and methodologies for fungal sensor fabrication and testing is equally pivotal, ensuring reproducibility and comparability of results and facilitating their widespread adoption and commercialization (Roth et al. 2023).

Nevertheless, despite these challenges, fungal sensors hold immense potential across various applications and are poised for significant advancements. In the domain of disease diagnostics, these sensors can be further tuned to detect fungal infections and monitor disease progression, ultimately enhancing diagnostic accuracy and enabling personalized treatment approaches (Naeem et al. 2022). Moreover, fungal sensors can assume a pivotal role in environmental monitoring, encompassing air and water quality sensing, pollutant detection, and ecological health monitoring. These sensors, coupled with advancements in technology, can substantially augment our capacity to assess and mitigate environmental risks (Toussaint et al. 2016).

To propel the field of fungal sensors forward and surmount existing challenges, several research directions and emerging technologies warrant exploration. Integrating multi-omics technologies offers the promise of comprehensively understanding fungal sensing mechanisms and the development of more sophisticated and accurate fungal sensors (Ball et al. 2020). Concurrently, the incorporation of nanomaterials like graphene, carbon dots, and metal nanoparticles presents opportunities to enhance sensitivity, selectivity, and stability, potentially enabling the creation of compact and portable fungal sensors (Gaikwad et al. 2019). Moreover, leveraging advanced data analysis techniques such as machine learning and artificial intelligence, in tandem with fungal sensor data, holds the potential to refine data interpretation and utilization, leading to more precise real-time monitoring and decision-making (Roshani et al. 2022). Interdisciplinary collaborations, encompassing mycology, biotechnology, materials science, and engineering, are indispensable in fostering innovation and addressing the challenges inherent in fungal sensor development. These cross-disciplinary approaches hold the promise of yielding novel sensor designs, innovative fabrication techniques, and diverse applications (Schouteden et al. 2015). Ultimately, the development of advanced fungal sensors with heightened sensitivity, selectivity, and reliability has the potential to revolutionize a plethora of domains, including healthcare, environmental monitoring, and technology.

20.9 Emerging Trends in Fungal Biotechnology and Future Opportunities

The landscape of emerging trends in fungal biotechnology presents a promising trajectory towards sustainable production through the utilization of endophytes and engineered microorganisms. Driven by cultivation-independent techniques like metagenomics and empowered by synthetic biology tools such as CRISPR-CAS systems, prospects are focused on the RNA editing for genetic improvement. These methodologies not only promise amplified yields but also hold the potential for the creation of novel variants with therapeutic applications. Unveiling the latent potential within fungal genomes, especially in t RiPP gene clusters. Future bioinformatic tools should encompass the distinctive characteristics outlined in this discourse to facilitate a more comprehensive identification of RiPP gene clusters within fungal genomes. As genome sequencing continues to advance and bioinformatic predictions align more effectively with our evolving understanding of fungal RiPP biosynthesis, coupled with high-throughput strategies to link gene clusters to compounds, the wealth of biosynthetic potential could become more accessible than ever before (Chu et al. 2023; Kessler and Chooi 2022; Punja 2010; Yapa et al. 2022).

In agricultural biotechnology, the endeavor to bolster plant resistance against fungal pathogens gains momentum through genome surveys, unveiling the intricate mechanisms residing within plant genomes to counter these threats. Harnessing the vast genetic information through the development of bioinformatic tools enables the decryption of plant-fungal interactions, setting the stage for precision breeding that capitalizes on the untapped genetic arsenal to reinforce fungal pathogen resistance in plants. The foliar endophytic fungi gradually unfurls, exposing a rich and diverse community thriving within plant leaves. Recent strides in DNA sequencing have revolutionized the exploration of this diversity. As genome surveys illuminate the abundance of foliar endophytic fungi, the vista of biotechnology broadens. Future research endeavors could tap into the multifaceted potential of these fungi across a spectrum of applications, ranging from agricultural enhancements to environmental remediation (Chu et al. 2023; Punja 2010).

Within the sphere of fungal biotechnology, genome surveys have cast a spotlight on fungal diversity and their inherent genetic capabilities. Notably, they have revealed species harboring traits like enzyme production for bioremediation and biofuel generation. As biotechnological methodologies continue to evolve, the intrinsic capacity of fungi to convert organic matter into valuable resources gains prominence, propelling us towards the ideals of a circular economy (Niego et al. 2023; Chu et al. 2023).

In the era of omics technologies, fungal biotechnology is undergoing a paradigm shift, with a comprehensive exploration of fungal genomes and their functional elements. As we delve deeper into fungal genomics, a plethora of future opportunities awaits—from identifying novel therapeutic compounds to engineering fungal strains tailored for industrial bioprocessing (Wijayawardene et al. 2023; Punja 2010).

20.10 Conclusion

Although traditional fungal biotechnology still has a place in the industry, the novel applications have begun to take over with the advent of new technologies such as the integrated omics, which solves the gray areas by the simultaneous analysis at all levels i.e. genomic, transcriptomic, proteoimic, and metabolomic. The future of fungal biotechnology may see an unprecedent number of novel products in all fields of Biotechnology.

References

Adeel S, Rehman FU, Rafi S, Mahmood Zia K, Zuber M (2019) Environmentally friendly plant-based natural dyes: extraction methodology and applications. In: Ozturk M, Hakeem K (eds) Plant and human health: phytochemistry and molecular aspects, vol 2. Springer International Publishing, pp 383–416

Aguiar TQ, Silva R, Domingues L (2016) New biotechnological applications for *Ashbya gossypii*: challenges and perspectives 8:309–315. https://doi.org/10.1080/21655979.2016.1234543

Akilandeswari P, Pradeep BV (2016) Exploration of industrially important pigments from soil fungi. Appl Microbiol Biotechnol 100:1631–1643. https://doi.org/10.1007/S00253-015-7231-8/FIGURES/2

Aman Mohammadi M, Ahangari H, Mousazadeh S, Hosseini SM, Dufossé L (2022) Microbial pigments as an alternative to synthetic dyes and food additives: a brief review of recent studies. Bioprocess Biosyst Eng 45:1–12. https://doi.org/10.1007/S00449-021-02621-8/TABLES/3

Arnold AE (2007) Understanding the diversity of foliar endophytic fungi: progress, challenges, and frontiers. Fungal Biol Rev 21:51–66. https://doi.org/10.1016/J.FBR.2007.05.003

Ashfaq A, Sharma P (1970) Environmental effects of air pollution and application of engineered methods to combat the problem. J Ind Control Pollut 29

Bălă GP, Râjnoveanu RM, Tudorache E, Motişan R, Oancea C (2021) Air pollution exposure—the (in)visible risk factor for respiratory diseases. Environ Sci Pollut Res 28:19615–19628. https://doi.org/10.1007/S11356-021-13208-X/FIGURES/1

Ball B, Langille M, MBio JGM (2020) Fun(gi)omics: advanced and diverse technologies to explore emerging fungal pathogens and define mechanisms of antifungal resistance. Am Soc Microbiol 11:1–18. https://doi.org/10.1128/mBio.01020-20

Buhaienko T, Popovych D, Bondarchuk V, Novakova L, Hevko U, Vayda O, Myndziv K (2022) Health monitoring by functional indicators with the help of sensors at the phase of patient rehabilitation in today's conditions. J Educ Health Sport 12:86–93. https://doi.org/10.12775/JEHS.2022.12.10.011

Caro Y, Venkatachalam M, Lebeau J, Fouillaud M, Dufossé L (2017) Pigments and colorants from filamentous fungi. Fungal Metabolites:499–568. https://doi.org/10.1007/978-3-319-25001-4_26

Cerimi K, Akkaya KC, Pohl C, Schmidt B, Neubauer P (2019) Fungi as source for new bio-based materials: a patent review. Fungal Biol Biotechnol 6:1–10. https://doi.org/10.1186/S40694-019-0080-Y/TABLES/3

Chambergo FS, Valencia EY (2016) Fungal biodiversity to biotechnology. Appl Microbiol Biotechnol 100(6):2567–2577. https://doi.org/10.1007/S00253-016-7305-2

Chu LL, Huy NQ, Tung NH (2023) Microorganisms for ginsenosides biosynthesis: recent progress, challenges, and perspectives. Molecules 28(3):1437, 1–23. https://doi.org/10.3390/MOLECULES28031437

Chungkrang L, Bhuyan S (2020) Natural dye sources and its applications in textiles: a brief review. Int J Curr Microbiol Appl Sci 9:261–269. https://doi.org/10.20546/ijcmas.2020.910.034

Dehghani MH, Norouzian A, Fazlzadeh M, Ghaffari HR (2019) Exposure and risk assessment of BTEX in indoor air of gyms in Tehran, Iran. Microchem J 150:1–6. https://doi.org/10.1016/J.MICROC.2019.104135

Delvendahl N, Dienel HL, Meyer V, Langen N, Zimmermann J, Schlecht M (2022) Narratives of fungal-based materials for a new bioeconomy era 36:96–106. https://doi.org/10.1080/13511610.2022.2110453

Devi R, Kaur T, Kour D, Rana KL, Yadav A, Yadav AN (2020) Beneficial fungal communities from different habitats and their roles in plant growth promotion and soil health. Microb Biosyst 5:21–47. https://doi.org/10.21608/MB.2020.32802.1016

Dey S, Nagababu BH (2022) Applications of food color and bio-preservatives in the food and its effect on the human health. Food Chem Adv 1:1–13. https://doi.org/10.1016/j.focha.2022.100019

Domingo JL, Rovira J (2020) Effects of air pollutants on the transmission and severity of respiratory viral infections. Environ Res 187:1–7. https://doi.org/10.1016/J.ENVRES.2020.109650

Dominski FH, Lorenzetti Branco JH, Buonanno G, Stabile L, Gameiro da Silva M, Andrade A (2021) Effects of air pollution on health: a mapping review of systematic reviews and meta-analyses. Environ Res 201:1–11. https://doi.org/10.1016/J.ENVRES.2021.111487

Fugiel A, Burchart-Korol D, Czaplicka-Kolarz K, Smoliński A (2017) Environmental impact and damage categories caused by air pollution emissions from mining and quarrying sectors of European countries. J Clean Prod 143:159–168. https://doi.org/10.1016/J.JCLEPRO.2016.12.136

Gaikwad KN, Sonawane MD (2012) Fungi as bio-indicators of air quality. Int J Life Sci Pharma Res 2:25–28

Gaikwad A, Joshi M, Patil K, Sathaye S, Rode C (2019) Fluorescent carbon-dots thin film for fungal detection and bio-labeling applications. ACS Appl Bio Mater 2:5829–5840. https://doi.org/10.1021/acsabm.9b00795

Ghasemi R, Golbabaei F, Rezaei S, Pourmand MR, Nabizadeh R, Jafari MJ, Masoorian E (2020) A comparison of biofiltration performance based on bacteria and fungi for treating toluene vapors from airflow. AMB Express 10:1–9. https://doi.org/10.1186/S13568-019-0941-Z/FIGURES/5

Gospodarek M, Rybarczyk P, Brillowska-Dąbrowska A, Gębicki J (2019) The use of various species of fungi in biofiltration of air contaminated with odorous volatile organic compounds. E3S Web Conf 100:1–8. https://doi.org/10.1051/E3SCONF/201910000021

Jain S, Rustagi A, Kumar D, Yusuf MA, Shekhar S, Sarin NB (2019) Meeting the challenge of developing food crops with improved nutritional quality and food safety: leveraging proteomics and related omics techniques. Biotechnol Lett 41:471–481. https://doi.org/10.1007/S10529-019-02655-9/TABLES/1

Janiszewska-Turak E, Pisarska A, Królczyk JB, Naturalne B, Żywności W (2017) Natural food pigments application in food products. Nauka Przyroda Technologie 10:1–13. https://doi.org/10.17306/J.NPT.2016.4.51

Keller NP (2018) Fungal secondary metabolism: regulation, function and drug discovery. Nat Rev Microbiol 17(3):167–180. https://doi.org/10.1038/s41579-018-0121-1

Kessler SC, Chooi YH (2022) Out for a RiPP: challenges and advances in genome mining of ribosomal peptides from fungi. Nat Prod Rep 39:222–230. https://doi.org/10.1039/D1NP00048A

Khan MS, Gao J, Munir I, Zhang M, Liu Y, Moe TS, Xue J, Zhang X (2021) Characterization of endophytic fungi, acremonium sp., from lilium davidii and analysis of its antifungal and plant growth-promoting effects. Biomed Res Int:1–14. https://doi.org/10.1155/2021/9930210

Kumar A, Singh BP, Punia M, Singh D, Kumar K, Jain VK (2014) Assessment of indoor air concentrations of VOCs and their associated health risks in the library of Jawaharlal Nehru University, New Delhi. Environ Sci Pollut Res 21:2240–2248. https://doi.org/10.1007/S11356-013-2150-7/TABLES/7

Kumar V, Sarma VV, Thambugala KM, Huang JJ, Li XY, Hao GF (2021) Ecology and evolution of marine fungi with their adaptation to climate change. Front Microbiol 12:1–12. https://doi.org/10.3389/FMICB.2021.719000/BIBTEX

Kumar A, Verma V, Dubey VK, Srivastava A, Garg SK, Singh VP, Arora PK (2023) Industrial applications of fungal lipases: a review. Front Microbiol 14:1–18. https://doi.org/10.3389/FMICB.2023.1142536/BIBTEX

Lagashetti AC, Dufossé L, Singh SK, Singh PN (2019) Fungal pigments and their prospects in different industries. Microorganisms 7:604, 1–36 (2019). https://doi.org/10.3390/MICROORGANISMS7120604

Li F, Xue F, Yu XH (2017) GC–MS, FTIR and raman analysis of antioxidant components of red pigments from *Stemphylium lycopersici*. Curr Microbiol 74:532–539. https://doi.org/10.1007/S00284-017-1220-3/TABLES/1

Liang B, Du XJ, Li P, Sun CC, Wang S (2018) Investigation of citrinin and pigment biosynthesis mechanisms in *Monascus purpureus* by transcriptomic analysis. Front Microbiol 9:1–11. https://doi.org/10.3389/FMICB.2018.01374/BIBTEX

Manan S, Ullah MW, Ul-Islam M, Atta OM, Yang G (2021) Synthesis and applications of fungal mycelium-based advanced functional materials. J Bioresour Bioprod 6:1–10. https://doi.org/10.1016/J.JOBAB.2021.01.001

Mani M, Mani VM, Shanmuga Priya M, Dhaylini S, Preethi K (2015) Antioxidant and antimicrobial evaluation of bioactive pigment from *Fusarium* sp. isolated from stressed environment. Int J Curr Microbiol Appl Sci 4:1147–1158

Manisalidis I, Stavropoulou E, Stavropoulos A, Bezirtzoglou E (2020) Environmental and health impacts of air pollution: a review. Front Public Health 8:1–13. https://doi.org/10.3389/FPUBH.2020.00014/BIBTEX

Marycz M, Brillowska-Dąbrowska A, Muñoz R, Gębicki J (2022) A state of the art review on the use of fungi in biofiltration to remove volatile hydrophobic pollutants. Rev Environ Sci Biotechnol 21:225–246. https://doi.org/10.1007/S11157-021-09608-7/FIGURES/2

Marzena B, Paulina M, Katarzyna B, Baran M, Miziak P, Bonio K (2020) Characteristics of carotenoids and their use in the cosmetics industry. J Educ Health Sport 10:192–196. https://doi.org/10.12775/JEHS.2020.10.07.020

Muro Urista C, Gracida Rodríguez J, Abreu Corona A, Arana Cuenca A, Téllez Jurado A (2016) Pigments from fungi, an opportunity of production for diverse applications. Biologia (poland) 71:1067–1079. https://doi.org/10.1515/BIOLOG-2016-0144/MACHINEREADABLECITATION/ENDNOTE

Muthusamy C, Murugaiyan K (2023) Bioprospecting Marine fungal enzymes-scope and challenges. In: Muthusamy C, Murugaiyan K (eds) fungiindia.co.in., vol 59, pp 33–47. https://doi.org/10.36460/Kavaka/59/1/2023/33-47

Naeem M, Manzoor S, Abid M-U-H, Burhan M, Tareen K, Asad M, Mushtaq S, Ehsan N, Amna D, Xu B, Hazafa A (2022) Fungal proteases as emerging biocatalysts to meet the current challenges and recent developments in biomedical therapies: an updated review. J Fungi 2022:1–32. https://doi.org/10.3390/jof8020109

Narsing Rao MP, Xiao M, Li WJ (2017) Fungal and bacterial pigments: Secondary metabolites with wide applications. Front Microbiol 8:1–13. https://doi.org/10.3389/FMICB.2017.01113/BIBTEX

Ngamwonglumlert L, Devahastin S, Chiewchan N (2017) Natural colorants: pigment stability and extraction yield enhancement via utilization of appropriate pretreatment and extraction methods. Crit Rev Food Sci Nutr 57:3243–3259. https://doi.org/10.1080/10408398.2015.1109498

Niego AGT, Lambert C, Mortimer P, Thongklang N, Rapior S, Grosse M, Schrey H, Charria-Girón E, Walker A, Hyde KD, Stadler M (2023) The contribution of fungi to the global economy. Fungal Divers 2023:1–43. https://doi.org/10.1007/S13225-023-00520-9

Oladipo OG, Awotoye OO, Olayinka A, Bezuidenhout CC, Maboeta MS (2018) Heavy metal tolerance traits of filamentous fungi isolated from gold and gemstone mining sites. Braz J Microbiol 49:29–37. https://doi.org/10.1016/J.BJM.2017.06.003

Pascoalino LA, Pires TCSP, Taofiq O, Ferreira ICFR, Barros L, Reis FS (2023) Biochemistry of secondary metabolism of fungi. In: Carocho M, Heleno SA, Barros L (eds) Natural secondary metabolites: from nature, through science, to industry. Springer International Publishing, pp 437–474

Ponnaiyan D, Anitha CM, Prakash PSG, Subramanian S, Rughwani RR, Kumar G, Nandipati SR (2023) Mucormycosis diagnosis revisited: current and emerging diagnostic methodologies for the invasive fungal infection (review). Exp Ther Med 25:1–6. https://doi.org/10.3892/ETM.2022.11746

Poorniammal R, Prabhu S, Sakthi AR (2019) Evaluation of In vitro antioxidant activity of fungal pigments. Pharma Innov J 8:326–330

Poorniammal R, Prabhu S (2023) Exploitation of Thermomyces fungal pigment for the synthesis of silver nanoparticles for textile application. Emergent Life Sci Res 09:40–48. https://doi.org/10.31783/ELSR.2023.914048

Punja ZK (2010) Genetic engineering of plants to enhance resistance to fungal pathogens—a review of progress and future prospects 23:216–23. https://doi.org/10.1080/07060660109506935

Tian X, Ding H, Ke W, Wang L (2021) Quorum sensing in fungal species. Annu Rev Microbiol 75:449–469. https://doi.org/10.1146/ANNUREV-MICRO-060321-045510

Ramesh C, Vinithkumar NV, Kirubagaran R, Venil CK, Dufossé L (2019) Multifaceted applications of microbial pigments: current knowledge, challenges and future directions for public health implications. Microorganisms 7:1–46. https://doi.org/10.3390/MICROORGANISMS7070186

Riza MA, Go YI, Harun SW, Maier RRJ (2020) FBG sensors for environmental and biochemical applications—a review. IEEE Sens J 20:7614–7627. https://doi.org/10.1109/JSEN.2020.2982446

Roshani S, Coccia M, Mosleh M (2022) Sensor technology for opening new pathways in diagnosis and therapeutics of breast, lung, colorectal and prostate cancer. medRxiv:1–31. https://doi.org/10.1101/2022.02.18.22271186

Roth MG, Westrick NM, Baldwin TT (2023) Fungal biotechnology: from yesterday to tomorrow. Front Fungal Biol 4:1–11. https://doi.org/10.3389/FFUNB.2023.1135263/FULL

Sajid S, Akbar N (2018) Applications of fungal pigments in biotechnology. Pesqui Agropecu Bras 7:922–930. https://doi.org/10.19045/BSPAB.2018.700111

Sarsaiya S, Jain A, Shi J, Chen J (2021) Fungi endophytes for biofactory of secondary metabolites: Genomics and metabolism. In: Sudisha Jogaiah (ed) Biocontrol agents and secondary metabolites: applications and immunization for plant growth and protection. Woodhead Publishing, pp 1–21

Schouteden C, Serwas D, Palfy M, Dammermann A (2015) The ciliary transition zone functions in cell adhesion but is dispensable for axoneme assembly in C. elegans. J Cell Biol 210:35–44. https://doi.org/10.1083/JCB.201501013/VIDEO-3

Soares DMM, Procópio DP, Zamuner CK, Nóbrega BB, Bettim MR, de Rezende G, Lopes PM, Pereira ABD, Bechara EJH, Oliveira AG, Freire RS, Stevani CV (2022) Fungal bioassays for environmental monitoring. Front Bioeng Biotechnol 10:1–6. https://doi.org/10.3389/FBIOE.2022.954579/BIBTEX

Gupta C, Aggarwal S (2016) Dyeing wet blue goat nappa skin with a microbial colorant obtained from *Penicillium minioluteum*. J Clean Prod 127:585–590. https://doi.org/10.1016/J.JCLEPRO.2016.03.043

Szutowski D, Szułczyńska J (2017) Product innovation in cosmetic industry—case study of major cosmetic companies. GISAP Econ Jurisprud Manage:1–4. https://doi.org/10.18007/GISAP:EJM.V0I12.1620

Toma MA, Rahman MH, Rahman MS, Arif M, Nazmul KHM, Nazir H, Dufossé L (2023) Fungal pigments: carotenoids, riboflavin, and polyketides with diverse applications. J Fungi:1–28. https://doi.org/10.3390/jof9040454

Toussaint M, Bontemps C, Besserer A, Hotel L, Gérardin P, Leblond P (2016) Whole-cell biosensor of cellobiose and application to wood decay detection. J Biotechnol 239:39–46. https://doi.org/10.1016/J.JBIOTEC.2016.10.003

Umar A, Darwish DBE, Alenezi MA (2024) Fungal pigments: Secondary metabolites and their application. In: Abd-Elsalam KA, Mohamed HI (eds) Fungal secondary metabolites. Elsevier, pp 173–195

van der Wielen PWJJ, van der Kooij D (2013) Nontuberculous mycobacteria, fungi, and opportunistic pathogens in unchlorinated drinking water in the Netherlands. Appl Environ Microbiol 79:825–834. https://doi.org/10.1128/AEM.02748-12/SUPPL_FILE/ZAM999104048SO1.PDF

Velmurugan P, Kamala-Kannan S, Balachandar V, Lakshmanaperumalsamy P, Chae JC, Oh BT (2010) Natural pigment extraction from five filamentous fungi for industrial applications and dyeing of leather. Carbohydr Polym 79:262–268. https://doi.org/10.1016/J.CARBPOL.2009.07.058

Venil CK, Velmurugan P, Dufossé L, Devi PR, Ravi AV (2020) Fungal pigments: potential coloring compounds for wide ranging applications in textile dyeing. J Fungi 6:68, 1–23. https://doi.org/10.3390/JOF6020068

Venkatachalam M, Magalon H, Dufossé L, Fouillaud M (2018) Production of pigments from the tropical marine-derived fungi *alaromyces albobiverticillius*: New resources for natural red-colored metabolites. J Food Compos Anal 70:35–48. https://doi.org/10.1016/J.JFCA.2018.03.007

Vivek-Ananth RP, Sahoo AK, Kumaravel K, Mohanraj K, Samal A (2021) MeFSAT: a curated natural product database specific to secondary metabolites of medicinal fungi. RSC Adv 11:2596–2607. https://doi.org/10.1039/D0RA10322E

Wainaina S, Taherzadeh MJ (2023) Automation and artificial intelligence in filamentous fungi-based bioprocesses: a review. Bioresour Technol 369:1–12. https://doi.org/10.1016/J.BIORTECH.2022.128421

Wijayawardene NN, Boonyuen N, Ranaweera CB, de Zoysa HKS, Padmathilake RE, Nifla F, Dai DQ, Liu Y, Suwannarach N, Kumla J, Bamunuarachchige TC, Chen HH (2023) OMICS and other advanced technologies in mycological applications. J Fungi 9(1–27):688. https://doi.org/10.3390/JOF9060688

Xu YJ (2017) Foodomics: a novel approach for food microbiology. TrAC, Trends Anal Chem 96:14–21. https://doi.org/10.1016/J.TRAC.2017.05.012

Yapa N, Lakmali D, De Zoysa KS, Silva S, Manawadu C, Herath BM, Madhushan A, Perera G, Ratnayakae O, Kapilan R, Rathnayake A, Sirisena AI, Asad S, Karunarathna SC, Bamunuarachchige C (2022) Biofertilizers: an emerging trend in agricultural sustainability. Chiang Mai J Sci 49:608–640. https://doi.org/10.12982/CMJS.2022.050

Zuhara S, Isaifan R (2018) The impact of criteria air pollutants on soil and water: a review. J Environ Sci Pollut Res Visit J 4:278–284. https://doi.org/10.30799/jespr.133.18040205

Chapter 21
Cellulases from Psychrophilic and Psychrotrophic Microorganisms and their Potential Applications

L. S. Mamatha Bhanu and Soumya Chatterjee

21.1 Introduction

Microbes have been living since time immemorial and have continued to survive through many harsh environmental changes, and they hold promising applications in various fields. Almost the Earth's entire atmosphere is covered by cold ecosystems, and the temperature is likely to be <5°C. Their diversity enables them to tolerate extremely cold or hot environments, known as extremophiles, which include archaea, bacteria, yeast, and viruses. About 80% of the earth is dominated by low-temperature environments (permafrost, polar surface, ice-cap regions, and deep oceans) (Moyer and Morita 2007). Cold environments like the polar surface and tundra are located in high-latitude areas in the mountain regions of the globe and have unique soil types, plants, and animals. Almost 12% of the earth is completely covered forever by snow and ice, which are found in Polar Regions, and about 33% is covered by seasonal snow and ice that is variable according to the seasons (Campbell et al. 2018). Cold-loving microorganisms or cold-adapted microorganisms successfully colonise cold regions and play a major role in nutrient succession and renewable energy production in the cold ecosphere. Cold-loving microorganisms are categorised as psychrophiles and psychrotrophs. (Morita 1975) defined psychrophiles as organisms that grow at an optimal temperature of about 15–20 °C. (Eddy 1960) proposed the use of the term 'psychrotroph' (psychrotolerant); their growth temperatures range between 15 °C and 20 °C. Psychrophiles use several biochemical pathways, including chemoautotrophy, heterotrophy, and photosynthesis, from diverse communities. Aerobic and

L. S. M. Bhanu (✉)
Department of Biotechnology, Yuvaraja's College, University of Mysore, Manasagangotri, Mysuru, Karnataka 570006, India
e-mail: ls.mamatha@gmail.com

S. Chatterjee
Biodegradation Technology Division, Defence Research Laboratory, DRDO, Tezpur, Assam, India

K. Bala et al. (eds.), *Harnessing Microbial Potential for Multifarious Applications*, Energy, Environment, and Sustainability, https://doi.org/10.1007/978-981-97-1152-9_21

anaerobic bacteria that inhabit this cold environment are affected by several factors, like high salt concentration, osmotic pressure, oxidative stress, cosmic radiation, the availability of nutrients, exopolymeric substances, and particulate matter; all of these tend to be maintained by temperature fluctuations (D'Amico et al. 2006). Consequently, the unique property of psychrophiles is to adapt to this extreme condition and remain metabolically active. Generally, microbial diversity includes the phylum of Proteobacteria (gram-negative α, β, and γ bacteria) and the species of *Pseudomonas* and *Vibrio* spp., whereas in gram-positive bacteria, *Bacteriodes* phylum*tophaga-Flavobacterium-Coryneforms, Ar Arthrobacter* sp., and *Micrococcus* sp. are found. There are also reports of the *Methanogenium* and *Methanococcus* genera. Among them, cyanobacteria, *Oscillatoria, Nostoc commune,* and *Phormidium* have been identified (Pandey et al. 2004). Low temperatures often affect the cell membrane fluidity, but cold-adapted bacteria have the ability to alter the fatty acid composition of the membranes, like changes in lipid head group, protein content, synthesis of different types of carotenoids, cis-to-trans fatty acid modulation, and altering chain length (Chintalapati et al. 2004). Further, low temperatures also affect the transcription and translation processes through reduced activity of enzymes, protein folding, and the instability of DNA and RNA structures. Evidently, the reports of methanogen, *Methanococcoidesburtonii,* from the Antarctic region show that dead-box RNA helicases (dead) involved in the hydrolysis of ATP and RNA binding help in tolerating cold shock and survive at low temperatures (Lim et al. 2000). Despite cellular activities, these organisms emerged by producing different cold-shock proteins (CSPs) or cold-active enzymes (Kuddus 2015; Saavedra et al. 2018). Cold-active enzymes catalyse the reaction efficiently in 0–25 °C temperature ranges (Stark et al. 2022). Indeed, these enzymes have several characteristic properties and multiple functions, such as efficient catalytic activity, lower activation energies, relative ease of large-scale production, and thermolabile (Kuddus 2015; Santiago et al. 2016; Nikolaivits et al. 2017). The proteins trehalose and extracellular polysaccharides, or exopolysaccharides (EPSs), present on the cell surface act as defences against adverse environmental conditions caused by freezing. It also has a significant role in ice-recrystallisation, or restraining ice growth (Deming and Young 2017). It also prevents protein denaturation and aggregation. Similarly, some cold-adapted microorganisms' production of chaperons helps promote protein folding and stability. (Romero-Romero et al. 2011; Piette et al. 2011a) (Fig. 21.1)

Cold-adapted enzymes have wide applications in biotechnological industries, including food processing (Kumar et al. 2021), pharmaceuticals (Ge et al. 2020), leather (Furhan et al. 2019), brewing and wine, paper and pulp industries (Kuhad et al. 2011a), and detergent industry (Kuddus and Ahmad 2012; Sarmiento et al. 2015), bioremediation and environmental management (Liu et al. 2023; Wiśniewska et al. 2021), straw resourcing (Wang et al. 2021), molecular biology research (Kumar et al. 2021; Berg et al. 2019) and as phosphate solubilizing bacteria (PSB) in the field of agriculture (Rizvi et al. 2451). Cold-adapted enzymes also facilitate plant growth by increasing growth hormones and siderophores (Tapia-Vázquez et al. 2020). Several other hydrolytic enzymes are also secreted that degrade cell walls, for example, chitinases, proteases, and cellulases. However, psychrophiles are widely

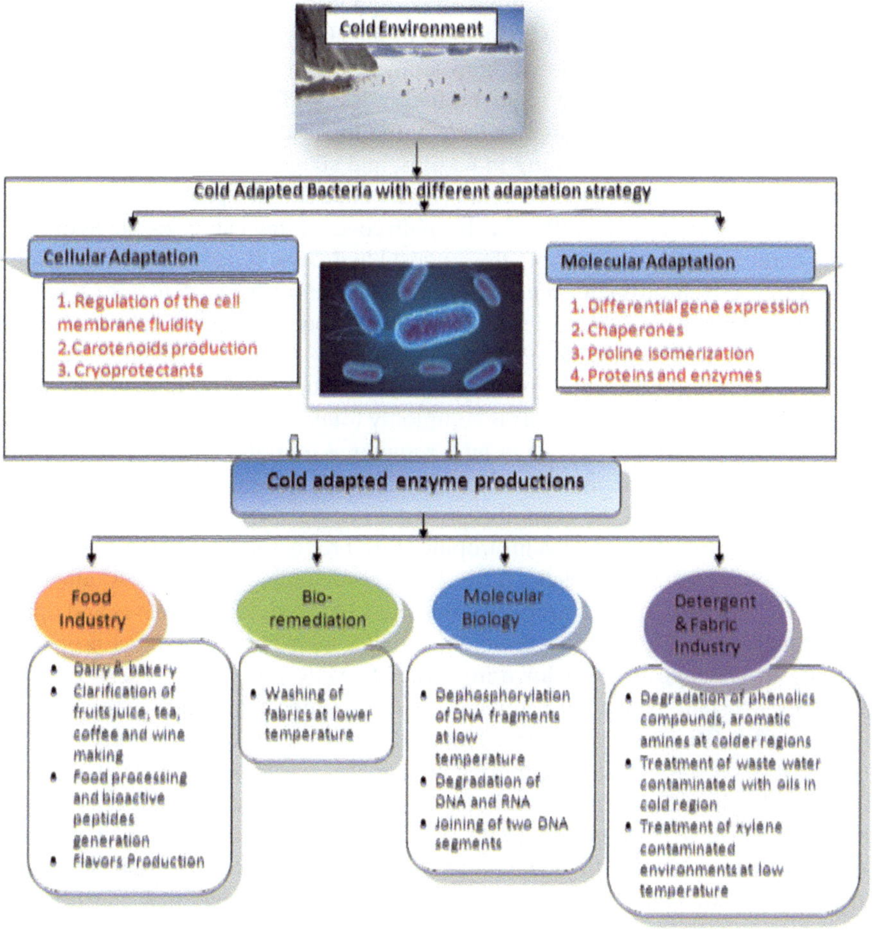

Fig. 21.1 An overview of Cold-adaptation strategy of Psychrophiles and Psychrotrophs in cold environment and the applications of cold-adapted enzymes in different fields

known for producing active enzymes at low temperatures and being stable under basic pH conditions. Cellulases are hydrolytic enzymes that hydrolyse cellulose into simpler sugars. Cellulases are produced from other groups (nonpsychrophiles) of microbes and animals and are also used as laundry additives. The cellulase enzyme from psychrophiles has applications mainly in agriculture, bioremediation, the food and pulp industry, and some in basic molecular biology (Kasana and Gulati 2011). Three types of cellulases, such as endogluconase, exogluconase, and glucosidase, are involved in the breakdown of the cellulose complex (Medie et al. 2011). The cold-active cellulase enzyme from *Acremonium alcalophilum* shows activity even at 40 °C (Hayashi et al. 1996), but the optimum temperature is 20 to 40 °C. This enzyme has greater potential to breakdown lignocellulosic biomass and is beneficial in the bioenergy sector (Cavicchioli et al. 2011). Alternatively, cellulose is used to convert into bioethanol at high temperatures (50–60 °C) for commercial purposes, and this can be used in the biofuel industry (Tiwari et al. 2015). Furthermore, cellulases from psychrophiles have a role in various sectors, such as pulping in the paper industry, biopolishing in textiles, and silage in food and feed formulations (Kasana and Gulati 2011). Recently, a new cold-adapted protein (RBcel1) from *Pseudomonas stutzeri* (Pst_2494) has been discovered that catalyses the hydrolysis of only carboxymethyl cellulose in the glycosyl hydrolase (GH5) family (Berlemont et al. 2009). Psychrotrophic microbes are able to grow from -1 to 10°C, but the optimal temperature is 21 °C. They are involved in food spoilage, like stored refrigerated food items (Yuan et al. 2018). The majority (90%) of psychrotrophs are isolated from stored vegetables, poultry, and meats, and most of them belong to the genus *Pseudomonas. P. fluorescens* is commonly found along with coliform bacteria, *Flavobacterium, and alcaligenes in milk samples. Klebsiella and Enterobacter* are also sometimes found in pasteurised milk. In milk, the main psychrotrophs are *Pseudomonas*, which is the commonest, but *Flavobacterium* and some *coliforms* have also been found to be psychrotrophic. Egg product spoilage was observed in psychrotrophic bacteria in the *Bacillus cereus* group collected in cold weather (Jan et al. 2011). These organisms produce cold-active enzymes like proteases and lipases that breakdown milk proteins and lipids and alter their physiochemical properties, resulting in spoilage without lowering the pH (McKellar 1989). These enzymes are highly resistant, as they are active even at ultra-high temperatures in superpasteurized milk (Oliveira et al. 2015). The cold-active enzymes that are produced by psychrotropes are effective in degrading organic waste substances (Hou et al. 2017). Additionally, they degrade urea and convert nitrate to nitrite (Kumari et al. 2022a) and are also used as plant growth promoter (PGPR) biofertilizers (Patni et al. 2018). Due to their unique climatic adaptations, they are used as model organisms in the study of cold-adaptation mechanisms and functional characteristics. Furthermore, they are economically important in industries for the production of peptides, bioactive compounds, antibiotics, and enzymes (Kumari et al. 2022b). Therefore, this chapter focuses on the adaptation, production, and potential biotechnological applications of cold-adapted cellulase enzymes from psychrophilic and psychrotrophic microorganisms.

21.1.1 Biodiversity of Psychrophilic Microorganisms

Schmidt-Nielsen (1902) coined the term ‘psychrophilic’ for the first time to indicate microorganisms that are capable of survival at 0 °C (Nogi and Kato 1999). Predominantly, psychrophiles and psychrotrophs represent all three domains: bacteria, archaea, and eukarya (Feller 2017a). The psychrotrophic microbes include different genera and phylums such as *Bacillus*, *Pseudomonas*, *Arthrobacter*, *Exiguobacterium*, *Serratia*, *Paenibacillus,* and *Providencia.* The phylum includes *Cyanobacteria, Euryarchaeota, Actinobacteria, Ascomycota, Basidiomycota, Bacteroidetes, Chlamydiae, Chloroflexi, Planctomycetes,* Firmicutes, *Gemmatimonadetes, Mucoromycota, Proteobacteria, Spirochaetes, Nitrospirae, Thaumarchaeota,* and *Verrucomicrobia.* Recent, new groups of psychrophilic cyanobacteria, fungi, and viruses have been reported (Margesin et al. 2008). The psychrophile species are isolated and characterised according to phylogenetic classifications based on 16S ribosomal RNA sequence. In 1964, the first taxonomical psychrophilic bacteria reported were *Vibrio* (*Moritella* gen. nov.) *marinus* MP-1 and *Vibrio* (*Colwellia* gen. nov.) psychroerythrus. However, the only genus composed exclusively of psychrophilic bacteria is *Moritella.* Other genera include both mesophilic and psychrophilic bacteria.

Class Gammaproteobacteria: Phylogenetically and physiologically, it belongs to the phylum Proteobacteria, which includes the Colwellia and Thalassomonas genera (Ivanova et al. 2004). It is composed of all gram-negative microbes. These microorganisms can live in several terrestrial and marine environments, and they play important roles in extreme climatic conditions such as hydrothermal vents. Genus Glaciecola was comprised of two species, *Glaciecolapunicea* and *G. pallidula* (Bowman et al. 1998), isolated from Arctic Ocean seawater, coastal surface seawater, deep-sea sediment, and East Sea sediment. In Genus Moritella, four psychropiezophilic species were identified: *M. abyssi, M. japonica, M. profunda, and M. yayanosii,* and they were isolated from the Japan Trench, the Mariana Trench, and the deep eastern tropical Atlantic Ocean (Nogi and Kato 1999; Xu et al. 2003). In the Genus Psychrobacter (Juni and Heym 1986), 14 species have been identified from Antarctic sea ice, krill Euphausia, ornithogenic soil and sediments, the deep sea, and marine ascidians (Maruyama et al. 2000; Romanenko et al. 2002). In Genus *Psychromonas* (Nogi et al. 2002, 2007) 14 species have been recognised by marine environments, of which most are true psychrophiles. Genus Shewanella has 51 species from marine, freshwater bodies, and industrial waste that have the ability to breakdown manganese and iron oxides (Bowman et al. 1997).

Class Deltaproteobacteria contains two genera, Desulfuromusa and *Desulfuromonas* (Pfennig and Biebl 1976; Liesack and Finster 1994) and three psychrophilic species, *Desulfofrigusoceanense, Desulfofabagelida,* and *Desulfotaleapsychrophila,* are isolated from marine Arctic sediments (Knoblauch et al. 1999).

Phylum Cytophaga-Flavobacterium: ***Bacteriodes*** contain two genera:

Genus *Flavobacterium:* This genus contains eight species of true psychrophiles (Humphry et al. 2001; Tamaki et al. 2003). Genus *Psychroflexus:* 4 species are recognised, but only one species, *P. torques,* from Antarctic sea ice, is psychrophilic (Bowman et al. 1998).

ClassDeinococci This includes gram-positive radiation-resistant Cocci. Almost 39 species have been recognised as psychrophilic from the Antarctic, Alpine regions, and industrial waste (Hirsch et al. 2004; Callegan et al. 2008)

In studies of psychrophilic bacteria in Antarctic regions, categories like obligate psychrophilic bacteria, psychrotolerant bacteria, and psychrotrophs were classified based on their ability to grow. The Antarctic soils consist of diverse groups of microbes such as Actinobacteria, Proteobacteria, and Bacteroidetes (Bottos et al. 2014). The microbial populations are influenced by several factors, such as the soil (moisture, pH, nutrients, carbon, copper), total water, organic content, and nitrogen (Aislabie et al. 2009). Chong et al. (2015) have studied the prokaryotes biodiversity from soil and cold-adapted extremophiles from the regions of Signy Island, Maritime Antarctic, and revealed that the dominant species are from the phylum Actinobacteria, Bacteroidetes, and Proteobacteria; other unusual phyla belong to Firmicutes, Planctomycetes, and Spirochetes. A comparative study of all these diverse habitats shows that water, algal blooms, and soli have more diversity compared to ice and deep-sea sediments. Therefore, the sea ice ecosystem has an excellent source of psychrophilic bacteria with low G+C content and low temperatures.

Bacterial cultures isolated from the Antarctic and Arctic regions have yielded new genera and species of the Proteobacteria phylum. The proteobacteria phylum consists of the α-proteobacteria, β-proteobacteria, and γ-proteobacteria classes, the Bacteriodetes phylum, and the Actinobacteria phylum. In the Baltic Sea ice, prototrophic sulphur bacteria have been found (Petri et al. 2001) that are anaerobic and contain high exopolysaccharides and mucopolysaccharides (Mock and Thomas 2005). There are also reports of denitrification at nitrite-accumulated ice (Kaartokallio 2001). In many sea ice habitats, including Arctic sea ice, anaerobic bacteria denitration and ammonium ion oxidation are found (Thomas and Dieckmann 2002). In 1989, Irgens reported a marine gas-vacuolated bacterium from euphotic waters. Gas-vacuolated cyanobacteria are commonly seen in freshwater. They belong to the Bacteroidetes genus *Polaribacter* (Gosink et al. 1998), and proteobacterial genera have *Octadecabacter* (Gosink et al. 1997) and *Psychromonas* (Auman et al. 2006). The cyanobacterial species detected in Antarctic Peninsula soils are *Phormidiumautumnale, Pseudoanabaenacatenata, and* the diatom *Pinnularia borealis.* There are a few populations of *Planktospheaerellaterrestris, Netrium* sp., Achnanthes *lapponica, Cylindrocystisrebissonii, Clamydomonaschlorostellata, Nostoc* sp., and *Cosmariumundulatum.* Additionally, Antarctic Peninsula soils dominate mixed microbial populations: algae, mosses, lichens, and liverworts. The most commonly observed ascomycetes are yeasts, and the genera include *Rhodotorula, Candida, Cryptococcus,* and *Aureobasidium.* In the continent Antarctica, the psychrotrophs identified in the oasis regions are mostly *Pseudomonas* that produced green fluorescent pigment and named a *Pseudomonas fluoresens* species capable of growing at 4–30°C and being halotolerant. Shivaji et al. (1989) reported two more species as *P. putida* or *P. syringae.* Achromobacteria and alcaligenes are gram-negative bacilli isolated from continental soils (Vincent 1988). The microbial populations in Adelie penguin ornithogenic soils include Gram-negative bacteria. Cell wall composition reveals that this bacterium belongs to the Psychrobacter genus, as members of this genus are halotolerant, psychrotolerant, and psychrophilic. *Pseudomonas*, A. psychophenolicus, and *Rhodococcus* species isolated from polluted Alpine soil are able to degrade phenol. Arctic soils are contaminated due to petroleum leakage with psychrotrophic Urediniomycete species with hydrocarbon degrading activity (Margesin and Schinner 2005). Similarly, cold-tolerant and psychrophilic alpine yeasts were isolated to breakdown phenol and other phenolic compounds (Bergauer et al. 2005). The fungal species identified include *Penicillium* species, *Aspergillus versicolor, Alternaria alternate, Chrysosporiumpannorum, Arthrobotrysoligospora, Epicoccum purpurascens, Botrytis cinera, and Sclerotium* sp. *Cladosporium cladosporoides, C. herbarum, Cylindrocarponmagnusianum, Phialophorafastigiata, Mortierella alpine, M. minutissim, Phomaexigua* var *exigua, Mycelia sterilia, Oidodendroncerealis,* and *Trichosporellacerebriformis.* The psychrophilic fungi present in Arctic permafrost are known to form ericoid mycorrhizae anchored to the alpine Ericales species roots and that belong to the genus *Geomyces.* Nevertheless, metagenomic phylogenetic studies on psychrophilic organisms help in understanding their functions in these environments.

21.1.2 Adaptations of Psychrophiles and Psychrotrophs

Microbes do not exhibit thermoregulatory mechanisms; when they encounter higher growth temperatures, they expire as the capacity of the cell to take up oxygen decreases in some psychrophiles at 10 and 20 °C. To overcome this kind of thermal cell damage, the microbes physiologically adapt to low temperatures (Moyer et al. 2017). The study on psychrophile growth conditions by Bhakoo and Herbert (1979) shows that at optimum growth temperatures, oxygen uptake is minimal, whereas it gradually increases at suboptimal and supraoptimal growth temperatures. It also proved that psychrophiles synthesise more rRNA and protein under suboptimal conditions. The post-translational modification of dihydrouridine represents the conformational flexibility and stability of RNA, which differ in the case of thermophiles. Thermophilic microbes have reduced RNA flexibility and stability and have adapted to higher temperatures. Psychrophiles have hemophasic adaptations that allow the transport of nutrients to take place even at low temperatures; hence, they retain membrane fluidity. In addition, they produce certain proteins and enzymes like cold-acclimation proteins, cryoprotectors, exopolysaccharides, and antifreeze molecules to lower ice crystallisation and grow normally at low temperatures (Bahar et al. 2017). Furthermore, there are physiological factors that affect the growth of psychrophiles, like salinity, nutrient content, water, pressure, and UV irradiation (Valdespino-Castillo et al. 2018). The adaptation strategies of cold-adapted microorganisms are discussed in detail both at the molecular and cellular levels.

21.1.3 Adaptations at the Molecular Level

Membrane fluidity and integrity:

The cell envelope of psychrophilic bacteria is complex and plays a major role in adaptation. The reduced fluidity of the membrane and the slower rate of diffusion across the envelope lead to high turgor pressure and later cell ruptures due to ice formation. The adaptation strategies of cold-adapted microorganisms, their envelope components, exterior coatings, and peptidoglycan layer are not completely understood. A decrease in membrane fluidity is the preliminary signal to sense variation in the external temperature (Jin et al. 2022). Adaptations to low temperatures by the cell membrane are accomplished by modifications in the lipid bilayer composition (D'Amico et al. 2001; Siddiqui et al. 2013) The cell membrane acts as a boundary for nutrient transport in and out of the cell by becoming permeable, semi-permeable, and non-permeable, which activates sensor mechanisms leading to upregulation of genes involved in membrane fluidity for the exchange of nutrients and minerals in and out of the cell (Shivaji and Prakash 2010). The physical property of the membrane depends on its fatty acid composition. An increase in unsaturated fatty acids disrupts packing order in the phospholipid bilayer, resulting in a gel-phase transition for membrane fluidity and rigidity, followed by a loss of function (Siliakus et al. 2017; Benforte et al. 2018). Steric alternation in the membrane rigidity due to increased methyl chains, the transition from trans to cis conformation of fatty acids, and short-chain fatty acids are also reported (D'Souza-Ault et al. 1993; Guan et al. 2013). For all these variations in the cell membrane compositions at cold temperatures, archaea, Gram-positive, and Gram-negative bacteria have adopted different strategies (D'Amico et al. 2001). Cell wall adaptation in psychrophiles is noticed by the thick peptidoglycan layer because of the upregulation of peptidoglycan biosynthesis genes (Rodrigues et al. 2009). Mykytczuk et al. (2013) observed a thick cell wall in *Planococcushalocryophilus* Or1, which is made up of peptidoglycan, choline, and calcium carbonate and protects against osmotic pressure and ice formation at low temperatures (Mykytczuk et al. 2013). Only a few cold-adapted bacteria were analysed for other cell envelope components. Gram-negative bacteria contain the lipopolysaccharide layer (LPS) as the outermost layer of the cell wall. Transcriptome analysis reveals the upregulation of LPS biosynthesis genes (glycosyltransferases) at low temperatures (Fig. 21.2). (Gao et al. 2007; De Maayer et al. 2014). Benforte et al. (2018) reported a mutation in the glycosyltransferases gene (wapH) that damaged the growth of bacteria in the Antarctic region at low temperatures. Interestingly, membrane pigments, exclusively carotenoids, play a role in membrane fluidity. Psychrophilic microorganisms isolated from high-altitude soils (Pandey et al. 2019), ice cores and glaciers (Shen et al. 2018), and marine surface waters (Dieser et al. 2010) have reported increased production of polar carotenoids known to increase cell rigidity. Moreover, these membrane pigments also act as photoprotectants against UV radiation. They act as antioxidants against ROS (reactive oxygen species), photosynthetic components, antimicrobials, and cryoprotectants, conferring resistance to low temperatures [meo-viscosity in temperature fluctuations] (Rodrigues et al. 2009; Chattopadhyay 2006; De Maayer et al. 2014)

Cryoprotectants: Low temperatures influence ice formation, resulting in osmotic stress and cell ruptures. In cold environments, ice formation is delayed or almost restricted inside the bacterial cell compared to the cell's outer surface (Fonseca et al. 2016). Ice formation on the outer surface may lead to membrane damage with the exit of minerals and nutrients and the dehydration of cells. This increases the solute concentration outside of the cell, creating an osmotic imbalance. Cold-adapted microbes overcome freezing by producing ice-binding proteins (antifreeze

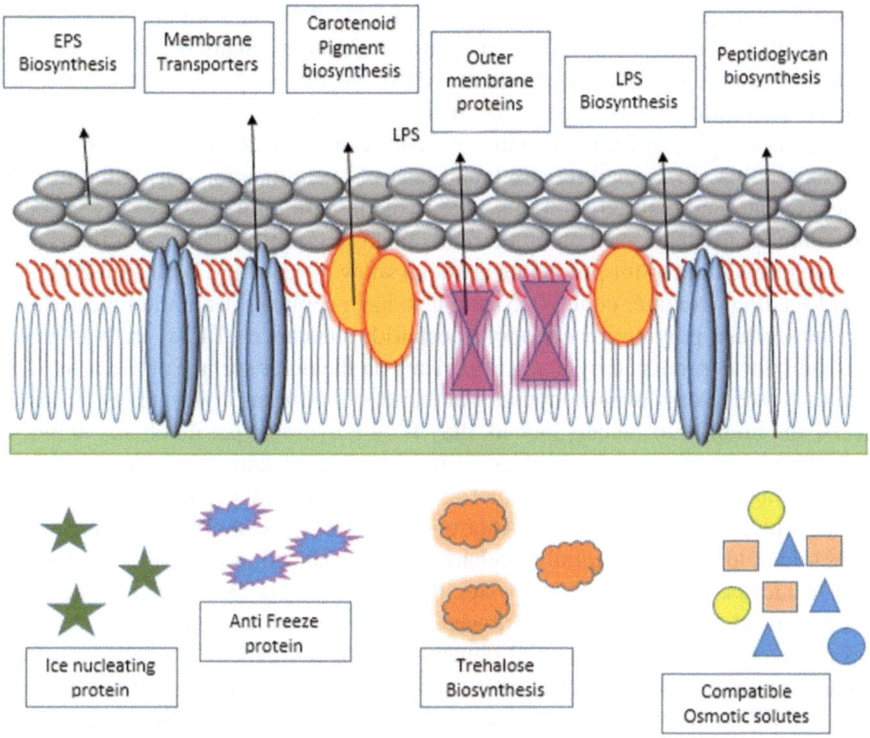

Fig. 21.2 Cold-adaptation mechanism of low-temperature microorganisms from the aspects of cell membrane structure, antifreeze protein, metabolic level, and low-temperature protective agent

and ice-nucleating proteins), compatible solutes, extracellular polymeric substances or exopolysaccharides (EPS), and/or biosurfactants.

Compatible solutes: Many cold-adapted microorganisms accumulate a variety of compatible solutes, like glycerol, glycine betaine, mannitol, sorbitol, trehalose, and N-Acetylglutaminylglutamine amide (D'Souza-Ault et al. 1993). These compatible solutes are low molecular weight, less toxic, organic, and low molecular weight, and less toxic organic solutes increase due to the upregulation of biosynthesis genes (Ghobakhlou et al. 2015; Goordial et al. 2016). This retains water content inside the cell, avoiding cell shrinkage and creating osmotic balance (Fonseca et al. 2016).

Ice-binding proteins: antifreeze proteins (AFPs): Antifreeze proteins with a molecular weight of ~2 to ~50 kDa bind to ice and inhibit ice growth and recrystallisation (Voets 2017). It was first identified in the fish blood of the Antarctic region; later, it was studied by all the populations of bacteria, fungi, diatoms, plants, insects, and crustaceans. *Marinomonasprimoryensis*, an Antarctic bacterium, produces AFP that specifically binds to ice and inhibits nucleation events (Bar Dolev et al. 2016). AFPs are known to have applications in the food and petroleum industries, aquaculture, and developing freeze-tolerant crops.

Ice-binding proteins: ice-nucleating proteins (INPs): They are membrane-bound proteins that favour ice formation. They are produced by many organisms as large, multimeric aggregates (Bar Dolev et al. 2016). In cold-adaptive microorganisms, it helps in preventing the damage caused by low temperatures, initiating ice nucleation at the exterior space by removing interior water (Lorv et al. 2014). Small extracellular aggregates that are formed release latent heat of crystallisation during the freezing process (Pummer et al. 2015). INPs have a role in agriculture, where they influence frost damage to plants and help in nutrient uptake. Currently, INPs are commercially available and are used for producing artificial snow (Cochet and Widehem 2000) and cloud seeding (Pummer et al. 2015) and also have applications in the food industry.

Extracellular polymeric or exopolysaccharide substances (EPSs) are high molecular substances produced by many organisms on the outer surface of the cells. They are composed mainly of carbohydrates and traces of proteins, lipids, phenols, nucleic acids, and humic substances. Indeed, they form a hydrated gel that helps in biofilm formation; they are mainly involved in osmoprotection and cryoprotection (Ewert and Deming 2013). Cold-adapted microorganisms have many genes for EPS biosynthesis, which was reported from Antarctic and Arctic ice shelf ponds (Varin et al. 2012). In cold habitats, exopolysaccharides are produced to prevent crystal formation inside the cell by directing water flow via osmosis (Kawahara et al. 1991). Cold-induced trehalose biosynthesis genes (otsA and otsB) are reported in E. coli, which favour the aggregation and denaturation of proteins by its collative effects (Delorge et al. 2014; Chen et al. 2017). *Colwelliapsychrerythraea*, a cold-tolerant bacterial EPS, alters the microstructure of ice crystals by raising pore density (Ewert and Deming 2013; Krembs et al. 2011). EPS counteracts as cryoprotectants by modifying cell adhesion, nutrient uptake, water retention, extracellular enzyme activity, and ice-nucleating proteins in cold conditions (Gerday 2013; Feller 2017b; Tribelli and López 2018) to maintain osmolality and prevent freezing (Casanueva et al. 2010).

Exopolysaccharides can be used as emulsifiers to improve the texture and stability of foods, which has a wide application in the food industry. Recently, Colliec-Jouault et al. (2019) and De Vuyst et al. (2016) reported beneficial effects as immunomodulatory, anti-inflammatory, and antioxidant. Microbial exopolysaccharides such as xanthan gum, dextran, and gellan gum have already been commercialised for various applications (Kumar et al. 2018).

21.1.4 Adaptations at the Molecular Level

Bacterial growth in freezing environments is influenced by modifications in their genome and protein levels. Protein misfolding affects membranes, movements, cold-shock proteins and enzymes, nutrient uptake, energy-generating systems, and other biodegradative enzymes (Russell 1998).

Cold temperatures reduce the stability of secondary structures of DNA and RNA, decrease the activity of enzymes involved in transcription and translation, and down-regulate gene expressions (Maayer et al. 2014). The proteomic analysis of *Pseudoalteromonashaloplanktis* TAC125, a psychrophile, reported a suppression of cold-induced proteins (Piette et al. 2011b) at a temperature change of 18–4 °C. Differential gene expression is generally observed in psychrotolerant cellulase microorganisms (Feller 2017b; De Maayer et al. 2014).

Psychrophiles and psychrotrophic microbes offer resistance to cryopreservation by producing cold-shock proteins (CSPs) (Sanchez et al. 1992). Morris et al. (1988) study on the growth phase observes that cells in the stationary phase are more resistant to freezing than in the exponential phase. CSPs are well studied in prokaryotes, especially in *Escherichia coli* (*E. coli*). During cold stress, *E. coli* accumulates nucleic acid-binding CSPs (Yamanaka et al. 1998; Bae et al. 2000). CspA is predominantly synthesised during cold stress, which has glycine-rich regions that account for about 10% of the total proteins. This is known to enhance translation at cold temperatures by preventing RNA structure formation (Karlson and Imai 2003).

The RbfA protein (ribosome-binding factor A) is a CSP recognised as a translation initiation factor that enhances translation in cold-shock mRNAs (Barria et al. 2013). RNase R and Csps, recognised as members of the mRNA surveillance system, help in translation under cold conditions (Yang et al. 2018).

Chaperones are transient proteins produced as a result of the cold-shock response in mesophilic and thermophilic microorganisms. They are often continuously overexpressed as cold-acclimation proteins at low temperatures in psychrophiles.

DNA and RNA chaperones efficiently undergo transcription, translation, and DNA replication. The protein and RNA/DNA chaperones favour efficient protein and RNA/DNA folding, respectively. It acts through protein misfolding and aggregation as well as the low-temperature stabilisation of RNA and DNA secondary structures. *E. coli* is used to produce recombinant protein chaperones from an Antarctic bacterium that is commercially available (Ferrer et al. 2003).

Proline isomerisation is a rate-limiting step in protein folding in many psychrophilic microorganisms. Generally, the cis–trans isomerisation (Fig. 21.3) of the peptidyl-prolyl bond is a slow process during protein folding at cold temperatures (Gerday 2013). The rate of the proline isomerisation reaction should also be slow at low temperatures for stability and substrate accessibility to the catalytic site. The majority of cold-adapted proteins have a reduced proline content (Baldwin 2008), which should diminish the negative effect of proline isomerisation on protein folding. 3-D structural analysis of psychrophilic proteins shows longer external loops with reduced proline content and flexible catalytic sites compared to their mesophilic potencies (Feller 2013a). The prolyl isomerases (PPiases) catalyse the isomerisation reaction. Interestingly, PPiases are overexpressed at low temperatures in the proteomes of most psychrophilic bacteria (Ting et al. 2010).

Proteins and enzymes: *P. haloplanktis, an*antarctic bacterium, synthesises proteins with 30% upregulation at 4 °C (Feller 2013a; Ricciardelli et al. 2019). Similarly, in other cold-adapted bacteria, overexpression of proteins is linked with the translation and availability of chaperones. Hence, this may be a rate-limiting

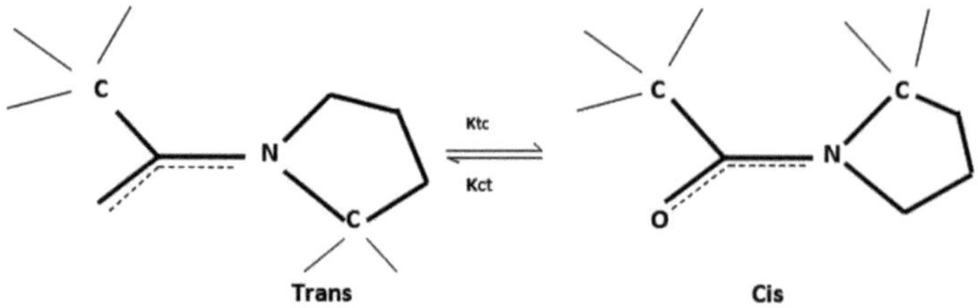

Fig. 21.3 Trans- and cis-isomers of a peptidyl-prolyl peptide bond

factor. In *M. burtonii*, genes are upregulated to retain RNA in a translational active state (Reed et al. 2013). In cold-active bacteria, *P. haloplanktis*, *P. ingrahamii*, and *C. psychrerythraea* reported higher amounts of tRNA and rRNA genes for higher translational rates (Feller 2013a). The psychrophilic bacteria, notably *M. burtonii*, *E. sibiricum*, *S. alaskensis*, *P. arcticus*, *S. denitrifycans*, and *P. haloplanktis*, overproduce RNA helicases that help the translation process at low temperatures. Psychrophilic archaea such as *Methanogeniumfrigidum* and *Methanococcoidesburtonii* produce a high amount of glycine and reduced proline residues in loop regions of psychrophilic proteins, which influences conformational flexibility, while the inclusion of glutamine and threonine residues helps reduce the charge on the protein (Reed et al. 2013). The cold-adapted archaea *Halorubumlacusprofundi*, tryptophan, and glutamic acid residues were reduced (Reed et al. 2013). A hydrophobic b-galactosidase surface protein in *H. lacusprofundi* was found to be similar to other proteins, the elongation factor 2 proteins of psychrophilic methanogens (Cavicchioli and Siddiqui 2006). Interestingly, psychrophilic enzymes have less thermostability but good structural flexibility at cold temperatures (Isaksen et al. 2016), resulting in effective binding between the active site and the substrate. Reducing the energy of activation leads to increased product turnover. It is known that the active site of a cold-active enzyme is more heat-labile as the activity is lost before unfolding of the protein, which might be due to flexibility (Struvay and Feller 2012; Zhang et al. 2017). Evidently, higher stability and flexibility in catalytic sites are reported in the isocitrate dehydrogenase and carbonic anhydrase enzymes of psychrophiles (Chiuri et al. 2009).

Metabolic adjustments: Various metabolic reprogrammings at low temperatures have been adapted, including a down-regulation of metabolic pathways, namely glycolysis, the pentose phosphate pathway (PPP), Kreb's (TCA) cycle, and the electron transport chain (ETC), as well as molybdopterin metabolism, which is reported to be downregulated at low temperatures in some psychrophiles (Piette et al. 2011; Tribelli and López 2018). The substitution of the primary metabolic pathway with secondary alternative pathways like ethanol oxidation, glyoxylate, methyglyoxal, or 2-methylcitrate cycles, and Propionyl-CoA catabolism (Tribelli and López 2018) depends on ecological habitat at low temperatures for energy conservation and survival.

21.2 Psychrophilic Cellulases and Their Properties

Cellulose (C6H10O5) is the most abundant carbohydrate in the plant cell wall. It's a polysaccharide composed of glucose monomers joined by b-1, 4 linkages. It provides the maximum renewable energy; therefore, it plays a vital role in the carbon cycle for all living organisms. About 80% of the biosphere and more than 90% of the marine environments have temperatures less than 5 °C. Cellulose synthesis and recycling in the cold environments account for a large proportion of the carbon cycle. Degradation of cellulose requires the combined action of the endoglucanases (EC 3.2.1.4) and the exoglucanases, or cellobiohydrolases (EC 3.2.1.91), which release the cellobiose and cellooligosaccharides. The process of cellulose hydrolysis is completed through the action of β-glucosidase (EC 3.2.1.21), which cleaves cellobiose, releasing glucose units. It belongs to the glycosyl hydrolase (GH) family and was the first low-temperature-active cellulase produced by the fungus *Acremonium alcalophilum* from the soil in 1996 (Hayashi et al. 1996). Later, the endo-β-glucanase was purified from the *Rhodotorulaglutinis* (psychrophilic yeast) of the soil surface (Oikawa et al. 1998). Cellulase-producing psychrophiles have been reported from different bacterial and fungal genera (Singh and Hayashi 1995). Cellulases generally contain three groups of cellulolytic activities, including endoglucanase, β-1, 4-cellobiohydrolase, and β-glucosidase. Due to their potential application in industries (Gupta et al. 2020), the cold-active enzymes are more efficient compared to their mesophilic microorganisms, as they represent protein stability. These enzymes are useful in studying protein folding (Feller et al. 1996). Crennell et al. reported that according to the Carbohydrate Active Enzymes database (CAZy), the cellulose enzymes belong to different GH families and are designated using two enzyme commission numbers (2002). Three main categories of cellulose enzymes have been known so far based on the site of action of the enzyme, that is, Endo-b-1, 4-glucanase or Endo-1, 4-b-Dglucan 4-glucanohydrolase (EG)EC 3.2.1.4: This binds to the noncrystalline part of cellulose. Cellobiohydrolase or 1,4-b-Dglucan Cellobiohydrolase (CBH) EC 3.2.1.91. This binds to the crystalline part of cellulose. β-glucosidase (BG) EC 3.2.1.21, and this breaks the cellobiose molecule (Kuhad et al. 2011b). Cel12A cellulase from *Rhodothermus marinus* from the GH12 family (Crennell et al. 2002) and CbhI from *Fusicoccum sp.* from the GH7 family (Kanokratana et al. 2011) have been reported. Bacterial cellulase is composed of one or more exocellobiohydrolases and endoglucanases that act in combination to hydrolyse cellulose into oligosaccharides (cellobiose) and β-glucosidases to glucose residues (Sukumaran et al. 2015).

Psychrophiles are cold-loving microorganisms. Psychrophilic bacteria produce enzymes (Table 21.1) that act on organic pollutants and xenobiotic compounds and degrade these toxic compounds into nontoxic substances; they have potential bioremediation applications. Subsequently, cellulases isolated and characterised from different psychrophilic microorganisms have varying optimal temperatures. Generally, the optimum temperature ranges from 20 to 40 °C; besides, there is also a study reporting a temperature of 50 °C by a cellulase from the psychrotrophic yeast *Rhodotorulaglutinis*. The optimum pH range is 4.5–7.0 (acidic to neutral pH), where the cellulases from *Paenibacillus*sp., *Shewanella*sp., and *Pseudoalteromonas* sp. MB-1 have optimum activity in the neutral to alkaline pH range. Psychrophilic cellulose has a molecular weight of 40–93 kDa. Endoglucanase (Cel9P) gene cloned *Paenibacillus* sp. BME-14 from marine bacteria has an optimum pH of 6.5 and a temperature of 35 °C with biotechnological application and is almost 52% similar to the natural cellulase enzyme (Fu et al. 2010). Another study is on the cellulosome-like enzyme purified from *Eisenia foetida*, CMCase, a multienzyme complex with β-glucosidase, β-1,3-glucanase, and β-xylosidase, with a molecular weight of 150 kDa active at 15 °C temperature and an optimum pH of 5.0 (Ueda et al. 2010). In bacteria, cellulases may be produced as a single enzyme or in a multienzyme complex or cellulosome complex. The catalytic domain is composed of cohesin, scaffoldin, and dockerin and binds to complimentary substrates through the carbohydrate binding module (CBM) (Chatterjee et al. 2015). In the early 1990s, cellulases were integrated into detergents (Ito 1997). It is used as a colour-brightening and softening agent in the detergent industry. Low-temperature cellulase is beneficial for lower (cooler) (30–40 °C) washing purposes with less water consumption. Psychrozymes, or psychrophilic enzymes, are adapted to allow for structural flexibility at the active site region, are active at lower temperatures, have less affinity for substrates, and have less enthalpy. Genome editing studies of the cold-adapted bacteria *Alteromonas*sp. SN2 were demonstrated through clustered regularly interspaced short palindromic repeats (CRISPRs). CRISPR identified 22 spacer genes that are linked to reverse transcriptase (Sharada et al. 2014). Industrial applications of psychrozymes are accomplished by increased catalytic activity, high stability, yield, diversity, and economic feasibility. Amylases, alkaline phosphatases, cellulases, catalase, nucleases, proteases, polygalacturonase, lipases, glycosylases, mannanases, pectate lyases, endo-arabinase, DNA ligase pectinase, and others are just a few of the enzymes produced by psychrophilic bacteria. Psychrozymes have high biocatalytic efficiency at mild to moderate temperatures.

Immobilised Cellulase enzyme: Cellulase exhibits better structural stability with maximum activity at alkaline pH; for large-scale applications, enzymes are immobilised on a gel-like matrix usually silica to retain stability and reusability (Bano et al. 2019). Immobilised cellulase has a high affinity to substrates and acts as a biocatalyst for cellulose bioconversion (Podrepšek et al. 2019). Glutaraldehyde is incorporated with immobilised cellulase to form cross-linked aggregates as it provides high stability and activity (Jia et al. 2017). Further, vinyl-cubic mesoporous silica used as a matrix for immobilised cellulase provides greater stability. Immobilised cellulose is combined with nanoparticles and used for bamboo biomass hydrolysis (Jesionowski et al. 2014; Hartono et al. 2010). Increased saccharification efficiency of immobilised cellulases in bioconversion was tested (Jia et al. 2017). Saccharification yield of holocellulase from Aspergillus niger immobilised with nanoparticle was investigated with free enzyme, where increased saccharification yield was observed (Kumar et al. 2017). *Trichoderma reesei* cellulase, immobilised

Table 21.1 Cellulase-producing psychrophiles (Kasana and Gulati 2011)

Psychrophilic microorganism	Cellulolytic enzyme	Temp. optimum (°C)	pH optimum	Molecular weight (kDa)	Reference
Acremonium alcalophilum	CMCase –(Piette et al. 2011a)	40	7.0		Hayashi et al. (1996)
Arthrobacter sp.	β-Glucosidase	35		93	Benesova et al. 2005)
Clostridium sp.	Endoglucanase, β-glucosidase, filter paper cellulase	20	5–6		Akila and Chandra 2003)
Fibrobacter succinogenes	Endoglucanase	25	5.5	55	You and Wang 2005)
Paenibacillus sp. strain C7	β-Glucosidase	30	35	7–8	Cristóbal et al. (2008)
Paenibacillus sp. BME-14	Endoglucanase	35	6.5	60	Fu et al. (2010)
Pseudoalteromonas sp. MB-1	Endoglucanase	35	7.2		You and Wang (2005)
Rhodotorulaglutinis	Endoglucanase	50	4.5	40	Oikawa et al. (1998)
Shewanella sp. G5	β-Glucosidase	37	8.0		Yadav et al. 2018)

on silica matrix (76%) and magnetic nanoparticles (85%), shows maximum immobilisation efficiency, high thermal stability, pH, and maximum velocity compared to non-immobilised enzymes (Grewal et al. 2017; Khoshnevisan et al. 2019).

21.2.1 Cold-Active Enzyme Activities for Different Applications

At present, research efforts are being made to develop substantially novel cold-adapted enzymes. The characteristic properties of cold-adapted enzymes at low temperatures and thermolability are highly valuable in different biotechnological fields in various industries, from molecular biology to food and beverage preparation (Cavicchioli and Siddiqui 2006). As a result, psychrophilic and psychrotrophic enzymes are gaining utmost importance and replacing mesophilic enzymes in industries. Due to their high catalytic activity and faster biochemical reactions, using less concentration of cold-adapted enzymes was found economically significant compared to other mesophiles analogous.

Agro-residual waste degradation: The optimisation of a cold-adapted bacterium, *Exiguobacteriumsibiricum* K1, with cellulase activity at 15 °C and pH 5, was used to degrade several agro-chemical wastes like sugar cane bagasse, corn stover, oil-extracted lemongrass, and wheat bran. This proved to be effective in degrading agro-based residual waste in cold regions (Kumari et al. 2022).

Plant growth promotion: The psychrotrophic microbes exhibit sustainable plant growth promotion with 1-aminocyclopropane-1-carboxylate (ACC) deaminase and biological nitrogen fixation; potassium, zinc, and phosphorussolubilisation; and production of biologically active compounds and phytoharmones such as ammonia, hydrogen cyanide, Fe-chelating compounds, cytokinins, gibberellic acids, and indole-3-acetic acid, which contribute to sustain environment and soil health (Yadav et al. 2018).

Biofilm Formation: Usually biofilms are composed of one or consortium of species associated on the surface of the bacterial populations submerged in a matrix of polymeric substances (EPSs) consisting of lipids, nucleic acids, polysaccharides, and proteins. Psychrotrophic milk-spoiling *Pseudomonas* isolated from biofilms have the ability to alter phenotypes by phase variation. During environmental stresses like temperature fluctuations, high-frequency phenotypic switching is mediated for the survival of the biofilm population.

Quorum Sensing and Metabolic Regulation: Bacteria communicate with each other using chemical signalling molecules called quorum sensing. N-acyl-homoserine lactones (AHLs), signalling molecules that are produced by a Gram-negative bacteria, psychrotrophic *P. fluorescens*, and other *Pseudomonas* spp, indicate spoilage of milk and milk-derived products.

Biofuels and Biorefineries: It is involved in the degradation and bioconversion of cellulosic materials. Renewable lignocellulosic biomasses are converted into pentose, hexose, or simple sugars; these sugars are later fermented to biofuel or bioethanol. This bioconversion includes three steps: pre-treatment, saccharification, and fermentation. Currently, many countries have been using bioethanol from plant biomass (Cristóbal et al. 2008). It is also involved in the production of organic acids, single-cell proteins, lipids, and animal feed (Oikawa et al. 1998; Ahmed and Bibi 2018; Chatterjee et al. 2015).

Textile and Detergent Industry: Alkaline cellulase helps improve colour brightness and dirt removal. In the textile industry, it is used for jeans biostoning, to improve the appearance of cotton fabrics (biopolishing), and to remove protuberances in cotton fabrics. Cellulase is incorporated in detergents to remove blood stains in cold, optimal washing. Anti-re-deposition of ink stains (Hayashi et al. 1996).

Paper and Pulp Industries: Biochemical pulping is done by cellulase to obtain bulk, high stiffness, and fines; this consumes 20–40% of energy. It is also used for the production of cardboard, sanitary paper, and paper towels. Cellulase is used as a co-additive in pulp bleaching, improving brightness. Cellulase from *Aspergillus niger* and *Trichodermreseei is* used to enhance the bleachability of softwood pulp kraft. CelB bacterial cellulase improves paper qualities like brightness, strength, pulp freeness, cleanliness, and improved drainage (Kuhad et al. 2011).

Animal Feed Industry: Agricultural waste and grain feed are treated with cellulase to improve the nutritional value of animal feed (Kuhad et al. 2011). It degrades anti-nutritional substances like cellulose, dextrins, lignin, β-glucan, pectins, inulin, and arabinoxylans, which in turn improve the nutritional value of animal feed (Kanokratana et al. 2011).

Food and Fermentation Industry: Improve the flavour, aroma, and volatile properties of fruits and vegetables for easy packing, transportation, and packing. Cellulase from *Paenibacillus* sp., *Bacillus* sp., *Trichoderma,* and *Aspergillus* is potentially used in the clarification of fruit juices and the extraction of antioxidants, starch, essential oils, flavouring agents, and proteins. Cellulases from *Aspergillus niger, Trichoderma reesi,* and *Humicolainsolens have been* reported to reduce dough time, tenderising meat (Oikawa et al. 1998; Singh and Hayashi 1995; Gupta et al. 2020; Feller et al. 1996) used in wine, bread making, stability of beer, clarification, to improve viscosity of wort and filterability, reduce dough time in fermentation, and to retain moisture, aroma, and shelf life (Fu et al. 2010; Feller 2013b; Kumari et al. 2022). β-glucosidases hydrolyse glycosylated oligosaccharide into glucose, and aglycones improve the aroma of the wine. Cellulase reduces the viscosity of fruit puree (Raveendran et al. 2018). It is used to extract carotenoids from tomato, carrot, sweet potato, and orange peels by enzymatic hydrolysis and has the potential to be used as colouring agent in the beverage and food industries (Yadav et al. 2018). Cellulases from *Aspergillus aculeatus are* used in the extraction of oil from olives (Oikawa et al. 1998). Polyphenols and flavonoids are extracted from grape pomace and tea leaves (Fu et al. 2010; Ueda et al. 2010)

21.3 Conclusions

Psychrophiles and psychrotrophs are ideal candidates that produce the hydrolytic enzyme cellulase, which have world-wide biotechnological applications due to their low thermostability at high temperatures and high catalytic effectiveness at cold temperatures. The quantitative economic overproduction of cold-adapted cellulase can be achieved by strain improvement and optimisation of various parameters, along with the identification of cost-effective nutrients. Furthermore, solving more crystal structures can provide in-depth knowledge and a better understanding of the structure–function relationship of cold-adapted enzymes. With advances in modern biotechnology, particularly in metagenomic and protein engineering, exploiting the properties of novel cold-active enzymes helps to overcome demands in industries and the agriculture sector, as well as remove environmental pollutants.

Conflict of Interest The authors declare no conflict of interest.

References

Ahmed A, Bibi A (2018) Fungal cellulase; production and applications: minireview. Int J Health Life Sci 4(1):19–36

Aislabie J, Jordan S, Ayton J, Klassen JL, Barker GM, Turner S (2009) Bacterial diversity associated with ornithogenic soil of the Ross Sea region, Antarctica. Can J Microbiol 55(1):21–36

Akila G, Chandra TS (2003) A novel cold-tolerant *Clostridium* strain PXYL1 isolated from a psychrophilic cattle manure digester that secretes thermolabile xylanase and cellulase. FEMS Microbiol Lett 219(1):63–67

Auman AJ, Breezee JL, Gosink JJ, Kämpfer P, Staley JT (2006) Psychromonasingrahamii sp. nov., a novel gas vacuolate, psychrophilic bacterium isolated from Arctic polar sea ice. Int J Syst Evol Microbiol 56(5):1001–1007

Bae W, Xia B, Inouye M, Severinov K (2000) Escherichia coli CspA-family RNA chaperones are transcription antiterminators. Proc Natl Acad Sci 97(14):7784–7789

Bahar S, Razavi S, Liu Yakov K (2017) Hot experience for cold-adapted microorganisms: temperature sensitivity of soil enzymes. Soil Biol Biochem 105:236e243

Baldwin RL (2008) The search for folding intermediates and the mechanism of protein folding. Annu Rev Biophys:371–21

Bano A, Chen X, Prasongsuk S, Akbar A, Lotrakul P, Punnapayak H, Anwar M, Sajid S, Ali I (2019) Purification and characterization of cellulase from obligate halophilic Aspergillus flavus (TISTR 3637) and its prospects for bioethanol production. Appl Biochem Biotechnol 189:1327–1337

Barria C, Malecki M, Arraiano CM (2013) Bacterial adaptation to cold. Microbiology 159(12):2437–2443

Benesova E, Markova M, Kralova B (2005) α-Glucosi-dase and β-glucosidase from psychrotrophic strain Arthro-bacter sp. C2–2. Czech J Food Sci 23:116–120

Benforte FC, Colonnella MA, Ricardi MM, Solar Venero EC, Lizarraga L, López NI, Tribelli PM (2018) Novel role of the LPS core glycosyltransferase WapH for cold adaptation in the Antarctic bacterium Pseudomonas extremaustralis. PLoS One 13(2):e0192559

Berg K, Leiros I, Williamson A (2019) Temperature adaptation of DNA ligases from psychrophilic organisms. Extremophiles 23:305–317

Bergauer UG, Janknecht S, Karsai S, Raulin C (2005) Praxis der Stimmtherapie. Springer, Berlin Heidelberg

Berlemont R, Delsaute M, Pipers D, D'amico S, Feller G, Galleni M, Power P (2009) Insights into bacterial cellulose biosynthesis by functional metagenomics on Antarctic soil samples. ISME J 3(9):1070–1081

Bhakoo M, Herbert RA (1979) The effects of temperature on the fatty acid and phospholipid composition of four obligately psychrophilic Vibrio spp. Arch Microbiol 121:121–127

Bottos EM, Scarrow JW, Archer SDJ, McDonald IR, Cary SC (2014) Bacterial community structures of Antarctic soils. In: Antarctic terrestrial microbiology: physical and biological properties of Antarctic soils, pp 9–33

Bowman JP, McCammon SA, Nichols DS, Skerratt JH, Rea SM, Nichols PD, McMeekin TA (1997) Shewanellagelidimarina sp. nov. and Shewanellafrigidimarina sp. nov., novel Antarctic species with the ability to produce eicosapentaenoic acid (20: 5ω3) and grow anaerobically by dissimilatory Fe (III) reduction. Int J Syst Evol Microbiol 47(4):1040–1047

Bowman JP, McCammon SA, Lewis T, Skerratt JH, Brown JL, Nichols DS, McMeekin TA (1998) Psychroflexustorquis gen. nov., sp. nov. a psychrophilic species from Antarctic sea ice, and reclassification of Flavobacterium gondwanense (Dobson et al. 1993) as Psychroflexusgondwanense gen. nov., comb. nov. Microbiology 144(6):1601–1609

Callegan RP, Nobre MF, McTernan PM, Battista JR, Navarro-Gonzalez R, McKay CP, da Costa MS, Rainey FA (2008) Description of four novel psychrophilic, ionizing radiation-sensitive *Deinococcus* species from alpine environments. Int J Syst Evol Microbiol 58(5):1252–1258

Campbell S, Affleck RT, Sinclair S (2018) Ground-penetrating radar studies of permafrost, periglacial, and near-surface geology at McMurdo Station, Antarctica. Cold Reg Sci Technol 148:38–49

Casanueva A, Tuffin M, Cary C, Cowan DA (2010) Molecular adaptations to psychrophily: the impact of 'omic'technologies. Trends Microbiol 18(8):374–381

Cavicchioli R, Siddiqui KS (2006) Cold-adapted enzymes. Enzyme Technol:615–638

Cavicchioli R, Charlton T, Ertan H, Mohd Omar S, Siddiqui KS, Williams TJ (2011) Biotechnological uses of enzymes from psychrophiles. MicrobBiotechnol 4:449–460

Chatterjee S, Sharma S, Prasad R, Datta S, Dubey D, Meghvansi M, Vairale M, Veer V (2015) Cellulase enzyme based biodegradation of cellulosic materials: an overview. S Asian J Exp Biol 5(6):271e282

Chattopadhyay MK (2006) Mechanism of bacterial adaptation to low temperature. J Biosci 31:157–165

Xi, Chen, Yang H, Wang Y (2017) Parametric study of passive design strategies for high-rise residential buildings in hot and humid climates: miscellaneous impact factors. Renew Sustain Energy Rev 69:442–460

Chintalapati S, Kiran MD, Shivaji S (2004) Role of membrane lipid fatty acids in cold adaptation. Cell Mol Biol (Noisy-le-Grand, France) 50(5):631–642

Chiuri R, Maiorano G, Rizzello A, Del Mercato LL, Cingolani R, Rinaldi R, Maffia M, Pompa PP (2009) Exploring local flexibility/rigidity in psychrophilic and mesophilic carbonic anhydrases. Biophys J 96(4):1586–1596

Chong C-W, Pearce DA, Convey P (2015) Emerging spatial patterns in Antarctic prokaryotes. Front Microbiol 6:1058

Cochet N, Widehem P (2000) Ice crystallization by Pseudomonas syringae. Appl Microbiol Biotechnol 54:153–161

Colliec-Jouault S, Sinquin C, Zykwinska A, Delbarre-Ladrat C (2019) Marine biodiversity as a new source of promising polysaccharides. Enzym Technol Mar Polysaccharides 3

Crennell SJ, Hreggvidsson GO, Karlsson EN (2002) The structure of rhodothermus marinus Cel12A, a highly thermostable family 12 endoglucanase, at 1.8 Å resolution. J Mol Biol 320(4):883–897

Cristóbal HA, Breccia JD, Abate CM (2008) Isolation and molecular characterization of Shewanella sp. G5, a producer of cold-active β-D-glucosidases. J Basic Microbiol 48 (1):16–24

D'Amico S, Gerday C, Feller G (2001) Structural determinants of cold adaptation and stability in a large protein. J Biol Chem 276(28):25791–25796

D'Amico S, Collins T, Marx J-C, Feller G, Gerday C, Gerday C (2006) Psychrophilic microorganisms: challenges for life. EMBO Rep 7(4):385–389

Delorge I, Janiak M, Carpentier S, Van Dijck P (2014) Fine tuning of trehalose biosynthesis and hydrolysis as novel tools for the generation of abiotic stress tolerant plants. Front Plant Sci 5:147

Deming JW, Young JN (2017) The role of exopolysaccharides in microbial adaptation to cold habitats. In: Psychrophiles: from biodiversity to biotechnology, pp 259–284

Dieser M, Greenwood M, Foreman CM (2010) Carotenoid pigmentation in Antarctic heterotrophic bacteria as a strategy to withstand environmental stresses. Arct Antarct Alp Res 42(4):396–405

Bar Dolev M, Braslavsky I, Davies PL (2016) Ice-binding proteins and their function. Annu Rev Biochem 85:515–542

D'Souza-Ault MR, Smith LT, Smith GM (1993) Roles of N-acetylglutaminylglutamine amide and glycine betaine in adaptation of Pseudomonas aeruginosa to osmotic stress. Appl Environ Microbiol 59(2):473–478

Eddy BP (1960) The use and meaning of the term "psychrophilic." J Appl Bacteriol 23:189–190

Ewert M, Deming JW (2013) Sea ice microorganisms: Environmental constraints and extracellular responses. Biology 2(2):603–628

Fonseca F, Meneghel J, Cenard S, Passot S, Morris GJ (2016) Determination of intracellular vitrification temperatures for unicellular micro organisms under conditions relevant for cryopreservation. PLoS One 11(4):e0152939

Feller G (2013a) Psychrophilic enzymes: from folding to function and biotechnology. Scientifica 2013

Feller G (2013b) Psychrophilic enzymes: from folding to function and biotechnology. Scientifica (Cairo) 5:12840

Feller G (2017a) Cryosphere and psychrophiles: insights into a cold origin of life? Life 7(2):25

Feller G (2017b) Cryosphere and psychrophiles: Insights into a cold origin of life. Life 7(2):25

Ferrer M, Chernikova TN, Yakimov MM, Golyshin PN, Timmis KN (2003) Chaperonins govern growth of Escherichia coli at low temperatures. Nat Biotechnol 21(11):1266–1267

Fu X, Liu P, Lin L, Hong Y, Huang X, Meng X, Liu Z (2010) A novel endoglucanase (Cel9P) from a marine bacterium Paenibacillus sp. BME-14. Appl Biochem Biotechnol 160:1627–1636

Furhan J, Awasthi P, Sharma S (2019) Biochemical characterization and homology modelling of cold-active alkophilic protease from Northwestern Himalayas and its application in detergent industry. Biocat Agric Biotechnol 17:726e735

Gao J-P, Chao D-Y, Lin H-X (2007) Understanding abiotic stress tolerance mechanisms: recent studies on stress response in rice. J Integr Plant Biol 49(6):742–750

Ge J, Jiang X, Liu W, Wang Y, Huang H, Bai Y, Su X, Yao B, Luo H (2020) Characterization, stability improvement, and bread baking applications of a novel cold-adapted glucose oxidase from Cladosporium neopsychrotolerans SL16. Food Chem 310:125970

Gerday C (2013) Psychrophily and catalysis. Biology 2(2):719–741

Feller G, Narinx E, Arpigny JL, Aittaleb M, Baise E, Genicot S, Gerday G (1996) Enzymes from psychrophilic organisms. FEMS Microbiol Rev 18(2–3):189–202

Ghobakhlou A-F, Johnston A, Harris L, Antoun H, Laberge S (2015) Microarray transcriptional profiling of Arctic Mesorhizobium strain N33 at low temperature provides insights into cold adaption strategies. BMC Genomics 16:1–14

Gosink JJ, Herwig RP, Staley JT (1997) Octadecabacter arcticus gen. nov., sp. nov., and O. antarcticus, sp. nov., nonpigmented, psychrophilic gas vacuolate bacteria from polar sea ice and water. Syst Appl Microbiol 20(3):356–365

Gosink JJ, Woese CR, Staley JT (1998) Polaribacter gen. nov., with three new species, P. irgensii sp. nov., P. franzmannii sp. nov. and P. filamentus sp. nov., gas vacuolate polar marine bacteria of the Cytophaga-Flavobacterium-Bacteroides group and reclassification of 'Flectobacillusglomeratus' as Polaribacterglomeratus comb. nov. Int J Syst Evol Microbiol 48(1):223–235

Goordial J, Raymond-Bouchard I, Zolotarov Y, de Bethencourt L, Ronholm J, Shapiro N, Woyke T et al (2016) Cold adaptive traits revealed by comparative genomic analysis of the eurypsychrophileRhodococcus sp. JG3 isolated from high elevation McMurdo Dry Valley permafrost, Antarctica. FEMS Microbiol Ecol 92(2):fiv154

Grewal J, Ahmad R, Khare SK (2017) Development of cellulase-nanoconjugates with enhanced ionic liquid and thermal stability for in situ lignocellulose saccharification. Bioresour Technol 242:236–243

Guan Q, Wu J, Zhang Y, Jiang C, Liu R, Chai C, Zhu J (2013) A DEAD box RNA helicase is critical for pre-mRNA splicing, cold-responsive gene regulation, and cold tolerance in Arabidopsis. Plant Cell 25(1):342–356

Gupta SK, Kataki S, Chatterjee S, Prasad RK, Datta S, Vairale MG, Sharma S, Dwivedi SK, Gupta DK (2020) Cold adaptation in bacteria with special focus on cellulase production and its potential application. J Clean Prod 258:120351

Hayashi K, Nimura Y, Ohara N, Uchimura T, Suzuki H, Komacata K, Kozaki M (1996) Low-temperature-active cellulase produced by Acremonium alcalophilum-JCM 7366-Note. Seibutsu-Kogaku Kaishi-J Soc Ferment Bioeng 74(1):7–10

Hirsch P, Gallikowski CA, Siebert J, Peissl K, Kroppenstedt R, Schumann P, Stackebrandt E, Anderson R (2004) Deinococcusfrigens sp. nov., Deinococcussaxicola sp. nov., and Deinococcusmarmoris sp. nov., low temperature and draught-tolerating, UV-resistant bacteria from continental Antarctica. Syst Appl Microbiol 27(6):636–645

Hou N, Wen L, Cao H, Liu K, An X, Li D, Wang H, Du X, Li C (2017) Role of psychrotrophic bacteria in organic domestic waste composting in cold regions of China. Biores Technol 236:20–28

Humphry DR, George A, Black GW, Cummings SP (2001) Flavobacterium frigidarium sp. nov., an aerobic, psychrophilic, xylanolytic and laminarinolytic bacterium from Antarctica. Int J Syst Evol Microbiol 51(4):1235–1243

Isaksen GV, Åqvist J, Brandsdal BO (2016) Enzyme surface rigidity tunes the temperature dependence of catalytic rates. Proc Natl Acad Sci 113(28):7822–7827

Ito S (1997) Alkaline cellulases from alkaliphilic Bacillus: enzymatic properties, genetics, and application to detergents. Extremophiles 1:61–66

Ivanova EP, Flavier S, Christen R (2004) Phylogenetic relationships among marine Alteromonas-like proteobacteria: emended description of the family Alteromonadaceae and proposal of Pseudoalteromonadaceae fam. nov., Colwelliaceae fam. nov., Shewanellaceae fam. nov., Moritellaceae fam. nov., Ferrimonadaceae fam. nov., Idiomarinaceae fam. nov. and Psychromonadaceae fam. nov. Int J Syst Evol Microbiol 54(5):1773–1788

Jan S, Brunet N, Techer C, Le Maréchal C, Koné AZ, Grosset N, Cochet MF et al (2011) Biodiversity of psychrotrophic bacteria of the Bacillus cereus group collected on farm and in egg product industry. Food Microbiol 28(2):261–265

Jesionowski T, Zdarta J, Krajewska B (2014) Enzyme immobilization by adsorption: a review. Adsorption 20:801–821

Jia J, Zhang W, Yang Z, Yang X, Wang N, Yu X (2017) Novel magnetic cross-linked cellulase aggregates with a potential application in lignocellulosic biomass bioconversion. Molecules 22(2):269

Jin S, Wang Y, Zhao X (2022) Cold-adaptive mechanism of psychrophilic bacteria in food and its application. Microb Pathog 169:105652

Juni E, Heym GA (1986) Psychrobacterimmobilis gen. nov., sp. nov.: genospecies composed of gram-negative, aerobic, oxidase-positive coccobacilli. Int J Syst Evol Microbiol 36(3):388–391

Kaartokallio H (2001) Evidence for active microbial nitrogen transformations in sea ice (Gulf of Bothnia, Baltic Sea) in midwinter. Polar Biol 24:21–28

Kanokratana P, Uengwetwanit T, Rattanachomsri U, Bunterngsook B, Nimchua T, Tangphatsornruang S, Plengvidhya V, Champreda V, Eurwilaichitr L (2011) Insights into the phylogeny and metabolic potential of a primary tropical peat swamp forest microbial community by metagenomic analysis. Microbial Ecol 61:518–528

Karlson D, Imai R (2003) Conservation of the cold shock domain protein family in plants. Plant Physiol 131(1):12–15

Kasana RC, Gulati A (2011) Cellulases from psychrophilic microorganisms: a review. J Basic Microbiol 51(6):572–579

Kawahara K, Seydel U, Matsuura M, Danbara H, Rietschel ET (1991) Zahringer U Chemical structure of glycosphingolipids isolated from *Sphingomonaspaucimobilis*. FEBS Lett 292:107–110

Khoshnevisan K, Poorakbar E, Baharifar H, Barkhi M (2019) Recent advances of cellulase immobilization onto magnetic nanoparticles: an update review. Magnetochemistry 5(2):36

Knoblauch C, Sahm K, Jørgensen BB (1999) Psychrophilic sulfate-reducing bacteria isolated from permanently cold Arctic marine sediments: description of Desulfofrigusoceanense gen. nov., sp. nov., Desulfofrigus fragile sp. nov., Desulfofabagelida gen. nov., sp. nov., Desulfotaleapsychrophila gen. nov., sp. nov. and Desulfotaleaarctica sp. nov. Int J Syst Evol Microbiol 49(4):1631–1643

Krembs C, Eicken H, Deming JW (2011) Exopolymer alteration of physical properties of sea ice and implications for ice habitability and biogeochemistry in a warmer Arctic. Proc US Natl Acad Sci 108:3653–3658

Kuddus M (2015) Cold-active microbial enzymes. Biochem Physiol 4:e132

Kuddus M, Ahmad IZ (2012) Cold-active extracellular α-amylase production from novel bacteria Microbacteriumfoliorum GA2 and Bacillus cereus GA6 isolated from Gangotri glacier, Western Himalaya. J Genet Eng Biotechnol 10 (1):151–159

Kuhad RC, Gupta R, Singh A (2011) Microbial cellulases and their industrial applications. Enzyme Res:280696

Kuhad RC, Gupta R, Singh A (2011) Microbial cellulases and their industrial applications. Enzyme Res

Kumar A, Singh S, Tiwari R, Goel R, Nain L (2017) Immobilization of indigenous holocellulase on iron oxide (Fe_2O_3) nanoparticles enhanced hydrolysis of alkali pretreated paddy straw. Int J Biol Macromol 96:538–549

Kumar A, Marcolli C, Luo B, Peter T (2018) Ice nucleation activity of silicates and aluminosilicates in pure water and aqueous solutions–Part 1: the K-feldspar microcline. Atmos Chem Phys 18(10):7057–7079

Kumari A, Upadhyay J, Joshi R (2021) Psychrotrophic microbes: biodiversity, adaptation, and implications. Microb Metatranscriptomics Belowground:273–293

Kumar R, Kumar P, Giri A (2019) Regional impact of psychrophilic bacteria on bioremediation. In: Smart bioremediation technologies. Academic Press, pp 119–135

Kumari S, Manyapu V, Kumar R (2022a) Recent advances in composting and vermicomposting techniques in the cold region: resource recovery, challenges, and way forward. Adv Org Waste Manage:131–154

Kumari S, Kumar A, Kumar R (2022b) A cold-active cellulase produced from Exiguobacteriumsibiricum K1 for the valorization of agro-residual resources. Biomass Convers Biorefinery:1–11

Liesack W, Finster K (1994) Phylogenetic analysis of five strains of gram-negative, obligately anaerobic, sulfur-reducing bacteria and description of Desulfuromusa gen. nov., including Desulfuromusakysingii sp. nov., Desulfuromusabakii sp. nov., and Desulfuromusasuccinoxidans sp. nov. Int J Syst Evol Microbiol 44(4):753–758

Lim J, Thomas T, Cavicchioli R (2000) Low temperature regulated DEAD-box RNA helicase from the Antarctic archaeon, Methanococcoidesburtonii. J Mol Biol 297(3):553–567

Liu Y, Zhang N, Ma J, Zhou Y, Wei Q, Tian C, Fang Y, Zhong R, Chen G, Zhang S (2023) Advances in cold-adapted enzymes derived from microorganisms. Front Microbiol 14:1152847

Lorv JSH, Rose DR, Glick BR (2014) Bacterial ice crystal controlling proteins. Scientifica 20

Hartono SB, Qiao SZ, Liu J, Jack K, Ladewig BP, Hao Z, Lu GQM (2010) Functionalized mesoporous silica with very large pores for cellulase immobilization. J Phys Chem C 114(18)

De Maayer P, Anderson D, Cary C, Cowan DA (2014) Some like it cold: understanding the survival strategies of psychrophiles. EMBO Rep 15(5):508–517

Margesin R, Schinner F, Marx J-C, Gerday C (eds) (2008) Psychrophiles: from biodiversity to biotechnology. Springer, Berlin

Margesin R, Schinner F (eds) (2005) Manual for soil analysis-monitoring and assessing soil bioremediation, vol 5. Springer Science & Business Media

Maruyama A, Honda D, Yamamoto H, Kitamura K, Higashihara T (2000) Phylogenetic analysis of psychrophilic bacteria isolated from the Japan Trench, including a description of the deep-sea species *Psychrobacterpacificensis* sp. nov. Int J Syst Evol Microbiol 50(2):835–846

McKellar RC (1989) Enzymes of psychrotrophs in raw food. CRC Press

Medie FM, Vincentelli R, Drancourt M, Henrissat B (2011) *Mycobacterium tuberculosis* Rv1090 and Rv1987 encode functional β-glucan-targeting proteins. Protein Expr Purif 75(2):172–176

Mock T, Thomas DN (2005) Recent advances in sea-ice microbiology. Environ Microbiol 7(5):605–619

Morita RY (1975) Psychrophilic bacteria. Bacteriol Rev 39(2):144–167

Morris GJ, Coulson GE, Clarke KJ (1988) Freezing injury in *Saccharomyces cerevisiae*: the effect of growth conditions. Cryobiology 25(5):471–482

Moyer CL, Morita RY (2007) Psychrophiles and psychrotrophs. Encycl Life Sci 1(6)

Moyer C, Collins R, Morita R (2017) En: Psychrophiles and psychrotrophs. Reference Module in Life Sciences. Chichester (Reino Unido):298–303

Mykytczuk N, Foote SJ, Omelon CR, Southam G, Greer CW, Whyte LG (2013) Bacterial growth at−15°C; molecular insights from the permafrost bacterium Planococcushalocryophilus Or1. ISME J 7(6):1211–1226

Nikolaivits E, Dimarogona M, Fokialakis N, Topakas E (2017) Marine-derived biocatalysts: importance, accessing, and application in aromatic pollutant bioremediation. Front Microbiol 8:265
Nogi Y, Kato C (1999) Taxonomic studies of extremely barophilic bacteria isolated from the Mariana Trench and description of Moritellayayanosii sp. nov., a new barophilic bacterial isolate. Extremophiles 3:71–77
Nogi Y, Kato C, Horikoshi K (2002) Psychromonaskaikoae sp. nov., a novel from the deepest piezophilic bacterium cold-seep sediments in the Japan Trench. Int J Syst Evol Microbiol 52(5):1527–1532
Nogi Y, Hosoya S, Kato C, Horikoshi K (2007) Psychromonashadalis sp. nov., a novel piezophilic bacterium isolated from the bottom of the Japan Trench. Int J Syst Evol Microbiol 57(6):1360–1364
Oikawa T, TsUKAGAwA Y, SoDA K (1998) Endo-β-glucanase secreted by a psychrotrophic yeast: purification and characterization. Biosci Biotechnol Biochem 62(9):1751–1756
Oliveira GB, Favarin L, Luchese RH, McIntosh D (2015) Psychrotrophic bacteria in milk: how much do we really know? Braz J Microbiol 46:313–321
Pandey KD, Shukla SP, Shukla PN, Giri DD, Singh JS, Singh P, Kashyap AK (2004) Cyanobacteria in Antarctica: ecology, physiology and cold adaptation. Cell Mol Biol Paris-Wegmann 50(5):575–584
Pandey A, Jain R, Sharma A, Dhakar K, Kaira GS, Rahi P, Dhyani A, Pandey N, Adhikari P, Shouche YS (2019) 16S rRNA gene sequencing and MALDI-TOF mass spectrometry based comparative assessment and bioprospection of psychrotolerant bacteria isolated from high altitudes under mountain ecosystem. SN Appl Sci 1:1–12
Patni B, Panwar AS, Negi P, Joshi GK (2018) Plant growth promoting traits of psychrotolerant bacteria: a boon for agriculture in hilly terrains. Plant Sci Today 5(1):24–28
Petri R, Podgorsek L, Imhoff JF (2001) Phylogeny and distribution of the sox B gene among thiosulfate-oxidizing bacteria. FEMS Microbiol Lett 197(2):171–178
Pfennig N, Biebl H (1976) Desulfuromonasacetoxidans gen. nov. and sp. nov., a new anaerobic, sulfur-reducing, acetate-oxidizing bacterium. Arch Microbiol 110:3–12
Piette F, Struvay C, Feller G (2011a) The protein folding challenge in psychrophiles: facts and current issues. Environ Microbiol 13(8):1924–1933
Piette F, D'Amico S, Mazzucchelli G, Danchin A, Leprince P, Feller G (2011b) Life in the cold: a proteomic study of cold-repressed proteins in the Antarctic bacterium Pseudoalteromonashaloplanktis TAC125. Appl Environ Microbiol 77(11):3881–3883
Podrepšek GH, Knez Z, Leitgeb M (2019) Activation of cellulase cross-linked enzyme aggregates (CLEAs) in scCO2. J Supercrit Fluids 154:104629
Pummer BG, Budke C, Augustin-Bauditz S, Niedermeier D, Felgitsch L, Kampf CJ, Huber RG et al (2015) Ice nucleation by water-soluble macromolecules. Atmos Chem Phys 15(8):4077–4091
Raveendran S, Parameswaran B, Ummalyma SB, Abraham A, Mathew AK, Madhavan A, Rebello S, Pandey A (2018) Applications of microbial enzymes in food industry. Food Technol Biotechnol 56(1):16
Reed CJ, Lewis H, Trejo E, Winston V, Evilia C (2013) Protein adaptations in archaeal extremophiles. Archaea 2013
Ricciardelli A, Casillo A, Vergara A, Balasco N, Corsaro MM, Tutino ML, Parrilli E (2019) Environmental conditions shape the biofilm of the Antarctic bacterium Pseudoalteromonashaloplanktis TAC125. Microbiol Res 218:66–75
Rizvi A, Ahmed B, Khan MS, Umar S, Lee J (2021) Psychrophilic bacterial phosphate-biofertilizers: a novel extremophile for sustainable crop production under cold environment. Microorganisms 9(12):2451
Rodrigues DF, da C Jesus E, Ayala-del-Rfo HL, Pellizari VH, Gilichinsky D, Sepulveda-Torres L, Tiedje JM (2009) Biogeography of two cold-adapted genera: psychrobacter and exiguobacterium. ISME J (6):658–665

Romanenko LA, Schumann P, Rohde M, Lysenko AM, Mikhailov VV, Stackebrandt E (2002) Psychrobactersubmarinus sp. nov. and Psychrobactermarincola sp. nov., psychrophilic halophiles from marine environments. Int J Syst Evol Microbiol 52(4):1291–1297

Romero-Romero ML, Ingles-Prieto A, Ibarra-Molero B, Sanchez-Ruiz JM (2011) Highly anomalous energetics of protein cold denaturation linked to folding-unfolding kinetics. PLoS One 6(7):e23050

Russell NJ (1998) Molecular adaptations in psychrophilic bacteria: potential for biotechnological applications. Biotechnol Extremophiles:1–21

Saavedra HG, Wrabl JO, Anderson JA, Li J, Hilser VJ (2018) Dynamic allostery can drive cold adaptation in enzymes. Nature 558(7709):324

Sanchez Y, Taulien J, Borkovich KA, Lindquist S (1992) Hsp104 is required for tolerance to many forms of stress. EMBO J 11(6):2357–2364

Santiago M, Ramírez-Sarmiento CA, Zamora RA, Parra LP (2016) Discovery, molecular mechanisms, and industrial applications of cold-active enzymes. Front Microbiol 7:1408

Sarmiento F, Peralta R, Blamey JM (2015) Cold and hot extremozymes: industrial relevance and current trends. Front Bioeng Biotechnol 20(3):148

Schmidt-Nielsen S (1902) UebereinigepsychorphileMikroorganismen und ihrVor- kommen. Centr Bakteriol Parasitenk Abt II 9:145–147

Sharada R, Venkateswarlu G, Venkateswar S, Anand Rao M (2014) Applications of cellulases-review. Int J Pharm Chem Biol Sci 4(2)

Shen L, Liu Y, Wang N, Jiao N, Xu B, Liu X (2018) Variation with depth of the abundance, diversity and pigmentation of culturable bacteria in a deep ice core from the Yuzhufeng Glacier, Tibetan Plateau. Extremophiles 22:29–38

Shivaji S, Prakash JSS (2010) How do bacteria sense and respond to low temperature? Arch Microbiol 192:85–95

Shivaji SSS, Rao NS, Saisree L, Vipula Sheth VS, Reddy GS, Bhargava PM (1989) Isolation and identification of Pseudomonas spp. from Schirmacher Oasis, Antarctica. Appl Environ Microbiol 55(3):767–770

Siddiqui KS, Williams TJ, Wilkins D, Yau S, Allen MA, Brown MV, Lauro FM, Cavicchioli R (2013) Psychrophiles. Annu Rev Earth Planet Sci 41:87–115

Siliakus MF, van der Oost J, Kengen SWM (2017) Adaptations of archaeal and bacterial membranes to variations in temperature, pH and pressure. Extremophiles 21:651–670

Singh A, Hayashi K (1995) Microbial cellulases: protein architecture, molecular properties, and biosynthesis. Adv Appl Microbiol 40:1–44

Stark C, Bautista-Leung T, Siegfried J, Herschlag D (2022) Systematic investigation of the link between enzyme catalysis and cold adaptation. Elife 11:e72884

Struvay C, Feller G (2012) Optimization to low temperature activity in psychrophilic enzymes. Int J Mol Sci 13(9):11643–11665

Sukumaran RK, Sankar V, Madhavan A, Sankar M, Satheesh V, Idris ASO, Beevi US (2015) Enzyme technologies: current and emerging technologies for development of novel enzyme catalysts. Enzym Food Beverage Process:39–66

Tapia-Vázquez I, Sánchez-Cruz R, Arroyo-Domínguez M, Lira-Ruan V, Sánchez-Reyes A, del Rayo Sánchez-Carbente M, Padilla-Chacón D, Batista-García RA, Folch-Mallol JL (2020) Isolation and characterization of psychrophilic and psychrotolerant plant-growth promoting microorganisms from a high-altitude volcano crater in Mexico. Microbiol Res 232:126394

Tamaki H, Hanada S, Kamagata Y, Nakamura K, Nomura N, Nakano K, Matsumura M (2003) *Flavobacterium limicola* sp. nov., a psychrophilic, organic-polymer-degrading bacterium isolated from freshwater sediments. Int J Syst Evol Microbiol 53(2):519–526

Thomas DN, Dieckmann GS (2002) Antarctic sea ice–a habitat for extremophiles. Science 295(5555):641–644

Ting L, Williams TJ, Cowley MJ, Lauro FM, Guilhaus M, Raftery MJ, Cavicchioli R (2010) Cold adaptation in the marine bacterium, Sphingopyxisalaskensis, assessed using quantitative proteomics. Environ Microbiol 12(10):2658–2676

Tiwari R, Nain PKS, Singh S, Adak A, Saritha M, Rana S, Sharma A, Nain L (2015) Cold active holocellulase cocktail from Aspergillus niger SH3: process optimization for production and biomass hydrolysis. J Taiwan Instit Chem Eng 56:57–66

Tribelli PM, López NI (2018) Reporting key features in cold-adapted bacteria. Life 8(1):8

Ueda M, Goto T, Nakazawa M, Miyatake K, Sakaguchi M, Inouye K (2010) A novel cold-adapted cellulase complex from Eisenia foetida: characterization of a multienzyme complex with carboxymethylcellulase, β-glucosidase, β-1, 3 glucanase, and β-xylosidase. Comp Biochem Physiol B Biochem Mol Biol 157(1):26–32

Valdespino-Castillo PM, Cerqueda-García D, Espinosa AC, Batista S, MerinoIbarra M, Taş N, Alcántara-Hernández RJ, Falcón LI (2018) Microbial distribution and turnover in Antarctic microbial mats highlight the relevance of heterotrophic bacteria in low-nutrient environments. FEMS Microbiol Ecol 94 (9):129

Varin T, Lovejoy C, Jungblut AD, Vincent WF, Corbeil J (2012) Metagenomic analysis of stress genes in microbial mat communities from Antarctica and the High Arctic. Appl Environ Microbiol 78(2):549–559

Vincent WF (1988) Microbial ecosystems of Antarctica, 1105. Cambridge University Press, Cambridge, UK

Voets IK (2017) From ice-binding proteins to bio-inspired antifreeze materials. Soft Matter 13(28):4808–4823

De Vuyst L, Harth H, Van Kerrebroeck S, Leroy F (2016) Yeast diversity of sourdoughs and associated metabolic properties and functionalities. Int J Food Microbiol 239:26–34

Wang Z, Wu S, Fan C, Zheng X, Zhang W, Wu D, Wang X, Kong H (2021) Optimisation of enzymatic saccharification of wheat straw pre-treated with sodium hydroxide. Sci Rep 11(1):23234

Wiśniewska KM, Twarda-Clapa A, Białkowska AM (2021) Screening of novel laccase producers—isolation and characterization of cold-adapted laccase from Kabatiellabupleuri G3 capable of synthetic dye decolorization. Biomolecules 11(6):828

Xu Y, Nogi Y, Kato C, Liang Z, Ruger HJ, De Kegel D, Glansdorff N (2003) Moritella profunda sp. nov. and Moritellaabyssi sp. nov., two psychropiezophilic organisms isolated from deep Atlantic sediments. Int J Syst Evol Microbiol 53(2):533–538

Yadav AN, Verma P, Sachan SG, Kaushik R, Saxena AK (2018) Psychrotrophic microbiomes: molecular diversity and beneficial role in plant growth promotion and soil health. In: Microorganisms for green revolution: vol 2. Microbes for sustainable agro-ecosystem, pp 197–240

Yamanaka K, Fang L, Inouye M (1998) The CspA family in Escherichia coli: multiple gene duplication for stress adaptation. Mol Microbiol 27(2):247–255

Yang Q, Zhang M, Zhang M, Wang C, Liu Y, Fan X, Li H (2018) Characterization of a novel, cold-adapted, and thermostable laccase-like enzyme with high tolerance for organic solvents and salt and potent dye decolorization ability, derived from a marine metagenomic library. Front Microbiol 9:2998

You YW, Wang TH (2005) Cloning and expression of endoglucanase of marine cold-adapted bacteria Pseudoalteromonas sp. MB-1. Wei Sheng wuxue bao=Acta MicrobiologicaSinica 45 (1):142–144

Yuan L, Sadiq FA, Liu TJ, Li Y, Gu J, Yang H, He G (2018) Spoilage potential of psychrotrophic bacteria isolated from raw milk and the thermo-stability of their enzymes. J Zhejiang Univ Sci 19(8):630

Zhang Z, Li J, Pan Y, Li J, Zhou L, Shi H, Zeng Y et al (2017) Natural variation in CTB4a enhances rice adaptation to cold habitats. Nat Commun 8(1):14788

MIX
Papier aus verantwortungsvollen Quellen
Paper from responsible sources
FSC® C105338

If you have any concerns about our products,
you can contact us on
ProductSafety@springernature.com

In case Publisher is established outside the EU,
the EU authorized representative is:
Springer Nature Customer Service Center GmbH
Europaplatz 3, 69115 Heidelberg, Germany

Printed by Libri Plureos GmbH
in Hamburg, Germany